Computational Microelectronics

Edited by S. Selberherr

MOSFET Models for VLSI Circuit Simulation

Theory and Practice

N. Arora

Springer-Verlag Wien New York

Dr. Narain Arora
Digital Equipment Corporation
Hudson, MA 01749, USA

Softcover reprint of the hardcover 1st edition 1993

Typesetting: Thomson Press (India) Ltd., New Delhi 110001

Printed on acid-free paper

With 270 Figures

ISSN 0179-0307
ISBN-13: 978-3-7091-9249-8   e-ISBN-13: 978-3-7091-9247-4
DOI: 10.1007/ 978-3-7091-9247-4

In the loving memory of my parents
Hukamdevi and Guranditta Arora

Preface

Metal Oxide Semiconductor (MOS) transistors are the basic building block of MOS integrated circuits (IC). Very Large Scale Integrated (VLSI) circuits using MOS technology have emerged as the dominant technology in the semiconductor industry. Over the past decade, the complexity of MOS IC's has increased at an astonishing rate. This is realized mainly through the reduction of MOS transistor dimensions in addition to the improvements in processing. Today VLSI circuits with over 3 million transistors on a chip, with effective or electrical channel lengths of 0.5 microns, are in volume production. Designing such complex chips is virtually impossible without simulation tools which help to predict circuit behavior before actual circuits are fabricated. However, the utility of simulators as a tool for the design and analysis of circuits depends on the adequacy of the device models used in the simulator. This problem is further aggravated by the technology trend towards smaller and smaller device dimensions which increases the complexity of the models.

There is extensive literature available on modeling these short channel devices. However, there is a lot of confusion too. Often it is not clear what model to use and which model parameter values are important and how to determine them. After working over 15 years in the field of semiconductor device modeling, I have felt the need for a book which can fill the gap between the theory and the practice of MOS transistor modeling. This book is an attempt in that direction.

The book deals with the MOS Field Effect transistor (MOSFET) models that are derived from basic semiconductor theory. Various models are developed ranging from simple to more sophisticated models that take into account new physical effects observed in submicron devices used in today's MOS VLSI technology. The assumptions used to arrive at the models are emphasized so that the accuracy of the model in describing the device characteristics are clearly understood. Due to the importance of designing reliable circuits, device reliability models have also been covered. Understanding these models is essential when designing circuits for state of the art MOS IC's.

Extracting the device model parameter values from device data is a very important part of device modeling which is often ignored. In this book the first detailed presentation of model parameter determination for MOS models is given. Since the device parameters vary due to inherent processing variations, how to arrive at worst case design parameters which ensure maximum yield is covered in some detail.

Presentation of the material is such that even an undergraduate student not well familiar with semiconductor device physics can understand the intricacies of MOSFET modeling. Chapter 1 deals with the overview of various aspects of device modeling for circuit simulators. Chapter 2 is a brief but complete (for understanding MOSFET models) review of semiconductor device physics and *pn* junction theory. The MOS transistor characteristics as applied to current MOS technologies are discussed in Chapter 3. The theory of MOS capacitors that is essential for the understanding of MOS models are covered in Chapter 4. Different MOSFET models, such as threshold voltage, DC (steady-state), AC and reliability models are the topic of discussion in Chapters 5, 6, 7 and 8, respectively. Chapters 9 and 10 deal with data measurements and model parameter extraction. The diode and MOSFET models implemented in Berkeley SPICE, a defacto industry standard circuit simulator, are covered in Chapter 11. Finally, the statistical variation of model parameters due to process variations are covered in Chapter 12.

It is my sincere hope that this book will serve as a technical source in the area of MOSFET modeling for state of the art MOS technology for both practicing device and circuit engineers and engineering students interested in the said area.

During the writing of this book I have received much help and encouragement, directly or indirectly, from my colleagues. First I would like to express my gratitude to the management of Digital Equipment Corporation, namely Dr. Rich Hollingsworth (Corporate Consultant) and Dr. Llanda Richardson (Consultant) for their encouragement and assistance in writing this book. I am deeply indebted to Dr. F. Fox, Dr. D. Ramey, and Mr. K. Mistry for their excellent work in careful reading of many of the chapters in the first draft of the manuscript and giving their critical comments. I am also indebted to Drs. R. Rios, J. Huang and Mr. K. Roal for this invaluable help during completion of this work. I would like to express my thanks to the large number of my colleagues, within and outside my organization, who helped in preparing the manuscript in one way or the other, namely Drs. A. Bose, D. Bell, B. Doyle, J. Faricelli, A. Enver, K. L. Kodandpani, L. Richardson, A. R. Shanker, Messers L. Bair, N. Khalil, L. Gruber, Prof. S. C. Jain (former Director SPL), Prof. D. Antoniadis (MIT), Prof. G. Gildenblat (Penn. State), Prof. D. J. Roulston (UW), Dr. R. Chadha (AT & T), and Dr. M. Sharma (Motorola). This acknowledgment will not be complete without the name of Dr. Risal Singh, my old colleague

and close friend, who in spite of his busy schedule, spent many many hours to help me bring this book to the present form.
Finally, I would like to express my deep gratitude to my family. This book would not have been possible without their support. The understanding of my wife Suprabha, and the cooperation of my son Surendra and daughter Shilpa all were indispensable in making this book a reality.

April 11, 1992
Shrewsbury, MA.

Contents

List of Symbols

The following is list of symbols used in the text. This list excludes those symbols which are used locally in a particular chapter.

Symbol	Description	Unit
A_i	Ionization constants	cm^{-1}
B_i	Ionization constants	V/cm
C_d	Depletion region capacitance per unit area	F/cm^2
C_{fb}	Flat band capacitance per unit area	F/cm^2
C_g	Capacitance per unit area of a MOS capacitor	F/cm^2
C_{gc}	Gate-to-channel capacitance per unit area	F/cm^2
C_{gb}	Gate to bulk capacitance per unit area	F/cm^2
C_{gso}	Gate to source overlap capacitance per unit length	F/cm
C_{gdo}	Gate to drain overlap capacitance per unit length	F/cm
C_{gbo}	Gate to bulk overlap capacitance per unit length	F/cm
C_{GS}	Intrinsic gate to source capacitance	F
C_{GD}	Intrinsic gate to drain capacitance	F
C_{GB}	Intrinsic gate to bulk capacitance	F
C_j	*pn* junction depletion capacitance	F/cm^2
C_{ox}	Gate oxide capacitance per unit area	F/cm^2
C_{oxt}	Total gate oxide capacitance ($C_{ox}WL$)	F
C_{sc}	Space charge capacitance per unit area	F/cm^2
D_{it}	Interface state density	$charges/cm^2$
D_n	Electron Diffusivity or diffusion constant	cm^2/s
D_p	Hole Diffusivity or diffusion constant	cm^2/s
E_a	Ionized acceptor energy level	eV
E_d	Ionized donor energy level	eV
E_c	Energy level for the lower edge of the conduction band	eV

Symbol	Description	Unit
E_v	Energy level for the upper edge of the valance band	eV
E_g	Energy gap of semiconductor	eV
E_i	Intrinsic energy level	eV
E_f	Fermi-energy or Fermi level in (*n* or *p*-type) bulk silicon	eV
E_{fn}	Fermi-energy level in *n*-type Silicon	eV
E_{fp}	Fermi-energy level in *p*-type Silicon	eV
$\mathscr{E}$	Electric field in the space-charge region	V/cm
$\mathscr{E}_x$	Vertical or normal electric field in the channel	V/cm
$\mathscr{E}_y$	Lateral electric field in the channel	V/cm
$\mathscr{E}_c$	Critical field for the carrier velocity saturation	V/cm
$\mathscr{E}_{eff}$	Effective vertical field	V/cm
$\mathscr{E}_{ox}$	Electric field in the oxide	V/cm
$\mathscr{F}_n$	Electron quasi-Fermi energy	eV
$\mathscr{F}_p$	Hole quasi-Fermi energy	eV
G	Carrier generation rate	s^{-1}
g_d	Diode small-signal conductance	A/V
g_{ds}	MOSFET small signal output conductance	A/V
g_m	MOSFET small signal transconductance	A/V
g_{mbs}	MOSFET small signal substrate transconductance	A/V
I_d	Current in a diode or drain current in a MOSFET	A
I_g	Gate current in a MOSFET	A
I_b	Substrate current in a MOSFET	A
I_s	Leakage current in a diode or source current in a MOSFET	A
I_{ds}	Drain to source current	A
J_n	Electron current density	A/cm^2
J_p	Hole current density	A/cm^2
J_l	Junction leakage current	A
k	Boltzmann constant	J/K
L	Effective or electrical channel length	μm
L_m	Mask or drawn channel length	μm
L_d	Extrinsic Debye length	cm
l_d	Length near the drain end due to channel length modulation	cm
m_n^*	Electron effective mass	g
m_p^*	Hole effective mass	g
m_0	Electron rest mass	g

Symbol	Description	Unit
N_a	Impurity (accepter) concentration in *p*-type silicon	cm^{-3}
N_d	Impurity (donor) concentration in *n*-type silicon	cm^{-3}
N_b	Impurity concentration in (*n* or *p*-type) bulk silicon	cm^{-3}
N_a^-	Ionized acceptor impurity concentration	cm^{-3}
N_d^+	Ionized donor impurity concentration	cm^{-3}
n	Free electrons concentration	cm^{-3}
n_i	Intrinsic carrier concentration	cm^{-3}
n_p	Electrons concentration in *p*-type silicon	cm^{-3}
p	Free holes concentration	cm^{-3}
p_n	Hole concentration in *n*-type silicon	cm^{-3}
q	Magnitude of Electronic Charge	C
Q_0	Oxide charge density at the interface	$charges/cm^2$
Q_b	Bulk (depletion) charge per unit area	C/cm^2
Q_i	Mobile (inversion) charge per unit area	C/cm^2
Q_g	Gate charge per unit area	C/cm^2
Q_s	Charge per unit area induced in the silicon	C/cm^2
Q_I	Total inversion charge	C
Q_G	Total gate charge	C
Q_S	Total source charge	C
Q_D	Total drain charge	C
R_{ch}	Intrinsic channel resistance	Ω
R_d	Drain resistancc	Ω
R_s	Source resistance	Ω
R_t	Sum of the source and drain resistance	Ω
S	Subthreshold slope	V/decade
T	Absolute temperature	K
t_{ox}	Gate oxide thickness	Å
V_g	Gate voltage	V
V_{cb}	Channel to bulk potential	V
V_{bs}	Bulk to source voltage	V
V_{ds}	Drain to source voltage	V
V_{gs}	Gate to source voltage	V
V_{gb}	gate to bulk voltage ($V_{gs} - V_{bs}$)	V
V_{gd}	gate to drain voltage ($V_{gs} - V_{ds}$)	V
V_{dsat}	Drain saturation voltage	V
V_{fb}	Flat-band voltage	V
V_{th}	Threshold voltage	V
V_t	Thermal voltage (kT/q)	V

Symbol	Description	Unit
V_{sb}	Substrate bias	V
υ	Carrier velocity	cm/sec
υ_{sat}	Carrier saturation velocity	cm/sec
W	Effective or electrical channel width	μm
W_m	Mask or drawn channel width	μm
x	Distance from Si–SiO_2 interface into silicon	cm
t_{ch}	Inversion layer thickness	cm
X_j	Junction depth	cm
X_n	Depletion width on *n*-side of a *pn* junction	cm
X_p	Depletion width on *p*-side of a *pn* junction	cm
X_d	Bulk depletion width in a MOS capacitor or MOSFET	cm
X_{dm}	Maximum depletion width in a MOS capacitor or MOSFET	cm
α	Body factor term $(1 + \delta\gamma)$	—
β_0	Device gain $= \mu_0 C_{ox} W/L$	V^{-1}
γ	Body factor $\sqrt{2q\epsilon_{si}N_b}/C_{ox}$	$V^{1/2}$
δ	Square-root approximation factor in the bulk charge Q_b term	V^{-1}
ΔL	Difference between drawn and electrical channel length $(L_m - L)$	μm
ΔW	Difference between drawn and electrical channel width $(W_m - L)$	μm
τ	Carrier life-time	sec
ϵ_{ox}	Dielectric permittivity of SiO_2	F/cm
ϵ_{si}	Dielectric permittivity of silicon	F/cm
θ	The mobility degradation factor resulting from the vertical field	V^{-1}
μ_0	Low field channel mobility	$cm^2/V \cdot s$
μ_{eff}	Effective channel mobility	$cm^2/V \cdot s$
μ_n	Electron mobility	$cm^2/V \cdot s$
μ_p	Hole mobility	$cm^2/V \cdot s$
μ_s	MOSFET surface mobility	$cm^2/V \cdot s$
μ_{eff}	Effective mobility due to gate and drain field	$cm^2/V \cdot s$
χ_s	Electron affinity for silicon	eV
Φ_m	Metal (gate) work function	eV
Φ_{ms}	Gate to substrate work function difference	eV
Φ_b	Electron potential barrier height at Si–SiO_2 interface	eV
μ_{sr}	Surface roughness scattering	$cm^2/V \cdot s$
ρ	Space charge density	C/cm^3

Symbol	Description	Unit
ρ_s	Sheet resistance	$\Omega/\square$
λ	channel length modulation factor	V^{-1}
ϕ	Electrostatic potential with respect to intrinsic level E_i	V
ϕ_s	Surface potential with respect to intrinsic level E_i	V
ϕ_{bi}	Built-in potential of a *pn* junction	V
ϕ_f	Fermi potential in (*n* or *p*-type) bulk silicon	eV
φ_n	Electron quasi-Fermi potential (imref)	V
φ_p	Hole quasi-Fermi potential (imref)	V

Acronyms

Symbol	Description
CHE	Channel Hot Electron
CLM	Channel Length Modulation
CMOS	Complementary Metal Oxide Semiconductor
C–V	Device Capacitance–Voltage Characteristics
DIBL	Drain Induced Barrier Lowering
GCA	Gradual Channel Approximation
IGFET	Insulated Gate Field Effect Transistor
LDD	Lightly Doped Drain
LOCOS	Localized Oxidation of Silicon
MOSFET	Metal Oxide Semiconductor Field-Effect Transistor
nMOST	*n*-channel MOSFET
pMOST	*p*-channel MOSFET
S/D	MOSFET Source/Drain
ZTC	Zero Temperature Coefficient

Overview 1

Even though the operation of the modern Metal-Oxide-Semiconductor (MOS) transistor was first described by Lilienfield in 1930 [1], it was not until 1960 that the first MOS transistor using silicon as the semiconductor material was reported by Kang and Atalla [2]. The MOS technology became viable only after methods of routinely growing reliable oxides were developed and reported by Snow, Grove, Deal and Sah in 1964 [3]. Since that time the MOS industry has expanded very quickly. Today MOS integrated circuits (ICs) have emerged as the dominant technology in the semiconductor industry. The exponential growth in the number of components per chip and projections for the future are shown in Figure 1.1 [4]. Also shown is the minimum feature size that can be produced on a chip. The dotted lines are projections for the future. Clearly with this technology it is now possible to have more than a million transistors on a single chip. All this has been possible due to the fact that the basic MOS transistor size has shrunk by a factor of about 20 during the last two decades, from a feature size of 20 μm to less than a micron. Much of this shrinkage can be attributed to advances in lithography, the use of ion implantation, and low temperature annealing [4].

During the early days of MOS technology, aluminum (Al) gate *p*-channel MOS transistors were the workhorse technology. In the late sixties polysilicon replaced Al as the material for the MOS transistor gate [5]. The next major milestone was the LOCOS (LOCalized Oxidation of Silicon) isolation technique [6]. Commercially successful products using the NMOS process (all *n*-channel MOS transistors) with LOCOS isolation were developed in the mid seventies. NMOS device technology became the driving force of the 1970s because of its reliability, reasonable manufacturing cost, and scalability. During the last decade, MOS transistors have been scaled down in dimensions both vertically and horizontally. Rules of this scaling were originally formulated by Dennard et al. [7] in 1974 and subsequently other schemes of scaling were proposed [8] (see Section 3.3).

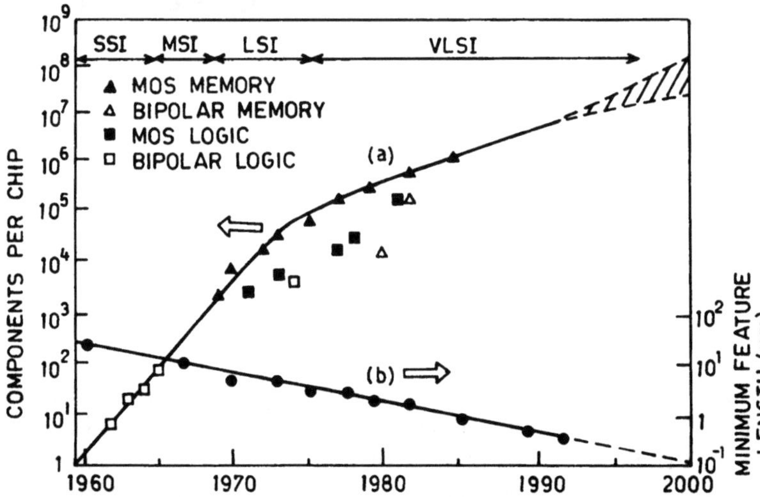

Fig. 1.1 (a) Exponential growth of the number of components on the chip (SSI = small-scale integration; MSI = medium-scale integration; LSI = large-scale integration; VLSI = very large-scale integration); (b) Exponential decrease of the minimum device dimensions. Dotted lines are projections. (From Sze [4, p. 3], slightly modified)

Unfortunately not all device parameters can be scaled proportionately. *These limits on scaling have increased the importance of device and circuit modeling.* The CMOS (Complementary MOS, with both *p*- and *n*-channel transistors) technology has revolutionized the state of the art of IC design due to its inherent noise immunity and reduced static power dissipation. CMOS technology became the technology of choice for the VLSI (Very Large Scale Integration) chips of the 1980s [9]. Although there has been considerable recent interest in incorporating bipolar transistors into CMOS processes, resulting in a BiCMOS technology [10, 11], we will restrict ourselves to device modeling for NMOS and CMOS technologies.

Although the MOS transistor (also called MOSFET) is the most important device for VLSI chips such as microprocessors and semiconductor memories, it is also becoming an important power device. MOS transistors based on DMOS (Double-diffused MOS) and VMOS (Vertical grooved MOS) technology have highly asymmetrical characteristics which makes these technologies unsuitable for integrated circuit applications [12]. Nevertheless, excellent discrete power devices are built with these technologies. The modeling of power MOSFETs is not covered in this book [13, 14].

1.1 Circuit Design with MOSFETs

For today's circuit design, computer-aided simulation [15]–[17] has become an indispensable tool because:

- Manual techniques traditionally used for circuit analysis and design are simply inadequate because of the complexity of today's circuits.
- Simulation allows designers to design their chips under worst case conditions so that manufacturing tolerances can be incorporated into the design. It thus greatly increases the likelihood that the circuit (chip) will work as desired and have good production yield.
- Simulation allows designers to predict and optimize circuit performance.

At the lower end of the hierarchy of VLSI design tools, *circuit simulators* offer the most detailed level of simulation normally used for circuit design. Some of the most successful circuit simulators of the early 1970s are still used extensively in the design and verification of VLSI chips; most notably are ASTAP(Advanced STatistical Analysis Program) from IBM [18] and SPICE2(Simulation Program with Integrated Circuit Emphasis) from the University of California, Berkeley[1] [19]. These simulators are typically used to analyze circuits with up to several hundred nodes. SPICE2 is the defacto industry standard and is used in many universities all over the world. Most of the circuit simulators which are available commercially are derived from SPICE. The commercial vendors claim to provide improved convergence, graphics capabilities, improved user interfaces, and often special analysis modes. Simulators that are not derived from SPICE differ from it in their choice of integration methods or in some aspect of modeling methodology. A very good survey of different commercially available circuit simulators was reported by Beresford and Domitrowich [20]. The capabilities of these simulators includes three basic types of analysis, e.g. nonlinear DC, nonlinear transient and linear AC analysis and several special options such as sensitivity analysis, noise and distortion analysis, worst case analysis, and Fourier analysis. In recent years relaxation based circuit simulators for special classes of MOS circuits have emerged that could speedup the simulation of big circuits by at least two orders of magnitude [21].

In general a circuit simulation program consists of the following four subprograms [22]–[24]: (1) the *input* subprogram that reads the input file, constructs the data structure for the original circuit description and checks

[1] SPICE3 is a redesigned implementation of SPICE2 program written in the C programming language and is designed to be modular. In terms of algorithms it is no improvement over SPICE2 which is written in Fortran. The SPICE software package is in public domain and can be obtained by writing to Ms. Cindy Manly, EECS/ERL Industrial Liaison Program, 497 Cory Hall, University of California, Berkeley, California, 94720.

it for user errors; (2) the *setup* subprogram that sets up data structures required for the circuit analysis; (3) the *analysis* subprogram which performs the desired circuit analysis; and (4) the *output* subprogram which generates the output specified by the user. It is the analysis part where the system of equations describing the complete circuit are solved numerically to give the desired analysis results. This system of equations is formed for each element in the circuit and the topological constraint connecting them. In general, it is of the form

$$\mathbf{f}(\mathbf{x}, \mathbf{x}', t) = 0 \tag{1.1}$$

where $\mathbf{x}$ is the vector of the unknown circuit variables, $\mathbf{x}'$ is the time derivative of $\mathbf{x}, t$ is the time, and $\mathbf{f}$ in general is a highly nonlinear function vector. A solution of Eq. (1.1) can be obtained by first converting nonlinear differential equations into nonlinear algebraic equations using numerical integration methods. The resulting nonlinear algebraic equations are then solved using the Newton–Raphson iterative algorithm. At each Newton–Raphson iteration the nonlinear equations are linearized around the operating point. The linear representation of nonlinear circuit elements like diodes and transistors is called the *companion model*. The latter describes the linearized characteristics of the nonlinear element as a function of its controlling voltage and current. The differential equation characterizing a capacitor or inductor is also approximated using a companion model (resistive circuit) that depends upon the integration algorithm [22, 23]. Thus a *companion model reduces a dynamic network into a resistive network*. It should be pointed out that the linearization process is approximate and therefore its accuracy depends on the error tolerance allowed.

Numerical errors are unavoidable in the simulation process. However, by choosing suitable variables these errors may be reduced. From the computational point of view, when the circuit contains nonlinear capacitors it is advantageous to use charge Q as the state variable as this has been shown to result in less propagation of numerical error [22]. For MOSFET capacitances this choice of Q as a state variable becomes essential. Otherwise charge nonconservation problems can arise, as will be discussed in detail in Chapter 7.

The utility of the circuit simulators as a tool for the design and analysis of VLSI circuits depends on the adequacy of the device model being used in the simulator. In particular, the accuracy and simplicity (computational efficiency) of the model directly affects the corresponding accuracy and speed of simulation. It has been found that for large circuits the MOSFET model evaluation accounts for a large percentage (up to 80%) of the total analysis time [15]. This problem is further aggravated by the technology trend towards smaller and smaller device dimensions which increases the complexities of the models. Thus realistic circuit modeling requires an understanding of the accuracy and limitations of the various device models and the computational techniques used for performing the analysis of the model.

1.2 MOSFET Modeling

The device models describe the terminal behavior of a device in terms of current-voltage (I-V), capacitance-voltage (C-V) characteristics, and the carrier transport process which takes place within the device. These models thus reflect device behavior in all regions of operation of the device. It is convenient to divide these models into two categories: (1) physical device models, and (2) equivalent circuit models. Physical device models are based on a careful definition of device geometry, doping profile, carrier transport equations (*semiconductor equations*) and material characteristics. These models can be used to predict both terminal characteristics and transport phenomenon. Modern MOS VLSI devices, due to their small size (micron and submicron), require two- or three-dimensional solutions of the coupled semiconductor equations which can be solved only by numerical methods [25, 26]. These so called numerical *device simulators* provide detailed insight into the physical aspect of device operation and can predict the characteristics of new devices. For this reason they are mostly used to study device physics and device design [27]. Several public domain and commercial software packages are now available for device analysis and simulation; the most well known among them are MINIMOS [28], PISCES [29], FIELDAY [30], CADDETH [31]. Since device simulators are computationally intensive and require large amount of computer memory, they are not suitable for circuit simulation.
Due to the 2-D and 3-D nature of the physical effects governing electrical behavior of VLSI MOS transistors, it is very difficult to obtain a closed form analytical formulation which is valid in all operating regions of interest. However, *one can still obtain closed form analytical models, based on device physics, that are generally valid only over a limited region of device operation.* Despite this limitation, such models are frequently used for circuit simulators because of ease of computation.
Equivalent circuit models describe electrical properties of the device by connecting electrical circuit elements in such a way that the model emulates the electrical terminal behavior of the device. These models are thus based on the device characteristics; the circuit elements of this model are derived either from closed form analytical function or using an empirical approach. These models are often used in circuit simulators to represent device characteristics because of the ease of evaluation; the circuit simulator SPICE exclusively uses equivalent circuit models. For semiconductor devices the equivalent circuit model elements are highly nonlinear and element values are strongly dependent on DC bias, frequency, signal level and temperature.

Therefore, in addition to having separate DC and AC circuit models, it is generally necessary to distinguish between the small-signal and large-signal (transient) models. Thus in general we require three types of circuit models—DC, transient and AC—corresponding to three basic types of circuit analysis:

- A DC model is a *static model* that evaluates the device current for a fixed voltage, not varying with time.Thus in a DC model dynamic effects such as time delay arising from the presence of energy-storage elements (capacitors and inductors) are ignored. This model is used to calculate quiescent operating points of a circuit.[2]
- A transient model is a *large-signal dynamic model* which evaluates the device current when the applied voltage is varying with time. It is called a large-signal model because no restrictions are placed on the magnitude of the applied voltage. This model is required for the time domain analysis. In this case current is the sum of both DC and transient currents arising from the charging or discharging of device storage elements, usually capacitances.
- An AC model is a *small-signal model* which evaluates the current when the variation in the applied voltage is so small that the resulting small current variations can be expressed using linear relations. The small-signal linear model can usually be obtained very easily and systematically from the DC model of the device. Since AC model is used for the frequency-domain analysis, it should take into account energy-storage elements and the frequency dependent effects of the transistor.

The MOSFET model we will be concerned with contains only capacitances as the storage elements and not the inductors. The latter are important only at very high frequencies (GHz range). For the transistor model to be used in a circuit simulator, the following requirements should be satisfied:

- The model should be *accurate* so that it simulates actual transistor behavior over all regions of operation of interest. An accuracy of about 5% between the experimental device current (and capacitances) and the model is generally sufficient for circuit modeling work
- During transient analysis, calculation of transistor current is carried out thousands of time, therefore, it is imperative that the model be both *computationally efficient*, and accurate. Thus, the model needs not only to be accurate but *simple* too; there is always a trade-off between accuracy and simplicity.
- In order to avoid any nonconvergence problems in the simulator, the

---

[2] The points (nodal voltages and branch currents) about which the circuit operates are termed *quiescent points* (Q-points) or *bias points*. Accurate Q-points are critical for the design and simulation of transient and AC response.

mathematical equations representing the device model must be continuous, with continuous first derivatives (which are required by the Newton–Raphson algorithm), although not necessarily in a strict mathematical sense. The degree of discontinuity, if present, must be so small that the resulting errors can be absorbed by the overall simulation program error tolerances.
- In MOS VLSI circuit design devices of different lengths and widths are used, therefore, it is desirable that a single model should fit all device sizes used in actual design practice.

Clearly, any choice of the model must be based on compromises between the accuracy of the model in predicting device characteristics over the operating range of interest and the computational efficiency of simulating large circuits. As the size and the complexity of modern circuits increases, the choice of appropriate models becomes more critical. For this reason a hierarchy of models of different levels of accuracy are normally available in a circuit simulator so that designers can choose a model best suited to their potential application. For example, Berkeley SPICE has four different levels of MOSFET models. The combined requirements of computational efficiency and available memory restrict the device models for circuit simulators into the following three categories.

Analytical Models. There are basically two types of analytical models where model equations are directly derived from device physics. One type of model is based on surface potential analysis, often called *charge sheet models* [32, 33]. These models are inherently continuous in all regions of operation of the device. The current can be accurately determined using these models, but the equations themselves are complex, involving transcendental expressions, and often require iterations just to compute the surface potential for a given bias condition. They are thus not very suitable for VLSI circuit simulation, although recently they have been used for simulation of small circuits [34, 35]. The second type of analytical model is the result of applying various approximations to the semiconductor equations, based upon decisions as to which physical phenomena dominate [8], [33], [36]–[38]. Thus, different equations are required to represent different regions of operation of the device. Such models represent first order device behavior fairly accurately, and higher order effects are normally accounted for through the introduction of physical and empirical parameters. These models are usually referred to as *semi-empirical analytical models. Practically all the models used in today's circuit simulators fall into this category, and range from simple to more complex models.* These are the type of models which are covered in this book.
The advantage of these models are that they do describe the relationship between the physical process and geometry structure on the one hand and

electrical behavior on the other, so that with some minor changes in the process, electrical behavior can still be predicted. However, the disadvantage is that they are technology dependent and takes considerable time to develop the model. Furthermore, effects resulting from new device structures often require minor or major modification of the existing model and may even require a new model.

Table Lookup Model. In a table lookup model the device current data are stored for different bias points and device geometries in a tabular form [39]–[41]. Generally some sort of interpolation scheme is used to obtain the current values which are not stored. This data base is collected from experimental devices or generated from device level simulators like MINIMOS/PISCES. In another approach, instead of directly storing the device current I_{ds}, the coefficient of some mathematical functions like cubic splines are precalculated from original I_{ds} data for different bias and geometry. It is the coefficients of this function which are then stored in a tabular form, and are later used to compute the currents and conductances required by the simulator [40]–[41]. This approach increases model evaluation speed and reduces storage. *These types of models have the advantage that they are technology independent and can be developed in a shorter time compared to physical models.* The disadvantage of this approach is that it gives no physical insight into device behavior. The model validity outside the data range is uncertain, and if accuracy is required storage is still a problem.

It should be pointed out that table lookup models are generally used for device DC models. For transient and AC models, we still use analytical models because the charges associated with different device terminals are difficult to measure. The charges can be calculated from terminal capacitances [42], but even the capacitance measurements for VLSI devices are difficult to carryout. We will not cover the table lookup model and the interested reader should look into the references cited.

Empirical Model. In an empirical model, the model equations representing device characteristics are purely of the curve fitting type and are thus not based on device physics [42a]. The only advantage of this type of model is that it requires small data storage as compared to table look models and model development time is shorter compared to other modeling approaches. The disadvantage is that this approach is not technology independent. *Purely empirical models are seldom used in circuit simulators*, although empirical (or curve fitting) parameters are often included in physical models to describe 2 or 3-D device behavior.

1.3 Model Parameter Determination

The accuracy of a device model in predicting device characteristics is fully dependent on the accuracy of the model parameter values being used. With ever decreasing device dimensions, the complexity of the models used in circuit simulators have increased significantly. Further, most circuit models are semi-empirical analytical models containing various fitting parameters that do not have a well defined physical meaning, and the number of these fitting parameters increases with the complexity of the model. Very often some of these fitting parameters become redundant,[3] and no unique value can be determined for those parameters. Therefore, care must be taken in extracting model parameter values from device data so that physical meaning of the parameter is retained as far as possible. The device data required for extracting model parameters may either be obtained from a device level simulator or from electrical measurements on a number of test devices of different geometries (different widths and lengths). For MOS VLSI it is common practice to fabricate MOS transistors of different widths and lengths on special test chips, also called *test patterns*, along with other test structures required for process development and characterization [43]. The electrical measurements on test transistors are then normally performed using an automatic wafer prober and measurement system as discussed in Chapter 9.

Various general purpose curve fitting programs called *optimizers* have been developed which extract model parameters by curve fitting model equations to the experimental device data using non-linear least square optimization techniques [44]–[46]. One such optimizer which is in the public domain is SUXES (Stanford University eXtractor modEl parameterS) from Stanford University.[4] Other similar packages are commercially available from different companies and universities.

To support designs that yield well across the full range of random variations in a process, the statistical behavior of the model parameters must be known. Since I-V and/or C-V characteristics of the devices represent the joint distribution of all the process variations, by extracting the model parameters from these curves for different size MOSFETs, and studying the observed distribution of extracted parameters, worst case design parameters can be created. This process spread information is essential for chip design for good manufacturing yield (see Chapter 12).

---

[3] If a physical effect can be described partially by two parameters or one parameter has much smaller influence than the other parameter, then one of the parameter becomes redundant.

[4] SUXES can be obtained by writing to Office of Technical Licensing, Stanford University, 105 Encina Hall, Stanford, California 94305.

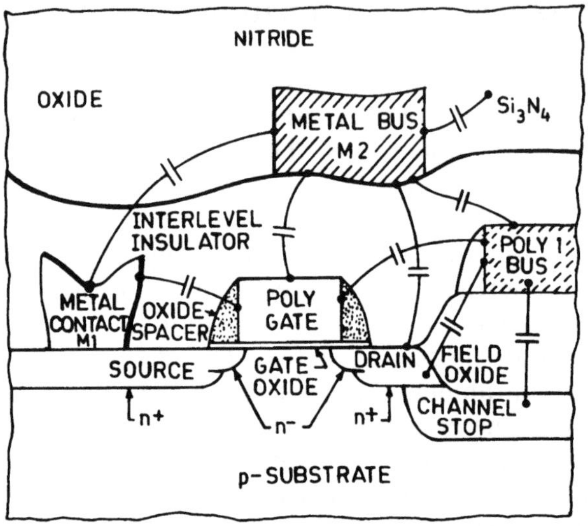

Fig. 1.2 Parasitic capacitances in 1 μm CMOS integrated circuits. The dimensions are approximately to scale. (From Yang and Chatterjee [47], slightly modified)

1.4 Interconnect Modeling

MOS VLSI circuits consist almost entirely of MOS transistors and their interconnections. In a typical MOS VLSI chip, active device area is 10% while the physical area occupied by interconnect and isolation regions is 6 to 10 times the active device area. For this reason the role of interconnect is becoming increasingly important as the feature size is scaled down to submicron dimensions and device density is increased on the chip. Figure 1.2 shows a vertical cross-section of a 1 μm design rule CMOS technology [47]. From this figure it is reasonable to expect that capacitive coupling between the metal lines and from metal lines to devices will play a significant role in the circuit response. In fact, interconnect capacitances are becoming dominant in determining the performance of VLSI circuits. Therefore, these parasitic capacitances must be taken into account during the chip design. The distributed resistance and capacitance of long signal lines form a low pass filter circuit which can affect signal timing. The switching power necessary to drive this interconnect loading is a significant part of the total chip power dissipation. Modeling of these interconnect properties is thus important and must be included by the designer when checking circuit performance through circuit simulation tools. The models for the parasitic capacitances and resistances, outside the device but part of the chip, are outside the scope of this book and interested readers are referred to the references cited [27], [48].

1.5 Subjects Covered

In this book we will cover analytical models for MOS VLSI devices and their model parameter determination. Emphasis will be on models that are suitable for VLSI circuit simulation. Although models discussed will be based on device physics, these models will often include empirical factors in order to account for the second order effects essential to model short geometry device behavior.

The basic semiconductor and *pn* junction theory essential for the development of the MOS transistor models are reviewed in Chapter 2. The overview of MOS transistor operation and characteristics are discussed in Chapter 3. Also included in this chapter is the overview of VLSI MOSFET characteristics such as MOSFET scaling, hot-electron effects, and MOSFET parasitic elements. The MOS capacitor, which is used for the characterization of MOS process and is basic to understanding MOSFET operation, is the topic for Chapter 4. From a circuit modeling point of view, MOSFET threshold voltage is the single most important device parameter. The threshold voltage models for large and small geometry MOSFETs are developed in Chapter 5. The device DC models are discussed in Chapter 6 while AC models, both small and large signal, are covered in Chapter 7. Models for hot-electron effect, particularly substrate and gate current models, and device life-time models are covered in Chapter 8.

The experimental setup, required for taking device data for different geometries and as a function of bias, is discussed in Chapter 9. Methods of determining some basic parameters such as threshold voltage, mobility of the carriers in the inversion region, doping profile, MOSFET capacitance measurements etc. are also covered in this chapter. The general purpose nonlinear optimization techniques for model parameter extraction are discussed in Chapter 10. The MOSFET model parameter extraction in general are also covered in this chapter. Since SPICE is used extensively through out the industry and at various universities, we have devoted Chapter 11 to the Diode and MOSFET models and their parameters as implemented in Berkeley SPICE. Finally the statistical variations of the MOSFET parameters due to the process variations are covered in Chapter 12.

References

[1] J. E. Lilienfield, US Patent 17, 45175 issued Jan. 28, 1930.

[2] D. Kahng and M. M. Atalla, 'Silicon-silicon dioxide field induced surface devices', IRE Solid State Device Research Conference, Pittsburgh, PA 1960. Also see references such as, D. Kahng, 'A historical perspective on the development of MOS transistors and related devices', IEEE Trans. Electron Devices, ED-23, pp. 655–660 (1976); J. D. Meindl, 'Ultralarge scale integration', ibid, ED-31, pp. 1555–1561 (1984).

[3] E. H. Snow, A. S. Grove, B. E. Deal, and C. T. Sah, 'Ion transport phenomena in insulating films', J. Appl. Phys., 36, pp. 1665–1673 (1965).

[4] S. M. Sze, Ed., *VLSI Technology*, 2nd Ed. McGraw-Hill Book Company, New York, 1988.

[5] F. Faggin and T. Klein, 'Silicon gate technology', Solid-State Electron., 13, pp. 1125–1144 (1970).

[6] J. A. Appels, E. Kooi, M. M. Paffen, J. J. H. Schiorje, and W. H. C. G. Verkuylen, 'Local oxidation of Silicon and its application in semiconductor technology', Philips Res. Rep., 25, pp. 118–132 (1970).

[7] R. H. Dennard, F. H. Gaensslen, H. N. Yu, V. L. Rideout, E. Bassous, and A. R. LeBlanc, 'Design of ion-implanted MOSFETs with very small physical dimensions', IEEE J. Solid-State Circuits. SC-9, pp. 256–268 (1974).

[8] N. G. Einspruch and G. Gildenblat, Eds., *Advanced MOS Device Physics*, VLSI Electronics: Microstructure Science, Vol. 18, Academic Press Inc., New York, 1989.

[9] J. Y. Chen, 'CMOS—The emerging VLSI Technology', IEEE Circuits and Device Magazine, 2, pp. 16–331 (1986).

[10] P. Ashburn, *Design and Realization of Bipolar Transistors*, John Wiley & Sons, New York, 1988.

[11] A. R. Alvarez, Ed., *BICMOS Technology and Application*, Kluwer Academic Publisher, Boston, 1989.

[12] S. K. Gandhi, *VLSI Fabrication Principles*, John Wiley & Sons, New York, 1983.

[13] A. Blicher, *Field-Effect and Bipolar Power Transistor Physics*, Academic Press, Inc., New York 1981.

[14] D. A. Grant and J. Gowar, *Power MOSFETS—Theory and Application*, John Wiley & Sons, New York, 1989.

[15] P. Antognetti, D. O. Pederson, and H. De Man, Eds., '*Computer Design Aids for VLSI Circuits*', NATO Advanced Institute 1980, Sigthoff & Noordhoff, Alphen aan den Rijn, The Netherlands, 1981.

[16] A. E. Ruehli, Ed., *Circuit Analysis, Simulation and Design*, North-Holland, New York, 1986.

[17] A. F. Schwarz, *Computer-Aided Design of Microelectronic Circuits and Systems*, Vols. I and II, Academic Press, New York, '1987.

[18] W. T. Weeks, A. J. Jimenez, G. W. Mahoney, D. Mehta, H. Qassemzadeh, and T. R. Scott, 'Algorithms for ASTAP—A network-analysis program', IEEE Trans. on Circuit Theory, CT-20, pp. 628–634 (1973). Also see, Program Reference Manual, Pub. no. SH20-1118-0, IBM Corp., Data Process Division, White Plains, NY 10604.

[19] L. W. Nagel, 'SPICE2: A computer program to simulate semiconductor circuits', Memorandum No. UCB/ERL-M520, Electronic Res. Lab., University of California, Berkeley, May 1975.

[20] R. Beresford and J. Domitrowich, 'Survey of circuit simulators', VLSI Design, Vol. 8, pp. 70–80, July 1987.

[21] J. K. White and A. Sangiovanni-Vincentelli, *Relaxation Techniques for the Simulation of VLSI Circuits*, Kluwer Academic Publisher, Boston, 1987.

[22] D. A. Calhan, *Computer-Aided Network Design*, revised Ed., McGraw-Hill Book Company, New York, 1972.

[23] L. O. Chua and P. M. Lin, *Computer-Aided Analysis of Electronic Circuits: Algorithms & Computational Techniques*, Prentice Hall, Englewood Cliffs, NJ, 1975.

[24] W. J. McCalla, *Fundamentals of CAD Simulation*, Kluwer Academic Publisher, Boston, 1988.

[25] S. Selberherr, *Analysis and Simulation of Semiconductor Devices*, Springer-Verlag, Wien, New-York, 1984.

[26] C. M. Snowden, *Semiconductor Device Modeling*, Peter Peregrinus Ltd., London, 1988.

[27] K. M. Cham, S. Y. Oh, D. Chin, J. L. Moll, K. Lee, and P. V. Voorde, *Computer-Aided Design and VLSI Device Development*, 2nd Ed., Kluwer Academic Publisher, Boston, 1988.

[28] S. Selberherr, A. Schutz, and H. W. Potzel, 'MINIMOS—a two-dimensional MOS transistor analyzer, IEEE, ED-27, pp. 1540–1550 (1980); See also 'MINIMOS 3: A MOSFET simulator that includes energy balance', ibid, ED-34, pp. 1074–1078 (1987).

[29] M. R. Pinto, C. S. Rafferty, and R. W. Dutton, 'PISCES-II: Poisson and continuity equation solver', Stanford Electronic Lab. Tech. Rep., Sept. 1984.

[30] E. M. Buturla, P. E. Cottrell, B. M. Grossman, and K. A. Salsburg, 'Finite-element analysis of semiconductor devices: The FIELDAY program', IBM J. Res. Dev., 25, pp. 131–146 (1981).

[31] T. Toyabe, H. Masuda, Y. Aoki, H. Shukuri, and T. Hagiwara, 'Three-dimensional device simulator CADDETH with highly convergent matrix solution algorithm', IEEE Trans. Computer-Aided Design, CAD-4, pp. 482–488 (1985).

[32] J. R. Brews, 'Physics of the MOS transistor', in *Silicon Integrated Circuits* (D. Kahng, Ed.), pp. 1–120, Applied Solid-State Science Series, Supplement 2A, Academic Press, New York, 1981.

[33] Y. P. Tsividis, *Operation and Modeling of the MOS Transistor*, McGraw-Hill Book Company, New York, 1987.

[34] H. J. Park, P. K. Ko and C. Hu, 'A charge sheet capacitance model of short channel MOSFET's for SPICE', IEEE Trans. Compter-Aided Design, CAD-10, pp. 376–389 (1991).

[35] A. R. Boothroyd, S. W. Tarasewicz, and C. Slaby, 'MISNAN—A physically based continuous MOSFET model for CAD applications', IEEE Trans. Compter-Aided Design, CAD-10, pp. 1512–1529 (1991).

[36] P. Antognetti and G. Massobrio, Eds., *Semiconductor Device Modeling with SPICE*, McGraw-Hill Book Company, New York, 1988.

[37] D. A. Divekar, *FET Modeling for Circuit Simulation*, Kluwer Academic Publisher, Boston, 1988.

[38] H. C. de Graaff and F. M. Klaassen, *Compact Transistor Modelling for Circuit Design*, Springer-Verlag Wien, New York, 1990.

[39] T. Shima, H. Yamada, and R. L. M. Dang, 'Table look-up MOSFET modeling system using 2-D device simulator and monotonic piecewise cubic interpolation', IEEE Trans. Computer-Aided Design, CAD-2, pp. 121–126 (1983).

[40] G. Bischoff and J. P. Krusius, 'Technology independent device modeling for simulation of integrated circuits for FET technologies', IEEE Trans. Computer-Aided Design, CAD-4, pp. 99–110 (1985).

[41] J. A. Barby, J. Vlach, and K. Singhal, 'Polynomial splines for MOSFET model approximation', IEEE Trans. Computer-Aided Design, CAD-7, pp. 557–565 (1988).

[42] T. Shima 'Table look-up MOSFET capacitance model for short channel devices', IEEE Trans. Computer-Aided Design, CAD-5, pp. 624–632 (1986).

[42a] R. F. Vogel, 'Analytical MOSFET model with easily extracted parameters', IEEE Trans. Computer-Aided Design, CAD-4, pp. 127–134 (1985).

[43] M. G. Buchler, 'Microelectronic test chips for VLSI electronics', in 'VLSI Electronics: Microstructure Science' (N. G. Einspruch, Ed.), Vol. 6, Chap. 9, pp. 529–576, Academic Press Inc., New York, 1986.

[44] K. Doganis and D. L. Scharfetter, 'General optimization and extraction of IC device model parameters', IEEE Trans. Electron Devices, ED-30, pp. 1219–1228 (1983).

[45] W. Maes, K. M. De Meyer, and L. H. Dupas, 'SIMPAR: A versatile technology independent parameter extraction program using new optimized fit strategy', IEEE Trans. Computer-Aided Design, CAD-5, pp. 320–325 (1986).

[46] M. S. Sharma and N. D. Arora, 'OPTIMA: A nonlinear model parameter extraction

program with statistical confidence region algorithms', IEEE Trans. Computer-Aided Design, CAD-12, May (1993).

[47] P. Yang and P. K. Chatterjee, 'SPICE modeling for small geometry MOSFET circuits', IEEE Trans. Computer-Aided Design, CAD-1, pp. 169–182 (1982).

[48] H. B. Bakoglu, *Circuits, Interconnects and Packaging for VLSI*, Addison-Wesley Publishing Co., Reading MA, 1990.

Review of Basic Semiconductor and *pn* Junction Theory 2

This chapter reviews some of the basics of semiconductor theory that are necessary for an understanding of the development of the device models which follows. Also reviewed is *pn* junction theory as its behavior is basic to the operation of transistors. The review is brief and covers only those topics which have direct relevance to MOS VLSI circuits. For more exhaustive treatments, the reader is referred to textbooks on the subject [1]–[12].

2.1 Energy Band Model

The starting material in the fabrication of MOS devices and integrated circuits (IC) is silicon in the crystalline form. The silicon wafers are cut parallel to either the $\langle 111 \rangle$ or $\langle 100 \rangle$ planes with $\langle 100 \rangle$ material being the most commonly used. This is largely due to the fact that $\langle 100 \rangle$ wafers, during processing, produce the lowest charges at the oxide-silicon interface and higher mobility [13]. Polycrystalline silicon (polysilicon) is also extensively used in IC technology as a conductor, contacts or gate in MOS devices [14]. This material is structurally more complex than single crystal silicon. It consists of many small regions, each having well defined structure but differing from its neighboring regions. For circuit model purposes we can treat polycrystalline silicon as being crystalline in nature.

In a silicon crystal each atom has four valence electrons to share with its four nearest neighboring atoms. The valence electrons are shared in a paired configuration called a *covalent bond*. It is predicted by quantum mechanics that the electrons of an isolated atom may exist only in certain discrete energy levels or orbitals which are characterized by specific values of the quantum numbers. When two atoms approach one another the levels must split so that there will be energy levels to accommodate all the electrons of the system. When the system has a large number of atoms, as in the case of crystalline material, the higher energy levels tend to merge into two

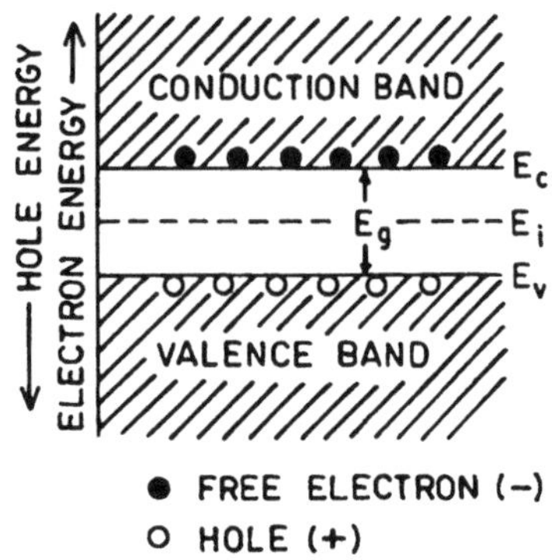

Fig. 2.1 Energy band diagram of a semiconductor (silicon)

separate bands of allowed energies, called the *valence band* and the *conduction band*. The energy levels in the valence bands are mostly filled with electrons forming the covalent bonds. The energy levels in the conduction bands are nearly empty. Electrons which occupy the energy levels in the conduction band are called *free electrons*, (or *conduction electrons*).

The very closely spaced energy levels in the valence and conduction bands are often separated by a energy range where there are no allowed quantum states or energy levels, known as the *energy gap* E_g (or *band gap*). This energy gap between the two allowed energy bands is often referred to as the *forbidden band* or *forbidden gap*. Although the energy is a complex function of momentum in three dimensions and there are so many energy levels, and so many electrons, the energy band picture will be tedious to draw out if all the energy levels are shown. Thus, only the edge levels of each of the allowed energy bands are shown in the energy band diagram (see Figure 2.1). *The electron energy is considered a positive quantity and is plotted upward on the energy-band diagram.* If E_c and E_v are the energy levels for the lower edge of the conduction band and upper edge of the valence band respectively, then the band gap E_g is[1]

$$E_g = E_c - E_v \quad \text{(eV)}. \tag{2.1}$$

[1] Note that the unit of E_g is electron-Volt eV (or qV). By definition, an electron-Volt is the energy an electron of charge q acquires when it moves through a potential difference of 1 Volt (V). When E_g in eV is divided by q (in units of electron charge, not Coulombs), the charge q of the electron cancels out, and the result is Volts (V). Thus E_g and E_g/q have *the same numerical value but different units.* In general any physical quantity expressed in eV can be converted into Volts by simply dividing the quantity by the charge q and vice versa. If E_g is in Joules, then E_g/q is in Volts where $q = 1.602 \times 10^{-19}$ C.

When a valence electron is given sufficient additional energy ($\geq E_g$), it can break out of the chemical bonding state and become a free electron that moves about freely in the lattice. A *hole* (the absence of an electron) is left where the electron was bonded. Since net positive charge is now associated with the atom from which the electron broke away, a hole is associated with a positive charge. Note that *both the electron and the hole are generated simultaneously*. In the energy band diagram holes are normally represented as circles and electrons as dots.

Experimental results show that the *band gap energy E_g of most semiconductors, including silicon, decreases as the temperature increases*, because the crystal lattice spacing increases due to the thermal expansion. The temperature dependence of the band gap E_g for silicon can be modeled using the following polynomial equation [15]

$$E_g(T)=\begin{cases} 1.206-2.73\times 10^{-4}T & \text{(2.2a)} \\ \quad (\text{for } T\geq 250\,\text{K}) & \\ 1.1785-9.025\times 10^{-5}T-3.05\times 10^{-7}T^2 & \text{(2.2b)} \\ \quad (\text{for } 300\,\text{K}>T>170\,\text{K}) & \\ 1.17+1.059\times 10^{-5}T-6.05\times 10^{-7}T^2 & \text{(2.2c)} \\ \quad (\text{for } T\leq 170\,\text{K}) & \end{cases}$$

where T is the temperature in Kelvin (K). Equation (2.2a) fits the experimental data to within 1 meV in the temperature 250–415 K, while at temperatures below 250 K, E_g is modeled to within 2 meV [15]. Note that both Eqs. (2.2a) and (2.2b) are accurate for $250 < T < 300$ K. Although Eq. (2.2) has been used for low-temperature device modeling work [16, 17], the circuit simulator SPICE uses the following equation for E_g [18]

$$E_g = 1.160 - \frac{7.02\times 10^{-4}T^2}{1108+T} \quad \text{(eV)} \quad \text{(SPICE)} \tag{2.3}$$

giving $E_g = 1.115$ eV at 300 K, which is somewhat lower than the more accurate value of 1.124 eV predicted by Eq. (2.2).

2.2 Intrinsic Semiconductor

A pure, single crystal semiconductor in which all the electrons in the conduction band are thermally excited from the valence band is called an *intrinsic semiconductor*. In other words, in an intrinsic semiconductor, at a given temperature, the number of holes in the valence band equals the number of electrons in the conduction band. Thus if n and p are the free electron and hole concentrations (per cm^3) respectively, then

$$n = p = n_i, \quad (\text{cm}^{-3}) \tag{2.4a}$$

Table 2.1. *Effective mass ratios for silicon at 300 K*

	Density-of-states effective mass m_n^*/m	Conduction effective mass m_p^*/m
Electrons	1.08	0.26
Holes	0.81	0.386

or,

$$np = n_i^2 \tag{2.4b}$$

where n_i by definition is the free electron (or hole) concentration in intrinsic silicon, often called the *intrinsic carrier concentration.*

Effective Mass of Electron and Hole. The electrons in the conduction band and the holes in the valence bands move freely throughout the crystal as if they were free particles, suffering only occasional scattering by impurities and defects present in the crystal. Thus electrons and holes are analogous to electrons moving in a vacuum. The difference is the presence of the potential or the Coulomb force experienced by the electrons due to the charged atomic cores of host atoms. These charged atomic cores are located on a regular lattice, giving rise to a periodic potential energy; in vacuum there is no such periodic potential.

The effect of the periodic potential on the motion of electrons in the conduction band and holes in the valence band is represented by the *effective masses* of the *electrons* (m_n^*) and *holes* (m_p^*) respectively, and by the equivalent positive charge of a hole. It should be pointed out that effective mass is not a simple scalar quantity. For a given material and carrier type there are several effective masses encountered in practice [1]–[10]. Further, the effective mass required to calculate carrier (electron and hole) concentration, called the *density of states*[2] effective mass, is different from the conductivity effective mass required to calculate carrier mobility. These effective masses are function of temperature as well. There is large variation in the values of m_n^* and m_p^* reported in the literature [15]. The commonly used values for the effective mass for electrons and holes at room temperature are summarized in the following Table 2.1 [6].

Intrinsic Carrier Concentration. According to Barber [19], who has reviewed and correlated the theoretical and experimental data on n_i for

[2] Density of states is the total number of energy levels per unit volume which are available for possible occupation by electrons.

silicon we have

$$n_i = 3.1 \times 10^{16} T^{3/2} \exp\left(-\frac{1.206}{2kT}\right) \quad (\mathrm{cm}^{-3}) \tag{2.5}$$

where 1.206 is the extrapolated zero-degree band gap energy $E_g(0)$ [cf. Eq. (2.2a)], k is the Boltzmann constant ($= 8.62 \times 10^{-5}$ eV/K) and T is the temperature in K. The term kT is called *Boltzmann factor*. Since it has the dimension of energy, it is often called the *thermal energy* and the corresponding factor kT/q (V), the *thermal voltage* which we will denote by V_t. The value of thermal energy at 300 K is 0.02586 eV. At T = 300 K, Eq. (2.5) yields $n_i = 1.19 \times 10^{10}\,\mathrm{cm}^{-3}$. Recently Green [15] has reported that at 300 K a more accurate value is $n_i = 1.08 \times 10^{10}\,\mathrm{cm}^{-3}$. Note that n_i *increases rapidly with temperature, doubling roughly every* 8 °C.[3]
Another expression which has been often used for n_i calculation as a function of temperature is

$$n_i = 3.9 \times 10^{16} T^{3/2} \exp\left[-\frac{E_g(T)}{2kT}\right] \quad (\mathrm{cm}^{-3}). \tag{2.6}$$

If $n_i(T_0)$ is the value of n_i at the nominal or reference temperature T_0 (say 300 K), then using (2.6), n_i at any other temperature T could be written as

$$n_i(T) = n_i(T_0)\left(\frac{T}{T_0}\right)^{3/2} \exp\left[-\frac{E_g(T)}{2kT} + \frac{E_g(T_0)}{2kT_0}\right] \quad (\mathrm{cm}^{-3}) \tag{2.7}$$

where $E_g(T)$ is given by Eq. (2.2). The above equation is used in SPICE for calculating n_i at any temperature T with $n_i = 1.45 \times 10^{10}\,\mathrm{cm}^{-3}$ at T = 300 K.

2.2.1 *Fermi Level*

The number of carriers available for conduction determines the electrical properties of a semiconductor. This number is found from the density of allowed states and the probability that these states are occupied. The probability that an available state with energy E is occupied by an electron under thermal equilibrium conditions is given by the Fermi–Dirac

[3] At 77 K n_i for silicon is $\sim 10^{-20}\,\mathrm{cm}^{-3}$, while at 400 K its value is $\sim 10^{12}\,\mathrm{cm}^{-3}$.

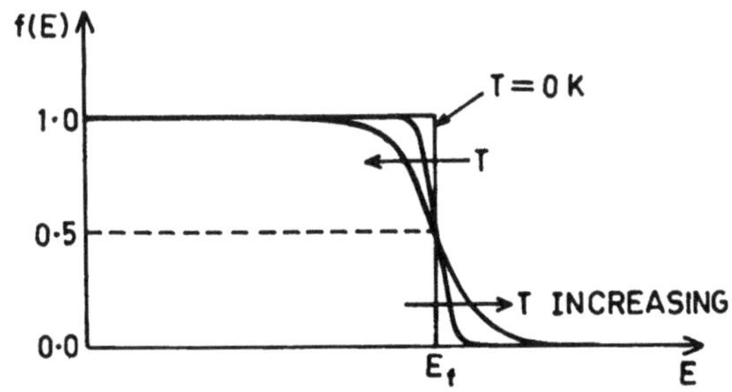

Fig. 2.2 A Fermi-Dirac distribution function

probability density function $f(E)$, also called the Fermi function [1]–[10]

$$f(E) = \frac{1}{1 + \exp\left(\dfrac{E - E_f}{kT}\right)} \tag{2.8}$$

where E_f is the *Fermi energy* or *Fermi level* defined as *the energy level at which the probability of finding an electron, for $T > 0$ K, is exactly one-half.* Note that the Fermi level is a purely mathematical parameter and provides a reference with which other energies can be compared. When $E = E_f$ we have $f(E) = 1/2$ which means that the electron is equally likely to have an energy above the Fermi level as below it. At $T = 0$ K, $f(E) = 1$ indicating thereby that the probability of finding an electron below E_f is unity and above E_f is zero. In other words all energy levels below E_f are filled and all energy levels above E_f are empty. As T is increased above zero, the function $f(E)$ changes as shown in Figure 2.2. Thus, the probability that energy levels above E_f are filled increases with temperature. It is important to note that the Fermi function (or Fermi energy) *applies only under equilibrium conditions.*[4]

The Fermi level can be considered to be the *chemical potential* for electrons and holes. Since the condition for any system in equilibrium is that the chemical potential must be constant through out the system, it follows that the *Fermi level must be constant throughout a semiconductor in equilibrium.* The Fermi level in intrinsic silicon, often referred to as *intrinsic Fermi level* E_i, is only 0.0073 eV below midgap at T = 300 K. Thus for all practical purposes it can be assumed that E_i is in the middle of the energy gap.

---

[4] Here 'equilibrium' means no applied voltage, no applied external fields or thermal gradients.

For all energy levels higher than $3kT$ above E_f the function $f(E)$ can be approximated by

$$f(E) = \exp\left(-\frac{E - E_f}{kT}\right) \tag{2.9}$$

which is identical to the Maxwell–Boltzmann density function for classical gas particles. For most device applications, the function $f(E)$ given by Eq. (2.9) is a good approximation.

2.3 Extrinsic or Doped Semiconductor

When elemental impurities called *dopants* are added to silicon,[5] free carrier concentration of intrinsic silicon changes and the resulting silicon is called *doped* or *extrinsic* silicon. The most commonly used dopants in integrated circuit technology are boron(B), phosphorous(P), and arsenic(As). If the dopants are phosphorous or arsenic they are called *donor atoms*, since they donate an electron to the crystal lattice, and the doped silicon is called n-type material that contains excess electrons. However, if the dopant is boron, it is called an *acceptor atom*, since it can be thought of as accepting an electron from the valence band, and the doped silicon is called p-type that contains excess holes. In terms of energy band diagrams, donors add allowed electron states in the band-gap close to the conduction band edge as pictured in Figure 2.3a; acceptors add allowed states just above the valence band edge as shown in Figure 2.3b. Also shown in this figure are positions of the Fermi level due to donors (Fig. 2.3c) and acceptors (Fig. 2.3d).

It is possible to dope silicon so that $p = n$. Material of this type is called *compensated* silicon. In practice, however, one impurity dominates so that semiconductor is either n-type or p-type. A semiconductor[6] is said to be *nondegenerate*, if the Fermi level lies in the band gap more than a few kT ($\sim 3kT$) from either band edge. Conversely, if the Fermi level is within a few kT ($\sim 3kT$) of either band edge, the semiconductor is said to be *degenerate*. In the nondegenerate case, the carrier concentration obeys Maxwell–Boltzmann statistics (2.9). However, for the degenerate case where the dopant concentration is in excess of approximately $10^{18}\,\text{cm}^{-3}$ (heavy

[5] The silicon crystal contains 5×10^{22} atoms/cm^3 [= Avogadro number (6.02×10^{23}) × Density (2.33)/Gram Molecular Weight (28.09)]. The doping concentration used in the devices ranges from 10^{14}–$10^{20}\,\text{cm}^{-3}$ or from less than one atom in hundred million to a fraction of a percent.

[6] Through out this book the word semiconductor and silicon are used interchangeably.

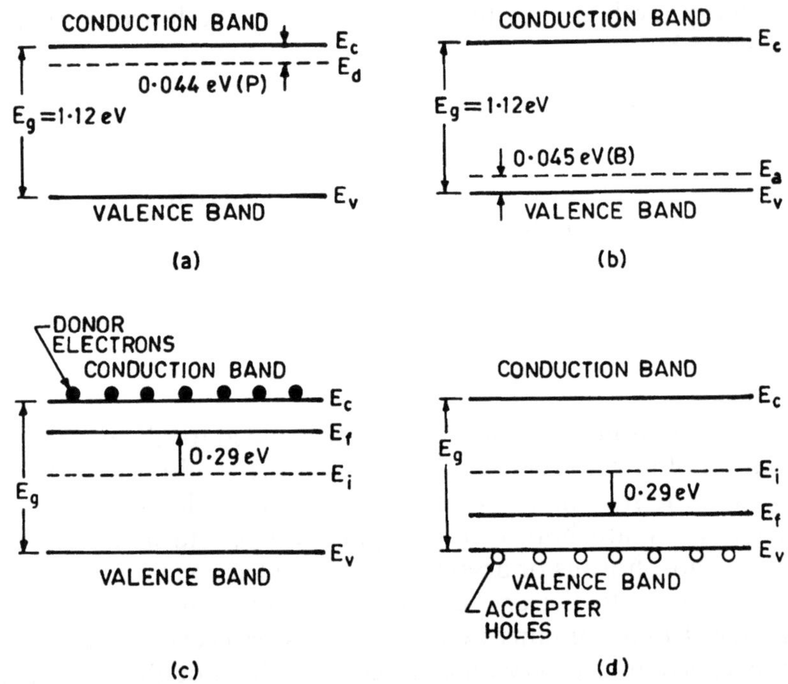

Fig. 2.3 Energy-band diagram representation of (a) donor level E_d (phosphorous, P) in silicon (b) acceptor level E_a (boron, B) (c) Fermi level with phosphorous doping concentration of $10^{15}\,\text{cm}^{-3}$ and (d) Fermi level with boron doping concentration of $10^{15}\,\text{cm}^{-3}$

doping)[7] one must use the Fermi–Dirac distribution function given by Eq. (2.8). In what follows, unless otherwise specified, we will assume the semiconductor to be nondegenerate.

In an *n*-type nondegenerate semiconductor the Fermi level E_f (or Fermi potential $\phi_f = -E_f/q$) lies *above* the intrinsic level E_i (or potential $\phi = -E_i/q$)[8] by an amount given by the following equation (see Figure 2.3c),

$$n = n_i \exp\left(\frac{E_f - E_i}{kT}\right) = n_i \exp\left[\frac{q}{kT}(\phi - \phi_f)\right] \quad (\text{cm}^{-3}) \tag{2.10a}$$

[7] The carrier concentration greater than $10^{18}\,\text{cm}^{-3}$ (heavy doping) is normally represented as n^+(electrons) or p^+(holes).

[8] Note the negative sign in the $\phi = -E/q$ relation; when the electron energy plotted upwards is positive, the *positive potential must be plotted downwards because of the negative electron charge.*

while in a p-type semiconductor the Fermi level E_f (Fermi potential ϕ_f) lies *below* the intrinsic level E_i (potential ϕ) by an amount (see Figure 2.3d)

$$\boxed{p = n_i \exp\left(\frac{E_i - E_f}{kT}\right) = n_i \exp\left[\frac{q}{kT}(\phi_f - \phi)\right] \quad (\mathrm{cm}^{-3}).} \tag{2.10b}$$

The Eqs. (2.10a) and (2.10b) for n and p, respectively, are often referred to as *Boltzmann's relations.*

At room temperature, the available thermal energy is sufficient to ionize nearly all acceptor and donor atoms due to their low ionization energies. Hence it is a safe approximation to say that *in nondegenerate silicon at room temperature*

$$n \approx N_d \quad (n\text{-type}) \tag{2.11a}$$

$$p \approx N_a \quad (p\text{-type}) \tag{2.11b}$$

where N_d is the concentration of donor atoms (cm^{-3}) and N_a is the concentration of acceptor atoms (cm^{-3}). In an n-type material, where $N_d \gg n_i$, electrons are *majority carriers* whose concentration is given by Eq. (2.11a), while the hole concentration p_n is[9]

$$p_n \approx \frac{n_i^2}{N_d} \quad (\mathrm{cm}^{-3}) \tag{2.12}$$

remembering that $pn = n_i^2$ [cf. Eq. (2.4b)]. The hole concentration p_n is much smaller than n_n. Thus holes are *minority carriers* in an n-type semiconductor. Similarly, in a p-type semiconductor where $N_a \gg n_i$, holes are majority carriers given by Eq. (2.11b), while electron concentration n_p is given by

$$n_p \approx \frac{n_i^2}{N_a} \quad (\mathrm{cm}^{-3}). \tag{2.13}$$

Since $n_p \ll p$, electrons are minority carriers in p-type semiconductor. Consequently, we often use the terminology of majority carriers and minority carriers.

Taking E_i as the zero reference level and making use of Eq. (2.11a) in (2.10a) we can write the electron concentration n in terms of ϕ_f, for an n-type semiconductor, as

$$n = N_d = n_i \exp\left(-\frac{q\phi_f}{kT}\right). \tag{2.14a}$$

---

[9] It is common practice to represent carrier concentrations with the subscript denoting the type of semiconductor. Thus p_n denotes hole concentration p in an n-type semiconductor and likewise n_n (or simply n) denotes electron concentration n in an n-type semiconductor.

Similarly, the hole concentration p in a p-type semiconductor becomes

$$p = N_a = n_i \exp\left(\frac{q\phi_f}{kT}\right). \tag{2.14b}$$

Rearranging Eq. (2.14a) or (2.14b) for ϕ_f we get

$$\boxed{\phi_f = \pm \frac{kT}{q} \ln\left(\frac{N_b}{n_i}\right) \quad \text{(V)}} \tag{2.15}$$

where the (+) sign is for p-type semiconductors ($N_b = N_a$) and the (−) sign is for n-type semiconductors ($N_b = N_d$). Note that ϕ_f *is not only a function of carrier concentration, but is also dependent on temperature.* The variation of Fermi potential $|\phi_f|$ at $T = 300$ K as a function of substrate concentration $N_b (= N_a$ or $N_d)$ is shown in Figure 2.4. As the temperature increases, n_i increases [cf. Eq. (2.5)] and therefore ϕ_f decreases (the increase in n_i with temperature is much faster than the increase in thermal energy kT). Thus, with an increase of temperature, the Fermi level approaches the midgap position i.e. the intrinsic Fermi level; showing thereby that the semiconductor becomes intrinsic at high temperature. Thus *doped or extrinsic silicon will become intrinsic if the temperature is high enough.* The temperature at which this happens depends upon the dopant concentration. *When the material becomes intrinsic, the device can no longer function* and therefore the intrinsic region is avoided in device operation.

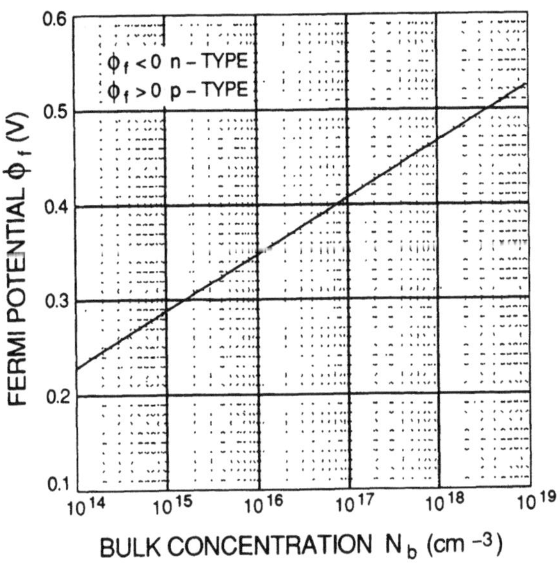

Fig. 2.4 Fermi potential ϕ_f in silicon as a function of substrate concentration N_b

The temperature coefficient of ϕ_f can be obtained by differentiating Eq. (2.15) giving

$$\frac{d\phi_f}{dT} = \frac{1}{T}\left[\phi_f - \left(0.603 + \frac{3}{2}\frac{kT}{q}\right)\right] \tag{2.16}$$

where we have made use of Eq. (2.5) for n_i. This gives $d\phi_f/dT \sim -1\,\mathrm{mV/K}$. If we use Eq. (2.7) for n_i, ϕ_f at any temperature T can be written in terms of its value at a nominal temperature T_0 as

$$\boxed{\phi_f(T) = \left(\frac{T}{T_0}\right)\phi_f(T_0) - \frac{kT}{q}\left[1.5\ln\left(\frac{T}{T_0}\right) + \left\{-\frac{E_g(T)}{2kT} + \frac{E_g(T_0)}{2kT_0}\right\}\right] \quad (\mathrm{V}).} \tag{2.17}$$

This is the equation used in SPICE for the temperature dependence of ϕ_f.

2.3.1 Generation-Recombination

Under thermal equilibrium, the condition $pn = n_i^2$ is maintained. This condition may be disturbed by the introduction of free carriers (only electrons, only holes, or electron-hole pairs) in the semiconductor. This process of introducing additional carriers (*excess carriers*) is called carrier *injection* and can occur in different ways (optical, electrical, etc). Note that the injection refers to any increments of carriers due to a nonthermal source, irrespective of the nature of this source. If the injected carrier density is small compared to the majority carrier density at equilibrium, so that the latter remains essentially unchanged while the minority carrier density is equal to the excess carrier density, then the process is called *low-level injection.* If the injected carrier density is comparable to or exceeds the majority carrier density before injection, then it is called *high level injection.* Although in semiconductor device operation it is generally low level injection which is important, one also encounters high level injection.
Let us take an example. Suppose in n-type silicon, $N_d = 10^{16}\,\mathrm{cm}^{-3}$ then from Eq. (2.11) the majority carrier concentration $n_{n0} = 10^{16}\,\mathrm{cm}^{-3}$, while from Eq. (2.12) minority carrier concentration[10] $p_{n0} = 2 \times 10^4\,\mathrm{cm}^{-3}$. Now

---

[10] The majority and minority carrier concentrations calculated using Eqs. (2.11)–(2.13) are thermal equilibrium values of the carrier concentration and are often denoted by adding the subscript 0 to any other subscript in order to distinguish them from the new carrier concentration after injection. Thus, thermal equilibrium values of the majority electrons and minority holes in a n-type silicon will be represented by the symbols n_{n0} and p_{n0} respectively.

suppose $10^8\,cm^{-3}$ electron-hole pairs are injected into the material, then the new n_n is still $10^{16}\,cm^{-3}$ while p_{n0} becomes $10^8\,cm^{-3}$. Thus, while the majority carrier concentration n_n remains unchanged by carrier injection, the minority carrier concentration p_n has risen significantly. This is low level injection.

While generation represents the process whereby electrons and holes are created, recombination is the process whereby electron and holes are annihilated or destroyed. If G and R represents generation and recombination rate ($cm^{-3}\cdot s^{-1}$), respectively, then at thermal equilibrium $R = G$. Crystalline defects due to dislocation in the crystal lattice, which could be the result of imperfect fabrication techniques, and impurities that form electronic states deep in the energy gap, assist recombination of electrons and holes. Here the word *deep* indicates that the states are away from the band edges and near the center of the energy gap. These deep states are commonly referred to as *recombination-generation* (or simply *recombination*) *centers* or *traps*[11]. Such recombination centers are usually unintentional impurities, which are not necessarily ionized at room temperature. These deep level impurities have concentrations far below the concentration of donor and acceptor impurities, which have shallow energy levels (see Figure 2.3a,b). Gold is a deep level impurity intentially used in silicon to increase the recombination rate. This recombination via deep level impurity or trap is often referred to as *indirect recombination* in order to distinguish it from *direct recombination* wherein carriers make a direct transition between the conduction and valence bands. The probability of direct recombination is very low in silicon due to the nature of the silicon band gap structure. This indirect recombination process was originally proposed by Shockley and Read [22] and independently suggested by Hall [23] and therefore is often referred to as *Shockley-Read-Hall* (SHR) recombination.

The other recombination process in silicon that does not depend on deep level impurities and which sets an upper limit on lifetime is *Auger recombination*, in which electrons and holes recombine without trap levels and the released energy (of the order of the energy gap) is transferred to another majority carrier (a hole in a *p*-type and electron in a *n*-type silicon). Usually Auger recombination is important when the carrier concentration is very large ($> 5 \times 10^{18}$) as a result of high doping or high level injection as is the case in bipolar transistors [12].

Recombination is not instantaneous. It takes a finite and measurable time, the average of which is called the *lifetime* denoted by the symbol τ. Physically τ represents the average time an electron remains free before it recombines

[11] Strictly speaking recombination centres and traps are not the same. Traps implies that the carrier is captured for a definite time after which it is either re-emitted back to where it came from or recombines with other carries. However, for our discussion we will treat them as being the same.

with a hole. If N_t is the density of the recombination centers corresponding to the energy level E_t then it can be shown that [1]–[10]

$$\tau = \frac{1}{CN_t} = \frac{1}{c_0 v_{th} N_t} \quad \text{(sec)} \tag{2.18}$$

where C is the proportionality constant for the capturing process and describes the effectiveness of the deep state in capturing an electron or hole. The constant C is often expressed in terms of *capture cross-section* c_0 such that $C = c_0 v_{th}$, where $v_{th} = \sqrt{3kT/m^*}$ is the thermal velocity. Here m^* is effective mass of the carriers. At room temperature (300 K) $v_{th} \sim 10^7$ cm/s. From Eq. (2.18) it is evident that the higher the N_t and/or larger the capture cross-section, the shorter the lifetime.

The lifetime of most interest in semiconductor devices is the *minority carrier lifetime* τ_p for holes in n-type silicon and τ_n for electrons in p-type silicon. Thus τ_p(or τ_n) represents the average time an excess minority carrier hole(or electron) will remain free before combining with a majority carrier electron(or hole). Experimentally, it is found that minority carrier lifetime generally decreases with increasing majority carrier concentration as there would be more opportunities for recombinations, apart from being governed by Eq. (2.18). The lifetime generally encountered in silicon is in the range 10^{-4} to 10^{-9} sec depending on the density of the recombination centers and doping concentration.

Lifetime is one of the important parameters that characterizes a semiconductor. Short lifetime generally enhances the undesired leakage currents. Long lifetimes are generally desirable in most cases, including MOSFETs. The carrier lifetime is temperature dependent [17], [21]. To a first approximation it can be assumed to vary as $T^{-1/2}$ through v_{th} dependence.

2.3.2 Quasi-Fermi Level

When carriers are injected into the semiconductor, the equilibrium condition is disturbed. The carrier densities can no longer be described by a constant Fermi level throughout the system. On the energy diagram this nonequilibrium condition is shown by using two different Fermi levels, $\mathscr{F}_n$ for electrons and $\mathscr{F}_p$ for holes, called *quasi-Fermi levels* (also sometime referred to *imref*). Thus quasi-Fermi levels apply when the equilibrium condition is disturbed and expresses the increased probability of finding minority carriers in an otherwise empty level. The quasi-Fermi levels $\mathscr{F}_n$ and $\mathscr{F}_p$ are related to the nonequilibrium carrier concentration in the same way as the Fermi level E_f is related to equilibrium carrier concentration. If φ_n and φ_p are the quasi-Fermi potentials for electrons and holes corresponding to the quasi-Fermi level $\mathscr{F}_n$ and $\mathscr{F}_p$ respectively, then under nonequilibrium condition we have

$$n = n_i \exp\left(\frac{\mathscr{F}_n - E_i}{kT}\right) = n_i \exp\left(-\frac{q\varphi_n}{kT}\right) \quad (\text{cm}^{-3}) \tag{2.19a}$$

$$p = n_i \exp\left(\frac{E_i - \mathscr{F}_p}{kT}\right) = n_i \exp\left(\frac{q\varphi_p}{kT}\right) \quad (\text{cm}^{-3}) \tag{2.19b}$$

so that in equilibrium $\mathscr{F}_n = \mathscr{F}_p = E_f$ and $\varphi_n = \varphi_p = \phi_f$. Note that under nonequilibrium conditions the *pn* product is not equal to the thermal equilibrium value of n_i^2 but is a function of the separation of the two quasi-Fermi levels. This separation is a measure of the deviation from the equilibrium values of free carrier concentrations.

2.4 Electrical Conduction

The current flow in a semiconductor is the result of two different phenomena, namely: (1) the *drift* of carriers (electron and holes), which is caused by the presence of an electric field, and (2) the *diffusion* of carriers which is caused by the concentration gradient in the semiconductor. We will now consider factors involved in both phenomena.

2.4.1 Carrier Mobility

When an electron (or hole) in free space moves under the influence of an electric field, $\mathscr{E}$, it will be accelerated (indefinitely) in proportion to the force of the field $\mathscr{E}$. However, an electron within a semiconductor (in fact any material) *will not* accelerate indefinitely in response to the field. This is because the carriers will collide with various scattering centers such as the atoms of the host lattice (lattice scattering), the impurity atoms (impurity scattering), other carriers (carrier-carrier scattering), etc.[12] These different scatterings tend to redirect the electron momentum, and in many cases tends to dissipate the electron energy gained from the electric field. Under the influence of a uniform electric field, the process of energy gain (from the field) and energy loss (due to the scattering) balance each other and

[12] The lattice scattering is due to the collision between carries and the thermally agitated lattice atoms. It involves transfer of kinetic energy from the carriers to vibrational energy of the host lattice. This energy transfer occurs in discrete quanta of vibrational energy known as *phonons*. For this reason lattice scattering is also known as phonon scattering. The *impurity scattering* is the result of Coulombic attraction or repulsion between charged carriers and the ionized donors and/or acceptors, and therefore also known as Coulombic scattering.

carriers attain a constant average velocity, called *drift velocity. Except for very large fields, drift velocity v of the carriers is directly proportional to the applied field* $\mathscr{E}$. Thus

$$v = \mu\mathscr{E} \quad \text{(cm/s)}. \tag{2.20}$$

The proportionality constant μ is called mobility of the carriers. Physically speaking μ represents the ease with which carriers move in the semiconductor crystal and is given by

$$\mu = \frac{q\tau}{m^*} \quad (\text{cm}^2\,\text{V}^{-1}\,\text{s}^{-1}) \tag{2.21}$$

where m^* is the carrier effective mass (cf. Table 2.1) and τ is the mean free time between the collisions which depends upon different scattering mechanisms. Assuming different scattering mechanisms are independent, the total mobility is determined by combining the mobilities for different scattering mechanisms: μ_L (mobility due to lattice scattering), μ_I (mobility due to ionized impurity scattering) etc, using Mathiessen's rule shown below

$$\frac{1}{\mu} = \frac{1}{\mu_L} + \frac{1}{\mu_I} + \cdots \tag{2.22}$$

Measurements show that electron mobility (μ_n) in n-type silicon is about 3 times greater than the hole mobility (μ_p) in p-type silicon. This difference is partly due to the difference in the effective mass of the electrons and holes and partly because of the difference in their scattering mechanisms. Carrier mobility in bulk silicon is a function of the doping concentrations as would be expected from Eq. (2.21) due to the presence of τ. Mobility as a function of concentration for silicon at 300 K is shown in Figure 2.5. This figure represents the 'best fit' to the measured data and can be generated from the following empirical relation

$$\boxed{\mu = \mu_{\min} + \frac{\mu_{\max} - \mu_{\min}}{1 + (N/N_{\text{ref}})^{\alpha}} \quad (\text{cm}^2/\text{V}\cdot\text{s})} \tag{2.23}$$

where N is the total dopant concentration in the silicon and the four parameters $\mu_{\max}$, $\mu_{\min}$, N_{ref} and α have different values for different types of impurities. Values of these parameters for different dopants are shown in Table 2.2 [6]. Note from Figure 2.5 that there is a monotonic decrease in mobility with increasing concentration. The decrease in μ with increasing N is caused by increased ionized impurity scattering. At low doping, μ approaches the doping independent limiting value set by lattice scattering.

The carrier mobility discussed above is called the *bulk mobility* in order to distinguish it from the so-called *surface mobility* in the channel region

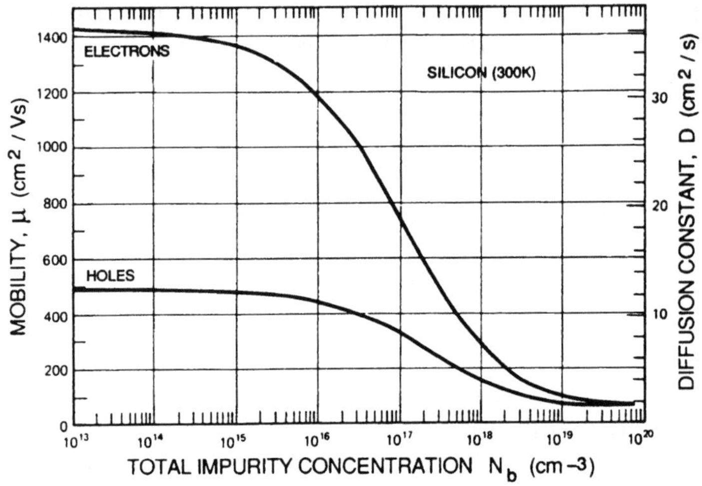

Fig. 2.5 Electron and hole mobilities in silicon at 300 K as a function of total dopant concentration

Table 2.2. *Mobility parameters for different dopants in silicon at 300 K [6]*

Parameter	Arsenic	Phosphorus	Boron
μ_{min}	52.2	68.5	44.9
μ_{max}	1417	1414	470.5
N_{ref}	9.68×10^{16}	9.20×10^{16}	2.23×10^{17}
α	0.680	0.711	0.719

(inversion layer) of a MOSFET. The surface mobility is much lower than the bulk mobility because of the existence of the high electric field normal to the channel and increased scattering due to "surface roughness" at the $Si\text{-}SiO_2$ interface. We will discuss surface mobility in more detail in Section 6.6.

It should be pointed out that the data in Figure 2.5 are applicable when the value of the electric field $\mathscr{E}$ is less than $1\text{–}5 \times 10^4$ V/cm. For higher fields, μ actually decreases from its low field value resulting eventually in the saturation of the drift velocity. This *limiting high-field drift velocity is referred to as the saturation velocity* v_{sat}. For silicon a typical value of $v_{sat} = 1.07 \times 10^7$ cm/s for electron and occurs at an electric field of $\sim 2.0 \times 10^4$ V/cm. The corresponding values for hole is $v_{sat} = 8.34 \times 10^6$ cm/s and $\mathscr{E} \sim 5.0 \times 10^4$ V/cm.

Figure 2.6 shows measured value of the drift velocity for electrons and holes at 300 K in silicon as a function of the applied field $\mathscr{E}$. At low fields, carrier velocity increases linearly with the electric field indicating constant mobility. When the field exceeds about 20 KV/cm, carriers begin to loose energy by scattering with optical phonons and their velocity saturates. As

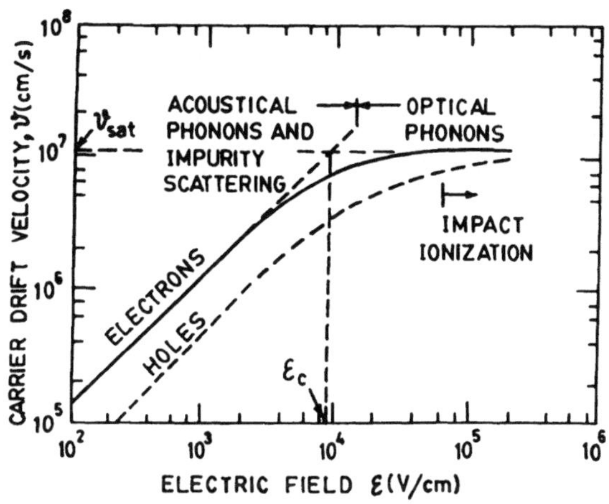

Fig. 2.6 Drift velocities of electrons and holes as a function of applied field showing velocity saturation at high fields

the field exceeds 100 KV/cm carriers gain more energy from the field than they can loose by scattering. Consequently their energy with respect to the bottom of the conduction band (for electrons) or top of the valence band (for holes) begins to increase. The carriers are no longer in thermal equilibrium with the lattice. Since they acquire energy higher than thermal energy (kT) they are called *hot-carriers*. It is these hot-carriers which are responsible for reducing the mobility at high fields. Figure 2.6 can be approximated by the following empirical expression [24]

$$v = v_{sat} \frac{\mathcal{E}/\mathcal{E}_c}{[1 + (\mathcal{E}/\mathcal{E}_c)^{\nu}]^{1/\nu}} \quad \text{(cm/sec)} \tag{2.24}$$

where $\mathcal{E}_c$ is the critical field at which carrier velocity saturates. The parameter v_{sat}, $\mathcal{E}_c$ and ν in the above equation are given in Table 2.3 [24].

When hot-carrier acquire energy in excess of the binding energy of an electron to its parent atom, it can break the covalent bond in collision with the atom and produce an electron-hole pair. *This process of creating electron-hole pairs by collision of energetic electrons (or holes) with parent atoms is called impact ionization.* It is characterized by the *ionization energy*

Table 2.3. *Parameters for field dependence of drift velocity for silicon at 300 K*

Parameter	v_{sat}(cm/sec)	$\mathcal{E}_c$(V/cm)	ν
Electrons	1.07×10^7	6.91×10^3	1.11
Holes	8.34×10^6	1.45×10^4	2.637

E_i (generally $E_i > E_g$, the band gap energy) which is the minimum energy required by the electron or hole to ionize an atom. The electron-hole pair generation rate G due to impact ionization depends upon the *ionization coefficients* for electrons and holes respectively and is *defined as the number of electron-hole pairs generated by an energetic carrier per unit distance traveled.* The ionization coefficients are strong functions of the electric field because the energy necessary for an ionizing collision is imparted to the carrier by the electric field. *The ionization rate for electrons is higher than that for holes by 2–3 orders of magnitude* [17].

Carrier mobilities exhibit a strong and somewhat complex dependence on temperature. This is because the scattering mechanisms are temperature dependent. For design and analysis purposes, the dependence of mobility on the temperature and dopant concentration are described fairly accurately by the following equations [7], [25]

$$\mu_n = 88T_n^{-0.57} + \frac{7.4 \times 10^3 T^{-2.33}}{1 + [(N/1.26 \times 10^{17})T_n^{2.4}]^\alpha} \tag{2.25a}$$

$$\mu_p = 54.3T_n^{-0.57} + \frac{1.36 \times 10^3 T^{-2.23}}{1 + [(N/2.35 \times 10^{17})T_n^{2.4}]^\alpha} \tag{2.25b}$$

where

$$\alpha = 0.88T_n^{-0.146} \text{ and } T_n = T/300$$

T being measured in Kelvin and N is the total dopant density (cm^{-3}) in the silicon.[13] Based on Eq. (2.25) temperature dependence of mobility is shown in Figure 2.7.

The temperature dependence of v_{sat} has been reported for electrons in silicon [24]

$$v_{sat} = \frac{2.4 \times 10^7}{1 + 0.8\exp(T/600)} \quad \text{(cm/sec)} \tag{2.26}$$

which predicts about 20% increase in v_{sat} at 77 K compared to its value at 300 K. However, no such expression exists for holes, due to the fact that it is difficult to access the high-field region where, if any, hole velocity saturation occurs. From a global best fit to the velocity-field curve, it turns out that in the temperature range 245–430 K we may assume [20], [21]

$$v_{sat} = \begin{cases} 10^7 T_n^{-0.87} & \text{(for electrons)} \\ 8.37 \times 10^6 T_n^{-0.52} & \text{(for holes)} \end{cases} \tag{2.26}$$

---

[13] There is a typo in Eqs. (8) and (13) of the original paper [25] where $\alpha = 0.88T_n^{-0.146}$ appears to be a multiplying factor while in fact it is in an exponent factor.

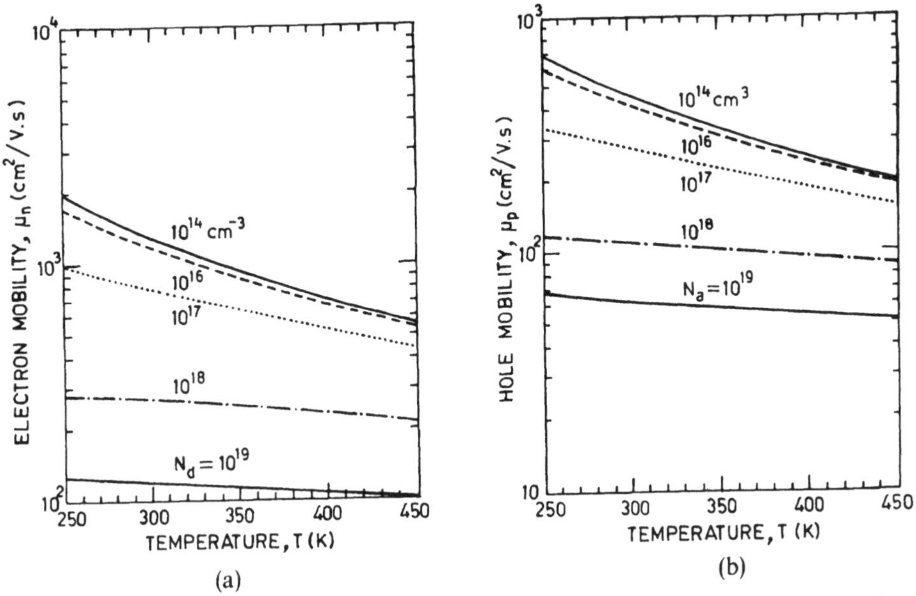

Fig. 2.7 Mobility as a function of temperature in silicon for doping concentration in the range 10^{14} to $10^{19}\,\text{cm}^{-3}$ (a) electrons and (b) holes

Other formulations for temperature dependence of v_{sat} has also been proposed [17].

2.4.2 *Resistivity and Sheet Resistance*

The flow of electrons carrying a charge q at drift velocity v constitute a current, called the *electron drift current*. If there are n free electrons per cm^3 then the drift electron current density[14] $J_{n,\text{drift}}$ is given by

$$J_{n,\text{drift}} = qnv = qn\mu_n\mathscr{E} = \sigma\mathscr{E} \quad (\text{A/cm}^2) \tag{2.28}$$

where we have made use of Eq. (2.20) for v. Since, for a given material μ_n and n are constants, we have lumped the product $q\mu_n n$ into a single parameter called *electron conductivity*[15] σ_n. The *resistivity* ρ of a material is the inverse of conductivity; thus electron resistivity ρ_n is

$$\boxed{\rho_n = \frac{1}{\sigma} = \frac{1}{q\mu_n n} \quad (\Omega\,\text{cm}).} \tag{2.29}$$

[14] Current density, represented by the symbol J, is the current I per unit cross-sectional area A of a conducting medium i.e. $J = I/A$ (A/cm^2). The area A is perpendicular to the current path.

[15] Equation $J = \sigma\mathscr{E}$ is another form of Ohm's law which states that the current density is proportional to the voltage gradient.

Similarly, one can write the hole current density $J_{p,\mathrm{drift}}$ and hole resistivity ρ_p by replacing n and μ_n by p and μ_p respectively in Eqs. (2.28) and (2.29). If the silicon is doped with both donor and acceptors, then resistivity is expressed as

$$\rho = \frac{1}{qn\mu_n + qp\mu_p}. \tag{2.30}$$

Thus the resistivity of the semiconductor depends on the electron and hole concentrations and their mobilities. Empirical curves of resistivity versus bulk impurity concentration are shown in Figure 2.8 for uniformly doped silicon at 300 K based on recent experimental data [26]. These curves differ from Irvin's curves (the most used curves prior to 1982 and given in many textbooks) by as much as 50% for boron doped p-type silicon. For phosphorous doped n-type silicon the difference is only 15%. Curves for n-type are lower than p-doped silicon because electron mobility is higher than hole mobility.

Sheet Resistance. The resistance of a uniform structure of width w, thickness t, and length l is given by

$$R = \rho \frac{l}{tw} \quad (\Omega) \tag{2.31}$$

where ρ is resistivity of the material in Ωcm. In integrated circuits, the diffusion lines are normally uniform in thickness, therefore, we can absorb t into resistivity ρ and define a new variable ρ_s, called *sheet resistance*, which has dimensions of Ohm (Ω). Thus, Eq. (2.31) becomes

$$R = \rho_s \frac{l}{w} \quad (\Omega). \tag{2.32}$$

When $l = w$, the structure becomes square with $R = \rho_s$. Thus *sheet resistance of a layer is the resistance measured between the opposite sides of a square of that layer, regardless of its actual dimensions.* Hence ρ_s is often expressed as $\Omega/\square$ (Ohm per square). Note that the sheet resistance does not depend on the size of the square. Typical sheet resistance of n^+ and p^+ polysilicon layer are 15 and 25 $\Omega/\square$. The resistance of any layer is simply ρ_s times the number of squares in the path of the current. The process parameters which determine the sheet resistance of a layer are the resistivity ρ and thickness (or depth) t of the layer.

The concept of sheet resistance is important in a integrated circuits as it can simplify handling resistance calculations even when the thickness is

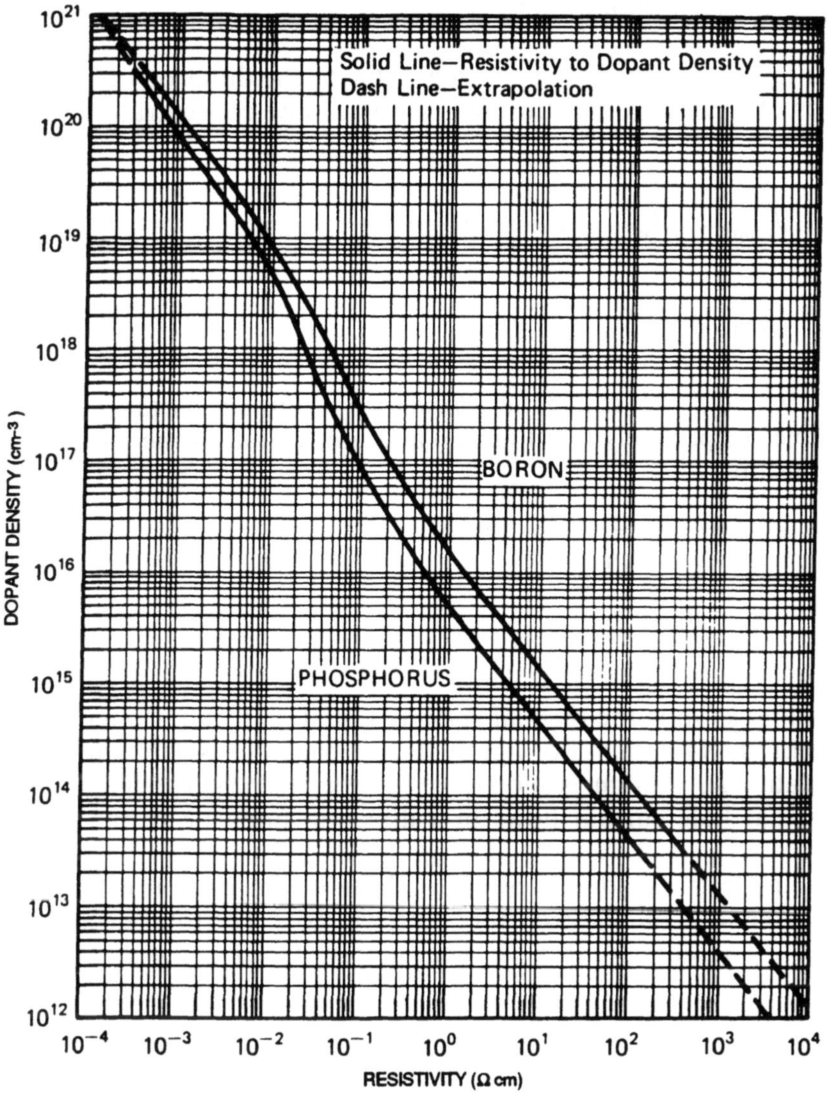

Fig. 2.8 Dopant density versus resistivity at 23°C for silicon doped with phosphorous and boron. (After Muller and Kamins [6].)

not uniform, such as would occur for ion implanted[16] or diffused layers. Since resistivity ρ is a function of carrier concentration and mobility, both of which are functions of temperature, ρ will be temperature dependent.

[16] Ion implantation is a technique of introducing a precise quantity of dopants (impurities) into the substrate.

2.4.3 Transport Equations

The diffusion current density, which is due to the diffusion of carriers from regions of high concentration to regions of low concentration,[17] is given by

$$J_{n,\text{diff}} = qD_n \frac{dn}{dx} \quad \text{(electrons)} \tag{2.33a}$$

$$J_{p,\text{diff}} = -qD_p \frac{dp}{dx} \quad \text{(holes)} \tag{2.33b}$$

where D_n and D_p are called the *diffusivity* or *diffusion constants* for electrons and holes respectively (see Fig. 2.5) and are related to the mobility by the relationship

$$\boxed{\frac{D_n}{\mu_n} = \frac{D_p}{\mu_p} = \frac{kT}{q} \equiv V_t} \tag{2.34}$$

where $V_t = kT/q$ is the *thermal voltage*. Equation (2.34) is often referred to as the *Einstein relation*. For lightly doped silicon ($N \sim 10^{14}\,\text{cm}^{-3}$) at room temperature $D_n = 38\,\text{cm}^2/\text{s}$ and $D_p = 13\,\text{cm}^2/\text{s}$. The negative sign in Eq. (2.33b) shows that hole current flows in a direction opposite to the gradient of holes.

When an electric field is present in addition to a concentration gradient, drift current and diffusion current both will flow. The total electron current density J_n at any point x is then simply $J_n = J_{n,\text{drift}} + J_{n,\text{diff}}$. In other words,

$$\boxed{J_n = qn\mu_n\mathscr{E} + qD_n \frac{dn}{dx} \quad (\text{A/cm}^2).} \tag{2.35a}$$

---

[17] Diffusion current Eq. (2.33) is a direct consequence of Fick's first law of diffusion, which states that the flux of carriers (numbers per sec. passing through a unit area) F is proportional to the concentration gradient dn/dx. Thus

$$F = -D\frac{dn}{dx}$$

where the proportionality constant D is called the diffusivity.

Similarly, the hole current density $J_p(=J_{p,\text{drift}}+J_{p,\text{diff}})$ is given by

$$\boxed{J_p = qp\mu_p\mathscr{E} - qD_p\frac{dp}{dx} \quad (\text{A/cm}^2)} \tag{2.35b}$$

so that the total current density $J = J_n + J_p$. The current equations (2.35a) and (2.35b) are often referred to as *transport equations.*
Under thermal equilibrium no current flows inside the semiconductor and therefore $J_n = J_p = 0$. However, under nonequilibrium conditions J_n and J_p can be written as

$$J_n = -qn\mu_n\frac{d\varphi_n}{dx} \tag{2.36a}$$

$$J_p = -qp\mu_p\frac{d\varphi_p}{dx} \tag{2.36b}$$

and are easily obtained by combining Eqs. (2.35) and (2.19) for n and p respectively (quasi-Fermi potentials).

2.4.4 Continuity Equation

When carriers diffuse through a certain volume of semiconductor, the current density leaving the volume may be smaller or larger depending upon the recombination or generation taking place inside the volume. Let us consider a small length Δx of a semiconductor with cross-sectional area A in the yz plane. The electron current density entering the volume $A\cdot\Delta x$ is $J_n(x)$ while that leaving is $J_n\cdot(x+\Delta x)$. The net increase in the electron concentration per unit time, $\partial n/\partial t$, is the difference between the electron flux per unit volume entering and leaving minus the recombination rate R_n plus the generation rate G_n. That is

$$\frac{\partial n}{\partial t} = -\frac{1}{q}\frac{J_n(x) - J_n(x+\Delta x)}{\Delta x} - R_n + G_n \tag{2.37}$$

in the limit of $\Delta x \to 0$, we get

$$\boxed{\frac{\partial n}{\partial t} = \frac{1}{q}\frac{\partial J_n}{\partial x} - R_n + G_n.} \tag{2.38a}$$

Similarly for holes we have

$$\boxed{\frac{\partial p}{\partial t} = -\frac{1}{q}\frac{\partial J_p}{\partial x} - R_p + G_p} \tag{2.38b}$$

where R_p and G_p are recombination and generation rates for holes. These equations are called *continuity equations* for electrons and holes respectively and describe the time dependent relationship between current density, recombination and generation rates and distance. They are used for solving transient phenomena and diffusion with recombination-generation of carriers.

2.4.5 Poisson's Equation

Poisson's equation is a very general differential equation, based on Maxwell's field equations, that relates the charge density ρ to the electric field $\mathscr{E}$ (or potential ϕ). When the semiconductor as a whole is charge neutral i.e. it exhibits no net charge, ρ must be zero. However, when space charge neutrality does not apply Poisson's equation must be invoked. Mathematically Poisson's equation is stated as follows:

$$\frac{d\mathscr{E}}{dx} = \frac{\rho(x)}{\epsilon_0 \epsilon_{si}} \tag{2.39}$$

where ρ is net space charge density (Coul/cm^3), $\epsilon_0 (= 8.854 \times 10^{-14}\ \mathrm{F/cm^2})$ is permittivity of free space, and $\epsilon_{si} (= 11.7)$ is relative permittivity of silicon. If n and p are free electron and hole concentrations and N_a^- and N_d^+ are concentrations of ionized acceptors and donors respectively in the space charge region, we have

$$\frac{d\mathscr{E}}{dx} = \frac{q}{\epsilon_0 \epsilon_{si}} [p(x) - n(x) + N_d^+(x) - N_a^-(x)]. \tag{2.40}$$

Remembering that $\mathscr{E} = -d\phi/dx$ and since at room temperature $N_a^- \approx N_a$ and $N_d^+ \approx N_d$, the Poisson equation in terms of potential ϕ can be written as

$$\boxed{\frac{d^2\phi}{dx^2} = -\frac{q}{\epsilon_0 \epsilon_{si}} [p(x) - n(x) + N_d(x) - N_a(x)].} \tag{2.41}$$

The current equation (2.35), the continuity equation (2.38) and the Poisson equation (2.41) are one-dimensional equations; however, they can easily be extended to three-dimensional equations (see section 6.1).

2.5 *pn* Junction at Equilibrium

The most remarkable property of the *pn* junction is that it *rectifies*, i.e. it allows current to flow in one direction but not in the opposite direction. When the *p*-side of the junction is made positive with respect to the *n*-side by applying an external voltage V_f, as shown in Figure 2.9a, the junction is said to be *forward biased*, giving rise to current that increases rapidly as the voltage increases. However when we reverse the polarity, i.e. a negative voltage is applied to the *p*-side with respect to the *n*-side, the junction is said to be *reverse biased* and in this case virtually no current flows initially (see Fig. 2.9b). As the reverse bias is increased the current remains negligible until a critical voltage is reached when the current suddenly increases. This

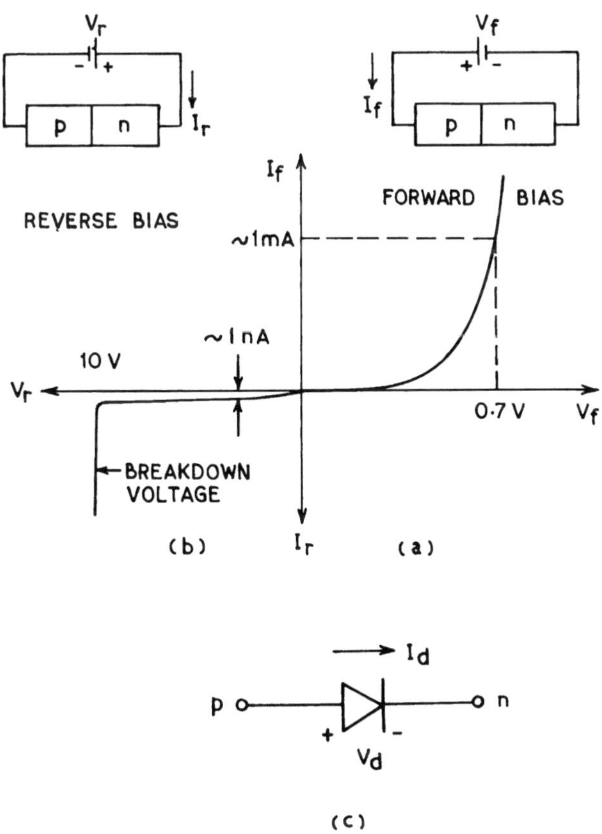

Fig. 2.9 Current voltage characteristic of an ideal diode under (a) forward bias and (b) reverse bias; note the change in the scale. (c) Circuit representation of a diode

critical voltage is called the *junction breakdown voltage*. The applied forward voltage is usually less than 1V while the reverse breakdown voltage could be tens of volts depending upon the structure of the *pn* junction and the dopant concentration in the *p*- and *n*-sides. The junction symbol for a diode is shown in Figure 2.9c. The arrowhead points from the *p*-type region towards the *n*-type region in the direction of forward-biased current flow.

The most common method of forming a *pn* junction is by diffusion or ion implantation of one type of impurity into a background of the opposite type, so that in one region donor impurity ions are in majority while in the other region acceptor impurities are prevalent. This is illustrated in Figure 2.10 for a planar type *pn* structure that also shows the doping profile.[18] The *metallurgical junction depth* X_j is indicated as the point where the net impurity concentrations of donors and acceptors are equal. Following the implantation process, devices go through various high-temperature fabrication steps which change the final profile. For circuit models, the actual impurity profile (dotted line) is often approximated by a *step* or *abrupt profile* (solid line) or sometime a *linearly graded profile* so that tractable circuit equations can be developed. Since most junctions encountered in MOS technology are, to a good approximation, step junctions, we will focus our analysis on step junctions (step doping profiles).
As shown in Figure 2.10b, a step doping profile is characterized by a constant *p*-type dopant concentration N_a that changes with position in a

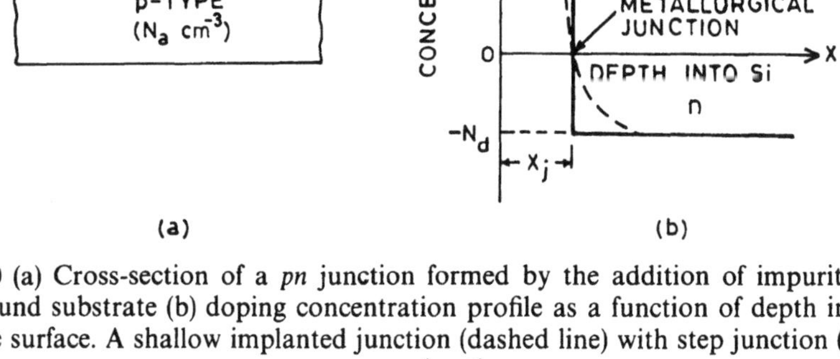

Fig. 2.10 (a) Cross-section of a *pn* junction formed by the addition of impurities into a background substrate (b) doping concentration profile as a function of depth into silicon from the surface. A shallow implanted junction (dashed line) with step junction (solid line) approximation

[18] The peak of the doping profile due to ion implantation is generally inside the silicon while that due to diffusion is at the surface.

stepwise fashion to a constant *n*-type dopant concentration N_d. Thus there is a large carrier concentration gradient at the junction resulting in carrier diffusion. Holes in the *p*-side diffuse to the *n*-side leaving behind negatively charged acceptor ions (N_a^-) and electrons from the *n*-side diffuse to the *p*-side leaving behind positively charged donor ions (N_d^+). Consequently, a *space charge region* is formed (negative charge on the *p*-side and positive charge on the *n*-side) creating thereby an electric field $\mathscr{E}$, and hence, a potential difference as shown in Figure 2.11. The direction of the field (*n*-region to *p*-region) is such that it opposes further diffusion of the carriers so that, in thermal equilibrium, the net flow of carriers is zero. The "internal" potential difference between the two sides of the junction is called the

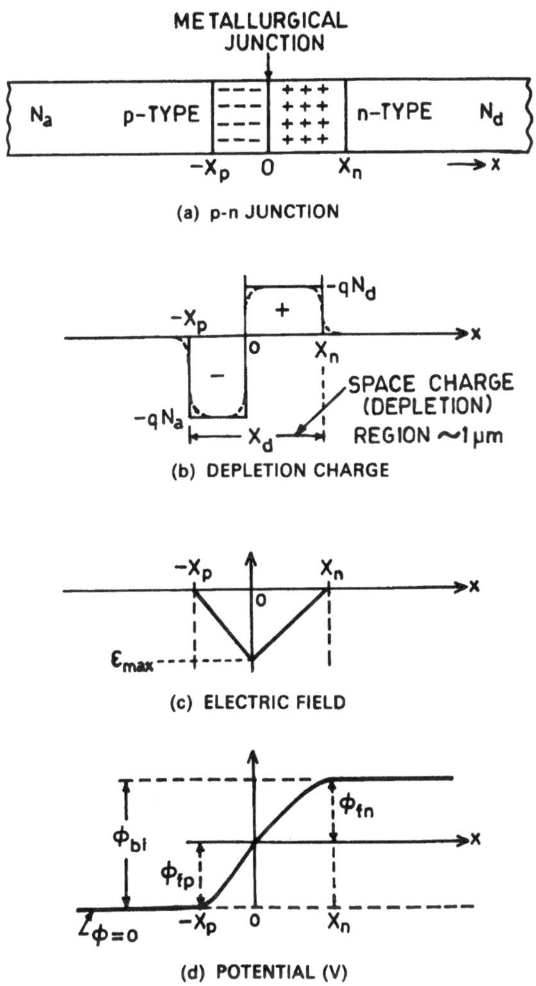

Fig. 2.11 Electric charge, field and potential relationships in the depletion region associated with a *pn* junction

built-in potential or barrier height ϕ_{bi}. The space charge region on the two sides of the metallurgical junction is often called the *depletion region*, because the region is depleted of the free (majority) carriers[19], or *transition region*.[20]

The boundary of the depletion region is labeled as X_p on the *p*-region side and X_n on the *n*-region side; the sum of X_p and X_n is called the *depletion width* X_d. The region outside the depletion width, on the two sides of the junction, is called the *quasi-neutral* region as in this region the majority carrier distribution does not differ much from the impurity distribution. This quasi-neutral region is also referred to as the *ohmic* or *bulk* region. The ϕ_{bi} and X_d are two important physical parameters that are used in developing the diode model and are discussed below.

2.5.1 Built-in Potential

The built-in potential ϕ_{bi} of a *pn* junction in equilibrium can easily be calculated using current equations (2.35). With no external voltage applied no current can flow (equilibrium condition), i.e. $J_n = J_p = 0$ and therefore the electric field $\mathscr{E}$ across the junction becomes

$$\mathscr{E} = -\left(\frac{kT}{q}\right)\frac{1}{n}\frac{dn}{dx} = \left(\frac{kT}{q}\right)\frac{1}{p}\frac{dp}{dx}. \tag{2.42}$$

Remembering $\mathscr{E} = -d\phi/dx$ and integrating the above equation from the *n*-type to *p*-type in the direction of the electric field, with equilibrium electron concentration n_{n0} (majority electrons in the *n*-region) and n_{p0} (minority electrons in the *p*-region), or equilibrium hole concentration p_{n0} and p_{p0}, respectively at the edges of the depletion layer, we get

$$\phi_{bi} \equiv \phi_n - \phi_p = V_t \ln\left(\frac{n_{n0}}{n_{p0}}\right) = V_t \ln\left(\frac{p_{p0}}{p_{n0}}\right) \tag{2.43}$$

where $V_t = kT/q$ and $\phi_n - \phi_p$ is the voltage difference between the two sides of the junction at thermal equilibrium which by definition is the built-in voltage ϕ_{bi}.

For an abrupt junction with uniform dopant concentration on the two sides we have $n_{n0} = N_d$ and $p_{p0} = N_a$. Making use of Eqs. (2.12) and (2.13) for

---

[19] Strictly speaking the depletion region is not devoid of free carriers but the number is so small compared to N_a and N_d that for all practical purposes it can be assumed to be depleted of free carriers.

[20] The three terms, *space-charge region, depletion region* or *transition region* are generally used synonymously.

minority carrier concentration, Eq. (2.43) becomes

$$\boxed{\phi_{bi} = V_t \ln\left(\frac{N_a N_d}{n_i^2}\right) \quad \text{(V)}.} \tag{2.44}$$

This is the potential difference which exists in a pn junction at thermal equilibrium and is a function of carrier concentration on the two sides of the junction. Its value is typically between 0.6–0.7 V for silicon junctions and is strongly temperature dependent due to its dependence on n_i. Note that the built-in voltage ϕ_{bi} can not be measured directly using a voltmeter. This is because the built-in voltage can be thought of as similar to contact potential between two dissimilar metals and the sum of all contact potential in a loop is zero.

The value of ϕ_{bi} given by Eq. (2.44) is valid under equilibrium conditions when no external voltage is applied to the diode. However, when voltage V_d is applied to the diode[21] (nonequilibrium situation) the potential barrier height becomes $(\phi_{bi} - V_d)$; *V_d is positive for forward bias and negative for reverse bias.* If the applied forward voltage is exactly equal to the built-in voltage, there will be no barrier and therefore, there will be copious flow of forward current. This is the maximum voltage which can be applied across the *pn* junction (provided there is no external resistance in series with the diode).

2.5.2 Depletion Width

The width X_d of the depletion region can be obtained by solving Poisson's equation (2.41). Let us assume that the free carrier concentrations *n* and *p* are negligibly small compared to the fixed ionized impurities $N_a^- \simeq N_a$ and $N_d^+ \simeq N_d$ over the entire region defined by the depletion width bounded by $-X_p$ and X_n i.e. $N_d \gg n_n$ or p_n and $N_a \gg n_p$ or p_p. This assumption is often referred to as the *depletion approximation.* It is an excellent approximation for most engineering purposes which will often be used during the development of analytical device models.

Assuming a step *pn* junction so that N_a and N_d are uniform in *p*- and *n*-regions respectively, the Poisson's equation (2.41) under the depletion approximation can easily be integrated using the boundary condition

[21] Note that voltage applied across the terminals of the diode will be the same as that appearing across the depletion region provided the diode contact resistance and *p*- and *n*- bulk (or neutral) regions resistance is negligible (see discussion in section 3.2.1, Eq. 2.64).

$\mathscr{E}(-X_p)=\mathscr{E}(X_n)=0$ (see Figure 2.11c) to give [2]–[12]

$$\mathscr{E}(x)=-\frac{qN_a}{\epsilon_0\epsilon_{si}}(X_p+x) \qquad -X_p\le x\le 0$$
$$\mathscr{E}(x)=-\frac{qN_d}{\epsilon_0\epsilon_{si}}(X_n-x) \qquad 0\le x\le X_n. \tag{2.45}$$

Since the field must be continuous at $x=0$, we get from Eq. (2.45) the maximum field $\mathscr{E}_{\max}$ as

$$\mathscr{E}_{\max}=-\frac{qN_aX_p}{\epsilon_0\epsilon_{si}}=-\frac{qN_dX_n}{\epsilon_0\epsilon_{si}} \tag{2.46}$$

or

$$\boxed{qN_aX_p=qN_dX_n} \tag{2.47}$$

which gives the distribution of the charge on either side of the junction and shows that the negative charge on the *p*-side exactly equals the positive charge on the *n*-side. Equation (2.47) also shows that the *width of the depletion region on each side of the junction varies inversely with the dopant concentration; the higher the dopant concentration, the narrower the depletion region.*

If we integrate Eq. (2.45) once again, remembering that $\mathscr{E}=-dV/dx$, and the potential difference between the *p* and *n* sides is ϕ_{bi} it can be seen that [2]–[12]

$$X_n=\sqrt{\frac{2\epsilon_0\epsilon_{si}}{q}\frac{N_a}{N_d(N_a+N_d)}\phi_{bi}}\quad \text{(cm)} \tag{2.48}$$

and

$$X_p=\sqrt{\frac{2\epsilon_0\epsilon_{si}}{q}\frac{N_d}{N_a(N_a+N_d)}\phi_{bi}}\quad \text{(cm)} \tag{2.49}$$

so that total depletion width $X_d\ (=X_p+X_n)$ becomes

$$X_d=\sqrt{\frac{2\epsilon_0\epsilon_{si}}{q}\left(\frac{1}{N_a}+\frac{1}{N_d}\right)\phi_{bi}}\quad \text{(cm).} \tag{2.50}$$

The value of X_d given above is at thermal equilibrium with no external voltage applied to the diode. However, when the diode is in a nonequilibrium condition, with voltage V_d applied to it, then as was stated earlier, the potential barrier height becomes $(\phi_{bi} - V_d)$, so that depletion width as a function of voltage becomes

$$X_d = \sqrt{\frac{2\epsilon_0\epsilon_{si}}{q}\left(\frac{1}{N_a} + \frac{1}{N_d}\right)(\phi_{bi} - V_d)} \quad \text{(cm)}. \tag{2.51}$$

This shows that forward bias $V_d(= V_f)$ will result in a decrease in the depletion width due to the decrease in barrier height, while reverse bias $-V_d(= V_r)$, will result in an increase in the depletion width due to a higher barrier height.

Using Eq. (2.48) for X_n or (2.49) for X_p in Eq. (2.46), the maximum electric field, $\mathscr{E}_{\max}$ in the depletion region becomes

$$\mathscr{E}_{\max} = \sqrt{\frac{2q}{\epsilon_0\epsilon_{si}} \frac{N_a N_d}{N_a + N_d}(\phi_{bi} - V_d)} \quad \text{(cm)}. \tag{2.52}$$

The higher the reverse voltage the higher is the field.

If the impurity concentration on one side of the junction is much higher than the other side, the junction is called a *one-sided step junction*; it is an excellent approximation for diffused junctions having shallow junction depths. In this case it can be seen that *the depletion region expands almost totally into the lighter doped side.* For example, in the case of an n^+p junction ($N_d \gg N_a$ and $X_p \gg X_n$) the depletion width X_d is almost entirely in the *p*-side. Thus from Eq. (2.51) it is easy to see that X_d for a one sided step junction becomes

$$X_d = \sqrt{\frac{2\epsilon_0\epsilon_{si}}{qN_b}(\phi_{bi} - V_d)} \quad \text{(cm)} \tag{2.53}$$

where $N_b = N_a$ for n^+p junction and $N_b = N_d$ for p^+n junction. A more accurate result for the depletion width can be obtained by considering majority carrier distribution tails (electrons in the *n*-side and holes in the *p*-side) as shown by dashed lines in Figure 2.11b. Each contributes a correction factor V_t to ϕ_{bi} [5]. Thus the depletion width is still given by Eq. (2.53) except that ϕ_{bi} is replaced by $\phi_{bi} - 2V_t$ so that using this more accurate expression, X_d for a one-sided step junction becomes

$$X_d = \sqrt{\frac{2\epsilon_0\epsilon_{si}}{qN_b}(\phi_{bi} - 2V_t - V_d)}. \tag{2.54}$$

However, Eq. (2.53) is accurate to within about 3% for the biases normally encountered in MOS circuitry.

2.6 Diode Current-Voltage Characteristics

When an external forward voltage V_d is applied to the junction, the holes that are injected from the *p*-region move across the depletion region and diffuse into the *n*-region. This is shown in Figure 2.12 where, for the condition of forward bias, the minority carrier concentration in the neutral region is plotted on a linear scale. As these holes (minority carriers in *n*-region) diffuse away from the junction they recombine with free electrons, which are majority carriers, so the free hole density decreases with distance x from the junction. If the *n*-region is long enough, the hole density approaches the equilibrium hole density $p_{n0} = n_i^2/N_d$ [cf. Eq. (2.12)] which is due to thermal generation. Similarly electrons injected into the *p*-region approach the equilibrium electron density $n_{p0} = n_i^2/N_a$. The complete minority carrier concentration distribution in the bulk of the semiconductor near the depletion region can be obtained using the continuity Eq. (2.38). Under the assumptions that:

1. The step junction profile is applicable.
2. The depletion approximation is valid.
3. Low level injection is maintained in the bulk region.
4. No generation-recombination takes place in the depletion region.
5. There is no voltage drop in the bulk region so that V_d is sustained entirely across the depletion region.
6. The width of the *p*- and *n*-regions outside the depletion region is much greater than minority carrier diffusion length for holes and electrons L_p and L_n respectively.[22]

The diode current I_d is given by [2]–[12]

$$\boxed{I_d = I_s\left[\exp\left(\frac{V_d}{V_t}\right) - 1\right] \quad \text{(A)}} \tag{2.55}$$

where I_s is called the *reverse saturation* current given by

$$\boxed{I_s = qA_d n_i^2\left[\frac{D_p}{L_p N_d} + \frac{D_n}{L_n N_a}\right] \quad \text{(A).}} \tag{2.56}$$

---

[22] The minority carrier *diffusion lenghts* L_n and L_p are defined as

$$L_p = \sqrt{D_p \tau_p} \quad \text{and} \quad L_n = \sqrt{D_n \tau_n} \quad \text{(cm)}$$

Physically, $L_p(L_n)$ is the mean distance traveled by the injected hole (electron) before it recombines with an electron (hole) and may range from 1 to 100 μm, depending upon the silicon purity and doping concentration.

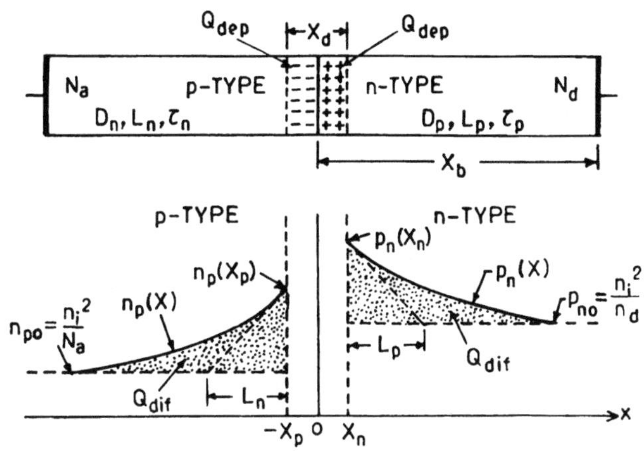

Fig. 2.12 Excess minority carrier distribution in the bulk region when the diode is forward bias

Clearly, I_s *may be considered as arising from thermal generation of minority carriers in the bulk region.* At $T = 300\,\text{K}$, $V_t \sim 26\,\text{mV}$ and for a forward bias $V_d(= V_f)$ greater than 60 mV, the exponential term is greater than ten. Therefore, the exponential term is dominant at practical levels of forward bias and the net forward current I_f is

$$I_f \equiv I_d \approx I_s \exp\left(\frac{V_d}{V_t}\right) \quad \text{(A)}. \tag{2.57}$$

Thus, *under forward bias with* $V_f/V_t \gg 1$, *the current varies exponentially with the applied voltage.*

When the diode is reverse biased by a voltage $-V_d(= V_r)$, the potential barrier increases resulting in an increase in the depletion width. Under reverse bias there is very little current flow as the minority carriers (electrons from *p*-side and holes from *n*-side) which constitute the current have very low density. The flow of minority carriers constitutes the so called *leakage current* of the junction. It turns out that Eq. (2.55) is applicable for both forward and reverse bias voltages. In reverse bias, the exponential term drops out at a relatively small value of V_r, the current approaches I_s and becomes independent of bias. For this reason I_s is called the reverse saturation current. Thus, in reverse bias for $V_d \leq 5V_t$

$$I_d = -I_s. \tag{2.58}$$

In practical diodes (step junction) the saturation current is mainly influenced by the lighter doped side, i.e. the side where minority carrier concentration is greatest. For example, for p^+n junctions, $N_a \gg N_d$, and therefore Eq. (2.56) becomes

$$I_s = qA_d n_i^2 \frac{D_p}{L_p N_d} \quad \text{(A)}. \tag{2.59}$$

The above equation for I_s is based on the assumption that the length X_b of the n-region from the junction to the ohmic contact is much larger than the diffusion length L_p (see Fig. 2.12). In this case ($X_b \gg L_p$) the minority carrier concentration decreases to its thermal equilibrium value in a distance less than X_b. When this happens it is referred to as a *long-base diode*. However in IC technology we often encounter situations when $X_b \ll L_p$. Such diodes are called *short-base diodes*. In this case ($X_b \ll L_p$), recombination is completed at the ohmic contact that makes the connection to the n-region; I_s for short base diodes simply can be obtained by replacing L_p with X_b in Eq. (2.56) [12].

2.6.1 *Limitation of the Diode Current Model*

The diode current Eq. (2.55), often called the ideal-diode equation, describes fairly accurately *pn* junction devices over a range of applied voltages. However, there exists a significant range of useful biases where the ideal diode equation becomes inaccurate. We will briefly discuss those regions of operations.

Forward Bias. The current voltage characteristics of a forward biased silicon *pn* junction diode is shown in Figure 2.13, where the ideal diode

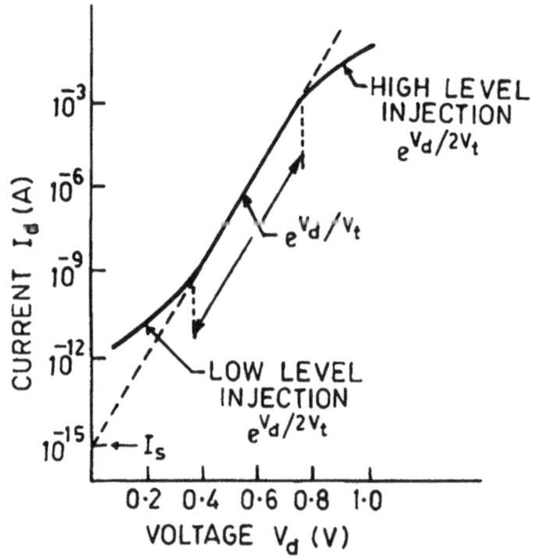

Fig. 2.13 Forward bias characteristics of a real diode showing low-level and high-level injections. Saturation current I_s is obtained by extrapolating current from the mid range (dotted line)

current is shown by the dotted line. Two different regions of nonideal behavior are illustrated in this figure. At a very small value of the forward bias ($V_d < 0.2$–0.3 V) the injected carrier densities are relatively small. When these carriers move through the depletion region, some of them may be lost by recombination in this region thereby forming a recombination current I_{rec}, which is added to the ideal diode diffusion current. The result is a larger total current than that predicted by the ideal diode Eq. (2.55). This recombination current dominates in the silicon diode at very small current levels and violates assumption 4. Note that the recombination current is present at large current levels also, but then it is only a small fraction of the total current. Using the Shockley-Read-Hall (SHR) theory of generation and recombination, it can be shown that the space-charge recombination current I_{rec} is [11, 12]

$$I_{rec} = \frac{qA_d n_i X_d}{\tau_r} \exp\left(\frac{V_d}{2V_t}\right) \quad \text{(A)} \tag{2.60}$$

where τ_r is the lifetime associated with the recombination of excess carriers in the depletion region. The lifetime τ_r is analogous to, but usually greater than τ_n or τ_p for the neutral regions and is generally approximately equal to $2\sqrt{\tau_p \tau_n}$ [12]. Thus total diode current is the sum of Eqs. (2.60) and (2.55). In general until V_d reaches a value of about 0.4 V, the neutral region diffusion current will be less than I_{rec}.

At high current levels, the injected minority carrier density is comparable to the majority carrier concentration (high-level injection)[23], and therefore assumption 3 is invalid. For high level injection, majority carrier concentration increases significantly above its equilibrium value, giving rise to an electric field. Thus in such cases *both drift and diffusion components must be considered.* The presence of the electric field results in a voltage drop across this region, and thus reduces the applied voltage that appears across the junction resulting in a lower current than expected. It can be shown that under high level injection the diode current I_d is [12]

$$I_d = \frac{2qA_d n_i D_p}{X_b} \exp\left(\frac{V_d}{2V_t}\right) \quad \text{(high-level injection)} \tag{2.61}$$

which indicates that high level current depends on $1/2V_t$ rather than $1/V_t$.

---

[23] Since the injected minority carrier is much greater than the background doping concentration, the conductivity of this region is increased. Therefore this region becomes conductivity modulated. For this reason the terms *conductivity modulation* and *high level injection* are often used interchangeably to denote this situation in different devices.

Thus, depending upon the magnitude of the applied forward voltage, the diode current can be represented by an empirical form

$$I_f \equiv I_d = I_s \left[\exp\left(\frac{V_d}{\eta V_t}\right) - 1 \right] \tag{2.62}$$

where the factor η is called the *ideality factor* and is a measure of how close to ideal the real diode curve is. When recombination current dominates or there is high level injection $\eta = 2$ and when diffusion current dominates $\eta = 1$.

Reverse Bias. The current of a reverse bias diode is shown in Figure 2.14 where the dotted line shows current I_s due to an ideal diode Eq. (2.56). Clearly, the current in a real diode does not saturate at $-I_s$ as predicted by Eq. (2.56). This is because when the diode is reverse biased, generation of electron-hole pairs in the depletion region takes place, which was neglected in the ideal diode equation. In fact, the generation current dominates because carrier concentrations are smaller than their thermal equilibrium values. Again, using SHR theory [5]–[12], it can be shown that the generation current I_{gen} is

$$I_{\text{gen}} = \frac{q A_d n_i X_d}{\tau_g} \tag{2.63}$$

where τ_g is the generation life time of the carriers in the depletion region and is approximately equal to $2\tau_p$ if we assume $\tau_p \sim \tau_n$ [12]. Note that while I_s is proportional to n_i^2, I_{gen} is proportional to n_i only. Thus I_{gen} will be dominant when n_i is small as is the case at room and low temperatures. Further, since the space charge width X_d increases as the square root of the reverse bias [cf. Eq. (2.51)], the generation current increases with reverse bias voltage as shown in Figure 2.14 as a solid line. Thus taking into account I_{gen} the total reverse current I_r becomes $I_r \equiv -I_d = -(I_s + I_{\text{gen}})$. This value of I_r not only approximately agrees with measured value of reverse current but in addition, it provides proper voltage dependence of the reverse current in properly constructed silicon planar *pn* junctions.

In real devices there is a third component of leakage current, called the surface leakage current I_{sl}. This current can be treated as a special case of I_{gen} modeled at the surface where a high concentration of dislocations at the oxide-silicon interface, often referred to as *fast surface states* (see section 4.1.2), provide additional generation centers over those present in the bulk. It is very much process dependent and is responsible for large variation in the leakage current. *Both process and electrically induced defects at the surface generally increase the charge generation rate by an order of magnitude compared with the bulk recombination-generation rate.* In that

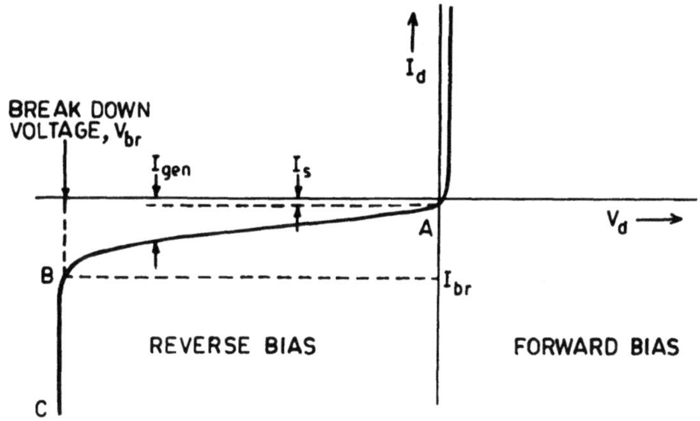

Fig. 2.14 Reverse bias characteristics of a real diode

case I_{sl} dominates over the other components of I_r and is thus responsible for higher leakage current for a diode compared to that predicted by the sum of I_{gen} and I_s. Leakage current is highly temperature dependent due to the presence of the n_i term. Also note that the generation limited leakage current is proportional to n_i while diffusion limited leakage current is proportional to n_i^2.

2.6.2 *Bulk Resistance*

At high current levels, bulk resistance and the metal-silicon contact resistance can produce a significant voltage drop (assumption 5) resulting in a smaller voltage across the junction and thus a lower current. Usually the bulk resistance and contact resistance are combined into one resistor called the *series resistance* r_s. Thus if V_d is the applied voltage to the diode terminals and V'_d is the voltage across the diode junction resulting in the current I_d as shown in Figure 2.15, we have

$$V_d = V'_d + r_s I_d. \tag{2.64}$$

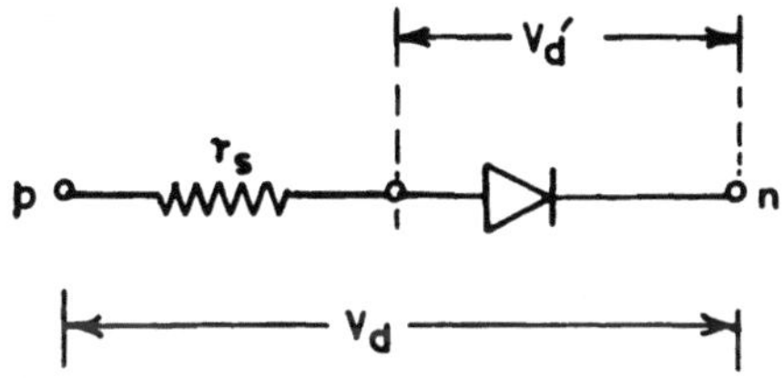

Fig. 2.15 Diode model at high level current; r_s is the diode resistance due to contact and bulk region resistivity

Under the ideal conditions when $r_s = 0$, $V_d = V'_d$ that is related to I_d by Eqs. (2.55) or (2.62). Thus, in the presence of the series resistance, the diode current-voltage expression becomes

$$I_d = I_s\left[\exp\frac{(V_d - I_d r_s)}{\eta V_t} - 1\right] \quad \text{(A).} \tag{2.65}$$

Rearranging this equation yields

$$V_d = \eta V_t \ln\left(1 + \frac{I_d}{I_s}\right) + r_s I_d. \tag{2.66}$$

Clearly, when I_d is large, the terminal voltage V_d will increase linearly with I_d because $I_d r_s$ increases faster than the logarithmic term.

2.6.3 Junction Breakdown Voltage

We have seen that the diode reverse (or leakage) current increases only as the square root of the reverse bias V_r [cf. Eq. (2.63)]. But as V_r increases so does the electric field in the depletion region [cf. Eq. (2.52)]. When the field reaches a certain critical field $\mathscr{E}_c$ corresponding to the reverse voltage $V_r = V_{br}$, called the *breakdown voltage*, a slight increase of reverse voltage causes a very large increase of current as shown in Figure 2.14 (region BC). This condition is often called the *breakdown condition* and is a most important consideration in device design.[24] The breakdown occurs because carriers, while moving through the depletion region, acquire sufficient energy to create new electron-hole pairs through impact ionization. The newly generated electron-hole pairs can also acquire sufficient energy from the field to create additional electron-hole pairs. Since electrons and holes travel in opposite directions, the carriers can multiply a few times in the depletion region before they reach the electrodes. This multiplicative process results in an avalanche effect. The resulting breakdown voltage, V_{br}, is called the *avalanche breakdown* and can be obtained using Eq. (2.52):

$$V_{br} = \frac{\epsilon_0 \epsilon_{si} \mathscr{E}_c^2}{2q}\left(\frac{1}{N_a} + \frac{1}{N_d}\right) \quad \text{(V).} \tag{2.67}$$

---

[24] The breakdown process is not inherently destructive so long as the current is limited. For very large currents, however, it will be destructive if sufficient heat is generated in the junction by the current. This is true whether the diode is forward biased or reverse biased and so, to limit the maximum current flow, an external resistance might be added in series with the diode.

The above equation shows that *any increase in the doping, either of n or p, results in a decrease in the breakdown voltage* V_{br}. Further, it shows that V_{br} is controlled by the concentration N of the lightly doped region and is proportional to N^{-1}. In practical diodes, V_{br} generally varies as $N^{-2/3}$ [12]. For moderately doped silicon (10^{14}–$10^{16}\,\mathrm{cm}^{-3}$) the value of the critical field is $\mathscr{E}_c \sim 4 \cdot 10^5$ V/cm and to a first approximation is independent of doping [11].

If the *pn* junction is heavily doped (concentration $> 10^{18}\,\mathrm{cm}^{-3}$) on both sides, the depletion layer is very narrow. Carriers cannot gain enough energy within the depletion region so that avalanche breakdown is not possible. However, in the depletion region the electric field is high; $\mathscr{E}_{\max}$ can be close to 10^6 V/cm. The field becomes so high that it exerts sufficient force to free the covalent bond electrons. This creates electron-hole pairs. This type of electron-hole pair generation is called *tunneling* and contributes to the current resulting in the breakdown of the junction. No carrier acceleration or collision is required for this type of breakdown. This mechanism of breakdown is called the *Zener breakdown.* In the source/drain *pn* junction of a MOSFET, it is the avalanche breakdown that is important.

2.7 Diode Dynamic Behavior

So far we have considered diode characteristics under constant applied voltage, not varying with time. In circuits, however, diodes are often subject to varying voltages. Such dynamic operation causes charges in the diode to vary resulting in an extra current not predicted by the DC current Eq. (2.55). There are two types of stored charge in a diode: (1) the charge Q_{dep} due to the depletion or space-charge region on each side of the junction [cf. Eq. (2.47)], and (2) the charge Q_{dif} due to minority carrier injection. Remember that it is these injected (excess) mobile carriers which generate current I_d and also represent a stored charge Q_{dif} in the diode. The latter is given by the area between the curve representing p_n(or n_p) and the steady state level p_{n0}(or n_{p0}) (see Figure 2.12). These two types of stored charges result in two types of diode capacitances, the junction capacitance C_j due to Q_{dep} and the diffusion capacitance due to Q_{diff}, and are discussed below.

2.7.1 Junction Capacitance

When the voltage applied to the *pn* junction is changed by a small amount (incremental change), there will be an incremental change in the depletion region charge Q_{dep} because the depletion width changes (see section 2.5.2). If the applied voltage is returned to its original value, carriers flow in such a direction that the previous increment of charge gets neutralized. The response of the *pn* junction to the incremental voltage thus results in generation of an effective capacitance C_j referred to as the *transition capacitance, junction capacitance or depletion layer capacitance.*[25] Recalling the definition of capacitance per unit area in terms of an incremental charge dQ_{dep} per unit area induced by an incremental change in applied voltage dV_d, we have

$$C_j = \frac{dQ_{\text{dep}}}{dV_d} \quad (\text{F/cm}^2). \tag{2.68}$$

Remembering that $Q_{\text{dep}} = qN_aX_p = qN_dX_n$ [cf. Eq. (2.47)], it is easy to see that

$$C_j = \sqrt{\frac{\epsilon_0\epsilon_{si}q}{2(\phi_{bi} - V_d)}\left(\frac{N_aN_d}{N_a + N_d}\right)} \quad (\text{F/cm}^2) \tag{2.69}$$

where we have made use of Eqs. (2.48)–(2.49).

This is the equation for the diode junction capacitance for a step profile in terms of the physical parameters of the device. Remember that *this equation is valid for* $V_d < \phi_{bi}$. Comparing Eqs. (2.69) and (2.51) it is easy to see that

$$C_j = \frac{\epsilon_0\epsilon_{si}}{X_d} \quad (\text{F/cm}^2). \tag{2.70}$$

This states that the junction capacitance is equivalent to that of a parallel plate capacitor with silicon as the dielectric and separated by a distance X_d, the depletion width. Though the derivation of Eq. (2.70) is based on a step profile, it can be shown [2] that *the relationship is valid for any arbitrary doping profile.*

It should be pointed out that although the *pn* junction capacitance can be calculated using the parallel plate capacitor formula, there are differences between the two types of capacitors. While true parallel plate capacitance is independent of applied voltage, *pn* junction capacitance given by Eq. (2.70) becomes voltage dependent through X_d. Therefore, the total charge in a *pn* junction cannot be obtained by simply multiplying the capacitance by the applied voltage, although a small variation in the charge

---

[25] The classical definition of the linear capacitance of a parallel plate capacitor structure is given by $C = Q/V$, where Q is the charge on the positively charged plate and V is the potential of that plate relative to the other plate. Since in many cases, capacitance may be bias dependent, more generally we define the capacitance C of any geometry as $C(V) = dQ/dV$ which is usually called the *incremental capacitance* and can easily be derived from the linear relation $C = Q/V$ remembering that the (displacement) current i_c through a capacitor is $i_c = dQ/dt$.

can still be obtained by multiplying a small variation in the voltage by the instantaneous capacitance value (see Eq. 2.68). Another difference is that in a *pn* junction the dipoles in the transition region have their positive charge in the *n*-side depletion region and negative charge in the *p*-side depletion region, while in a parallel plate capacitor the separation between the charges in the dipoles is much less and they are distributed homogeneously throughout the dielectric.
For a one-sided step junction, say n^+p diode with $N_d \gg N_a$, Eq. (2.69) becomes

$$C_j = \sqrt{\frac{\epsilon_0 \epsilon_{si} q N_a}{2(\phi_{bi} - V_d)}} \quad (\mathrm{F/cm^2}). \tag{2.71}$$

For the circuit designer it is more convenient to express capacitance in terms of electrical parameters. If C_{j0} is the junction capacitance at equilibrium i.e. at $V_d = 0$ V, then from Eq. (2.69) we obtain

$$C_{j0} = \sqrt{\frac{\epsilon_0 \epsilon_{si} q}{2\phi_{bi}}\left(\frac{N_a N_d}{N_a + N_d}\right)} \quad (\mathrm{F/cm^2}). \tag{2.72}$$

The junction capacitance C_j can be written in terms of C_{j0} as

$$C_j = \frac{C_{j0}}{\sqrt{1 - V_d/\phi_{bi}}} \quad (\mathrm{F/cm^2}). \tag{2.73}$$

In practical diodes, the doping profile is neither abrupt nor linearly graded as assumed in the derivation for C_j and therefore, to calculate the capacitance for real devices, we replace the one-half power in (2.73) by m, called the *grading coefficient*, resulting in the following equation for C_j

$$\boxed{C_j = \frac{C_{j0}}{\left(1 - \dfrac{V_d}{\phi_{bi}}\right)^m} \quad (\mathrm{F/cm^2}).} \tag{2.74}$$

For real devices m ranges between 0.2 and 0.6. Figure 2.16 shows a plot of the junction capacitance C_j as a function of junction voltage V_d. Note that the capacitance C_j decreases as the reverse biased $|V_d|$ increases (V_d is negative). When the diode is forward bias (V_d is positive) the capacitance C_j increases and becomes infinite at $V_d = \phi_{bi}$ as shown in Figure 2.16 (continuous line, curve A). This is because Eq. (2.74) no longer applies due to the depletion approximation becoming invalid [27]. A more exact analysis of the behavior of C_j as a function of the forward bias V_d is shown by the dotted line (curve b) [28]. However, in SPICE a straight line

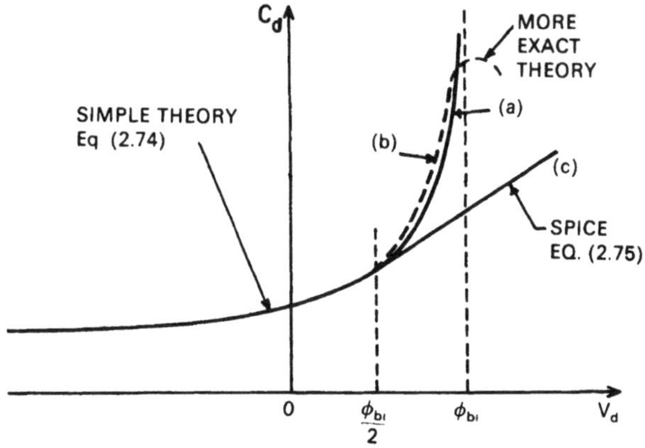

Fig. 2.16 Behavior of a *pn* junction depletion capacitance C_j as a function of the voltage V_d across the diode

approximation is used instead (see curve c in Figure 2.16). In this case we define a parameter $F_c (0 < F_c < 1)$ such that when the diode is forward bias and $V_d \geq F_c \phi_{bi}$ the following equation for C_j is used

$$C_j = \frac{C_{j0}}{(1-F_c)^{1+m}} \left[\frac{m}{\phi_{bi}} V_d + 1 - F_c(1+m) \right], \quad V_d \geq F_c \phi_{bi} \tag{2.75}$$

that is obtained by matching slopes at $F_c \phi_{bi}$. Thus, F_c determines how depletion capacitance is calculated when the junction is forward biased. Normally F_c is taken as 0.5. The above approximation avoids infinite capacitance and, though not accurate, is acceptable for circuit design work. This is because, under forward bias conditions, diffusion capacitance, as discussed below, dominates.

It should be pointed out that for circuit models ϕ_{bi} and m become fitting parameters and are obtained by fitting Eq. (2.74) to experimental capacitance data, as is discussed in detail in section 9.14.2.

2.7.2 Diffusion Capacitance

The variation in the stored charge Q_{dif}, associated with excess minority carrier injection in the bulk region under forward bias, is modeled by another capacitance C_{df}. The capacitance C_{df} is called *diffusion capacitance*, because the minority carriers move across the bulk region by diffusion. Since Q_{dif} is proportional to the current I_d, for an n^+p diode we can write [5]–[12]

$$Q_{\mathrm{dif}} = \frac{1}{A_d} \tau_p I_d \quad (\mathrm{C/cm^2}). \tag{2.76}$$

For a short base diode, τ_p is replaced by τ_t, the transit time of the diode. For the case of a long base diode the transit time τ_t is the excess minority carrier lifetime. Differentiating Eq. (2.76) gives

$$C_{df} = \frac{dQ_{\mathrm{dif}}}{dV_d} = \frac{\tau_p I_s}{A_d V_t} \exp\left(\frac{V_d}{V_t}\right) \quad (\mathrm{F/cm^2}). \tag{2.77}$$

A more accurate derivation results in a C_{df} half of that shown in Eq. (2.77) [11, 12].[26]
Let us compare the magnitude of the two capacitances at a forward bias of say 0.3 V; assume we have a n^+p diode with $N_a = 10^{15}\,\mathrm{cm}^{-3}$ and $N_d = 10^{19}\,\mathrm{cm}^{-3}$, then Eq. (2.44) gives $\phi_{bi} = 0.814$ V. For a forward bias of 0.3 V, Eq. (2.50) gives $X_d = 8.15 \times 10^{-5}$ cm and Eq. (2.70) gives $C_j = 1.27 \times 10^{-8}\,\mathrm{F/cm^2}$. Assuming $\tau_t = 10^{-7}$ sec, and $I_s = 4 \times 10^{-12}$ A for a junction area of $20 \times 20\,\mu\mathrm{m}^2$ gives $C_{df} = 4 \times 10^{-7}\,\mathrm{F/cm^2}$, which is much larger than C_j. It should be noted that under forward bias C_{df} increases much faster with increasing $V_d(= V_f)$, due to the exponential dependence on V_d, as compared to C_j. However, under reverse bias C_j decreases much more slowly with increasing $V_d(= -V_r)$, as compared to C_{df}. Therefore, *C_j is the dominant capacitance for reverse bias and small forward bias ($V_d < \phi_{bi}/2$), while diffusion capacitance C_{df} is dominant for forward bias ($V_d > \phi_{bi}/2$).*

2.7.3 Small Signal Conductance

In the large signal model discussed in the previous section we did not place any restriction on the allowed voltage variations. However, in some circuit situations, voltage variations are sufficiently small so that the resulting small current variations can be expressed using linear relationships. This is the so called *small signal behavior* of the diode. An example of linear relations are the capacitances C_j and C_{df} in Eqs (2.74) and (2.77), respectively, as they represent an overall nonlinear charge storage effect in terms of linear circuit elements (capacitors), although we did not label them as such.
For small variations about the operating point, which is set by the DC condition, the nonlinear diode current can be linearized so that the incremental diode current is proportional to the incremental applied voltage. This linear relationship is used to calculate the small signal conductance g_d,

$$g_d = \frac{dI_d}{dV_d} \quad (\mathrm{mho}). \tag{2.78}$$

[26] C_{df} is independent of frequency $\omega(= 2\pi \times \text{frequency})$ for $\omega\tau_p \ll 1$.

Using Eq. (2.55) we have

$$\boxed{g_d = \frac{I_s}{V_t}\exp\left(\frac{V_d}{V_t}\right) = \frac{1}{V_t}(I_d + I_s) \quad \text{(mho)}.} \tag{2.79}$$

Clearly g_d is proportional to the slope of the DC characteristics at the operating point. When the diode is forward biased, I_d is much larger than I_s and therefore g_d is proportional to I_d. However, when the diode is reverse biased $I_d = -I_s$ and therefore from the above equation g_d becomes zero. But in real diodes, $g_d \neq 0$ in the reverse bias condition due to the fact that the generation current, I_{gen}, [cf. Eq. (2.63)] is the dominant conduction mechanism.

2.8 Real *pn* Junction

In the discussion so far we have assumed that the junction is planar. However, real junctions fabricated by IC technology depart from true planarity as shown in Figure 2.17. When the junction is formed by diffusion through a window in the oxide mask, the impurities will diffuse downward (depth X_j) and sideways (L_{dif}) resulting in a planar region with nearly cylindrical edges (see Figure 2.17). Thus, in reality, the junction boundary consists of the flat planar bottom and its rounded sides and corners. Typically, the radius of the cylindrical sides of the junction is 0.6–0.8 times the junction depth X_j. Clearly the width X_d of the depletion region

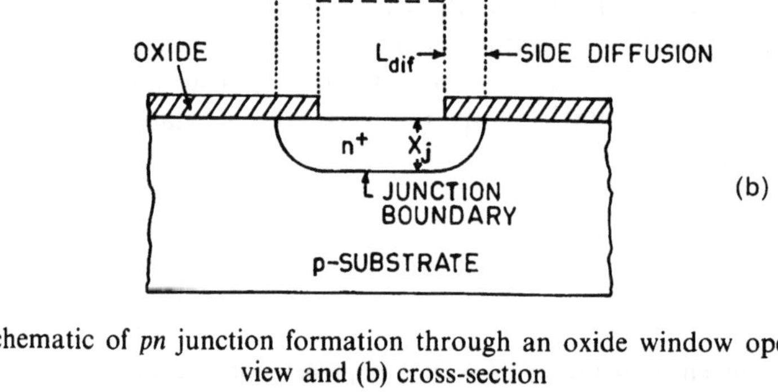

Fig. 2.17 Schematic of *pn* junction formation through an oxide window opening (a) top view and (b) cross-section

will not be uniform along the boundary of the junction. The depletion width will be narrower at the cylindrical edges due to the charge crowding at the edges resulting in a larger electric field than in the plane part of the junction. The higher electric field at the corners will result in a lower breakdown voltage of the real diode as compared to the true planar diode. The reduction in the *breakdown voltage for a shallower junctions with smaller radius of curvature will be more severe as compared to true planar junctions.* This is because the lines of force will concentrate more on the corners where the electric field is higher as compared to the planar region, resulting in a lower breakdown voltage at the corners.

Due to the smaller depletion width at the edges (because of high fields), the junction capacitance will be larger at the edges compared to the plane portion of the junction. Thus, capacitance in a real junction can be thought of as consisting of two components:

- the *area component*, C_{area}; it is the capacitance per unit area due to the area A defined by the opening in the oxide mask through which impurities have been diffused. This is also called the *bottom-wall capacitance.*
- the *periphery component*, C_{peri}; it is the capacitance per unit length due to the periphery P of the oxide window opening, also known as the *side-wall capacitance.*

so that the total capacitance C_J becomes[27] the sum of $C_{\text{area}} \times A$ and $C_{\text{peri}} \times P$. *Traditionally, the measured junction capacitance of discrete diodes is the area capacitance which submerges the periphery component.* However, if the junctions are shallower, as is usually the case with source/drain junctions of VLSI MOSFETs, the periphery component is often larger than the area component. Both these capacitances follow the model described in section 2.7.1 with the model parameter (C_{j0}, ϕ_{bi}, and m) values being different in the two cases [cf. Eq. (2.74)].
In order to separate the two components of the junction capacitance, measurements are made on special *test structures* with extreme area to periphery ratios [29]. One such test structure which maximizes the area is shown in Figure 2.18a (structure 'a') and other which maximizes perimeter is shown in Figure 2.18b (structure 'p'). Another structure which is often used for perimeter maximization is the "serpentine" structure. If C_A is the total capacitance for structure 'a' and C_P is the total capacitance for

[27] Throughout the text, the lower case subscript for charge Q and capacitance C denote per unit quantity while upper case subscript represent total quantity. Thus, for example, C_j represents junction capacitance per unit area while C_J denotes total junction capacitance. Similarly, charge Q_{dep} represents depletion charge per unit area while Q_{DEP} will represent total depletion charge.

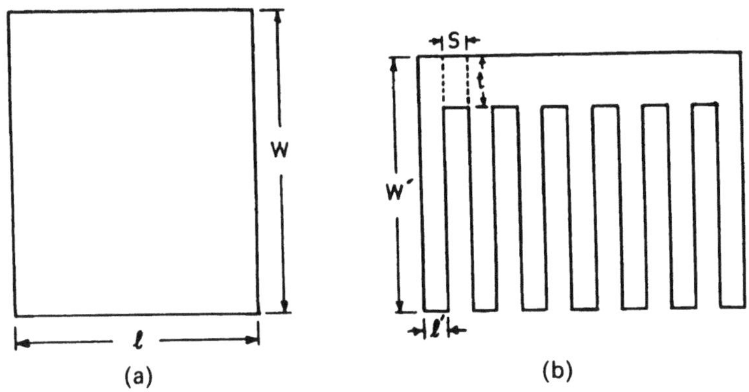

Fig. 2.18 Test structures for separating area and periphery capacitance components of a junction diode. (a) maximum area structure (b) maximum perimeter structure

structure 'p', then we can write

$$C_A = C_{\text{area}} A_a + C_{\text{peri}} P_a \quad \text{(F)} \tag{2.80a}$$

$$C_P = C_{\text{area}} A_p + C_{\text{peri}} P_p \quad \text{(F)} \tag{2.80b}$$

where

A_a = Area of the structure 'a' = $l \times w$ (see Figure 2.18a)
P_a = Perimeter of the structure 'a' = $2\,(l + w)$
A_p = Area of the structure 'p' $\approx$ m $(l' \times w')$, (see Figure 2.18b)
P_p = Perimeter of the structure 'p' ≈ 2m $(l' + w')$
C_{area} = capacitance per unit area (F/cm^2)
C_{peri} = capacitance per unit perimeter (F/cm)
m = number of fingers in structure 'p'

Note that Eqs. (2.80) are based on the assumption that C_{area} and C_{peri} are the same for the two structures at a given voltage and temperature.This is normally the case when the test structures are side by side on a chip. Given measured data for C_A and C_P and knowing A_a, P_a, A_p and P_p for the two structures, we can calculate C_{area} and C_{peri} at each reverse voltage point using Eq. (2.80). In order to ensure that C_{area} and C_{peri} are the true area and perimeter capacitances, respectively, we must exclude any additional parasitic effects such as overlap capacitance between the junction and crossing conductors. The area and periphery capacitances, C_{area} and C_{peri}, respectively, as a function of reverse bias are shown in Figure 2.19 where dots are measured data (calculated from Eq. (2.80) using measured C_A and C_P), while the solid lines are the fit to the data (dots) using Eq. (2.74).

Similar to junction capacitances, the reverse leakage current will also be different in the plane portion and corners of the junction resulting in the *area* (*or bottom-wall*) and *periphery* (*or side-wall*) components. The area

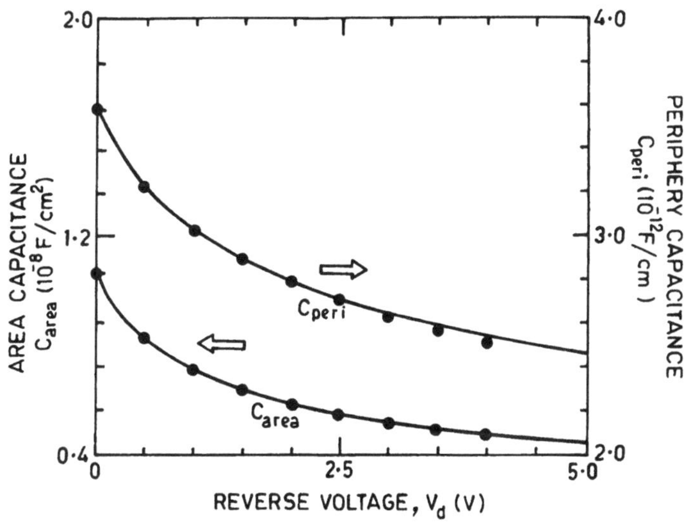

Fig. 2.19 Area capacitance C_{area} and periphery capacitance C_{peri} as a function of reverse bias $V_d(= V_r)$. Dots are experimental points (see text), while continuous lines are obtained by fitting the data to the Eq. (2.74)

component is the current crossing the area defined by the opening in the oxide mask through which impurities have been diffused. The periphery component is the current crossing the periphery of the oxide window opening and is usually dominated by the surface generation. The two components of I_R are again separated by doing measurements on two different test structures, one that maximizes area and another which maximizes perimeter, similar to the structures shown in Figure 2.18. If I_A is total current for structure a and I_p is the total current for structure p, then we can write

$$I_A = I_{\text{area}} A_a + I_{\text{peri}} P_a \quad \text{(A)} \tag{2.81a}$$

$$I_P = I_{\text{area}} A_p + I_{\text{peri}} P_p \quad \text{(A)} \tag{2.81b}$$

where I_{area} and I_{peri} are the currents per unit area (A/cm^2) and per unit perimeter (A/cm), respectively. Measuring the diode current I_A and I_P for the two different structures as a function of voltage and knowing A_a, P_a, A_p and P_p for the two structures, we can calculate I_{area} and I_{peri} using Eq. (2.81) respectively for a given voltage V_d.

2.9 Diode Circuit Model

The DC equivalent circuit model for a *pn* junction diode is shown schematically in Figure 2.20a, which establishes dependence of the diode current I_d on the diode voltage V_d. The rhombic symbol for I_d simply

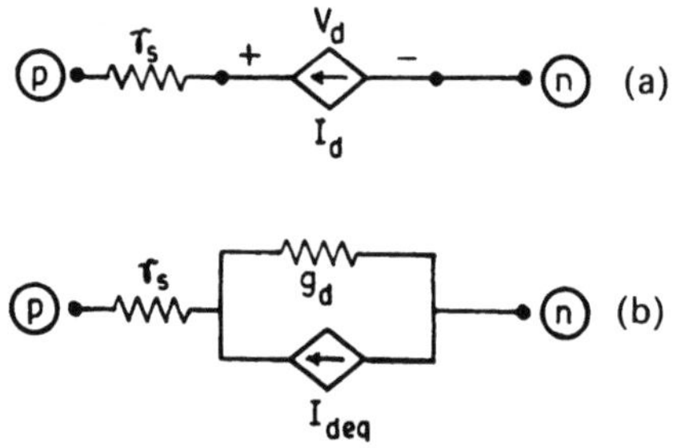

Fig. 2.20 Diode (a) equivalent circuit model for the DC analysis (b) linearized equivalent circuit model

represents a controlled current source. In this figure r_s is the diode series resistance and p and n are the nodes as specified in a SPICE input file. The value of I_d is determined by the following equations

$$I_d = \begin{cases} I_s\left[\exp\left(\dfrac{V_d - I_d r_s}{\eta V_t}\right) - 1\right] & V_d \geq -5\eta V_t \\ -I_s & V_{br} < V_d < -5\eta V_t \\ -I_{br} & V_d = -V_{br} \end{cases} \tag{2.82}$$

where I_s is the ideal saturation or leakage current defined by Eq. (2.56) and η is the ideality factor defined in section 2.6.1 and lies in the range 1–2. Note that η *is constant for the whole DC current computation.* The SPICE diode model is not capable of simulating diode characteristics that allows η to vary depending upon regions of operation. Therefore, for a fixed η (say $\eta = 1$) the model becomes inaccurate at low and high current level as discussed earlier.

Since I_d is a nonlinear function of V_d, in order to solve nonlinear circuit equations, the equivalent circuit model of Figure 2.20a is converted into its companion model (linearization of the nonlinear current) as shown in Figure 2.20b. In this figure g_d is the conductance of the pn junction given by Eq. (2.79) while the corresponding equivalent current I_{deq} is given by

$$I_{deq} = I_d - g_d \cdot V_d. \tag{2.83}$$

The small signal g_d is related to the large signal model by the following equation

$$g_d = \left.\frac{\Delta I_{ds}}{\Delta V_{gs}}\right|_{op} \tag{2.84}$$

where the subscript op denotes that the relation is evaluated at the operating-point bias value. Thus, to describe the DC behavior of the diode, we need four parameters: I_s, η, r_s and breakdown voltage V_{br} (or current I_{br} corresponding to the breakdown voltage V_{br}).

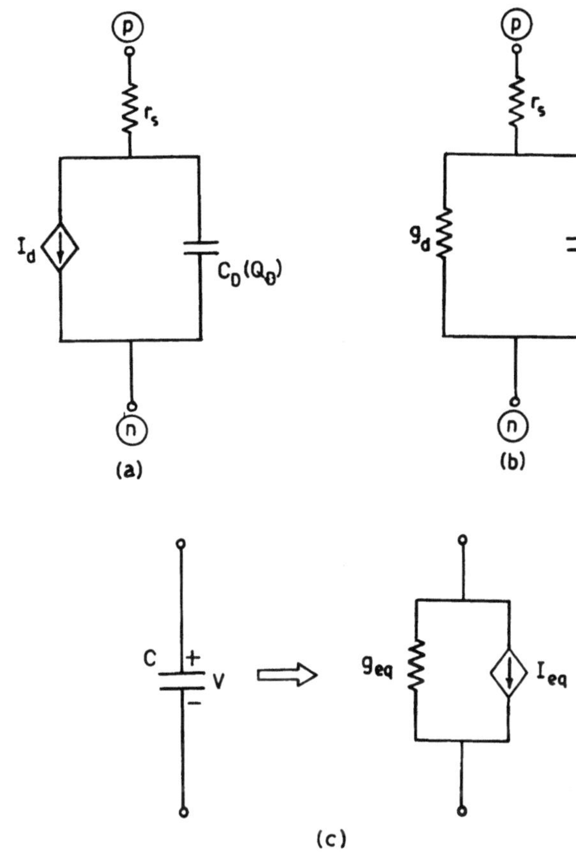

Fig. 2.21 Diode (a) large signal model for the transient analysis (b) linearized small signal model. (c) Companion model for the nonlinear capacitance

The large signal equivalent circuit model for the diode transient analysis is shown in Figure 2.21a. The total stored charge Q_D is given by

$$Q_D = A_d(Q_{\text{dif}} + Q_{\text{dep}}) = \tau_t I_d + A_d \int_0^{V_d} C_j dV \tag{2.85}$$

where we have made use of (2.76) for Q_{dif} and I_d is given by Eq. (2.82). Using Eqs. (2.74) and (2.75) for C_j we get

$$Q_D = \begin{cases} \tau_t I_d + A_d C_{j0} \dfrac{\phi_{bi}}{(1-m)} \left[1 - \left(1 - \dfrac{V_d}{\phi_{bi}} \right)^{1-m} \right], & V_d < F_c \phi_{bi} \\ \tau_t I_d + A_d C_{j0} F_1 + A_d \dfrac{C_{j0}}{F_2} \left[F_3 (V_d - F_c \phi_{bi}) + \dfrac{(V_d^2 - (F_c \phi_{bi})^2) m}{2\phi_{bi}} \right], & V_d > F_c \phi_{bi}. \end{cases} \tag{2.86}$$

The variables F_1, F_2 and F_3 are

$$F_1 = \frac{\phi_{bi}}{(1-m)}[1-(1-F_c)^{1-m}]$$
$$F_2 = (1-F_c)^{1+m} \tag{2.87}$$
$$F_3 = 1 - F_c(1+m)$$

where F_c is normally taken as 0.5 (cf. section 2.7.1) and is not a fitting parameter. The charge Q_D can be defined equivalently by the capacitance C_D as

$$C_D = \frac{dQ_D}{dV_d} = \begin{cases} \tau_t \dfrac{dI_d}{dV_d} + A_d C_{j0}\left(1 - \dfrac{V_d}{\phi_{bi}}\right)^{-m} & V_d < F_c\phi_{bi} \\ \tau_t \dfrac{dI_d}{dV_d} + \dfrac{A_d C_{j0}}{F_2}\left(F_3 + \dfrac{mV_d}{\phi_{bi}}\right) & V_d > F_c\phi_{bi}. \end{cases} \tag{2.88}$$

Again, C_D is first linearized using the companion model for the capacitance (see Figure 2.21c), which is nothing but a parallel combination of equivalent current and equivalent conductance whose value depends upon the integration method used [30]. Thus, to describe the large signal behavior of the diode, we need four parameters namely C_{j0}, m, ϕ_{bi} and τ_t.
The small signal equivalent circuit model for the diode AC analysis is shown in Figure 2.21b. The model requires small signal conductance g_d which is obtained from Eq. (2.78). Methods of determining diode model parameters are discussed in section 9.14 and 11.1.

2.10 Temperature Dependent Diode Model Parameters

Of the eight diode model parameters discussed in the previous section, those which change with temperature are I_s, τ_t, C_{j0} and ϕ_{bi}. The transit time τ_t varies rather weakly with temperature and therefore, its temperature dependence is not modeled in SPICE. Thus, the temperature dependence of only three parameters is considered.

2.10.1 *Temperature Dependence of I_s*

The saturation current I_s depends on temperature T through n_i^2 (Eq. 2.56) and hence, it increases strongly with temperature. Using Eq. (2.5) for n_i we can write I_s as

$$\boxed{I_s = CT^3 \exp\left[-\frac{E_g(T)}{kT}\right]} \tag{2.89}$$

where C includes all terms which are approximately independent of T. Note that we are ignoring any temperature dependence of D_p, D_n, L_p and L_n, although strictly speaking all these terms are temperature dependent. The *temperature coefficient* of I_s (fractional change in I_s per unit change in temperature) can be obtained by differentiating Eq. (2.89) as

$$\frac{1}{I_s}\frac{dI_s}{dT} = \frac{3}{T} + \frac{E_g(T)}{kT^2}. \tag{2.90}$$

The first term is ~1%/K at $T = 300\,\text{K}$ but the second term is ~14%/K. In other words, I_s *approximately doubles every* 5°C. However, experimentally it has been observed that the I_s reverse current doubles every 8°C. This is because Eq. (2.90) assumes that I_s is governed by n_i^2 while in reality, as was pointed out earlier (section 2.6.1), leakage current is governed by n_i rather than n_i^2.

A relation similar to (2.89) holds for other types of diodes, like Schottkey Barrier Diodes (SBD), and in general

$$I_s = CT^p \exp\left(\frac{-E_g(T)}{qV_t}\right) \tag{2.91}$$

where p is the saturation-current temperature exponent and E_g is the band gap energy, which is a function of temperature. SPICE assumes $E_g = 1.11\,\text{eV}$ for silicon, 0.67 eV for Germanium, and 0.69 eV for SBD. The temperature exponent factor p equals 3 for silicon and germanium while for SBD its value is 2. From Eq. (2.91), I_s at any temperature T can be calculated in terms of its value $I_s(T_0)$ at a known temperature T_0 (say room temperature) from the relation

$$\boxed{I_s(T) = I_s(T_0)\left(\frac{T}{T_0}\right)^p \exp\left[-\frac{E_g(T)}{kT} + \frac{E_g(T_0)}{kT_0}\right].} \tag{2.92}$$

This is the equation used in SPICE for temperature dependence of I_s. The temperature coefficient of diode forward current for a fixed forward bias is given by

$$\begin{aligned}\frac{1}{I_d}\frac{dI_d}{dT} &\approx \frac{1}{T}\frac{d}{dT}(I_s e^{V_d/V_t}) \\ &= \left[\frac{1}{I_s}\right] - \left[\frac{1}{T}\frac{V_d}{V_t}\right].\end{aligned} \tag{2.93}$$

This shows that the fractional change in the forward current is less than the fractional change in the saturation current.

2.10.2 Temperature Dependence of ϕ_{bi}

According to the Eq. (2.44), the temperature dependence of ϕ_{bi} is through $V_t(=kT/q)$ and n_i, i.e.

$$\phi_{bi} = \frac{2kT}{q} \ln\left(\frac{C}{n_i}\right) \tag{2.94}$$

where $C=\sqrt{N_a N_d}$ is a constant independent of temperature. The temperature dependence of ϕ_{bi} is obtained in a way similar to that for the temperature dependence of ϕ_f [see Eq. (2.17)] and is given by the following equation

$$\phi_{bi}(T) = \left(\frac{T}{T_0}\right)\phi_{bi}(T_0) - \frac{2kT}{q}\left[1.5\ln\left(\frac{T}{T_0}\right) + \left\{-\frac{E_g(T)}{2kT} + \frac{E_g(T_0)}{2kT_0}\right\}\right] \quad \text{(V)} \tag{2.95}$$

where $\phi_{bi}(T_0)$ is the value of ϕ_{bi} at reference temperature T_0. This is the equation used in SPICE for temperature dependence of ϕ_{bi}.

2.10.3 Temperature Dependence of C_{j0}

The temperature dependence of zero-bias depletion layer capacitance C_{j0} is due to the temperature dependence of the dielectric constant of silicon (ϵ_{si}) and that of ϕ_{bi}. Generalizing Eq. (2.72) for any doping profile we have

$$C_{j0} = B\left(\frac{\epsilon_{si}^2}{\phi_{bi}}\right)^m \tag{2.96}$$

where B is a constant. Differentiating Eq. (2.96) with respect to T and remembering that

$$\frac{1}{\epsilon_{si}}\frac{\partial \epsilon_{si}}{\partial T} = 2\cdot 10^{-4}/°\text{C}$$

for silicon, we obtain C_{j0} at any temperature T in terms of a known temperature T_0 as

$$C_{j0}(T) = C_{j0}(T_0)\left[1 + m\left\{4\cdot 10^{-4}(T-T_0) - \frac{\phi_{bi}(T)-\phi_{bi}(T_0)}{\phi_{bi}(T_0)}\right\}\right]. \tag{2.97}$$

Measured C_{j0} in the temperature range 0–120 °C agrees fairly well with Eq. (2.97). This can be seen from Figure 2.22 where measured diode junction

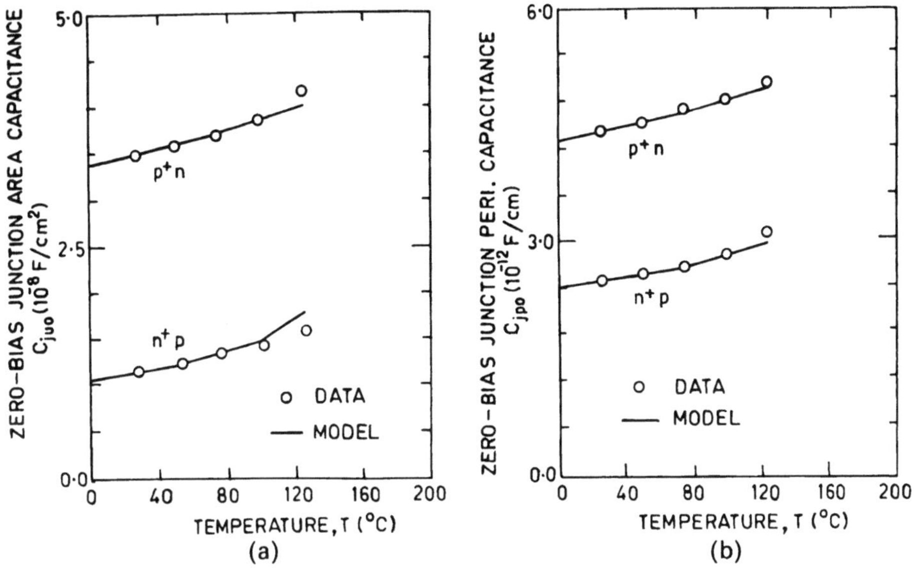

Fig. 2.22 Diode zero-biased junction capacitances as a function of temperature; (a) area capacitance and (b) periphery capacitance. Circles are measured data while lines are based on Eq. (2.97)

bottom-wall (area) and side-wall (periphery) capacitances at zero bias, C_{j0} and C_{jsw0}, respectively, are plotted as a function of temperature for a n^+p and p^+n diodes. The capacitances were measured using the test structures shown in Figure 2.18.

References

[1] R. A. Smith, *Semiconductors*, 2nd Ed., Cambridge University Press, London, 1978.

[2] A. S. Grove, *Physics and Technology of Semiconductor Devices*, John Wiley & Sons, New York, 1965.

[3] B. G. Streetman, *Solid State Electronic Devices*, 2nd ed., Prentice Hall, Englewood Cliffs, NJ, 1981.

[4] R. M. Warner Jr. and B. L. Grung, *Transistors—Fundamentals for the Integrated-Circuit Engineer*, John Wiley & Sons, New York, 1983.

[5] S. M. Sze, *Physics and Technology of Semiconductor Devices*, John Wiley & Sons, New York, 1985.

[6] R. S. Muller and T. I. Kamins, *Device Electronics for Integrated Circuits*, John Wiley & Sons, New York, 1986.

[7] R. F. Pierret, *Advanced Semiconductor Fundamentals*, Vol. VI, Modular Series on Solid-State Devices, Addison-Wesley Publishing Co., Reading MA, 1987.

[8] S. Wang, *Fundamentals of Semiconductor Theory and Devices*, Prentice Hall, N.J., 1989.

[9] M. Zambuto, *Semiconductor Devices*, McGraw-Hill Book Company, New York, 1989.

[10] M. Shur, *Physics of Semiconductor Devices*, Prentice Hall, Englewood Cliffs, N.J., 1990.

[11] G. W. Neudeck, *The PN Junction Diode*, Vol. II, 2nd Ed., Modular Series on Solid-State Devices, Addison-Wesley Publishing Co., Reading MA, 1987.

[12] D. J. Roulston, *Bipolar Semiconductor Devices*, McGraw-Hill Publishing Company, New York, 1990.

[13] M. Aoki, K. Yano, T. Masuhara, S. Ikeda, and S. Meguro, 'Optimum crystallographic orientation of submicron CMOS devices', 1985 IEDM Technical Digest, pp. 577–579.

[14] T. Kamins, *Polycrystalline Silicon for IC Application*, Kluwer Academic Publisher, Boston, 1988.

[15] M. A. Green, 'Intrinsic concentration, effective density of states, and effective mass in silicon', J. Appl. Phys., 67, pp. 2944–2954 (1990).

[16] F. H. Gaensslen and R. C. Jaeger, 'Temperature dependent threshold voltage behavior of depletion-model MOSFETS—characterization and simulation', Solid-State Electron., 22, pp. 423–430 (1979).

[17] S. Selberherr, 'MOS device modeling at 77K', IEEE Trans. Electron. Devices, ED-36, pp. 1464–1474 (1989).

[18] Y. P. Varshni, 'Temperature dependence of the energy gap in semiconductors', Physica (Amsterdam), 34, p. 149 (1967).

[19] H. D. Barber, 'Effective mass and intrinsic concentration in silicon', Solid-State Electronic, Vol. 10, pp. 1039–1051 (1967).

[20] S. Selberherr, *Analysis and Simulation of Semiconductor Devices*, Springer-Verlag, Wien, New-York, 1984.

[21] M. Chrzanowska-Jeske and R. C. Jaeger, 'BILOW—simulation of low temperature bipolar device behavior', IEEE Trans. Electron. Devices, ED-36, pp. 1475–1488 (1989).

[22] W. Shockley and W. T. Read, 'Statistics of the recombination of holes and electrons', Phys. Rev., Vol. 87, p. 835 (1952)

[23] R. N. Hall, 'Electron-hole recombination in germanium', Phys. Rev., Vol. 87, pp. 387–392 (1952)

[24] C. Jacoboni, C. Canalo, G. Ottaviani, and A. Quaranta, 'A review of some charge transport properties of silicon', Solid State Electron., 20, pp. 77–89 (1977).

[25] N. D. Arora, J. R. Hauser, D. J. Roulston, 'Electron and hole mobilities in silicon as a function of concentration and temperature', IEEE Trans. Electron. Devices, ED-29, pp. 292–295 (1982).

[26] W. R. Thurber and J. R. Lowney, 'Electrical transport properties of silicon', in *VLSI Handbook*, Ed. N. G. Einspruch, Academic Press, New York, 1985.

[27] B. R. Chwala and H. K. Gummel, 'Transition region capacitance of diffused *pn* junctions', IEEE Trans. Electron. Devices, ED-18, pp. 178–195 (1971).

[28] H. G. Poon and H. K. Gummel, 'Modeling of emitter capacitance', Proc. IEEE (Lett.), 57, pp. 2181–2182 (1969)

[29] B. A. Freese and G. L. Buller, 'A method of extracting SPICE2 Junction capacitance parameters from measured data', IEEE Electron Devices Lett., EDL-5, pp. 261–263, (1984).

[30] L. O. Chua and P. M. Lin, *Computer Aided Analysis of Electronic Circuits: Algorithms & Computational Techniques*, Prentice Hall, Englewood Cliffs, NJ, 1975.

MOS Transistor Structure and Operation 3

In this chapter we will give an overview of the MOS transistor as used in VLSI technology, and its behavior under operating biases will be explained qualitatively. First we will describe the basic MOSFET structure and then qualitatively discuss its current-voltage characteristics. During the last two decades, device lengths have been reduced from 20 μm to less than a micron, which has resulted in high fields in the device. The rules of device scaling are first discussed followed by the impact of high field effects on device characteristics. Although there are various high field effects, the one which is of most concern for VLSI design is the so called hot-carrier effects. Only an overview is covered in this chapter, the detailed hot-carrier modeling is the subject of discussion in Chapter 8. Finally, a brief description of device structures specifically for VLSI design, that is important from a device modeling point of view, will be covered.

3.1 MOSFET Structure

As the name metal-oxide-semiconductor (MOS) suggests, the MOS transistor consists of a semiconductor substrate (usually silicon) on which is grown a thin layer of insulating oxide (SiO_2) of thickness t_{ox} (80–1000 Å).[1] A conducting layer (a metal or heavily doped polysilicon) called the *gate* electrode is deposited on top of the oxide. Two heavily doped regions of depth X_j (0.1–1.0 μm), called the *source* and the *drain* are formed in the substrate on either side of the gate. The source and the drain regions overlap slightly with the gate (see Fig. 3.1). The source-to-drain electrodes are equivalent to two *pn* junctions back to back. This region between the source and drain junctions is called the *channel region.* Thus a *MOS transistor is essentially a MOS structure, called the MOS capacitor, with two pn junctions on either side of the gate.* The field oxide (FOX) shown in Figure 3.1 is for

[1] In the future, with higher package density chips, t_{ox} will be less than 80 Å.

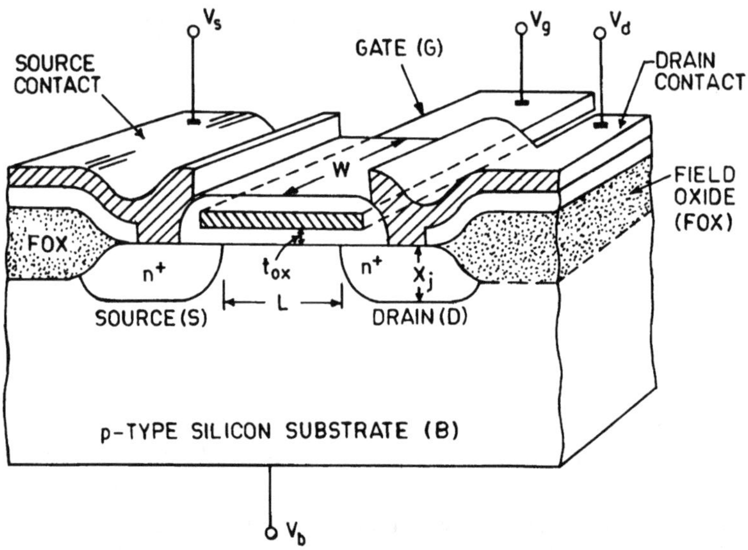

Fig. 3.1 MOS transistor structure showing three-dimensional view

isolating various devices on the same substrate as will be discussed in section 3.5.4. From the circuit model point of view, a MOS transistor is a four terminal device, the four terminals are designated as *gate g*, *source s*, *drain d*, and *substrate or bulk b*. Note that the structure is symmetrical. Because of this symmetry one cannot distinguish between the source and drain of an unbiased device; the roles of the source and the drain are defined only after the terminal voltages are applied.

Under normal operating conditions, a voltage V_g applied to the gate terminal creates an *electric field* that controls the flow of the charge carriers in the channel region between the source and the drain. Since the device current is controlled by the electric field (vertical field due to the gate voltage and lateral field due to the source to drain voltage) the device is known as a *MOS Field-Effect-Transistor* (MOSFET). Because the gate is electrically isolated from the other electrodes, this device is also called an *Insulated-Gate Field-Effect Transistor* (IGFET). Another acronym sometime used is MOST for the MOS Transistor. The bulk of the semiconductor region, shown as substrate in Figure 3.1, is normally inactive, since the current flow is confined to a thin channel (10–100 Å thick) at the surface of the semiconductor. It is for this reason the substrate region is also referred to as the *body* or *bulk* of the MOSFET.

MOSFETs may be either *n*-channel or *p*-channel depending upon the type of the carriers in the channel region. An *n*-channel MOS transistor (nMOST) has heavily doped n^+ source and drain regions with a *p*-type substrate and has electrons as the carriers in the channel region. While a *p*-channel MOS transistor (pMOST) has heavily doped p^+ source and drain regions with an n-type substrate and has holes as the carriers in the channel region.[2] Since a single type of charge carrier is involved for normal device operation (electrons for n-channel and holes for p-channel), these devices are also called *unipolar transistors* in contrast with the bipolar transistors whose operation depends on both type of carriers (electrons and holes). In addition to the type of the channel, MOSFETs are also classified according to the mode of operation.

The MOSFET which has no conducting channel between the source and drain at zero gate voltage is termed a *normally-off* device or more commonly an *enhancement-mode* device (E-device). In such devices a certain minimum gate voltage, called the *threshold* or *turn-on voltage* V_{th} is required to induce a conducting channel. In other words, the channel must be "enhanced" to cause conduction and hence the name enhancement mode device. If a conducting channel exists between the source and the drain so that the device is conducting even at zero gate voltage (i.e. the device is *normally-on*) then it is called a *depletion-mode* device (D-device) as a gate voltage is required to "deplete" the channel so as to turn the device off. The depletion-mode device is sometimes referred to as a *buried channel* device, because current flow is not exactly at the surface, as in the case of the enhancement-mode device, but some what away from the surface in the bulk of the silicon. Table 3.1 gives conditions on the gate electrode for turning 'on' or turning 'off' the four types of MOSFETs.

Since the gate is isolated from other electrodes by the insulating oxide layer, there is effectively no DC path between the gate and other electrodes. This results in a very *high DC input impedance* of the order of 10^{13}–$10^{15}\,\Omega$ and is primarily capacitive. Because of its high input impedance, a MOSFET requires very low steady state input power. This means that one transistor can conceptually drive many other transistors similar to it, i.e. it has a high *fan-out* capability[3].

The MOSFET shown in Figure 3.1 is an n-channel device. The distance L between the n^+ source-drain edges is called *channel length*. The distance

Table 3.1. *Four different types of MOSFETs*

Device Type	Normal State	Gate Voltage n-channel	Gate Voltage p-channel
Enhancement mode	OFF	$+V_g$ turns on	$-V_g$ turns on
Depletion mode	ON	$-V_g$ turns off	$+V_g$ turns off

[2] Throughout the book we will use the acronym nMOST and pMOST for n-channel and p-channel MOSFET respectively.

[3] Note that switching time will be affected by the capacitive loading and requires careful attention in real circuit design.

W to which the device is extended in the lateral direction (i.e. into and out of the page) is called *channel width*. The device width to length ratio (W/L) is called the *aspect ratio* and is normally used as a design parameter that can be varied to set the desired drain-source conduction properties of the MOSFET.

MOSFET Circuit Symbols. Circuit symbols for MOSFETs are shown in Figure 3.2. The symbols for enhancement mode devices are shown in Figure 3.2a while those for depletion mode devices are shown in Figure 3.2b. In fact these symbols reflect the basic structural features of the device, that is, gate to be physically isolated from the source and drain regions. The type of the MOSFET (i.e. either *n*- or *p*-type) is designated by the direction of the arrow on the body or the substrate terminal. This arrow designates the polarity of the *pn* junction formed between the source/drain and the substrate, and is in the same direction as a forward biased diode, that is, it points from the *p*-side to the *n*-side of the junction. The depletion mode MOSFETs are shown with a thick line across the source-drain regions to show the existence of a conducting channel under the gate. In many circuit drawings where the body or substrate connection is not shown explicitly, a slightly different set of symbols are used for enhancement devices as shown in Figures 3.2c and 3.2d.
In circuit design it is customary to define voltages at different terminals of the device with respect to the source as the reference potential. Thus, if V_g, V_s, V_d and V_b are the gate, source, drain and bulk (substrate) voltages respectively to some arbitrary ground reference, then we normally define

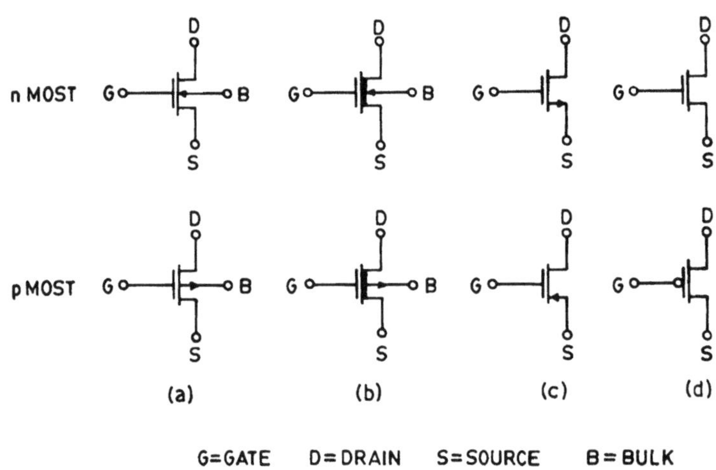

Fig. 3.2 Set of commonly used circuit symbols for *n*-channel and *p*-channel (a) enhancement mode MOSFET, (b) depletion mode MOSFET, (c) and (d) alternate symbols for enhancement mode devices

Table 3.2. *Operating voltages for nMOST and pMOST*

Device type	V_{ds}	V_{gs}	V_{bs}	I_{ds}
nMOST	+	+	−	+
pMOST	−	−	+	−

terminal voltages as drain-source voltage $V_{ds}(=V_d - V_s)$, gate-source voltage $V_{gs}(=V_g - V_s)$, and bulk-source voltage $V_{bs}(=V_b - V_s)$. *For normal DC operation of the device it is implicitly assumed that the only current that flows through the device is the drain-source current or simply drain current* I_{ds} and is defined to be positive flowing into the drain terminal; it is the terminal current of the MOSFET. The current-voltage (I-V) relation assumes the general form

$$I_{ds} = f(V_{gs}, V_{ds}, V_{bs}) \tag{3.1}$$

indicating that all of the device voltages are important in controlling the drain current. Note that the controlling parameters of a MOSFET are voltages as opposed to currents in a bipolar transistor. Polarities of voltages and currents for nMOST and pMOST are reversed as shown in the Table 3.2.

3.2 MOSFET Characteristics

This section gives an informal, qualitative description of the classical long channel enhancement n-type MOSFET (nMOST) shown in Figure 3.3. Although continuous shrinking of MOSFET size and technology improvements have resulted in a more complicated structure, which has its effect on modeling, the essential structure remains the same as that shown in Figure 3.3. Under normal operating conditions, the *source and drain voltages are always such that the source and drain-to-substrate pn junctions are reverse biased.* The simplest bias arrangement that can be used to illustrate the operation of a MOSFET is when both the source and the bulk are at ground potential i.e. $V_b = V_s = V_{sb} = 0$. Even at $V_{gs} = V_{ds} = 0$ a depletion region is formed around n^+ source and drain regions (see dashed lines Figure 3.3) due to the n^+p junction formed with the p-type substrate of concentration N_b (cm^{-3}). The width X_{sd} and X_{dd} of this depletion region under the source and drain, respectively, based on the one-dimensional abrupt junction approximation [cf. Eq. (2.53)], is given by the following equation

$$X_{sd} = X_{dd} = \sqrt{\frac{2\epsilon_0 \epsilon_{si} \phi_{bi}}{qN_b}} \quad \text{(cm)} \quad \text{at } V_{ds} = V_{bs} = 0 \tag{3.2}$$

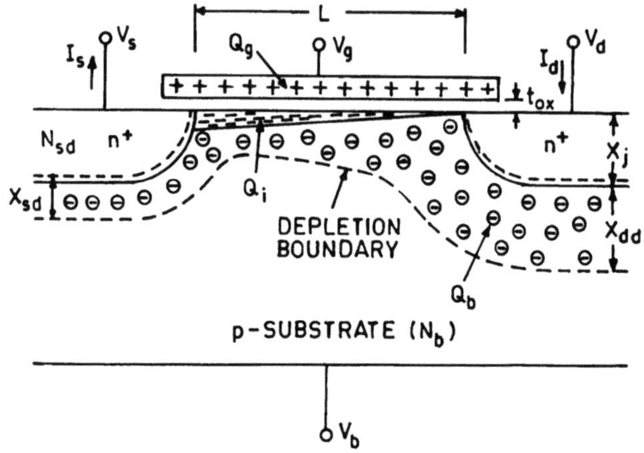

Fig. 3.3 Cross-section of a *n*-channel MOSFET showing voltages, currents, and charge symbols. Dotted lines show depletion boundaries

where ϕ_{bi} is the built-in potential between the source/drain to substrate *pn* junction given by [cf. Eq. (2.44)]

$$\phi_{bi} = V_t \ln\left(\frac{N_{sd}N_b}{n_i^2}\right) \quad \text{(V)} \tag{3.3}$$

where $V_t = kT/q$ is the thermal voltage, $N_{sd}(\sim 10^{20}\,\text{cm}^{-3})$ is the concentration of the source/drain region, and n_i is the intrinsic carrier concentration.
Let us assume the drain terminal is at a certain positive voltage V_{ds}. When a positive V_{gs} that is less than a certain minimum gate voltage, called the *threshold voltage* V_{th}, is applied to the gate, the *p*-type surface region is depleted of holes underneath the gate oxide.[4] Because holes are pushed away from the surface leaving behind the immobile negatively ionized atoms, a negative charge is built up at the silicon surface. This charge is called the *depletion or bulk charge* Q_b. Under this condition the only current that flows is the leakage current.
If V_{gs} is now increased so that $V_{gs} > V_{th}$ is applied to the gate, a conducting channel with a mobile negative charge Q_i is formed at the surface. This channel at the surface is also called an *inversion layer* because the *surface layer is inverted from p-type, before conduction, to n-type after the conducting channel is formed.* The thickness of this inversion layer is 10–100 Å and depends upon the applied bias (see section 4.2.3). At $V_{gs} = V_{th}$, the concentration of the minority carriers (electrons) at the surface equals the majority carrier holes (*p*-type substrate). The higher the $V_{gs}(> V_{th})$, the higher the minority carrier charge density Q_i. The mobile charge Q_i is also called

[4] This can be understood more clearly after we study the MOS capacitor in Chapter 4.

inversion charge. From the charge conservation principle, the sum of Q_i and Q_b equals the gate charge Q_g. Now if there is a voltage difference between the source and drain, a current I_{ds} will flow, *due to the diffusion of the carriers* (electrons in nMOST) from the drain to source.[5] Note that *pn* junction leakage current still flows and adds to the current due to channel formation. However, it is so small in magnitude compared to the current due to the channel formation that it can be neglected. Since the inversion charge Q_i depends heavily on the applied gate voltage, the gate can be used to control the current through the channel. Thus, an amplifying function can be realized. By biasing the structure in the cutoff region ($V_{gs} < V_{th}$) current is prevented to flow between the source and drain. Hence the transistor can be used as a switch.
For a fixed $V_{gs}(> V_{th})$, the drain current I_{ds} increases linearly with increasing drain voltage V_{ds}. The rate of increase decreases until I_{ds} saturates to a constant value. In this region the MOSFET is operating as a variable resistor which varies with the gate voltage; the channel resistance decreases with increasing V_{gs}. For this reason the MOSFET is said to be a *voltage controlled device.* The relationship between I_{ds} and V_{ds} for various values of V_{gs} for an experimental polysilicon gate nMOST with $L = 10\,\mu$m is shown in Figure 3.4. It shows four distinct operating regions [1]–[3]:

Linear Region. It is the region in which I_{ds} increases linearly with V_{ds} for a given $V_{gs}(> V_{th})$. To a first approximation, I_{ds} in the linear region is given by (see section 6.4.1)

$$I_{ds} = \mu C_{ox}\left(\frac{W}{L}\right)(V_{gs} - V_{th} - 0.5V_{ds})V_{ds} \quad \text{(linear-region)} \tag{3.4}$$

where μ is mobility of the carriers (electrons for nMOST) in the channel (inversion) region, C_{ox} is the gate oxide capacitance per unit area,[6] W/L is device width to length ratio, and V_{th} is threshold voltage. As we shall see later in section 5.1, to the first order V_{th} depends upon the gate oxide thickness t_{ox}, substrate doping concentration N_b, and type of the gate

[5] An alternative view point of the current flow from source to drain that does not require the concept of surface inversion layer is as follows. There exists an energy barrier across the source which inhibits the flow of electrons from the source. Application of gate voltage reduces this energy barrier and when V_{gs} becomes greater than V_{th}, electrons are emitted from the source. For nonzero V_{ds}, emitted electrons are collected by the drain resulting in a flow of current from drain to source. The name "source" and "drain", probably have been derived from this concept where source is emmitter of the carriers and drain is collector of the carriers.

[6] In MOS modeling it is common practice to express the gate oxide capacitance as per unit area rather than the total capacitance. If t_{ox} is the gate oxide thickness then $C_{ox} = \epsilon_0\epsilon_{ox}/t_{ox}$, where $\epsilon_0(= 8.854 \cdot 10^{-14}$ F/cm) is the permittivity of free space and $\epsilon_{ox}(= 3.9)$ is the relative permittivity (or dielectric constant) of gate oxide material.

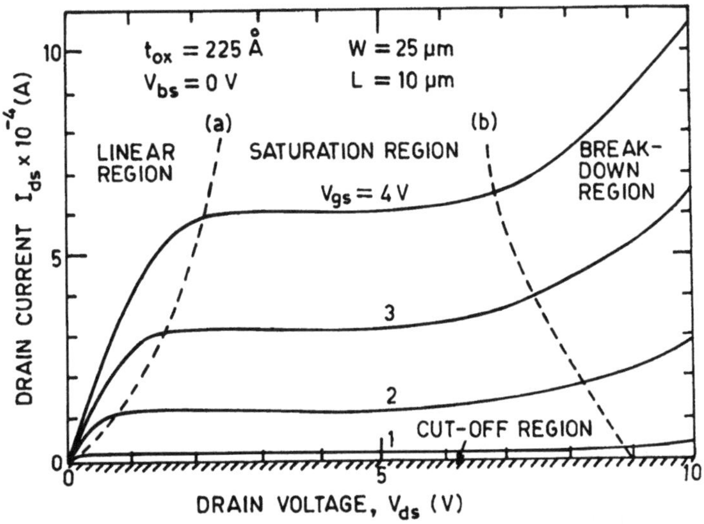

Fig. 3.4 Typical enhancement MOSFET (L = 10 µm) drain current (I_{ds}) drain voltage (V_{ds}) characteristics with gate voltage (V_{gs}) as a parameter, showing different regions of device operation; (a) linear, (b) saturation, (c) cut-off, and (d) breakdown regions

material (Al or polysilicon); V_{th} increases with increase of t_{ox} or N_b. Although to a first approximation μ is assumed constant in Eq. (3.4), in reality it is a function of gate and drain fields as we shall see in section 6.6.

Saturation Region. In this region I_{ds} no longer increases as V_{ds} increases, i.e. it saturates. Again to a first approximation, I_{ds} in the saturation region is given by

$$I_{ds} = \frac{1}{2}\mu C_{ox}\left(\frac{W}{L}\right)(V_{gs} - V_{th})^2 \quad \text{(saturation-region)} \tag{3.5}$$

showing that I_{ds} does not depend on V_{ds}. This is evident from Figure 3.4. It should be pointed out that this complete current saturation occurs only for MOSFET's with long channel lengths ($L = 10\,\mu$m, for example). As L decreases the saturation behavior degrades rapidly, causing an increase in I_{ds} when V_{ds} is increased (for details see section 6.7). Different mechanisms contribute to the degradation of saturation behavior for short channel devices, like mobility degradation due to carrier velocity saturation, source-drain resistance, etc. Dashed line 'a' (Figure 3.4), shows an approximate boundary between the linear and saturation region.

Breakdown Region. With further increase of V_{ds} beyond saturation, the transistor enters a region in which I_{ds} suddenly increases until breakdown of the drain-to-substrate *pn* junction occurs (cf. section 2.7) and is caused by the high electric field at the drain end. This *increase is quite sharp in aluminum gate technology but is much softer for polysilicon gate technology.* In short-channel devices, particularly in nMOST, the so called *hot-carrier effect*, due to the high electric field at the drain end, can also result in device breakdown [4]–[6]. Dashed line 'b' (Figure 3.4) shows the boundary between the saturation and breakdown region.

Cut-Off Region. This is the region in which $V_{gs} < V_{th}$ so that no channel exist between the source and the drain, resulting in $I_{ds} = 0$. However, in real devices, behavior is different as shown in Figure 3.5. In fact for $V_{gs} < V_{th}$, drain current follows an exponential decay (see Figure 3.5; note the log scale in the current axis) and is referred to as *weak inversion* or *subthreshold* (or leakage) current. *This leakage current is over and above the leakage current due to the reverse biased source-drain junction diode which is of the order of* 10^{-12} A *or lower.* In weak inversion, the *p*-type silicon surface has changed to *n*-type, but the inversion is weak meaning electrons concentration at the surface is still lower than the holes concentration. The low electron concentration results in low electric field along the channel and hence the *subthreshold current is mainly due to diffusion of carriers.* The current in subthreshold region is approximated as (see section 6.4.5)

$$I_{ds} = \frac{\mu C_{ox} W}{L}(\eta - 1)V_t^2 \exp\left[\frac{V_{gs} - V_{th}}{\eta V_t}\right](1 - e^{-V_{ds}/V_t}), \qquad V_{gs} < V_{th} \tag{3.6}$$

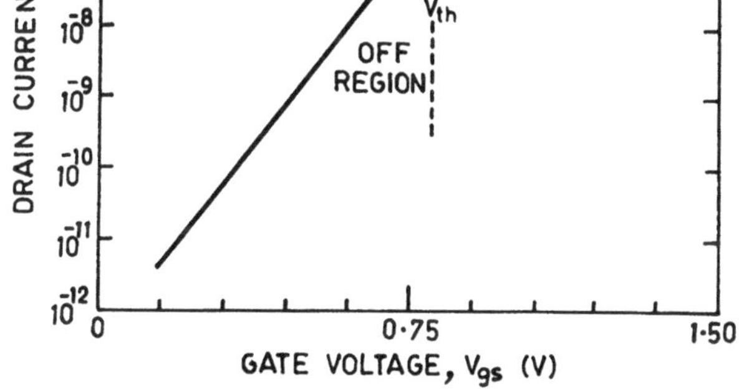

Fig. 3.5 Typical enhancement MOSFET $I_{ds} - V_{gs}$ characteristics at a fixed $V_{ds} = 0.1$ *V*. The drain current I_{ds} follows an exponential decay for gate voltage V_{gs} less than the threshold voltage V_{th}.

where η is a factor between 1 and 3 and signifies capacitive coupling between the gate and silicon surface. Note that I_{ds} in the subthreshold region is almost independent of V_{ds}, for V_{ds} larger than a few kT/q (i.e. $V_{ds} > 0.1$ V). However, for short channel devices I_{ds} does depends upon V_{ds}.
Normally leakage currents are minimized in device design. Leakage currents are sometimes ignored for circuit simulation models. Cut-off is important in digital switching circuits since it allows the drain current to be switched off by setting $V_{gs} < V_{th}$. The slope of the I_{ds}–V_{gs} curve in the subthreshold or cut-off region, often referred to as *subthreshold slope*, is an important parameter in device design as we shall see in section 6.4.5.

Effect of Substrate Bias. In the discussion so for we have assumed that $V_{bs} = 0$. However, when bias is applied to the substrate, modulation of the channel conductance results. Thus the substrate can act as a second gate and in fact is sometimes referred to as the *back gate* and V_{bs} is referred as *backgate bias.*[7] The input impedance of this control electrode is much lower than that of the front gate. Leakage currents are also higher in this case. Increasing the magnitude of V_{bs} will lower I_{ds} for a given value of V_{gs} and V_{ds}. This is because as $|V_{bs}|$ increases, the depletion region extends further into the substrate as more holes are depleted from the surface and immobile acceptor ions are generated—just as in *pn* junction theory. This results in higher bulk charge Q_b. A portion of the electric field lines originating from the inversion layer charge Q_i will now be diverted to the new immobile fixed charge Q_b. For a fixed V_{gs} and V_{ds}, the gate charge Q_g is fixed; therefore from the charge conservation principle ($Q_g = Q_i + Q_b$), there will now be less inversion charge Q_i and hence less conduction. Mathematically, as V_{bs} increases so does V_{th} (as we shall see in section 5.1), therefore according to Eqs. (3.1) and (3.4) I_{ds} decreases. This indeed is the case as shown in Figure 3.6, where experimental $I_{ds} - V_{ds}$ characteristics at two V_{bs} values (0V and -3V) are shown for a long channel MOSFET.
The MOSFET characteristics shown in Figures 3.4 and 3.6 are often called *output* characteristics while those shown in Figure 3.5 are called *transfer* characteristics. The transfer characteristics showing all regions of device operation are shown in Figure 3.7. The above considerations for *n*-channel enhancement MOSFETs also applies to *p*-channel MOSFETs, except that channel conduction is then due to holes and hence, hole mobility μ_p must be used. Also, all polarities of voltages and currents are reversed (see Table 3.2).

As was pointed out earlier, unlike enhancement devices, the depletion devices conduct even at zero V_{gs}. In this case, the gate voltage which completely turns the device off (no conduction) is called 'pinch-off voltage'

[7] The voltage V_{bs} is also called the *body* or *back bias.* The three terms *body bias, backgate bias* or simply *back bias* are often used synonymously.

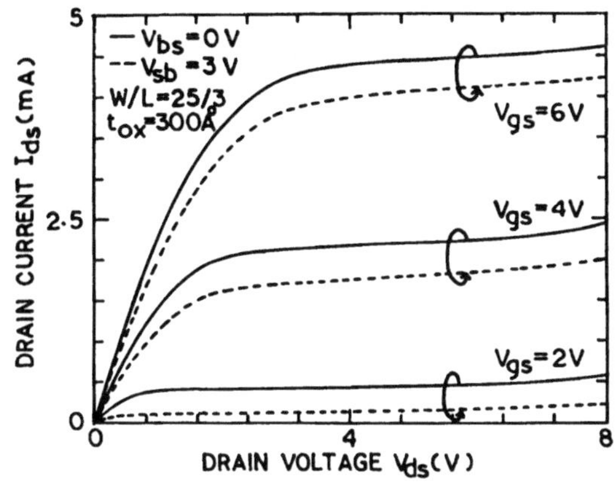

Fig. 3.6 Typical enhancement MOSFET output characteristics with gate voltage V_{gs} as parameter at two back bias V_{sb}

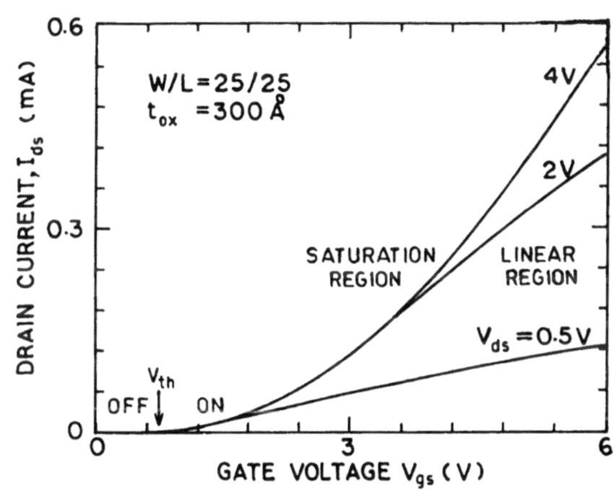

Fig. 3.7 Typical MOSFET transfer characteristics showing all regions of device operation

V_p, also referred to as threshold voltage V_{th}. Thus, in depletion devices V_p and V_{th} are used synonymously. While the *n*-channel enhancement device has a positive threshold voltage, the corresponding *n*-channel depletion device will have a negative threshold voltage. *The depletion device has similar structure to that of the enhancement device except that a conducting channel region is diffused into the substrate surface so that the device is normally on.* The transfer characterstics of an *n*-channel depletion device are shown in Figure 3.8a; compare these with the transfer characterstics of an enhancement device (Figure 3.7). For $V_{gs} = 0$, application of V_{ds} causes I_{ds} to flow through the conducting channel. As V_{ds} increases saturation occurs at the drain end similar to the enhancement device. Considering an *n*-channel depletion device, negative V_{gs} causes electrons to be repelled from

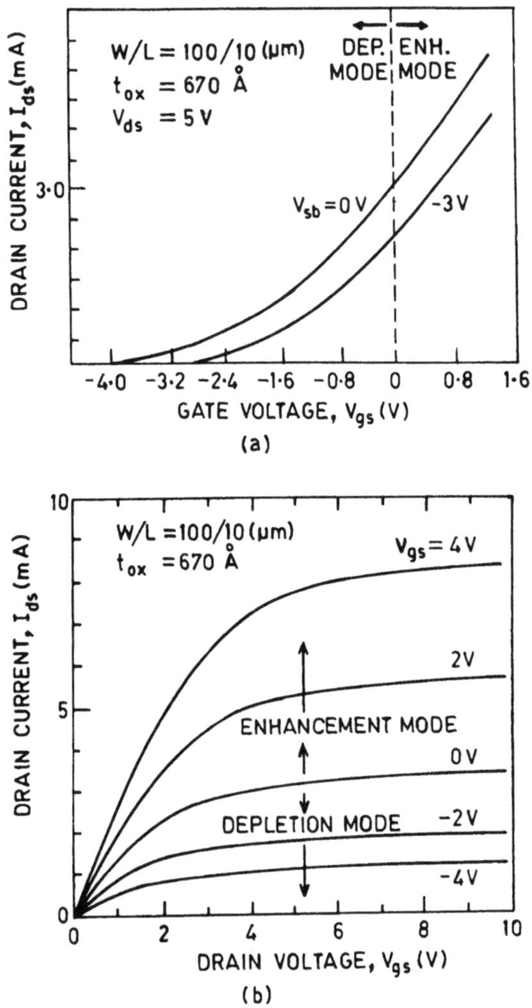

Fig. 3.8 Typical *n*-channel depletion MOSFET characteristics: (a) transfer, (b) output

the surface of the channel thereby reducing the conductivity and hence the drain current (see Figure 3.8b). This is the *depletion mode* of operation. If V_{gs} is positive, electrons are attracted to the channel region thus increasing the conductivity and hence the drain current increses. This is the *enhancement mode* of operation.

To a first order the drain current in a depletion device can be modeled using Eqs. (3.4) and (3.5) with negative V_{th}. Since in depletion devices the channel is formed away from the surface, the inversion layer mobility μ of the cariers is higher (10–20%), due to less scattering, than the enhancement devices where channel is formed at the surface. Note that depletion devices can be operated in both enhancement and depletion mode depending upon the applied voltage. In fact these devices have more than two modes of operations as discussed in section 6.5.

3.2.1 Punchthrough

Let us consider a MOSFET to which a small $V_{gs}(< V_{th})$ is applied and $V_{ds} = 0$ so that the device is off. An energy barrier exists between the source and the region under the gate, it is this barrier that holds the electrons in the source. The only current that flows is the drain leakage current of the reversed biased drain-substrate junction. The space-charge depletion width at the source and drain ends are symmetrical (dashed line *a* in Figure 3.9a). Applying a positive $V_{gs}(> V_{th})$ lowers the energy barrier, thus allowing electrons to move from the source and form a conducting channel connecting the source and the drain. Now if V_{ds} is increased with V_{gs} held constant, then the depletion width of the drain becomes closer to the source (dashed line *b* in Figure 3.9a). If V_{ds} is increased still further, eventually *at a certain drain voltage the drain depletion width touches the source depletion width resulting in lowering of the energy barrier of the source* (dashed line *c* in Figure 3.9a). When this happens a large amount of current flows even though the gate has been biased to turn the device off. The gate loses control over the drain and the device fails to operate normally. This phenomena is called *punchthrough* and V_{ds} induced current is called the *punchthrough current*. The corresponding drain voltage V_{ds} that causes a small but finite amount of drain current (usually 1nA-1pA) at or near zero V_{gs} is called punchthrough voltage V_{pt}. The amount of drain current used to define V_{pt} varies depending on the circuit requirements. For example, for DRAM's it is 1 pA while for SRAM's it is 1 nA. The punchthrough voltage V_{pt} can be calculated from the following approximate relation [2]

$$V_{pt} \approx \frac{qN_b}{2\varepsilon_0\varepsilon_{si}}(L - X_{dd})^2 - \phi_{bi}. \tag{3.7}$$

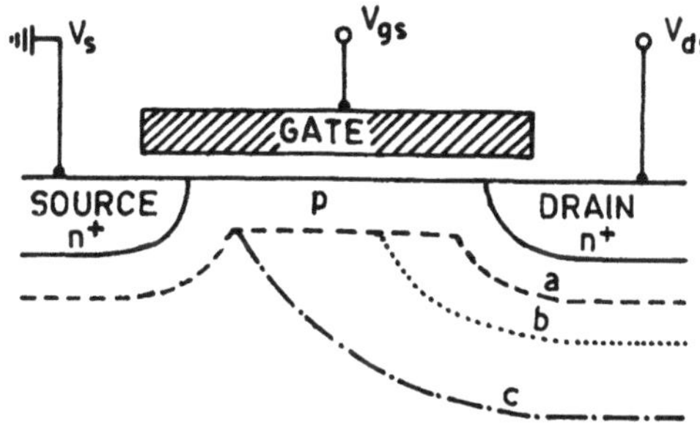

Fig. 3.9a Illustration of punchthrough phenomena in a MOSFET. The drain voltage V_{ds} increases from (a) to (c). At (c) drain depletion width touches source depletion width resulting in the punchthrough

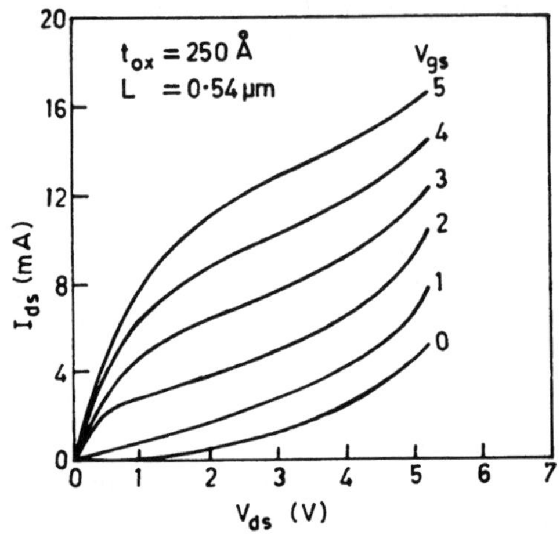

Fig. 3.9b MOSFET output characteristics showing punchthrough phenomena

In short-channel devices, punchthrough causes breakdown of the source/drain junctions. The effect of the punchthrough on the MOSFET current-voltage characteristics is shown in Figure 3.9b for a device with $L = 0.54\ \mu m$ and $t_{ox} = 225$ Å. When the device is in punchthrough, the drain current increases superlinearly with the drain voltage even at gate voltages below an expected threshold voltage (0.65 V, in this case). The punchthrough is normally avoided during device design and operation [7].

3.2.2 *MOSFET Capacitances*

The I–V characteristics shown in Figures 3.4–3.7 are steady-state or DC characteristics of a typical MOSFET. The transient behavior of a MOSFET is due to the device capacitive effects, which in fact are the results of the charges stored in the device. The stored charges are (1) the inversion charge Q_i in the inversion or channel region, (2) the bulk charge Q_b in the depletion region underlying the channel, (3) the gate charge $Q_g(-Q_i + Q_b)$ at the gate terminal, and (4) the charges due to the source/drain *pn* junctions. These charges give rise to the device capacitances shown in Figure 3.10. From the point of developing MOSFET dynamic or transient models, it is instructive to divide the device into two parts:

- The *intrinsic* part which forms the channel region of the device, and is mainly responsible for the transistor action (see Figure 3.10, dashed line). The charges which are responsible for the transistor action are the gate charge Q_g, the depletion or bulk charge Q_b and the inversion charge Q_i. The capacitances arising from these charges are called *intrinsic capacitances.* Thus, the capacitances C_{GS}, C_{GD} and C_{GB}, shown in Figure 3.10, are the

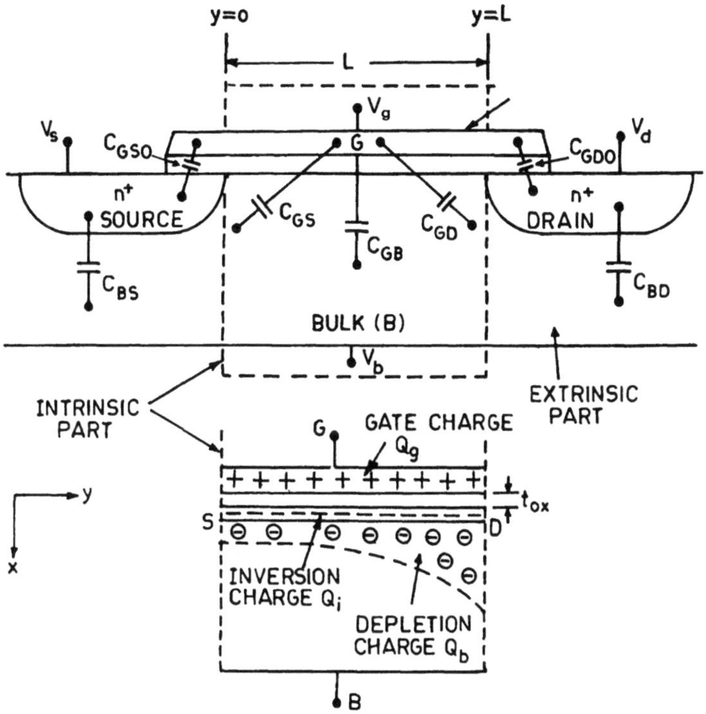

Fig. 3.10 MOSFET capacitances showing intrinsic and extrinsic part

intrinsic capacitances. The simple first order MOSFET capacitance model assumes only these three intrinsic capacitance, but in fact there are many more as we shall see in Chapter 7. These intrinsic capacitances are normally derived from the charges which are used to calculate the steady-state current I_{ds}. Thus, the capacitance expressions do not involve any new parameters other than those required for I_{ds} calculations (see Chapter 7).

- The *extrinsic part* which includes the source and drain *pn* junction portion of the MOSFET. Note that in the extrinsic part there is an inevitable overlap of the gate over the source and drain region. The capacitance arising from this overlap are called *gate overlap capacitances* shown as C_{GSO} and C_{GDO} in Figure 3.10. The source and drain *pn* junction capacitances (C_{BS} and C_{BD}) plus the gate overlap capacitances (C_{GSO} and C_{GDO}) are called *extrinsic capacitances.* These extrinsic capacitances are often called *parasitic capacitances* of the MOSFET (see section 3.6).

The capacitive characteristics of a MOSFET are then the sum of the intrinsic and extrinsic capacitances. It is these capacitances which are responsible for limiting overall device performance in terms of device switching speed.

3.2.3 Small-Signal Behavior

Under normal operation, a voltage applied to the gate, drain or substrate results in a change in the drain current. The ratio of change (increase) in the drain current (ΔI_{ds}) to the change (increase) in the gate voltage (ΔV_{gs}) while keeping drain and substrate voltages (V_{ds} and V_{bs}) constant is called *gate transconductance* or simply *transconductance* g_m,

$$g_m = \left.\frac{\Delta I_{ds}}{\Delta V_{gs}}\right|_{V_{ds}, V_{bs}} \quad \text{(A/V)}. \tag{3.8}$$

The transconductance g_m is one of the important device parameters as it is a measure of device gain. From Eqs. (3.4) and (3.5) it can be seen that

$$g_m = \begin{cases} \mu C_{ox}\left(\dfrac{W}{L}\right) V_{ds} & \text{(linear region)} \qquad (3.9a) \\[2ex] \mu C_{ox}\left(\dfrac{W}{L}\right)(V_{gs} - V_{th}) & \text{(saturation region).} \qquad (3.9b) \end{cases}$$

Thus, the gain of a MOSFET can be increased by

- Increasing C_{ox}, that is, using a MOSFET with a thinner gate oxide (lower t_{ox}).
- Using devices with higher carrier mobility μ. Since μ of the electrons is higher than that of holes, an nMOST has higher gain compared to pMOST.
- Using devices with larger channel width W and shorter channel length *L*. While decreasing *L*, scaling considerations must be taken into account as discussed in section 3.3.

It should be pointed out that though Eq. (3.9a) shows that g_m in the linear region is constant independent of the gate voltage V_{gs}, in real devices g_m varies with V_{gs}, being maximum at low V_{gs}. This discrepancy is due to the assumption of constant μ (independent of V_{gs}) in Eq. (3.90), while in reality μ is V_{gs} and V_{ds} dependent as we shall see later in section 6.6. For the same reason, in real devices g_m in saturation is V_{ds} dependent while Eq. (3.9b) predicts constant g_m (see Figure 3.11).

Since g_m is sensitive to the quality of the gate oxide (through μ and V_{th} parameters), it is frequently used to monitor the effect of hot-carrier stress and accelerated aging on device reliability as discussed in section 8.5.

In addition to g_m, the MOSFET has two other conductances. The ratio of change (decrease) in the drain current due to change (increase) in $|V_{bs}|$ for a fixed V_{gs}, and V_{ds} is called *substrate transconductance* g_{mbs},

$$g_{mbs} = \left.\frac{\Delta I_{ds}}{\Delta V_{bs}}\right|_{V_{gs}, V_{ds}} \quad \text{(A/V)}. \tag{3.10}$$

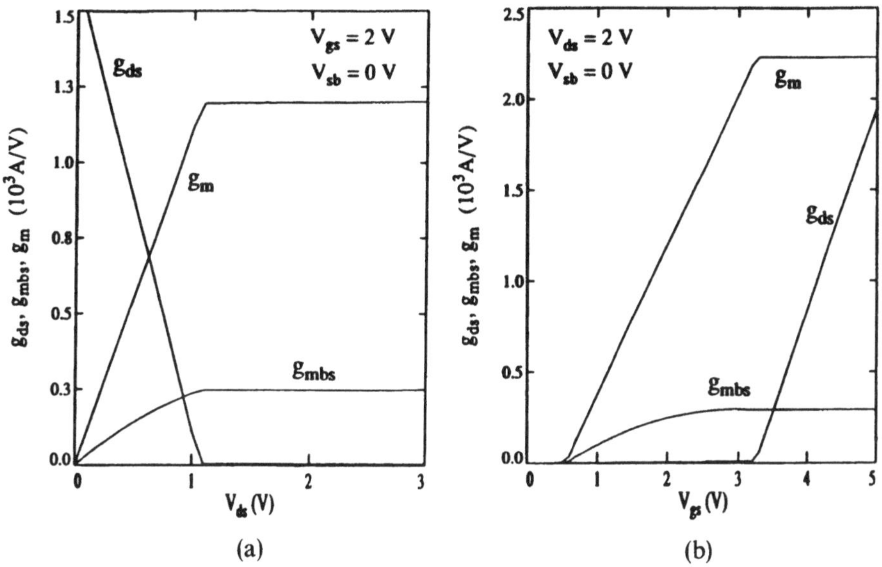

Fig. 3.11 Small-signal parameters g_m, g_{mbs} and g_{ds} as a function of (a) V_{ds}, and (b) V_{gs} based on Eqs (3.5)–(3.7)

Finally, the ratio of change (increase) in the drain current (ΔI_{ds}) to the change (increase) in the drain voltage (ΔV_{ds}) is called *drain conductance* or simply *conductance* g_{ds},

$$g_{ds} = \left.\frac{\Delta I_{ds}}{\Delta V_{ds}}\right|_{V_{gs},V_{bs}} \quad \text{(A/V)}. \tag{3.11}$$

Figure 3.11 shows plot of g_m, g_{mbs} and g_{ds} as a function of V_{gs} and V_{ds} obtained using Eqs. (3.4) and (3.5) that are based on first order MOSFET model.

If all the voltages are changed simultaneously, then the corresponding total change in the drain current is

$$\begin{aligned}\Delta I_{ds} &= \frac{\partial I_{ds}}{\partial V_{gs}}\cdot\Delta V_{gs} + \frac{\partial I_{ds}}{\partial V_{ds}}\cdot\Delta V_{ds} + \frac{\partial I_{ds}}{\partial V_{bs}}\cdot\Delta V_{bs} \\ &= g_m\cdot\Delta V_{gs} + g_d\cdot\Delta V_{ds} + g_{mbs}\cdot\Delta V_{bs}.\end{aligned} \tag{3.12}$$

If the change in the voltages is small, approaching zero, then these (trans)conductances are called *small signal* (*trans*)*conductances*. The small-signal equivalent circuit of a MOSFET is shown in Figure 3.12, where rhomic symbols represents controlled current sources. Note that the linear model can be easily derived from the device DC model using Eqs. (3.9)–(3.11).

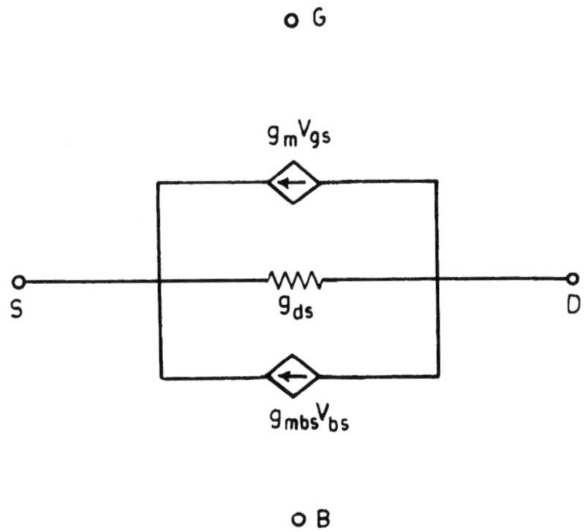

Fig. 3.12 A low frequency, usually less than 1 KHz, small-signal model for a MOSFET

3.2.4 Device Speed

The ratio of transconductance g_m to the gate input capacitance C_G gives a relative measure of the *switching speed* of the device. This ratio g_m/C_G is roughly the 3-dB bandwidth of the device itself [1]. To a first order the intrinsic gate capacitance C_G is simply that of a parallel plate capacitance of area WL and thickness t_{ox}, i.e $C_G = WLC_{ox}$, so that

$$\frac{g_m}{C_G} = \frac{\mu}{L^2}(V_{gs} - V_{th}) \tag{3.13}$$

where we have made use of Eq. (3.9b) for the saturation region transconductance g_m. Note that Eq. (3.13) depends only upon the length of the channel and not on the width. Thus increasing the channel width increases the gate capacitance as much as the transconductance and no increase in the speed is achieved. *Thus, to increase device speed, the channel length L should be as short as possible.* Assuming $L = 1\ \mu$m and $\mu = 600\ \text{cm}^2/V\cdot s$ and $V_{gs} - V_{th} = 5$ V gives a device with cutoff frequency of

$$f_c = \frac{g_m}{2\pi C_G} \approx 10\,\text{GHz}.$$

However, due to the device extrinsic capacitances (S/D junction and gate overlap capacitances), the switching speeds of the actual devices are much less than that predicted by the above equation.

3.3 MOSFET Scaling

During the last two decades, MOS transistors have been systematically scaled down in dimensions in order to achieve increased circuit density (more circuit functions in a given silicon area) and higher performance (higher switching speed, lower power dissipation, etc). Rules of scaling were first proposed by Dennard et al. [8] with the idea of reducing the device dimensions while still maintaining the current-voltage behavior of a large device. According to this rule, all horizontal and vertical device dimensions (i.e., device length L, width W, gate-oxide thickness t_{ox} and source/drain junction depth X_j) as well as voltages are scaled down by a factor $\mathscr{K} > 1$, called the *scaling factor*, while the doping concentration N_b is increased by the same factor. This scaling rule, often known as *classical or constant field scaling*, results in electric fields inside the device that are unchanged compared to the unscaled or original device. The effect of keeping the electric field unchanged in the scaled device is to avoid undesirable high field effects such as mobility degradation, impact ionization, hot-carrier effect etc. Let us see how the drain current I'_{ds} of the scaled device changes compared to the current I_{ds} of the original device. According to Eq. (3.5) the drain current I'_{ds} after scaling becomes

$$I'_{ds} = \frac{\mu C'_{ox} W'}{2L'}(V'_{gs} - V'_{th})^2 = \frac{1}{2}\mu(C_{ox}\mathscr{K})\left(\frac{W/\mathscr{K}}{L/\mathscr{K}}\right)\left(\frac{V_{gs}}{\mathscr{K}} - \frac{V_{th}}{\mathscr{K}}\right)^2 = \frac{I_{ds}}{\mathscr{K}}$$

where prime represents a scaled parameter and remembering that

$$C'_{ox} = \frac{\epsilon_0 \epsilon_{ox}}{t_{ox}/\mathscr{K}} = C_{ox}\mathscr{K}.$$

Note that we have assumed V_{th} scaled by $\mathscr{K}$ which is only approximately true [8]. The scaled total gate capacitance C_G is

$$C'_G = W'L'C'_{ox} = \frac{W}{\mathscr{K}}\frac{L}{\mathscr{K}}\frac{\epsilon_0\epsilon_{ox}}{t_{ox}/\mathscr{K}} = \frac{C_G}{\mathscr{K}}$$

and, the scaled gate delay τ is

$$\tau' = \frac{C_G}{\mathscr{K}}\frac{V_{dd}/\mathscr{K}}{I_{ds}/\mathscr{K}} = \frac{\tau}{\mathscr{K}}.$$

The results of this scaling, are summarized in Table 3.3 [8].
As is evident from this table, when the devices are properly scaled for increased packing density, it also increases the performance in terms of power dissipation and speed. This simple MOS scaling rule provides a useful starting point for the design of small-geometry MOS transistors. For example, according to these rules a 1 μm gate length MOS transistors

Table 3.3. *MOSFET scaling laws [8]*

Scaled Parameters:	
Device dimensions, W, L, t_{ox}, X_j	$1/\mathscr{K}$
Substrate doping N_b	$\mathscr{K}$
Supply voltage V_{dd}	$1/\mathscr{K}$
Affected Parameters:	
Gate capacitance C_G	$1/\mathscr{K}$
Drain current I_{ds}	$1/\mathscr{K}$
Gate delay $V_{dd}C_G/I_{ds}$	$1/\mathscr{K}$
Power dissipation $I_{ds}V_{dd}$	$1/\mathscr{K}^2$
Speed-power product	1

should be designed to have $t_{ox} \approx 200$ Å, $X_j = 0.2$–$0.25\,\mu$m and $N_b \approx 2-3 \times 10^{16}\,\text{cm}^{-3}$. These values are also consistent with the following empirical relationship between these parameters and the minimum channel length $L_{\min}$ above which a MOSFET will behave like a long channel device [9]

$$L_{\min} = 0.4[X_j t_{ox}(X_{sd} + X_{dd})^2]^{1/3} \quad (\mu\text{m}) \tag{3.14}$$

where X_j is the junction depth in μm, t_{ox} is gate oxide thickness in Å and X_{sd} and X_{dd} are the depletion widths under the source and drain respectively in μm [cf. Eq. (3.2)].

Figure 3.13 shows Eq. (3.14) compared with experimental and 2-D simulation results. Note from this figure that if for a given technology $\chi = X_j t_{ox}(X_{sd} + X_{dd})^2 = 10^5$ (μm^3 Å) then a device with $L = 10\,\mu$m is a short device. However, if $\chi = 1$ (μm^3 Å) then a device with $L = 0.5\,\mu$m is a long

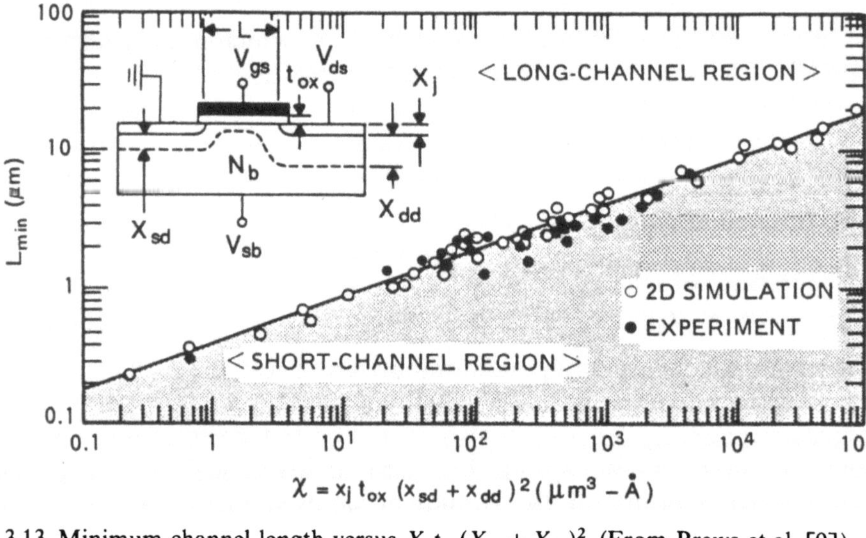

Fig. 3.13 Minimum channel length versus $X_j t_{ox}(X_{sd} + X_{dd})^2$. (From Brews et al. [9])

device. Thus, *electrically, whether a device is long or short really depends upon the process conditions rather than strictly on the physical dimensions.*

It is worth pointing out that not all device parameters scale proportionally. For example, subthreshold slope does not scale [8]. This means that subthreshold current in scaled devices becomes larger while current above threshold is reduced. This is undesirable for digital circuit design because it means that it will be difficult to turn off the device. In addition to this, for practical and standardization reasons, the supply voltage was not scaled as per the scaling rules.[8] In order to minimize the undesirable high electric field in the devices, due to unscaled voltage, the gate oxide was scaled by a factor v which is less than $\mathscr{K}$ $(1 < v < \mathscr{K})$ [12]. Nonetheless, in this case of the so called *constant voltage scaling*, the electric fields can still be very large, resulting in undesirable effects. This has lead to alternative schemes of scaling in which supply voltage does not scale as fast as the device dimensions [12]–[14]. The *quasi-constant voltage scaling* is the same as constant voltage scaling except that the voltage is scaled by a factor v. This means that the electric field will be smaller compared to constant voltage scaling. Baccarani et al. [14] have proposed a more general scaling law in which scaled devices perform somewhere between constant field and constant voltage scaling. These different schemes are summarized in Table 3.4 [3], [15, pp. 1–37].

Although these scaling rules have provided some useful guidelines for reducing the size of a MOSFET, practical considerations like processing capability and device reliability tend to limit strict observance of these scaling rules.

Table 3.4. *MOSFET scaling rules*

Parameters	Constant field scaling	Constant voltage scaling $1 < v < \mathscr{K}$	Quasi-constant voltage scaling $1 < v < \mathscr{K}$	Generalized scaling $1 < v < \mathscr{K}$
W, L, X_J	$1/\mathscr{K}$	$1/\mathscr{K}$	$1/\mathscr{K}$	$1/\mathscr{K}$
t_{ox}	$1/\mathscr{K}$	$1/v$	$1/\mathscr{K}$	$1/\mathscr{K}$
N_b	$\mathscr{K}$	$\mathscr{K}$	$\mathscr{K}$	$\mathscr{K}^2/v$
V_{dd}	$1/\mathscr{K}$	1	$1/v$	$1/v$

[8] The standard supply voltage is 5 V. This was not scaled until very recently when technology moved into submicron feature size. To reduce hot-carrier effects it is important that supply voltage be reduced. The suggested standard supply voltage for these device is 3.3 V [10]–[11].

3.4 Hot-Carrier Effects

When the channel length L is reduced, while keeping the supply voltage constant, the maximum electric field experienced by the carriers in the channel region near the drain end is increased. As the carriers move from the source to drain they can acquire enough kinetic energy in the high field region of the drain junction so as to cause impact ionization. Some of them can even surmount the Si-SiO_2 interface barrier and enter into the oxide. These high energy carriers that are no longer in thermal equilibrium with the lattice, and have energy higher than the thermal energy (kT), are called *hot-carriers*. Effects arising from heating of carriers from the channel under normal MOSFET operation are called *channel hot-carrier effects.*

When the impact ionization occurs, a primary hot-carrier will generate a secondary electron-hole pair. While the electrons (primary and secondary for the case of nMOST) continue to constitute drain-to-source current, the secondary holes generated by the impact ionization drift through the substrate to the substrate contact resulting in a so called the *substrate current* I_b (see Figure 3.14). *Measurement of I_b provides a good monitor as to the heating of the channel carriers and to the electric field in the drain region.* A low level of I_b usually will cause no undesirable effect. However, if the substrate current from a single MOSFET or the sum of the substrate

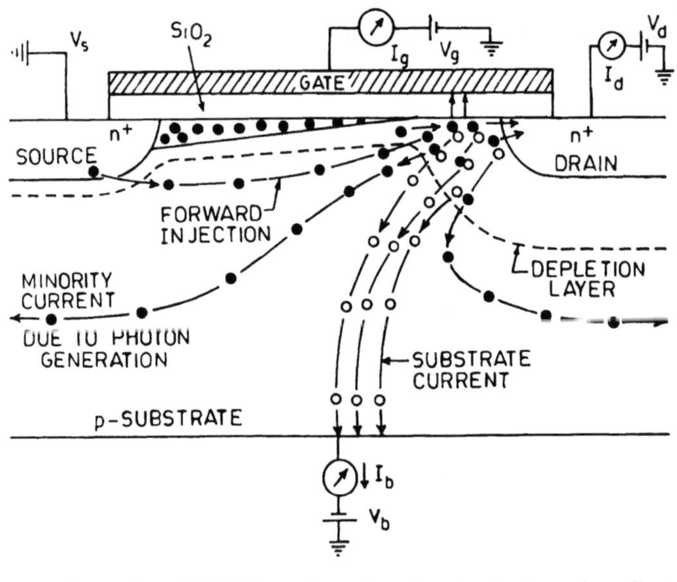

Fig. 3.14 Cross-section of a nMOST in saturation showing hot-carrier effects (1) substrate hole current I_b due to impact ionization, (2) electron injection into the oxide resulting in the gate current I_g, (3) electron forward injection, and (4) secondary electrons leading to photogeneration

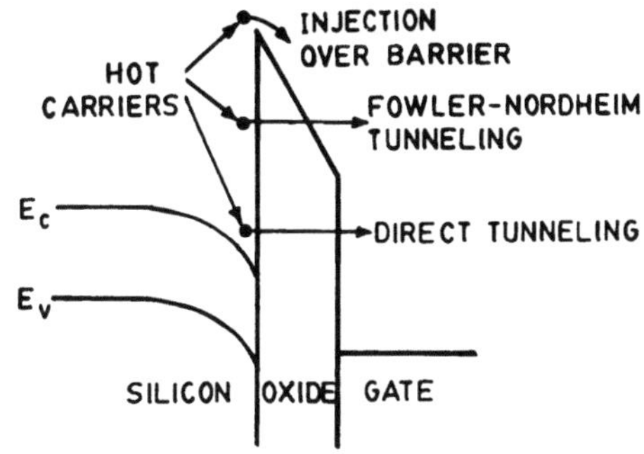

Fig. 3.15 Three different types of carrier injection into the gate resulting in hot-carrier effects

currents from a large number of MOSFETs (for example, in a RAM chip) is excessively high, the substrate current will potentially saturate an on-chip substrate bias generator [16] and lead to circuit malfunction. An excessive substrate current flowing through the substrate contact will result in an ohmic voltage drop in the substrate. Since the source of the MOSFET is usually grounded, this ohmic drop in the substrate can forward bias the source-to-substrate junction [17]. When coupled with the drain, a parasitic bipolar transistor exists in parallel with the MOSFET. This composite structure (MOSFET and parasitic bipolar transistor) is the cause of most drain-to-source breakdown in short-channel MOSFETs [4]–[6] and results in snap back characteristics of the $I-V$ curves. In CMOS circuits the same mechanism can trigger a similar phenomena resulting in a latchup (see section 3.5.5).

Some hot-carriers, though small in numbers, can acquire enough energy (higher than that required for impact ionization) so as to surmount the $Si\text{-}SiO_2$ interface barrier ($\sim 3.2\,eV$ for electrons and $\sim 4.9\,eV$ for holes) and thus move (get injected) into the gate oxide (see Figure 3.15). Most of the injected carriers will be collected by the gate electrode resulting in a so called the *gate current* I_g. Since the energy barrier for this process is very high, the number of hot-carriers injected into the gate will be much smaller compared to those which cause impact ionization. Therefore, the gate current will be smaller than the substrate current by a few orders of magnitude. Figure 3.16 shows gate, substrate and drain current for a typical nMOST with $L = 1.3\,\mu m$. For a fixed drain bias I_b peaks when $V_{gs} \approx V_{ds}/2$ while I_g peaks at $V_{gs} \approx V_{ds}$. The reduction in the I_b or I_g beyond the maximum is due to decreasing channel electric field with increasing gate bias. It should be pointed out that carriers can also enter the gate oxide by tunneling (see Figure 3.15). For direct tunneling the oxide has to be very thin ($< 100\,Å$) and the field be high ($\mathscr{E}_{ox} > 10^6$ V/cm). Even for thicker oxide the carrier with energy close to but less than the energy barrier can tunnel

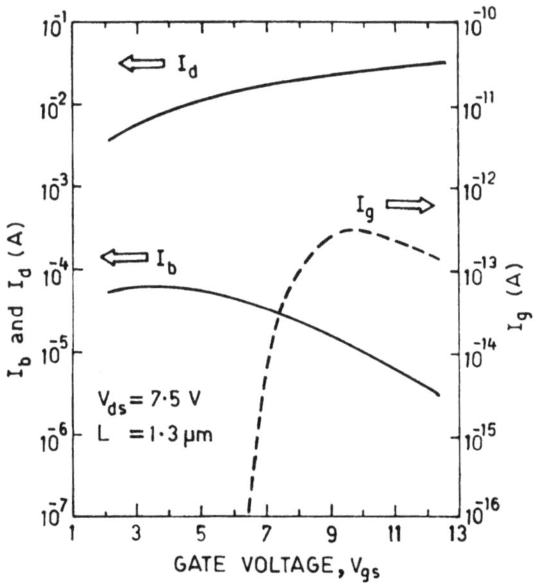

Fig. 3.16 Typical gate and substrate currents I_g and I_b respectively as a function of gate voltages V_{gs} at drain voltages $V_{ds} = 7.5$ V. The device has $t_{ox} = 200$ Å, $L = 1.3\,\mu$m

through the barrier. This effect is called *field assisted* or *Fowler-Nordheim tunneling*.

Another result of hot-carrier injection into the gate oxide is that some carriers are either (1) captured by electrically active defects in SiO_2 (trapping) thus modifying fixed oxide charge density Q_{ot} in the oxide and/or (2) create fast interface state density Q_{it} at the Si-SiO_2 interface (see section 4.1). The charge accumulation due to either Q_{ot} or Q_{it} leads to the creation of a potential barrier to the carriers in the channel. The interface charge Q_{it} also increases the Coulombic scattering degradation of electron mobility at the interface. Either one of the effects or both combined brings about significant device degradation over a period of time. The degraded region is located near the drain and stretches towards the source with increasing stressing time [18]. *This degradation of device characteristics manifests itself as shift in the threshold voltage ΔV_{th}, transconductance degradation Δg_m, and reduced subthreshold slope, thereby affecting the circuit performance.* Hence, channel hot-carrier injection poses a long-term reliability issue in VLSI MOS integrated circuits. Obviously this is not desirable and should be avoided. From this discussion it is clear that *the cause of device degradation is the high channel electric field near the drain junction that accelerates or heats inversion layer charge carriers* [15, pp. 119–160], [19, 20]. Although special source/drain structures such as the lightly doped drain (LDD) are fabricated (see section 3.5.3) to reduce this field thereby minimizing the hot-carrier effect [21], it remains an increasingly important issue. This is because as device are scaled down, the power supply is not scaled proportionally.

From the device modeling point of view, the *substrate current* I_b, and *gate current* I_g are two basic monitors for assessing the overall effect of hot carriers on device performance. The detailed models for these currents are discussed in chapter 8. It should be pointed out that at 1 μm, the hot-carrier problems are more severe in nMOST as compared to pMOST. This is because at a given channel field the impact ionization rate of holes is 2 to 3 orders of magnitude lower than that of electrons. Also the potential barrier heights for holes is higher compared to electrons. However, for sub-half micron pMOST devices hot-carrier effects can begin to cause reliability problems.

Besides these undesirable effects, a beneficial application of channel hot-carrier (electron) injection into the gate oxide is the programming (writing) mechanism in Erasable Programmable Read Only Memories (EPROM) [22]. By the injection of hot electrons into the floating gate of EPROMs, the threshold voltage of the transistor is changed and thus the 'on' or 'off' state of the memory cell is set.

3.5 VLSI Device Structures

Although the MOSFET structure shown in Figure 3.3 is the basic device structure, it has become more complicated as device dimensions have been reduced [23]–[25]. We will now discuss some of the important features of these devices in order to better model them.

3.5.1 Gate Material

In present day MOS technology the gate electrode is invariably made of degenerate polysilicon[9] (doping concentration $> 5 \times 10^{19}\,cm^{-3}$). Typical thickness of the polysilicon gate is about 0.35 μm. The polysilicon gate technology has many advantages over the aluminum (Al) gate technology. Some of these advantages are (1) the polysilicon functions as a gate mask during the high temperature source/drain diffusion step so as to avoid alignment difficulties, (2) the stability of the polysilicon-SiO_2 interface, and (3) the ability to shift the threshold voltage of MOSFETs by about 1 V by varying polysilicon doping from degenerate *p*-type to degenerate *n*-type. A disadvantage of the polysilicon gate is its high resistivity compared to

---

[9] The gate has to be degenerately doped so that it can be treated as an equipotential surface, otherwise one needs to consider the so called polydepletion effect while modeling current and capacitances in a MOSFET (see section 4.5).

the Al gate; typical sheet resistance of n^+ and p^+ polysilicon layers are 15 and 25 Ω/□ respectively compared to Al sheet resistance of few mΩ/□. The problem of high resistivity has been solved by using a combination of polysilicon with refractory metal silicides (such as $CoSi_2$) [26]–[28]. The combination is called polycide and has a sheet resistance of 2–5 Ω/□. For submicron technology, gates are generally polycides.

3.5.2 Nonuniform Channel Doping

To avoid the problem of device punchthrough, VLSI devices are fabricated with channel implants which change the doping at the surface; higher doping concentration near the surface reduces the extension of the source and drain depletion layers into the channel. In fact the *channel implant is also used to achieve a desired threshold voltage.* The implant used to adjust the threshold voltage is often referred to as *threshold adjust implant.* Thus in VLSI devices more than one implant is often used in the channel region, one to adjust the threshold voltage and the other to avoid the punchthrough effect. The latter implant is normally of high energy and high dose and thus is deep (extends close to source/drain depletion widths) compared to the threshold adjust implant, which is of low energy and is thus closer to the surface.

Generally in 1 μm and higher CMOS technologies both nMOST and pMOST are fabricated with n^+ polysilicon gates. In this technology (n^+ polysilicon gates) the nMOST has channel implant dopants (boron) which are of the same type as that of the substrate (*p*-type). However, in order to obtain an appropriate threshold voltage for pMOST, a shallow channel implant of dopants which are of the opposite type to that of the substrate (*p*-type implant in *n*-type substrate) has to be performed, thereby forming a *pn* junction under the gate. Without this extra implant, V_{th} of the *p*-device will be too negative a value in a CMOS process. These *p*-channel devices in n^+ polysilicon gate technology are sometime called *compensated devices.* In a recent submicron CMOS technology, *p*-channel MOSFETs are being fabricated with p^+ polysilicon gates, while *n*-channel MOSFETs are with n^+ polysilicon gates [29]. With p^+ polysilicon gate the pMOST has the same type of channel implant as that of the substrate. The pMOST with p^+ polysilicon gate has various advantages over n^+ polysilicon gate devices (compensated devices), such as (1) lower gate current and hence more resistant to hot-carrier effects, (2) smaller subthreshold slope and hence better turn-off characteristics, and (3) smaller off-state drain leakage current. However, a disadvantage is lower drain current due to lower hole mobility at the surface inversion layer. In a CMOS technology both *n*- and *p*-channel MOSFETs are *enhancement* type devices and as such they are *normally-off* at zero V_{gs}.

In the *depletion type* MOSFET, the channel implant is opposite to that of the substrate and is deep instead of shallow (unlike in the compensated devices) so that significant current flows even at $V_{gs} = 0\,\text{V}$. Since these devices are *normally-on* they are also referred to as *normally-on buried channel* (BC) MOSFETs. The term buried channel is used because the current flows away from the surface in the area confined between the surface and bulk depletion regions. The depletion devices are used as a load in high performance enhancement/depletion (E/D) logic circuits and their advantage over enhancement devices include increased carrier mobility, low noise, reduced hot electron injection and improved breakdown voltage.
Because compensated devices also have channel implant opposite to that of the substrate, these devices are some time referred to as the *normally-off buried channel* (BC) MOSFETs. Thus, a buried channel can result in a MOSFET which is normally-on or normally-off depending on the surface doping in the channel region. These two types of devices have entirely different behaviors as discussed in Chapter 5 and 6. It should be pointed out that in practice *p*-channel depletion devices are not made. It is the *n*-channel depletion devices, with negative threshold voltage, which are more important.

3.5.3 Source-Drain Structures

As was discussed in section 3.4, high channel electric fields at the drain end are the main cause of hot-carrier effects in short channel devices. To reduce the channel fields, VLSI *n*-channel devices almost universally use *graded*

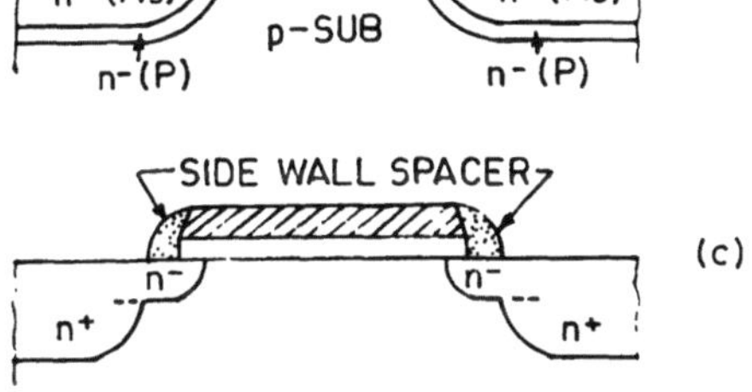

Fig. 3.17 Cross-section of different source/drain MOSFET structures (a) conventional (single diffused) drain structure (b) Double diffused drain (DDD) structure, and (c) Lightly doped drain (LDD) structure

drain structures formed by using two donor type implants. The two most commonly used graded junctions are *double diffused drain* (DDD) [30] and *lightly doped drain* (LDD) [31, 32] (Figure 3.17). A DDD structure for *n*-channel MOSFET is formed by implanting phosphorous (P) and arsenic (As) into the source-drain region. A lightly doped *n*-region (n^-) is first formed using *P* and then a heavily doped n^+ region using As, remembering that *P* is lighter than As and therefore diffuses faster. Thus, in a DDD structure a lightly doped n^- region encloses the n^+ region as shown in Figure 3.17b; the doping level drops by 2–3 orders of magnitude from the n^+ to n^- region. Note that in practice the source is also modified due to the symmetrical nature of the MOSFET although it is the drain side where the maximum field is to be reduced. The DDD structure, though simple, is normally used to reduce the hot-carrier effects for channel lengths down to 1.5–2 μm devices. However, this structure is not suitable for submicron devices due to the fact that it results in deeper junctions and hence increased shortchannel effects and more gate-to-source/drain overlap capacitance.

For submicron devices, the most commonly used S/D structure is the LDD. In this structure a lightly doped *n*-region (n^-) is first created by implanting low energy P or As and then oxide spacers are formed at the side wall of the polysilicon gate (see Figure 3.17c). The oxide spacers then serve as a mask for the standard n^+ As implant. The n^+ implants do not diffuse laterally under the gate but diffuse under the spacers to the edges of the gate. The lateral doping profile of the LDD structure is shown in Figure 3.18a; also shown (Figure 3.18b) is a conventional nMOST with its doping profile [31]. By introducing an n^- region between the drain and channel, the peak channel field is not only shifted towards the drain, but is also

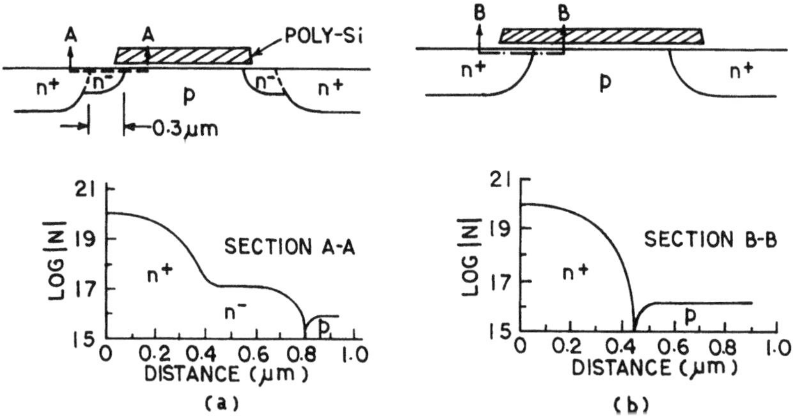

Fig. 3.18 MOSFET cross-section and doping profile for (a) lightly doped drain (LDD), and (b) conventional source and drain. (After Ogura et al. [31])

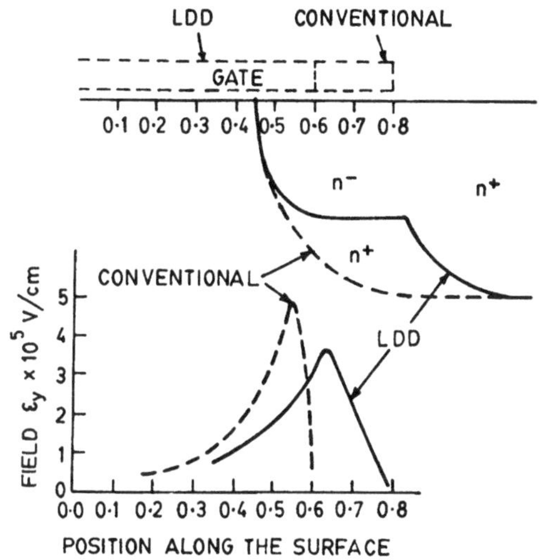

Fig. 3.19 Magnitude of the electric field at the Si-SiO$_2$ interface as a function of distance; L = 1.2 μm, $V_{ds} = 8.5$ V, $V_{gs} = V_{th}$ in a conventional S/D (dashed line and LDD (continuous line). The physical geometries are shown above the plots. (After Ogura et al. [31])

reduced to about 80% of the value for a conventional device (see Figure 3.19). Since the peak field is now reduced and shifted inside the drain region, carrier injection into the oxide is reduced resulting in a more reliable device. This structure results in a higher breakdown voltage and substrate current I_b is reduced considerably. Note that the overlap capacitance is also reduced resulting in a lower gate capacitance and hence higher speed. This improvement is not without cost. Apart from having additional fabrication steps as compared to the standard source drain structure, performance is slightly reduced (4–8%) due to the higher series resistance of the n^- region [23, 24].

As junction depths are scaled down, the resistivity of the source/drain diffusion region becomes higher which again results in higher source/drain resistance and hence lower transconductance as we shall see in section 3.6.1. Low resistivity materials such as refractory metal silicides are often used to reduce this resistance [26]. In fact *self-aligned silicides (called salicides) have become essential ingredients in present day submicron VLSI technology* [27]–[28]. In this process Titanium (Ti) or Cobalt (Co) film is first deposited on the wafer after the formation of source/drain and polysilicon gate. The metal is then reacted with silicon at ~600 °C to form $TiSi_2$($CoSi_2$). The silicide is formed only on the silicon surface (source/drain and polysilicon gate) and not on the oxide. There can be some variation in this process.

3.5.4 Device Isolation

In MOS integrated circuits, all active devices are built on a common silicon substrate and therefore, it is important that they should be adequately isolated from each other. This isolation becomes more important in a VLSI chip because of increased numbers of transistors and decreased isolation space on the chip. If the isolation is not adequate, a leakage current will flow through the substrate resulting in a DC power dissipation and crosstalk among different transistors, which ultimately can destroy the logic state (on-off) of each device.

The most commonly used isolation technique is the so called LOCOS (LOCalised Oxidation of Silicon) scheme which depends upon the local oxidation of silicon using a silicon-nitride mask. In this scheme a thick oxide is grown over heavily doped silicon regions except where it is actually intended to form active transistors. The thick oxide is often called the *isolation* or *field oxide* and the heavily doped region under the field oxide is called the *channel stop* (see Figure 3.20). The implant used to create the heavily doped region under the field oxide is called the *field implant* or *channel stop implant*. Typical thickness of the field oxide t_{fox} is of the order of 3000 Å as compared to 200 Å for t_{ox}, the gate oxide thickness in a typical 1 μm CMOS technology.

In the LOCOS isolation technique a parasitic MOSFET is also formed because the metal or polysilicon lines used to interconnect transistors acts as a parasitic gate with two n^+ diffusion areas adjacent to it acting as source/drain. It is necessary to keep the threshold voltage V_{tfox} of this parasitic MOSFET high compared to that of the active MOSFET in order to avoid formation of a channel region under the field oxide so as not to create any leakage paths. Normally V_{tfox} is around 10 V or more compared to V_{th} of 1 V for an active MOSFET. As we shall see in Chapter 5, threshold

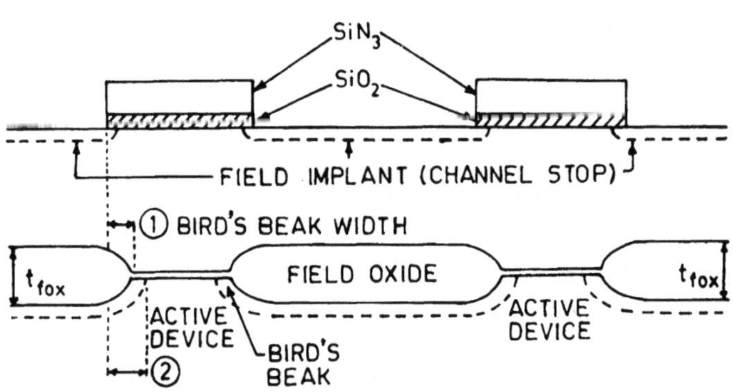

Fig. 3.20 Active area width reduction during LOCOS (top) Nitride/oxide stack (bottom) field implant after LOCOS field oxide (1) encroachment of the field oxide (2) lateral diffusion of the field dopants

voltage is directly proportional to the oxide thickness, therefore, higher t_{fox} is used to achieve high V_{tfox}. This explains why $t_{fox} \gg t_{ox}$.

The LOCOS scheme for VLSI isolation is limited by the field oxide encroachment and lateral diffusion of the field implant dopants into the active device area (see Figure 3.20). The lateral oxidation encroachment makes the edges of LOCOS oxide resemble a bird's beak. The "bird's beak" width usually ranges from 0.5 to 1 μm per side. The LOCOS surface area represents a significant overhead in surface wafer utilization and thus hinders achieving higher packing density.

Several other isolation approaches have been used which are either improvements over LOCOS isolation in terms of reducing the bird's beak or creating a fully recessed isoplaner bird's beak free configuration [23]–[25], [33]. The one most promising technology is the *trench isolation* technique, where lateral encroachment is all but eliminated and isolation can be achieved with very narrow n^+ to p^+ spacing, thus resulting in very high packing density [33]. A typical trench is shown in Figure 3.21. A deep grove (more than twice the S/D junction depths) is first etched into the silicon by reactive ion etching (RIE) and then the side walls are oxidized. The oxide on the walls blocks the diffusion of impurities in subsequent process steps. Next the trench is filled with SiO_2 or polysilicon and is capped with SiO_2. All this is achieved at the cost of a more complex process, resulting in a higher cost and probably lower yield. The trench isolation technique will most likely replace LOCOS for future sub-half micron MOSFET technologies.

3.5.5 *CMOS Process*

In a CMOS process both *p*- and *n*-channel transistors have to be on the same substrate. This is normally achieved by creating a secondary substrate, called the *well or tub*, in the main (primary) substrate. Thus, depending upon the primary substrate type, the process is called an *n-well process*

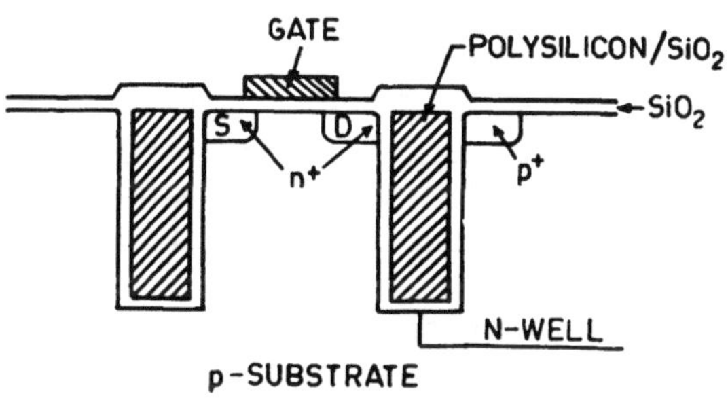

Fig. 3.21 Cross-section of a CMOS trench isolation process

(primary substrate *p*-type for nMOST and *n*-well for pMOST) or a *p-well process* (primary substrate *n*-type for pMOST and *p*-well for nMOST). Another alternative is to form two separate wells (*n*-well for pMOST and *p*-well for nMOST) in the primary substrate so that *n*- and *p*-device characteristics can be adjusted independently. This is called a *twin tub* (*well*) CMOS process. *It is the n-well process which is most commonly used.* This is because when technology transitioned from NMOS to CMOS, the then existing *n*-channel MOSFET designs could easily be exploited for CMOS circuit designs. Moreover, at micron and submicron channel lengths hot-carrier effects in nMOST become very severe. It is easier to ensure a low resistance path for nMOST channel to substrate contact if nMOSTs are formed in *p*-substrate than if they are formed in a *p*-well.

The cross-section of a CMOS *n*-well device structure is shown in Figure 3.22. Note from this figure, that the CMOS process creates two parasitic bipolar transistors, *lateral and vertical.* In a *n*-well process the p^+ source, *n*-well and the *p*-substrate constitute a vertical *pnp* transistor while the *n*-well, *p*-substrate and n^+ source form a lateral *npn* transistor. These are parasitic transistors intrinsic to the process, not required for MOS operation. Notice that the base of each parasitic transistor (*npn* and *pnp*) is driven by the collector of the other thereby forming a feedback loop. The loop gain of this *pnpn* switch, called silicon controlled rectifier (SCR), is equal to the product of the common-emitter current gains, β_{npn} and β_{pnp}, of the *npn* and *pnp* transistors respectively. When the loop gain is greater than one, the SCR can be switched to a low impedance state with large current conduction (often many milliamperes). This condition is called *latchup* [34]. It is a very important effect in CMOS technology as it can easily destroy a chip. Under certain

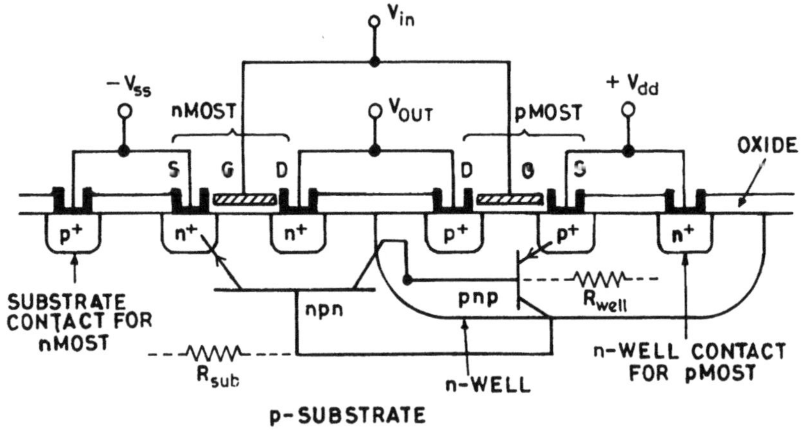

Fig. 3.22 Cross-section of a *n*-well CMOS process showing both *n*- and *p*-channel MOSFET with isolation

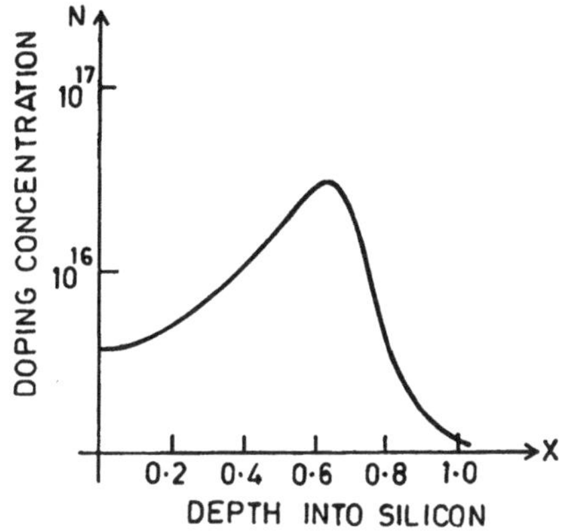

Fig. 3.23 A typical retrograde doping profile in a CMOS process

conditions such as transient currents, ionizing radiations, etc. lateral currents in the well and substrate can forward bias emitter-base junctions of the bipolar transistors, thus activating the switch resulting in the latchup. Latchup is a problem inherent to CMOS technology. The critical parameters that affect latchup are well and substrate resistance, R_{well} and R_{sub}, respectively, and parasitic transistor current gains β_{npn} and β_{pnp}. By reducing R_{well} and R_{sub}, the gain $\beta_{npn}\beta_{pnp}$ can be kept below one thus avoiding latchup. The well resistance is normally reduced by forming *a retrograde well* with a doping profile somewhat similar to that shown in Figure 3.23. The high doping concentration in the bulk provides a low resistivity path for lateral current, while relatively low doping at the surface maintains high breakdown voltage of the S/D junctions. To reduce substrate resistance R_{sub} often a lightly doped epitaxial layer is formed on a heavily doped substrate of the same type. For a *n*-well process a p^- epitaxial layer (concentration $\sim 10^{14}\,cm^{-3}$) is grown on a p^+ substrate (concentration $\sim 10^{18}\,cm^{-3}$) as shown in Figure 3.24 and the process is called *epi-CMOS process.* The heavily doped substrate provides a low resistivity path for lateral substrate currents. The effectiveness of this approach depends on the thickness and bias voltage of the epitaxial layer. A twin tub CMOS process is far less

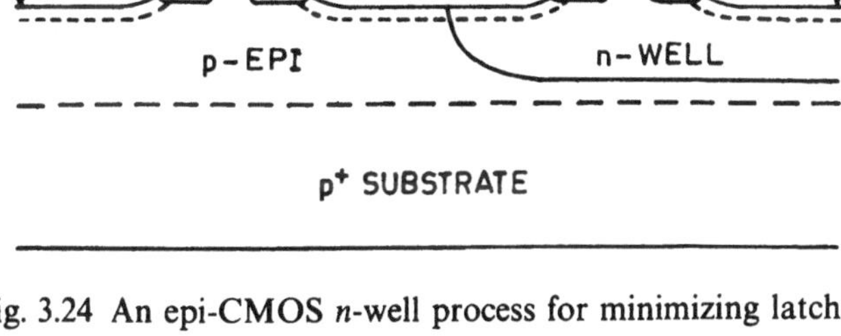

Fig. 3.24 An epi-CMOS *n*-well process for minimizing latchup

prone to latchup compared to a n-well process [34]. Thus *using suitable fabrication and appropriate layout techniques, latchup is generally minimized, although it can never be eliminated.*

3.6 MOSFET Parasitic Elements

As was pointed out earlier, the source/drain junction portion of a MOSFET is a parasitic component. These junctions have resistance and capacitances (S/D *pn* junction capacitance and gate-to-source/drain overlap capacitance). These parasitic elements (resistance and capacitance) limit the drive capability and switching speed of the device and therefore should be minimized. However, S/D is an essential part of the device, therefore, these effects can not be eliminated. It is therefore important to model these elements in order to simulate the device switching behavior accurately.

3.6.1 Source-Drain Resistance

The first order drain current Eqs. (3.4)–(3.6) implicitly assume that the voltages applied at the device terminals are the same as those across the channel region. In other words, voltage drops across the intrinsic resistances R_s and R_d associated with the source and drain regions, respectively, are negligible compared to the applied voltages. Stated another way, the series resistance R_s and R_d are negligible compared to the channel resistance R_{ch}. This indeed is true for long channel devices as R_{ch} is directly proportional to channel length L; the higher the L the higher the R_{ch} as can easily be seen from the following equation obtained by differentiating Eq. (3.4) with respect to V_{ds}

$$R_{ch} = \frac{1}{g_{ds}} = \left(\left. \frac{\partial I_{ds}}{\partial V_{ds}} \right|_{V_{gs}} \right)^{-1} = \left(\frac{L}{W} \right) \frac{1}{\mu C_{ox}(V_{gs} - V_{th} - V_{ds})}. \tag{3.15}$$

However, as the channel length L decreases the series resistance R_s and R_d become appreciable fractions of R_{ch} and thus can no longer be neglected. This can be seen from Figure 3.25 where the ratio of the series resistance $R_t(=R_s+R_d)$ to the total device resistance $R_0(=R_t+R_{ch})$ is plotted against channel length; R_t and R_0 were measured on a set of n-channel MOSFETs fabricated using a typical 2 μm CMOS process. Note that for $L=25\ \mu$m this ratio is 1.2% while it becomes 15% at $L=2\ \mu$m. The impact of R_t, particularly for short channel devices, is a reduction of the transconductance g_m and the device current drive capability. The fact that *series resistance* is less sensitive to scaling than the device itself, it *is one of the major factors limiting the performance of scaled MOS devices* [35]. In order to understand

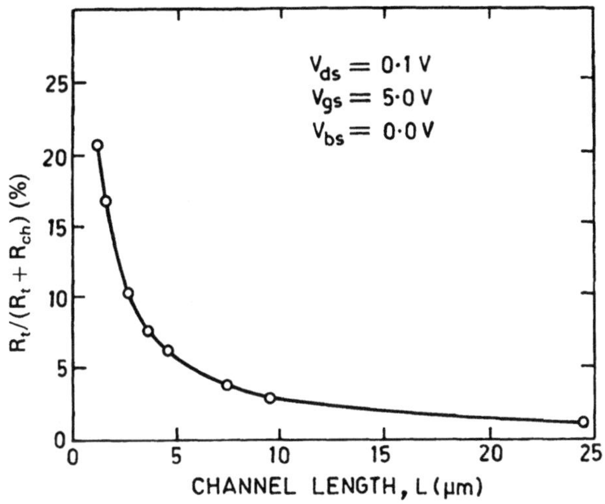

Fig. 3.25 Ratio of the source and drain series resistance R_t to the total device resistance ($R_t + R_{ch}$) as a function of channel length

this, it will be instructive to see what are the factors which influence R_s and R_d.

The schematic diagram of the current pattern in the source (drain) region is shown in Figure 3.26. The resistance $R_s(R_d)$, which is in series with the channel resistance R_{ch}, can be expressed as the sum of three terms[10] [36]

$$R_s = R_{sh} + R_{co} + R_{sp} \quad (\Omega) \tag{3.16}$$

where R_{sh} is the sheet resistance of the heavily doped source (drain) diffusion region where the current flows along the parallel lines, R_{co} is the contact resistance between the metal and the source (drain) diffusion region, and R_{sp} is the spreading resistance due to the current lines crowding near the channel end of the source (drain) (see Figure 3.26).

The sheet resistance R_{sh} is simply given by

$$R_{sh} = \frac{\rho_s S}{W} \tag{3.17}$$

where S is the distance between the contact via and the channel region, ρ_s is the sheet resistance per square ($\Omega/\square$) and W is device width. For a typical 1 μm CMOS technology, $\rho_s = 30$ and $60\,\Omega/\square$ for n^+ and p^+ regions, respectively (see Table 3.5). For LDD source/drain structures, commonly used

[10] The separation of the series resistance into three terms is of course only an approximation, which is convenient for qualitative discussions. Strictly speaking, R_s should be determined by matching the solution of the field and current continuity equation in the channel and the source/drain region using approximate boundary conditions at the metal semiconductor contact.

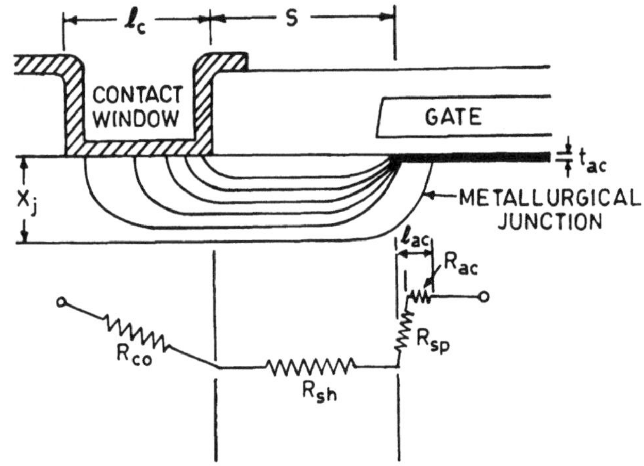

Fig. 3.26 Schematic diagram showing current pattern in a source/drain region and (b) their representative resistance components. (After Ng and Lynch [35])

in (sub)micron n-channel devices to reduce hot-electron effects, an additional sheet resistance due to the n^- region needs to be considered which results in a higher ρ_s.

Within the contact area, the voltage drop in the diffused region results in current crowding near the front end of the contact. This effect results in a contact resistance R_{co}. Based on the transmission line model of the interface between the metal and the diffused region, it has been shown that R_{co} for a rectangular contact of length l_c can be expressed as [36, 37]

$$R_{co} = \frac{\sqrt{\rho_c \rho_s}}{W} \cdot \coth\left(l_c \sqrt{\frac{\rho_s}{\rho_c}} \right) \tag{3.18}$$

where ρ_c is the interfacial specific contact resistivity ($\Omega\,\mathrm{cm}^2$) between the metal and the source (drain) region. The magnitude of ρ_c depends on charge transport mechanisms and is determined primarily by the surface impurity concentration N_s, potential barrier height ϕ and ambient temperature T [36]. In practice ρ_c is sensitive to the metal-silicon interface preparation procedure. In particular, the presence of an oxide in the contact hole strongly affects ρ_c. A good aluminum-diffusion contact should have ρ_c below $100\,\Omega\,\mu\mathrm{m}^2$. Equation (3.18) assumes that all channel width is used for the contact. However, this is not generally the case and multiple standard contacts of minimum size l_c separated by spacing d are used (see Figure 3.27), therefore, Eq. (3.18) is multiplied by a factor $(1 + d/l_c)$.

The spreading resistance R_{sp} arises from the radial pattern of current spreading from the MOSFET channel, which has a thickness of the order of 50 Å. A first order expression for R_{sp}, based on the assumption of uniform doping

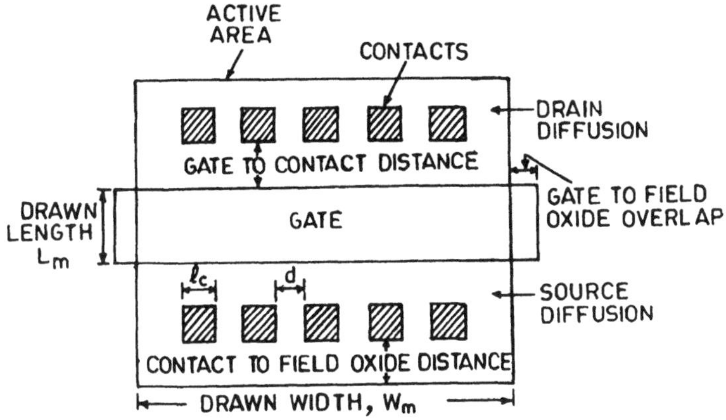

Fig. 3.27 Layout of a MOSFET showing relevant source/drain dimensions. l_c = S/D contact size and d = S/D contact space

in the source (drain) region, is given by [38]–[40]

$$R_{sp} = \left(\frac{2\rho_s X_j}{\pi W}\right) \cdot \ln\left(H \frac{X_j}{t_{ac}}\right) \tag{3.19}$$

where t_{ac} is the thickness of the surface accumulation layer of length l_{ac} in the gate-to-source/drain overlap region and H is a factor that has been found to have a value in the range 0.37–0.9 [38]–[40]. The exact value of H plays a minor role due to the logarithmic nature of the equation and the fact that the ratio X_j/t_{ac} is large (see Table 3.5). Since the current is first confined to the accumulation layer and then spreads into the bulk, the resistance R_{ac} of this accumulation layer must be added to R_{sp} [39]. Notice that R_{sh} and R_{sp} are invariant with scaling mainly due to increased ρ_s caused by using the solid solubility doping concentration at the surface of heavily doped shallow junctions and by the corresponding decrease in junction depth due to scaling. For a typical 1 μm CMOS technology, values

Table 3.5. *Typical 1 μm CMOS process parameters*

Parameter	n-channel n^+ S/D	p-channel p^+ S/D	Units
ρ_s	30	60	Ω/□
ρ_c	10	60	$\Omega\,\mu m^2$
X_j	0.25	0.35	μm
t_{ac}	50	50	Å
R_{sh}	30	60	Ω
R_{co}	18	78	Ω
R_{sp}	17	52	Ω

of the different parameters are shown in Table 3.5. While calculating R_{co} and R_{sh} it has been assumed that $S = W = 1\,\mu$m, $l_c = d = 1\,\mu$m.
Note that all three components of the series resistance have values of the same order of magnitude. Also note that the pMOST has higher S/D resistance as compared to the nMOST. Both ρ_s and ρ_c will increase by a factor of 2 for future 0.5 μm technology because of the decrease in junction depth X_j. However, self-aligned silicided S/D (a standard technique for 0.5 μm technology) will reduce the sheet resistance ρ_s to roughly 2–4 $\Omega/\square$ and contact resistivity ρ_c to below $20\,\Omega\mu^2$. Without the silicide technology, the g_m degradation will be over 10% for 0.5 μm technology.
Equations (3.17)–(3.19) give fairly good estimates of the series resistance for standard source/drain junctions. However, experimental results, particularly for LDD source/drain devices, show underestimation of the R_{sp} term. The series resistance of LDD source/drain structures, is much higher compared to devices with standard (conventional) source/drain structures. This is due to the n^- region which has lower X_j and lower doping compared to the n^+ region.
More realistic expressions for R_{sp} have been determined assuming a non-uniform doping profile of the source (drain) junction near the vicinity of the channel end. This results in R_{sp} being gate bias dependent. In addition, R_{ac} (accumulation layer resistance) also becomes gate bias dependent [39]. Both R_{sp} and R_{ac} have been shown to be strong function of the slope of the doping profile near the junction. A large slope minimizes both R_{ac} and R_{sp} and both these resistances decrease with increasing V_{gs}. The gate bias dependence of $R_t (= R_s + R_d)$ is shown in Figure 3.28 for standard and LDD devices [43]. Note that while the change in R_t due to a change in gate

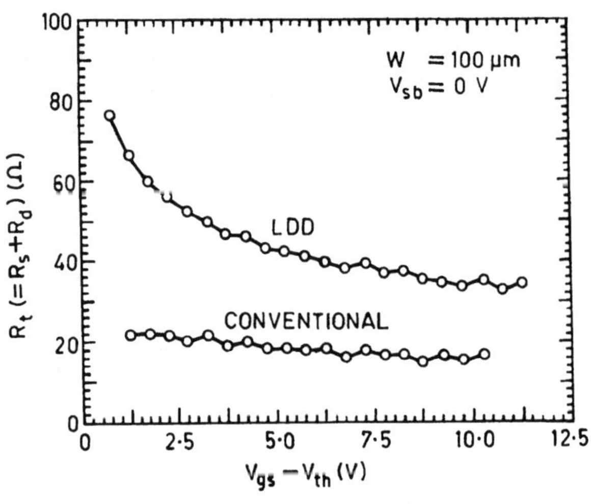

Fig. 3.28 Extracted R_t from conventional and LDD nMOST as a function of effective gate voltage ($V_{gs} - V_{th}$). (After Hu et al. [43])

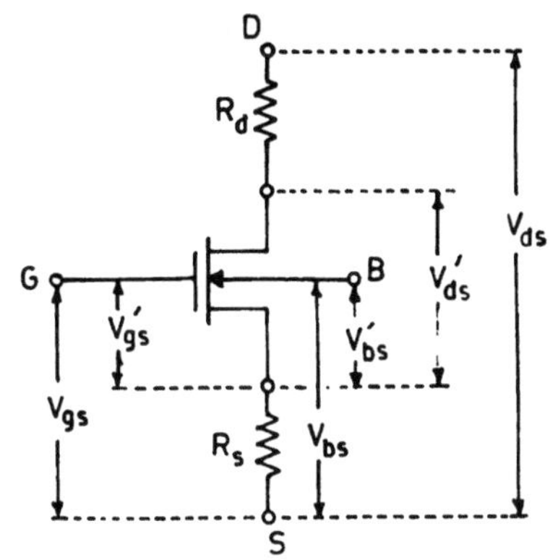

Fig. 3.29 MOSFET showing source and drain resistance R_s and R_d respectively

overdrive ($V_{gs} - V_{th}$) from 0.75 V to 11.25 V is only 6 Ω for a conventional device, the corresponding change in an LDD device is almost 45 Ω (78-33 Ω) with n^- concentration of $1 \times 10^{17}\,\text{cm}^{-3}$. It has been found that the bias dependence of R_t varies considerably with concentration of the n^- region. The bias dependence of R_t is small for n^- implant dose $\geq 10^{13}\,\text{cm}^{-2}$. At higher V_{ds}, R_{sp} decreases at the drain end because channel current has already spread into the bulk in the pinch-off region [42]. For MOSFET circuit models bias dependence of R_t is ignored in order to keep current equations simple.

Effect of Source/Drain Resistance on Device Transconductance. The effect of R_s and R_d in calculating I_{ds} can be understood from the equivalent circuit shown in Figure 3.29. We will assume that R_s and R_d are independent of the bias. If the effective or intrinsic drain and gate voltages are denoted by V'_{ds} and V'_{gs} respectively while V_{ds} and V_{gs} are the voltages applied at device external terminals, then from Kirchiff's law we get

$$\begin{aligned} V'_{gs} &= V_{gs} - I_{ds} \cdot R_s \\ V'_{ds} &= V_{ds} - I_{ds} \cdot (R_s + R_d). \end{aligned} \tag{3.20}$$

From above equations it follows that at constant V_{ds} (i.e, $dV_{ds} = 0$)

$$\begin{aligned} dV'_{gs} &= dV_{gs} - dI_{ds} \cdot R_s \\ dV'_{ds} &= -dI_{ds} \cdot (R_s + R_d). \end{aligned} \tag{3.21}$$

Assuming V_{sb} is constant and using Eq. (3.12) the differential of the drain current can be written as

$$\begin{aligned} dI_{ds} &= \left.\frac{\partial I_{ds}}{\partial V'_{gs}}\right|_{V'_{ds}} \cdot dV'_{gs} + \left.\frac{\partial I_{ds}}{\partial V'_{ds}}\right|_{V'_{gs}} \cdot dV'_{ds} \\ &= g'_m \cdot dV'_{gs} + g'_d \cdot dV'_{ds} \end{aligned} \tag{3.22}$$

where g'_m and g'_{ds} are the intrinsic transconductance and drain conductance respectively ($R_s = R_d = 0$). Substituting Eqs. (3.21) in (3.22) we get [44]

$$g_m = \frac{g'_m}{1 + (R_s + R_d)g'_{ds} + R_s g'_m}. \tag{3.23}$$

If however, we also include V_{bs} dependence on the drain current then Eq. (3.23) is modified as [45]

$$g_m = \frac{g'_m}{1 + (R_s + R_d)g'_{ds} + R_s(g'_m + g'_{mbs})}. \tag{3.24}$$

Both Eqs. (3.23) and (3.24) show that g_m reduces due to the presence of S/D resistance. In the saturation region, g'_{ds} is zero (strictly speaking, true only for long channel devices) and therefore

$$g_m = \frac{g'_m}{1 + R_s g'_m} \tag{3.25}$$

which shows that in the saturation region g_m is degraded by a factor $(1 + R_s g'_m)$. Since g'_{ds} is non-zero in the linear region, it is evident from Eq. (3.23) that *the impact of R_s and R_d on g_m in the linear region is more pronounced than in the saturation region.* This is true of drain current also, that is, the drain current reduction due to R_t is more pronounced in the linear region compared to the saturation region [46].

3.6.2 Source/Drain Junction Capacitance

The source/drain to substrate boundary is an n^+p (or p^+n) junction. Recall from our discussion in section 2.8, the capacitance of a *pn* junction consists of two components, the area component (or bottom-wall capacitance) and the periphery component (or side-wall capacitance). In a MOSFET the S/D doping concentration towards the outer side (field side) is different from the inner side (channel side), therefore a MOSFET junction capacitance is divided into the following three components as shown in Figure 3.30

1. bottom-wall capacitance C_{jw} (F/cm^2),
2. outer side-wall capacitance C_{jsw1} (F/cm),
3. inner side-wall capacitance C_{jsw2} (F/cm).

All three capacitances are generally modeled by Eq. (2.74). For submicron devices, due to thinner gate oxides, low junction depth and higher channel doping concentration (as per scaling laws), the inner side-wall capacitance becomes higher than the outer side-wall capacitance. For shallow junctions, the side-wall capacitance is much larger than the bottom-wall capacitance

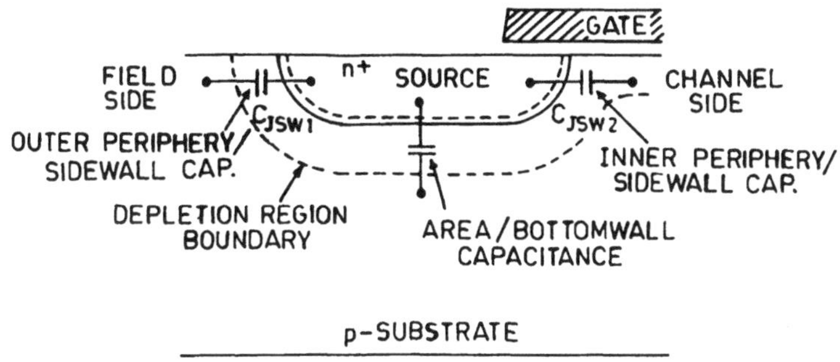

Fig. 3.30 MOSFET source or drain junction capacitances showing area (bottom-wall) and periphery (side-wall) components

(cf. section 2.8). Thus, the total junction capacitance (source or drain) can be written as

$$C_J = \frac{C_{jw0}A_{jw}}{(1 - V_d/\phi_{jw})^{mj}} + \frac{C_{jsw10}P_{sw1}}{(1 - V_d/\phi_{sw1})^{m1}} + \frac{C_{jsw20}P_{sw2}}{(1 - V_d/\phi_{sw2})^{m2}} \quad (\mathrm{F/cm^2}) \tag{3.26}$$

where C_{jw0}, C_{jsw10} and C_{jsw20} are bottom-wall, outer and inner side-wall capacitances at zero substrate bias. A_{jw}, P_{sw1} and P_{sw2} are bottom-wall area, outer and inner periphery respectively of the S/D opening. Special test structures are used to measure the three components of the capacitances. Since the values of ϕ and m depend upon the exact doping profile, they can vary considerably for the three capacitances.

In older technologies the difference between the inner and outer side-wall capacitances was insignificant, therefore, SPICE only allows for 'one' side-wall diffusion capacitance. Thus, SPICE has only two components of the S/D junction capacitances. However, in submicron technology the difference between the inner and outer side wall capacitance is significant and must be taken into account.

3.6.3 Gate Overlap Capacitances

The overlap capacitances are parasitic elements that originate from the basic fabrication steps. In the self-aligned process, the polysilicon gate is employed as the mask to define the source and drain regions. The overlaps occur because the remaining processing steps require heating of the wafer. This gives rise to *lateral diffusion* of the source/drain dopants so the polysilicon gate overlaps the source and drain regions of the final structure. Since in a MOSFET source and drain regions are normally symmetrical,

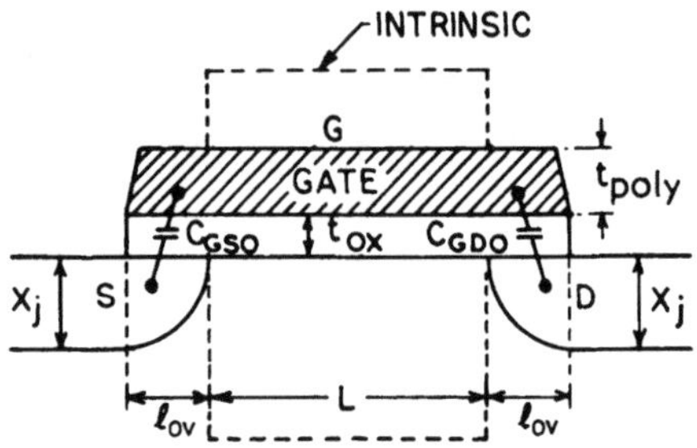

Fig. 3.31 Cross-section showing overlap capacitances between the source/drain and the gate which give rise to C_{GSO} and C_{GDO}

one can assume the source and drain overlap distance l_{ov} to be equal[11] (see Figure 3.31). Assuming the parallel plate formulation, the overlap capacitance C_{GSO} and C_{GDO} for the source and drain regions respectively may be approximated as

$$C_{GSO} = C_{GDO} = \frac{\epsilon_0 \epsilon_{si}}{t_{ox}} W l_{ov} = C_{ox} W l_{ov} \quad \text{(F)}. \tag{3.27}$$

If C_{gso} and C_{gdo} are the overlap capacitance per unit width (F/cm) for the gate-source or gate-drain overlap respectively, then

$$C_{gso} = C_{gdo} = C_{ox} l_{ov} \quad \text{(F/cm)}. \tag{3.28}$$

A third overlap capacitance that can be significant is due to the overlap between the gate and the bulk as shown in Figure 3.32a. This is the capacitance C_{GBO} that occurs due to the overhang of the transistor gate required at one end and is a function of the effective polysilicon width that is equivalent to the drawn channel length (Figure 3.32b). Thus, if C_{gbo} is the gate bulk overlap capacitance per unit length, then the total gate-to-bulk overlap capacitance becomes

$$C_{GBO} = C_{gbo} L_{poly} \tag{3.29}$$

where L_{poly} is the width of the polysilicon (defining gate length) after etch. Normally C_{GBO} is much smaller than C_{GSO}/C_{GDO} and therefore is often neglected.

Clearly circuit designers do not have control over the overlap distances, and hence the overlap capacitances, as these are the fabrication parameters that are defined by the processing steps.

For a typical 2 μm CMOS process the junction depth $X_j = 0.3$ μm and thus l_{ov} is approximately 0.21 μm (assuming 0.7 to be the side diffusion). Since

[11] Strictly speaking this is not necessarily the case for submicron devices with extremely shallow junctions ($X_j \sim 0.1$ μm) where off-axis S/D implants can cause implant shadowing, producing asymmetry in the source and drain overlap capacitances.

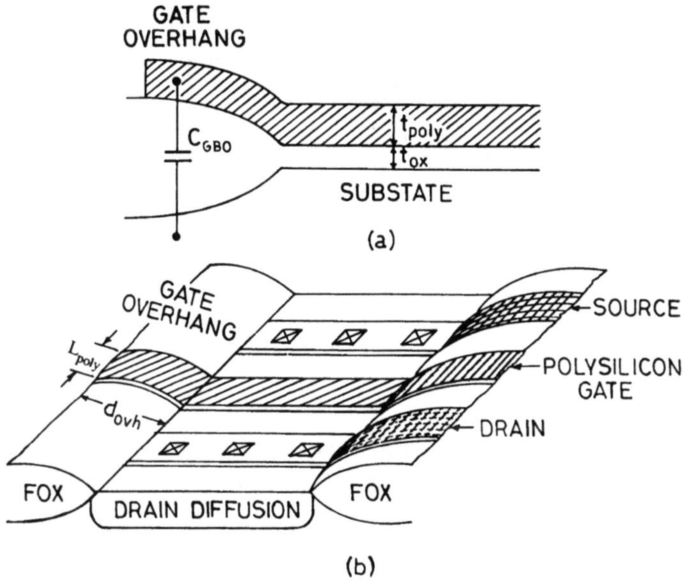

Fig. 3.32 Cross-section (a) 3-D representation (b) showing gate overlap capacitance C_{GBO} between the overlap of the polysilicon gate over the field oxide

l_{ov} is small, Eq. (3.28) underestimates the overlap capacitance [47, 48] due to the fringe capacitance being ignored, which could be significant percentage of the total capacitance. This can be seen from Figure 3.33 which is the plot of overlap capacitance C_{gxo} (x = s or d) as a function of overlap distance l_{ov} for a typical $2\,\mu$m CMOS technology with gate oxide thickness $t_{ox} = 300$ Å and gate polysilicon thickness $t_{poly} = 0.4\,\mu$m, based on 2-D

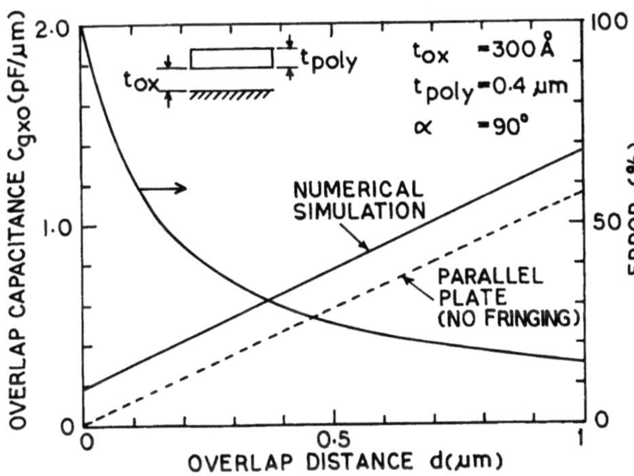

Fig. 3.33 Plot of the overlap capacitance (including fringing) versus overlap distance using parallel plate capacitance formula and exact numerical solution

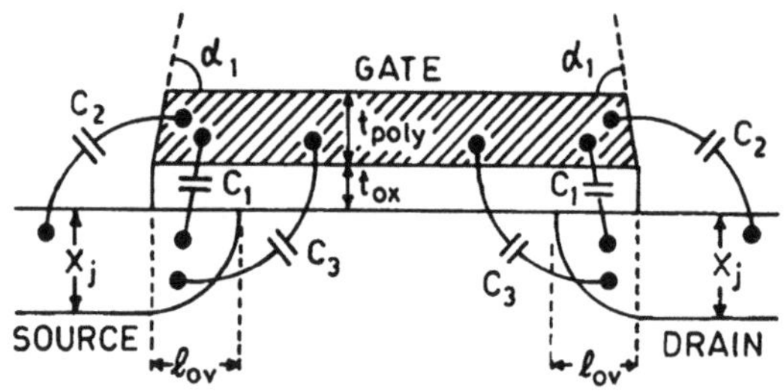

Fig. 3.34 Cross-sectional view of the geometrical parameters of a MOSFET and three components of the overlap capacitances, C_1 (parallel plate), C_2 (outer fringing) and C_3 (inner fringing)

numerical solution of the field equations. Also shown is a plot of the parallel plate component (dashed line) which would be the value of C_{gxo} if one used Eq. (3.28) to estimate the overlap capacitance. For a typical case of $l_{ov} = 0.2\,\mu\text{m}$, the error using Eq. (3.28) will be greater than 40% in calculating the overlap capacitances.

In a MOSFET, in addition to the outer fringing field capacitance, there is another parasitic capacitance which must be taken into account while calculating the overlap capacitance (see Figure 3.34). Thus MOSFET overlap capacitance can be approximated by the parallel combination of the (1) direct overlap capacitance C_1 between the gate and the source/drain, (2) fringing capacitance C_2 on the outer side between the gate and source/drain and (3) fringing capacitance C_3 on the channel side (inner side) between the gate and side wall of the source/drain junction such that [47]

$$C_{gso} = C_{gdo} = \underbrace{C_{ox}(l_{ov} + \Delta)}_{C_1} + \underbrace{\frac{\epsilon_{ox}}{\alpha_1}\ln\left(1 + \frac{t_{poly}}{t_{ox}}\right)}_{C_2} + \underbrace{\frac{2\epsilon_{si}}{\pi}\ln\left[1 + \frac{X_j}{t_{ox}}\sin\alpha_1\right]}_{C_3}. \tag{3.30}$$

Note that C_1 is the parallel plate component of length $(l_{ov} + \Delta)$ where Δ accounts for the fact that polysilicon thickness has a slope of angle α_1. It is a correction factor of higher order and is given by

$$\Delta = \frac{t_{ox}}{2}\left[\frac{1 - \cos\alpha_1}{\sin\alpha_1} + \frac{1 - \cos\alpha_2}{\sin\alpha_2}\right]$$

where $\alpha_2 = (\pi/2\cdot\epsilon_{ox}/\epsilon_{si})$. It is interesting to note that the fringing component C_3 (channel side) is much larger than C_2 (outer side) because ϵ_{si} is roughly 3 times ϵ_{ox} and also quite often $\alpha_1 \geq \pi/2$. From Eq. (3.30) it is clear that even if the overlap distance l_{ov} is reduced to zero, there will be an "overlap" capacitance present due to the fringing components still.

Although in Eq. (3.30) the channel side fringing capacitance C_3 is assumed to be bias independent, in reality it is gate and drain voltage dependent. C_3 in Eq. (3.30) gives the maximum value of the inner fringing capacitance. Note that *when the device is in inversion* C_3 *is zero.*
The overlap capacitance is bias dependent particularly for LDD and thin gate oxide devices [49, 50]. The bias dependence of the overlap capacitance in accumulation has been modeled using an equation similar to the junction capacitance Eq. (2.74) [51]. However, most of the circuit simulators, including SPICE, use only one value of the bias independent overlap capacitance.

3.7 MOSFET Length and Width Definitions

At the device level, the circuit designer has control over only two parameters, the device channel length L and width W, that have important effects on device behavior. For increased current drive and hence circuit speed a large W and small L is required. It is important to understand what device L and W stand for from the modeling point of view.

3.7.1 *Effective or Electrical Channel Length*

We define the *channel length* L *as the distance between the source-drain junctions* as shown in Figure 3.35a. This distance is in reality process dependent and hence, we call it *effective channel length* in order to distinguish it from the drawn gate length (physical mask dimensions) L_m. Due to the manufacturing tolerances, L_m is slightly different from the final gate length L_{poly}. The difference between the drawn and final gate length[12] is $L_{var}(= L_m - L_{poly})$. During the high temperature fabrication steps the source and drain junctions not only diffuse vertically but also move laterally under the gate. This lateral diffusion L_{dif} is typically 0.6–0.8 of the source-drain junction depth X_j, depending upon the type of dopants. Thus we define the effective channel length L as

$$L = L_m - L_{var} - 2L_{dif} \quad \text{(effective channel length).} \tag{3.31}$$

These definitions are shown in Figure 3.35a. For circuit models, L_{var} and L_{dif} are combined together and therefore

$$L = L_m - \Delta L \quad \text{(effective channel length).} \tag{3.32}$$

---

[12] Note that L_{var} is not necessarily a positive number, it could be negative too depending upon the tolerance.

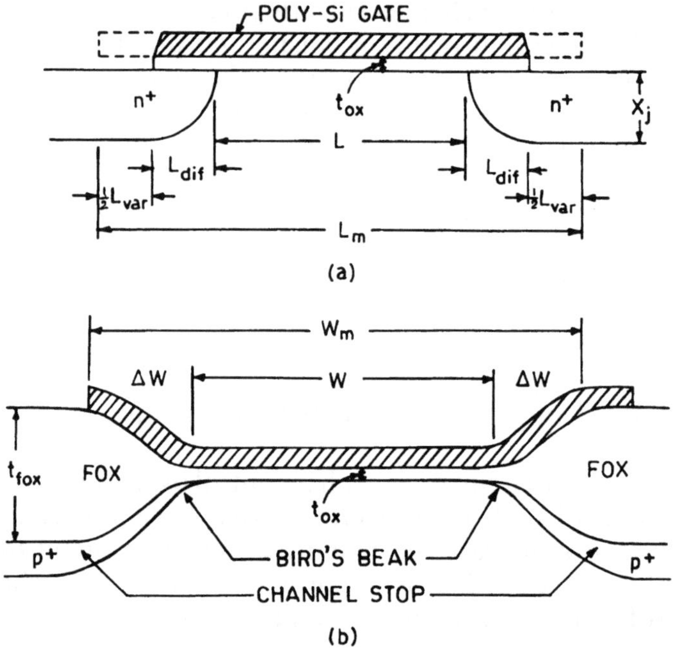

Fig. 3.35 MOSFET (a) channel length definition and (b) channel width definition

Thus ΔL contains both L_{dif} and L_{var}. It is this ΔL which we measure electrically, and therefore L is also called the *electrical channel length.*

3.7.2 Effective or Electrical Channel Width

As was pointed out earlier, in present day MOSFET isoplanar processes different devices are isolated from the neighboring devices by the so called field oxide whose thickness $t_{fox} \gg t_{ox}$. This is typical of devices made from LOCOS technology. During high temperature processing steps, the heavily doped region under the field oxide will encroach into the channel, which when combined with some fabrication process, causes tapering of the thin oxide (active) to thick oxide (field) resulting in a structure that looks like a bird's beak. This causes the *effective, or electrical, device width* W to be smaller than the drawn device width W_m (physical mask dimension) by a factor ΔW (see Figure 3.35b). Thus

$$W = W_m - \Delta W \quad \text{(effective channel width)} \tag{3.33}$$

Thus both ΔL and ΔW depend on mask fabrication techniques, photolithographic process and equipment, production quality control, S/D junction depths, and the size of the minimum dimensions. In most processes L is less than L_m because lateral diffusion is dominant compared to other photolithographic variations. In the older Al gate process when X_j was around 2 μm, ΔL was $\sim 2.5\ \mu$m. In present day polysilicon gate processes with $X_j \sim 0.3\ \mu$m or even less, ΔL is about $0.8 \sim 1\ \mu$m. In the Al gate process W is always larger than W_m because over-etching of either or both S/D diffusion and gate oxide is used to compensate for the misalignment in order to improve yield. *Polysilicon gate processes result in a W which is almost always smaller than W_m because of the bird's beak and because gate and junction regions combined on the same pattern makes the active area pattern larger.*

Remember that it is the L and W, defined by Eqs. (3.32) and (3.33) respectively, that are used in MOSFET model equations. *Unless otherwise stated, through out this book we will use the symbols L and W for the effective or electrical channel length and width, respectively.* The detailed methods of determining ΔL and ΔW and hence L and W are discussed in section 10.6.

It is worth pointing out here that methods of extracting ΔL and ΔW resulting in effective L and W are purely electrical parameters. Thus for example the effective channel length L is not necessarily the same as the physical parameter L, which is the distance between the metallurgical source and drain (S/D) junctions. The difference between the physical and electrical L is small for conventional S/D junctions. However, it can be quite significant for LDD devices, especially when the n^- implant dose is low or n^- region is long. For LDD junctions, the electrical methods generally give L that represents the distance between points somewhere within the n^- S/D regions. This is because, in this case current is confined to the silicon surface even beyond the metallurgical junctions in the n^- regions [39], [46]. The higher the n^- dose, the closer these points are to the metallurgical n^-/p-substrate (n-channel device) junction; the lower the dose the closer they are to high-low n^+/n^- transition point.

While the effective channel length L depends upon the S/D structure, the effective channel width W depends upon the isolation structure. Thus MOSFETs with LOCOS isolation structure have higher ΔW compared to trench isolated structures.

3.8 MOSFET Circuit Models

The equivalent circuit model for the DC operation of a MOSFET is shown in Figure 3.36 which establishes the dependence of the drain current I_{ds} on the source, drain, gate and bulk voltages [52]. In Figure 3.36 S and D are the source and drain nodes as specified in a SPICE input file (wirelist); S' and D' are the corresponding internal node respectively representing channel portion of the device (intrinsic part). g_s and g_d are conductances of the source and drain series resistance, respectively. The gate and bulk nodes

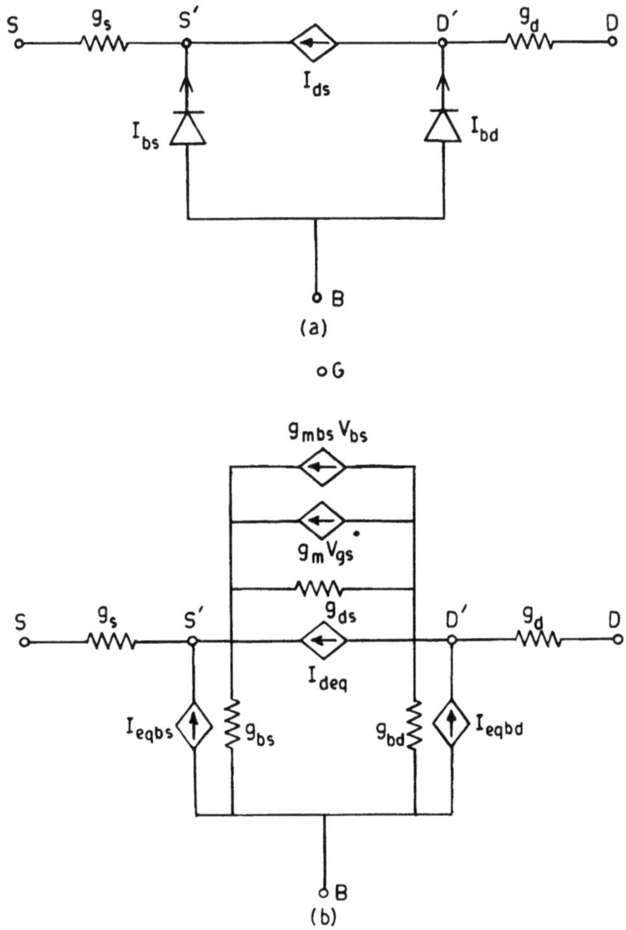

Fig. 3.36 MOSFET (a) equivalent circuit model and (b) linearized equivalent circuit model for DC analysis.

are represented by G and B respectively. Since the gate is separated from the source and drain by an insulator (gate oxide), the gate is assumed to be a DC open circuit. The drain-source current is represented by a voltage controlled current source I_{ds} which is given by Eqs. (3.4)–(3.6) for the first order model. The I_{bs} and I_{bd} are the DC source and drain *pn* junction currents respectively.

The I_{ds}, I_{bs} and I_{bd} are all nonlinear functions of the node or terminal voltages. To solve the nonlinear circuit equation by the nonlinear Newton Raphson method, the equivalent circuit model in Figure 3.36a is converted into its companion model as shown in Figure 3.36b. In this figure g_m, g_{mbs} and g_{ds} are small signal MOSFET intrinsic conductances defined by

Eqs. (3.8)–(3.11) and are related to the large signal model by the equations

$$g_m = \left.\frac{\Delta I_{ds}}{\Delta V_{gs}}\right|_{op} \qquad g_{ds} = \left.\frac{\Delta I_{ds}}{\Delta V_{ds}}\right|_{op} \qquad g_{mbs} = \left.\frac{\Delta I_{ds}}{\Delta V_{gs}}\right|_{op} \tag{3.34}$$

where the subscript *op* denotes that the independent variables V_{gs}, V_{ds} and V_{bs} assume the values at the operating-point bias. These conductances can easily be obtained from the DC drain current model. The equivalent current source for the intrinsic MOSFET is calculated from the following equation

$$I_{\text{deq}} = \pm(I_{ds} - g_m \cdot V_{gs} - g_{mbs} \cdot V_{bs}) - g_{ds} \cdot V_{ds} \tag{3.35}$$

where the + sign indicates that S and D nodes specified in SPICE input file is for normal operation ($V_{ds} > 0$), while the − sign is for inverted operation ($V_{ds} < 0$), such as when the source and drain are interchanged in the SPICE input or voltage polarity across the device changes, as is typical in a pass gate.

The conductances for the source and drain junctions are represented by g_{bs} and g_{bd} respectively, while the corresponding equivalent currents are shown as I_{eqbs} and I_{eqbd} respectively and are given by

$$I_{eqbs} = I_{bs} - g_{bs} \cdot V_{bs} \tag{3.36a}$$

$$I_{eqbd} = I_{bd} - g_{bd} \cdot V_{bd}. \tag{3.36b}$$

The equivalent current source I_{deq}, I_{eqbs} and I_{eqbd} are multiplied by (−1) for *p*-channel devices.

For transient and small-signal analysis, we need capacitance components as well as the previously described DC model. Figure 3.37 shows the complete equivalent circuit for a MOSFET. C_{bs} and C_{bd} are the source and

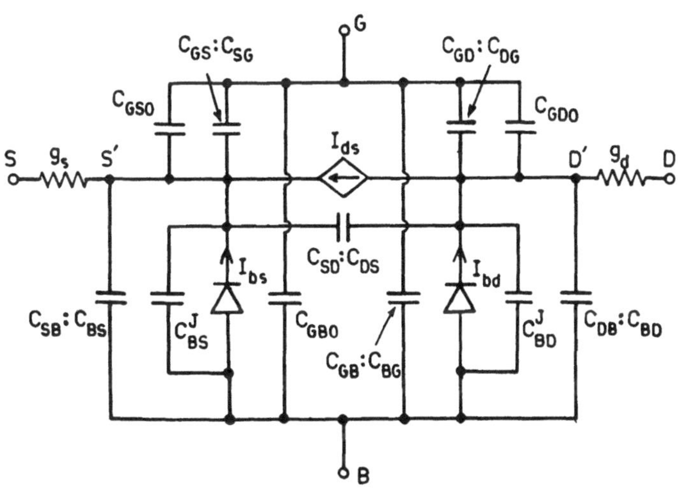

Fig. 3.37 MOSFET large signal equivalent circuit model for transient analysis

drain junction capacitance respectively while Q_{bs} and Q_{bd} are the corresponding charges. The gate overlap capacitances are shown as C_{GSO}, C_{GDO} and C_{GBO}. The twelve intrinsic capacitances are shown as C_{GS}, C_{GD}, C_{GB}, C_{BS}, C_{BD}, C_{BG}, C_{SD}, C_{SB}, C_{SG}, C_{DS}, C_{DG}, C_{DB}. These nonlinear and nonreciprocal capacitances are required to conserve the charge in the device during the transient analysis and are discussed in Chapter 7. In the first order equivalent circuit model, only 3 nonlinear reciprocal capacitances C_{GS}, C_{GD} and C_{GB} are normally considered as was discussed in section 3.2. However, this simple model is inadequate for many circuits as discussed in chapter 7. For the transient analysis companion models are first formed for these capacitances similar to the case of the diode model discussed earlier in section 2.9.

References

[1] C. T. Sah, 'Characteristics of the metal-oxide-semiconductor transistors', IEEE Trans. Electron Devices, ED-11, pp. 324–345 (1964).

[2] P. Richman, *MOS Field-Effect Transistors and Integrated Circuits*, John Wiley & Sons, New York, 1973.

[3] Y. P. Tsividis, *Operation and Modeling of the MOS Transistor*, McGraw-Hill Book Company, New York, 1987.

[4] F. C. Hsu, P. K. Ko, S. Tam, C. Hu, and R. S. Muller, 'An analytical breakdown model for short-channel MOSFETs', IEEE Trans. Electron Devices, ED-29, pp. 1735–1740 (1982).

[5] F. C. Hsu, R. S. Muller, and C. Hu, 'A simplified model of short-channel MOSFET characteristics in the breakdown mode', IEEE Trans. Electron Devices, ED-30, pp. 571–576 (1983).

[6] M. Pinto-Guedes and P. C. Chan, 'A circuit model for bipolar-induced breakdown in MOSFET', IEEE Trans. Compter-Aided Design, CAD-7, pp. 289–294 (1988).

[7] S. Vaidya, E. N. Fuls, and R. L. Johnston, 'NMOS ring oscillators with cobalt-silicided-diffused shallow junctions formed during poly-plug contact doping cycle', IEEE Trans. Electron Devices, ED-33, pp. 1321–1328 (1986).

[8] R. H. Dennard, F. H. Gaensslen, H. N. Yu, V. L. Rideout, E. Bassous, and A. R. LeBlanc, 'Design of ion-implanted MOSFETs with very small physical dimensions', IEEE J. Solid-State Circuits, SC-9, pp. 256–268 (1974).

[9] J. R. Brews, W. Fichtner, E. H. Nicollian, and S. M. Sze, 'Generalized guide for MOSFET miniaturization', IEEE Electron Devices Lett., EDL-1, pp. 2–5 (1980).

[10] J. W. Mathews and C. K. Erdelyi, 'Power supply voltages for future VLSI', IEEE Proc. CICC, pp. 149–152 (1986).

[11] M. Kakumu, M. Kinugawa, K. Hashimoto, and J. Matsunaga, 'Power supply voltage for future CMOS VLSI in half and sum-micrometer', IEEE IEDM-86, *Tech. Dig.*, pp. 399–402 (1986).

[12] P. K. Chatterjee, W. R. Hunter, T. C. Holloway, and Y. T. Lin, 'The impact of scaling laws on the choice of *n*-channel and *p*-channel for MOS VLSI', IEEE Trans. Electron Device Lett., EDL-1, pp. 220–223 (1980).

[13] H. Shichijo, 'A re-examination of practical performance limits of scaled n-channel and p-channel MOS devices for VLSI', Solid-State Electron., 26, pp. 969–986 (1983).

[14] G. Baccarani, M. R. Wordeman, and R. H. Dennard, 'Generalized scaling theory

and its application to 1/4 micron MOSFET design', IEEE Trans. Electron Devices, ED-31, pp. 452–462 (1984).
[15] N. G. Einspruch and G. Gildenblat, Eds., *Advanced MOS Device Physics*, VLSI Electronics Vol. 18, Academic Press Inc., New York, 1989.
[16] Y. W. Sing and B. Sudlow, 'Modeling and VLSI design constraints of substrate current', IEEE IEDM-80, *Dig. Tech. Papers*, pp. 732–735 (1980).
[17] B. Eitan, D. Frohman-Bentchkowsky, and J. Shappir, 'Holding time degradation in dynamic MOS RAM by injection-induced electron currents', IEEE Trans. Electron Devices, ED-28, pp. 1515–1519 (1981).
[18] E. Takeda, A. Shimizu, and T. Hagiwara, 'Role of hot-hole injection in hot-carrier effects and the small degraded channel region in MOSFETs', IEEE Electron Device Lett., EDL-4, pp. 329–331 (1983).
[19] E. Takeda, H. Kume, T. Toyabe, and S. Asai, 'A submicrometer MOSFET structure for minimizing hot-carrier generation', IEEE Trans. Electron Devices, ED-29, pp. 611–618 (1982).
[20] E. Takeda, 'Hot-carrier effects in submicrometer MOS VLSI', Proc. IEE, 131, Pt I, no. 5, pp. 153–164 (1984).
[21] J. J. Sanchez, K. K. Hsueh, and T. A. DeMassa, 'Drain-engineered hot-electron-resistant device structures—A Review', IEEE Trans. Electron Devices, ED-36, pp. 1125–1131 (1989).
[22] D. Frohman-Bentchkowsky, 'FAMOS—A new semiconductor charge storage device', Solid-State Electron., 17, pp. 517–529 (1974).
[23] K. M. Cham, S. Y. Oh, D. Chin, J. L. Moll, K. Lee, and P. V. Voorde, *Computer-Aided Design and VLSI Device Development*, 2nd Ed., Kluwer Academic Publisher, Boston, 1988.
[24] J. M. Pimbley, M. Ghezzo, H. G. Parks, and D. M. Brown, in: *Advanced CMOS Process Technology* (N. G. Einspruch, Ed.), VLSI Electronics: Microstructure Science, Vol. 19, Academic Press Inc., New York, 1989.
[25] J. Y. Chen, '*CMOS Devices and Technology for VLSI*', Prentice Hall, Englewood Cliffs, NJ, 1990.
[26] S. P. Murarka, *Silicide for VLSI Applications*, Academic Press, New York, 1983.
[27] D. L. Kwong, Y. H. Ku, S. K. Lee, E. Louis, N. J. Alvi, and P. Chu, 'Silicided shallow junction formation by ion implantation of impurity ions into silicide layers and subsequent drivein', J. Appl. Phys., 61, pp. 5084–5088 (1987).
[28] L. Van den Hove, R. Wolters, K. Maer, R. F. De Keersmaecker, and G. J. Declerack, 'A self-sligned $CoSiO_2$ interconnection and contact technology for VLSI applications', IEEE Trans. Electron Devices, ED-34, pp. 554–562 (1987).
[29] C. Y. Wong, J. Y.-C. Sun, Y. Taur, C. S. Oh, R. Angelucci, and B. Davari, 'Doping of N^+ and P^+ poly-Si in a dual-gate CMOS process', IEEE-IEDM88, *Tech. Dig.*, pp. 238–241 (1988).
[30] M. Koyanagi, H. Kaneko, and S. Shinizu, 'Optimum design of n^+–n^- double-diffused drain MOSFETS to reduce hot-carrier emission', IEEE Trans. Electron Devices, ED-32, pp. 562–570 (1985).
[31] S. Ogura, P. J. Tsang, W. W. Walker, D. L. Critchlow, and J. F. Shepard, 'Design and characteristics of the lightly doped drain-source (LDD) insulated gate field-effect transistor', IEEE Trans. Electron Devices, ED-27, pp. 1359–1367 (1980).
[32] P. J. Tsang, S. Ogura, W. W. Walker, J. F. Shepard, and D. L. Critchlow, 'Fabrication of high-performance LDD FETS with sidewall spacer technology', IEEE Trans. Electron Devices, ED-29, pp. 590–596 (1982).
[33] R. D. Rug, H. Momose, and Y. Nagakubo, 'Deep trench isolated CMOS devices', IEEE-IEDM82, *Tech. Dig.*, p. 62– (1982). See also R. D. Rung, 'Trench isolation prospects for application in CMOS VLSI', IEEE-IEDM84, *Tech. Dig.*, p. 574–578 (1984).

[34] R. R. Troutman, '*Latchup in CMOS Technology: The Problem and Its Cure*', Kluwer Academic Publisher, Boston, 1987.

[35] K. K. Ng and W. T. Lynch, 'The impact of intrinsic series resistance as MOSFET scaling', IEEE Trans. Electron Devices, ED-34, pp. 503–511 (1987).

[36] G. Sh. Gildenblat and S. S. Cohen, 'Contact metalization', in: N. G. Einspruch and G. Gildenblat, Eds., VLSI Electronics Vol. 15, Academic Press Inc., New York, 1987.

[37] J. M. Pimbley, E. Cumberbatch, and P. S. Hagan, 'Analytical treatment of MOSFET source-drain resistance', IEEE Trans. Electron Devices, ED-34, pp. 834–838 (1987).

[38] G. Baccarani and G. A. Sai-Halasz, 'Spreading resistance in submicron MOSFETs', IEEE Trans. Electron Device Lett., EDL-4, pp. 27–29 (1983).

[39] K. K. Ng and W. T. Lynch, 'Analysis of the gate-voltage dependent series resistance of MOSFETs', IEEE Trans. Electron Devices, ED-33, pp. 965–972 (1986).

[40] J. M. Pimbley, 'Two dimensional current flow in the MOSFET source-drain', IEEE Trans. Electron Devices, ED-33, pp. 986–996 (1986).

[41] W. M. Loh, S. E. Swirhun, T. A. Schreyer, R. M. Swanson, and K. C. Saraswat, 'Current crowding effects and determination of specific contact resistivity from contact end resistance (CER) measurements', IEEE Trans. Electron Devices, ED-34, pp. 512–524 (1987).

[42] F. M. Klaassen, P. T. J. Biermans, and R. M. D. Velghe, 'The series resistance of submicron MOSFETs and its effect on their characteristics', Proc. ESSDERC 1988, J. De Physique, pp. 257–260 (1988).

[43] G. J. Hu, C. Chang, and Y. T. Chia, 'Gate-voltage-dependent effective channel length and series resistance of LDD MOSFETs', IEEE Trans. Electron Devices, ED-34, pp. 2469–2475 (1987).

[44] S. Y. Chou and D. A. Antoniadis, 'Relationship between measured and intrinsic transconductances of FETs', IEEE Trans. Electron Devices, ED-34, pp. 448–450 (1987).

[45] S. Cserveny, 'Relationship between measured and intrinsic transconductances of MOSFETs', IEEE Trans. Electron Devices, ED-37, pp. 2413–2414 (1990).

[46] M. H. Seavey, 'Source and drain resistance determination for MOSFETs', IEEE Electron Device Lett., EDL-5, pp. 479–481 (1984).

[47] R. Shrivastava and K. Fitzpatrick, 'A simple model for the overlap capacitance of a VLSI MOS device', IEEE Trans. Electron Devices, ED-29, 1870–1875 (1982).

[48] E. W. Greeneich, 'An analytical model for the gate capacitance of small-geometry MOS structures', IEEE Trans. Electron Devices, ED-30, pp. 1838–1839 (1983).

[49] T. Smedes and F. M. Klaassen, Effects of the lightly doped drain configuration on capacitance characteristics of submicron MOSFETs', IEEE IEDM-90, *Technical Digest*, pp. 197–200 (1990).

[50] N. D. Arora, D. A. Bell, and L. A. Bair, 'An accurate method of determining MOSFET gate overlap capacitance', Solid-State Electron., 35, pp. 1817–1822 (1992).

[51] S. W. Lee and R. C. Rennick, 'A compact IGFET model—ASIM', IEEE Trans. Computer-Aided Design, CAD-7, pp. 952–975 (1988).

[52] S. Liu and L. W. Nagel, 'Small-signal MOSFET models for analog circuit design', IEEE J. Solid-State Circuits, SC-17, pp. 983–998 (1982).

MOS Capacitor 4

The metal–oxide–semiconductor (MOS) structure is the heart of MOS technology. When this structure, commonly referred as MOS capacitor, is connected as a two terminal device, with one electrode connected to the metal and the other electrode connected to the semiconductor, a *voltage dependent capacitance* results. The MOS capacitor is a very useful device both for evaluating the MOS IC fabrication process and for predicting the MOS transistor characteristics. For this reason MOS capacitors are often included on the chip test sites. Note that the term MOS is still used even if the top electrode is not a metal and the insulator is not an oxide. Sometimes the acronym MIS (Metal–Insulator–Semiconductor) is also used for the MOS structure.
In this chapter we first discuss the behavior of a MOS capacitor and then develop the charge–voltage (Q–V) and capacitance–voltage (C–V) relationships, which will be used later in the development of the MOS transistor model. We will conclude the chapter by discussing applications of the C–V curves in MOS technology.

4.1 MOS Capacitor with No Applied Voltage

The cross-section of an MOS capacitor is shown in Figure 4.1. Essentially it consist of a *p*- or *n*-type silicon substrate covered by an insulating layer of oxide.[1] The oxide is thermally grown silicon dioxide (SiO_2) with a thickness that usually lies between 100 and 1000 Å. On the top of the oxide is a conducting layer of metal (usually aluminum) or, more commonly, degenerately doped polysilicon or a combination of polysilicon and silicide (e.g. $TiSi_2, CoSi_2$). This conducting metal layer at the top is called *gate*. The

[1] One may use other insulating materials such Si_3N_4 (silicon nitride) or even sandwich structures, the most common being a combination of SiO_2, Si_3N_4 and SiO_xN_y.

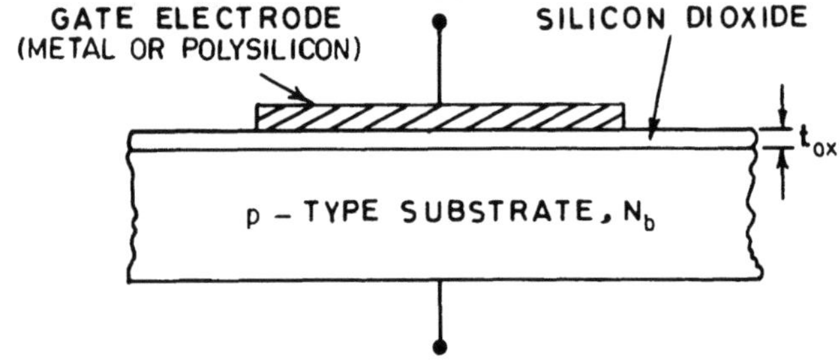

Fig. 4.1 Cross-section of a MOS capacitor fabricated on a p-type substrate of concentration N_b

gate structure is defined either by photolithography or the metal is evaporated on the wafer through a metal defining mask. The second electrode is the back (bulk) contact obtained by evaporating metal which is in intimate contact with silicon to make an ohmic contact. If the substrate conducts sufficiently to support displacement currents, this configuration results in a parallel plate capacitor with the gate as one electrode, the silicon as the other electrode, and SiO_2 as the dielectric. Such a system, called a *MOS capacitor, is in thermal equilibrium with the DC voltage applied, or if the voltage changes sufficiently slowly to be approximated as being constant.* Recalling the parallel plate capacitance formulation, we can write the oxide capacitance per unit area, C_{ox}, between the metal and the silicon surface as[2]

$$\boxed{C_{ox} = \frac{\epsilon_0 \epsilon_{ox}}{t_{ox}} \quad (\mathrm{F/cm^2})} \tag{4.1}$$

where ϵ_{ox} is the dielectric constant of the oxide (SiO_2) and t_{ox} is the oxide thickness. C_{ox} is a very important parameter in MOS technology.

In order to study the MOS capacitor, let us first consider the metal, oxide and semiconductor (p-type silicon) as three separate components. The energy band diagram of the three components is shown in Figure 4.2. In this Figure E_0 denotes a convenient reference potential energy level, which is the vacuum or free electron energy.[3] It is the level at which the $(1/r)$ Coulombic potential of an isolated positive charge becomes zero. Note that

[2] If a dual insulator is used, the capacitance per unit area of the structure could be obtained by defining an equivalent oxide thickness t_{eq} that is electrically equivalent to that of the two insulators. Thus, for SiO_2/Si_3N_4 composite system $t_{eq} = t_{ox} + (\epsilon_{ox}/\epsilon_{nit})t_{nit}$, so that $C_{eq} = \epsilon_0\epsilon_{ox}/t_{eq}$. Here t_{nit} and ϵ_{nit} are thickness and dielectric constant, respectively, of the nitride layer. Note that in MOS modeling it is common practice to express the gate oxide capacitance as per unit area rather than the total capacitance.

[3] The potential energy of an electron sufficiently far away from the material surface, to be considered at infinity, is called vacuum level, symbolized by E_0.

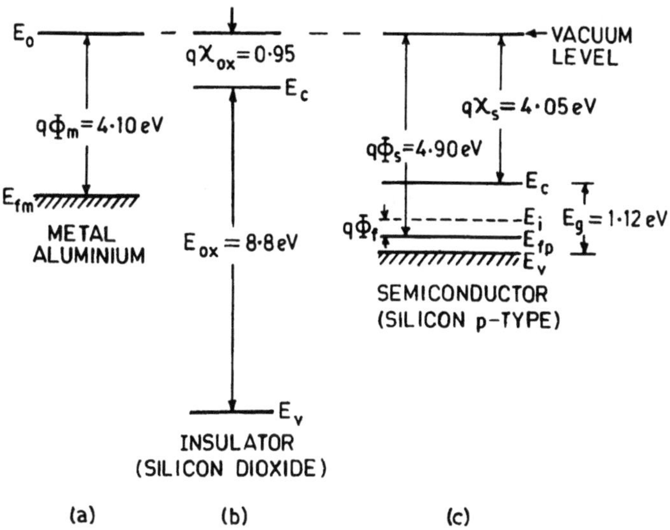

Fig. 4.2 The energy-band diagram of three separated components that form the MOS capacitor; (a) aluminum, (b) thermally grown SiO_2, and (c) *p*-type silicon ($N_b = 10^{15}\,cm^{-3}$)

the bandgap energy of SiO_2 is shown to be 8.8 eV [1] although others have used 8.0–9.0 eV [2]–[4]. The exact value of the energy band gap is debatable due to the amorphous nature of the oxide.

4.1.1 Work Function

The characteristic energy of a metal is its *work function* usually expressed by Φ_m in eV or Φ_m/q in Volts. It is the energy that must be supplied to an electron to take it across the surface energy barrier. In other words the *work function Φ_m is the energy difference between the vacuum level E_0 and the Fermi energy of the metal E_{fm}* ($\Phi_m = E_0 - E_{fm}$). Φ_m depends only on the charge distribution of the atomic core or the type of atom involved if the surface is free of foreign impurities and contamination. For aluminum $\Phi_m = 4.10\,eV$.

In a semiconductor and insulator, the height of the surface energy barrier is specified in terms of *electron affinity χ. It is the energy difference between the vacuum level E_0 and the conduction band edge E_c at the surface*, that is, $\chi = E_0 - E_c$. χ is a property of the material and is not affected by the presence of impurities or imperfections to any extent but only varies from one atomic type to another or is changed by alloy composition. In semiconductors electron affinity χ_s is used instead of $(E_0 - E_f)$, because the latter quantity is not a constant in the semiconductor but varies as a function of doping. From Figure 4.2, it is clear that the work function Φ_s

for *p*-type semiconductors is given by

$$\Phi_s = \chi_s + \frac{E_g}{2} + q\phi_{fp} \quad \text{(eV)} \quad \text{for } p\text{-type} \tag{4.2}$$

where E_g is the band gap energy given by Eq. (2.2) and ϕ_{fp} is the Fermi potential for *p*-type silicon. Similarly for an *n*-type semiconductor

$$\Phi_s = \chi_s + \frac{E_g}{2} - q\phi_{fn} \quad \text{(eV)} \quad \text{for } n\text{-type} \tag{4.3}$$

where ϕ_{fn} is the Fermi potential for *n*-type silicon. For the same doping concentration $|\phi_{fp}| = |\phi_{fn}| = \phi_f$, given by Eq. (2.15), repeated here for convenience

$$\phi_f = V_t \ln\left(\frac{N_b}{n_i}\right) \quad \text{(V)} \tag{4.4}$$

where $V_t(=kT/q)$ is the thermal voltage and N_b is the substrate dopant concentration. It is the magnitude of ϕ_f which is to be used in these and subsequent equations involving these quantities. For silicon, $\chi_s = 4.05\,\text{eV}$ and for *p*-type silicon with an acceptor concentration $N_b = 10^{15}\,\text{cm}^{-3}$, $\phi_f = 0.29\,\text{V}$, giving $\Phi_s = 4.90\,\text{eV}$.[4] This shows that the energy required for an electron to escape from *n*-type silicon ($\Phi_s = 4.9\,\text{eV}$) is higher than the energy required for an electron to escape from aluminum ($\Phi_m = 4.1\,\text{eV}$). The work functions for the commonly used gate materials for IC technology are shown in Table 4.1. When calculating Φ for degenerately doped polysilicon, it is assumed that the Fermi energy lies at the band edges.
We can visualize the work function difference between two materials as a contact potential between them. It can be easily shown that *when different*

Table 4.1. *Work function Φ of different materials as determined by photoresponse [4,5]*

Material	qΦ (eV)
Al	4.10
Au	5.27
$MoSi_2$	4.73
$TiSi_2$	3.95
n-type degenerate polysilicon	4.05
p-type degenerate polysilicon	5.17

[4] In the literature, to calculate Φ_s for silicon, some authors have used ϕ_{so}, the energy difference between vacuum level and valance band, rather than χ_s, resulting in Φ_s for *p*-type silicon as $\Phi_s = \phi_{so} - E_g/2q + q\phi_f$. The quoted value of $\phi_{so} = 5.35\,\text{eV}$ [1] results in Φ_s of 5.08 eV assuming $\phi_f = 0.29\,\text{V}$. This results in a difference of 0.18 V with those calculated using Eq. (4.2).

materials are in contact with each other, the work function between its two ends depends only on the first and the last material [6]. Thus, when three components of the MOS structure are brought into contact, it is only the work function difference between the metal and the semiconductor which matters. Since in general the metal and semiconductor will have different work functions, this difference will cause distortion in the band shape. This is because the Fermi levels E_f have to be aligned when the system is in equilibrium, and the vacuum energy level E_0 has to be continuous. When the metal of the MOS structure is shorted to the semiconductor, electrons will flow from the metal to the semiconductor or vice versa until a potential is built up between the two, that will counterbalance the difference in their work functions. Thus, there is a variation in electrostatic potential from one region to another resulting in $E_c(E_v)$ band bending in the interior of the structure. Since metal is an equipotential region, no band bending occurs there. Therefore the energy bands bend in the oxide and semiconductor showing upward slope when $\Phi_m > \Phi_s$ and downward slope when $\Phi_m < \Phi_s$.

For the case Al–SiO_2–Si(p-type) system $\Phi_m < \Phi_s$, the bands bend downwards as shown in Figure 4.3a. This of course assumes that *oxide is an ideal insulator with no charges of its own.* We can compensate for this band bending by applying an external voltage ΔV_{fb}, which is simply the work function difference that caused the band bending in the first place. Thus for the bands to become flat at the surface (see Figure 4.3b) we have

$$\boxed{\Delta V_{fb} = \Phi_m - \Phi_s = \Phi_{ms}} \tag{4.5}$$

where Φ_{ms} is the work function difference between the gate electrode and bulk silicon in Volts.

For an Al–SiO_2–Si(p-type) system

$$\Phi_{ms} = (\Phi_m - \Phi_s) = \Phi_m - (\chi_s + E_g + q\phi_f) \quad \text{(V)} \tag{4.6}$$

or

$$\boxed{\Phi_{ms} = -0.51 - \phi_f \quad \text{(V)} \quad p\text{-type Si.}} \tag{4.7}$$

Since Φ_f is typically 0.3 V, Φ_{ms} is a negative number. Similarly for an Al–SiO_2–Si(n-type) system,

$$\boxed{\Phi_{ms} = -0.51 + \phi_f \quad \text{(V)} \quad n\text{-type Si}} \tag{4.8}$$

which is again a negative number. Thus for an Al–SiO_2–Si(n- or p-type) system Φ_{ms} is always negative.

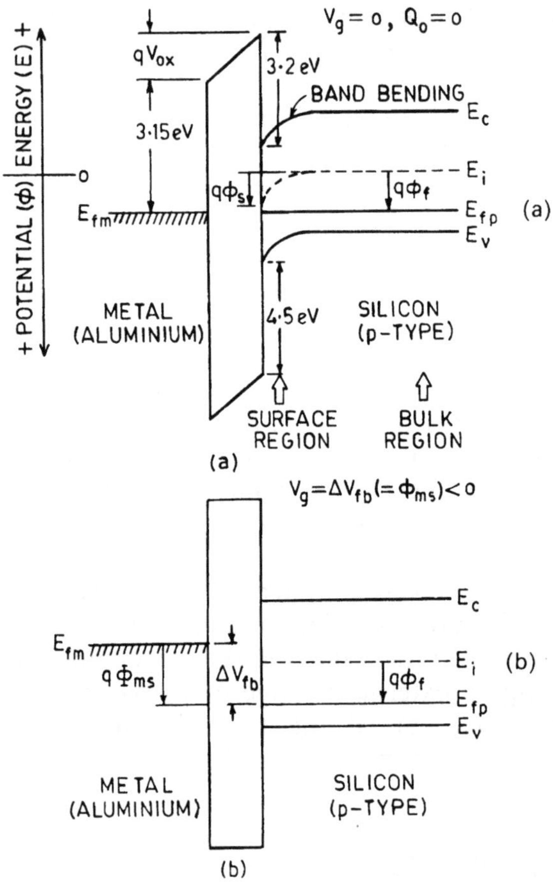

Fig. 4.3 MOS capacitor (a) showing band bending at the surface due to Φ_{ms}, the work function difference between the metal (Al) and semiconductor (*p*-type silicon), with no bias applied; (b) flat band condition for structure shown in (a). Oxide is assumed to be free of any charges

As was pointed out earlier (cf. section 3.5.1) in present day MOS technologies the gate electrode is invariably made of degenerate polysilicon. For the polysilicon gate electrode, the workfunction difference Φ_{ms} becomes

$$\Phi_{ms} = \phi_f(\text{polysilicon gate}) - \phi_f(\text{substrate}). \tag{4.9}$$

Figure 4.4 shows the energy band diagram for the *p*-type substrate with an n^+ polysilicon gate. In this case Φ_{ms} becomes

$$\boxed{\Phi_{ms} = -0.56 - \phi_f \quad (\text{V}) \quad p\text{-type Si}} \tag{4.10}$$

which is a negative number. Here 0.56 is half of E_g for silicon. Similarly

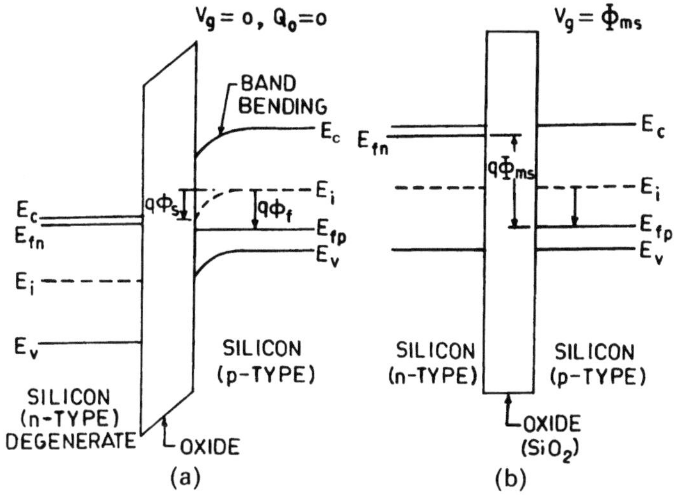

Fig. 4.4 MOS capacitor (a) showing band bending at the surface due to Φ_{ms}, the work function difference between the n^+ degenerately doped polysilicon and semiconductor (*p*-type silicon), with no bias applied; (b) flat band condition for structure shown in (a). Oxide is assumed to be free of any charges

for an *n*-type substrate with a n^+ polysilicon gate, we have

$$\boxed{\Phi_{ms} = -0.56 + \phi_f \quad (\text{V}) \quad n\text{-type Si}} \tag{4.11}$$

which shows that Φ_{ms} will still be negative but by a smaller amount as compared to the case for n^+ polysilicon with a *p*-substrate. Detailed investigation of n^+ polysilicon gate suggests that Φ_{ms} has a maximum magnitude at a dopant concentration of about $5 \times 10^{19}\,\text{cm}^{-3}$ and becomes less negative for higher dopant concentration; the behavior is different for As and P, suggesting that the polysilicon grain structure affects the work function difference [7, 8]. Note that for *n*-type substrate with a p^+ polysilicon gate, Φ_{ms} will be a positive quantity. The work function difference Φ_{ms} as a function of silicon doping concentration N_b for aluminum, p^+ and n^+ polysilicon gates are shown in Figure 4.5 and are based on Eqs. (4.7)–(4.11).

4.1.2 Oxide Charges

Some impurities or defects can be inadvertently incorporated into the oxide during oxide growth or subsequent processing steps. This results in the oxide being contaminated with various types of charges and traps. Four different types of charges have been identified in thermally grown oxide on a silicon surface [11]. These charges are shown schematically in

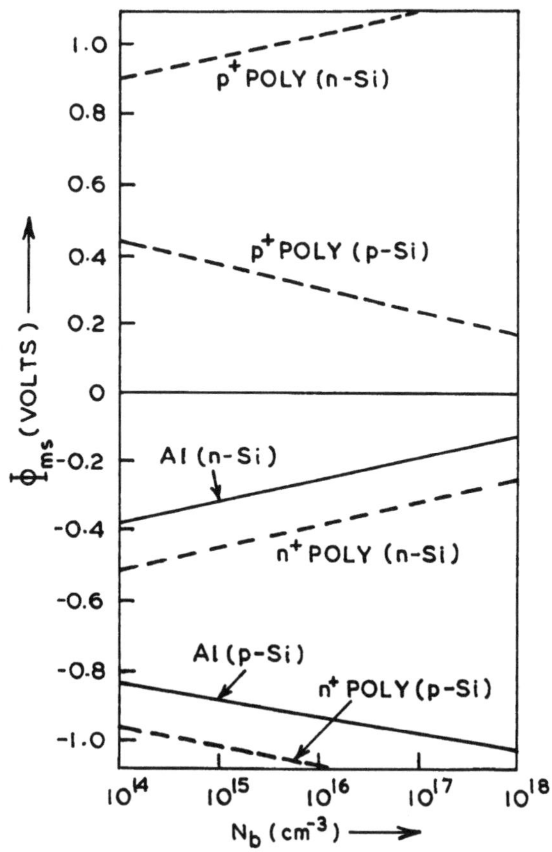

Fig. 4.5 Work function difference Φ_{ms} versus substrate doping N_b for degenerately doped polysilicon and aluminium. (From Kim et al. [10])

Figure 4.6. They are (1) interface-trapped charge Q_{it}, (2) fixed-oxide charge Q_f, (3) oxide trapped charge Q_{ot}, and (4) mobile ionic charge Q_m. All of these charges are very much dependent on the device fabrication process. Here we discuss basic properties of these charges, however for details of the origin and techniques of measurements of different oxide charges, the reader is referred to [1]–[3], [13].

Interface Trapped Charge Q_{it}. The interface trapped charge, Q_{it}, is the charge due to electronic energy levels located at the Si–SiO_2 interface with energy states in the silicon band gap that can capture or emit electrons (or holes). These electronic states arise because of the lattice mismatch at the interface, dangling (incomplete) bonds, the adsorption of foreign impurity atoms at the silicon surface, and other defects caused by radiation or similar bond-breaking processes. These are the most important type of charges because of their wide-ranging and degrading effect on device behavior.

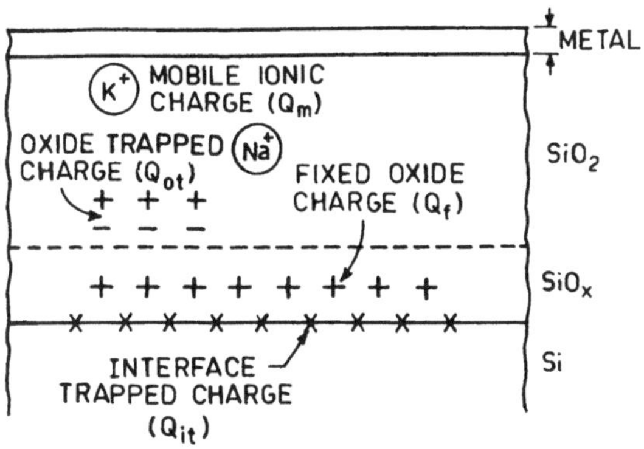

Fig. 4.6 Names and location of the charges associated with thermally grown SiO_2 on silicon (From Deal [11])

These types of charges have also been referred to as *surface states, fast states, or interface states* denoted by Q_{ss}. Under equilibrium condition, the occupancy of these interface states or traps is governed by the position of the Fermi level in the same way as for any other electron energy level. Because interface trap levels are distributed across the silicon energy band, we normally define an interface trap density D_{it} [1–4]

$$D_{it} = \frac{1}{q}\frac{dQ_{it}}{dE} \quad \text{number of charges/cm}^2\cdot\text{eV}.$$

D_{it} is extremely sensitive to even minor process details, varies significantly from process to process, and is orientation dependent. Present day MOS devices with thermally grown SiO_2 on Si have most of the interface trapped charge neutralized by low temperature ($\leq 500\,^\circ$C) hydrogen annealing. In $\langle 100 \rangle$ orientation, D_{it} is about an order of magnitude smaller than in $\langle 111 \rangle$; smaller D_{it} is correlated with the smaller density of the available bonds at the surface. The value of D_{it} at mid gap for $\langle 100 \rangle$ oriented silicon for modern MOS VLSI process can be as low as 5×10^9 cm^{-2}eV^{-1}. *An increase in D_{it} causes instabilities in the MOS transisfor threshold voltage and reduces the carrier mobility at the surface, which results in a reduction in the device transconductance.*

Fixed Oxide Charge Q_f. As the name suggests these are the immobile charges located within approximately 25 Å of the Si-SiO_2 interface (see Figure 4.6) and normally arise from structural damage associated with oxidation or various impurity atoms. Generally Q_f is positive and depends on the oxidation ambient, temperature and annealing conditions, and silicon orientation. Like Q_{it}, Q_f is low for $\langle 100 \rangle$ orientation. However, it is independent of the doping type and concentration in the silicon, oxide thickness and oxidation time. Q_f can be minimized by annealing in an inert ambient, such as, Argon at a temperature in excess of 900°C. Typical values of Q_f for a carefully treated Si–SiO_2 system is about $\sim 10^{10}\,cm^{-2}$ for the ⟨100⟩ surface. *Because of the low value of Q_{it} and Q_f, the ⟨100⟩ orientation is preferred for silicon MOSFET.*

Oxide Trapped Charge Q_{ot}. The oxide trapped charge Q_{ot} is associated with defects in SiO_2. The oxide traps are usually electrically neutral and are charged by introducing electrons and holes into the oxide through ionizing radiation such as implanted ions, X-rays, electron beams, etc. The magnitude of Q_{ot} depends on the amount of radiation dose and energy and the field across the oxide during irradiation. Like Q_{it} these charges could be positive (trapped hole) or negative (trapped electrons). Q_{ot} resembles Q_f in that its magnitude is not a function of silicon surface potential and there is no capacitance associated with it.

Mobile Ionic Charge Q_m. The mobile ionic charges Q_m are due to sodium (Na^+) or other alkali ions that gets into the oxide during cleaning, processing and handling of the MOS devices.[5] These ions move very slowly within the oxide; their transport depends strongly on the applied electric field ($\sim 1\,MV \cdot cm^{-1}$) and temperature (30–400°C). Positive voltages push the ions towards the interface, while negative voltages draw them towards the gate. A current is observed in the external circuit during ion drift. After the ion drift, the changed centroid of charge within the oxide layer results *in a shift of the flat band voltage or in a threshold voltage shift in the MOSFET, and may cause an unexpected device failure.* Device instabilities from mobile ions are minimized by avoiding contamination during processing. Chlorine neutralization is the most commonly accepted procedure for reducing mobile ion contamination in gate oxides. The other most common approach to counter the action of alkali ions is by gettering them far from the interface in a gettering layer. The most commonly used gettering medium is Borophospho-silicate glass (BPSG).
The result of these charges in the oxide and at the SiO_2 interface is to bend the bands at the silicon surface in a fashion similar to that shown in Figure 4.3 due to the work function difference Φ_{ms}. If $\rho(x)$ is the charge density/unit volume within the oxide, then the total voltage shift ΔV_{fb} or band bending due to the various charges is simply the sum of the voltage shifts due to the individual charges; that is [4]

$$\boxed{\Delta V_{fb} = -\frac{Q_f + Q_{it}}{C_{ox}} - \frac{1}{C_{ox}} \int_0^{t_{ox}} \frac{x}{t_{ox}} \rho(x) dx \quad (V)} \tag{4.12}$$

[5] In the early 1960s Na^+ ions were the main source of the instability in the threshold voltage of a MOSFET.

remembering that the charges Q_{it} and Q_f are located at or near the Si–SiO$_2$ interface ($x = t_{ox}$), while the mobile and the oxide trap charges Q_m and Q_{ot} respectively are distributed throughout the oxide.
For circuit simulation models, it is common practice to model these different types of charges by a sheet of charge Q_o which is assumed to be located at the Si–SiO$_2$ interface and causes the same effect as that of actual charges of unknown distribution. This charge Q_o is called the *equivalent interface charge and is always positive for both p- and n-type substrate.* The total band bending due to the effective oxide charge becomes

$$\Delta V_{fb} = -\frac{Q_o}{C_{ox}} \quad \text{(V)}. \tag{4.13}$$

This is the gate voltage that is needed to cause all the charge Q_o to be imaged in the gate electrode so that none is induced in the silicon. However, when the gate "floats"or the gate electrode is absent, the oxide charges will seek all their image charges in the silicon.

4.1.3 Flat Band Voltage

We have already seen (cf. section 4.1.1) that when different elements of a MOS system are brought into contact with each other forming a MOS capacitor, the oxide and the semiconductor bands bend as shown in Figure 4.3. This band bending is due to the work function difference Φ_{ms} between the metal and the semiconductor given by Eq. (4.5). Band bending also takes place due to the oxide charges and is given by Eqs. (4.12) or (4.13). The total band bending V_{fb} due to Φ_{ms} and Q_0 is obtained by adding Eqs. (4.5) and (4.13). That is

$$\boxed{V_{fb} = \Phi_{ms} - \frac{Q_o}{C_{ox}} \quad \text{(V)}.} \tag{4.14}$$

It should be pointed out that in present-day VLSI technology the contribution of the Q_0/C_{ox} term to V_{fb} is much smaller than the Φ_{ms} term, and therefore the value of V_{fb} is controlled mainly by Φ_{ms}. Thus, for nMOST with n^+ polysilicon gate, V_{fb} is a negative number because Φ_{ms} is negative. However, for pMOST V_{fb} is positive for p^+ polysilicon gate and near zero or negative for n^+ polysilicon gate (see section 4.1.1).
Let us now see how this band bending at the silicon surface affects surface behavior. The hole concentration in a p-type substrate is given by Eq. (2.10b), repeated here for convenience,

$$p = n_i \exp\left(\frac{E_i - E_f}{kT}\right) \quad \text{cm}^{-3}. \tag{4.15}$$

When the bands bend downwards, as is the case for the Al–SiO_2–Si system, the energy difference $(E_i - E_f)$ decreases at the surface. According to Eq. (4.15), this decrease in $(E_i - E_f)$ will result in a decrease in p, the hole concentration. In other words, holes are depleted at the surface giving rise to a space charge region. On the contrary if the band bend upwards, as happens in the case of an Au–SiO_2–Si system $(\Phi_m > \Phi_s)$, the energy difference $(E_i - E_f)$ increases which results in an increase in the hole concentration at the surface. Thus, we see that even when no external voltage is applied to a MOS capacitor, there will be a change in the concentration of carriers at the surface as compared to that in the bulk, due to the Φ_{ms} and/or charges Q_0 in the oxide. This change in the concentration creates an electric field at the surface and hence a voltage difference between the silicon surface and the silicon bulk. This voltage difference, referred to as the *surface potential* ϕ_s, *represents the electrostatic potential at the surface measured from the bulk intrinsic level* E_i. *It is a measure of the amount of band bending at the surface*, as shown in Figure 4.7. Note that ϕ_s is a positive quantity when bands bend downward and it is negative when bands bend upward. When ϕ_s is zero the bands are flat at the surface. Thus, at any point x from the surface towards the bulk the potential is $\phi(x)$.

Band bending can be compensated by applying an external voltage equal to V_{fb} [cf. Eq. (4.14)]. This condition under which bands become flat at the surface is called the *flat band condition* and the corresponding voltage which is required to bring about the flat band condition is called the *flat band voltage*, V_{fb}. Thus, the *flat band voltage is that gate voltage which must be applied so as to have zero surface potential* $(\phi_s = 0)$ with *flat energy bands*

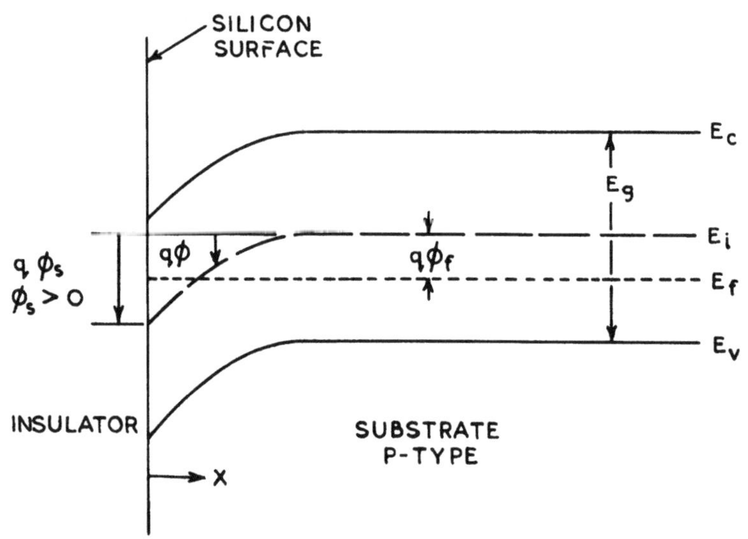

Fig. 4.7 Band bending showing surface potential ϕ_s at the surface of a p-type silicon

over the entire semiconductor surface. The flat band condition is often used as a reference state and the flat band voltage as a reference voltage and is thus an important MOS device parameter.

4.2 MOS Capacitor at Non-Zero Bias

The behavior of the MOS capacitor described in the previous section was under conditions when no external voltage was applied to the gate. Let us now discuss what happens when an external voltage V_g is applied between the gate and the substrate (bulk) as shown in Figure 4.8. The applied voltage V_g is shared between the voltage across the oxide V_{ox}, the surface potential ϕ_s and the work function difference Φ_{ms} (V) between the metal and the substrate. Thus,

$$V_g = V_{ox} + \phi_s + \Phi_{ms}. \tag{4.16}$$

When a gate voltage V_g is applied to a MOS capacitor, it induces a charge Q_s in the silicon. From the charge point of view, we will have three charges in the system:

- Charge Q_g on the gate due the voltage V_g applied to the gate.
- Effective interface charge Q_o in the oxide silicon interface because the oxide is not a perfect insulator (section 4.1.2).
- Charge Q_s induced in the silicon under the oxide.

The charge neutrality condition requires that

$$\boxed{Q_g + Q_o + Q_s = 0.} \tag{4.17}$$

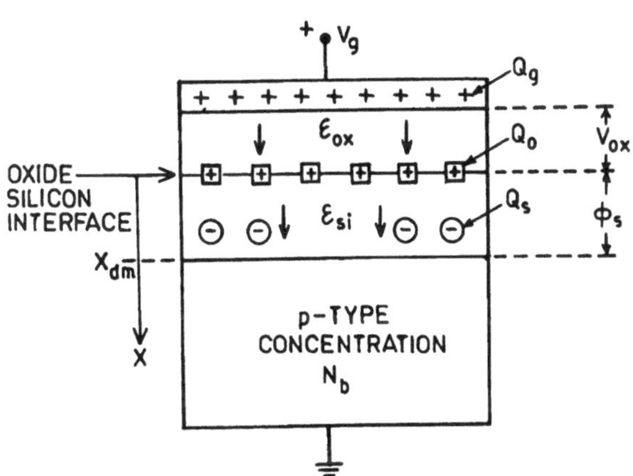

Fig. 4.8 MOS capacitor under applied gate voltage V_g showing charges, fields and potentials

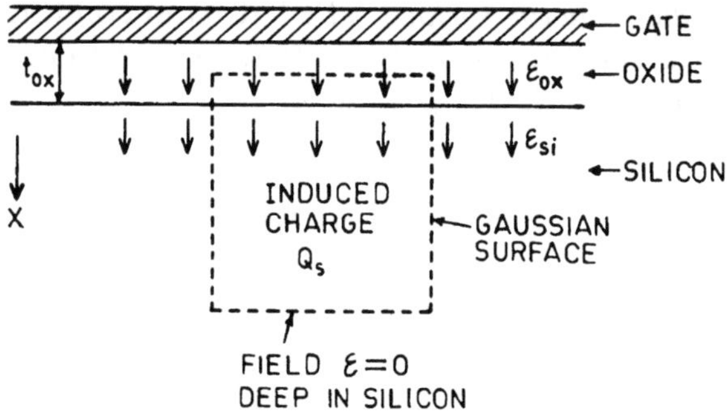

Fig. 4.9 Gaussian volume relating charge Q_s stored in the silicon to the field $\mathscr{E}_{ox}$ in the oxide

If the applied voltage V_g is positive, then the electric field $\mathscr{E}_{si}$ is directed *into* the silicon surface at the interface and will induces a charge Q_s in the silicon. The density of the induced charge Q_s per unit area can be calculated using Gauss' law.[6] Figure 4.9 shows an imaginary Gaussian surface enclosing part of the MOS capacitor in the silicon. Deep in the silicon the electric field $\mathscr{E}$ is zero, therefore there is no contribution to the flux through the Gaussian volume from the bottom surface. Also there is no flux contribution from the sides of the Gaussian volume, since the electric field lines run parallel to the sides. The only flux contributing to Q_s is from the oxide silicon side and as such Q_s becomes

$$Q_s = -\epsilon_0 \epsilon_{si} \mathscr{E}_{si} \quad (\text{F/cm}^2). \tag{4.18}$$

Similarly, applying Gauss' law at the metal-oxide interface gives

$$Q_g = \epsilon_0 \epsilon_{ox} \mathscr{E}_{ox} \equiv V_{ox} C_{ox} \quad (\text{F/cm}^2) \tag{4.19}$$

where $\mathscr{E}_{ox} = V_{ox}/t_{ox}$ is the electric field in the oxide. The field $\mathscr{E}_{ox}$ and $\mathscr{E}_{si}$ are related by Eq. (4.17). For the ideal case with $Q_0 = 0$ (no interface/oxide charge) we have $\epsilon_0 \epsilon_{ox} \mathscr{E}_{ox} = \epsilon_0 \epsilon_{si} \mathscr{E}_{si}$. Now combining Eqs. (4.14), (4.16)–(4.19) it is easy to see that

$$\boxed{V_g = V_{fb} + \phi_s - \frac{Q_s}{C_{ox}} \quad (\text{V}).} \tag{4.20}$$

This gives the relationship between the applied gate voltage V_g and the surface potential ϕ_s. Note that under the flat band condition $V_g = V_{fb}$,

[6] Gauss' theorem states that the outgoing flux of the electric field through a closed surface is equal to the sum of the charge located within the closed surface.

because $\phi_s = 0$ and $Q_s = 0$. Depending upon the value of V_g, different surface conditions will results, which will now be discussed.

4.2.1 Accumulation

Suppose we apply a gate voltage V_g such that $V_{go}(= V_g - V_{fb})$ is negative, that is $V_g < V_{fb}$. This negative voltage at the gate will create an electric field $\mathscr{E}_{ox}$ which points towards the gate electrode as shown in Figure 4.10a. Since the applied negative voltage *depresses* the electrostatic potential of the metal relative to the substrate, electron energies are *raised* in the metal relative to the substrate. As a result the Fermi level E_{fm} for the metal lies above its equilibrium position by qV_{go}. Moving E_{fm} up in energy relative to the substrate Fermi level E_f causes the oxide conduction band to bend upwards, consistent with the direction of the field $\mathscr{E}_{ox}$.
From the charge point of view, $V_{go} < 0$ results in a negative charge on the gate. This in turn will induce a positive charge Q_s at the silicon surface. Such a positive charge in the *p*-type silicon means excess hole concentration is created at the surface (see Figure 4.10a). Since excess holes are accumulated at the surface, this is referred to as the *accumulation* condition. As the hole concentration increases at the surface, $(E_i - E_f)$ must increase in accordance with Eq. (4.15), resulting in the bands bending upwards as shown in Figure 4.10a. Thus in

$$\text{accumulation} \begin{cases} V_g & < V_{fb}, \\ \phi_s & < 0, \\ Q_s & > 0. \end{cases} \tag{4.21}$$

Although this bias condition is extremely useful in measuring some basic MOS system characteristics, it is not of much importance from the MOS circuit point of view.

4.2.2 Depletion

Now consider the case when the gate voltage is such that $V_{go}(= V_g - V_{fb})$ is positive, that is $V_g > V_{fb}$. This positive gate voltage will create an electric field $\mathscr{E}_{ox}$ pointed from the gate towards the substrate as shown in Figure 4.10b. A positive gate voltage raises the potential of the gate, lowering the Fermi level E_{fm} by qV_{go}. Moving E_{fm} down in energy relative to the substrate Fermi level E_f causes band bending downward in the oxide conduction band, consistent with the direction of the field $\mathscr{E}_{ox}$.
A positive voltage at the gate places positive charge on it, which in turn will repel holes from the silicon surface and thus exposes the negatively

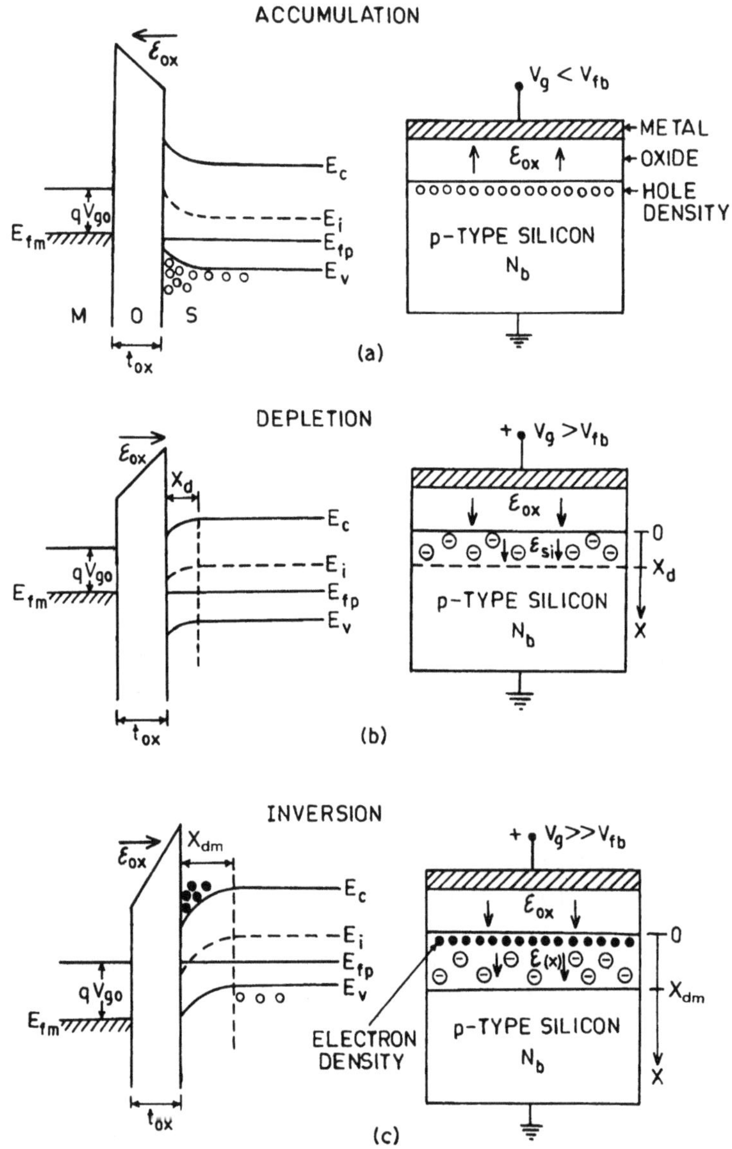

Fig. 4.10 Effect of applied voltage on a *p*-type MOS capacitor; (a) negative voltage $V_{go} = (V_g - V_{fb})$ causes hole accumulation at the surface; (b) positive voltage depletes holes from the silicon surface; and (c) a large positive V_{go} causes inversion, forming an *n*-type layer at the silicon surface

charged acceptor ions. In other words, a positive charge on the gate induces a negative charge Q_s at the silicon surface. Since holes are depleted at the surface it is referred to as the *depletion* condition. This is analogous to the depletion region in *pn* junctions discussed in section 2.5. Since hole concentration decreases at the surface, we see from Eq. (4.15) that $(E_i - E_f)$

must decrease resulting in E_i coming closer to E_f thereby bending the bands downward near the surface (Figure 4.10b). Thus in

$$\text{depletion}\begin{cases} V_g & > V_{fb}, \\ \phi_s & > 0, \\ Q_s & < 0. \end{cases} \tag{4.22}$$

Let us now calculate the depletion layer charge. The band bending potential $\phi(x)$ must satisfy Poisson's equation (2.41) and is used to calculate the induced charge Q_s within the space charge region of width X_d at the surface, also called the depletion width. We refer to this induced charge in the depletion region as the *bulk charge* denoted by Q_b. Applying Gauss' law we have [cf. Eq. (4.18)]

$$Q_b \equiv Q_s(\text{depletion}) = -\epsilon_0\epsilon_{si}\mathscr{E}_{si} \quad (\text{F/cm}^2). \tag{4.23}$$

Under the depletion approximation (cf. section 2.5.2) $n = p = 0$ (no free carriers) and the assumption that the substrate is p-type (uniform concentration $N_b\,\text{cm}^{-3}$) so that $N_a(=N_b) \gg N_d$, the Poisson equation (2.41) becomes

$$\frac{d^2\phi}{dx^2} = \frac{qN_b}{\epsilon_0\epsilon_{si}} \quad \text{for} \quad 0 \le x \le X_d. \tag{4.24}$$

Integrating the above equation twice from the interface ($x = 0$) to the depletion edge ($x = X_d$) and using the boundary conditions

$$\phi = \phi_s \quad \text{and} \quad \frac{d\phi}{dx} = -\mathscr{E}_{si} \quad \text{at} \quad x = 0 \tag{4.25a}$$

and

$$\phi = \frac{d\phi}{dx} = 0 \quad \text{at} \quad x = X_d \tag{4.25b}$$

we get

$$\phi(x) = \phi_s\left(1 - \frac{x}{X_d}\right)^2 \quad (\text{V}) \tag{4.26}$$

which gives a relationship between the band bending and the surface potential. The depletion width X_d in the above equation can easily be calculated by substituting Eq. (4.26) in (4.24) giving

$$\boxed{X_d = \sqrt{\frac{2\epsilon_0\epsilon_{si}\phi_s}{qN_b}} \quad (\text{cm}).} \tag{4.27}$$

Note that the depletion width given by the equation above is the same as that obtained for one sided step *pn* junction under the depletion approximation [cf. Eq. (2.53)]. This shows that *we can treat the silicon surface/silicon bulk system as a one sided step junction.*

The depletion or bulk charge Q_b can now be obtained from Eq. (4.23) using Eqs. (4.26) and (4.27) giving

$$Q_b = -\epsilon_0\epsilon_{si}\mathscr{E}_{si} = \epsilon_0\epsilon_{si}\frac{d\phi}{dx}\bigg|_{x=0} = -\sqrt{2\epsilon_0\epsilon_{si}qN_b\phi_s} \quad (\mathrm{F/cm^2}). \qquad (4.28)$$

Alternatively, Q_b can also be obtained by integrating the charge qN_b under the depletion width X_d giving

$$\boxed{Q_b = -q\int_0^{X_d} N_b dx = -qN_bX_d = -\sqrt{2\epsilon_0\epsilon_{si}qN_b\phi_s} \quad (\mathrm{F/cm^2})} \qquad (4.29)$$

where we have made use of Eq. (4.27) for X_d. For *n*-type silicon, Q_b, given by Eq. (4.29), is a positive quantity.

Note that Eq. (4.27) for X_d is in terms of surface potential ϕ_s. Since ϕ_s itself is a function of gate voltage V_g [cf. Eq. (4.20)], one can also write X_d in terms of V_g. Thus, by substituting ϕ_s from Eq. (4.27) and $Q_s(=Q_b)$ from Eq. (4.29) in Eq. (4.20) and solving the resulting quadratic equation in X_d we get, under the depletion approximation

$$X_d = \frac{\epsilon_0\epsilon_{si}}{C_{ox}}\left(\sqrt{1+\frac{2(V_g - V_{fb})C_{ox}^2}{\epsilon_0\epsilon_{si}qN_b}} - 1\right) \quad (\mathrm{cm}). \qquad (4.30)$$

Equation (4.30) shows that when $V_g = V_{fb}$, the depletion width $X_d = 0$, consistent with the definition of the flat band voltage.

4.2.3 Inversion

If we continue to increase the positive gate voltage $V_{go}(=V_g - V_{fb})$, the downward band bending would further increase. In fact, a sufficiently large voltage can cause so much band bending that it may cause the midgap energy E_i to cross over the constant Fermi level E_f i.e. $E_f > E_i$. When this happens the surface behaves like *n*-type material with an electron concentration given by Eq. (2.10a). Note that this *n*-type surface is formed not by doping but instead by *inversion* of the original *p*-type substrate due to the applied gate voltage. This is referred to as the *inversion* condition and is shown in Figure 4.10c. Thus in

$$\text{inversion} \begin{cases} V_g & \gg V_{fb}, \\ \phi_s & > 0, \\ Q_s & < 0. \end{cases} \tag{4.31}$$

The surface is inverted as soon as $E_f > E_i$. This is called *the weak inversion regime* because the electron concentration remains small until E_f is considerably above E_i. If we further increase V_{go}, the concentration of electrons at the surface will equal, and then exceed, the concentration of the holes in the substrate. This is called the *strong inversion* regime.

One may ask, where these electrons (minority carriers) in the p-substrate come from when inversion sets in. Physically speaking these electrons come from the electron-hole generation, within the space charge (depletion) region, caused by the thermal vibration of lattice phonons. The rate of thermal generation depends upon the minority carrier life time τ_0 which is typically in μsec (10^{-6}sec). This means that minority carriers are not immediately available when an inverting gate voltage is applied. The time t_{inv} required to form an inversion layer at the surface is approximated by [14].

$$\boxed{t_{inv} = \frac{2N_b\tau_0}{n_i} \quad \text{(sec.)}} \tag{4.32}$$

where n_i is the intrinsic carrier concentration. For a typical value of $\tau_0 = 1\,\mu$sec and $N_b = 10^{15}\,\text{cm}^{-3}$, $t_{inv} \sim 0.2$ sec. Thus the *formation of the inversion layer is a relatively slow process compared to the time required for the holes (majority carriers) to flow from or to the silicon surface* which is of the order of picoseconds (i.e. the dielectric relaxation time associated with the substrate.)

The inversion layer is important from the MOS transistor operation point of view. It is the nature of the inversion layer, that is, number of carriers in the inversion layer (i.e. inversion layer charge Q_i), the mobility of the carriers in the layer etc. which determines the current in the transistor. The inversion layer charge Q_i can be calculated by including the electron concentration n in Poisson's equation (4.24). Let us first calculate n. Rewriting Eq. (4.4) as $n_i = N_b e^{-\phi_f/V_t}$ and substituting n_i in Eq. (2.10) we get

$$n = N_b e^{(\phi - 2\phi_f)/V_t}. \tag{4.33a}$$

and

$$p = N_b e^{-\phi/V_t} \tag{4.33b}$$

At the surface $\phi = \phi_s$ and therefore, from Eq. (4.33a), the electron concentration n_s at the surface is given by

$$n_s = N_b e^{(\phi_s - 2\phi_f)/V_t}. \tag{4.34}$$

When $\phi_s = \phi_f$, i.e. $E_i = E_f$, we see that $n_s = n_i$. That is, the silicon becomes intrinsic. When $\phi_s > \phi_f$, we have $E_f > E_i$ and the surface is inverted. At the onset of weak inversion the surface potential ϕ_s is slightly larger than ϕ_f and in this case the depletion width is given by Eq. (4.27). As we further increase ϕ_s by increasing the gate voltage V_g, the depletion width X_d widens and the electron concentration n_s at the surface increases (see Eq. 4.34). When the gate voltage is such that $\phi_s = 2\phi_f$, $n_s = N_b$, i.e., *the electron concentration at the surface becomes equal to the hole concentration in the bulk.* When this happens the surface is said to be strongly inverted, and under this condition, the depletion width reaches its maximum value X_{dm}, which can be obtained by replacing $\phi_s = 2\phi_f$ in Eq. (4.27). Thus,

$$X_{dm} = \sqrt{\frac{4\epsilon_0 \epsilon_{si} \phi_f}{q N_b}} \quad \text{(cm)}. \tag{4.35}$$

The condition $\phi_s = 2\phi_f$ is often referred to as the *classical condition for strong inversion.* When $\phi_s > 2\phi_f$, the depletion width increases but very slowly. This is because the inversion charge immediately adjacent to the oxide-silicon interface shields the interior (bulk) of the semiconductor from any additional charge on the gate.

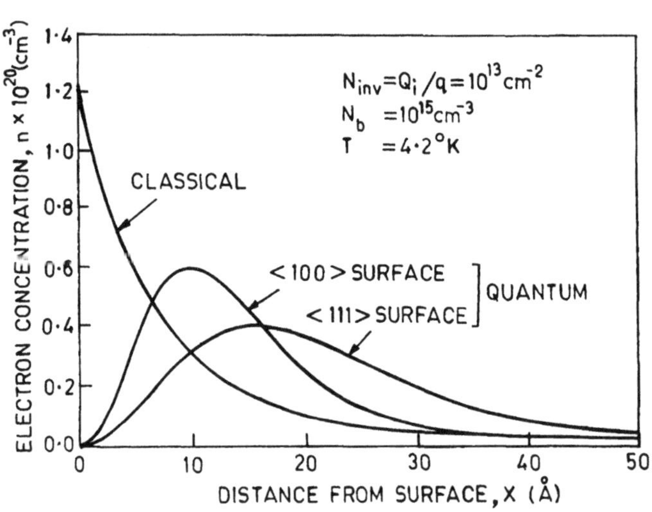

Fig. 4.11 Calculated electron concentration in silicon ⟨100⟩ and ⟨111⟩ surface as a function of distance from the surface for classical and quantum case. (From Stern and Howard [15])

The thickness of the inversion layer has been calculated using both quantum mechanical and classical approaches. These calculation show [15, 16] that the average "inversion layer thickness" at room temperature is about 50 Å, depending on the substrate doping concentration and gate voltage. Although not important from a circuit modeling point of view, it is interesting to consider the differences in the charge distributions calculated using the two approaches. The differences are, as shown in Figure 4.11, in two aspects.

- In the classical case, the electron density has its maximum value at the oxide-silicon interface, and it decreases steadily as we move from the surface into the bulk. In the quantum mechanical case, the electron density is zero at the interface, increases to its maximum value, and then decreases with the distance from the surface.
- In the classical case, the electron distribution is independent of the crystal orientation while it depends on the crystal orientation in the quantum mechanical case.

Figure 4.11 also shows that most of the electrons are confined in a layer 50 Å thick. For this reason, the motion of the electrons in the channel of a MOSFET can be regarded as two-dimensional, provided device width and length are not very small (cf. section 3.7.7).
We will now return to calculate the inversion layer charge density Q_i. Including the electron concentration n from Eq. (4.33) in Poisson's equation (4.24) yields

$$\frac{d^2\phi}{dx^2} = \frac{qN_b}{\epsilon_0\epsilon_{si}}\left[1 + e^{(\phi - 2\phi_f)/V_t}\right] \quad \text{for} \quad 0 \le x \le X_d. \tag{4.36}$$

Integrating once under the boundary conditions (4.25) we get[7]

$$-\left.\frac{d\phi}{dx}\right|_{x=0} = \mathscr{E}_{si} = \sqrt{\frac{2qN_b}{\epsilon_0\epsilon_{si}}}[\phi_s + V_t e^{-2\phi_f/V_t}(e^{\phi_s/V_t} - 1)]^{1/2}. \tag{4.37}$$

[7] Multiplying both sides of the Poisson's equation by $2(d\phi/dx)$ and using the identity,

$$\frac{1}{2}\frac{d}{dx}\left(\frac{d\phi}{dx}\right)^2 = \left(\frac{d\phi}{dx}\right)\left(\frac{d^2\phi}{dx^2}\right)$$

Eq. (4.36) becomes

$$\frac{d}{dx}\left(\frac{d\phi}{dx}\right)^2 = \frac{2qN_b}{\epsilon_0\epsilon_{si}}\frac{d}{dx}[\phi + V_t e^{(\phi - 2\phi_f)/V_t}]$$

which can easily be integrated to give (4.37).

Using Gauss' theorem, we get the induced charge density Q_s in the silicon as [cf. Eq. (4.18)]

$$Q_s = -\epsilon_0\epsilon_{si}\mathscr{E}_{si}$$
$$= -\sqrt{2\epsilon_0\epsilon_{si}qN_b}\,[\phi_s + V_t e^{(\phi_s - 2\phi_f)/V_t}]^{1/2} \quad (\mathrm{C/cm^2}) \tag{4.38}$$

which could further be simplified to

$$Q_s = -\sqrt{2\epsilon_0\epsilon_{si}qN_b}\,[\phi_s + V_t e^{(\phi_s - 2\phi_f)/V_t}]^{1/2} \quad (\mathrm{C/cm^2}) \tag{4.39}$$

where we have dropped -1 after the e^{ϕ_s/V_t} term because the exponential term is so large in strong inversion that -1 makes no difference, and in weak inversion the term $V_t e^{(\phi_s - 2\phi_f)/V_t}$ is so small that the entire minority carrier term can be neglected. Note that this induce charge Q_s is the sum of the inversion charge Q_i and depletion charge Q_b, that is

$$Q_s = Q_i + Q_b. \tag{4.40}$$

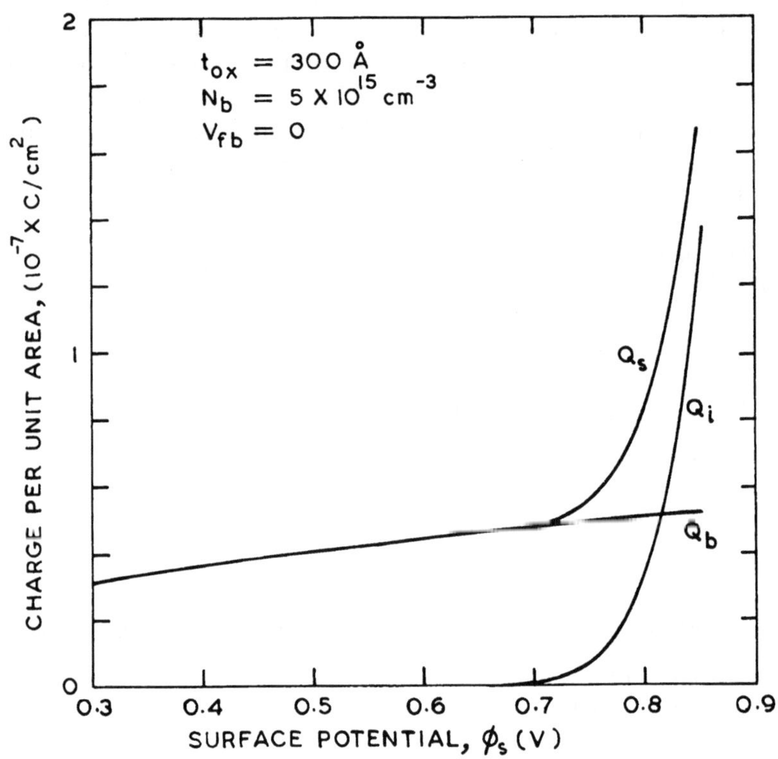

Fig. 4.12 Variation of inversion layer charge density Q_i [Eq. (4.41)], bulk charge density Q_b [Eq. (4.29)], and the total semiconductor charge density $Q_s(=Q_b+Q_i)$ [Eq. (4.39)] versus surface potential ϕ_s in all regimes of device operation for a p-type substrate, $N_b = 5\cdot10^{15}\,\mathrm{cm^{-3}}$, $t_{ox} = 300\,$Å, and $V_{fb} = 0\,$V

Using Eq. (4.29) for Q_b and Eq. (4.39) for Q_s, we get the inversion charge Q_i from Eq. (4.40) as

$$Q_i = -\sqrt{2\epsilon_0\epsilon_{si}qN_b}\left[\sqrt{\phi_s + V_t e^{(\phi_s - 2\phi_f)/V_t}} - \sqrt{\phi_s}\right] \quad (\text{C/cm}^2). \quad (4.41)$$

This gives the relationship between the inversion charge density Q_i and the surface potential ϕ_s. Figure 4.12 shows various charges as a function of ϕ_s. Note that the depletion charge Q_b does not vary appreciably. Also note that Q_i and Q_s have two distinct regions, which become more apparent when plotted on a logarithmic scale as shown in Figure 4.13a, where Q_i is plotted as a function of ϕ_s. These regions are (a) *weak inversion* and (b) *strong inversion.* Classically, the condition $\phi_s = 2\phi_f$ separates the region between the weak and strong inversion. Often, however, the inversion regime is divided into three regions; the third region which lies between the weak and strong inversion is called *moderate inversion,* defined as the region between $2\phi_f$ and $2\phi_f + 6V_t$ (see Figure 4.13b). In this scheme the region beyond $2\phi_f + 6V_t$ is the strong inversion region [6].

Weak Inversion. Weak inversion sets in when the surface band bending is ϕ_f and it extends to $2\phi_f$ (see Fig. 4.13). Within this region, the inversion-layer charge Q_i is small compared to the depletion-layer charge Q_b, that is

$$|Q_i| \ll |Q_b| \quad \text{(weak-inversion)}. \quad (4.42)$$

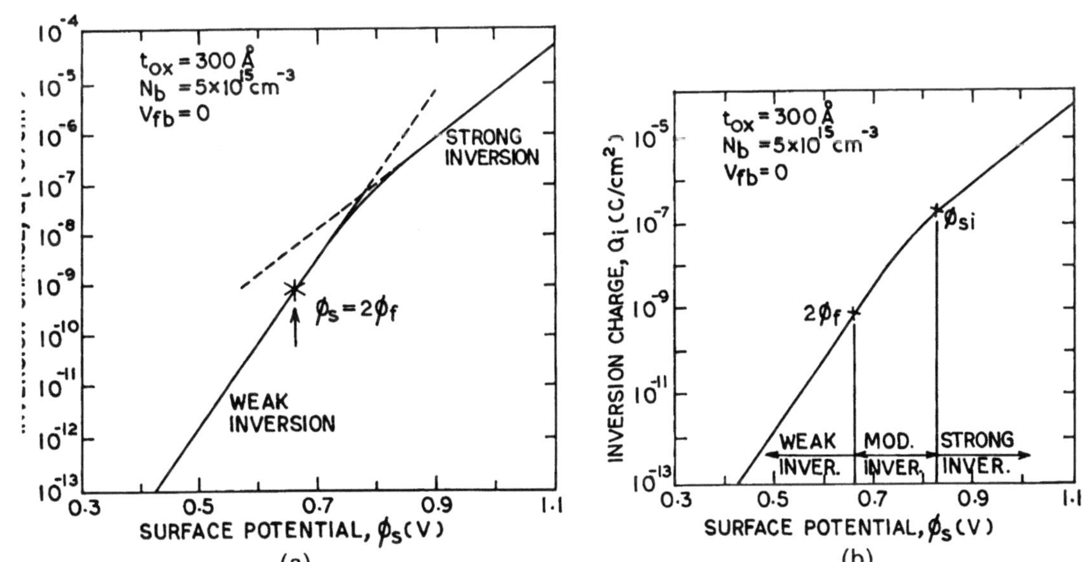

Fig. 4.13 Variation of inversion layer charge density Q_i versus surface potential ϕ_s for p-type substrate. (a) showing weak and strong region of operation (b) three different regimes of inversion; weak, moderate and strong inversion. $N_b = 5\cdot10^{15}\,\text{cm}^{-3}$, $t_{ox} = 300$ Å and $V_{fb} = 0$ V

For a small ϕ_s, Eq. (4.41) could be simplified[8] by assuming that the exponential term is small compared to ϕ_s, resulting in the following expression for Q_i

$$Q_i = -\sqrt{\frac{\epsilon_0 \epsilon_{si} q N_b}{2\phi_s}}\, V_t e^{(\phi_s - 2\phi_f)/V_t} \quad \text{(weak-inversion)} \quad (\text{C/cm}^2). \tag{4.43}$$

Thus, *in the weak inversion regime Q_i is essentially an exponential function of the surface potential ϕ_s*. This is plotted as a dashed line in Figure 4.13a.

Strong Inversion. Strong inversion is defined by the condition that the inversion layer charge Q_i is large compared to the depletion region charge Q_b, i.e.

$$|Q_i| > |Q_b| \quad \text{(strong-inversion)}. \tag{4.44}$$

Here the exponential term in Eq. (4.41) is large compared to ϕ_s and thus Q_i in strong inversion becomes

$$Q_i = \sqrt{2\epsilon_0 \epsilon_{si} q N_b V_t}\, e^{\phi_s/2V_t} \quad \text{(strong-inversion)} \quad (\text{C/cm}^2). \tag{4.45}$$

The inversion layer charge is an exponential function of the surface potential with a slope of $1/2V_t$ (on a log scale). Therefore, *a small increment of the surface potential induces a large change in the inversion layer charge.* Using Eq. (4.39) for Q_s in Eq. (4.20) we get a relationship between the gate voltage and surface potential as

$$V_g = V_{fb} + \phi_s - \frac{\sqrt{2\epsilon_0 \epsilon_{si} q N_b}}{C_{ox}}\left[\phi_s + V_t e^{(\phi_s - 2\phi_f)/V_t}\right]^{1/2}. \tag{4.46}$$

This is an implicit relation in ϕ_s and must be solved numerically (see Appendix E). The result of such simulations are shown in Figure 4.14. At low gate voltage ($> V_{fb}$) ϕ_s increases reasonably rapidly with gate bias and so does the depletion width X_d under the gate. This regime corresponds to the depletion and weak inversion regions of the device operation. At larger gate biases, ϕ_s hardly changes; ϕ_s has become pinned. The classical condition for the pinning is $\phi_s = 2\phi_f$. This pinning occurs when strong inversion sets in. The condition when this happens is often called the condition for *threshold* and the corresponding gate voltage is called *threshold voltage V_{th}*. It is one of the important device parameter which will be discussed in more details in Chapter 5.

[8] Using the Binomial theorem $\sqrt{1+x} = 1 + x/2 - x^2/8 + \cdots$ and retaining the first two terms we have $\sqrt{1+x} = 1 + x/2$

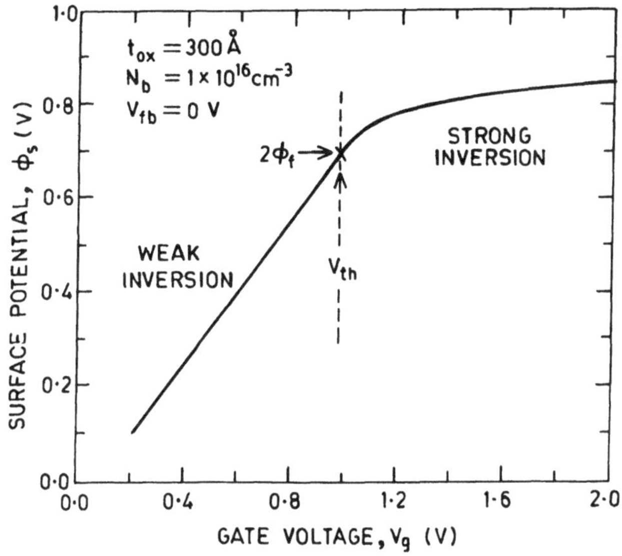

Fig. 4.14 Variation of surface potential ϕ_s with gate voltage V_g obtained using Eq. (4.46). $N_b = 1.0 \times 10^{16}$ cm^{-3}, $t_{ox} = 300$ Å, $V_{fb} = 0$ V. Cross indicates $\phi_s = 2\phi_f$ point which separate weak and strong inversion region

To summarize, we have calculated separate expressions for the induced charge Q_s that are valid in the depletion [Q_b, Eq. (4.29)] and inversion [Q_i, Eq. (4.41)] regime of MOS capacitor operation. However, one can easily derive a general expression for Q_s that is valid for all the regimes of device operation by including both the holes and electrons and thus solving the Poisson Eq. (2.41). Using Eqs. (4.33) for n and p and noting that in the bulk charge neutrality dictates that $N_d - N_a = n_{po} - p_{po}$, n_{po} and p_{po} being the carrier density in the bulk ($p_{po} \approx N_b$ and $n_{po} \approx N_b e^{-2\phi_f/V_t}$), the Poisson Eq. (2.41) becomes

$$\frac{d^2\phi}{dx^2} = \frac{qN_b}{\epsilon_0\epsilon_{si}}[1 + e^{(\phi - 2\phi_f)/V_t} - e^{-\phi/V_t} - e^{-2\phi_f/V_t}] \quad \text{for} \quad 0 \le x \le X_d. \tag{4.47}$$

Integrating the above equation, under the boundary condition (4.25), results in the following expression for Q_s which includes both holes and electrons [6], [12],

$$Q_s = -\sqrt{2\epsilon_0\epsilon_{si}qN_b}[\phi_s + e^{-2\phi_f/V_t}(V_t e^{\phi_s/V_t} - V_t - \phi_s) + V_t e^{-\phi_s/V_t} - V_t]^{1/2} \quad (\text{C/cm}^2). \tag{4.48}$$

The charge expression (4.48) is valid in all the regions of MOS capacitor operation—accumulation, depletion, and inversion. It should be pointed out that in the literature Eq. (4.48) is also written in terms of n_{po} and p_{po}

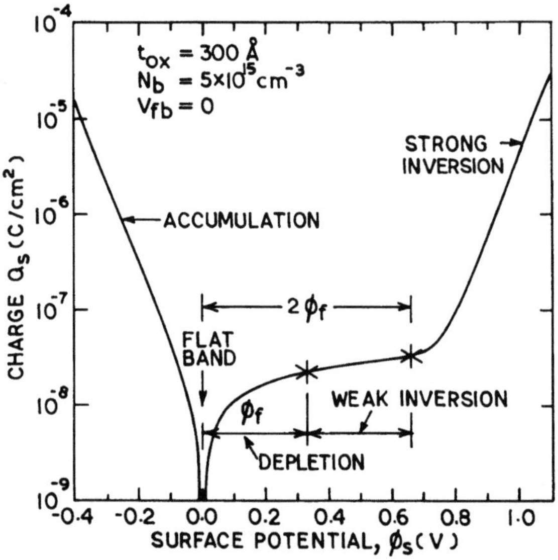

Fig. 4.15 Variation of the induced charge density Q_s in silicon versus surface potential ϕ_s for p-type substrate in all regimes of device operation obtained using Eq. (4.48). $N_b = 5 \times 10^{15}\,\text{cm}^{-3}$, $t_{ox} = 300\,\text{Å}$ and $V_{fb} = 0\,\text{V}$

as [4]

$$Q_s = -\frac{\sqrt{2}V_t\epsilon_0\epsilon_{si}}{L_d}\left[V_t e^{-\phi_s/V_t} + \phi_s - V_t + \frac{n_{p0}}{p_{p0}}(V_t e^{\phi_s/V_t} - \phi_s - V_t)\right]^{1/2} \quad (\text{C/cm}^2) \tag{4.49}$$

where L_d is the Debye length defined as

$$L_d = \sqrt{\frac{\epsilon_0\epsilon_{si}kT}{q^2N_b}} \quad (\text{cm}) \tag{4.50}$$

and the ratio $n_{p0}/p_{p0} = V_t e^{-2\phi_f/V_t}$. Equations (4.29) and (4.41) are special cases of Eq. (4.48). The variation of the induced charge Q_s as a function of ϕ_s using Eq. (4.48) for p-type substrate is illustrated in Figure 4.15 which clearly shows all the three regimes of operation. These regimes are easily identified:

- When $\phi_s < 0$, the MOS structure is in the accumulation mode. The term that predominates in Eq. (4.48) is $e^{-\phi_s/V_t}$ and therefore in this regime Q_s varies as

$$Q_s \sim e^{-\phi_s/2V_t} \quad (\text{accumulation}). \tag{4.51a}$$

Table 4.2. *Definition of silicon surface parameters*

Substrate type		ϕ_s	Q_s	$V_g(V_{fb}=0)$
p	Accumulation	$-$	$+$	$-$
	Depletion/Inversion	$+$	$-$	$+$
n	Accumulation	$+$	$-$	$+$
	Depletion/Inversion	$-$	$+$	$-$

- When $\phi_s > 0$ such that $0 < \phi_s < 2\phi_f$, the structure is in the depletion and weak inversion regime. In this case the term that predominates in Eq. (4.48) is $\sqrt{\phi_s}$ and therefore Q_s varies as

$$Q_s \sim \sqrt{\phi_s} \quad \text{(depletion and weak inversion).} \tag{4.51b}$$

- When $\phi_s > 2\phi_f$, the structure is in the strong inversion mode. The predominant term in Q_s varies as

$$Q_s \sim e^{\phi_s/2V_t} \quad \text{(strong inversion).} \tag{4.51c}$$

Note that the accumulation, depletion and inversion conditions described by Eqs. (4.21), (4.22), and (4.31) are for *p*-type substrates. However, for *n*-type substrates these conditions will be reversed as shown in Table 4.2.

4.3 Capacitance of MOS Structures

In the previous section we developed relationships between the charge and potential under different gate voltage conditions across a MOS capacitor. Now we will see how the capacitance of the MOS system varies with the applied voltage . The capacitance of any system is the ratio of the variation in charge to the corresponding variation in the small-signal applied voltage. Thus the capacitance C_g of a MOS structure is

$$C_g = \frac{dQ_g}{dV_g} \quad (\text{F/cm}^2). \tag{4.52}$$

Substituting the value of V_{ox} from Eq. (4.19) in Eq. (4.16) we get

$$V_g = \Phi_{ms} + \phi_s + \frac{Q_g}{C_{ox}}.$$

Since Φ_{ms} is a constant, it follows that

$$dV_g = d\phi_s + \frac{dQ_g}{C_{ox}}. \tag{4.53}$$

Combining Eqs. (4.52) and (4.53) gives the capacitance of an MOS system as

$$\frac{1}{C_g}=\frac{1}{C_{ox}}+\frac{1}{dQ_g/d\phi_s} \quad (\mathrm{F/cm^2}). \tag{4.54}$$

Rearranging Eq. (4.17) and differentiating with respect to ϕ_s we get

$$\frac{dQ_g}{d\phi_s}=-\frac{dQ_s}{d\phi_s}-\frac{dQ_o}{d\phi_s}. \tag{4.55}$$

The quantity $-dQ_s/d\phi_s$ *can be interpreted as the capacitance per unit area,* C_s *associated with the silicon depletion or space charge region, i.e.*

$$\boxed{C_s=-\frac{dQ_s}{d\phi_s}} \quad (\mathrm{F/cm^2}) \tag{4.56}$$

and, can easily be obtained by differentiating Eq. (4.48) giving the following expression for C_s

$$C_s=\frac{\sqrt{2V_t}\epsilon_0\epsilon_{si}}{L_d}\,\frac{[1-e^{-\phi_s/V_t}+e^{-2\phi_f/V_t}(e^{\phi_s/V_t}-1)]}{2[V_te^{-\phi_s/V_t}+\phi_s-V_t+e^{-2\phi_f/V_t}(V_te^{\phi_s/V_t}-\phi_s-V_t)]^{1/2}} \quad (\mathrm{F/cm^2}). \tag{4.57}$$

Similarly, the capacitance per unit area C_o associated with the interface charge density Q_o can be defined as

$$C_o=-\frac{dQ_o}{d\phi_s} \quad (\mathrm{F/cm^2}) \tag{4.58}$$

so that we have from Eqs. (4.55)–(4.58),

$$\frac{dQ_g}{d\phi_s}=C_s+C_o. \tag{4.59}$$

Combining Eqs. (4.54) with (4.59) we get

$$\frac{1}{C_g}=\frac{1}{C_{ox}}+\frac{1}{C_s+C_o}. \tag{4.60}$$

Thus the *MOS capacitor is the series combination of the oxide capacitor* C_{ox} *and the parallel combination of the silicon capacitor* C_s and interface charge capacitance C_0. For a given oxide thickness t_{ox}, the value of C_{ox} is constant and corresponds to the maximum capacitance of the system. This equivalent circuit of the MOS capacitor is shown in Figure 4.16, where R_o is the

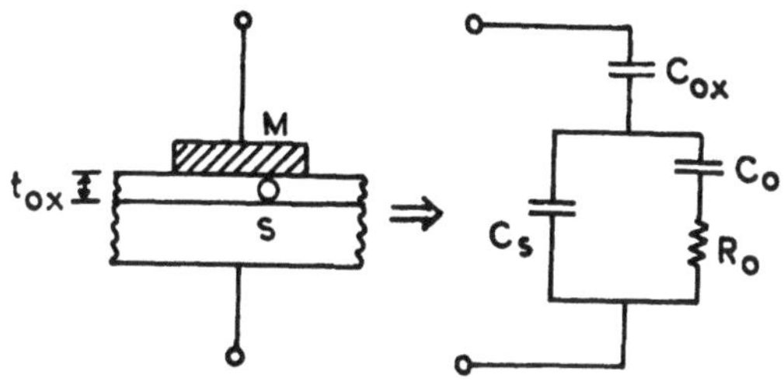

Fig. 4.16 Equivalent circuit of an MOS capacitor. R_0 is the resistance associated with the interface charge capacitance C_0

resistance associated with the interface charge capacitance C_o and is in parallel with the silicon capacitance C_s. The fixed positive interface charge density Q_o is independent of the surface potential and if it is assumed that no voltage dependent trapping mechanisms are occurring at the Si–SiO_2 interface, then C_o will be zero and C_g will be given by

$$\boxed{\frac{1}{C_g}=\frac{1}{C_{ox}}+\frac{1}{C_s}.} \tag{4.61}$$

Combining Eqs. (4.20), (4.57), and (4.61) gives a complete description of an MOS capacitor as a function of gate voltage V_g. Thus, to calculate the MOS Capacitance–Voltage (C–V) curve we first choose a set of ϕ_s values compatible with the silicon band gap. For each value of ϕ_s we in turn

- calculate C_s, the space charge region capacitance, using Eq. (4.57),
- calculate C_g, the total MOS structure capacitance, using Eq. (4.61),
- calculate Q_s, the charge contained in the silicon space charge region, using Eq. (4.49),
- finally determine the gate voltage V_g using Eq. (4.20) for a given value of V_{fb}.

For each value of ϕ_s chosen we can draw one point of coordinate (V_g, C_g). The set of chosen ϕ_s values allows us to plot the C_g–V_g curve point by point. Note that if we assume $V_{fb}=0$, the resulting C–V curve is called *ideal* C–V curve and is shown in Figure 4.17. Depending upon the applied voltage, the MOS capacitor will either be in accumulation, depletion or inversion. Let us now consider these cases.

Accumulation. We have already seen that for a *p*-type substrate in accumulation there are excess carriers (majority holes) at the surface. In this case, the applied voltage $V_g < V_{fb}$ and the surface potential ϕ_s is

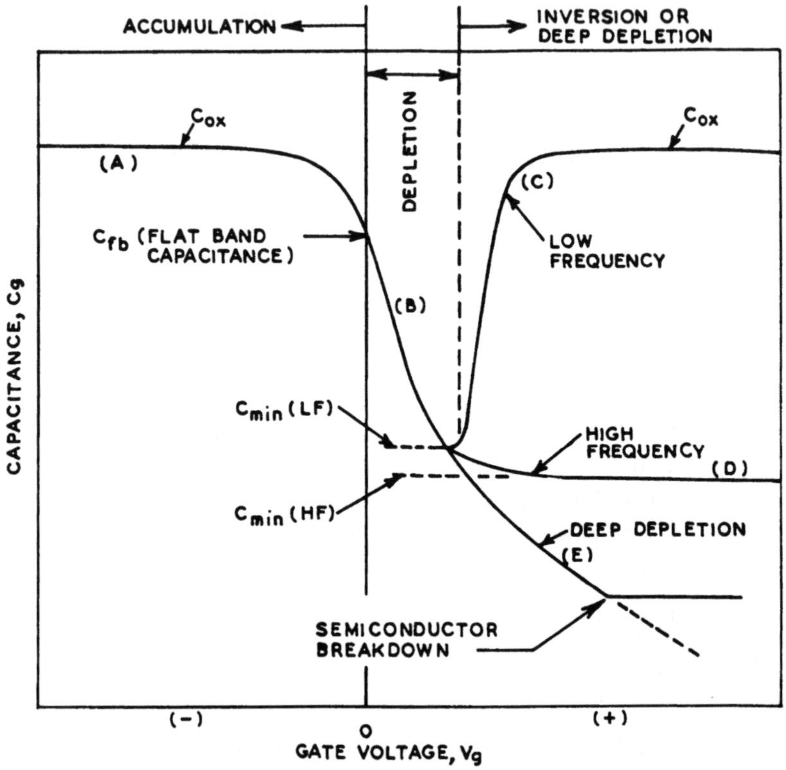

Fig. 4.17 Capacitance-voltage (C–V) curve of a MOS capacitor under (A) accumulation, (B) depletion, and (C)–(E) inversion. Curve (C) is at low frequency and (D) at high frequency. (After Sze [4], slightly modified.)

negative. Recall from Eq. (4.51a), Q_s and hence C_s in accumulation is proportional to $e^{-\phi_s/V_t}$, which means that for negative ϕ_s, C_s becomes very large. Therefore, as can be seen from Eq. (4.61), the total MOS capacitance C_g is approximately C_{ox}. Thus, in accumulation

$$\boxed{C_g \cong C_{ox} \quad \text{(accumulation).}} \tag{4.62}$$

This is plotted as curve A in Figure 4.17.

Depletion. As the negative voltage is reduced sufficiently so that $V_g > V_{fb}$, a depletion region of width X_d is formed near the silicon surface. This depletion width acts as a dielectric in series with the oxide. Consequently the silicon capacitance C_s decreases and according to Eq. (4.61) the total capacitance decreases resulting in the following expression for the capacitance in depletion

$$C_g = \left(\frac{1}{C_{ox}} + \frac{1}{C_s}\right)^{-1} \quad \text{(depletion)} \tag{4.63}$$

where C_s is the capacitance per unit area associated with the depletion region at the surface. General expression for C_s is given by Eq. (4.57). However, a much simpler expression for C_s can be obtained using the depletion approximation. As was mentioned earlier, the silicon-surface/silicon-bulk system may be approximated by a one sided step junction in depletion or inversion. Therefore, from Eq. (2.70) we have

$$C_s(\text{depletion}) = \frac{\epsilon_0 \epsilon_{si}}{X_d} \quad (\text{F/cm}^2) \tag{4.64}$$

where X_d is the depletion width at the surface and is given by Eq. (4.27) in terms of ϕ_s or Eq. (4.30) in terms of V_g. Substituting X_d from Eq. (4.30) in Eq. (4.64) we get space-charge capacitance $C_{s,d}$ in depletion mode as

$$C_{s,d} = C_{ox}\left[\sqrt{1 + \frac{2(V_g - V_{fb})C_{ox}^2}{\epsilon_o \epsilon_{si} q N_b}} - 1\right]^{-1} \tag{4.65}$$

so that the gate capacitance C_g in depletion becomes

$$\boxed{C_g = C_{ox}\left[1 + \frac{2(V_g - V_{fb})C_{ox}^2}{\epsilon_o \epsilon_{si} q N_b}\right]^{-1/2} \quad \text{(depletion).}} \tag{4.66}$$

This is plotted as curve B in Figure 4.17. The MOS capacitor follows this curve until inversion sets in. From Eq. (4.66) it is clear that *for a given voltage $V_g - V_{fb}$, the capacitance in the depletion region will be higher for higher N_b and/or low C_{ox}* (larger t_{ox}). Further, note that at $V_g = V_{fb}$ (flat band condition, $\phi = 0$), we have $C_g = C_{ox}$. However, in a real MOS capacitor at the flat band, C_g is less than C_{ox} (see Fig. 4.17). This is because the transition between the accumulation and depletion regions is not abrupt as is assumed in the depletion approximation on which Eq. (4.66) is based.

To solve for the MOS capacitance at flat band, called the *flat band capacitance C_{fb}*, we need to use Eq. (4.48) for Q_s or Eq. (4.57) for C_s. For $\phi < 0$ we have $e^{-\phi_s} \gg e^{-2\phi_f} > e^{-2\phi_f + \phi_s}$ and therefore, Eq. (4.48) can be approximated as

$$Q_s \approx -\frac{\sqrt{2V_t \epsilon_0 \epsilon_{si}}}{L_d}[V_t e^{-\phi_s/V_t} + \phi_s - V_t]^{1/2} \tag{4.67}$$

which on differentiation gives[9]

$$C_s(\text{flat band}) = -\left.\frac{dQ_s}{d\phi_s}\right|_{\phi_s \to 0} = \frac{\epsilon_0 \epsilon_{si}}{L_d} \quad (\text{F/cm}^2). \tag{4.68}$$

Combining Eqs. (4.68) and (4.61) we get the MOS capacitance at flat band as

$$\boxed{C_g \equiv C_{fb} = \left(\frac{1}{C_{ox}} + \frac{L_d}{\epsilon_0 \epsilon_{si}}\right)^{-1} \quad (\text{flat band}).} \tag{4.69}$$

Inversion. If the gate voltage $(V_g - V_{fb})$ is sufficiently positive such that $\phi_s > \phi_f$, an inversion layer is formed at the surface of the silicon. Recall that this inversion layer is formed from the generation of minority carriers (electrons in our example of a *p*-type substrate). The concentration of minority carriers can change only as fast as carriers can be generated within the depletion region near the surface. This limitation causes the MOS capacitance in inversion to be a function of frequency of the AC signal used to measure the capacitance. If the AC signal frequency is sufficiently low (typically less than 10 Hz) so that the inversion charge carriers (minority carriers) are able to follow the AC bias voltage and the DC sweeping voltage, then the resulting C–V curve is know as a *low-frequency* (LF) C–V plot.[10] However, if the AC bias signal frequency is too high (typically above

---

[9] Using the expansion $e^x = 1 + x + x^2/2 + x^3/6 + x^4/24 \cdots$ and retaining its first 5 terms we get from Eq. (4.67), after some algebra,

$$Q_s \approx -\frac{\epsilon_0 \epsilon_{si}}{L_d} \phi_s \left[1 - \frac{\phi_s}{3V_t} + \frac{\phi_s^2}{12V_t^2}\right]^{1/2}$$

which could further be simplified using Binomial expansion of $\sqrt{1+x} \approx 1 + x/2$, resulting in

$$Q_s \approx -\frac{\epsilon_0 \epsilon_{si}}{L_d} \phi_s \left[1 - \frac{\phi_s}{6V_t} + \frac{\phi_s^2}{24V_t^2}\right].$$

Above equation, on differentiation with respect to ϕ_s, in the limit ϕ_s tends to zero, leads to Eq. (4.68).

[10] For a typical value of $\tau_0 = 1\,\mu$sec and a dopant concentration of $10^{15}\,\text{cm}^{-3}$, the time required to form an inversion layer is roughly 0.2 sec [cf. Eq. (4.32)]. This explains why the AC signal voltage must be changed very slowly to observe the low-frequency C–V curve.

10^5 Hz) so that the inversion charge carriers do not follow the AC voltage, the measured C–V curve is called the *high-frequency* (HF) C–V plot.

It is worth noting here that the calculations leading to MOS capacitance Eq. (4.57) assumes that all charges in the depletion region follow the variation of ϕ_s. This means that Eq. (4.57) is valid only for low frequency C–V curve. In order to derive a general expression for the high frequency capacitance we first evaluate the charge contained in the space-charge layer by neglecting the contribution of minority carriers in Eq. (4.48). We then determine the equivalent thickness of the depletion layer possessing the same integrated charge and then calculate the corresponding capacitance using Eq. (4.64) [2, p. 247, Pt. I].
Note that *in the accumulation and depletion mode the MOS capacitance C_g is independent of the frequency for all frequencies of practical interest.* This is because in this region minority carriers are negligible and so do not contribute to the total charge, which is governed by majority carriers. The latter have transport time of the order of picoseconds (see section 4.2.3). Thus, depending upon the frequency of the AC signal and measurement conditions, three types of C–V plots are generally obtained as discussed below.
It should be pointed out that frequency dependence of capacitance in inversion is true only for MOS capacitor and not MOS transistors. In the case of MOS transistors, the source and drain diffusions can supply minority carriers to the inversion layer almost instantaneously.

4.3.1 *Low Frequency C–V Plots*

In this case the DC gate voltage and the AC signal voltage are changed very slowly so that the MOS capacitor always approaches equilibrium. This means that the signal frequency is low enough so that the inversion layer carriers can follow it. In this case the capacitance C_g is just that associated with the charge storage on either side of the oxide. Thus, in inversion at low frequency (LF)

$$\boxed{C_g \cong C_{ox} \quad \text{(inversion-LF signal).}} \tag{4.70}$$

Under these conditions a plot of measured capacitance versus gate voltage follows curve C in Figure 4.17. Starting from the value of C_{ox} in accumulation, the capacitance decreases (as the depletion region is formed) and goes through the minimum and then increases moving back to C_{ox} as the surface becomes strongly inverted. Note that the increase in the capacitance depends upon the ability of the minority carriers to follow the AC signal.

It can be shown that the space charge capacitance at low frequency $C_{s,lf}$ is[11]

$$C_{s,lf}(\text{Min, LF}) \approx \frac{\epsilon_0 \epsilon_{si}}{L_d} \frac{1}{2\sqrt{U_f - 1}} \tag{4.71}$$

where $U_f(=\phi_f/V_t)$ is the normalized Fermi potential so that LF minimum capacitance becomes

$$\boxed{C_{min} = \left(\frac{1}{C_{ox}} + \frac{2\sqrt{U_f - 1}}{\epsilon_0 \epsilon_{si}/L_d}\right)^{-1} \quad (\text{Min } C_g \text{ at LF}).} \tag{4.72}$$

The discussion above assumed that the MOS system was in a dark enclosure so that no external source of minority carriers generation other than thermal generation was available. However, if the surface is illuminated, the surface carrier generation rate will increase resulting in an increase in the low frequency capacitance.

4.3.2 High Frequency C–V Plot

If the AC measuring signal frequency is so high that the inversion layer charge density Q_i cannot follow the high frequency (HF) variation in the gate voltage, Q_i can be assumed to be constant for a given DC bias. Under this condition the depletion region charge density Q_b and the width X_d of the depletion region will fluctuate around the quiescent value Q_{bmax} and X_{dm} respectively. In this case the capacitance of the depletion region is given by

$$C_{s,hf}(\text{inversion}) = \frac{\epsilon_0 \epsilon_{si}}{X_{dm}} = \sqrt{\frac{\epsilon_0 \epsilon_{si} q N_b}{4\phi_f}} \tag{4.73}$$

---

[11] The minimum of $C_{s,lf}$ can be obtained by differentiating Eq. (4.57) with respect to ϕ_s, equating the resulting equation to zero and then solving for $\phi_s = \phi_{smin}$. The algebra could be simplified if one writes Eq. (4.57) in terms of normalized potentials $U_f(=\phi_f/V_t)$ and $U_s(=\phi_s/V_t)$ as

$$C_s = \frac{\epsilon_0 \epsilon_{si}}{\sqrt{2} L_i} \frac{\sinh(U_f) + \sinh(U_s - U_f)}{[\cosh(U_s - U_f) + U_s \sinh(U_f) - \cosh(U_f)]^{1/2}}.$$

At $U_s = U_{smin}$, $dC_s/d\phi_s = 0$ which gives $U_{smin} \approx 2U_f - \ln(4U_f - 4)$. Substituting value of $U_s = U_{smin}$ in C_s gives Eq. (4.71). It should be pointed out that an approximate expression for $C_{s,lf}$ (min) is given as [17]

$$C_s(\text{Min, LF}) \approx \frac{\sqrt{2}\epsilon_0 \epsilon_{si}}{5L_d}.$$

where we have made use of Eq. (4.35) for X_{dm}. Thus, the gate capacitance in inversion at HF becomes

$$C_g = \left(\frac{1}{C_{ox}} + \frac{X_{dm}}{\epsilon_0 \epsilon_{si}}\right)^{-1} \quad \text{(inversion-HF)} \tag{4.74}$$

and is constant independent of bias because X_{dm} is constant. This is shown in Figure 4.17 as curve D. Note that C_g given by the above equation is also C_{min} at HF.

Experimentally, at HF more rapid saturation of capacitance to its minimum is observed than is predicted by the above equation. This would be expected if minority carrier redistribution is taken into account. Since the inversion layer is not infinitesimally thin, a redistribution of the carriers within the inversion layer at the AC measurement frequency will cause capacitance to saturate more abruptly. As a further consequence, the saturation capacitance will be larger than predicted by the above equation. Several authors have taken into account the redistribution of minority carriers and have used a more accurate estimation of the space charge layer. However, an excellent approximation for the asymptotic behavior of $C_{s,hf}$ at inversion is due to Berman and Kerr [18] which gives

$$C_{s,hf}(\text{Min, HF}) = \sqrt{\frac{1}{2V_t} \frac{\epsilon_0 \epsilon_{si} q N_b}{(2U_f - 1 + \ln[1.15(U_f - 1)])}}. \tag{4.75}$$

It should be reiterated that though the inversion layer of a MOS capacitor cannot follow the high frequency signal, this is not the case with MOS transistors which are capable of operating at much higher frequencies. This is because the heavily doped source region of the MOS transistor will always be in contact with the inversion layer and thus can supply the charge required to follow the high frequency gate signal.

4.3.3 Deep Depletion C–V Plot

If a MOS capacitor is swept from the accumulation to the inversion region at a relatively fast rate (about 10 V/s and higher) so that there is not enough time for the thermal generation of the inversion charge carriers (minority carriers), the capacitance will continue to drop following the depletion curve. This is a non-equilibrium situation in which the depletion width widens (to balance the increased gate charge) past its maximum value X_{dm} and C_d does not reach a minimum. This is shown as curve E in Figure 4.17. This expansion of the depletion region deep in the silicon bulk is referred to as *deep depletion.* The capacitance in the deep depletion is given by Eq. (4.65). The deep depletion curve is obtained when the DC voltage sweep rate is high, independent of the frequency of the AC signal voltage

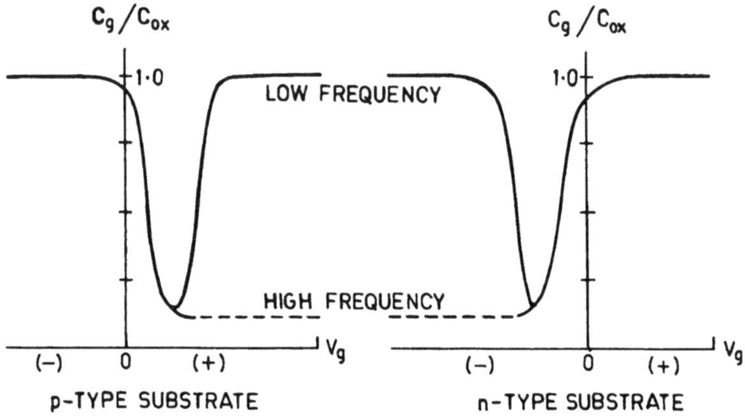

Fig. 4.18 Typical C–V relationship for an MOS capacitor with (a) *p*-type silicon substrate, and (b) *n*-type silicon substrate

(HF) and no inversion charge can form. *The easiest way to obtain deep depletion is to sweep the DC voltage by either applying a voltage step or using a fast voltage ramp on the gate.*
Deep depletion is a nonequilibrium condition. The generation rate of minority carriers increases as the depletion layer is widened, and the deep depletion curve is frequently observed to relax to the high frequency curve at higher biases. *A fast relaxation time indicates an excessive generation rate and hence excessive leakage in the device.* Thus the measurement of relaxation time (minority carrier lifetime) provides a tool for the detection of defects near the surface that may be induced during processing.
Which of the three curves (LF, HF or DD) are obtained during C–V measurement depends upon the frequency of the applied AC signal and the DC sweep rate. The shape of the low and high frequency curves vary as a function of doping concentration N_b and gate oxide thickness t_{ox} and can easily be computed as discussed earlier. In MOS C–V plots the value of gate-to-substrate capacitance C_g is often normalized to gate oxide capacitance C_{ox}. It is this normalized capacitance (C_g/C_{ox}) which is normally plotted against the gate voltage V_g. Figure 4.18 shows typical C–V relationships for MOS capacitors for *p*- and *n*-type silicon substrate. Solid curves are low frequency C–V plots while dashed curves indicates high frequency behavior after inversion sets in. Note that the C–V plots for an *n*-type substrate is obtained simply by changing the voltage axis of a *p*-type substrate.

4.4 Deviation from Ideal C–V Curves

The MOS capacitance plots shown in the Figure 4.17 are for the ideal case, wherein it was assumed that the gate oxide is a perfect insulator free of

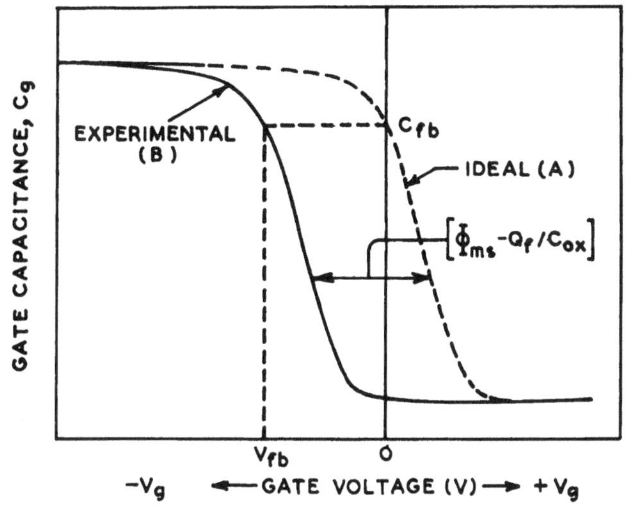

Fig. 4.19 Influence of the metal-semiconductor work function difference Φ_{ms} and fixed oxide charge Q_f on HF C–V curve for an MOS capacitor. Curve A is for ideal case with $\Phi_{ms} = 0$ and $Q_f = 0$, while curve B is experimental curve. Parallel shift of curve A to curve B is direct measure of V_{fb}

charges ($Q_0 = 0$) and $\Phi_{ms} = 0$, so that $V_{fb} = 0$. In reality, the gate oxide is not a perfect insulator and contains various type of charges as was discussed in section 4.1.2. Due to the nonideal nature of the real MOS structures, experimental C–V plots, both LF and HF, deviates from the ideal by one or more of the following parameters: (1) nonzero Φ_{ms}, the metal-silicon workfunction difference, (2) interface traps, (3) mobile ions in the oxide, and (4) fixed oxide charge.

In fact this deviation is used to study the properties of the silicon surfaces. Figure 4.19 illustrates an experimental HF C–V plot of an MOS capacitor on a *p*-type substrate (curve B) along with a ideal curve ($V_{fb} = 0$) shown as a dotted line (curve A). The *horizontal parallel voltage shift* between the curve A and the experimental curve B is a direct measure of the flat band voltage V_{fb}. A negative V_{fb} causes a shift to the left of the curve A for both *n*- and *p*-substrates. If V_{fb} is positive (*n*-substrate with p^+ polysilicon gate) the shift is to the right of curve A. Recall from Eq. (4.14) that V_{fb} is due to the work function difference Φ_{ms} and effective gate oxide charge Q_0, that is,

$$V_{fb} = \Phi_{ms} - \frac{Q_0}{C_{ox}} = \Phi_{ms} - \frac{Q_0 t_{ox}}{\epsilon_0 \epsilon_{ox}}. \tag{4.76}$$

If it is assumed that effective gate oxide charge Q_0 is independent of the oxide thickness, t_{ox}, and remains constant during processing, then from Eq. (4.76) the amount of the shift is directly proportional to the workfunction difference Φ_{ms}. Thus, using MOS capacitors of different t_{ox} and then measuring V_{fb} as a function of t_{ox} one can easily calculate Φ_{ms} [19].

As was mentioned earlier, mobile alkali ions move in the oxide under high temperature and high field resulting in the shift of C–V curve when a sample is heated during the applied bias. This is often referred to as a *bias-temperature* stress cycle or simply B–T cycle. Typically the device is heated around 200–300°C and a gate bias that results in the oxide field of around 10^6 V/cm is applied for about 10 minutes or so. The device is then cooled to room temperature and the C–V curve is plotted. The procedure is then repeated with opposite bias polarity. At any instant of time the effect of mobile ions on the C–V curve is the same as though they are fixed charges (curve B in Figure 4.19). A negative B–T stress will result in the C–V curve shifted towards the right of curve B due to mobile ions moving towards the gate. A positive B–T stress will result in the C–V curve being shifted towards left of curve B. The total shift in the C–V curve becomes

$$V_{fb}(+) - V_{fb}(-) = -\frac{Q_m}{C_{ox}} \tag{4.77}$$

which can be used to calculate the total mobile ionic charge per unit area Q_m. Although in modern MOS processing the contamination due to mobile alkali ions are minimized, routine C–V plots using B–T stress are used as one of the several monitors of checking oxide quality. Unexpected sources of contamination are plentiful resulting, for example, from a bad batch of a chemical, a leaky vacuum system, etc.

Unlike the nonideal C–V curves discussed so far, which results in a parallel shift of the C–V curve along the voltage axis, experimental C–V curves are sometime distorted or smeared out with a slope which is less than that of the ideal (theoretical) curve, as shown in Figure 4.20. This distortion or decrease in the slope can be directly attributed to the interface traps charge Q_{it} at the oxide–silicon interface, because the amount of charge trapped at the interface depends on the surface band bending, i.e., surface potential. One can see from Figure 4.20 that the influence of donor-like interface traps (positively charged) is to stretch the curve outwardly to the left (curve B) in accordance with Eq. (4.12). In this case the C–V curve shows a maximum negative shift in accumulation, gradually changing to almost no shift in inversion. This is because, in accumulation, all interface traps will be positively charged for donor like states and, in inversion, the amount of positive charge reduces to a minimum.[12] If interface traps are acceptor-like, then the trapped charge Q_{it} will be negative and the voltage shift will be in the positive direction resulting in a C–V curve which is shifted towards the right (curve C). This is due to the variation in the total number of

[12] Remember that this is the case for *p*-type substrate. However, for *n*-type substrate with donor-like traps C–V curve will show smallest shift in accumulation. The shift continues to increase as the device goes through flat band, depletion and inversion.

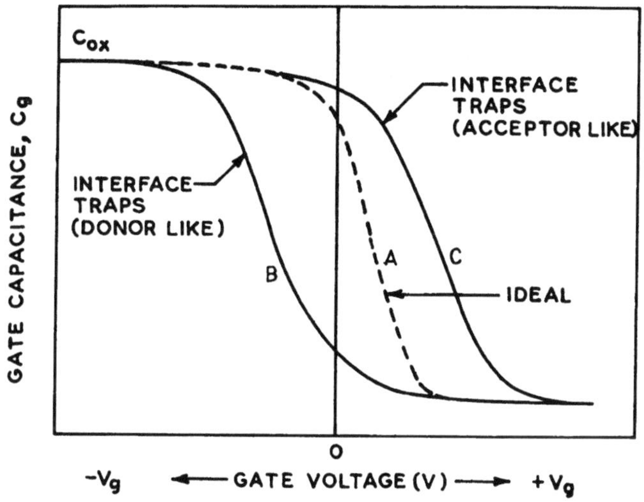

Fig. 4.20 Influence of the interface traps on the high frequency MOS C–V curve. Curve A is ideal C–V curve with no traps, curve B is with donor like traps, and curve C is with only acceptor like interface traps

empty or filled states as the Fermi level is swept across the band gap by changing the surface potential (or applied voltage).

Nonuniform Substrate. The discussion so far has assumed that the substrate is uniformly doped. Frequently, ions are implanted into the substrate, through the oxide, for example, to adjust the threshold voltage of a MOSFET. The ion implantation leads to nonuniform doping in the substrate, or may result in the formation of a layer of opposite doping buried in the substrate, thereby forming a *pn* junction underneath the Si–SiO_2 interface. Depending upon the energy and the dose of the implant, the C–V curve may look different. When the dose is low (say, 10^{10} Boron/cm^2 in an *n*-type substrate), the substrate doping is not fully compensated and the C–V curves obtained are almost identical to those of true *n*-type substrates. The same applies to large doses (10^{12} Boron/cm^2) if the peak concentration is sufficiently close to the Si–SiO_2 interface.

4.5 Anomalous C–V Curve (Polysilicon Depletion Effect)

In the discussion so far we had assumed that the polysilicon gate is degenerately doped (concentration $> 5 \times 10^{19}\,\mathrm{cm}^{-3}$). This is usually the case when gates are $POCl_3$ doped (for n^+ polysilicon gates). However, in submicron technologies the gates are ion implanted and may not be degenerately doped depending upon the process conditions. This is specially true when the gate oxide is thin, of the order of 100 Å and lower. If the gate is non-

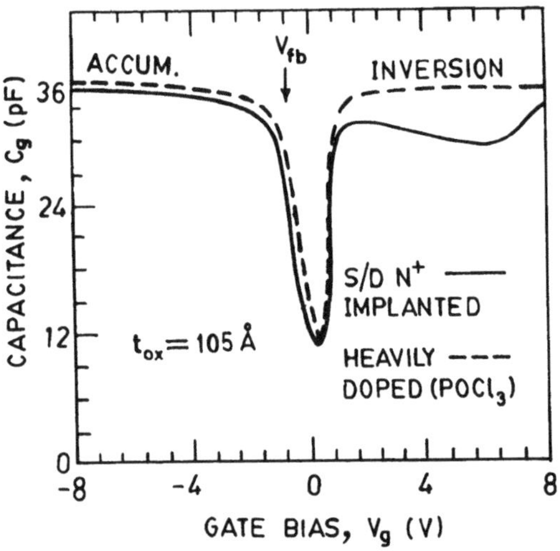

Fig. 4.21 MOSFET gate capacitance showing heavily doped polysilicon gate (dotted line) and source/drain implanted *n*-doped polyside gate (continuous line) showing polysilicon depletion effect. (From Chapman et al. [22])

degenerately doped, it can no longer be treated as an equipotential area. This means that the capacitance describing the MOS capacitor can no longer be given by Eq. (4.61) and one needs to include the capacitance C_{poly} due to the polysilicon gate. The resulting gate capacitance equation becomes [20]

$$\frac{1}{C_g} = \frac{1}{C_{poly}} + \frac{1}{C_{ox}} + \frac{1}{C_s}. \tag{4.78}$$

The result of the nondegenerate polysilicon is that the LF capacitance in inversion ($C_{g,inv}$) is much smaller than in accumulation, and $C_{g,inv}$ decreases slightly with gate bias. However, at gate bias larger than a certain voltage the $C_{g,inv}$ recovers to C_{ox} rather abruptly [21]–[24]. This is shown in Figure 4.21, where high frequency C–V curve of an MOS capacitor (using split-CV method[13]) is plotted for $POCl_3$ doped (dotted lines) and implanted gate (continuous line). Note that in addition to the lowering of the gate capacitance in inversion there is a slight shift in the flat band voltage resulting in an increase in the threshold voltage [24].
This anomalous C–V behavior has been explained assuming that there exists a layer of partially activated dopants near the polysilicon/SiO_2 interface [23] or by assuming that the polysilicon gate is non-degenerately doped (concentration $\sim 10^{19}\,\text{cm}^{-3}$) [24]. When the gate bias is swept positive (for the *As* implanted gate/SiO_2/*p*-substrate), the *p*-substrate is

[13] Split-CV method is discussed in section 9.9.1.

inverted and the electrons inverted at the SiO_2-Si interface may deplete the electrons in the polysilicon gate, if the activated concentration of electrons near the polysilicon-SiO_2 interface is not high enough. The total capacitance will be lowered by the series polysilicon depletion capacitance. In addition, the depletion width increases with gate bias and when it extends to the highly activated region, the holes will flood the polysilicon-SiO_2 interface and the apparent $C_{g,\mathrm{inv}}$ will recover to C_{ox} (see Figure 4.21).

Since this anomalous C–V behavior is due to the depletion in the polysilicon near the polysilicon-SiO_2 interface, it is also referred to as the *polysilicon depletion effect.* The decrease in capacitance due to polysilicon depletion effect can be expressed as an increase in the "effective" gate oxide thickness and thus a reduction in the MOSFET drain current. In other words, *polysilicon depletion effect results in a degraded MOSFET, that has higher threshold voltage and lower drain current, compared to the device with degenerate doped polysilicon gate (no poly-depletion effect).* Although in this case gate capacitance in inversion is reduced and thus should result in a higher device speed, but due to reduction in the device current the overall device speed is reduced. For this reason polysilicon depletion effect is avoided in an MOSFET.

4.6 MOS Capacitor Applications

As was said in the beginning of this chapter, the simple MOS capacitor structure is a very powerful tool for studying the silicon surface and the results obtained are directly applicable to MOS transistors. In fact C–V measurements are routinely used to monitor IC fabrication processes. For this reason, MOS capacitor structures are often placed within the scribe lanes between circuits in regular production wafers; or they could be part of stand alone test structures, where the entire wafer consists of different structures for process and device modeling work.

An MOS capacitor as a test device is used to measure HF and LF C–V plots, which in turn are normally used to measure gate oxide thickness t_{ox}, oxide charges and interface-state density D_{it}, flat band voltage V_{fb}, doping concentration in the silicon, etc. [1]–[3]. These test devices are also used to measure non-equilibrium properties such as generation and recombination lifetime τ and surface generation and recombination velocity. These characteristics are derived from the transient capacitance response (C–t), that is by measuring the speed at which an inversion layer forms from initial deep depletion to its final minimum capacitance due to the minority carriers generation. Thus, C–t curves provide an indication of the charge generation processes both in the bulk of the silicon and at the surface. Both these processes have a significant effect on device performance, particularly for charge coupled devices. In this book we will be only

concerned with C–V plots and not C–t plots; readers interested in studying the latter are referred to [1]–[3].
The general setup for measuring LF and HF MOS C–V curves are discussed in Chapter 9. Also discussed there are device parameter measurements using MOS capacitor C–V plots, that are routinely used for MOS process/device characterization.

4.7 Nonuniformly Doped Substrate and Flat Band Voltage

As was stated earlier, the flat band voltage is an important MOS parameter. It is a good reference potential as it represents that gate voltage at which surface potential $\phi_s = 0$; that is, energy bands are flat for the entire silicon surface. For uniformly doped substrate V_{fb} can easily be determined as the gate voltage corresponding to the theoretically computed flat band capacitance C_{fb} on C–V curve [cf. Eq. (4.69)]. Alternatively, V_{fb} can also be obtained by measuring the *parallel voltage shift* in the measured HF C–V curve relative to the *ideal HF C–V* curve (see Figure 4.19). The first method, however, has large uncertainty and therefore the second method is preferred [1, pp. 477–499]. Moreover, at low temperatures, particularly below 180 K, the first method gives inconsistent values of V_{fb} [25]. This is because Eq. (4.69) was derived assuming complete ionization of impurity energy levels, which is a fairly good assumption for temperatures above 250 K. However, at lower temperatures this assumption is no longer valid, and in this case one must take into account the impurity freezeout effect [26].

Real MOS devices are invariably nonuniformly doped. Strictly speaking, the concept of the flat band is not applicable for nonuniformly doped substrates. This is because non-uniform doping causes a built-in field at the surface resulting in band bending independent of that caused by the metal-semiconductor workfunction difference Φ_{ms} and oxide charge Q_0. Thus, with the implanted channel it is not possible to achieve flat band conditions throughout the silicon for any combination of voltages. However, since V_{fb} *is a good reference potential, it is still used in context with non-uniform doping but redefined as that gate voltage which causes overall space charge to be zero* [27]–[29]. With this definition, V_{fb} for implanted devices can be expressed as [29]

$$V_{fb} = V_{fb}^0 \pm V_t \ln\left(\frac{N_s}{N_b}\right) \tag{4.79}$$

where V_{fb}^0 represents the flat band voltage as defined in Eq. (4.22) for uniformly doped substrate, N_s is the concentration at the surface, and '+'

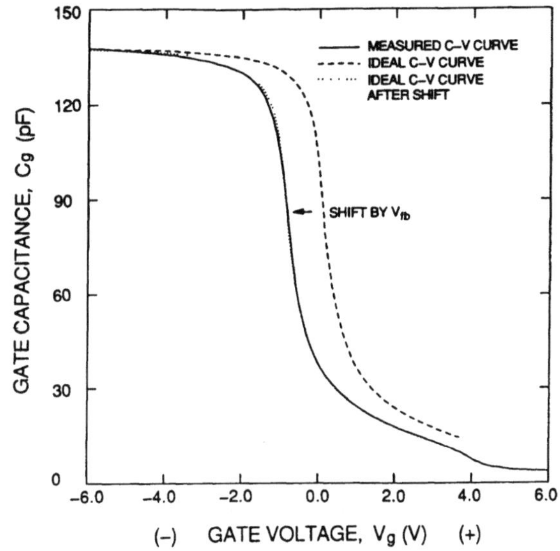

Fig. 4.22 Method showing flat-band voltage V_{fb} determination by measuring deep depletion MOS capacitance C–V curve and its comparison with the ideal C–V curve (dotted line)

and '−' signs are for *p*-type and *n*-type substrate, respectively. There are other definitions of V_{fb} which are not based on charge neutrality like the one given in [30].

It should be pointed out that differences in the definitions of V_{fb} is not important for circuit modeling work so long as one understands where it is coming from. This is because for circuit modeling work V_{fb} is considered as a model parameter to be determined for a given process. For a nonuniformly doped substrate, V_{fb} is obtained by comparing the experimental and ideal HF C–V curve, (see Figure 4.22), similar to the uniformly doped case. Generating an ideal C–V curve for a uniformly doped substrate is straightforward as discussed in section 4.3. However, for a nonuniformly doped substrate one needs to solve Poisson's equation numerically using the nonuniform doping concentration of the substrate. Program like MOSCAP could be used for this purpose [31].

4.7.1 Temperature Dependence of V_{fb}

The flat band voltage decreases with increasing temperature. This variation is caused by the temperature dependence of the work function difference Φ_{ms} between the gate material and the substrate, as Q_0 is independent of temperature to a first order approximation. Thus, one can write [cf.

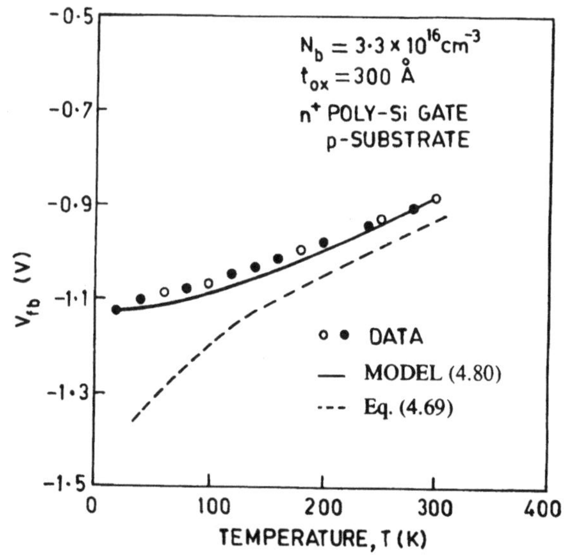

Fig. 4.23 Variation of the flat-band voltage V_{fb} with temperature for a p-type MOS capacitor. Symbols are V_{fb} measured using MOS C–V curve while continuous line is based on Eq. (4.80). (After Huang and Gildenblat [25])

Eq. (4.9)]

$$V_{fb}(T) = \Phi_{ms}(T) = \phi_{\text{poly}}(T) - \phi_{si}(T) \tag{4.80}$$

where ϕ_{poly} and ϕ_{si} are Fermi-potentials of the polysilicon gate and substrate, respectively. This simple model accurately predicts measured V_{fb} over a wide range of temperatures (77–300 K) for n-channel devices [25]. Figure 4.23 shows V_{fb} as a function of temperature for an MOS capacitor with n^+ polysilicon gate and p-substrate ($t_{ox} = 300$ Å). Circles are experimental data obtained by measuring the voltage shift in the measured HF C–V curve relative to the ideal C–V curve at each temperature, while the continuous line is calculated V_{fb} based on Eq. (4.80). Also shown in the figure (dashed line) is V_{fb} calculated using Eq. (4.69). This shows that *at low temperatures Eq. (4.69) becomes invalid for calculating* V_{fb}.

Depending upon the type of gate material and substrate concentration, the temperature coefficient of V_{fb} lies in the range 1–2 mV/K and can easily be calculated from the temperature coefficient of ϕ [cf. Eq. (2.17)]. Differentiating Eq. (4.80) and making use of Eq. (2.17) we have, for an MOS capacitor with n^+polysilicon gate and p-substrate,

$$\frac{\partial \Phi_{ms}}{\partial T} = \frac{1}{T}\left[\Phi_{ms} + 2\left(0.603 + \frac{3}{2}\frac{kT}{q}\right)\right]. \tag{4.81}$$

For an MOS capacitor with n^+ polysilicon gate and n-silicon surface we get

$$\frac{\partial \Phi_{ms}}{\partial T} = \frac{1}{T}\Phi_{ms}. \tag{4.82}$$

This explains why temperature coefficient of n-channel enhancement devices is higher than that of depletion devices (see section 5.4).

References

[1] E. H. Nicollian and J. R. Brews, *MOS (Metal Oxide Semiconductor) Physics and Technology*, John Wiley & Sons, New York, 1982.
[2] G. B. Barbottin and A. Vapaille, Eds., *Instabilities in Silicon Devices*, Vols I and II, North-Holland, New York, 1989.
[3] D. K. Schroder, *Semiconductor Material and Device Characterization*, John Wiley & Sons, New York, 1990.
[4] S. M. Sze, *Semiconductor Physics and Technology*, John Wiley & Sons, New York, 1982.
[5] T. P Chow and A. J. Steckl, 'Refractory metal silicides: Thin film properties and processing technology', IEEE Trans. Electron Devices, ED-30, pp. 1480–1497 (1983).
[6] Y. P. Tsividis, *Operation and Modeling of the MOS Transistor*, McGraw-Hill Book Company, New York, 1987.
[7] N. Lifshitz, 'Dependence of the work function difference between the polysilicon gate and silicon substrate on the doping level in polysilicon', IEEE Trans. Electron Devices, ED-32, pp. 617–621 (1985).
[8] T. Kamins, *Polycrystalline Silicon for IC Application*, Kluwer Academic Publisher, Boston, 1988.
[9] W. M. Werner, 'The work function difference of MOS system with aluminum field plates and polycrystalline silicon field plates', Solid-State Electron., 17, pp. 769–775 (1974).
[10] O. H. Kim and C. K. Kim, 'Threshold voltage shift due to change of impurity type of polysilicon in heavily doped polysilicon gate MOSFET', Proc. Intern. Symposium on VLSI Technology, System and Applications, pp. 170–173, March (1983).
[11] B. E. Deal, 'Standardized terminology for oxide charge associated with thermally oxidized silicon,' IEEE Trans. Electron Devices, ED-27, pp. 606–608 (1980).
[12] P. Richman, *MOS Field-Effect Transistors and Integrated Circuits*, John Wiley & Sons, New York, 1973.
[13] R. F. Pierret, *Field Effect Devices, Vol VI: Modular Series on Solid-State Devices*, Addison-Wesley Publishing Co., Reading MA, 1983.
[14] R. S. Muller and T. I. Kamins, *Device Electronics for Integrated Circuits*, John Wiley & Sons, New York, 1986.
[15] F. Stern and W. E. Howard, 'Properties of semiconductor surface inversion layers in quantum limits', Phys. Rev. 163, pp. 816–835 (1967).
[16] A. P. Gnadinger and H. E. Talley, 'Quantum mechanical calculation of the carrier distribution and thickness of the inversion layer of a MOS field-effect transistor', Solid-State Electron., 13, pp. 1301–1309 (1970).
[17] K. H. Zaininger and F. Heiman, 'The C–V technique as an analytical tool', Solid-State Techn., Part I, pp. 49–56, May (1970); Part II, pp. 46–55, June (1970).
[18] A. Berman and D. R. Kerr, 'Inversion charge distribution model of the high frequency MOS capacitance', Solid-State Electron., 17, pp. 735–742 (1974).

[19] T. W. Hickmott and R. D. Issac, 'Barrier heights at the polycrystalline Silicon–SiO_2 interface', J. Appl. Phys., 52, pp. 3464–3475 (1981).
[20] G. Yaron and D. Frohman-Bentchkowsky, 'Capacitance voltage characterization of poly Si–SiO_2–Si structures', Solid-State Electron., 23, pp. 433–439 (1980).
[21] C. Y. Wong, J. Y.-C. Sun, Y. T. Aur, C. S. Oh, R. Angelucci, and B. Davari, 'Doping of N^+ and P^+ poly-Si in a dual-gate CMOS process', IEEE IEDM, *Tech. Dig.*, pp. 238–241 (1988).
[22] R. A. Chapman, C. C. Wei, D. A. Bell, S. Aur, G. A. Brown, and R. A. Haken, '0.5 micron CMOS for high performance at 3.3 V', IEEE IEDM, *Tech. Dig.*, pp. 52–55 (1988).
[23] C. Y. Lu, J. M. Sung, H. C. Kirsch, S. J. Hollenius, T. E. Smith, and L. Manchanda, 'Anomalous C–V characteristics of implanted poly MOS structure in n^+/p^+ dual gate CMOS technology', IEEE Electron Devices Lett., EDL-10, pp. 192–194 (1989).
[24] P. Habas and S. Selberherr, 'On the effect of nodegenerate doping of polysilicon gate in thin oxide MOS devices—analytical modeling', Solid-State Electron., 33, pp. 1539–1544 (1990).
[25] C. L. Huang and G. Sh. Gildenblat, 'MOS flat-band capacitance method at low temperatures', IEEE Trans. Electron Devices, ED-36, pp. 1434–1439 (1989).
[26] R. C. Jaeger, F. H. Gaensslen, 'Simulation of impurity freezeout through numerical solution of Poisson's equation with application to MOS devices,' IEEE Trans. Electron Devices, ED-27, pp. 914–920 (1980).
[27] R. R. Troutman, 'Ion-implanted threshold voltage tailoring for insulated gate field-effect transistors', IEEE Trans. Electron Devices, ED-24, pp. 182–192 (1977).
[28] A. H. Marshak and R. Shrivastava, 'On threshold and flat-band voltages for MOS devices with polysilicon gate and nonuniformly doped substrate', Solid-State Electron., 26, pp. 361–364 (1983).
[29] F. Van de Weile, 'On the flat-band voltage of MOS structures on nonuniformly doped substrate, Solid-State Electron., 27, pp. 824–826 (1984).
[30] D. A. Antoniadis, 'Calculation of threshold voltage in nonuniformly doped MOSFETs', IEEE Trans. Electron Devices, ED-31, pp. 303–307 (1984).
[31] R. C. Jaeger, F. H. Gaensslen, and S. E. Diehl, 'An efficient algorithm for simulation of MOS capacitance', IEEE Trans. Computer-Aided Design, CAD-2, pp. 111–116 (1983).

Threshold Voltage 5

One of the most important physical parameters of a MOSFET is its threshold voltage V_{th}, defined as the gate voltage at which the device starts to turn on. The accurate modeling of threshold voltage is important to predict correct circuit behavior from a circuit simulator. Since V_{th} has profound effect on circuit operation, it is often used to monitor process variations. Present day MOS process invariably use ion implantation into the channel region, a step often called the *threshold adjust implant*, that alter the doping profile near the surface of silicon substrate. By changing dose and energy of the threshold implant a desired threshold voltage is achieved. The threshold voltage is by no means a constant quantity but varies with the back bias. With larger back bias, circuits have slower transitions due to decreased drain current and as a result, noise margins decrease.
In this chapter we develop models for the threshold voltage of MOSFETs. First, we will derive the expression for V_{th} for a uniformly doped substrate. We will modify the model for channel implanted devices. The effect of device channel length and width on V_{th} is then modeled from a circuit simulation point of view.

5.1 MOSFET with Uniformly Doped Substrate

In this section we will develop a basic threshold voltage expression for large and wide MOSFETs, neglecting edge effects due to short channel and narrow widths. Let us consider an n-channel device with a uniformly doped substrate of concentration N_b (cm^{-3}), the structure and dimensions of which are shown in Figure (5.1a). In our coordinate system, the x direction is the distance into the silicon measured from the oxide–silicon interface, the y direction is the distance along the length of the channel measured from the source end increasing towards the drain, and z direction is the distance along the width of the device.

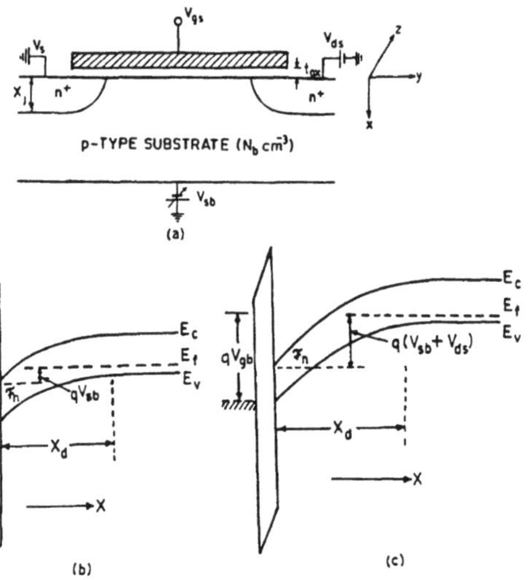

Fig. 5.1. (a) Schematic diagram of *n*-channel MOSFET (nMOST). Its energy band diagram at (b) the source end, and (c) the drain end of the channel

Strictly speaking, modeling a MOSFET is a three-dimensional (3-D) problem; however, for all practical purposes (unless the width W and length L are very small) we can treat the system as a 2-D problem in the x and y direction only. Even as a 2-D problem, calculation of charges in the system is quite complex. Therefore, in order to obtain a MOSFET model for use in circuit simulators, we generally make some simplifying assumptions. It is generally assumed that the variation of the electric field $\mathscr{E}_y$ in the y direction (along the channel) is much less than the corresponding variation of the field $\mathscr{E}_x$ in the x direction[1] (perpendicular to the channel). This is called the *gradual channel approximation* (GCA). Mathematically the GCA means

$$\frac{\partial \mathscr{E}_y}{\partial y} \ll \frac{\partial \mathscr{E}_x}{\partial x}, \tag{5.1}$$

which, in terms of electrostatic potential ϕ, is equivalent to

$$\left|\frac{\partial^2 \phi}{\partial y^2}\right| \ll \left|\frac{\partial^2 \phi}{\partial x^2}\right|.$$

Although GCA is valid for most of the channel length, it breaks down near the drain end. In spite of this shortcoming, it is used for MOSFET modeling because Poisson's equation becomes one-dimensional. This means that the charge expressions developed in chapter 4 for an MOS capacitor could be used for a MOS transistor. Recall that those charge expressions were

[1] The field $\mathscr{E}_x$ is also termed *normal or transverse field*, while the field $\mathscr{E}_y$ is also termed *longitudinal, tangential or lateral* field.

developed using Gauss' Law under the assumption that the electric field was perpendicular to the gate oxide. This means that implicitly we assumed the GCA to be valid. Also recall that the gate, bulk and inversion charges, Q_g, Q_b and Q_i, respectively, were assumed to be uniformly distributed along the y direction. This will not be the case now because the surface potential will vary along the y direction due to the applied source and drain voltages relative to the substrate (bulk). This means, the application of the drain voltage V_{ds} and substrate voltage V_{sb} in a MOSFET will result in *potentials and charges per unit area that will be position dependent along the y direction.* In normal operation of an n-channel MOSFET, the only major current results from the flow of electrons from the source to drain. Since the hole current is negligible everywhere, it is reasonable to assume that the electron quasi-Fermi energy (or level) $\mathscr{F}_n$ is constant in the surface region where electron concentration is large. Similarly, one can assume the hole quasi-Fermi energy (or level) $\mathscr{F}_p$ to be constant where hole concentration is significant. Application of a positive bias V_{sb} to the source relative to the substrate lowers $\mathscr{F}_n$ in the source relative to the Fermi level E_f in the substrate (bulk) by an amount qV_{sb} (see Figure 5.1b). Similarly, the positive drain voltage V_{ds} lowers $\mathscr{F}_n$ in the drain relative to the substrate by an amount $q(V_{ds}+V_{sb})$ (see Figure 5.1c). It is the difference in $\mathscr{F}_n$ (or quasi-Fermi potential φ_n) between the source and drain that drives the electrons down the channel. If we define $V_{cb}(y)$, called the *channel potential*, as the potential difference between $\mathscr{F}_n$ at the silicon surface and E_f in the bulk, such that

$$V_{cb}(y)=\begin{cases} V_{sb} & \text{at } y=0 \text{ (source end)} \\ V_{sb}+V_{ds} & \text{at } y=L \text{ (drain end)} \end{cases} \tag{5.2}$$

then, compared to the case of an MOS capacitor, the surface potential or band bending ϕ_s in a MOSFET becomes $\phi_s+V_{cb}(y)$. Strictly speaking, this relation is only approximate and breaks down near the source/drain boundaries. However, it is good enough for MOSFET circuit models. We will use this relationship for developing threshold voltage and drain current models of a MOSFET. Let us now discuss the threshold voltage model.

At equilibrium, there exists a depletion region between the pn junction formed by the p-substrate and the n^+ source and drain regions (see dotted lines in Figure 5.2a). With no applied voltages, the source and drain depletion widths represented by X_{sd} and X_{dd}, respectively, under the one-dimensional abrupt junction approximation, are given by Eq. (3.2). In general a back bias V_{sb} is applied to the source relative to the substrate. This V_{sb} is such that it reverse biases the source/drain junction for proper operation of the device. This results in an increase in the source/drain depletion widths [cf. Eq. (2.53)]. Let us assume that only a small voltage is applied at the drain, relative to the source (i.e. $V_{ds}<0.1\text{V}$). Then, to a good approximation, $V_{cb}(y)\cong V_{sb}$.

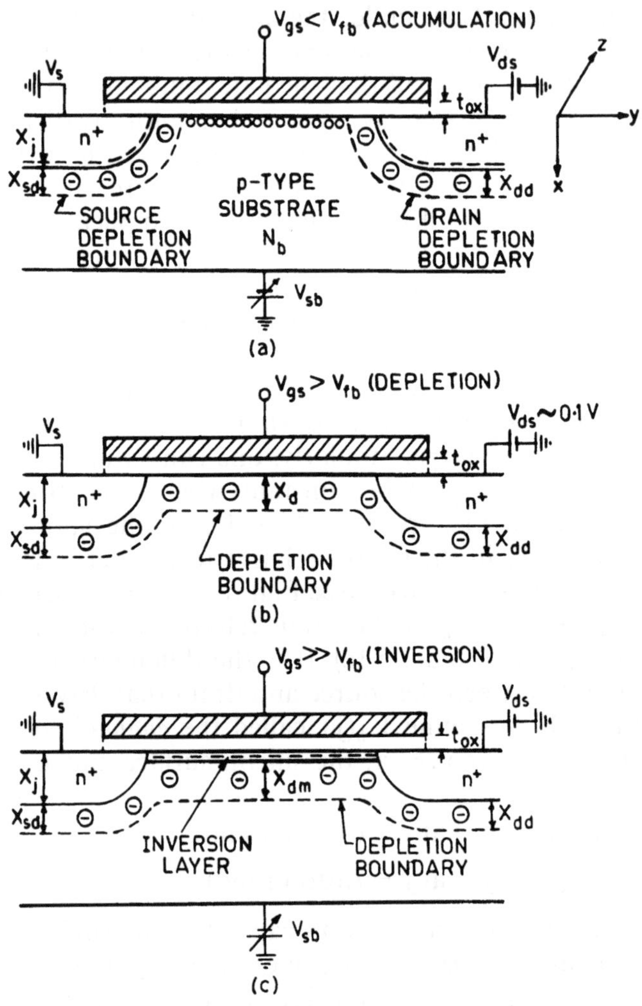

Fig. 5.2 Cross-section of an nMOST in (a) accumulation, (b) depletion, and (c) inversion. Dotted lines show depletion width under source and drain *pn* junction and the gate

From our earlier discussion on an MOS capacitor, recall that a positive gate voltage V_{gb} (for *p*-substrate) induces a charge Q_s in the silicon such that [cf. Eq. (4.20)]

$$V_{gb} = V_{fb} + \phi_s(y) - \frac{Q_s(y)}{C_{ox}} \tag{5.3}$$

where V_{fb} is the flat band voltage and ϕ_s is surface potential or band bending at the surface. Note that both ψ_s and Q_s are functions of distance along the y direction (length of the channel), due to the applied channel voltage V_{cb}. However, under the assumption that V_{ds} is very small (< 0.1V) we can replace $\phi_s(y)$ with $(\phi_s + V_{sb})$. Remembering $V_{gb} = (V_{gs} + V_{sb})$, Eq. (5.3) can be simplified as

$$V_{gs} = V_{fb} + \phi_s - \frac{Q_s}{C_{ox}}. \tag{5.4}$$

Similar to the case of an MOS capacitor, the channel region of a MOSFET exhibits three different regimes of operation—accumulation, depletion and inversion—that depends upon the values of the applied gate voltages.

Accumulation. When the gate voltage $V_{gs} < V_{fb}$, holes are accumulated at the silicon surface under the gate. No current flows between the source and the drain as the source and drain to substrate *pn* junctions are still separated by the accumulated *p*-region (see Figure 5.2a).

Depletion. When $V_{gs} > V_{fb}$, the silicon surface is depleted of holes thereby forming a depletion layer under the gate as shown by the dotted line in Figure 5.2b. The depth X_d of this depletion layer is easily obtained from the MOS capacitor Eq. (4.27) by replacing ϕ_s with $\phi_s + V_{sb}$. Thus,

$$X_d = \sqrt{\frac{2\epsilon_0\epsilon_{si}}{qN_b}(\phi_s + V_{sb})} \quad \text{(cm)} \tag{5.5}$$

and is constant along the length of the channel under our assumption of small $V_{ds}(< 0.1\text{V})$. Since there are no free carriers, only a small leakage current flows from the source to the drain.

Inversion. When $V_{gs} \gg V_{fb}$, an inversion layer is formed at the surface under the gate (see Figure 5.2c). This induced layer of electrons extends continuously from source to drain, thus forming a conduction path resulting in electron current flow between source and drain.[2] The higher the gate voltage, the larger will be the number of induced electrons and thus, the larger the current. Recall that in inversion, the induced charge Q_s consists of fixed bulk charge Q_b (due to the ionized acceptors in the depletion layer) and the mobile inversion charge Q_i (due to the electrons in the inversion layer) that is, $Q_s = Q_b + Q_i$ [cf. Eq. (4.40)]. Accordingly Eq. (5.4) becomes

$$V_{gs} = V_{fb} + \phi_s - \frac{Q_b + Q_i}{C_{ox}}. \tag{5.6}$$

As was discussed earlier for the case of an MOS capacitor, at low values of the gate voltages $(> V_{fb})$, ϕ_s increases reasonably rapidly with gate bias

[2] Recall that in an MOS capacitor under inversion there is no conduction path and hence no DC current flows.

and so does the gate depletion width X_d. This regime corresponds to the depletion and weak inversion regions of device operation (see Figure 4.14). At higher values of gate voltages ($\gg V_{fb}$), ϕ_s hardly changes with the gate bias; ϕ_s has become pinned. This pinning occurs when strong inversion sets in. The condition when this happens is often called the condition for *threshold* and the corresponding gate voltage is called the *threshold voltage* V_{th}. Thus, *threshold voltage is that gate voltage, relative to the source, at which the transistor enters the strong inversion regime, thereby forming a conducting channel at the surface from the source to the drain.* It must be emphasized that the formation of a conducting channel is not an abrupt process in which the channel is completely depleted of carriers at a given V_{gs} and then a small increase in V_{gs} causes channel to be formed. It is rather *a continuous process.*

At the edge of the strong inversion, Q_i is small compared to Q_b. Therefore, neglecting Q_i and replacing ϕ_s in Eq. (5.6) with ϕ_{si} (the value of ϕ_s at strong inversion) yields the following expression for $V_{gs}(=V_{th})$,

$$V_{th} = V_{fb} + \phi_{si} - \frac{Q_b(\phi_{si})}{C_{ox}} \quad \text{(V)} \tag{5.7}$$

where $Q_b(\phi_{si})$ is the value of Q_b at strong inversion. Note that *neglecting Q_i in Eq. (5.7) results in an abrupt transition to the conducting state, although the exact transition is exponential and occurs over several tenths of a volt* (see Figure 4.12). Despite this inaccuracy, V_{th} as defined by Eq. (5.7) is a very useful parameter for describing MOSFET characteristics.

When strong inversion sets in, the gate depletion width X_d reaches its maximum value X_{dm} and is obtained by replacing ϕ_s in Eq. (5.5) with ϕ_{si}, that is,

$$X_{dm} = \sqrt{\frac{2\epsilon_0\epsilon_{si}}{qN_b}(\phi_{si} + V_{sb})} \quad \text{(strong inversion).} \tag{5.8}$$

Similarly, the bulk charge Q_b in the depletion region under the gate, when strong inversion sets in, is simply obtained by replacing X_d with X_{dm} in Eq. (4.29) for Q_b. Thus, under the depletion approximation we have

$$Q_b(\phi_{si}) = -qN_bX_{dm} \quad \text{(C/cm}^2\text{).} \tag{5.9}$$

Combining Eqs. (5.7)–(5.9) yields the following expression for V_{th}:

$$V_{th} = V_{fb} + \phi_{si} + \gamma\sqrt{\phi_{si} + V_{sb}} \quad \text{(V)} \tag{5.10}$$

where

$$\boxed{\gamma = \frac{\sqrt{2\epsilon_0\epsilon_{si}qN_b}}{C_{ox}}} \quad \text{(V}^{1/2}\text{).} \tag{5.11}$$

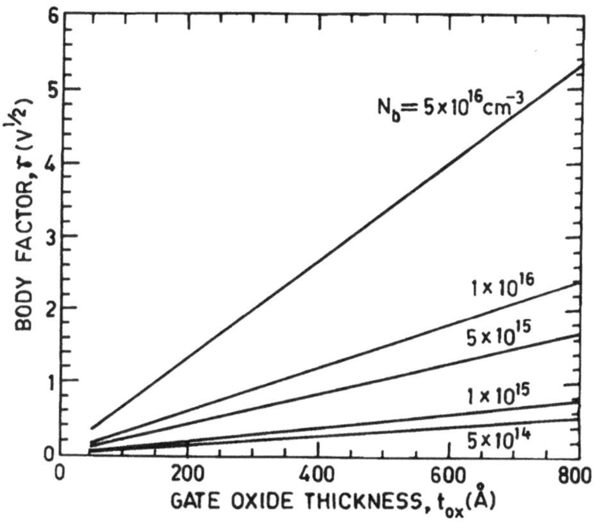

Fig. 5.3 Variation of the body factor γ as a function of gate oxide thickness t_{ox} for different substrate concentrations N_b

The factor γ defined by Eq. (5.11) is called the *body factor or body-effect coefficient* and depends upon both the bulk concentration N_b and the oxide thickness t_{ox} (see Figure 5.3). Since both these parameters are process dependent, knowing the dependence of γ on process parameters t_{ox} and N_b permits designers to determine the sensitivity of V_{th} to process control.

There are several criteria used in the literature to define the onset of strong inversion [1]–[4]. The two most commonly quoted criteria are:

1. In strong inversion, the minority carrier (electrons in our case of an nMOST) density at the surface is equal to the majority carrier (holes) density in the substrate. Stated another way, when the surface potential ϕ_s equals twice the Fermi potential ϕ_f with respect to the substrate [cf. Eq. (4.4)]

$$\boxed{\phi_{si} = 2\phi_f = 2V_t \ln\left(\frac{N_b}{n_i}\right).} \tag{5.12}$$

   This criterion is often referred to as the *classical criterion for strong inversion.*

2. Given the fact that pinning of ϕ_s occurs at somewhat higher potential than $2\phi_f$, it has been suggested that the criterion which gives better estimate of the measured threshold voltage is [3]

$$\phi_{si} = 2\phi_f + 6V_t. \tag{5.13}$$

   This definition of strong inversion is six times the thermal voltage ($6V_t$) higher than the classical criterion. The inversion regime between $\phi_{si} = 2\phi_f$ to $\phi_{si} = 2\phi_f + 6V_t$ is called moderate inversion by Tsividis [1].

Various different criteria reported in the literature [1]–[4] are not equivalent and hence lead to different values of V_{th}. However, almost all of the *threshold voltage models implemented in circuit simulators are based on the classical criterion* and hence it is this criterion we will use to develop the V_{th} model.

While it is convenient to model V_{th} as described above, it is difficult to measure the V_{gs} value at which a particular strong inversion condition occurs. Experimentally, V_{th} is obtained by measuring drain current I_{ds} as a function of gate voltage V_{gs}. Several definitions are in use for determining V_{th} from $I_{ds} - V_{gs}$ data, as discussed in section 9.4. However, the most commonly used definition is called the extrapolated V_{th}. In the latter method, V_{th} is obtained by extrapolating the $I_{ds} - V_{gs}$ curve to $I_{ds} = 0$. *Most often the extrapolated V_{th} is matched to the model equation that incorporates a particular strong inversion condition.*

Assuming the classical criterion for strong inversion, the threshold voltage Eq. (5.10) becomes

$$\boxed{V_{th} = V_{fb} + 2\phi_f + \gamma\sqrt{2\phi_f + V_{sb}}.} \tag{5.14}$$

This *is the basic equation for the threshold voltage of a MOSFET.* It shows that plot of V_{th} as a function $\sqrt{2\phi_f + V_{sb}}$ will result in a straight line, the slope of which gives the body factor γ. This indeed is the case as shown in Figure 5.4 for n-channel devices fabricated using NMOS technology

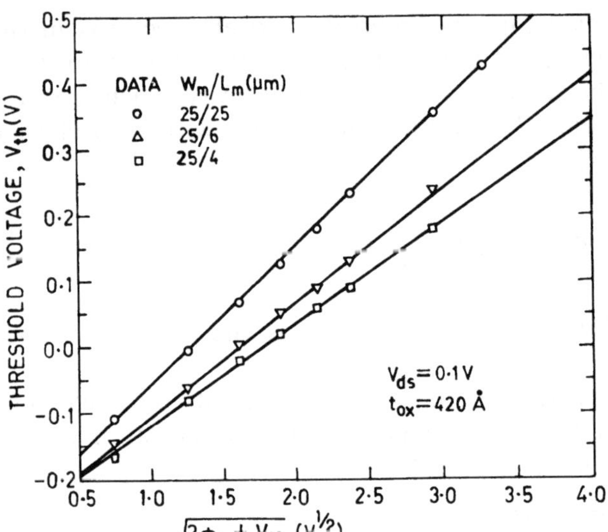

Fig. 5.4 Variation of threshold voltage V_{th} against $\sqrt{2\phi_f + V_{sb}}$ for experimental uniformly doped nMOSTs; $t_{ox} = 420\,\text{Å}$, $W_m/L_m = 25/25$ (○), 25/6 (△), 25/5 (◇)

Table 5.1. *Sign convention of different parameters in V_{th} Eq. (5.14)*

	Substrate	V_{fb}	ϕ_f	Q_b	γ	V_{sb}	V_{TO}
nMOST	p-type	− (for metals and n^+ polysilicon gate)	+	−	+	+	+
pMOST	n-type	− (for metals and n^+ polysilicon gate) + (for p^+ polysilicon gate)	−	+	−	−	−

($t_{ox} = 420$ Å), having width to length ratio $W_m/L_m(\mu m/\mu m) = 25/25, 25/6$, and 25/5. If the gate oxide thickness t_{ox} and hence C_{ox} is known, the intercept of the line could be used to calculate V_{fb} and N_b. In fact, this procedure is often used to calculate γ and V_{fb}, as will be discussed later in section 9.5. When the body bias $V_{sb} = 0$, Eq. (5.14) yields

$$\boxed{V_{TO} \equiv V_{th} = V_{fb} + 2\phi_f + \gamma\sqrt{2\phi_f}} \tag{5.15}$$

and is called the *zero body bias threshold voltage* V_{TO}. For circuit modeling purpose, Eq. (5.14) is often written as

$$V_{th} = V_{TO} + \gamma(\sqrt{2\phi_f + V_{sb}} - \sqrt{2\phi_f}) \quad \text{(V)}. \tag{5.16}$$

Equation (5.16) for V_{th} is used in the SPICE Level 1 MOSFET model. Although Eq. (5.14) has been derived for n-channel MOSFETs, it is also valid for p-channel MOSFETs (pMOSTs), provided proper signs are used for different parameters, as shown in Table 5.1.

Based on the sign conventions shown in Table 5.1, V_{th} for nMOSTs become a positive number while for pMOSTs it is a negative number.

From Eq. (5.14) it is clear that the threshold voltage depends upon the following parameters:

- The gate oxide thickness t_{ox}, through C_{ox} [cf. Eq. (4.1)]; *the thicker the oxide, the higher the threshold voltage.* This explains why field oxides in LOCOS isolation technique are very thick (0.2–0.4 μm) because field transistors need to have higher threshold voltage than the active transistors (cf section 3.5.4).
- The flat band voltage V_{fb} that depends upon the work function difference Φ_{ms} between the gate material and the substrate, and charge Q_0 at the oxide–silicon interface [cf. Eq. (4.14)].
- The substrate doping concentration N_b that enters through the ϕ_f and γ terms; *the higher the N_b, the higher the V_{th}*. A plot of V_{th} versus N_b, for both nMOST and pMOST with n^+ and p^+ polysilicon gates for three different t_{ox} (150 Å, 250 Å and 650 Å), is shown in Figure 5.5 [5]. These

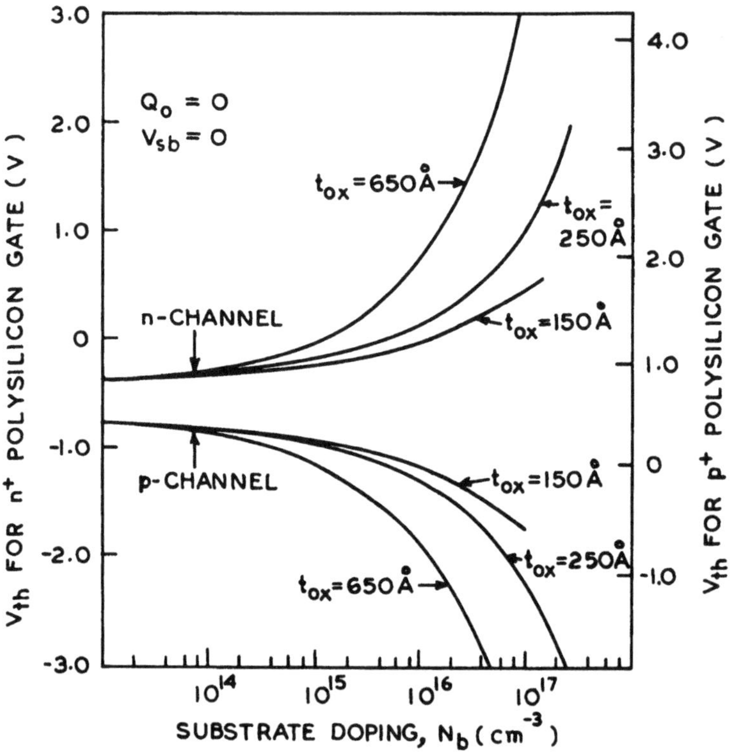

Fig. 5.5 Calculated threshold voltage V_{th} for *n*- and *p*-channel MOSFETs as a function of substrate doping N_b for n^+ polysilicon gate (left scale) and p^+ polysilicon gate (right scale) for three different oxide thickness. (After Sze [5])

curves are based on the assumption that $Q_0 = 0$, a reasonable assumption for modern VLSI processes.

- The temperature through the ϕ_f term, *the higher the temperature, the lower the* V_{th} (for details see section 5.4)
- The body bias V_{sb}; *the higher the* V_{sb}, *the higher the* V_{th}. The increase in V_{th} due to an increase in V_{sb} can be obtained from Eq. (5.16) as

$$\Delta V_{th} = V_{th} - V_{TO} = \gamma[\sqrt{2\phi_f + V_{sb}} - \sqrt{2\phi_f}]. \tag{5.17}$$

Body-Effect. The variation of V_{th} with V_{sb} is often called the *substrate bias sensitivity* or *body-effect*. Differentiating Eq. (5.14) with respect to V_{sb} we get

$$\frac{dV_{th}}{dV_{sb}} = \pm \frac{\gamma}{2\sqrt{2\phi_f + V_{sb}}} \tag{5.18}$$

where the + and the − signs are for *n*- and *p*-channel devices, respectively. This equation shows that the body-effect increases as the body factor γ

increases and body bias V_{sb} decreases. For circuit design, it is often desirable to lower the body effect,[3] which means the body factor γ should be reduced. From Eq. (5.11) it is evident that γ can be reduced with lower doping concentration N_b and/or lower oxide thickness t_{ox}. However, lowering N_b, for example, conflicts with the scaling rule (cf. section 3.4). In fact, the choice of process or circuit parameters is a trade off between various parameters involved in device design.

SPICE Implementation. Note that Eq. (5.14) becomes invalid for $V_{sb} \leq -2\phi_f$, i.e., when the S/D diodes become forward biased by an amount $2\phi_f$. Although during normal operation of the device the S/D diodes will not be forward biased; however, in SPICE, during Newton–Raphson iterations, it is possible to encounter $V_{sb} < -2\phi_f$. This is just an artifact of the iteration solution process, and convergence to a proper solution requires the model to behave well even in such invalid operating regions. Therefore, to use Eq. (5.14) or (5.16) in the forward biased region of S/D junction, some sort of smoothing function is used to limit the value of $(V_{sb} + 2\phi_f)$ such that it is always positive. The smoothing function assures a smooth transition without any discontinuity. In SPICE, the transition point is chosen as $V_{sb} = -\phi_f$. Thus, when $V_{sb} + \phi_f \geq 0$, Eq. (5.14) is used and when $V_{sb} + \phi_f < 0$, the term $\sqrt{V_{sb} + 2\phi_f}$ is replaced by $2\sqrt{\phi_f}/(1 - V_{sb}/\phi_f)$ such that V_{th} and its first derivative are continuous at $V_{sb} = -\phi_f$ in the forward biased S/D region.

5.2 Nonuniformly Doped MOSFET

In the previous section we have seen that for a given gate material the threshold voltage V_{th} of a MOSFET depends upon the substrate doping concentration N_b and the gate oxide thickness t_{ox}. Therefore, in principle, V_{th} could be set to any value by proper choice of N_b and t_{ox} (see Figure 5.5). However, considerations like the body-effect, source-drain junction capacitances and breakdown voltages often dictate desirable values of these parameters. In practice this is achieved by ion implanting a shallow layer of dopant atoms into the substrate in the channel region. Thus, by adjusting the channel surface concentration (using ion implantation) any desired value of V_{TO} can be achieved. In fact, in VLSI devices, more than one implant is often used in the channel region—one to adjust the threshold voltage and another to avoid the punchthrough effect—as was discussed

[3] During circuit operation, in NMOS circuits, the MOSFET source voltage often increases which results in higher V_{sb} thereby causing V_{th} to increase. This results in a decrease in the drain current I_{ds} [see Eq. (3.4)], consequently the circuit runs at a lower speed and might not even function properly. For this reason, it is desirable to reduce the change in V_{th} due to increase in V_{sb}, that is reduce body-effect.

in section 3.5.2. The fact that the surface is no longer uniformly doped, due to the channel implant, means Eq. (5.14) is generally not valid.

Recall that in n^+ polysilicon gate CMOS technology, an nMOST has channel implant dopants (boron) which are of the same type as that of the substrate (*p*-type), while pMOST (compensated *p*-device) has shallow channel implant dopants, which are of the type opposite to that of the substrate (cf. section 3.5.2). Since compensated pMOST has shallow channel implant, the surface layer is depleted at zero gate bias. When $V_{gs} > V_{th}$, the current flows at the surface. Therefore, these compensated devices are usually modeled in the same way as nMOST, so far as the drain current modeling is concerned; however they have a different threshold voltage model as we will see later. In a recent submicron CMOS technology, pMOSTs are being fabricated with p^+ polysilicon gate while nMOSTs are with n^+ polysilicon gates. With p^+ polysilicon gate, pMOST has channel implant of the same type as substrate and therefore, from a modeling point of view, these devices are similar to nMOST.

In the depletion type devices the channel implant, which is of opposite type to that of the substrate, is deep so that significant current flows even at $V_{gs} = 0$V. These (depletion) devices are referred to as *normally-on buried channel* (BC) MOSFET as against the compensated devices, which are also referred to as *normally-off buried channel* (BC) MOSFETs. The two BC MOSFETs result in entirely different V_{th} behavior due to the different potential distributions associated with the built-in *pn* junctions in the channel region. This can easily be seen from their energy band diagrams as shown in Figure 5.6 for a *p*-channel device with a *p*-type buried layer in *n*-type substrate and an n^+ polysilicon gate. While for the normally-off BC MOSFET (Figure 5.6a) the energy band bending of the bulk junction extends to the channel surface, the depletion device (normally-on) has an (hole) energy minimum (Figure 5.6b). It was pointed out earlier (cf. section 3.5.2) that in practice *p*-channel depletion devices are not usually made. It

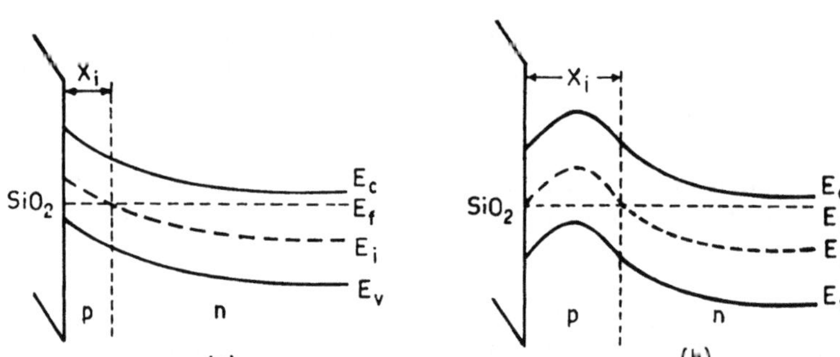

Fig. 5.6 Energy band diagram for a *p*-type buried-channel MOSFET in (a) the surface channel mode and (b) the buried channel mode (depletion device)

is the *n*-channel depletion device, with negative threshold voltage, which is more important and thus modeled here.

There is an extensive literature on threshold voltage models for ion implanted devices [1]–[36]. However, we will discuss and develop only those models which are suitable for circuit simulators. We will first discuss enhancement mode devices and then depletion mode devices.

5.2.1 Enhancement Type Device

When ions are implanted into the channel, the implanted profile can be fairly accurately approximated by the following Gaussian distribution function (see Figure 5.7, also see Appendix H, Eq. (H.5))

$$N(x) = N_0 \exp\left[-\frac{(x - R_p)^2}{2\Delta R_p^2} \right] \quad (\mathrm{cm}^{-3}) \tag{5.19}$$

where

$N_0 = D_i/(\Delta R_p \sqrt{2\pi})$ is the maximum concentration and occurs at $x = R_p$,
x = the depth measured from the oxide–silicon interface,
R_p = *projected range* (average penetration depth),
ΔR_p = *straggle* (standard deviation)
D_i = *dose*, i.e., number of implanted ions per unit area.

The channel implant dose D_i is typically of the order of $10^{10} - 10^{12}\,\mathrm{cm}^{-2}$ while the implant energy varies from 10–200 KeV. Following the implantation process, devices go through various high-temperature fabrication steps, which change the final profile. Figure 5.8 shows the final channel implant profiles for nMOST and pMOST for a typical $2\,\mu$m CMOS technology with n^+ polysilicon gate.

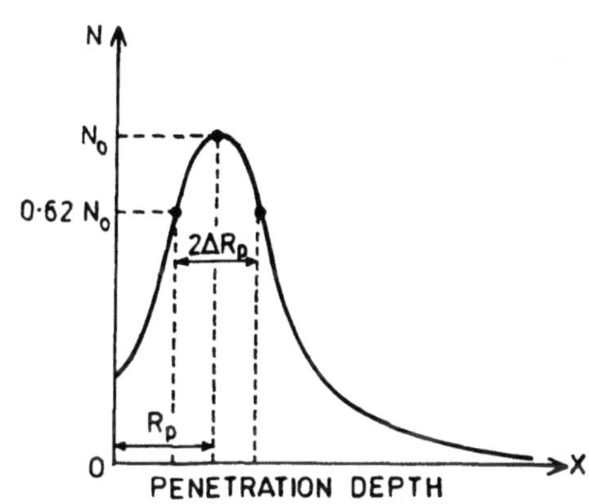

Fig. 5.7 Gaussian doping profile in the channel region of a VLSI MOSFET

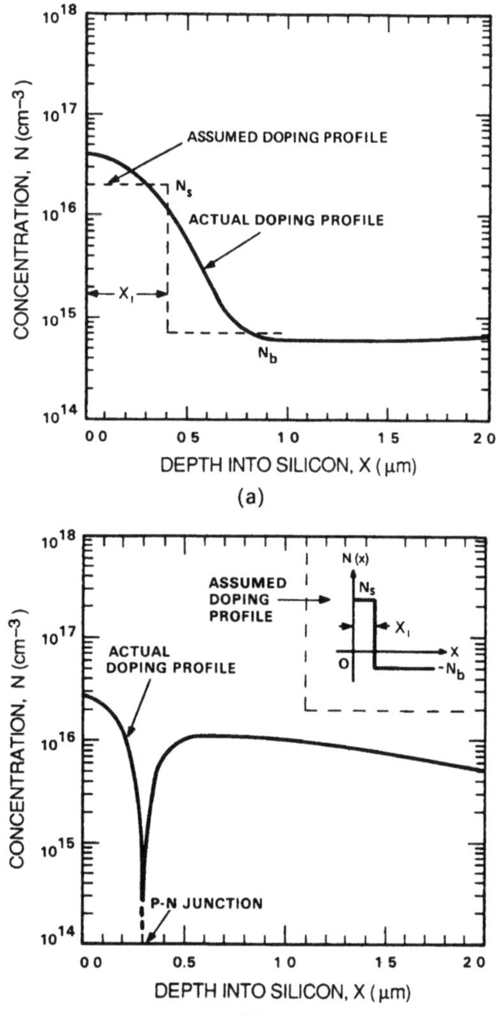

Fig. 5.8 Vertical doping profile of channel implanted region under the gate for a typical $2\,\mu$m CMOS process for (a) nMOST and (b) pMOST

The result of the channel implant in an otherwise uniformly doped substrate is to change the threshold voltage. The extrapolated threshold voltage V_{th} as a function of $\sqrt{2\phi_f + V_{sb}}$ for different channel implant dose is shown in Figure 5.9 [6]. One can see from this figure that *the slope of the V_{th} versus V_{sb} curve changes from single slope at low doses to two distinct slopes at higher doses.* This shows that a simple square root dependence, which relates V_{th} to V_{sb}, is not correct for channel implanted devices with high doses. However, for these devices Eq. (5.7) is still valid provided we use appropriate value of V_{fb}, ϕ_{si}, and $Q_b(\phi_{si})$. We will now consider each of these terms and see how they are modified for implanted devices.

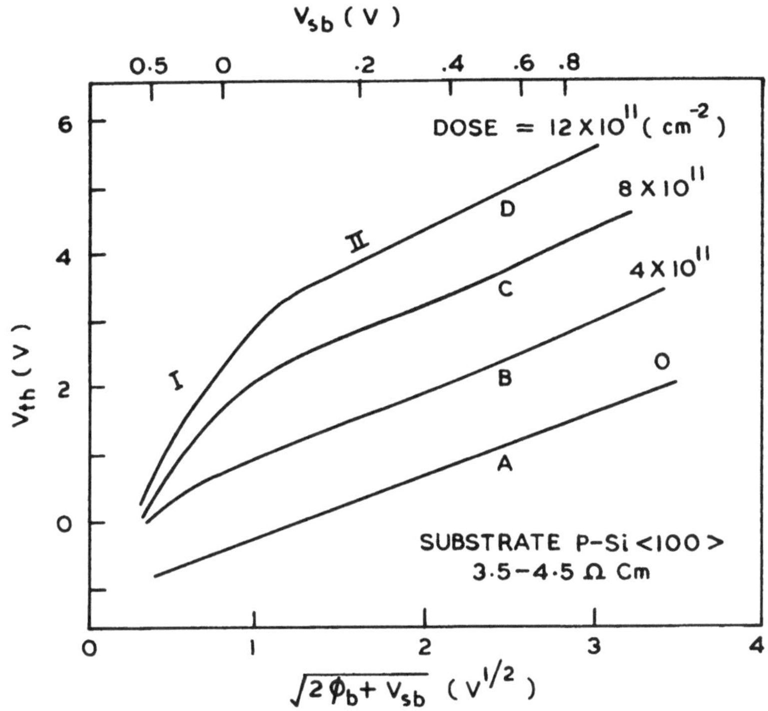

Fig. 5.9 Threshold voltage dependence on body bias for different channel implants. (After Kamoshida [6])

Flat Band Voltage V_{fb}. As has been discussed earlier (cf. section 4.7), the concept of the flat band voltage is strictly applicable to a uniformly doped substrate. However, being an important reference voltage, it has been redefined for nonuniformly doped substrates as that gate voltage which causes the overall space charge to be zero [cf. Eq. (4.79)]. Whatever definition is used, for circuit modeling V_{fb} is treated as a model parameter to be determined for a given process.

Surface Potential at Strong Inversion (ϕ_{si}). Like the uniformly doped case, different criteria for strong inversion have been suggested for non-uniformly doped substrates [2], [33]–[36]. Some of these are:

1. The classical criterion given by Eq. (5.12) is still used for non-uniformly doped substrate [33], although strictly speaking it is valid only for implanted channels with low dose. Others [11] have used this criterion by replacing N_b (bulk concentration) with N_s (surface concentration) in the ϕ_f term, that is,

$$\phi_{si} = 2\phi_f(\text{surface}) = 2V_t \ln\left(\frac{N_s}{n_i}\right). \qquad (5.20)$$

Compare Eq. (5.20) with the corresponding Eq. (5.12) for uniform doped substrate.

2. The minority carrier concentration at the surface is equal to majority carrier density at the boundary of the depletion region [7], [9], [31] that is,

$$\phi_{si} = V_t \ln\left[\frac{N(X_{dm})N_b}{n_i^2}\right] \tag{5.21}$$

where $N(X_{dm})$ is the dopant density at the edge of the depletion region of width X_{dm}. Note that, the condition defined by Eq. (5.21) reduces to Eq. (5.12) when $N(X_{dm}) = N_b$, i.e. when the boundary of the depletion region is located in the uniformly doped part of the profile. In real devices this will be the case for shallow implants or higher values of the back bias.

3. The variation in the inversion and depletion charge densities Q_i and Q_b, respectively, with respect to the surface potential ϕ_s are equal [34]–[36], that is,

$$\frac{dQ_i}{d\phi_s} = \frac{dQ_b}{d\phi_s}.$$

This criterion is equivalent to

$$\phi_{si} = V_t \ln\left(\frac{\bar{N}N_b}{n_i^2}\right) \tag{5.22}$$

where $\bar{N}$ is the average concentration given by

$$\bar{N} = \frac{1}{X_{dm}}\int_0^{X_{dm}} N(x)dx.$$

It can easily be seen that this criterion is equivalent to the classical criterion for uniformly doped substrate.

Again, these different criteria result in slightly different values of threshold voltages. In fact, the above three criteria lead to threshold voltages that are about 0.2 V apart. A detailed comparison of the threshold voltage shift as a function of implant dose for boron implanted MOS structures, based on both 2-D numerical solution and depletion approximation, has been studied by Demoulin and Van De Wiele [2]. It has been found that agreement between the criterion 3 and experimentally measured V_{th} is fairly good, while the classical condition 1 is not valid for heavy implant doses. *In spite of this inadequacy of the classical criterion* [cf. Eq. (5.20)], *it is still used for circuit models because of its simplicity.*

Bulk Charge Q_b. Under the depletion approximation, the bulk charge Q_b for implanted channels can be obtained from the following equation[4] [cf. Eq. (4.29)]

$$Q_b = -q \int_0^{X_{dm}} N(x)dx. \tag{5.23}$$

Therefore, Eq. (5.14) for implanted devices becomes

$$V_{th} = V_{fb} + \phi_{si} + \frac{q}{C_{ox}} \int_0^{X_{dm}} N(x)dx. \tag{5.24}$$

Assuming that the implanted profile is Gaussian as given by Eq. (5.19), many authors [9], [27] have calculated Q_b using Eq. (5.23). The resulting expression for Q_b is fairly complex, involving error functions. These expressions have predictive capabilities so that, for example, one can know how the change in the implant dose D_i will affect the bulk charge and hence threshold voltage. However, such complex models are not suitable for use in circuit simulators. For this reason they are not discussed here and details of the equations for Q_b and V_{th} are left to the interested reader.

The fact that the threshold voltage is determined by the integral of the doping profile rather than by its actual shape, and the desire to get tractable equations for V_{th} have led to the replacement of the exact profiles by idealized step profiles of concentration N_s and width X_i, as shown by dotted lines in Figure 5.8a, such that

$$(N_s - N_b)X_i = \int_0^{\infty} [N(x) - N_b]dx = D_i. \tag{5.25}$$

We choose N_s and X_i such that the total charge under the exact profile is the same as that under the step profile. Rearranging Eq. (5.25) yields the following expression for the surface concentration N_s of the step profile

$$N_s = \frac{D_i}{X_i} + N_b \quad (\text{cm}^{-3}). \tag{5.26}$$

Although one can express N_s and X_i in terms of implant parameters

---

[4] Equation (5.23) assumes that quasi-neutrality holds at every point outside the depletion region of width X_{dm}. This in general is not true and concentration gradient causes a built-in field which has to be taken into account when integrating Poisson's equation. Therefore, strictly speaking Eq. (5.23) needs to be modified as [9]

$$Q_b = -q \int_0^{x_{dm}} N(x)dx - \epsilon_0 \epsilon_{si} \mathscr{E}(X_{dm})$$

where $\mathscr{E}(X_{dm})$ is the electric field at the boundary X_{dm} of the depletion region.

(R_p, ΔR_p and D_i as given in Eq. (5.19)) [29], for circuit models it is more appropriate to use N_s and X_i as model parameters. These parameters are then chosen to make the resulting threshold voltage model match the experimental data.

Shallow Implant Model. In many devices, a very shallow implant is used to modify V_{th}. The limiting case would be an infinitely thin sheet, approximately a delta function, of ionized charge qD_i localized at the Si–SiO$_2$ interface. This is equivalent to modifying the flat band voltage by an amount qD_i/C_{ox} resulting in the following equation for V_{th} [8]

$$V_{th} = V_{fb} + 2\phi_f + \frac{qD_i}{C_{ox}} + \gamma\sqrt{2\phi_f + V_{sb}} \quad \text{(shallow implant).} \tag{5.27}$$

Thus, a shallow implant increases V_{th} without increasing the depletion width X_{dm}.

Deep Implant Model. The threshold model described by Eq. (5.27) is fairly good for shallow channel implants. However, the model becomes inaccurate when the implant becomes deep. In such cases the channel doping profile is often replaced by an idealized step profile (see Figure 5.10a). Depending upon the depth of the channel depletion width X_{dm}, in relation to the depth X_i of the step profile, two cases will arise:

Case I. When the back gate bias V_{sb} is such that the depletion depth X_{dm} is less than the depth of the implant X_i (i.e. $X_{dm} < X_i$), the surface can be considered to be uniformly doped with concentration N_s given by Eq. (5.26). In this case V_{th} is obtained simply by replacing N_b in Eq. (5.14) with N_s, that is,

$$\boxed{V_{th1} = V_{fb} + \phi_{si} + \gamma_1\sqrt{\phi_{si} + V_{sb}}} \tag{5.28}$$

where ϕ_{si} is given by equation Eq. (5.20) and

$$\gamma_1 = \frac{\sqrt{2\epsilon_0\epsilon_{si}qN_s}}{C_{ox}}. \tag{5.29}$$

In fact, for low values of V_{sb} (0–1 V), the slope of the V_{th} versus V_{sb} curve could be used to calculate N_s.

Case II. When V_{sb} is such that X_{dm} lies outside X_i (i.e. $X_{dm} > X_i$), V_{th} is no longer given by Eq. (5.28) because X_{dm} has now to be determined from the high-low step doping profile. In this case the bulk charge Q_b is given by (see Figure 5.10a, shaded area)

$$-Q_b = qN_sX_i + qN_b(X_{dm} - X_i). \tag{5.30}$$

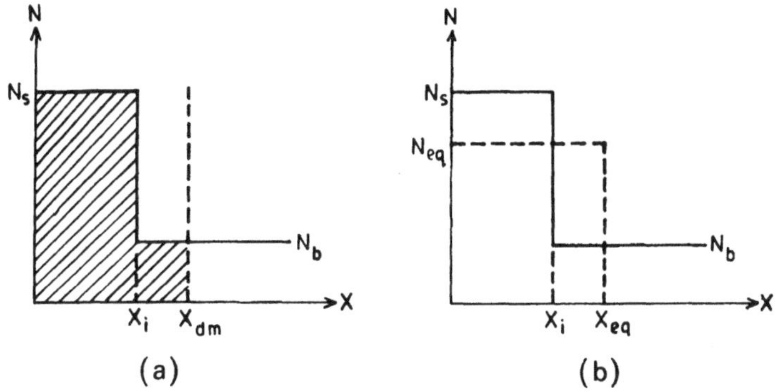

Fig. 5.10 (a) Step doping profile for an n-channel MOSFET, (b) Doping transformation procedure for calculating the equivalent concentration N_{eq} and width X_{eq} of the transformed box

Thus, Q_b can be determined provided X_{dm} is known. The latter can be obtained by solving the Poisson's equation (2.41) under the depletion approximation in the two regions subject to the following doping distribution:

$$N(x)=\begin{cases} N_s & \text{for } x \le X_i \\ N_b & \text{for } x > X_i \end{cases} \tag{5.31}$$

and satisfying the following two boundary conditions:

- the electric field $\mathscr{E}(x)$ is continuous at $x = X_i$,
- the field $\mathscr{E}(x) = 0$ at $x = X_{dm}$.

This yields, after some algebraic manipulation,

$$X_{dm} = \sqrt{\frac{2\epsilon_0\epsilon_{si}}{qN_b}}\left[\phi_{si} + V_{sb} - \frac{qX_i^2}{2\epsilon_0\epsilon_{si}}(N_s - N_b)\right]^{1/2}. \tag{5.32}$$

Combining Eqs. (5.30) and (5.32) and using the resulting value of Q_b in Eq. (5.7) yields the following expression for the threshold voltage[5]

$$\boxed{V_{th2} = V_{fb} + \phi_{si} + \frac{q(N_s - N_b)X_i}{C_{ox}} + \gamma\sqrt{\phi_{si} + V_{sb} - V_0}} \tag{5.33}$$

[5] Often Eq. (5.33) is written in terms of dose D_i as

$$V_{th2} = V_{fb} + \phi_{si} + \frac{qD_i}{C_{ox}} + \gamma\sqrt{\phi_{si} + V_{sb} - \frac{qX_i}{2\epsilon_0\epsilon_{si}}D_i}$$

where we have made use of Eq. (5.25). Compare this with Eq. (5.27) for shallow implants.

where

$$V_0 = \frac{qX_i^2}{2\epsilon_0\epsilon_{si}}(N_s - N_b) \tag{5.34}$$

and γ is given by Eq. (5.29) with N_s replaced by N_b [cf. Eq. (5.11)]. Note that Eq. (5.33) has the same functional dependence on V_{sb} as Eq. (5.28) and the two become the same for uniformly doped substrate ($N_s = N_b$). Thus, V_{th} of an *implanted device could be modeled using* Eqs. (5.28) and (5.33) depending upon the substrate bias.

This model is often referred to as the two sections model. In practice, in order to implement this two sections V_{th} model [Eqs. (5.28) and (5.33)], we normally define a potential ϕ_i such that it results in a depletion width X_{dm} which is exactly equal to the depth of the implant X_i, that is,

$$\phi_i = \frac{qN_sX_i^2}{2\epsilon_0\epsilon_{si}}, \tag{5.35}$$

called the *critical voltage*. In fact ϕ_i is the intersection of slopes I and II (see Figure 5.9), and is a function of the ion implant parameters and the surface concentration. In terms of ϕ_i, Eq. (5.32) for X_{dm} could be written as

$$X_{dm} = X_i\sqrt{1 + \frac{2\phi_0\epsilon_{si}}{qN_bX_i}(\phi_s - \phi_i)} \tag{5.36}$$

where

$$\phi_s = \phi_{si} + V_{sb}. \tag{5.37}$$

When $\phi_i \leq \phi_s$ we use Eq. (5.28) for V_{th} while when $\phi_i > \phi_s$, we use Eq. (5.33). It should be pointed out that for two sections models, not only must V_{th} be continuous at the boundary but its first derivative must also be continuous, a convergence requirement for the model to be used in a circuit simulator as discussed in Chapter 1.

Doping Transformation Model. Very often in VLSI devices, we need deep channel implants such that the resulting implant depth X_i is comparable to depletion region depth X_{dm} in the back bias range of interest. In such cases the two-sections V_{th} model [cf. Eqs. (5.28) and (5.33)] becomes inaccurate for $V_{sb} > 1$ V. Accurate results have been obtained using a method called the *doping transformation* procedure [11], [13], [30]. In this method we transform the doping (actual or step) profile into another step profile of equivalent doping concentration N_{eq} and width X_{eq} (see Figure 5.10b). While the method of calculating N_{eq} proposed by Ratnam and Salama [13] is an improvement over that of Chatterjee et al. [11], it has the drawback that the doping transformation procedure must be done for every different channel length device fabricated by the same channel implant.

Another procedure which is device independent and is applicable for step profiles was proposed by Arora [30] and is based on the following conditions:

1. the total induced charge Q_s under the channel is conserved, and
2. the surface potential ϕ_s is constant.

If X_{eq} is the width of the new transformed profile of concentration N_{eq} as shown in Figure 5.10b, then condition (1) leads to the following equation

$$qN_{eq}X_{eq} = qN_sX_i + qN_b(X_{dm} - X_i), \tag{5.38}$$

while condition (2) leads to

$$X_{eq} = \sqrt{\frac{2\epsilon_{si}\phi_s}{qN_{eq}}} \tag{5.39}$$

where ϕ_s is given by Eq. (5.37). Solving Eqs. (5.38) and (5.39) for N_{eq} and using Eq. (5.36) for X_{dm} yields

$$N_{eq} = N_s\frac{\phi_i}{\phi_s}\left[1 - \frac{N_b}{N_s} + \frac{N_b}{N_s}\sqrt{1 + \frac{N_s}{N_b}\left(\frac{\phi_s}{\phi_i} - 1\right)}\right]^2 \tag{5.40}$$

where ϕ_i is given by Eq. (5.35). This value of N_{eq} is used for N_s in Eq. (5.29) for the body factor term γ_1 when $\phi_s > \phi_i$. We thus see that in this procedure N_{eq} becomes a function of back bias, and therefore γ is no longer a constant but is bias dependent.

For a uniformly doped substrate N_s equals N_b, and therefore Eq. (5.40) gives $N_{eq} = N_b$. Thus, using either N_s (when $\phi_i \le \phi_s$) or N_{eq} (when $\phi_i > \phi_s$) in Eq. (5.29), one can calculate the threshold voltage V_{th} from Eq. (5.28) for a large geometry enhancement MOSFET having a nonuniformly doped substrate. *This procedure of calculating V_{th} has been found to work well with different generations of VLSI technologies* [30], [59]. This doping transformation model is also a two-sections model, since one needs to use either N_s or N_{eq} depending upon values of V_{sb}. The calculated threshold voltages (continuous lines) as a function of back bias shown in Figure (5.4) are based on Eq. (5.40).

Compensated Devices. The threshold voltage models developed so far assumed that the channel implant is of the same type as that of the substrate. Although, the model equations were developed for n-channel devices, these models are also valid for p-channel devices with p^+ polysilicon gate and with appropriate sign changes (see Table 5.1). However, as was pointed out earlier, p-channel CMOS devices with n^+ polysilicon gate need shallow channel implant of the type opposite to that of the substrate or well (which is n-type). Therefore, strictly speaking, the model developed earlier for n-channel implanted devices are not valid for p-channel compensated devices. Since these p-devices are normally-off at $V_{gs} = 0\,\text{V}$, the shallow

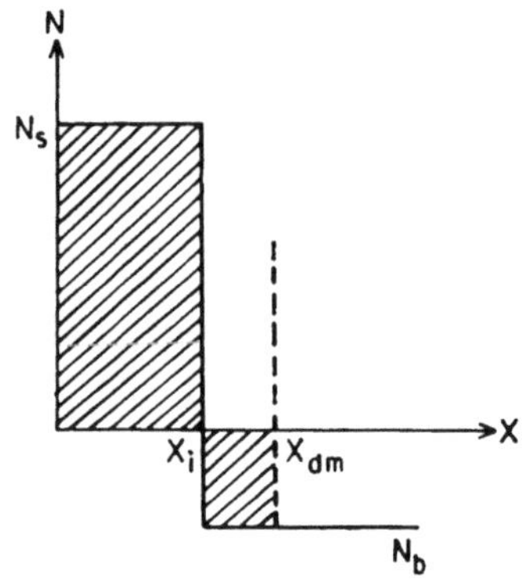

Fig. 5.11 Step doping profile for a compensated p-channel MOSFET

implanted layer is completely depleted and therefore, a sufficiently negative voltage is required for an inversion layer to form. Again, approximating the actual doping profile by a step profile of concentration N_s and width X_i (see Figure 5.8b), the bulk charge Q_b is given by (see shaded area in Figure 5.11)

$$Q_b = qN_b(X_{dm} - X_i) - qN_sX_i. \tag{5.41}$$

The channel depletion width X_{dm} can be obtained as usual by solving Poisson's equation under the boundary conditions given by Eq. (5.31) resulting in the following expression

$$X_{dm} = \sqrt{\frac{2\epsilon_0\epsilon_{si}}{qN_b}}\left[\phi_{si} - V_{sb} + \frac{qX_i^2}{2\epsilon_0\epsilon_{si}}(N_s + N_b)\right]^{1/2}. \tag{5.42}$$

Combining Eqs. (5.41) and (5.42) and using the resulting equation for Q_b in Eq. (5.7), with appropriate sign changes, yields the following equation for p-channel V_{th}[6] [15–16]

$$\boxed{V_{th} = V_{fb} - \phi_{si} + \frac{q(N_s + N_b)X_i}{C_{ox}} - \gamma\sqrt{\phi_{si} - V_{sb} + V_0} \quad \text{(V)}} \tag{5.43}$$

where

$$V_0 = (N_s + N_b)\frac{qX_i^2}{2\epsilon_0\epsilon_{si}} \tag{5.44}$$

---

[6] Note that when $N_sX_i = N_b(X_{dm} - X_i)$, the depletion charge at the surface (p-type) just balances the depletion charge in the substrate (n-type). Under this so called compensation condition, $V_{th} = V_{fb} - \phi_{si} \equiv V_{thc}$. When $V_{th} < V_{thc}$ we have a surface channel device, however, when $V_{th} > V_{thc}$ we have a buried channel device. For n-well CMOS p-channel devices, $V_{thc} \sim -1.0$ V.

and γ is given by Eq. (5.29) with N_s replaced by N_b. Note that for p-channel devices, V_{sb} is negative and the dose $D_i = (N_s + N_b)X_i$. Note also that Eq. (5.43) is similar to Eq. (5.33) for n-channel devices except that the term V_0 is now added to ϕ_{si} term. It should be pointed out that for compensated devices, N_s and N_b are usually of the same order of magnitude which yields $V_0 \sim 0.1$ V. Therefore, to a first order, one can still use Eq. (5.14) for V_{th} of p-channel compensated devices with appropriate sign changes. Due to the positive value of V_0 the back bias dependence of V_{th} for compensated devices is smaller compared to n-channel devices with n^+ polysilicon gates or p-channel with p^+ polysilicon gates.

Empirical Models. Various empirical approaches have been suggested to model V_{th} for implanted devices [37]–[39]. Note from Figure 5.9 that for channel implanted devices the slope of the V_{th} curve decreases as back bias increases. This change in slope can be accounted for in the V_{th} expression (5.14) with replacing the voltage V' corresponding to the depletion charge Q_b [cf. Eq. (5.7)] with a polynomial of the form [37]

$$V' = \sum_{k=1}^{n} \gamma_k (2\phi_f + V_{sb})^{k/2}. \tag{5.45}$$

In practice, it is quite sufficient to add only one more term to the classical body factor term so that Eq. (5.14) for implanted devices become

$$\boxed{V_{th} = V_{fb} + 2\phi_f + \gamma\sqrt{2\phi_f + V_{sb}} + \gamma_0(2\phi_f + V_{sb}).} \tag{5.46}$$

The parameter γ_0 adjusts the body-effect relationship for nonuniformity of the doping. In general, γ_0 will be a negative number. This is because in general the doping concentration decreases as we move away from the surface into the bulk (see Fig. 5.8) and thus offsets an initially high value of γ. Note that ϕ_f in Eq. (5.46) is now determined using Eq. (5.20) with N_s replaced by some average value of the substrate concentration N_{avg}. It is N_{avg} which in turn is used to calculate γ from Eq. (5.11). The N_{avg} and γ_0 are normally obtained by curve fitting Eq. (5.46) to the experimental data. Equation (5.46) for V_{th} is used in the SPICE Level 4 MOSFET model (BSIM model) [40]. Comparing Eq. (5.46) with Eq. (5.14), it is easy to see that the modified γ for implanted devices becomes

$$\gamma_{im} = \gamma + \gamma_0\sqrt{2\phi_f + V_{sb}} \tag{5.47}$$

and is bias dependent, similar to the doping transformation model.
In another approach, threshold behavior of implanted devices has been modeled by the following relationship [39]

$$V_{th} = V_{fb} + 2\phi_f + \gamma\sqrt{2\phi_f} - \frac{G_{11}V_{bs}}{1 - G_{12}V_{bs}} \tag{5.48}$$

where G_{l1} and G_{l2} are fitting parameters and are obtained by curve fitting the experimental data with Eq. (5.48).
The advantage of using empirical relations in V_{th} models is that they can be used for both *p*- and *n*-channel devices. Note that not all V_{th} models discussed above will work for a given technology, because of the semi-empirical nature of these models. It has been found that the doping transformation procedure of modeling *n*-channel threshold voltage works very well for present day MOS technologies. On the other hand Eq. (5.46) seems to work well for *p*-channel devices.

5.2.2 Depletion Type Device

As was pointed out earlier, depletion type MOSFETs (normally-on BC MOSFETs) conduct even at $V_{gs} = 0$ V. A cross-section of an *n*-channel depletion mode MOSFET is shown schematically in Figure 5.12. When $V_{gs} < V_{fb}$, a surface space charge region develops under the gate in the

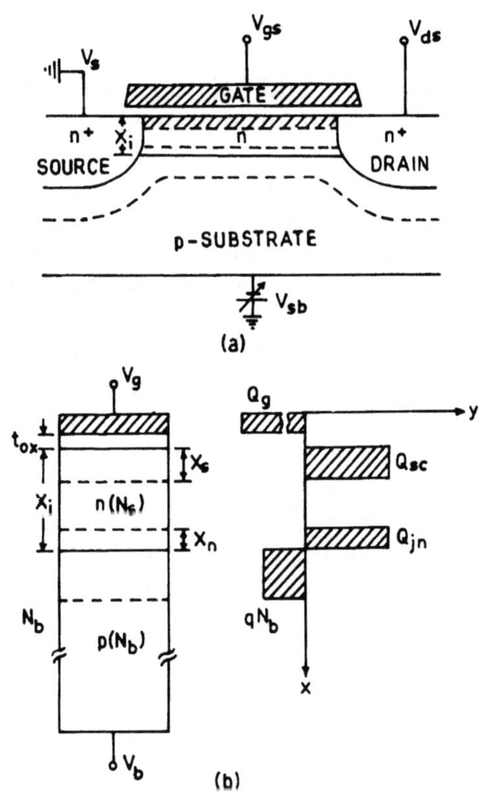

Fig. 5.12 (a) Cross-section of an *n*-channel depletion type device and (b) its charge distribution

channel region. The depletion width X_s of this surface space charge region is due to the combined effects of the gate voltage V_{gs} and channel voltage $V_{cb}(y)$ [cf. Eq. (5.2)]. Another space charge region is formed along the channel and substrate *pn* junction. The depletion width of this *pn* junction in the channel region is controlled by the channel voltage V_{cb}. *A conducting channel is thus formed between the boundaries of the two space charge regions.* With decreasing values of the gate voltages (more negative V_{gs}), the surface depletion region penetrates deeper into the channel until the depleted region at the surface reaches the depleted region of the *pn* junction. When this happens at the source end of the channel the device is turned off. The gate voltage which "pinches off" the channel is called the *pinch-off voltage* V_p or threshold voltage[7] V_{th}. Under pinch off condition, the surface space charge Q_{sc} under the gate and the charge Q_{jn} stored in *n*-side of the substrate must balance the charge Q_{im} in the implanted region. That is [17]–[22]

$$-Q_{im} + Q_{jn} + Q_{sc} = 0. \tag{5.49}$$

Under these conditions the charge distribution is shown in Figure 5.12b. Approximating the channel doping profile by a step profile of width X_i and concentration N_s (see Figure 5.8b), the implanted layer charge density Q_{im} can be written as

$$\boxed{Q_{im} = qN_sX_i \quad (\mathrm{C/cm^2}).} \tag{5.50}$$

The *pn* junction space charge density Q_{jn} is given by

$$Q_{jn} = qN_sX_n \quad (\mathrm{C/cm^2}) \tag{5.51}$$

where X_n is the depletion width on the *n*-side of the *pn* junction in the channel region. Recall from *pn* junction theory that X_n under the depletion approximation is [cf. Eq. (2.51)]

$$X_n = \sqrt{\frac{2\epsilon_0\epsilon_{si}}{q}\frac{N_b}{N_s(N_b + N_s)}(\phi_j + V_{sb})} \quad (n\text{-side depletion width}) \tag{5.52}$$

where we have assumed $V_{cb} \approx V_{sb}$ because V_{ds} is small (< 0.1 V), and ϕ_j is the built-in voltage of the *pn* junction in the channel region, given by [cf. Eq. (2.44)]

$$\phi_j = V_t \ln\left(\frac{N_bN_s}{n_i^2}\right). \tag{5.53}$$

[7] For depletion devices, the two terms *threshold voltage* and *pinch-off voltage* are generally used synonymously.

If we define γ_e as the effective body-factor term

$$\gamma_e = \frac{\sqrt{2\epsilon_0 \epsilon_{si} q N_e}}{C_{ox}} \tag{5.54}$$

where

$$N_e = \frac{N_s N_b}{N_s + N_b}$$

then combining Eqs. (5.51), (5.52) and (5.54) we get

$$\boxed{Q_{jn} = \gamma_e C_{ox} \sqrt{\phi_j + V_{sb}} \quad (\mathrm{C/cm^2}).} \tag{5.55}$$

The surface charge density Q_{sc} is given by

$$Q_{sc} = q N_s X_s \tag{5.56}$$

where X_s is the surface depletion width. Recall from the MOS capacitor theory that X_s under the depletion approximation is given by Eq. (4.30). However, in a MOSFET, the effective V_{gb} varies from the source to the drain end. Therefore, to calculate X_s for a MOSFET one should replace V_{gb} in Eq. (4.30) with $V_{gb} - V_{cb}(y) \approx V_{gs}$ (assuming $V_{cb} \approx V_{sb}$) and N_b with N_s. This results in the following expression for X_s,

$$X_s = \frac{\gamma_s C_{ox}}{q N_s}\left(-\frac{\gamma_s}{2} + \sqrt{\frac{\gamma_s^2}{4} + V_{gs} - V_{fb}}\right) \tag{5.57}$$

where γ_s is defined as

$$\gamma_s = \frac{\sqrt{2\epsilon_0 \epsilon_{si} q N_s}}{C_{ox}}.$$

Substituting Eq. (5.57) in (5.56) yields

$$\boxed{Q_{sc} = \gamma_s C_{ox}\left(-\frac{\gamma_s}{2} + \sqrt{\frac{\gamma_s^2}{4} + V_{gs} - V_{fb}}\right).} \tag{5.58}$$

At the pinch-off, i.e. when device is turned off, $V_{gs} = V_{th}$, and $X_i = X_s + X_n$. Thus, combining Eqs. (5.49), (5.50), (5.55) and (5.58) yields, after some algebraic manipulation, the following expression for the threshold voltage of a depletion MOSFET

$$\boxed{V_{th} = V_{fb} - q N_s X_i\left(\frac{1}{C_{ox}} + \frac{X_i}{2\epsilon_0 \epsilon_{si}}\right) - \frac{N_b}{N_b + N_s}(\phi_i + V_{sb}) + \gamma_d \sqrt{\phi_j + V_{sb}}} \tag{5.59}$$

where

$$\gamma_d = \left(\frac{1}{C_{ox}} + \frac{X_i}{\epsilon_0 \epsilon_{si}}\right)\sqrt{2\epsilon_0 \epsilon_{si} q N_e}$$

is the body-factor for depletion devices. For $N_s \gg N_b$, V_{th} can be approximated as

$$V_{th} = V_{fb} - qN_s X_i\left(\frac{1}{C_{ox}} + \frac{1}{2C_i}\right) + \left(\frac{1}{C_{ox}} + \frac{1}{C_i}\right)\sqrt{2\epsilon_0 \epsilon_{si} q N_b(\phi_j + V_{sb})} \tag{5.60}$$

where $C_i = \epsilon_0 \epsilon_{si}/X_i$. It is interesting to note the following

- The threshold voltage equation defined this way has exactly the same form as for an enhancement mode device.
- The body factor for depletion devices is higher than the enhancement devices and depends on the width X_i of the implant.

The threshold voltage Eq. (5.59) is based on approximating the channel doping profile by a step junction. However, it has been suggested that a linearly graded profile would approximate the actual profile more closely, thus resulting in a more accurate threshold voltage model, although at the expense of more complexity of the model [23, 24].

If the substrate doping is high or the ion implanted dose is low (lightly doped layer) the depleted region of the channel *pn* junction on the *n*-side can reach the interface. This of course can happen much more readily when the substrate is reverse biased. Under these conditions, free charge carriers can only be accumulated at the interface (as in the enhancement devices), so that in this case we have

$$Q_{sc} = -C_{ox}(V_{gs} - V_{fb}) \tag{5.61}$$

instead of Eq. (5.58). In this case the V_{th} equation will be different because the gate controlled charge is either a depletion one or an accumulation one.

Another threshold voltage, called the *threshold for inversion* at the source end, is also defined for depletion devices. It is the gate voltage that causes channel surface inversion, denoted by V_{thi}. When inversion occurs at the surface, the surface space charge region X_s attains a maximum value X_{sm} given by

$$X_{sm} = \frac{\gamma_s C_{ox}}{qN_s}\sqrt{V(y)}$$

and results in the following value of V_{thi},

$$V_{thi} = V_{fb} - \gamma_s\sqrt{\phi_j + V_{sb}} - (\phi_j + V_{sb}). \tag{5.62}$$

If $V_{th} > V_{thi}$, then the device cannot be completely turned off, because inversion will occur at the surface first, resulting in a constant drain current. It should be pointed out that in the Berkeley SPICE, depletion mode MOSFETs are treated as enhancement mode devices with a negative threshold voltage corresponding to the charge introduced to form the built-in channel. This zero order model ignores the channel depth and assumes the channel charge to exist as a thin sheet at the Si–SiO_2 interface. If the device is used simply as load then this model is good enough. However, if it is to be used in other applications, then it requires a separate model.
Considering both pMOST and nMOST devices, a general expression for the threshold voltage can be written as

$$\boxed{V_{th} = V_{fb} \pm 2\phi_f + \Delta V_{th} \pm \gamma\sqrt{|2\phi_f \pm V_{sb} \mp V_0|}} \tag{5.63}$$

where the + and − signs are for *n*- and *p*-channel devices respectively, and ΔV_{th} is the threshold voltage shift due to the channel implant of depth X_i. The term $V_0(N_s, N_b, X_i)$ is a correction term due to the threshold voltage implant. For a uniformly doped substrate (unimplanted channels), $\Delta V_{th} = V_0 = 0$. For channel implanted enhancement devices, with a channel implant of the same type as that of the substrate, V_0 has a sign opposite to that of ϕ_{si} $(=2\phi_f)$ for classical criterion). Therefore, $\phi_{si} + V_0$ can approach zero. For depletion devices or unimplanted (uniformly doped) devices, V_0 has a value of zero. For compensated *p*-channel devices with a channel implant of the opposite type to that of the substrate, V_0 has the same sign as ϕ_{si}, therefore, $\phi_{si} + V_0$ may take values in excess of 1 V.

5.3 Threshold Voltage Variations with Device Length and Width

The threshold voltage models presented in the previous sections indicate that V_{th} is independent of the device length L and width W. This is true only for large geometry MOSFETs, but not when L and W become small as is evident from Figure 5.4 which shows different values of V_{th} for different W/L devices from same technology. Experimentally it has been found that when L and W become small, V_{th} changes from its long channel value. This is shown in Figure 5.13 where curve A shows variation of V_{th} with L for a fixed W, while curve B shows variation of V_{th} with W for a fixed L [41]. Clearly *for a fixed* W, V_{th} *decreases with decreasing* L, *while for a fixed* L *decreasing* W *increases* V_{th}. This reduction in V_{th} with decreasing L becomes noticeable when L becomes comparable to X_{sd} and X_{dd} the source and drain depletion widths, respectively. When this happens the MOSFET is considered a *short channel* device. Similarly, when W becomes comparable to X_{dm}, the depletion width in the channel region, then the MOSFET is

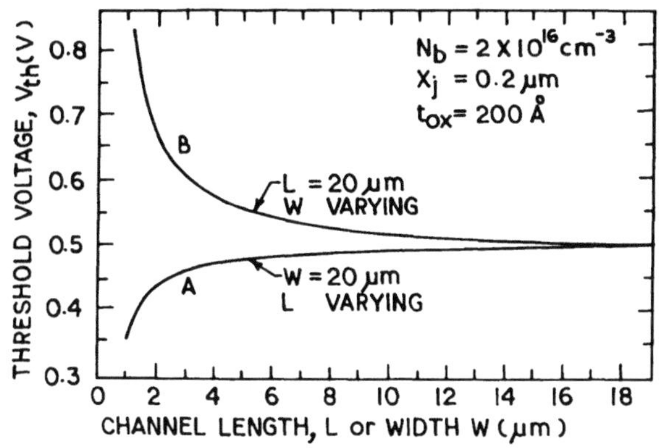

Fig. 5.13 Threshold voltage variation with channel length L (curve A) and width W (curve B) based on 2-D device simulation. (From Akers and Sanchez [41])

called *narrow width device*. Indeed, these variations in V_{th} are not predicted by the model developed in the previous sections.

5.3.1 Short-Channel Effect

Recall that while deriving Eq. (5.9) for Q_b we implicitly assumed that the depletion region due to the gate field was rectangular in shape with charge $|Q_b| = qN_bX_{dm}$. This approximation neglects the charges near the source and drain ends that terminate the built-in field from the source and drain junctions. In fact, the depletion regions in the channel due to the gate overlap with that due to the source/drain junctions. Due to the overlapping of the fields, the effective gate controlled charge Q_b' becomes smaller than Q_b. In other words, as the channel length reduces, the gate controls less charge by an amount $\Delta Q_l (= Q_b - Q_b')$, resulting in a decrease in the V_{th}. Because of the two dimensional nature of the charge and electric field distribution, this decrease in V_{th} (short-channel effect) could best be analyzed by solving a 2-D Poisson's equation either numerically or analytically [46]–[49]. However, *for reasons of simplicity, the most widely used* V_{th} *models for circuit simulators are based on either charge sharing concepts or empirical relationships.*

Charge sharing models account for the reduction in V_{th} through the sharing of the channel depletion region charge between the gate and source-drain junctions. These models assume *a priori* geometrical forms for the source and drain depletion regions and their boundaries. The channel depletion width is then geometrically divided into two parts, one associated with the gate and the other associated with the junctions. It is the gate controlled charge Q_b' which is then used as Q_b in Eq. (5.7). The accuracy of the models obviously is dependent on how Q_b is geometrically divided to get Q_b'. Based on charge division and geometric shapes, various V_{th} models, ranging from simple to more complex models, have been developed. The most simple of many geometrically based models is that of Yau [51], shown in Figure 5.14a, and is based on the following assumptions:

- the substrate is uniformly doped with concentration N_b,
- the source and drain are at zero potential, i.e., $V_{ds}=0$,
- the source/drain junctions (depth X_j) are cylindrical in shape with radius X_j,
- the charges at the source/drain end of the channel are shared equally between the gate and the source/drain junctions resulting in a trapezoidal shape for the gate controlled depletion charge,
- the channel depletion width is equal to that of the source/drain depletion widths, that is, $X_{sd}=X_{dd}=X_{dm}=\sqrt{2\varepsilon_0\varepsilon_{si}(2\phi_f+V_{sb})/qN_b}$ [cf. Eq. (5.11)].

From Figure 5.14a, the gate controlled depletion charge Q_b' is in a trapezoidal area of depth X_{dm}, length L at the surface, and length L' at the bottom of the depletion region, and is given by

$$Q_b' = qN_bX_{dm}\left(\frac{L+L'}{2L}\right) \tag{5.64}$$

where X_{dm} is given by Eq. (5.8). From Figure 5.14b it can easily be seen, using triangle ABC, that

$$X_c = X_j\left(\sqrt{1+\frac{2X_{dm}}{X_j}}-1\right) \tag{5.65}$$

which leads to

$$\frac{L+L'}{2L}=\frac{L+(L-2X_c)}{2L}=1-\frac{X_j}{L}\left(\sqrt{1+\frac{2X_{dm}}{X_j}}-1\right).$$

This equation when combined with Eq. (5.64) yields

$$Q_b' = qN_bX_{dm}\left[1-\frac{X_j}{L}\left(\sqrt{1+\frac{2X_{dm}}{X_j}}-1\right)\right]. \tag{5.66}$$

If we define

$$\boxed{F_l = 1-\frac{X_j}{L}\left(\sqrt{1+\frac{2X_{dm}}{X_j}}-1\right)} \tag{5.67}$$

then Eq. (5.66) reduces to

$$Q_b' = qN_bX_{dm}F_l = \gamma C_{ox}F_l\sqrt{2\phi_f+V_{sb}} \tag{5.68}$$

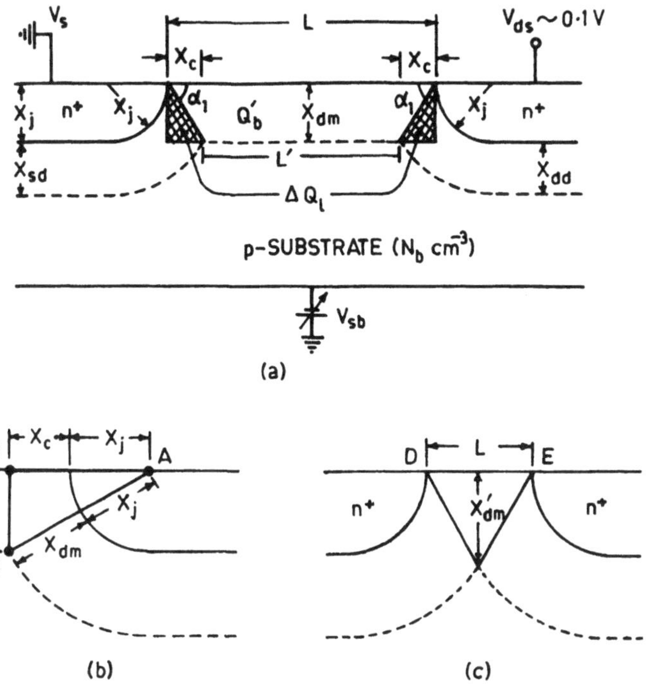

Fig. 5.14 Yau charge sharing model (a) for calculating threshold voltage V_{th} in a short channel MOSFET and (b) calculation of X_c from the triangle ABC, (c) condition when source and drain depletion boundaries meet and depletion width X_{dm} reaches maximum value X'_{dm}

where we have made use of Eqs. (5.8) and (5.11). Now substituting Q'_b for Q_b in Eq. (5.14) yields the following equation for the threshold voltage of short channel MOSFETs

$$\boxed{V_{th} = V_{fb} + 2\phi_f + \gamma F_l \sqrt{2\phi_f + V_{sb}} \quad \text{(V)} \quad \text{(short-channel).}} \tag{5.69}$$

The factor F_l is called the *charge sharing factor*. It is a means of describing the fraction of the total depletion charge in the channel that is terminated on the gate; its value being always less than one. Clearly for long channel devices F_l approaches unity, so that, Q'_b approaches Q_b.

Equation (5.69) remains valid as long as the substrate bias V_{sb} is less than the voltage needed to cause the source and drain depletion regions to meet. As the substrate bias is increased to the point where both regions touch, the charge enclosed is represented by the triangular region shown in Figure 5.14c. If X'_{dm} denotes the channel depth where both the source and drain regions meet, then

$$X'_{dm} = \frac{L}{2X_j}\left[X_j + \frac{L}{4}\right]. \tag{5.70}$$

For $X_{dm} > X'_{dm}$, we assume $X_{dm} = X'_{dm}$. Comparing Eq. (5.69) with (5.14) we get the change $\Delta V_{th,l}$ in V_{th} due to the short-channel effect as

$$\Delta V_{th,l} = V_{th}(\text{long channel}) - V_{th}(\text{short channel}) = \frac{\Delta Q_l}{C_{ox}}$$

$$= \frac{Q_b}{C_{ox}} \frac{X_j}{L}\left(\sqrt{1 + \frac{2X_{dm}}{X_j}} - 1\right). \tag{5.71}$$

This simple model predicts most short-channel effects and the results, in general, are in agreement with the experimental data, although the exact amount of the change in V_{th} may not be represented by Eq. (5.71). For a given channel length, $\Delta V_{th,l}$ depends upon the following device parameters:

- The gate oxide thickness t_{ox}; the higher the t_{ox} (or lower the C_{ox}) the higher the $\Delta V_{th,l}$ and hence the higher the short-channel effect. To reduce the short-channel effect, VLSI/ULSI devices need to have thinner gate oxides.
- The substrate doping concentration N_b; the lower the N_b, the higher the X_{dm}, and therefore the higher is the short-channel effect. This is why (sub)micron devices have higher substrate doping at the surface, obtained using a channel implant.
- The junction depth X_j; the higher the X_j, the higher the short-channel effect.

Dependence of the short-channel effects on process parameters are evident not only from Eq. (5.71) but also from Eq. (3.14). As was pointed out earlier, whether or not a device is short channel depends not so much on the physical mask length of the channel, but rather more on t_{ox}, N_b and X_j. A 4 μm device with higher X_j, higher t_{ox} and/or lower N_b can evidence more severe short-channel effect than a 2 μm device with lower X_j, lower t_{ox} and/or higher N_b. The short-channel effect becomes higher with back bias V_{sb}; the higher the V_{sb}, the higher the X_{dm} thus resulting in an increased short-channel effect.

The simple geometrical approximation of Yau has been modified by many others, resulting in different expressions for F_l. Some of these are reviewed by Akers et al. [41] and Fichtner et al. [52]. For example, Dang [53] assumed the boundary of the space charge shared by the S/D to bulk to take the form of an ellipse with the center at the gate, and axes as follows:

$$\begin{aligned} &\text{minor axis, } 2a_1 = 2(X_l + aX_j) \\ &\text{major axis, } 2b_1 = 2(X_{dm} + X_j) \end{aligned} \tag{5.72}$$

where the factor $a(= 0.6\text{–}0.8)$ is the side diffusion factor. In this case it can easily be seen that the factor F_l is [30], [53]

$$F_l = 1 - \frac{X_c}{L}, \tag{5.73}$$

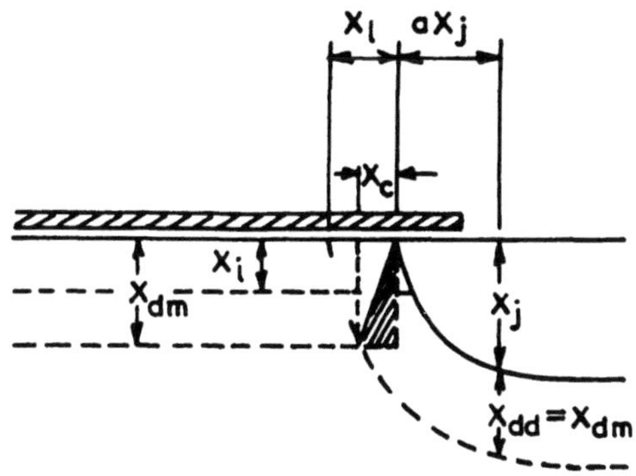

Fig. 5.15 Diagram illustrating the partioning of the depletion charge for calculating charge sharing factor F_l assuming cylindrical junctions

where (see Figure 5.15)[8]

$$X_c = \left[(X_l + aX_j)\sqrt{1 - \frac{X_{dm}^2}{(X_{dm} + X_j)^2}} - aX_j \right], \tag{5.74}$$

and

$$X_l = X_j \left[C_0 + C_1 \left(\frac{X_{dm}}{X_j} \right) + C_2 \left(\frac{X_{dm}}{X_j} \right)^2 \right]. \tag{5.75}$$

Here $C_0(= 0.0631353)$, $C_1(= 0.8013292)$ and $C_2(= -0.01110777)$ are constants that relate the depletion width of a cylindrical junction to that of a planar junction through Eq. (5.75). This model for F_l is used in the SPICE Level 3 MOSFET model.

Using 2-D device simulators, it has been shown that the charge sharing scheme [Eqs. (5.67) or (5.73)] in general, overpredicts the reduction ΔQ_l in the charge Q_b. In order to correct for this overprediction we multiply X_c by a fitting parameter G_l, whose value is less than unity and is technology dependent [30]. Otherwise, one would have to use more complicated expression for F_l, which is not desirable for circuit models.

In order to use Eq. (5.69) for the implanted devices γ will be replaced by γ_{im}. The variation of V_{th} with the channel length L for nMOST, fabricated using an NMOS process, is shown in Figure 5.16a. The corresponding variation for devices fabricated using a CMOS process is shown in Figure 5.16b. Dots are experimental data while continuous lines are calculated based on Eqs. (5.28) and (5.40) with F_l given by Eqs. (5.73)–(5.75). In order to obtain the best fit between the experimental data and calculated V_{th}, a

[8] To arrive at Eq. (5.74), we use the equation of an ellipse $x^2/a_1^2 + y^2/b_1^2 = 1$, where $2a_1$ and $2b_1$ are given by Eq. (5.72).

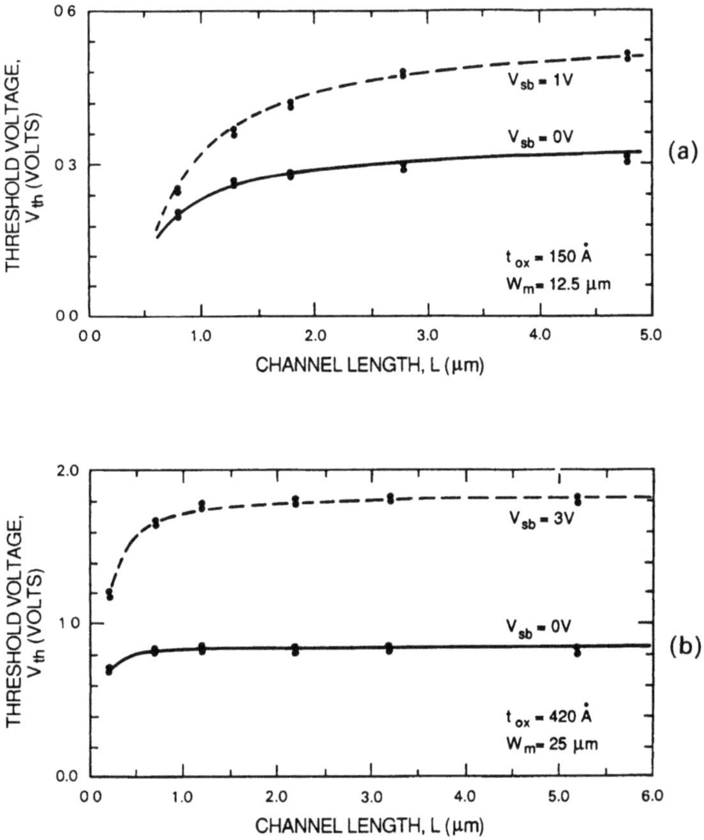

Fig. 5.16 Threshold voltage of n-channel devices as a function of device length at $V_{ds} = 0.1$ V for two different substrate bias V_{sb} for devices fabricated using (a) NMOS process and (b) CMOS process. (After Arora [30])

nonlinear least-square curve fitting routine such as SUXES was used (see Chapter 10).

An approximate expression for Q_b' based on Eq. (5.66), has also been suggested for CAD applications. It is based on the assumption that the angle α_1 (see Figure 5.14a) does not change over the bias range in which the device is used. By fixing $\tan\alpha_1$, which is defined as [37]

$$\tan\alpha_1 = \left[\frac{X_j}{X_{dm}}\left(\sqrt{1+\frac{2X_{dm}}{X_j}}-1\right)\right]^{-1}$$

and substituting it into Eq. (5.71) we get

$$\Delta V_{th,l} = \frac{\Delta Q_l}{C_{ox}} = \frac{Q_b}{C_{ox}}\frac{1}{L}\frac{X_{dm}}{\tan\alpha_1} = \frac{\epsilon_{si}}{\epsilon_{ox}}\frac{1}{L}\frac{t_{ox}}{\tan\alpha_1}(2\phi_f + V_{sb}) = \frac{G_l}{L}(2\phi_f + V_{sb}) \tag{5.76}$$

where $G_l = \epsilon_0 \epsilon_{si}/(C_{ox} \tan \alpha_1)$ is a fitting parameter. Although the assumption that the angle α_1 is constant is not strictly correct, the resulting error in V_{th} for a properly scaled process is very small ($\sim 4\%$), because G_l is insensitive to the angle α_1. However, in a poorly scaled process, if short-channel effects change the V_{th} by more than 30%, this would imply that the initial assumption of charge partitioning illustrated in Figure 5.14 might not be correct. In that case more complicated two-dimensional analysis is required. Comparing Eqs. (5.69) and (5.14), one can easily see that the *effective body factor* for short channel devices reduces to γF_l from its long channel value of γ. This is also evident from Figure 5.4 where the slope γ is smaller for shorter devices. For a given t_{ox}, this reduction in the effective body factor will result in the lowering of effective substrate doping. Thus, *the smaller the channel length, the lower will be the effective substrate doping and hence, the lower the threshold voltage.* Based on this reasoning, the SPICE BSIM model uses the following simple empirical formula for F_l

$$F_l = 1 - \frac{K_l}{L\sqrt{N_b}} \tag{5.77}$$

where K_l is another fitting parameter. Note that in this simple expression for F_l there is no V_{sb} dependence.

Buried Channel Devices. The short-channel effect in buried channel devices is relatively small compared to the enhancement type devices. This is because the *p*-implanted layer (*p*-channel device) under the gate suppresses charge sharing between the gate and the source/drain. The small dependence of V_{th} on L can be understood from Figure 5.17a, where one needs to subtract the hatched depletion charge area from the charge balance equation at the threshold condition. Since this shaded area is relatively small compared to an *n*-channel device, the effect of roll-off due to short-channel effects is small. Using Yau's approach, one can easily calculate the charge sharing factor F_l for compensated *p*-devices as

$$F_l = 1 - \frac{X_c}{L} = 1 - \frac{X_j}{L}\left(\sqrt{\left(1 + \frac{X_{sd}}{X_j}\right)^2 - \left(\frac{X_{dm}}{X_j}\right)^2} - 1\right). \tag{5.78}$$

Comparing this equation with F_l calculated for an enhancement device [cf. Eq. (5.67)] it is easy to see that the correction factor for buried channel devices is small. Basically the same result may be obtained using more complicated expressions for F_l [20]. This can be seen from Figure 5.17b where *p*-channel V_{th} has been plotted against channel length L (continuous lines). For comparison, V_{th} values of *n*-channel devices (dotted lines) from the same CMOS process have also been shown [16]. Remember that for depletion mode devices, X_n (Eq. (5.52), and Figure 5.12) will be changed to $F_l X_n$.

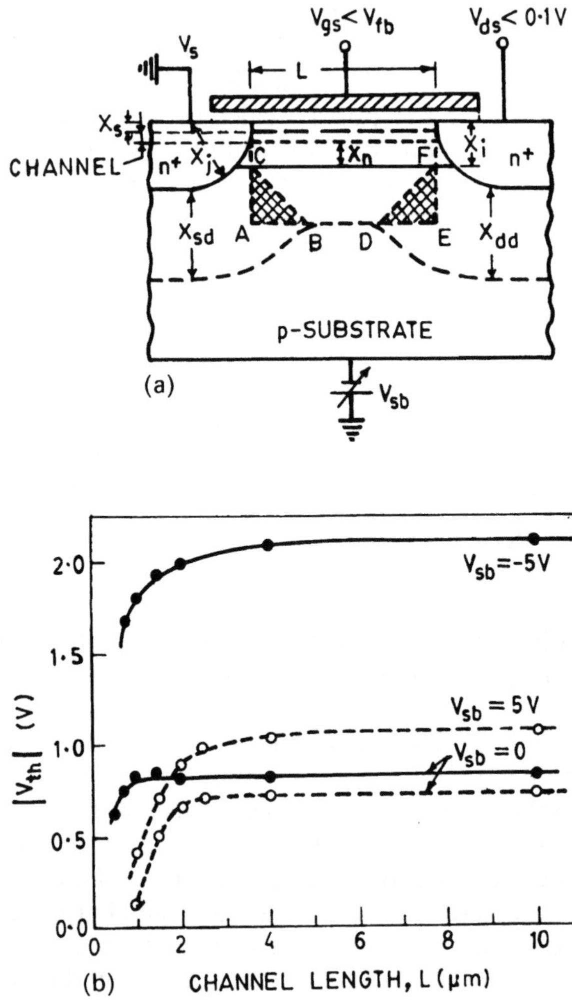

Fig. 5.17 (a) Schematic of a buried channel MOSFET for use in calculation of the charge sharing factor. (b) Threshold voltage as a function of channel length at two values of back bias for compensated *p*-channel (continuous line) and *n*-channel (dotted line) devices. (After Klaassen and Hes [16])

Note that for *all charge sharing models* $\Delta V_{th,l}$ *is proportional to* $1/L$. In terms of ΔQ_l the expression for short channel V_{th} becomes

$$V_{th}(\text{short channel}) = V_{fb} + 2\phi_f + \gamma\sqrt{2\phi_f + V_{sb}} - \frac{\Delta Q_l}{C_{ox}}. \tag{5.79}$$

For implanted devices γ is replaced by γ_{impl}. Although this charge sharing concept does explain the reduction in threshold voltage, a more physical explanation of the short-channel effect is due to the penetration of the junction electric field into the channel region [14], [43]–[50], as we will discuss later in section 5.3.3.

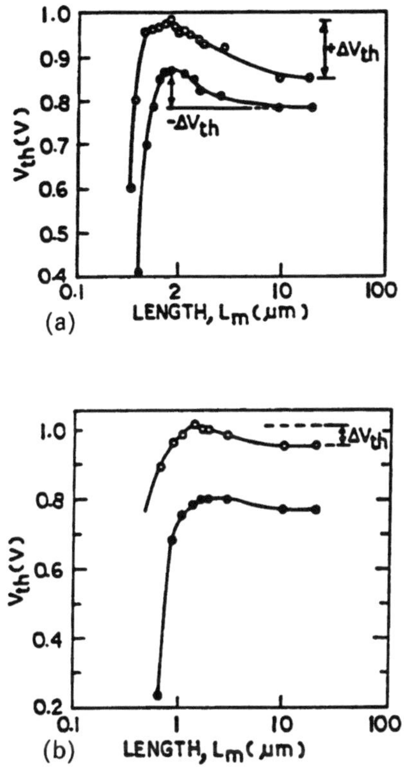

Fig. 5.18 Experimental results for threshold voltage versus channel length showing the threshold hump in CMOS (a) *n*-channel (b) compensated *p*-channel MOSFET. (After Orlowski et al. [56])

Anomalous Threshold Voltage Characteristics. Normally, the threshold voltage decreases monotonically with decreasing channel length. However, in some situations, it has been observed that V_{th} will initially increase with decreasing channel length. After it reaches a maximum value, it will start to decrease at even shorter channel lengths resulting in a "hump" in the threshold voltage versus channel length characteristics. This hump or anomalous behavior has been observed experimentally for both enhancement [56], [59] and buried channel devices [25, 26] as shown in Figure 5.18. It has been found that V_{th} enhancement depends on the energy and dose of the punchthrough implant and reoxidation (REOX) time.[9]

This threshold enhancement has been explained on the basis of nonuniform channel dopant distribution along the channel region. During the gate REOX step oxidation enhanced diffusion (OED) occurs. As a consequence,

[9] As was pointed out earlier, at sub-micron channel lengths high doping concentration in the channel region is required. This is achieved by a double channel implant. After the gate patterning etch a reoxidation (REOX) step follows. This reoxidation step in fact is responsible for the hump in V_{th} [56].

diffusion of the channel dopants is enhanced not only under the oxidizing regions but also in the adjacent channel regions. The result is a laterally nonuniform increase of N_s. The concentration gradient at the surface dN_s/dy is the driving force for the dopant redistribution. The magnitude of the V_{th} enhancement has been modeled by assuming that there is a surface charge density at the source/drain ends, which varies as [59]

$$Q_{fs}(y) = Q_{fs0} \exp\left(-\frac{y}{G_0}\right) \tag{5.80}$$

where Q_{fs0} (C/cm^2) represents the peak charge density at the source/drain ends, y is the distance along the channel and G_0 (cm) is the characteristic length over which this charge distribution is spread. This charge could be generated, during processing, because of the lateral distribution of the impurities in silicon at the surface near the source and drain ends. Integrating this equation from the source ($y = 0$) to the drain end ($y = L$) yields the total surface charge Q_{FS}, so that the shift (increase) in the threshold voltage due to Q_{FS} becomes [59]

$$\Delta V_{th} = \frac{Q_{FS}}{C_{ox}LW} = \frac{2Q_{fs0}G_0}{C_{ox}L}\left[1 - \exp\left(-\frac{L}{G_0}\right)\right] \tag{5.81}$$

where the factor 2 accounts for the charge at both the source and drain ends. This shift in the threshold voltage due to the fixed charge is independent of the back bias, and therefore, can be absorbed in the flat-band voltage V_{fb}. This means that V_{fb} becomes length dependent. Thus, modeling

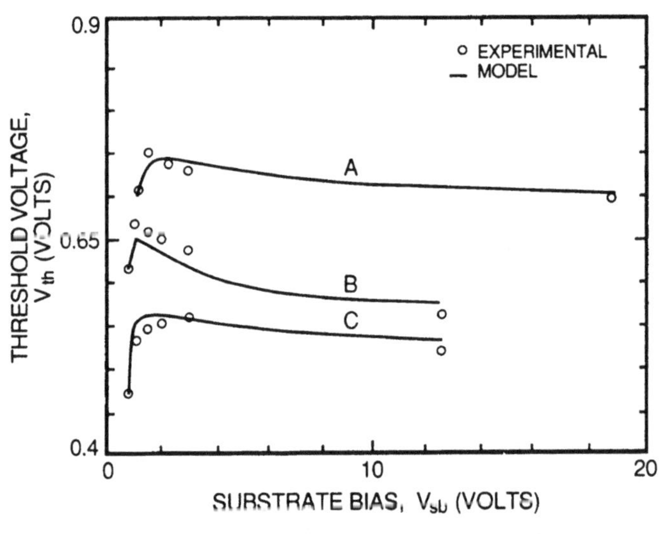

Fig. 5.19 Measured and modeled threshold voltage V_{th} versus channel L showing the threshold hump in n-channel CMOS devices. (After Arora and Sharma [59])

the anomalous V_{th} needs two extra parameters G_0 and Q_{fs0}. Figure 5.19 shows measured (symbols) and calculated (continuous lines) V_{th} based on Eq. (5.81).

For buried channel devices, the enhanced V_{th} has also been explained on the basis of a majority carrier spill over effect [20], [25]. For example, in the case of p-channel compensated devices a high-low p^+p junction is formed between the source/drain and the channel. Therefore, majority carriers from the source/drain tend to diffuse into the channel region resulting in an increase in the carrier concentration. As the carrier concentration increases, V_{th} increases and as the channel length becomes shorter, the ratio of the diffused charge to the implanted charge increases further, thereby increasing the threshold voltage.

5.3.2 Narrow-Width Effect

In a modern MOSFET technology there is a tapering of the oxide from thin to thick oxide as was discussed earlier in section 3.5.4. This results in a gate controlled depletion region at the edges of the device where thin oxide at the center under the gate transitions to the thick field oxide that is part of the isolation between the devices (LOCOS isolation). This oxide transition region is the bird beaks discussed earlier and shown in Figure 5.20. As a result of the gate inducing fringing field around the device edges, there is an extra depletion charge ΔQ_w under the gate with $\Delta Q_w/2$ on each side. This charge ΔQ_w must also be supported by the gate voltage. So long as the device width W is large compared to the gate depletion width X_{dm}, the charge ΔQ_w can be neglected compared to the total depletion charge Q_b. However, when W becomes comparable to X_{dm}, the charge ΔQ_w becomes significant with respect to Q_b. Since this additional charge must also be supported by the gate, it causes an increase in V_{th} by an amount

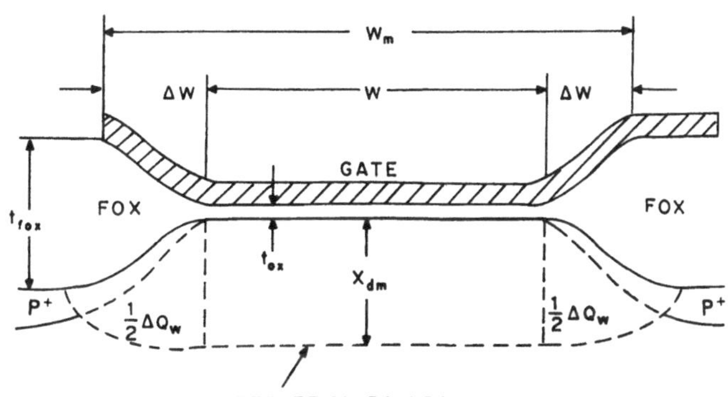

Fig. 5.20 Cross-section of a narrow-width device showing depletion charge ΔQ_w

$\Delta V_{th,w}(=\Delta Q_w/C_{ox})$. Thus, decreasing W increases V_{th}, assuming of course that the device L is large. Similar to the short channel modeling, the threshold voltage for narrow width devices, with large L, becomes

$$V_{th}(\text{narrow-width}) = V_{fb} + 2\phi_f + \gamma\sqrt{2\phi_f + V_{sb}} + \frac{\Delta Q_w}{C_{ox}}. \tag{5.82}$$

Different circuit models have been proposed to calculate the value of ΔQ_w. However, these models differ primarily in their treatment of the transition region from thin to thick oxide and their accounting for the lateral diffusion of the channel stop implant [60]–[64]. One model proposed by Akers et al. [61] is shown in Figure 5.21. It includes the effect of tapered oxide, field dopant encroachment at the channel edges, and gate overlap at the thick recessed field oxide edge, so that the net change in

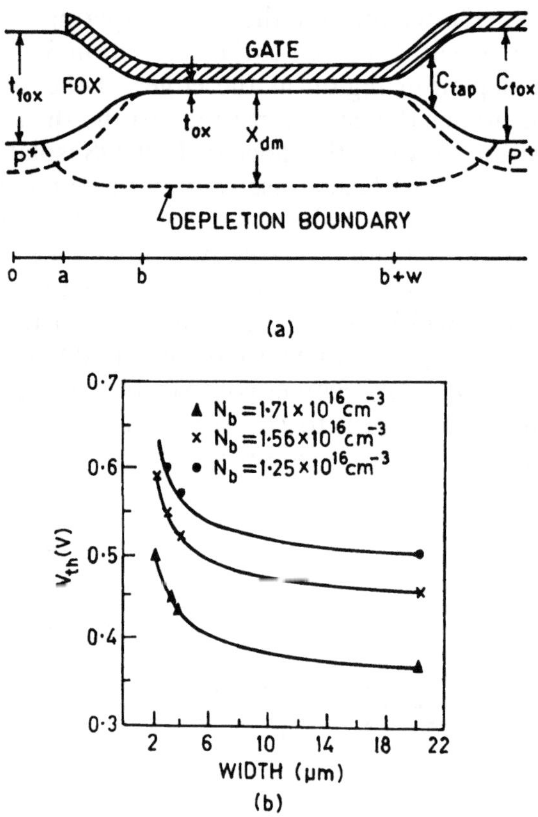

Fig. 5.21 (a) Aker's model for calculating threshold voltage of narrow width MOSFETs fabricated using the LOCOS isolation. (b) Threshold voltage versus width for various channel doping. Continuous lines are model based on Eq. (5.85) while symbols are data points. (After Akers et al. [61])

$V_{th,w}$ due to the narrow-width effect is given by

$$\Delta V_{th,w} = \frac{Q_T}{C_T} - \frac{Q_b}{C_{ox}} \tag{5.83}$$

where Q_T is the total gate controlled charge and is the sum of the charge Q_b in the channel depletion region, the charge Q_{tap} in the tapered region, and the charge Q_{fox} in the field oxide depletion region, i.e.

$$Q_T = Q_b + 2Q_{fox} + 2Q_{tap}. \tag{5.84}$$

Similarly, C_T is determined from the parallel combination of the gate capacitance C_{ox}, the capacitance C_{tap} due to the tapered oxide, and the capacitance C_{fox} due to the overlap of the gate to the field oxide, i.e.

$$C_T = C_{ox} + 2n_1 C_{fox} + 2n_2 C_{tap} \tag{5.85}$$

where the factors n_1 and n_2 account for the edge fringing of the electric field and the potential difference between the surface potential in the channel and the potential under the tapered and thick oxide regions. The oxide transition region is determined by fitting the linear variation over a distance $(b-a)$, which is the width of the tapered region. Figure 5.21b illustrates comparison of the experimental V_{th} due to change in the device width W, for different substrate concentration N_b, and the model Eq. (5.83).
This model gives fairly accurate results, but the final expression for $\Delta V_{th,w}$ is fairly complex. Normally, for circuit models, a more simplified approach is used which assumes a step like transition from the gate to the field oxide resulting in a planer silicon surface. Let A denote the cross-sectional area of the additional bulk charge ΔQ_w (see Figure 5.22) so that

$$\frac{1}{2}\Delta Q_w = qN_b\frac{A}{W} = Q_b\frac{A}{X_{dm}W} \quad (\mathrm{C/cm^2}) \tag{5.86}$$

where Q_b is the bulk charge given by Eq. (5.9). Different approximations have been used to calculate A. If the depletion edge is modeled as a quarter circle arc with radius X_{dm}, then $A = \frac{1}{4}\pi X_{dm}^2$ so that

$$\Delta Q_w = Q_b\frac{\pi X_{dm}}{2W}. \tag{5.87}$$

If a triangular cross-section is assumed, then $A = \frac{1}{2}\cdot(G_w X_{dm})\cdot X_{dm}$ (see Figure 5.22b) so that

$$\Delta Q_w = Q_b\frac{G_w X_{dm}}{W} \tag{5.88}$$

where G_w is a fitting parameter. For circuit models, the above variations

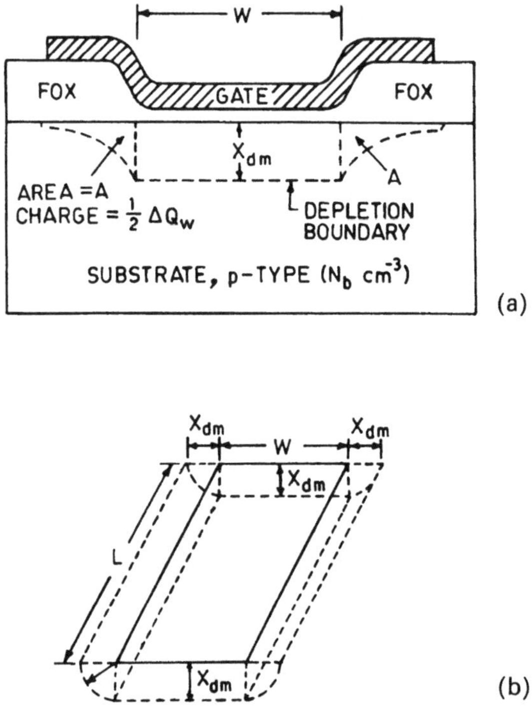

Fig. 5.22 (a) MOSFET structure in width direction assuming step like transition for isolation. (b) 3-D model for V_{th} calculation of the MOSFET shown in (a).

in general can be written as [30]

$$\boxed{\Delta Q_w = \frac{G_w(2\phi_{fld} + V_{sb})}{W}} \tag{5.89}$$

where G_w is a fitting parameter to account for the shape of the transition region and ϕ_{fld} is the surface field potential, given by

$$\phi_{fld} = 2V_t \ln\left(\frac{N_{fld}}{n_i}\right). \tag{5.90}$$

Equation (5.89) is based on the assumption that the field implant concentration N_{fld} is much larger than the channel surface concentration N_s, and there is no short-channel effect. It should be pointed out that taking ϕ_{fld} as a fitting parameter rather than calculating it from (5.90) gives a better fit to the experimental data. The fitting parameter G_w helps in getting accurate V_{th} for channel widths down to 1 μm; its value is always less than unity and is technology dependent [30]. For circuit models N_{fld}

often is assumed to be equal to N_b so that ϕ_{fld} equals $2\phi_f$. The SPICE level 2 model uses a similar form

$$\Delta Q_w = \frac{\pi \epsilon_0 \epsilon_{si}}{W}(2\phi_f + V_{sb}). \tag{5.91}$$

Note that the models that use geometrical shapes for the additional charge ΔQ_w have $\Delta V_{th,w}$ proportional to $1/W$ due to the narrow-width effect.
It is important to note that change in V_{th} due to the width effect is strongly dependent on the device isolation geometry [65]. In the LOCOS method, tapering of the oxide from thin to thick oxide resulted in extra charge ΔQ_w which has to be supported by the gate. This results in an increase in V_{th} with decreasing W. Recently, a new device isolation structure, called the *fully recessed* or *trench isolation* structure, has been developed in which the field oxide is buried in the substrate to a give relatively flat oxide surface (cf. section 3.5.4). In this case, no bird beaks are formed and therefore, this structure results in a high packaging density. Such devices behave somewhat differently than LOCOS isolated devices [66, 67]. It has been found that V_{th} in such devices decreases with decreasing W as shown in Figure 5.23 [66]. This is known as the *inverse width effect* [67].
In the trench isolated (fully recessed isolation oxide) MOSFET structure (see Figure 5.23a), the gate overlaps the field oxide on both sides of the channel. Based on 3-D device simulation, it has been shown that the potential at the edges of the channel are enhanced by fringing gate fields terminating on the side wall of the channel. The contribution of the fringing capacitance C_f is significant as device width is reduced below 3 μm and is given by [67]–[70]

$$C_f = \left(\frac{2\epsilon_0 \epsilon_{ox} L}{\pi}\right) \ln\left(\frac{2t_d}{t_{ox}}\right)$$

where t_d is the depth of the trench oxide (Figure 5.23a). Due to this fringing field the total effective gate oxide capacitance becomes

$$C_{ox}(\text{effective}) = C_{ox} WL\left(1 + \frac{F}{W}\right)$$

resulting in a net gate controlled charge Q_b' as

$$Q_b' = Q_b\left(\frac{W}{W+F}\right) \tag{5.92}$$

where F is fringing factor defined as

$$F = \left(\frac{4t_{ox}}{\pi}\right) \ln\left(\frac{2t_d}{t_{ox}}\right).$$

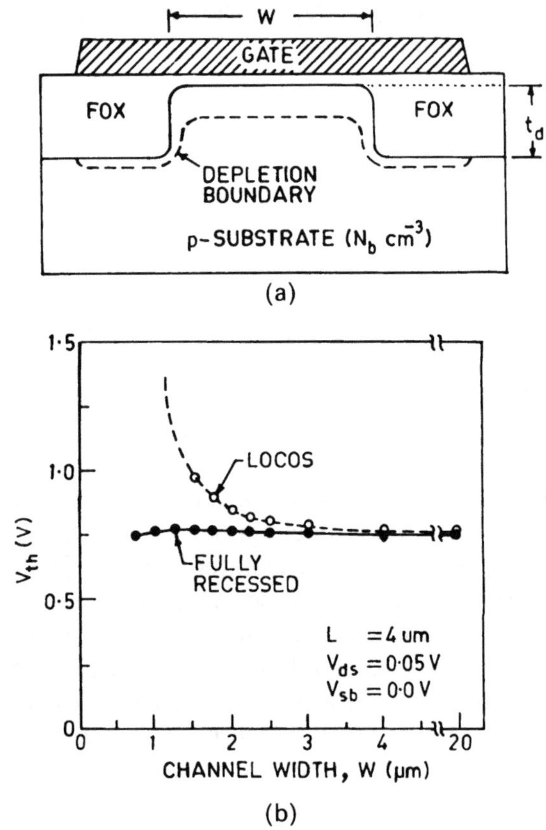

Fig. 5.23 A fully recessed (bird-beak free isolation) MOSFET (a) structure in width direction and (b) V_{th} variation with width W. (After Kurosawa et al. [66])

It is the Q_b' given by Eq. (5.92) that is used for Q_b in Eq. (5.7). Thus, for *nonrecessed oxide V_{th} increases, while for fully recessed oxide V_{th} decreases with decreasing width* (see Figure 5.23b).

5.3.3 Drain Induced Barrier Lowering (DIBL) Effect

The preceding discussion of short-channel and narrow-width effects on the threshold voltage assumed that source to drain voltage V_{ds} is very small (<0.1 V). However, when higher V_{ds} is applied to the device, the channel depletion width X_{dm} is no longer constant along the length of the device as was assumed earlier, but varies from the source to the drain as shown in Figure 5.24. From Eq. (5.8) it follows that for an applied voltage V_{ds}, the channel depletion width becomes a function of y in the form

$$X_{dm}(y) = \sqrt{\frac{2\epsilon_0 \epsilon_{si}}{qN_b}[\phi_{si} + V_{sb} + V(y)]}. \tag{5.93}$$

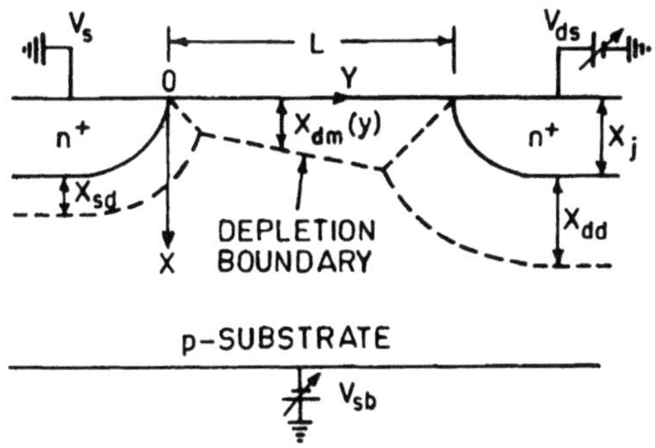

Fig. 5.24 Diagram illustrating charge sharing model with applied drain voltage

This shows that X_{dm} varies along the length of the channel, being maximum at the drain end, where $V(y) = V_{ds}$ and minimum at the source end, where $V(y) = 0$. Since V_{ds} reverse-biases the drain-substrate junction, it increases the depletion charge near the drain end due to the increase in X_{dm}. This further reduces the gate controlled charge thereby lowering V_{th} compared to when $V_{ds} \sim 0$. The decrease in V_{th} due to increasing V_{ds} can be modeled using Yau's approach where X_{dm} at the source end will be X_{sd} while that at the drain end will be X_{dd}, so that

$$F_l = 1 - \left[\frac{X_j}{2L}\left(\sqrt{1 + \frac{2X_{sd}}{X_j}} - 1\right) + \frac{X_j}{2L}\left(\sqrt{1 + \frac{2X_{dd}}{X_j}} - 1\right)\right] \tag{5.94}$$

where X_{sd} and X_{dd} are now given by [cf. Eq. (2.53)]

$$X_{sd} = \sqrt{\frac{2\epsilon_0\epsilon_{si}}{qN_b}(\phi_{bi} + V_{sb})} \tag{5.95a}$$

$$X_{dd} = \sqrt{\frac{2\epsilon_0\epsilon_{si}}{qN_b}(\phi_{bi} + V_{sb} + V_{ds})}. \tag{5.95b}$$

In this case, V_{th} is still given by Eq. (5.69) with F_l replaced by Eq. (5.94). It is through this F_l term that the dependence of V_{th} on V_{ds} is accounted for and is used in the SPICE Level 2 model. This procedure of generalizing Yau's approach imposing no restriction on the possible value of V_{ds} has been used by many others [41], [71]. Due to the simplifying assumptions made in the charge sharing model to account for the drain and back bias dependences, such models often include fitting parameters in order to better fit the experimental data.

The decrease in V_{th} due to an increase in V_{ds} could be looked at in another way. As L is reduced and V_{ds} is increased, the drain depletion region moves closer to the source depletion region, resulting in a significant field penetration from the drain to the source. Due to this field penetration, the

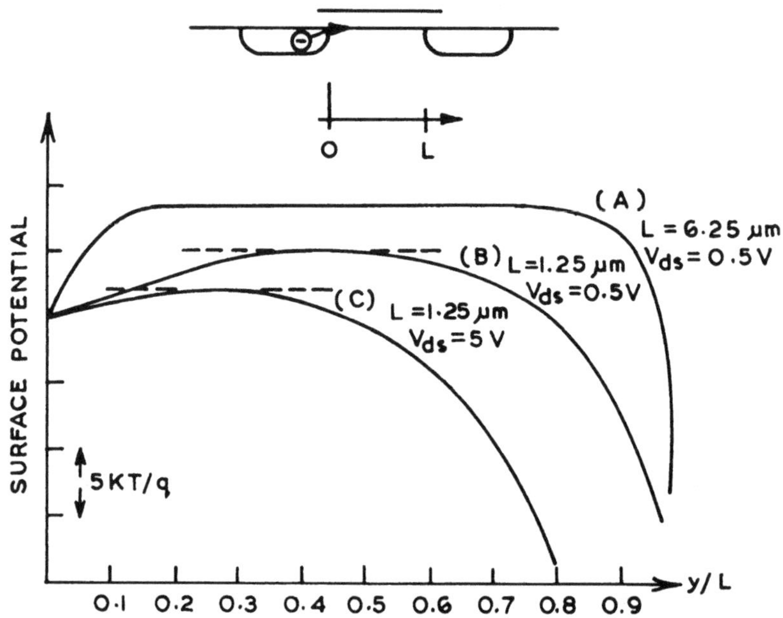

Fig. 5.25 Surface potential distribution for a constant gate voltage. Only channel length and drain voltage are varied. (After Troutman [72])

potential barrier at the source is lowered, resulting in increased injection of electrons by the source over the reduced channel barrier, giving rise to increased drain current. This process is called *drain induced barrier lowering* or simply DIBL [72]–[77].

Figure 5.25 illustrates the surface potential distribution along the channel between the source and drain for long (curve A) and short channel (curve B and C) devices. For long channel devices, the surface potential is constant over most of the channel length. As the channel length is reduced (curve B), keeping all other parameters constant, the peak of the surface potential is reduced and is constant only over a small fraction of the channel length. Since the peak potential is reduced, the current will increase. If V_{ds} is increased (curve C) the peak is further reduced and the region of constant potential is also reduced. From these observations, the short-channel effect has been attributed to the penetration of the junction electric fields into the channel region, causing barrier lowering, which in turn leads to V_{th} reduction. This reduction in V_{th} depends linearly on V_{ds} (see Fig. 5.26), so that

$$V_{th}(V_{ds}) \equiv V_{thd} = V_{th} - \sigma V_{ds} \tag{5.96}$$

where V_{th} is threshold voltage at low V_{ds} (< 0.1 V) as discussed in previous sections and σ is called the DIBL parameter. *It is interesting to note that since the drain modulates the potential in the channel, it is sometimes called*

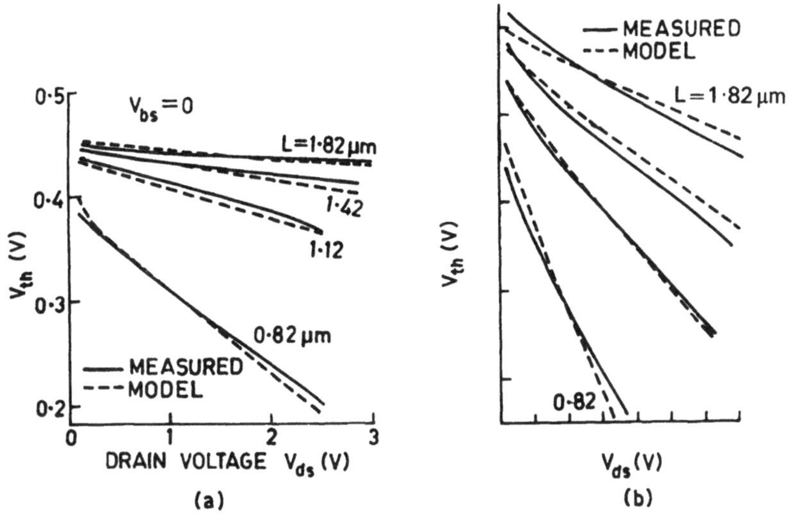

Fig. 5.26 Threshold voltage versus drain bias at different back bias (a) $V_{sb}=0$ and (b) $V_{sb}=-2\,\text{V}$ with channel length as a parameter. Measurements were taken on implanted devices. Parameters of doping box approximation are: $N_s=1.6\cdot 10^{16}\,\text{cm}^{-3}$, $N_b=4\cdot 10^{14}\,\text{cm}^{-3}$ and $X_i=0.29\,\mu\text{m}$. Other parameters are: $t_{ox}=250\,\text{Å}$, $X_j=0.27\,\mu\text{m}$, $V_{fb}=-0.706$ and $\eta=0.942$. (After Skotnicki et al. [50])

a second gate and σ is called the coefficient of static feedback. DIBL is a strong effect for short channel devices operated near threshold. For such devices operating in saturation, the DIBL parameter σ is the principal factor determining the output conductance.

Based on 2-D device simulations it has been found that the degree of lateral field penetration or the DIBL, which results in reduction in V_{th}, depends on the following factors:

- The channel length L; the lower the L, the higher the DIBL, i.e., lower V_{thd}. This is evident from Figure 5.26, which is a plot of experimental V_{th} as a function of V_{ds} for different channel lengths [72].
- The gate oxide thickness t_{ox}; the higher the t_{ox}, the higher the DIBL. This can be seen from Figure 5.27 which shows that as drain voltage is increased from 0.1 to 5 V and channel length is reduced the DIBL effect increases ($-\Delta V_{th}$ reduces) unless t_{ox} is decreased. In general this effect adds about 0.1 V to V_{th} [73].
- The source/drain junction depth X_j; the higher the X_j, the higher the DIBL. Simulated results showing V_{th} variation with two X_j are shown in Figure 5.28.
- The channel doping concentration N_b; the higher the N_b, the higher the DIBL.
- The back gate bias V_{sb}; the higher the V_{sb}, the lower the DIBL effect (see Figure 5.26b).

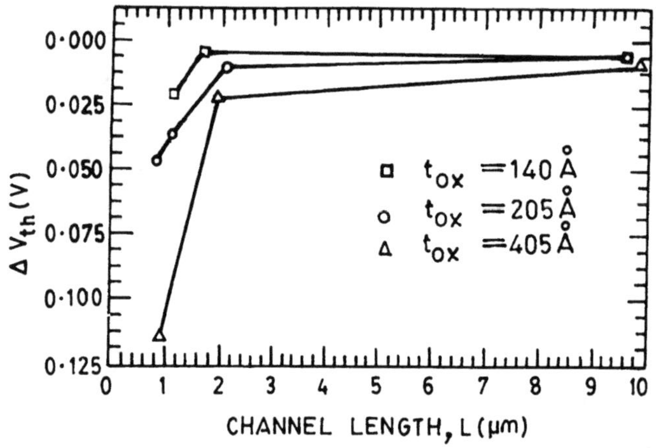

Fig. 5.27 Measured reduction of the threshold voltage versus channel length with oxide thickness as a parameter at $V_{ds} = 5$ V compared to $V_{ds} = 0.1$ V. (After Sodini et al. [73])

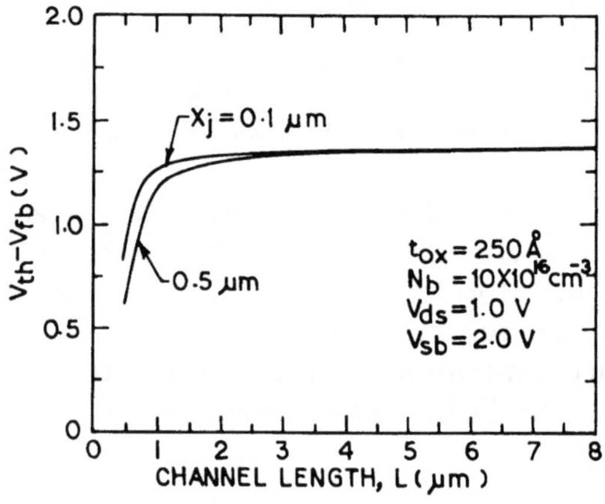

Fig. 5.28 Simulated results for threshold voltage versus channel length for two different junction depths X_j

Note that the above are the same parameters on which the short channel V_{th}, at low V_{ds}, depends (cf. section 5.3.1). Thus, higher V_{ds} increases the short-channel effects due to DIBL effect.

Based on varying surface potential along the channel, various authors have calculated DIBL parameter σ using either a charge sharing model [78] or solving analytically 2-D Poisson's equation [43]–[49]. In their *2-D solution of Poisson's equation different authors have made different assumptions, but all these results show an exponential dependence of V_{th} on channel length.* Specifically, the change from the long channel V_{th} value is proportional to $\exp(-L/L_0)$, L_0 being the long channel length. For example, in one approach by Toyabe and Asai [43] and Wu et al. [44], the silicon potential ϕ in the x direction (depth) is assumed to be a cubic function as

$$\phi(x, y) = a_0 + a_1 x + a_2 x^2 + a_3 x^3 \tag{5.97}$$

where the coefficients a_0, a_1 etc. (which may be functions of y) are found from the assumed boundary conditions for the potential and electric field at the silicon surface and depletion layer edge:

$$\phi(x, y) = \phi_s(y) \quad \text{and} \quad \frac{d\phi}{dx} = \frac{\epsilon_{ox}}{\epsilon_{si}} \frac{\phi_s(y) - V_{gs}}{t_{ox}} \quad \text{at } x = 0 \text{ (silicon surface)}$$

$$\phi(x, y) = V_{sb} \quad \text{and} \quad \frac{d\phi}{dx} = 0 \quad \text{at } x = X_{dm} \text{ (depletion edge).} \tag{5.98}$$

The assumed potential distribution ϕ [cf. (Eq. (5.97] then allows the Poisson equation for ϕ_s to be reduced to a differential equation in one variable (y), which is then solved using the above boundary conditions. The final equation for V_{th} becomes fairly complex. In other approaches such as that of Ratnakumar and Meindel [45], the 2-D nature of the Poisson equation is explicitly maintained. By solving the Poisson equation in terms of a Fourier series, it has been shown that for short channel devices the V_{th} is lowered by an amount

$$\Delta V_{th} = \frac{6t_{ox}}{d_1}[2(\phi_{bi} + V_{sb}) + V_{ds}]\exp\left(-\frac{\pi L}{4d_1}\right) \tag{5.99}$$

where d_1 is the depth of the equivalent box profile. Although this approach is physically sound, results are still approximate due to various assumptions made in arriving at Eq. (5.99). Other models using this methodology differ mainly in the approximation used to determine boundary conditions and the assumptions made for device doping [46]–[48].

In an approach called the voltage-doping transformation, a 2-D Poisson's equation is reduced to 1-D by assuming that the potential distribution along the channel (y direction) varies quadratically with distance. The resulting equation is then solved for only the potential at the center of the effective channel length, which is the minimum potential point (for implanted channels). In this case the equation to be solved is [49]

$$\frac{d^2\phi}{dx^2} = -\frac{q}{\epsilon_0\epsilon_{si}}\left[N(x) - \frac{2\epsilon_0\epsilon_{si}V'_{ds}}{qL^2}\right] \tag{5.100}$$

where

$$V'_{ds} = V_{ds} + 2(\phi_{bi} + V_{sb} - \eta\phi_{si}) + 2\sqrt{(\phi_{bi} + V_{sb} - \eta\phi_{si})(\phi_{bi} + V_{ds} - \eta\phi_{si})} \tag{5.101}$$

and η is a spreading parameter, which accounts for the transverse distribution of the potential ϕ. Physically speaking, Eq. (5.100) means that the influence of the lateral drain-source field on the potential barrier height is equivalent to and can be replaced by the reduction in the doping concentration [50]. Although Eq. (5.100) could be solved for any doping profile, assuming a step doping profile one obtains [49]

$$V_{th1} = V_{fb} + \phi_{si} + \frac{1}{C_{ox}}\sqrt{2\epsilon_0\epsilon_{si}qN_{eff}(\phi_{si} - V_{sb})} \quad \text{when} \quad X_{dm} \leq X_i \tag{5.102a}$$

while

$$V_{th2} = V_{fb} + \phi_{si} + \frac{qX_i}{C_{ox}}(N_{eff} - N_b) + \gamma\sqrt{\phi_{si} - X_i^2(N_{eff} - N_b)} \quad \text{when} \quad X_{dm} > X_i \tag{5.102b}$$

where

$$N_{eff} = N_s - \frac{2\epsilon_0\epsilon_{si}V'_{ds}}{qL^2}$$

and V'_{ds} is given by Eq. (5.101). The above model for V_{th} shows an approximately inverse quadratic dependence on channel length L and an inverse dependence on oxide capacitance C_{ox}. It should be pointed out that the weak dependence of the short-channel effect on the junction depth X_j has not been taken into account. In normal enhancement devices, this effect is small. Figure 5.26 shows the V_{th} variation with drain bias for two different back bias ($V_{sb} = 0\,\text{V}$ and $V_{sb} = -2\,\text{V}$). Continuous lines are experimental data while dashed lines are based on Eq. (5.102). As can be seen, agreement between the experimental data and the model is fairly good.

Empirical Approach. Very often in actual devices, the exponential dependence of the DIBL effect on L is not observed. In such cases an empirical approach is often used, assuming the surface potential to be constant along the length of the channel, even for short channel devices. This assumption results in a very simple expression for σ, which can be derived as follows [79]:

When V_{ds} is small (< 0.1 V), the substrate depletion region width X_{dm} may be calculated using the Poisson equation. When V_{ds} is large, an additional potential will be imposed in the region already depleted. Since no additional charge appears in Poisson's equation, this additional potential satisfies the Laplace equation:

$$\frac{d^2V_1}{dx^2} = 0. \tag{5.103}$$

Under the simplifying assumptions that (1) source/drain junction depths are small compared to the channel length, and (2) using the approximate boundary conditions that $V_1 = 0$ at the source region and $V_1 = V_{ds}$ at the drain region, we can solve Eq. (5.103), resulting in the following expression for the field E_1 at the source end [79]

$$E_1 = -\frac{V_{ds}}{\pi L}. \tag{5.104}$$

According to Gauss' law, this field is equivalent to a charge $\epsilon_0 \epsilon_{si} E_1$, which reduces the threshold voltage by an amount

$$\Delta V_{th} = \frac{\epsilon_0 \epsilon_{si}}{\pi C_{ox} L} V_{ds} = \sigma V_{ds} \tag{5.105}$$

where $\sigma = \epsilon_0 \epsilon_{si}/(\pi C_{ox} L)$. Note that the dependence of σ on V_{sb} has not been included, although depending on the process, the effect could be large (see Fig. 5.26). For circuit models the following empirical expression for σ is normally used [81]

$$\boxed{\sigma = \frac{\epsilon_0 \epsilon_{si}(\sigma_0 + \sigma_1 V_{sb})}{\pi C_{ox} L^m}} \tag{5.106}$$

where σ_0, σ_1 and m are constants that are used to better fit the model for the geometry dependence of the DIBL effect, for a given range of X_j and N_b. The exponent m of L varies in the range 1–3. The back bias dependence of m has also been proposed [83], but for circuit models, it is more appropriate to take m as constant as is done in almost all empirical models [80]. SPICE Level 3 model uses Eq. (5.106) with $m = 3$, $\sigma_1 = 0$ and σ_0 as a fixed constant value, not a fitting parameter. The threshold voltage as a function of drain bias for a typical n-channel 1 μm CMOS technology at two substrate bias is shown in Figure 5.29. Symbols are experimental data while dashed and continuous lines are based on Eq. (5.106).

Based on a 2-D solution, Masuda et al. [82] have proposed the following empirical formula

$$\sigma = \frac{\eta_0(2\phi_{bi} + V_{sb} + V_{ds})}{C_{ox} L^m} \tag{5.107}$$

which was fitted to the data for the following range of parameters:

$$N_b = 10^{15} - 10^{16}\,\mathrm{cm}^{-3}, \quad X_j = 0.15 - 0.41\,\mu\mathrm{m}$$

$$t_{ox} = 500\,\text{Å} \qquad \frac{\eta_0}{C_{ox}} = 4.1\cdot 10^{-2} - 8.8\cdot 10^{-2} \tag{5.108}$$

$$m = 2.6 - 2.3.$$

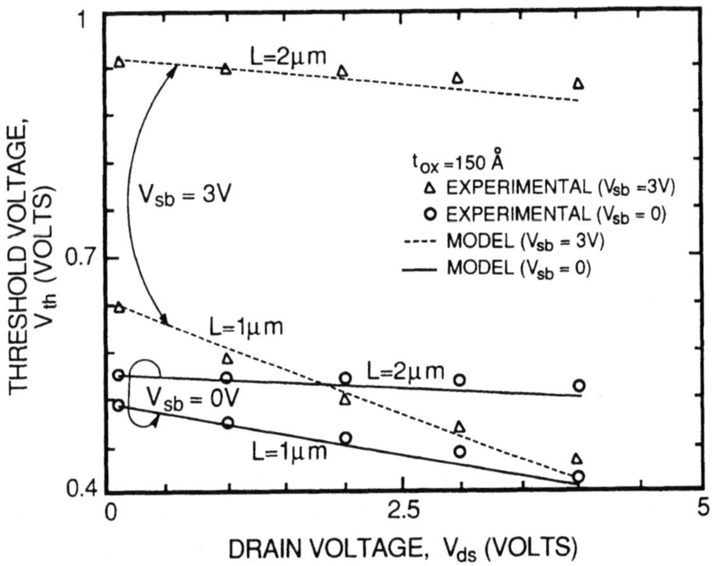

Fig. 5.29 Variation of threshold voltage with drain voltage at two different back bias $V_{sb} = 0$ and 3 V with channel length as a parameter for a typical n-channel 1 μm CMOS process

A slightly different form for the back bias V_{sb} (through C_d) dependence has also been proposed [84]

$$\sigma = \frac{\eta_0 \epsilon_0 \epsilon_{si}}{L(C_{ox} + C_d)} \tag{5.109}$$

where C_d is the depletion capacitance and is obtained by differentiating the bulk charge Q_b. For example, for a uniformly doped substrate, we can write

$$C_d = \frac{\gamma C_{ox}}{2\sqrt{\phi_f + V_{sb}}}. \tag{5.110}$$

On the other hand, Yang and Chaterjee's [85] model uses an effective body factor to account for the drain bias effect as

$$\gamma_{eff} = \gamma - G_l(\sqrt{\phi_{si} + V_{bs} + V_{ds}} - \sqrt{\phi_{si}}) \tag{5.111}$$

where γ is the long channel body factor, which becomes bias dependent for implanted devices, and G_l is a fitting parameter.
It is interesting to note that different short channel models show different functional dependences on the channel length L. Thus,

- the charge sharing models show a V_{th} dependence of $1/L$,
- the empirical models show V_{th} dependence of $1/L^n$ $(1 < n < 3)$,
- the 2-D models show V_{th} dependence of $\exp(-L/L_0)$.

The dependencies are quite different and do cause confusion as to which model is valid. Obviously, the model to choose depends upon the process technology. It has been found that for circuit models Eq. (5.106) is fairly general and fits most of the technology data the author has come across.

5.3.4 Small-Geometry Effect

When both the device width W and length L are small, that is, when both W and L are of the same order of magnitude as the depletion width X_{dm}, then the device is called a small geometry device.[10] For example, in a 2 μm CMOS technology a device with $W/L = 3/2\,(\mu\text{m})$ could be called a small geometry device. A first order estimate of the threshold voltage induced by small-geometry effect can be obtained by superposing the short-channel and narrow-width effects such that

$$\Delta V_{th} \approx \Delta V_{th,l} + \Delta V_{th,w}$$

so that the total threshold voltage at low V_{ds} for small-geometry devices becomes

$$V_{th} = V_{fb} + 2\phi_f + \gamma\sqrt{(2\phi_f + V_{sb})} - \frac{\Delta Q_l}{C_{ox}} + \frac{\Delta Q_w}{C_{ox}}. \tag{5.112}$$

This is the approach used in most of the circuit simulators, including SPICE. However, Eq. (5.112) overestimates ΔV_{th} due to the small-geometry effect. This is because short-channel and narrow-width effects are not really independent as assumed in Eq. (5.112). In fact, *there is a coupling between these two effects which results in a compensating effect.* This is because both W and L determine the gate controlled charge and the volume of the charge must be properly identified. Therefore, mere addition of the two effects does not accurately predict V_{th}.

We will develop a simple model for small small-geometry devices based on Yau's charge sharing approach which will illustrate this compensating effect on V_{th}. Figure 5.30 shows one side of the extra charge $\Delta Q_w/2$ in the depletion region due to the narrow width effect for an assumed triangular region. Since the short-channel effect reduces the bulk charge, the resulting geometry indicated by the dotted line shows the amount of the extra charge induced due to short-channel effect. The volume $\mathscr{V}$ of the extra charge that is responsible for $\Delta V_{th,w}$ is obtained by first finding the wedge volume and then subtracting the volume of the pyramid shape regions shown by

---

[10] A device with minimum L and W allowed by the process technology is referred to as the *minimum size device* for that technology.

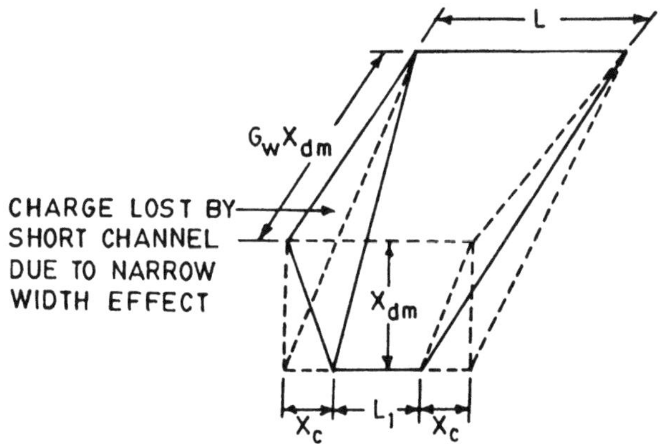

Fig. 5.30 Geometry for the computation of the threshold voltage of a small-geometry MOSFET

dotted lines, that is,

$$\mathscr{V} = 2(\tfrac{1}{2}G_w L X_{dm}^2 - \tfrac{1}{3}G_w X_c X_{dm}^2)$$

where the factor of 2 accounts for both sides of the device and X_c is given by Eq. (5.65). The compensated depletion charge due to width effect becomes

$$\Delta Q_w = \frac{qN_b\mathscr{V}}{WL} = Q_b\left[\frac{G_w X_{dm}}{W} - \frac{2}{3}\frac{G_w X_{dm}}{W}\frac{X_j}{L}\left(\sqrt{1+\frac{2X_{dm}}{X_j}} - 1\right)\right]. \tag{5.113}$$

Comparing this equation with Eq. (5.88) clearly shows the effect of the short channel on the extra charge in the width direction. From Yau's model [cf. Eq. (5.66)] we have

$$\Delta Q_l = Q_b - Q_b' = Q_b(1 - F_l) = Q_b\frac{X_j}{L}\left(\sqrt{1+\frac{2X_{dm}}{X_j}} - 1\right) \tag{5.114}$$

so that ΔV_{th} due to the small geometry effect is obtained by adding Eqs. (5.113) and (5.114) as

$$\Delta V_{th} = \frac{Q_b}{C_{ox}}\left[\frac{G_w X_{dm}}{W} - \frac{X_j}{L}\left(1 + \frac{2}{3}\frac{G_w X_{dm}}{W}\right)\left(\sqrt{1+\frac{2X_{dm}}{X_j}} - 1\right)\right]. \tag{5.115}$$

Clearly, the change in threshold voltage due to the small geometry effect is reduced by the extra term that originates due to the compensating effect. This is basically the model proposed by Merckel [87]. Note that the value of the fitting parameter G_w will be different when used with Eq. (5.88). Others

[89] have also proposed models for the small-geometry effect, but they are not very different from the above model.

The small geometry V_{th} model for n-channel MOSFET with fully recessed isolation oxide will be [67]

$$V_{th} = V_{fb} + 2\phi_f + \frac{Q_b}{C_{ox}} F_l \cdot \left(\frac{W}{W+F}\right). \tag{5.116}$$

5.4 Temperature Dependence of the Threshold Voltage

The threshold voltage of long channel implanted MOSFETs is given by Eq. (5.63) and is determined by device physical parameters, such as flat band voltage V_{fb}, bulk Fermi potential ϕ_f, and body factor γ. Since both ϕ_f (cf. section 2.4) and V_{fb} (cf. section 4.7) are temperature dependent, V_{th} is also temperature dependent. In fact temperature dependence of V_{th} is primarily governed by the temperature dependence of ϕ_f and V_{fb} [30], [90]–[93]. *Recall that the magnitude of both ϕ_f and V_{fb} decreases with increasing temperature, therefore the magnitude of V_{th} also decreases with increasing temperature for both n- and p-channel devices.* Typically, the

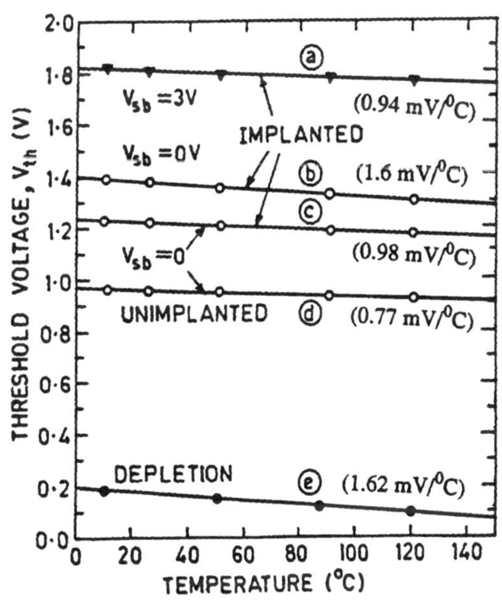

Fig. 5.31 Measured threshold voltage variation with temperature for different types of n-channel MOSFETs. Curves (a), (b), (c) and (d) are for enhancement type devices, while curve (e) is for depletion type device. All devices have n^+ polysilicon gates. The temperature coefficient of V_{th} (dV_{th}/dT) is shown on each curve

temperature coefficient of threshold voltage $|dV_{th}/dT|$ lies in the range 1–3 mV/degree depending upon the type of MOSFET and its physical parameters such as gate oxide thickness t_{ox}, substrate concentration N_b, etc. Measured values of the threshold voltage as a function of temperature for different types of long-channel MOSFETs are shown in Figures 5.31 and 5.32. The data shown as curves (a), (b), (c) and (d) in Figures 5.31 are for n-channel enhancement type devices, while curve (e) is for n-channel depletion type device; all these devices are with n^+ polysilicon gates. While curves (a), (b) and (c) are for channel implanted devices, curve (d) is for uniformaly doped device. Although both curves (b) and (c) are for enhancement devices, curve (b) has higher t_{ox}(300 Å), and lower surface concentration ($3 \times 10^{15}\,\mathrm{cm}^{-3}$) compared to curve (c), which has $t_{ox} = 105$ Å and $N_s = 3 \times 10^{16}\,\mathrm{cm}^{-3}$. The temperature dependence of V_{th} for p-channel devices is shown in Figures 5.32; curves (a) and (b) are for n^+ polysilicon gates while (c) is for p^+ polysilicon gate. The $|dV_{th}/dT|$ is shown on each curve. Note that in all these devices V_{th} *varies linearly with the temperature.* This linear dependence of V_{th} on temperature is valid down to 50 K [94, 95]. The temperature behavior of V_{th} can easily be predicted from Eq. (5.63). Remembering that the temperature dependence of V_{fb} is governed by the temperature dependence of Φ_{ms}, the work function difference between the gate material and the substrate [cf. section 4.7.1], differentiating Eq. (5.63)

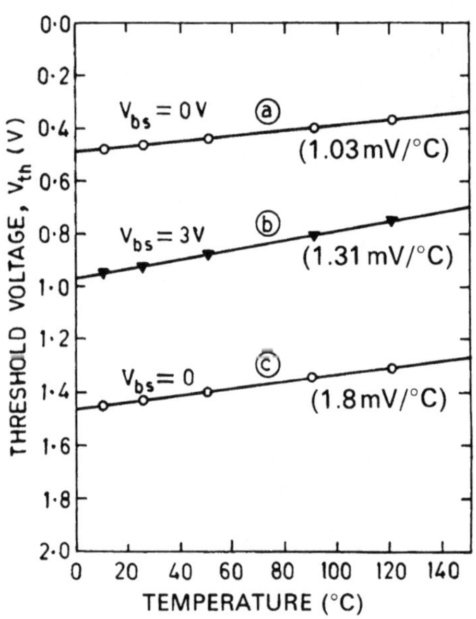

Fig. 5.32 Measured threshold voltage variation with temperature for different types of p-channel enhancement type MOSFETs. Curves (a) and (b) are with n^+ polysilicon gate, while curve (c) is with p^+ polysilicon gate

yields

$$\frac{\partial V_{th}}{\partial T} \equiv T_{vth} = \frac{\partial \Phi_{ms}}{\partial T} \pm 2\frac{\partial \phi_f}{\partial T} \pm 2\mathscr{A}\frac{\partial \phi_f}{\partial T} \tag{5.117}$$

where the + and the − signs are for n- and p-channel devices, respectively, and $\mathscr{A}$ is given by

$$\mathscr{A} = \frac{\gamma}{2\sqrt{|\pm 2\phi_f \pm V_{sb} \mp V_0|}}. \tag{5.118}$$

Recall that the temperature coefficient of ϕ_f is given by Eq. (2.31), repeated here for convenience,

$$\frac{\partial \phi_f}{\partial T} = \frac{1}{T}\left[\phi_f - \left(0.603 + \frac{3}{2}\frac{kT}{q}\right)\right] \tag{5.119}$$

while that for Φ_{ms} is given by Eq. (4.81) or (4.82) depending upon type of the gate and substrate materials; that is,

$$\frac{\partial \Phi_{ms}}{\partial T} = \begin{cases} \dfrac{1}{T}\left[\Phi_{ms} + 2\left(0.603 + \dfrac{3}{2}\dfrac{kT}{q}\right)\right] & (n^+ \text{ poly-Si gate and } p\text{-bulk}) \quad (5.120a) \\ \dfrac{1}{T}\Phi_{ms} & (n^+ \text{ poly-Si gate and } n\text{-bulk}). \quad (5.120b) \end{cases}$$

Equation (5.117) shows that the temperature coefficient of threshold voltage, T_{vth}, of a MOSFET depends upon the following parameters:

- The substrate concentration N_b; the higher the N_b, the higher the ϕ_f, and therefore, the higher the T_{vth}. At lower-temperatures ϕ_f increases; therefore, $\mathscr{A}$ will be reduced further and consequently, *substrate sensitivity (change in V_{th} due to change in V_{sb}) decreases at lower temperatures.* Thus, the lower the temperature, the lower the substrate sensitivity.
- The gate oxide thickness t_{ox}; the higher the t_{ox}, the higher the body factor γ and hence higher the $\mathscr{A}$ term. In other words, thicker gate oxides result in higher T_{vth}.
- The back bias V_{sb}; the higher the V_{sb}, the lower the $\mathscr{A}$ term, and consequently T_{vth} becomes smaller at higher V_{sb}. This is evident from curve (a) and (b) (Figs. 5.31 and 5.32) which are for $V_{sb} = 3$ V and 0 V, respectively.

n-Channel Devices (nMOST). For n-channel enhancement devices (n^+ polysilicon gate and p-substrate) $\partial\Phi_{ms}/\partial T$ is given by Eq. (5.120a). Since Eqs. (5.119) and (5.120) when used in Eq. (5.117) almost compensate each other, it is the last term in Eq. (5.117) that mainly contributes to the temperature coefficient of V_{th}. Furthermore, for implanted devices, V_0 and

$2\phi_f$ almost compensate each other, therefore $\mathscr{A}$ according to Eq. (5.118) has a relatively larger value compared to unimplanted devices where V_0 is zero. This explains why unimplanted devices have a lower temperature coefficient compared to the implanted devices (see curves (b) and (c) in Figure 5.31). Since V_{sb} increases $\mathscr{A}$, T_{vth} decreases (compare curves (a) and (b) in Figure 5.31). For *n*-channel depletion devices (n^+ polysilicon gate and *n*-silicon surface), Eqs. (5.120b) for $\partial\Phi_{ms}/\partial T$ and (5.119) for $\partial\Phi_{ms}/\partial T$ when used in Eq. (5.117), do not compensate each other, but rather add up. Therefore, all three terms of Eq. (5.117) contribute to T_{vth} resulting in a higher temperature coefficient compared to enhancement devices (curve e, Figure 5.31). For curve (d), higher T_{vth} is also due to higher t_{ox} (=420 Å) compared to 300 Å for curve b).

p-Channel Devices (pMOST). For *p*-channel devices with p^+ polysilicon gates, the situation is similar to that of *n*-channel enhancement devices with n^+ polysilicon gates and therefore, T_{vth} is almost the same as that of

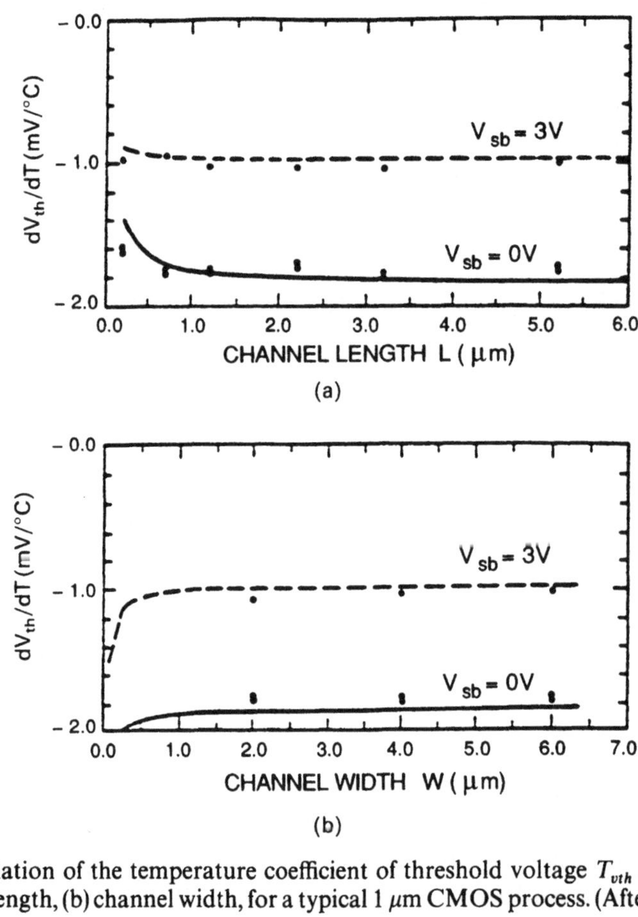

Fig. 5.33 Variation of the temperature coefficient of threshold voltage T_{vth} as a function of (a) channel length, (b) channel width, for a typical 1 μm CMOS process. (After Arora [30].)

n-channel enhancement devices. However, for p-channel compensated devices (n^+ polysilicon gate and n-type concentration at the silicon surface), the situation is similar to n-channel depletion devices and therefore, T_{vth} in this case is higher compared to that of surface channel p-devices (see curves (a) and (b) Figure 5.32).

For short channel devices, part of the gate induced charge is depleted from the source and drain resulting in the effective body factor (γF_l) being smaller compared to the long channel value; recall that F_l is less than unity for short-channel devices [cf. Eq. (5.67)]. Furthermore, since $\mathscr{A}$ is directly proportional to γ, Eq. (5.118) predicts that T_{vth} will decrease with decreasing channel length. This behavior is indeed observed experimentally as shown in Figure 5.33 where the dots are experimental points and the continuous lines are based on the Eq. (5.117).

References

[1] Y. P. Tsividis, *Operation and Modeling of the MOS Transistor*, McGraw-Hill Book Company, New York, 1987.

[2] E. Demoulin and F. Van De Wiele, 'Ion implanted MOS transistors', in *Process and Device Modeling for Integrated Circuits* (Wiele, Engl and Jespers, Eds.), pp. 617–676, NATO Advanced Study Institute, 1977, Noordhoff Publishing, Leyden, 1977.

[3] Y. Tsividis and G. Masetti, 'Problems in precision modeling of the MOS transistor for analog application', IEEE Trans. Computer-Aided Design, CAD-3, pp. 72–79 (1984).

[4] L. Lewyn and J. D. Meindel, 'An IGFET inversion charge model for VLSI system', IEEE Trans. Electron Devices, ED-32, pp. 434–440 (1985).

[5] S. M. Sze, Ed., *VLSI Technology*, 2nd Ed., McGraw-Hill Book Company, New York, 1988.

[6] M. Kamoshida, 'Threshold voltage and gain term β of ion implanted n-channel MOS transistors', J. Appl. Phys. Lett., 22, pp. 404–405 (1973).

[7] G. Doucet and F. Van de Weile, 'Threshold voltage of nonuniformly doped structures', Solid-State Electron., 16, pp. 417–423 (1973).

[8] V. L. Rideout, F. H. Gaensslen, and A. LeBlanc, 'Device design consideration for ion implanted n-channel MOSFETs', IBM J. Res. Dev., 19, pp. 50–59 (1975).

[9] R.R. Troutman, 'Ion-implanted threshold voltage tailoring for insulated gate field-effect transistors', IEEE Trans. Electron Devices, ED-24, pp. 182–192 (1977).

[10] J. R. Brews, 'Threshold shifts due to nonuniform doping profiles in surface channel MOSFETs', IEEE Trans. Electron Devices, ED-26, pp. 1696–1710 (1979).

[11] P. K. Chatterjee, J. E. Leiss, and G. W. Taylor, '1A dynamic average model for the body effect in the implanted short channel ($L = 1\,\mu$m) MOSFETs', IEEE Trans. Electron Devices, ED-28, pp. 606–607 (1981).

[12] D. A. Antoniadis, 'Calculation of threshold voltage in nonuniformly doped MOSFETs', IEEE Trans. Electron Devices, ED-31, pp. 303–307 (1984).

[13] P. Ratnam and C. A. T. Salama, 'A new approach to the modeling of nonuniformly doped short-channel MOSFETs', IEEE Trans. Electron Devices, ED-31, pp. 1289–1298 (1984).

[14] C. R. Viswanathan, B. C. Burkey, G. Lubberts, and T. J. Tredwell, 'Threshold voltage in short channel MOS devices', IEEE Trans. Electron Devices, ED-32, pp. 932–940 (1985).

[15] T. W. Sigmon and R. Swanson, 'MOS Threshold shifting by ion implantation', Solid-State Electron., 16, pp. 1217–1232 (1973).
[16] F. M. Klaassen and W. Hes, 'Compensated MOSFET devices', Solid-State Electron., 28, pp. 359–373 (1985).
[17] J. S. T. Huang and G. W. Taylor, 'Modeling of an ion-implanted silicon gate depletion-mode IGFET', IEEE Trans. Electron Devices, ED-22, pp. 995–1001 (1975).
[18] R. A. Haken, 'Analysis of the deep depletion MOSFET and the use of the DC characterstics for determining bulk-channel charge-coupled device parameters', Solid-State Electron., 21, pp. 753–761 (1978).
[19] Y. A. El-Mansy, 'Analysis and characterization of depletion mode MOSFET', IEEE J. Solid-State Circuits, SC-15, pp. 331–340 (1980).
[20] N. Ballay and B. Baylac, 'Analytical modeling for depletion-mode MOSFET with short and narrow-channel effects', IEE Proc. Part I, Solid-State & Electron Dev., 128, pp. 225–238 (1981).
[21] D. A. Divekar and R. I. Dowell, 'A depletion-mode MOSFET model for circuit simulation', IEEE Trans. Computer-Aided Design, IEEE CAD-3, pp. 80–87 (1984).
[22] M. J. Van de Tol and S. G. Chamberlain, 'Buried-channel MOSFET model for SPICE', IEEE Trans. Computer-Aided Design, CAD-10, pp. 1015–1035 (1991).
[23] G. R. Mohan Rao, 'An accurate model for a depletion mode IGFET used as a load device', Solid-State Electron., 21, pp. 711–714 (1978).
[24] S. W. Tarasewicz and C. A. T. Salama, 'Threshold voltage characteristics of ion-implanted depletion MOSFETs', Solid-State Electron., 31, pp. 1441–1446 (1988).
[25] J. S. T. Huang, J. W. Schrankler, and J. S. Kueng, 'Short-channel threshold model for buried-channel MOSFETs', IEEE Trans. Electron Devices, ED-31, pp. 1889–1895 (1984).
[26] K. C. K. Weng, P. Yang, and J. H. Chern, 'A predictor/CAD model for buried-channel MOS transistors', IEEE Trans. Computer-Aided Design, CAD-6, pp. 4–16 (1987).
[27] K. Shenai, 'Analytical solutions of threshold voltage calculations in ion-implanted IGFETs', Solid-State Electron., 26, pp. 761–766 (1983).
[28] A. Roychaudhuri, M. Jha, S. K. Sharma, P. A. Govindacharyulu, and M. J. Zarabi, 'Substrate bias dependence of short-channel MOSFET threshold voltage—a novel approach', IEEE Trans. Electron Devices, ED-35, pp. 167–172 (1988).
[29] S. Karmalkar and K. N. Bhat, 'A process-parameter-based circuit simulation model for ion-implanted MOSFETs and MESFETs', IEEE J. Solid-State Circuits, SC-24, pp. 139–145 (1989).
[30] N. D. Arora, 'Semi-empirical model for the threshold voltage of a double implanted MOSFET and its temperature dependence', Solid-State Electron., 30, pp. 559–569 (1987).
[31] A. H. Marshak and R. Shrivastava, 'On threshold and flat-band voltages for MOS devices with polysilicon gate and nonuniformly doped substrate', Solid-State Electron., 26, pp. 361–364 (1983).
[32] F. Van de Weile, 'On the flat-band voltage of MOS structures on nonuniformly doped substrate, Solid-State Electron., 27, pp. 824–826 (1984).
[33] M. R. MacPherson, 'Threshold shift calculations for ion implanted MOS devices', Solid-State Electron., 15, pp. 1319–1328 (1972).
[34] M. C. Tobey and N. Gordon, 'Concerning the onset of heavy inversion in MIS devices', IEEE Trans. Electron Devices, ED-21, pp. 649–650 (1974).
[35] H. Feltl, 'Onset of heavy inversion in MOS devices doped nonuniformly near surface', IEEE Trans. Electron Devices, ED-24, pp. 288–289 (1977).
[36] M. Nishida and M. Aoyane, 'An improved definition for the onset of heavy inversion in a MOS structure with nonuniformly doped semiconductors', IEEE Trans. Electron Devices ED-27, pp. 1222–1230 (1980).

[37] A. L. Silburt, R. C. Foss, and W. F. Petrie, 'An efficient MOS transistor model for computer-aided design', IEEE Trans. Computer-Aided Design, CAD-3, pp. 104–110 (1984).

[38] H. J. Park and C. K. Kim, 'An empirical model for the threshold voltage of enhancement nMOSFETs', IEEE Trans. Computer-Aided Design, CAD-4, pp. 629–635 (1985).

[39] G. T. Wright, 'Physical and CAD models for implanted-channel VLSI MOSFET', IEEE Trans. Electron Devices, ED-34, pp. 823–833 (1987).

[40] B. J. Sheu, D. L. Scharfetter, P. K. Ko, and M. C. Jeng, 'BSIM: Berkeley short-channel IGFET model for MOS transistors', IEEE Journal Solid-State Circuits, SC-22, pp. 558–565 (1987).

[41] L. A. Akers and J. J. Sanchez, 'Threshold voltage models of short, narrow and small geometry MOSFET's: a review', Solid-State Electron., 25, pp. 621–641 (1982).

[42] J. R. Brews, W. Fichtner, E. H. Nicollian, and S. M. Sze, 'Generalized guide for MOSFET miniaturization', IEEE Electron Devices Lett., EDL-1, pp. 2–5 (1980).

[43] T. Toyabe and S. Asai, 'Analytical models of threshold voltage and breakdown voltage of short-channel MOSFET's derived from two-dimensional analysis', IEEE Trans. Electron Devices, ED-26, pp. 453–460 (1979).

[44] C. Y. Wu, S. Y. Yang, H. H. Chen, F. C. Tseng, and C. T. Shih, 'An analytic and accurate model for the threshold voltage of short channel MOSFETs in VLSI', Solid-State Electron., 27, pp. 651–658 (1984).

[45] K. N. Rantnakumar and J. D. Meindel, 'Short-channel MOST threshold voltage model', IEEE J. Solid-State Circuits, SC-17, 937–948 (1982).

[46] T. N. Nguyen and J. D. Plummer, 'Physical mechanisms responsible for short channel effects in MOS devices', IEEE IEDM, *Dig. Tech. Papers*, pp. 596–599 (1981).

[47] D. R. Poole and D. L. Kwong, 'Two-dimensional analytical modeling of threshold voltage of short-channel MOSFETs', IEEE Electron Device Lett., EDL-5, pp. 443–446 (1984).

[48] V. Marsh, and R. W. Dutton, 'Submicron 2D MOS modeling', IEEE ICCAD, *Dig. Tech. Papers*, pp. 476–479 (1986).

[49] T. Skotnicki and W. Marciniak, 'A new approach to threshold voltage modeling of short-channel MOSFETs', Solid-State Electron., 29, pp. 1115–1127 (1986).

[50] T. Skotnicki, G. Merckel, and T. Pedron, 'The voltage doping transformation, a new approach to the modeling of MOSFET short channel effects', Proc. ESSERC 87, Bologna, pp. 543–546, North-Holland, Amsterdam (1987).

[51] L. D. Yau, 'A simple theory to predict the threshold voltage of a short-channel IGFET's', Solid-State Electron., 17, pp. 1059–1063 (1974).

[52] W. Fichtner and H. W. Potzl, 'MOS modeling by analytical approximation. I. Subthreshold current and threshold voltage', Int. J. Electron., 46, pp. 33–55 (1979).

[53] L. M. Dang, 'A simple current model for short-channel IGFET and its application to circuit simulation', IEEE Trans. Electron Devices, ED-26, pp. 436–445 (1979).

[54] M. Simard-Normandin, 'Channel length dependence of the body-factor effect in NMOS devices', IEEE Trans. Computer-Aided Design, CAD-2, pp. 2–4 (1983).

[55] M. Nishida and H. Onodera, 'An anomalous increase of threshold voltages with shortening the channel lengths for deeply boron-implanted n-channel MOSFETs', IEEE Trans. Electron Devices, ED-28, pp. 1101–1103 (1981).

[56] M. Orlowski, C. Mazure, and F. Lau, 'Submicron short channel effects due to gate reoxidation induced lateral interstitial diffusion', IEEE IEDM, *Dig. Tech. Papers*, pp. 632–635 (1987).

[57] C. Y. Lu and J. M. Sung, 'Reverse short-channel effects on threshold voltage in submicrometer salicide devices', IEEE Electron Devices Lett., EDL-10, pp. 446–448 (1989).

[58] J. S. T. Huang and J. W. Schrankler, 'On flat-band voltage dependence on channel length in short-channel threshold voltage model', IEEE Trans. Electron Devices, ED-36, pp. 1226, 1989. Also see ED-32, pp. 1001–1002 (1985).
[59] N.D. Arora and M. Sharma 'Modeling the anomalous threshold voltage behavior of submicron MOSFETs', IEEE Electron Devices Lett., EDL-13, pp. 92–94 (1992).
[60] K. E. Kroell and G. K. Ackermann, 'Threshold voltage of narrow channel field effect transistors', Solid-State Electron., 19, pp. 77–81 (1974).
[61] L. A. Akers, M. M. E. Beguwala, and F. Z. Custode, 'A model of a narrow width MOSFET including tapered oxide and doping encroachment', IEEE Trans. Electron Devices, ED-28, pp. 1490–1495 (1981).
[62] C. R. Ji and C. T. Sah, 'Two-dimensional numerical analysis of the narrow gate effect in MOSFET', IEEE Trans. Electron Devices, ED-30, pp. 635–647 (1983).
[63] P. T. Lai and Y. C. Cheng, 'An analytical model for the narrow-width effect in ion-implanted MOSFETs', Solid-State Electron., 27, pp. 639–649 (1984).
[64] S. S. Chuang and C. T. Sah, 'Threshold voltage models of the narrow-gate effect in micron and submicron MOSFETs', Solid-State Electron., 31, pp. 1009–1021 (1988).
[65] P. T. Lai and Y. C. Cheng, 'Comparison of threshold modulation in narrow MOSFETs with different isolation structure', Solid-State Electron., 28, pp. 551–554 (1985).
[66] A. Kurosawa, T. Shibata, and H. Iozuka, 'A new bird's beak free isolation technology for VLSI devices', IEEE IEDM, *Dig. Tech. Papers*, pp. 384–387 (1981).
[67] L. A. Akers, M. Sugino, and J. M. Ford, 'Characterization of the inverse-narrow-width effect', IEEE Trans. Electron Devices, ED-34, pp. 2476–2484 (1987).
[68] K. K. Hsueh, J. J. Sanchez, T. A. Demassa, and L. A. Akers, 'Inverse-narrow-width effects and small-geometry MOSFERT threshold voltage model', IEEE Trans. Electron Devices, ED-35, pp. 325–338 (1988).
[69] K. Ohe, S. Odanaka, K. Moriyama, T. Hori, and G. Fuse, 'Narrow-width effects of shallow trench-isolated CMOS with n^+-polysilicon gate', IEEE Trans. Electron Devices, ED-36, pp. 1110-1115 (1989).
[70] R. C. Vankemmel and K. M. De Meyer, 'A study of the corner effects in trench-like isolated structures', IEEE Trans. Electron Devices, ED-37, pp. 168–175 (1990).
[71] G. W. Taylor, 'The effects of two-dimensional charge sharing on the above threshold characteristics of short channel IGFETs', Solid-State Electron., 22, p. 701 (1979).
[72] R. Troutman, 'VLSI limitations from drain induced barrier lowering', IEEE Trans. Electron Devices, ED-26, pp. 461–469 (1979).
[73] C. G. Sodini, S. S. Wong, and P. K. Ko, 'A framework to evaluate technology and device design enhancements for MOS Ic's', IEEE J. Solid-State Circuits, SC-24, pp. 118–127 (1989).
[74] Y. Ohno, 'Short-channel MOSFET VT-VDS characteristics model based on a point charge and its mirror image', IEEE Trans. Electron Devices, ED-29, pp. 211–216 (1982).
[75] C. S. Chao, L. A. Akers, and D. N. Pattanayak, 'Drain voltage effects on the threshold voltage of a small geometry MOSFET', Solid-State Electron., 26, pp. 851–860 (1983).
[76] S. C. Chamberlain and S Ramanan, 'Drain-induced barrier-lowering analysis in VLSI MOSFET devices using two-dimensional numerical simulation', IEEE Trans. Electron Devices, ED-33, pp. 1745–1752 (1986).
[77] S. C. Jain and P. Balk, 'A unified analytical model for drain-induced barrierlowering and drain-induced high electric field in a short-channel MOSFET', Solid-State Electron., 30, pp. 503–511 (1987).
[78] Y. Omura and K. Ohwada, 'Threshold voltage theory for short-channel MOSFET model using a surface-potential distribution model', Solid-State Electron., 22, pp. 1045–1051 (1979).
[79] B. J. Sheu, D. L. Scharfetter, and H. C. Poon, 'Compact short channel IGFET model (CSIM)', Electronics Research Laboratory Memo No. UCB/ERL-M84/20, University of California, Berkeley, March 1984.

[80] F. M. Klaassen and W. C. J. de Groot, 'Modeling of scaled-down MOS transistors', Solid-State Electron., 23, pp. 237–242 (1980).
[81] N. D. Arora and L. M. Richardson, 'MOSFET modeling for circuit simulation' in: *Advanced MOS Device Physics* (N. G. Einspruch and G. Gildenblat, Eds.), VLSI Electronics: Microstructure Science, 18, pp. 236–276, Academic Press Inc., New York, 1989.
[82] H. Masuda, M. Nakai, and M. Kubo, 'Characteristics and limitation of scaled-down MOSFETs due to two-dimensional field effect', IEEE Trans. Electron Devices, ED-26, pp. 980–986 (1979).
[83] M. J. Deen and Z. X. Yan, 'Substrate bias effects on drain-induced barrier lowering in short-channel PMOS devices', IEEE Trans. Electron Devices, ED-37, pp. 1707–1713 (1990).
[84] T. Grotjohn and B. Hoefflinger, 'A parametric short-channel MOS transistor model for subthreshold and strong inversion current', IEEE Trans. Electron Devices, ED-31, pp. 234–246 (1984).
[85] P. Yang and P. K. Chatterjee, 'SPICE modeling for small geometry MOSFET circuits', IEEE Trans. Computer-Aided Design, CAD-1, pp. 169–182 (1982).
[86] P. P. Wang, 'Double boron implanted short channel MOSFET', IEEE Trans. Electron Devices, ED-24, pp. 196–204 (1977).
[87] G. Merckel, 'A simple model of the threshold voltage of short and narrow channel MOSFETs', Solid-State Electron., 23, pp. 1207–1213 (1980).
[88] O. Jantsch, 'A geometrical model of the threshold voltage of short and narrow channel MOSFETs', Solid-State Electron., 25, pp. 59–61 (1982).
[89] L. A. Akers and C. J. Chao, 'A closed form threshold voltage expression for small geometry MOSFET', IEEE Trans. Electron Devices, ED-29, pp. 776–778 (1982).
[90] R. C. Jager and F. H. Gaensslen, 'Simple analytical models for the temperature dependent threshold behavior of depletion-mode devices', IEEE Trans. Electron Devices, ED-26, pp. 501–507 (1979).
[91] B. S. Song and P. R. Gray, 'Threshold-voltage temperature drift in ion-implanted MOS transistors', IEEE Trans. Electron Devices, ED-29, pp. 661–668 (1982).
[92] J. J. Tzou, C. C. Yao, R. Cheung and H. Chan, 'The temperature dependence of threshold voltages in submicrometer CMOS', EDL-6, pp. 250–252 (1985).
[93] F. M. Klaassen and W. Hes, 'On the temperature coefficient of the MOSFET threshold voltage', Solid-State Electron., 29, pp. 787–789 (1986).
[94] S. K. Tewksbury, 'N-channel enhancement-mode MOSFET characteristics from 10–300K', IEEE Trans. Electron Devices, ED-28, pp. 1519–1529 (1981).
[95] R. M. Fox and R. C. Jager, 'MOSFET behavior and circuit considerations for analog applications at 77K', IEEE Trans. Electron Devices, ED-34, pp. 114–123 (1987).

6 MOSFET DC Model

The MOSFET model required for circuit simulation consists of two parts: (a) a steady-state or DC model, where the voltages applied at the terminals of the device remain constant, that is they do not vary with time; (b) a dynamic or AC model, where the device terminal voltages do not remain constant but vary with time. In this chapter we will discuss only DC MOS transistor models for different regions of device operation. In the next chapter we will deal with the dynamic models.

We will first develop a rigorous drain current model for long channel devices. We then simplify the model and derive a first order model based on various assumptions. This first order simple model is important in itself because it could be used for hand calculations of the drain current in a MOSFET circuit. This simple model will be improved upon as we remove some of the assumptions. The long channel model is then extended to short-geometry devices. This will be followed by studying the effect of temperature on the drain current characteristics.

6.1 Drain Current Calculations

Let us consider an *n*-channel device with uniformly doped substrate of concentration N_b (cm^{-3}), the structure and dimensions of which are shown in Figure 6.1. For the sake of simplicity we will assume this to be a large geometry device so that the short-channel and narrow-width effects can be neglected. The static and dynamic characteristics of a device under the influence of external fields in general can be described by the following three sets of coupled differential equations:

1. The Poisson equation (2.41) for the electrostatic potential ϕ,

$$\nabla^2\phi = -\frac{\rho}{\epsilon_0\epsilon_{si}}. \tag{6.1}$$

2. The current equation (2.35a) for electrons,

$$\mathbf{J}_n = \underbrace{q\mu_n n\mathscr{E}}_{\text{drift}} + \underbrace{qD_n \nabla n}_{\text{diffusion}} \quad (\text{A/cm}^2) \tag{6.2a}$$

which is the sum of two terms, *drift* due to the field $\mathscr{E}$ and *diffusion* due to the concentration gradient. Similarly, for holes, we have

$$\mathbf{J}_p = q\mu_p p\mathscr{E} - qD_p \nabla p \quad (\text{A/cm}^2). \tag{6.2b}$$

These two equations, under non-equilibrium condition, become [cf. Eq. (2.36)]

$$\mathbf{J}_n = -qn\mu_n \nabla\varphi_n \quad \text{(electrons)} \tag{6.3a}$$

$$\mathbf{J}_p = -qp\mu_p \nabla\varphi_p \quad \text{(holes)} \tag{6.3b}$$

where φ_n and φ_p are the electron and hole quasi-Fermi potentials, respectively. The total current density $J = J_n + J_p$.

3. The continuity equations (2.38)

$$\frac{\partial n}{\partial t} = \frac{1}{q}\nabla\cdot\mathbf{J}_n - R_n + G_n \quad \text{(electrons)} \tag{6.4a}$$

$$\frac{\partial p}{\partial t} = -\frac{1}{q}\nabla\cdot\mathbf{J}_p - R_p + G_p \quad \text{(holes).} \tag{6.4b}$$

As was pointed out earlier, modeling a MOSFET is a 3-dimensional (3-D) problem but for all practical purposes (unless the width W and length L are very small), we can treat the system as a 2-D problem in the x and y direction only (see Figure 6.1). Even as a 2-D problem, the equations above

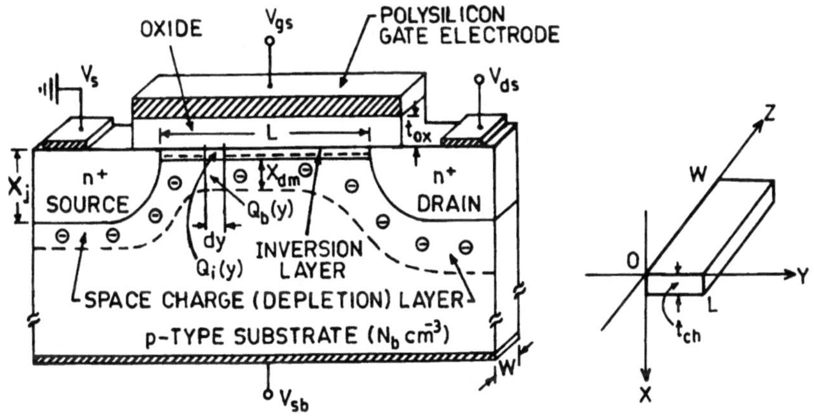

Fig. 6.1 Schematic of a n-channel MOSFET (nMOST) showing voltages and reference direction. The x, y and z directions are the distances into the silicon, along the channel and width of the device, respectively.

are fairly complex; one could solve exactly only using numerical techniques, as is done in 2-D device simulators such as MINIMOS [1], PISCES [2], etc. However, in order to obtain approximate analytical solutions for use in circuit simulators, we generally invoke assumptions that, though not rigorously true, do help to simplify these equations substantially.

Assumption 1. Let us assume that the variation of the electric field $\mathscr{E}_y$ in the y direction (along the channel) is much less than the corresponding variation of the field $\mathscr{E}_x$ in the x direction. That is, the *gradual channel approximation* (GCA) is valid. Recall it was on this assumption that we have developed threshold voltage models (cf. section 5.1). As was pointed out in section 5.1, with this approximation the Poisson equation (6.1) becomes one-dimensional, that is we need only to solve

$$\frac{d^2\phi}{dx^2} = -\frac{\rho(x)}{\epsilon_0\epsilon_{si}}. \tag{6.5}$$

Using 2-D numerical analysis, it can be verified that the GCA is valid for most of the channel length. However, it does fail near the drain region, where the longitudinal field $\mathscr{E}_y$ is comparable to the transverse field $\mathscr{E}_x$ even for long channel devices. In spite of its failure near the drain end, the GCA is used as it reduces the system to a 1-D current flow problem. The fact that we have to solve only a 1-D Poisson's equation means that the charge expressions developed in chapter 4 for an MOS capacitor could be used for a MOS transistor, with the modification that charge and potential will now be position dependent in the y direction.

Assumption 2. Hole current can be neglected.[1] This is because for normal operation of the n-channel device, $V_{ds} \geq 0$ and $V_{bs} \leq 0$. It should be pointed out that holes do become important in describing the device behavior in the avalanche or breakdown region, where impact ionization can create electrons and holes. Since the current equation we are going to derive will not include breakdown regions, *the drain current model needs to consider only the electron current density* J_n.

Assumption 3. Recombination and generation are neglected, that is $R_n = G_n = 0$; and if one considers only the static characteristics of the device, as is presently the case, then the continuity equation (6.4) reduces to

$$\nabla \cdot \mathbf{J}_n = 0. \tag{6.6}$$

This means that the drain current density is an electron current of vanishing

[1] Electrons are neglected for pMOST so that the current flow is assumed due to holes only, therefore, we need to consider only the hole current density J_p.

divergence, that is the total drain current I_{ds} *is constant at any point along the channel.*

Assumption 4. Current flows in the y direction only. This means that $d\varphi_n/dx = 0$, that is, φ_n is constant in the x direction. Such assumptions have been used even in 2-D simulators and the results are very satisfactory [3]. Thus, the current density J_n [cf. Eq. (6.3)], which consists of both drift and diffusion components, is given by

$$J_n(x, y) = -qn(x, y)\mu_n(x, y)\frac{\partial \varphi_n}{\partial y}. \tag{6.7}$$

Since the cross-sectional area of the channel in which the current flows is the channel width W times the channel depth, integrating the above equation across the channel depth (x direction) and width (z direction) gives the drain current I_{ds} at any point y in the channel as:

$$I_{ds}(y) = -W\int_0^\infty \left[qn(x, y)\mu_n(x, y)\frac{\partial \varphi_n}{\partial y}\right]dx = \text{constant}. \tag{6.8}$$

It is important to note that μ_n in the above equation is the electron surface channel mobility (for nMOST), often referred to as the *surface mobility* μ_s in order to distinguish it from the *bulk mobility*, the mobility far away from the surface that was discussed in section 2.5.1. The value of μ_s for electrons is in the range 400–700 cm^2/V·s, while for holes it is in the range 100–300 cm^2/V·s, and depends on both the gate and drain fields as discussed in section 6.6. Since the electron to hole mobility ratio is 2 to 3, this results in nMOST's being faster (higher current) than pMOST's. In the rest of the discussion we will replace μ_n by μ_s to emphasize that the mobility we are dealing with is the surface mobility.

In a MOSFET, the application of source and drain voltages relative to the substrate results in lowering of the quasi-Fermi level $\mathcal{F}_n$ (or potential φ_n) at the source end by an amount qV_{sb}, and at the drain end by an amount $q(V_{ds} + V_{sb})$, relative to the Fermi level E_f in the substrate (see Figure 5.1). It is the difference in φ_n between the source and drain that drives the electrons down the channel. If we define $V_{cb}(y)$ as the channel potential at any point y in the channel [cf. Eq. (5.2)], then (see Figure 6.2)

$$V_{cb}(y) = \varphi_n(y) - \varphi_n|_{\text{source}} = (E_f - \mathcal{F}_n)/q. \tag{6.9}$$

At the source end $V_{cb}(y = 0) = V_{sb}$ and at the drain end $V_{cb}(y = L) = V_{ds} + V_{sb}$. Thus, compared to the case of an MOS capacitor, $\mathcal{F}_n$ in the surface region of a MOSFET is lowered by an amount $qV_{cb}(y)$ thereby lowering the surface electron concentration n_s by a factor $e^{-V_{cb}(y)/V_t}$. Therefore, we can write for the electron concentration in a MOSFET as [cf. Eq. (4.34) for an MOS

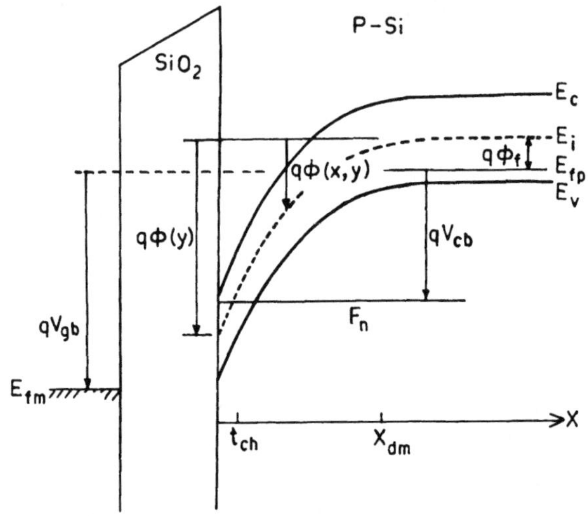

Fig. 6.2 Energy band diagram of a MOSFET shown in Figure 6.1. E_c and E_v represents the edges of the conduction and valence bands, respectively; $\mathscr{F}_n$ and $\mathscr{F}_p(=E_{fp})$ are quasi-Fermi level of the electrons and holes, respectively

capacitor]

$$n(y) = N_b e^{(\phi(y) - 2\phi_f)/V_t} e^{-V_{cb}(y)/V_t} \tag{6.10}$$

where $V_t(=kT/q)$ is the thermal voltage and ϕ_f is the bulk Fermi-potential given by Eq. (4.4). The hole concentration is unchanged and is given by Eq. (4.33b), i.e., $p = N_b e^{-\phi(y)/V_t}$.
Using Eq. (6.9), the I_{ds} Eq. (6.8) can be written as

$$I_{ds}(y) = -W\frac{dV_{cb}}{dy}\int_0^\infty qn(x, y)\mu_s(x, y)dx. \tag{6.11}$$

Assumption 5. Although μ_s depends on both $\mathscr{E}_x$ and $\mathscr{E}_y$, as we will see later in section 6.6, for now we will assume that *μ_s is constant*, taken at some average gate and drain field. With this assumption μ_s can be taken outside the integral in Eq. (6.11). If we define Q_i as the *mobile charge density*, that is,[2]

$$Q_i(y) = q\int_0^\infty n(x, y)dx \tag{6.12}$$

[2] All the charges discussed in this Chapter are charge per unit area, as was the case with the MOS capacitor, and are represented by Q with appropriate lower case subscript. For example, gate charge density will be represented by $Q_g(\mathrm{C/cm^2})$.

then Eq. (6.11) becomes

$$I_{ds}(y)\,dy = -\mu_s W Q_i(y) dV_{cb}. \tag{6.13}$$

Assuming the GCA is valid along the whole length of the channel, then integrating Eq. (6.13) along the length of the channel (y direction) we get

$$I_{ds} = -\mu_s \frac{W}{L}\int_{V_{sb}}^{V_{sb}+V_{ds}} Q_i(y) dV_{cb}. \tag{6.14}$$

Thus, to calculate the current I_{ds}, we need to calculate the mobile charge density Q_i. Different current models have been developed depending upon different estimations for $Q_i(y)$. We will now discuss these models.

6.2 Pao–Sah Model

In this model $Q_i(y)$ is calculated numerically by integrating the electron concentration in the x direction. Equation (6.12) can be rewritten as

$$Q_i(y) = q\int_0^{\infty} n(x, y)dx = q\int_{\phi_s}^{\phi_f} n(\phi, V_{cb})\frac{dx}{d\phi}d\phi = q\int_{\phi_f}^{\phi_s}\frac{n(\phi, V_{cb})}{\mathscr{E}_x}d\phi \tag{6.15}$$

where ϕ_s is the surface potential ($\phi = \phi_s$ at $x = 0$) and is position dependent due to the voltage applied between the source and drain terminals. Note the lower limit of integration is ϕ_f. This is because electron charge comes mostly from the area where electron concentration exceeds the hole concentration, the inversion layer therefore ends at a point where $\phi = \phi_f$. In the equation above $\mathscr{E}_x$ is the field in the x direction and is obtained by solving the Poisson equation (6.5). In analogy with an MOS capacitor [cf. Eq. (4.47)], the Poisson equation for a MOSFET can be written as

$$\frac{d^2\phi}{dx^2} = \frac{qN_b}{\epsilon_0\epsilon_{si}}[1 + e^{(\phi(y) - 2\phi_f - V_{cb}(y))/V_t} + e^{-2\phi_f/V_t} - e^{-\phi(y)/V_t}]. \tag{6.16}$$

The only difference between Eq. (6.16) and the corresponding Eq. (4.47) for the MOS capacitor is the presence of the potential $V_{cb}(y)$ in the exponent and the position dependence of $\phi(y)$ in the y direction [cf. Eq. (6.10)]. Integrating Eq. (6.16) in the x direction, and following the same procedure as was used in solving Eq. (4.47), we get the field $\mathscr{E}_x$ in silicon for the case of a MOSFET as

$$\mathscr{E}_x = -\frac{d\phi}{dx} = -\sqrt{\frac{2qN_b}{\epsilon_0\epsilon_{si}}}F(\phi, \phi_f, V_{cb}) \tag{6.17}$$

where

$$F(\phi, \phi_f, V_{cb}) = [\phi(y) + V_t e^{-\phi(y)/V_t} - V_t + e^{(-2\phi_f - V_{cb}(y))/V_t} \times \{V_t e^{\phi(y)/V_t} - \phi(y)e^{V_{cb}(y)/V_t} - V_t\}]^{1/2} \tag{6.18}$$

is essentially a function describing the electric field in which the term $V_t(e^{-\phi/V_t} - 1)$ is contributed by the majority carriers (holes), ϕ is contributed by the acceptors, $V_t e^{(-V_{cb}-2\phi_f)/V_t}(e^{\phi/V_t} - 1)$ is contributed by the minority carriers (electrons) and $\phi e^{-2\phi_f/V_t}$ is contributed by donors. Thus, knowing the field $\mathscr{E}_x$ [cf. Eq. (6.17)] and electron concentration n [cf. Eq. (6.10)], the mobile charge density Q_i can be calculated from Eq. (6.15), provided the surface potential ϕ_s is known. We will now calculate ϕ_s using the charge conservation principle. The application of the gate voltage induces a charge Q_s in silicon. The induced charge Q_s is related to the gate voltage V_{gb}, as in the case of an MOS capacitor [cf. Eq. (4.20)]

$$V_{gb} = V_{fb} + \phi_s(y) - \frac{Q_s}{C_{ox}} \tag{6.19}$$

where V_{fb} is the flat band voltage. Now from Gauss' theorem, the induced charge density $Q_s(y)$ in silicon is given by (see section 4.2)

$$\begin{aligned} Q_s(y) &= -\epsilon_0\epsilon_{si}\mathscr{E}_x|_{\phi=\phi_s} = -\sqrt{2\epsilon_0\epsilon_{si}qN_b}\,F(\phi_s, \phi_f, V_{cb}) \\ &= -\gamma C_{ox}F(\phi_s, \phi_f, V_{cb}) \end{aligned} \tag{6.20}$$

where $F(\phi_s, \phi_f, V_{cb})$ is obtained by replacing ϕ with ϕ_s in Eq. (6.18), and γ is the body factor, defined earlier as [cf. Eq. (5.11)],

$$\gamma = \frac{\sqrt{2\epsilon_0\epsilon_{si}qN_b}}{C_{ox}} \quad (V^{1/2}). \tag{6.21}$$

Equation (6.20) is valid in all regions (accumulation, depletion and inversion) of MOSFET operation. However, in the useful range of operation (depletion, weak inversion and strong inversion) ϕ_s is positive, and $\phi_s \gg V_t$, $2\phi_f \gg V_t$, and $2\phi_f + V_{cb} \gg V_t$. Therefore, the function F can be approximated as

$$F(\phi_s, \phi_f, V_{cb}) \approx [\phi_s(y) + V_t e^{(\phi_s(y) - 2\phi_f - V_{cb}(y))/V_t}]^{1/2}. \tag{6.22}$$

Combining Eqs. (6.19)–(6.22) yields

$$V_{gb} = V_{fb} + \phi_s(y) + \gamma[\phi_s(y) + V_t e^{(\phi_s(y) - 2\phi_f - V_{cb}(y))/V_t}]^{1/2}. \tag{6.23}$$

Equation (6.23) is a implicit equation in ϕ_s and can only be solved for given bias conditions using iterative procedures as discussed in Appendix E. Figure 6.3 shows a plot of ϕ_s as a function of $(V_{gb} - V_{fb})$ for different values of V_{cb}. Now combining Eqs. (6.10), (6.15), (6.17) and (6.23) with (6.14) yields

$$I_{ds} = \mu_s \frac{W}{L} C_{ox}\gamma \int_{V_{sb}}^{V_{sb}+V_{ds}} \int_{\phi_f}^{\phi_s} \frac{e^{(\phi - 2\phi_f - V_{cb})/V_t}}{F(\phi, \phi_f, V_{cb})} d\phi dV_{cb}. \tag{6.24}$$

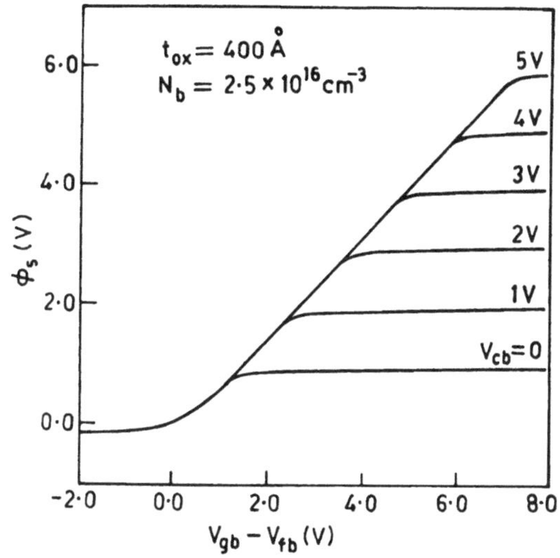

Fig. 6.3 The surface potential ϕ_s as a function of $(V_{gb} - V_{fb})$ for different values of V_{cb}

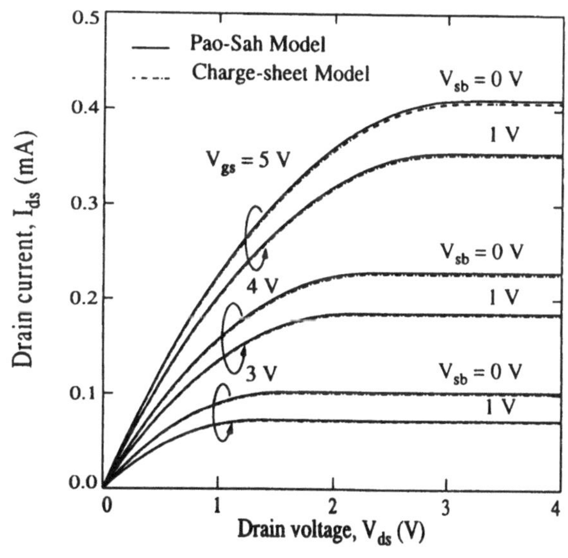

Fig. 6.4 The drain current-drain voltage characteristics of a MOSFET based on Pao–Sah model Eq. (6.24) (continuous lines) and charge-sheet model [Eqs. (6.35) and (6.36)] (dash-dot lines) at two V_{sb} using parameter values shown in Table 6.1

This double integral equation for I_{ds} is referred to as the Pao–Sah model [4] and can be only solved numerically. It takes into account both the drift and diffusion components of the drain current, and is valid in all regions of device operation including subthreshold and saturation. The different operation modes are not distinctly separated from each other and

Table 6.1. *nMOST parameter values used for Figures 6.4–6.6*

Parameter symbol	Parameter description	Parameter value	Units
L	Effective channel length	10	μm
W	Effective channel width	10	μm
t_{ox}	Gate oxide thickness	300	Å
μ_s	Channel mobility	600	cm^2/V.s
V_{fb}	Flat band voltage	-0.7	V
N_b	Substrate concentration	3×10^{16}	cm^{-3}

the transition from one mode to other is achieved asymptotically. Figure 6.4 shows complete I_{ds}–V_{ds} characteristics (continuous lines) obtained using Eq. (6.24). The MOSFET parameters used to calculate the current are described in Table 6.1 and are typical of a 2μm CMOS process.
The main disadvantage of this model is long computation time resulting from double numerical integration and multiple numerical solution of Eq. (6.22). Although the complexity of the Pao–Sah model makes it unsuitable for use in circuit simulators such as SPICE, it is the bench mark for model accuracy to which other simplified models are compared. Various simplifications of the Pao–Sah model have been proposed [5, 6], but the one which is still very accurate and does not require any numerical integration is called the *charge-sheet model* [7]–[12] and is derived in the next section.

6.3 Charge-Sheet Model

The analysis so far was very general. No assumption was made for the thickness of the inversion layer and it was assumed that both holes and electrons exist in the depletion region. Let us now assume the inversion layer to be of zero thickness (i.e., simply a sheet of charges) so that no potential is dropped across it.[3] Let us further assume that *the depletion region under the gate is practically free of mobile carriers so that the depletion approximation is valid.* This means that the mobile charge density Q_i is due to the *inversion* or *channel region* charge only, and is often referred to as inversion charge density.

[3] Recall that the inversion or channel region at the silicon surface is confined to a very thin layer of the order of 10–100 Å. Neglecting the inversion layer thickness is a fairly good approximation because this thickness is at least two orders of magnitude smaller than the depletion layer thickness.

The bulk charge density Q_b can be calculated as [see Eq. (6.15)]

$$Q_b(y) = -q\int_0^{\infty}(N_b - p)dx = -\sqrt{2\epsilon_0\epsilon_{si}qN_b}\int_0^{\phi_s}\frac{\sqrt{V_t}(e^{\phi(y)/V_t} - 1)}{F(\phi_s, \phi_f, V_{cb})}d\phi. \tag{6.25}$$

Assuming depletion approximation, the above integral can be approximated as

$$Q_b(y) = -\gamma C_{ox}\sqrt{\phi_s(y) - V_t} \quad \text{(depletion approximation)} \tag{6.26}$$

where we have made use of Eq. (6.21) for γ. Similar to the case of an MOS capacitor, one can also derive the following expression for Q_b, under the depletion approximation [cf. Eq. (4.29)],

$$Q_b(y) = -\gamma C_{ox}\sqrt{\phi_s(y)} \tag{6.27}$$

assuming an abrupt depletion layer boundary. The above equation is within 3% of Eq. (6.26) for the bias range of interest. It is this equation for Q_b which we will use in rest of this chapter. Noting that the induced charge Q_s in the channel is the sum of the inversion or channel charge Q_i and the bulk charge Q_b (i.e., $Q_s = Q_b + Q_i$) we can rearrange Eq. (6.19) in terms of Q_i as

$$Q_i(y) = -C_{ox}[V_{gb} - V_{fb} - \phi_s(y)] - Q_b(y)$$

or

$$Q_i(y) = -C_{ox}[V_{gb} - V_{fb} - \phi_s(y) - \gamma\sqrt{\phi_s(y)}]. \tag{6.28}$$

Now rewriting Eq. (6.13) in terms of the surface potential ϕ_s we have

$$I_{ds}(y) = -\mu_s W Q_i \frac{dV_{cb}(y)}{d\phi_s}\frac{d\phi_s}{dy}. \tag{6.29}$$

Rearranging Eq. (6.23) as

$$(V_{gb} - V_{fb} - \phi_s(y))^2 - \gamma^2[\phi_s(y) + V_t e^{(\phi_s(y) - 2\phi_f - V_{cb}(y))/V_t}] = 0 \tag{6.30}$$

and differentiating with respect to ϕ_s we get

$$\frac{dV_{cb}(y)}{d\phi_s} = 1 + V_t\left[\frac{2(V_{gb} - V_{fb} - \phi_s(y)) + \gamma^2}{(V_{gb} - V_{fb} - \phi_s(y))^2 - \gamma^2\phi_s(y)}\right]$$

which in terms of Q_b [Eq. (6.27)] and Q_i [Eq. (6.28)] becomes

$$\frac{dV_{cb}(y)}{d\phi_s} = 1 - \frac{V_t C_{ox}}{Q_i(y)}\left[1 + \frac{Q_i(y) - \gamma^2 C_{ox}}{Q_i(y) + 2Q_b(y)}\right]. \tag{6.31}$$

Combining Eqs. (6.31) and (6.29), and integrating the resulting equation from the source to drain yields I_{ds}. However, the final expression is rather complicated [11]. Equation (6.31) can be simplified by noting that the second term in the square bracket becomes important only when Q_i becomes small. Therefore, the above equation can be approximated by setting $Q_i = 0$ in the square bracket term giving [11]

$$\frac{dV_{cb}(y)}{d\phi} = 1 - \frac{V_t C_{ox}}{Q_i(y)}\left[1 - \frac{\gamma^2 C_{ox}}{2Q_b(y)}\right]. \tag{6.32}$$

Using this equation with Eq. (6.29) and simplifying we get

$$\begin{aligned} I_{ds}(y) &= -\mu_s W\left[Q_i(y)\frac{d\phi_s}{dy} - V_t\frac{dQ_i(y)}{dy}\right] \\ &= I_{ds1} + I_{ds2} \end{aligned} \tag{6.33}$$

where

$$I_{ds1} = -\mu_s W Q_i(y)\frac{d\phi_s}{dy} \quad \text{(drift component)}$$

$$I_{ds2} = \mu_s W V_t \frac{dQ_i(y)}{dy} \quad \text{(diffusion component).}$$

Equation (6.33), first derived by Brews, is known as the charge-sheet model [7], probably because it assumes the inversion layer to be simply a sheet of charges. This equation shows that the drain current is the sum of two components I_{ds1} and I_{ds2}, the drift and diffusion currents respectively. Note that although I_{ds} is constant, the two components, I_{ds1} and I_{ds2}, are functions of the distance y along the length of the channel. Also note that I_{ds1} and I_{ds2} are coupled differential equations and cannot be integrated separately. Assuming only the drift component is present, integrating the drift part of Eq. (6.33) under the boundary condition

$$\phi_s(y) = \begin{cases} \phi_{s0} & \text{at } y = 0 \\ \phi_{sL} \quad \text{and} & \text{at } y = L \end{cases} \tag{6.34}$$

and making use of Eq. (6.28) for Q_i yields

$$\int_0^L I_{ds1}\,dy = -\mu_s W \int_{\phi_{s0}}^{\phi_{sL}} Q_i\,d\phi_s$$

$$\text{or, } I_{ds1} = \mu_s C_{ox}\frac{W}{L}\left[(V_{gb} - V_{fb})(\phi_{sL} - \phi_{s0}) - \frac{1}{2}(\phi_{sL}^2 - \phi_{s0}^2) - \frac{2}{3}\gamma\{\phi_{sL}^{3/2} - \phi_{s0}^{3/2}\}\right] \tag{6.35}$$

where ϕ_{s0} and ϕ_{sL} are the values of ϕ_s at the source and drain ends of the channel, respectively. Similarly, if only diffusion current is present, the drain current again is obtained by integrating I_{ds2} from the source to drain

$$\int_0^L I_{ds2}dy = \mu_s W V_t \int_{\phi_{s0}}^{\phi_{sL}} dQ_i$$

$$\text{or, } I_{ds2} = \mu_s C_{ox}\frac{W}{L}V_t[\gamma(\phi_{sL}^{1/2} - \phi_{s0}^{1/2}) + (\phi_{sL} - \phi_0)]. \tag{6.36}$$

The total drain current is obtained by adding Eqs. (6.35) and (6.36). The values of ϕ_{s0} and ϕ_{sL}, required to calculate I_{ds}, are obtained numerically by solving the implicit Eq. (6.30) for $\phi_s(y)$ under the conditions

$$\phi_s(y) = \begin{cases} \phi_{s0} \quad \text{and} \quad V_{cb}(y) = V_{sb} & \text{at } y = 0 \\ \phi_{sL} \quad \text{and} \quad V_{cb}(y) = V_{sb} + V_{ds} & \text{at } y = L. \end{cases} \tag{6.37}$$

In weak inversion, where ϕ_{s0} is almost equal to ϕ_{sL}, even small errors in the values of ϕ_{s0} and ϕ_{sL} can lead to a large error in the current I_{ds1}, which depends on the difference $\phi_{sL} - \phi_{s0}$. Therefore, *an accurate solution is required for the surface potential, particularly for weak inversion current calculations.*

The implicit Eq. (6.30) in ϕ_s can be solved very efficiently by using the Schroder series method [13], which is based on a Taylor series expression of the inverse function, provided a good initial guess is used. Approximate solutions have also been suggested, but these are applicable only in certain regions of device operation. These are dicussed in Appendix E.

The procedure used to calculate the drain current I_{ds} as a function of drain voltage V_{ds} is to first choose V_{sb}, and then calculate ϕ_{s0}, for a given value of V_{gs}, using Eq. (6.30). Next, assuming a set of values of V_{ds}, calculate ϕ_{sL}, again using Eq. (6.30). The ϕ_{s0} and ϕ_{sL} are in turn used to calculate I_{ds} using Eqs. (6.35) and (6.36). Figure 6.5 shows the drain current I_{ds} and its components I_{ds1} and I_{ds2}, as a function of V_{gs} at $V_{ds} = 5$ V and $V_{sb} = 0$. These current values were obtained for a MOSFET with the parameters described in Table 6.1. As can be seen from this figure, *in strong inversion* $I_{ds} \approx I_{ds1}$ *so that the current is mainly due to drift. In weak inversion* $I_{ds} \approx I_{ds2}$ *and the current is mainly due to diffusion.* However, there is a region between the weak and strong inversion, called the *moderate inversion* by Tsividis [14, 15], where both drift and diffusion are important. The width of the moderate inversion in terms of the gate voltage is several tenths of a volt. A precise limit of this region has been given by Tsividis [15]. It turns out that the lower limit (ϕ_{mL}) of ϕ_s for moderate inversion is within a V_t of $2\phi_f$, while the upper limit (ϕ_{mH}) is 5 V_t to 6 V_t above ϕ_{mL}. The corresponding V_{gb} values are V_{gbL} and V_{gbH}, respectively and are obtained from Eq. (6.30) by using ϕ_s equals ϕ_{mL} and ϕ_{mH}, respectively.

The complete $I_{ds} - V_{ds}$ characteristics, obtained using the charge-sheet model (dashed-dot lines), are shown in Figure 6.4. Also shown in this figure are

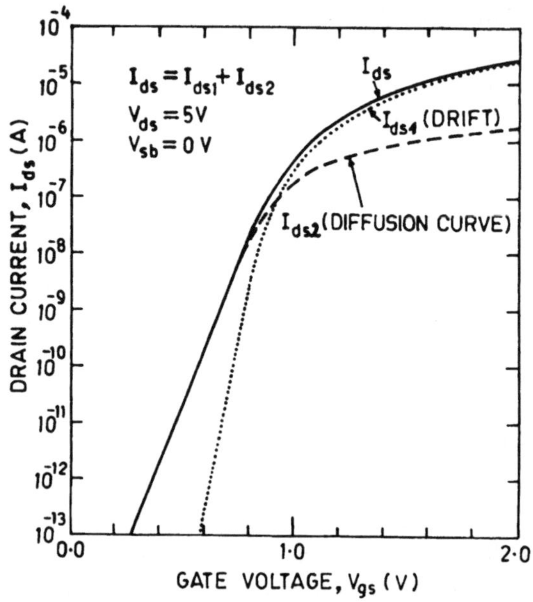

Fig. 6.5 The drain current as a function of V_{gs} based on charge sheet model. (a) diffusion current I_{ds2} (b) drift current I_{ds1} and (c) total current (continuous line) $I_{ds} = I_{ds1} + I_{ds2}$

I_{ds} calculated using the Pao-Sah model (continuous lines). Note that the charge-sheet model predicts I_{ds} within 1% of that calculated using the Pao–Sah model [cf. Eq. (6.24)] under most operating conditions [7]–[10]. Although the charge-sheet model is simpler compared to the Pao–Sah model, it still requires time consuming iterations to calculate ϕ_s at the source and drain ends. Therefore, it takes much longer computation time compared to the piece-wise multisection model to be discussed in the next section. Hence, in spite of its advantages, this model is not widely used in real circuit simulation programs [17]–[19].

In what follows we will use Eq. (6.14) to develop drain current models based on additional approximations to circumvent the implicit relation (6.30) for ϕ_s. This is often done by separately modeling distinct regions of device operation. One very commonly used boundary between the weak and strong inversion regions is the threshold voltage V_{th}. Thus, we will have one model for strong inversion and the other for weak inversion regions of device operation. Note that both models discussed so far have completely natural transitions between different regions. The model discussed next is a piece-wise multisection model. This type of model is the one most commonly used for circuit simulation because of its simplicity. Initially, we will develop the first order piece-wise model for a long and wide device. Subsequently, this first order model will be improved upon to develop more accurate models for short channel VLSI devices.

6.4 Piece-Wise Drain Current Model for Enhancement Devices

The drain current Eq. (6.14) includes both the drift and diffusion components of the current.

Assumption 6. Let us assume that *the diffusion current is negligible so that all the current flow is due to drift only.* This is a fairly good assumption provided the device is in strong inversion, that is, the gate voltage is greater than the threshold voltage ($V_{gs} > V_{th}$). If the diffusion current is neglected then, from Eq. (6.2), it is easy to see that

$$J_n \cong qn\mu_s\mathscr{E} = -qn\mu_s\frac{d\phi_s}{dy}. \tag{6.38}$$

Comparing this equation with Eq. (6.7) we find that for a device in strong inversion, we can write

$$\frac{\partial\varphi_n}{\partial y} \approx \frac{\partial\phi_s}{\partial y},$$

and therefore, from Eq. (6.9) it follows that

$$\phi_s(y) = \phi_s(0) + V_{cb}(y), \quad \text{(strong inversion)} \tag{6.39}$$

where $\phi_s(0)$ is the surface potential at $y = 0$ (source end). For the sake of algebraic manipulation, it is more convenient to write

$$V_{cb}(y) = V_{sb} + V(y)$$

where $V(y)$ now varies from 0 at the source end to V_{ds} at the drain end. With this assumption, Eq. (6.39) can be rewritten as

$$\phi_s(y) = 2\phi_f + V_{sb} + V(y), \quad \text{(strong inversion)} \tag{6.40}$$

where we have replaced $\phi_s(0)$ by $2\phi_f$, the classical criterion for strong inversion, while Eq. (6.14) reduces to

$$I_{ds} = -\mu_s\frac{W}{L}\int_0^{V_{ds}} Q_i(y)dV. \tag{6.41}$$

Thus, to calculate I_{ds} we need to calculate Q_i. We will now derive a simple and more useful expression for Q_i using a charge balance equation given by Eq. (6.28). It is interesting to note that Eq. (6.41) can also be written as

$$I_{ds}(y) = -WQ_i(y)v \tag{6.42}$$

where $v(=\mu_s\mathscr{E})$ is the velocity of the carriers in the channel region [cf. Eq. (2.20)].

6.4.1 First Order Model

Linear Region of the MOSFET Operation. Substituting the value of $\phi_s(y)$ from Eq. (6.40) in Eq. (6.28) results in the following expression for the inversion charge density Q_i

$$Q_i(y) = -C_{ox}[V_{gs} - V_{fb} - 2\phi_f - V(y)] - Q_b(y) \quad (\text{C/cm}^2). \tag{6.43}$$

Assumption 7. Let us assume that the bulk charge density Q_b *is constant along the length of the channel, independent of the applied drain voltage* V_{ds} so that $\phi_s(y) = 2\phi_f + V_{sb}$ is constant along length of the channel. With this assumption Eq. (6.27) becomes

$$Q_b(y) = -C_{ox}\gamma\sqrt{2\phi_f + V_{sb}} \quad (\text{C/cm}^2). \tag{6.44}$$

Combining Eq. (6.43) and (6.44) we get

$$Q_i(y) = -C_{ox}[V_{gs} - V_{th} - V(y)] \tag{6.45}$$

where V_{th} is the threshold voltage defined as [cf. Eq. (5.14)]

$$V_{th} = V_{fb} + 2\phi_f + \gamma\sqrt{2\phi_f + V_{sb}} \quad (\text{V}). \tag{6.46}$$

Using $Q_i(y)$ from Eq. (6.45) in Eq. (6.41) and carrying out the integration results in the following equation for the drain current I_{ds}

$$\boxed{I_{ds} = \frac{\mu_s C_{ox} W}{L}[V_{gs} - V_{th} - 0.5\,V_{ds}]V_{ds}, \quad V_{gs} > V_{th}.} \tag{6.47}$$

This is the current equation first derived by Sah [20] and later used by Schichman and Hodges [21] as a MOSFET model for circuit simulation. It is the SPICE MOSFET level 1 model. The factor $\mu_s C_{ox}$ is often referred as the *process transconductance parameter* κ, that is,

$$\kappa = \mu_s C_{ox} \quad (\text{A/V}^2) \tag{6.48}$$

and along with V_{th} accounts for the basic process variation in the current equation. Thus, Eq. (6.47) is often written as

$$I_{ds} = \kappa\frac{W}{L}[V_{gs} - V_{th} - 0.5\,V_{ds}]V_{ds}. \tag{6.49}$$

A typical value of κ (for $t_{ox} = 300$ Å) is about 25 μA/V^2 for nMOST and 10 μA/V^2 for pMOST. The factor $\kappa W/L$ is called the *gain factor* β, defined as

$$\boxed{\beta = \frac{\mu_s C_{ox} W}{L} \quad (\text{A/V}^2)} \tag{6.50}$$

If V_{ds} is small (< 0.1 V), Eq. (6.47) can be approximated as

$$I_{ds} \cong \beta(V_{gs} - V_{th})V_{ds}. \tag{6.51}$$

This shows that current varies linearly with the gate voltage. Consequently this region of MOSFET operation is often referred to as the *linear region of operation.* Rearranging the above equation we get,

$$R_{ch} = \frac{V_{ds}}{I_{ds}} \cong [\beta(V_{gs} - V_{th})]^{-1} \tag{6.52}$$

where R_{ch} is the effective resistance between the source and drain, often called the *channel resistance.* Note that R_{ch} varies linearly with the difference $(V_{gs} - V_{th})$, often referred to as the *effective gate drive* or *effective gate voltage.* This explains why MOSFETs are sometime referred to as voltage controlled variable resistors.

The plot of I_{ds} versus V_{ds}, derived from Eq. (6.47), for different values of V_{gs} is shown in Figure 6.6. For a given value of V_{gs}, the current I_{ds} initially increases with increasing V_{ds}. It reaches a peak value and then begins to decrease with further increase in V_{ds}. However, experimentally one does not observe a decrease in the current with increasing V_{ds}. The measured current follows Eq. (6.47) till the peak is reached and then saturates with further increase in V_{ds}. The reason why Eq. (6.47) does not predict the saturation is that this equation becomes invalid once the peak is reached. This can be seen as follows: The peak position of the current can be obtained by differentiating Eq. (6.47) with respect to V_{ds} and equating the resulting

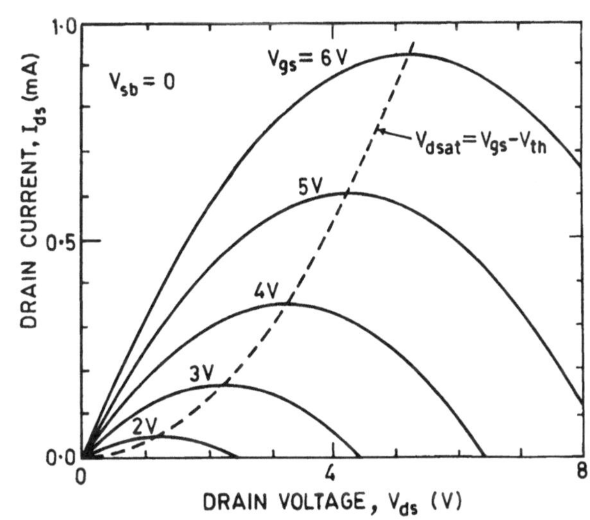

Fig. 6.6 The current-voltage characteristics of a nMOST based on Eq. (6.47) for the model parameter shown in Table 6.1. Dotted line shows saturation voltage for a given V_{gs}

equation to zero, that is,

$$\frac{dI_{ds}}{dV_{ds}} = \frac{\mu_s C_{ox} W}{L} [V_{gs} - V_{th} - V_{ds}] = 0. \tag{6.53}$$

This shows that the peak occurs when $V_{ds} = V_{gs} - V_{th}$. Also note from Eq. (6.45) that when $V(y) = V_{gs} - V_{th}$, the channel charge $Q_i = 0$; that means the channel does not exist. The maximum value of $V(y)$ will be at the drain end of the channel where $V(y) = V_{ds}$; therefore, when $V_{ds} \geq (V_{gs} - V_{th})$, we find that $Q_i = 0$. In other words, once the peak current is reached the GCA fails (see assumption 1) and therefore Eq. (6.47) is no longer valid.

Saturation Voltage. In deriving Eq. (6.47) it was assumed that an inversion layer exists along the channel from the source to drain. This indeed is true for $V_{gs} > V_{th}$ and small V_{ds} (see Figure 6.7a). However, for a given V_{gs}, when V_{ds} reaches a certain value, the Q_i at the drain end drops to zero as is

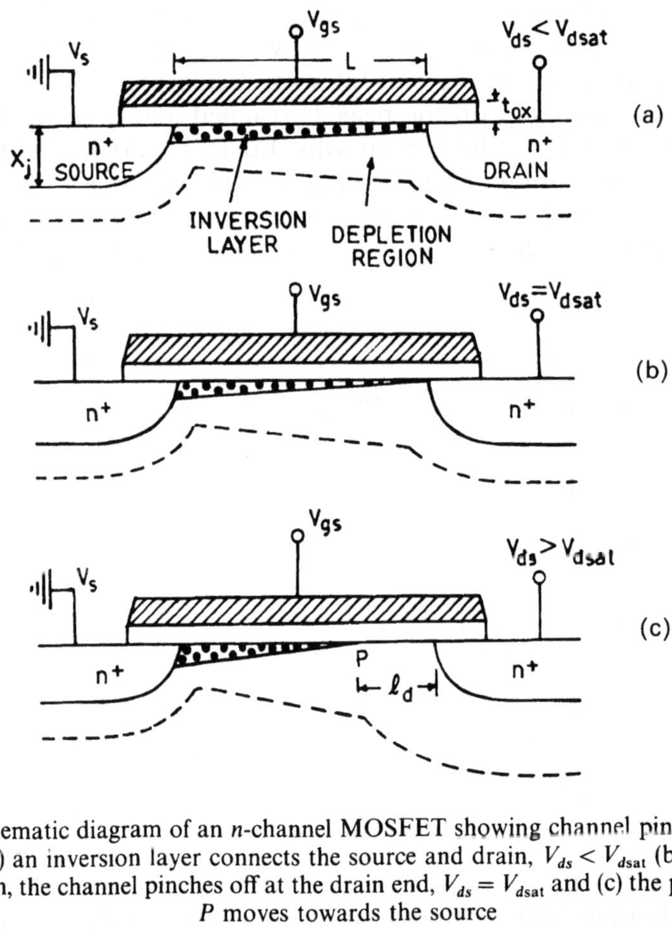

Fig. 6.7 Schematic diagram of an n-channel MOSFET showing channel pinch-off as V_{ds} is increased (a) an inversion layer connects the source and drain, $V_{ds} < V_{dsat}$ (b) at the on set of saturation, the channel pinches off at the drain end, $V_{ds} = V_{dsat}$ and (c) the pinch-off point P moves towards the source

evident from Eq. (6.45). That is, the channel is pinched-off near the drain end as shown in Figure 6.7b. The drain voltage that results in the disappearance of the channel (i.e., channel pinch-off) at the drain end is referred to as the *pinch-off or saturation voltage* V_{dsat}. The corresponding current at V_{dsat} is called the *saturation current* I_{dsat}. Note that the condition for pinch-off ($Q_i = 0$) is equivalent to the condition $dI_{ds}/dV_{ds} = 0$. In other words, *at pinch-off the slope of the* $I_{ds} - V_{ds}$ *characteristics becomes zero.* Physically, at the pinch-off point, the normal field $\mathscr{E}_x$ is inverted which pushes the mobile carriers (that are flowing to the drain area) away from the surface. By definition, the pinch-off voltage V_{dsat} is obtained either from Eq. (6.45) by equating Q_i to zero, or from Eq. (6.53) by replacing V_{ds} with V_{dsat}. Thus

$$\boxed{V_{dsat} = V_{gs} - V_{th}.} \tag{6.54}$$

This shows that the pinch-off voltage V_{dsat} (V_{ds} at which current saturation occurs) equals the effective gate voltage, and will increase with increasing V_{gs}. The behavior of V_{dsat} as a function of V_{gs} is shown as a dashed line in Figure 6.6.

Replacing V_{ds} with V_{dsat} in Eq. (6.47) results in the following equation for the drain current I_{dsat} at the pinch-off point

$$I_{dsat} = \frac{\beta}{2}(V_{gs} - V_{th})^2. \tag{6.55}$$

Note that Eqs. (6.54) and (6.55) are obtained on the assumption that $Q_i = 0$ in the pinch-off region. Physically speaking, this is incorrect as we can see from Eq. (6.13), which gives the field $\mathscr{E}_y$ along the channel as

$$\mathscr{E}_y(y) = \frac{I_{ds}(y)}{\mu_s W Q_i(y)}. \tag{6.56}$$

Since $Q_i = 0$ at the drain end, the field $\mathscr{E}_y$ becomes infinity. This means that carriers would have to move with infinite drift velocity in order for current to be nonzero, which obviously is not possible. Therefore, *it is incorrect to assume* $Q_i = 0$ at the pinch-off point. A more *correct statement should be that* Q_i *has a very small but finite value in the pinch-off region.*

Saturation Region of MOSFET Operation. In the discussion so far we have seen that as V_{ds} increases (for a given V_{gs}), the channel charge Q_i decreases near the drain end, and when $V_{ds} = V_{dsat} (= V_{gs} - V_{th})$, the channel is pinched-off. For $V_{ds} > V_{dsat}$, the pinched-off portion of the channel moves towards the source end due to the widening of the drain depletion region (see Figure 6.7c). In other words, as V_{ds} increases beyond pinch-off, the pinched-off region l_d between the channel pinch-off point P and the n^+

drain region causes the effective channel length to decrease from L to $L - l_d$. Since the channel can support only V_{dsat}, any voltage greater than V_{dsat} (i.e., $V_{ds} - V_{dsat}$) must be absorbed by the l_d portion of the channel. Clearly l_d is bias dependent and thus modulates the effective channel length. This is referred to as *channel length modulation* (CLM). If the channel length L is large enough such that $l_d \ll L$, then the drain current I_{ds} remains approximately constant at I_{dsat} for any V_{ds} in excess of V_{dsat}. Thus, to a first order, for V_{ds} beyond pinch-off, the current $I_{ds} = I_{dsat}$ and is given by Eq. (6.55), that is,

$$\boxed{I_{ds} = \frac{\beta}{2}(V_{gs} - V_{th})^2, \quad V_{ds} > V_{dsat}.} \tag{6.57}$$

The region of operation of the MOSFET beyond pinch-off ($V_{ds} > V_{dsat}$) is referred to as the *saturation region* because I_{ds} ideally does not increase in this region. For this reason the region below V_{dsat}, which we have called the linear region, is also referred to as the *nonsaturation region.*[4] Note that Eq. (6.57) predicts that I_{ds} in saturation varies as the square of the effective gate voltage and hence it is often referred to as the *square law model* of the MOSFET. Equations (6.47), (6.54) and (6.57) when plotted together give

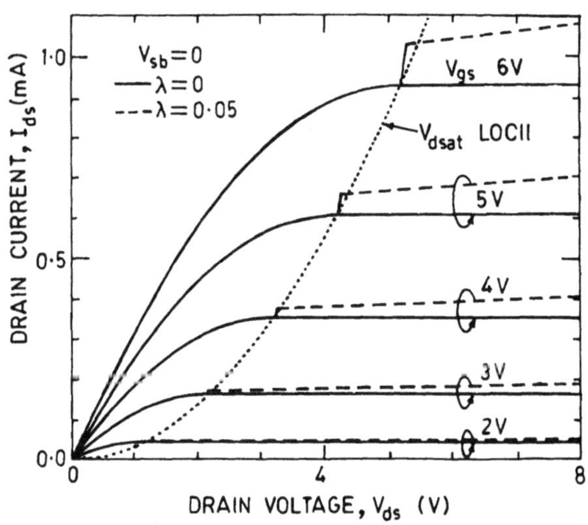

Fig. 6.8 Linear and saturation region of the device operation and the dividing line between the two based on Eqs. (6.47) (continuous lines) and (6.58) (dotted lines). The non zero slope is caused by channel length modulation effect

[4] The linear or nonsaturation region is sometime also referred to as the triode region, because the characteristics of this region are similar to the triode vacuum tube.

the MOSFET output characteristics as shown by the continuous lines in Figure 6.8.

It should be pointed out that Eq. (6.57) for the drain current in saturation is based on the assumption that the current is independent of V_{ds}. In practice this is true only for long channel devices. As L is reduced, the experimentally observed I_{ds} increases slowly with increasing V_{ds} for V_{ds} beyond pinch-off. This increase in the current can be modeled using the CLM effect through the l_d term. As V_{ds} increases beyond V_{dsat}, l_d also increases so that the effective channel length becomes $L_{eff} = L - l_d$. Using L_{eff} for L in Eq. (6.57) we get

$$I_{ds} = \frac{\mu_s C_{ox} W}{2(L - l_d)} (V_{gs} - V_{th})^2, \quad V_{ds} > V_{dsat} \tag{6.58}$$

which shows that as l_d increases, so does the drain current. Using Eqs. (6.55) and (6.58), it is easy to see that

$$I_{ds} = \frac{I_{dsat}}{1 - l_d/L}. \tag{6.59}$$

In general l_d is small compared to L, therefore to a first approximation

$$\left(1 - \frac{l_d}{L}\right)^{-1} \approx 1 + \frac{l_d}{L}.$$

With this approximation, Eq. (6.59) becomes

$$\boxed{I_{ds} = I_{dsat}\left(1 + \frac{l_d}{L}\right).} \tag{6.60}$$

It is this equation which is generally used for circuit simulation rather than Eq. (6.59) because it is more stable with respect to numerical problems. The computation of the exact value of l_d requires a two-dimensional solution as we will see later (section 6.7.3), but to a first approximation (first order model) we can write the following simple empirical relation for l_d [21]

$$1 + \frac{l_d}{L} \cong 1 + \lambda V_{ds}$$

so that Eq. (6.60) becomes

$$I_{ds} = I_{dsat}(1 + \lambda V_{ds}) \tag{6.61}$$

where λ is called the *channel length modulation parameter* and represents the small influence of V_{ds} on I_{ds}. Note that when $V_{ds} = -1/\lambda$, $I_{ds} = 0$. This means that when I_{ds} is extrapolated backward (from the saturation region)

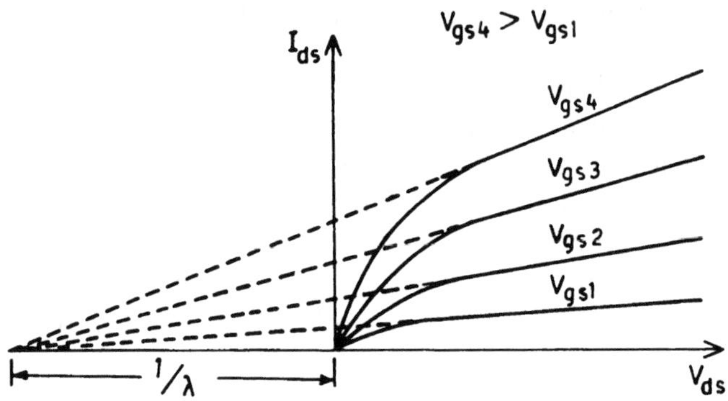

Fig. 6.9 Measurement of channel length modulation factor λ

it will cut the V_{ds} axis at a value of $1/\lambda$ as shown in Figure 6.9. However, this is an ideal case, and generally the value of λ is obtained by curve fitting the experimental I_{ds} data with that of Eq. (6.61) so as to minimize the error between the measured data and model. Typical values of λ fall in the range 0.05–0.001 V^{-1}.

Equation (6.61) is a first order approximation for the CLM effect. It provides the basic feature of nonzero slope for the saturated drain current as shown by the dotted lines in Figure 6.8. Note that the use of Eq. (6.61) for calculating I_{ds} in saturation results in a discontinuity of the current at $V_{ds} = V_{dsat}$. The SPICE MOS Level 1 model corrects for this discontinuity by multiplying the linear region current Eq. (6.47) by the factor $(1 + \lambda V_{ds})$. Other better approaches used to avoid this discontinuity and modeling the CLM effect are discussed later in section 6.7.3.

To summarize, we have developed a first order MOSFET model which can be described by the following equations:

$$I_{ds} = \begin{cases} 0 & V_{gs} \leq V_{th} \quad \text{(cutoff region)} \\ \beta(V_{gs} - V_{th} - 0.5V_{ds})V_{ds} & V_{gs} > V_{th},\ V_{ds} \leq V_{dsat} \quad \text{(linear region)} \\ \dfrac{\beta}{2}(V_{gs} - V_{th})^2(1 + \lambda V_{ds}). & V_{gs} > V_{th},\ V_{ds} > V_{dsat} \quad \text{(saturation region)} \end{cases} \tag{6.62}$$

and is based on the following assumptions:

1. the gradual channel approximation (GCA) is valid,
2. hole current can be neglected (for nMOST),
3. recombination and generation are neglected,
4. current flows in the y direction (along the length of the channel) only,
5. carrier mobility μ_s in the inversion layer is constant in the y direction,
6. current flow is due to drift only (diffusion current is neglected),
7. bulk charge Q_b is constant at any point in the y direction.

Table 6.2. *Typical SPICE Level 1 model parameters for 2 μm CMOS process*

Parameter symbol	Parameter name	Parameter description	Typical nMOST	values pMOST	Units
V_{To}	VTO	Zero-bias Threshold voltage	0.8	−0.8	V
κ	KP	Transconductance parameter	$2 \cdot 10^{-5}$	$1 \cdot 10^{-5}$	A/V^2
γ	GAMMA	Body factor	1.3	0.6	$V^{1/2}$
λ	LAMBDA	Channel length modulation factor	0.01	0.02	V^{-1}
$2\|\phi_f\|$	PHI	Bulk Fermi-potential	0.7	0.6	V

Although Eq. (6.62) is derived for a nMOST, the same equations apply for a pMOST once all polarities of voltages and current are reversed (see Table 3.2). Typical values of the model parameters of Eq. (6.62) are shown in Table 6.2 and are based upon a 2 μm CMOS *n*-well process. The accuracy of this simple model, represented by Eq. (6.62), is far from satisfactory even for relatively long channel devices (say 10 μm) by today's standards. However, because of its simplicity, it is *very useful model for performing basic circuit analysis and developing design equations for circuit performance.*

6.4.2 Bulk-Charge Model

The first order MOSFET current model developed in the previous section, though very useful for hand calculations, is not normally used for present day circuit simulation. This is because the model, in general, overestimates the current and the saturation drain voltage. This is not surprising since the model is based on various simplifying assumptions, some of which are not valid. Let us examine more carefully assumption 7, which states that Q_b is constant along the length of the channel. This means that the depletion width X_{dm} under the gate is constant from the source to drain even though $V_{ds} \neq 0$. In reality when $V_{ds} \neq 0$, X_{dm} will increase as we move from the source towards the drain (see Figure 5.24). Consequently, it is more appropriate to include the changing bulk charge in the depletion region from the source to the drain. In other words, we must use Eq. (6.27) for $Q_b(y)$ with $\phi_s(y)$ from Eq. (6.40), that is,

$$Q_b(y) = -C_{ox}\gamma\sqrt{\phi_s(y)} = -C_{ox}\gamma\sqrt{2\phi_f + V_{sb} + V(y)}. \tag{6.63}$$

This value of $Q_b(y)$ when used in Eq. (6.43) yields

$$Q_i(y) = -C_{ox}[V_{gs} - V_{fb} - 2\phi_f - V(y) - \gamma\sqrt{2\phi_f + V_{sb} + V(y)}]. \tag{6.64}$$

Using the above value of Q_i in Eq. (6.41) and integrating we get

$$I_{ds} = \frac{\mu_s W C_{ox}}{L}[(V_{gs} - V_{fb} - 2\phi_f - 0.5V_{ds})V_{ds} - \tfrac{2}{3}\gamma\{(V_{ds} + 2\phi_f + V_{sb})^{3/2} - (2\phi_f + V_{sb})^{3/2}\}]. \tag{6.65}$$

This is the current equation that takes into account varying bulk charge in the depletion region. Comparing Eq. (6.47) with (6.65) we find that the latter equation predicts lower current compared to the former. Physically speaking, this is understandable because the increasing bulk charge will reduce the inversion charge (for the same bias conditions) resulting in a lower current. However, Eq. (6.65) is more complex compared to Eq. (6.47). Indeed this is the price paid for achieving higher accuracy. The current Eq. (6.65) is sometimes referred to as the Ihantola–Moll model [22]. This is the current equation used by the SPICE Level 2 MOSFET model [23]. Figure 6.10 shows plots of I_{ds} versus V_{ds} for two values of V_{gs} using Eq. (6.65) (dotted line) and (6.47) (continuous line). The parameters used for calculating current are shown in Table 6.1. Note that Eq. (6.47) overestimates the current. The difference becomes much more pronounced at non zero V_{sb}. At low V_{ds} such that

$$V_{ds} \ll V_{gs} - V_{fb} - 2\phi_f,$$

and

$$V_{ds} \ll 2\phi_f + V_{sb},$$

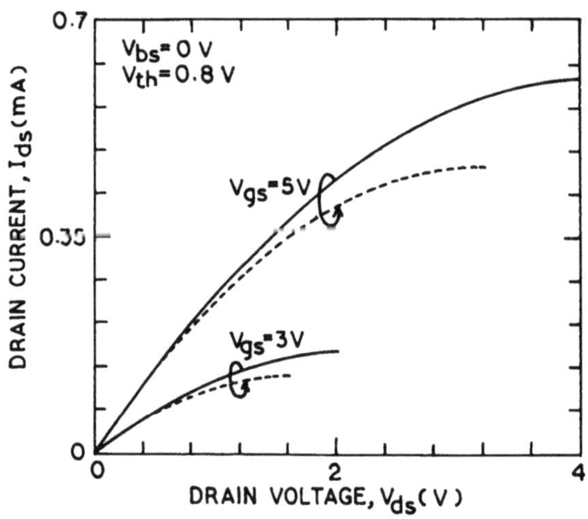

Fig. 6.10 Comparison of the $I_{ds} - V_{ds}$ characteristics in the linear region of device operation based on the first order theory (continuous lines, Eq. (6.47)) and bulk-charge theory (dotted lines, Eq. 6.65))

it follows (using the Binomial expansion) that

$$(V_{ds} + 2\phi_f + V_{sb})^{3/2} \cong (2\phi_f + V_{sb})^{3/2} + \tfrac{3}{2}(2\phi_f + V_{sb})^{1/2} V_{ds}$$

therefore, Eq. (6.65) reduces to

$$I_{ds} \cong \beta(V_{gs} - V_{th})V_{ds} \tag{6.66}$$

which is the same as Eq. (6.51). This makes sense because now V_{ds} is small, and therefore the depletion width X_{dm} under the gate can be assumed approximately constant along the length of the channel, the assumption upon which Eq. (6.51) is based. Note that Eq. (6.65) for I_{ds} is the linear region current equation, and therefore is valid only for $V_{ds} \leq V_{dsat}$. In this case V_{dsat} is obtained in the same way as discussed in the previous section, namely differentiating Eq. (6.65) with respect to V_{ds} and equating the resulting equation to zero. This results in the following expression for V_{dsat},

$$V_{dsat} = V_{gs} - V_{fb} - 2\phi_f + \frac{\gamma^2}{2} - \gamma\sqrt{V_{gs} - V_{fb} + V_{sb} + \frac{\gamma^2}{4}} \tag{6.67}$$

which again is more complicated compared to Eq. (6.54). Substituting the above value of V_{dsat} for V_{ds} in Eq. (6.65) gives the saturation region current.

6.4.3 Square-Root Approximation

If we look carefully at the derivation for the current Eq. (6.65), it is easy to see that the 3/2 power terms in the I_{ds} expression are due to the integration of the square root term in Q_b [cf. Eq. (6.63)]. This square-root term not only results in a complex I_{ds} equation, but it makes difficult, if not impossible, to arrive at closed form current and capacitance equations for the short-channel devices, as we shall see later. However, approximating the square root term by a linear function results in more tractable equations. For circuit models, the square-root term in Q_b, namely

$$F(V, V_0) \equiv \sqrt{2\phi_f + V_{sb} + V} \equiv F_{\mathbf{exact}} \tag{6.68}$$

is often approximated by the following expression with a linear dependence in $V(y)$

$$F(V, V_0) \cong \sqrt{2\phi_f + V_{sb}} + \delta \cdot V \equiv F_{\mathbf{approx}} \tag{6.69}$$

where $V_0 = 2\phi_f + V_{sb}$ and δ is a constant for a given value of V_0. Various different expressions for δ have been proposed in approximating the function $F(V, V_0)$ [24]–[28]. However, the first order approximation for δ can be obtained by retaining the first two terms in the Taylor series expansion of the function $F(V, V_0)$ resulting in[5]

$$F(V, V_0) \cong \sqrt{2\phi_f + V_{sb}} + \frac{0.5\,V}{\sqrt{2\phi_f + V_{sb}}}. \tag{6.70}$$

Comparing Eqs. (6.69) and (6.70) we see that

$$\delta = \frac{0.5}{\sqrt{2\phi_f + V_{sb}}}. \tag{6.71}$$

The plot of $F(V, V_0)$ versus V for different values of V_0 is shown in Figure 6.11. For a given V_0, dotted lines are the approximate values of $F(V, V_0)$ obtained from Eq. (6.70), while the continuous lines are the exact expression (6.68). Note that by taking the first two terms of the Taylor

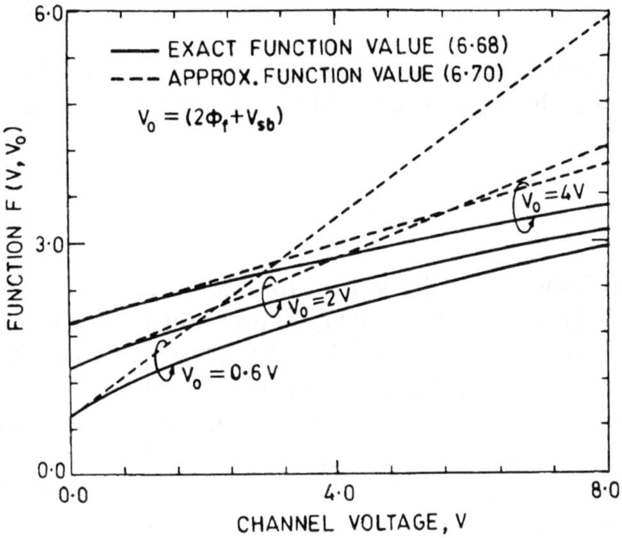

Fig. 6.11 Comparison of the exact and approximate values for the function $F(V, V_0)$ against channel voltage V for different values of V_0

---

[5] Assuming $V_0 = 2\phi_f + V_{sb}$, the Taylor series expansion of the function $F(V, V_0)$ results in

$$F(V, V_0) = \sqrt{V + V_0} = V_0^{1/2} + \frac{1}{2} V_0^{\,3/2} V - \frac{1}{8} V_0^{-3/2} V^2 + \cdots.$$

Neglecting second and higher order terms in V results in Eq. (6.70).

series, the value of δ becomes too large to provide a good approximation, particularly at low V_{sb} and high V. For this reason, a value of δ slightly smaller than that defined by Eq. (6.71) is often used, that is,

$$\delta = \frac{0.5}{\sqrt{2\phi_f + V_{sb}}} g \tag{6.72}$$

where g is a correction factor which varies between 0.5 and 0.8. In one approach $g = 0.8$ [27], while in some other approaches g is allowed to depend on V_{sb} for a better fit [24, 25]. In the latter case, the following semi-empirical expression for δ is suggested

$$\delta = \frac{0.5}{\sqrt{2\phi_f + V_{sb}}} \left[1 - \frac{1}{a_1 + a_2(2\phi_f + V_{sb})} \right] \tag{6.73}$$

where a_1 and a_2 are constant chosen such that the expressions (6.73) and (6.69) will give the best least square-fit to the function $F(V, V_0)$ over a range of V and V_0. Different values of a_1 and a_2 have been proposed. For example, Hanafi et al. [24] assume $a_1 = 1.41$ and $a_2 = 0.43$, while Sheu et al. [25] assume $a_1 = 1.744$ and $a = 0.8364$ using V from 0–10 V in 0.5 V increments and $(2\phi_f + V_{sb})$ in the range 0.7 to 20.7 V.
Another simple value proposed for δ is [27]

$$\delta = \frac{0.5}{\sqrt{1 + 2\phi_f + V_{sb}}}. \tag{6.74}$$

The most accurate approximation for δ is obtained by minimizing the least square function f [26]

$$f = \int_0^V [\sqrt{(V + 2\phi_f + V_{sb})} - (\delta V + \sqrt{2\phi_f + V_{sb}})]^2 \, dV \tag{6.75}$$

such that

$$\frac{df}{dV} = 0.$$

Differentiating Eq. (6.75) and equating it to zero yields, after a little algebra, the following expression for δ,

$$\delta = \frac{1}{V_r^3 \sqrt{2\phi_f + V_{sb}}} [0.8 - 1.5V_r^2 + (1 + V_r)^{3/2}(1.2V_r - 0.8)] \tag{6.76}$$

where

$$V_r = \frac{V}{2\phi_f + V_{sb}}.$$

Though accurate, this is a complicated expression and not suitable for CAD models. However, the following simplified form of Eq. (6.76) has been used in the drain current model [28]

$$\delta = \frac{0.5}{\sqrt{2\phi_f + V_{sb}}}\left[1 - \frac{V}{30(2\phi_f + V_{sb})}\right]. \tag{6.77}$$

This approximation, though accurate, has δ as a function of the variable V; so δ must be calculated for each V.

The effect of approximating the function $F(V, V_0)$ with different δ expressions is most sensitive at zero V_{sb}. Therefore, a comparison is made between the exact and approximate functions at zero V_{sb} by calculating the relative errors between them using different δ expressions. The results are shown in Figure 6.12 where the error E_r is defined as

$$E_r = 100 \times \frac{F_{\text{exact}} - F_{\text{approx}}}{F_{\text{exact}}}$$

where F_{exact} and F_{approx} are values of F given by Eqs. (6.68) and (6.69), respectively. Note from this figure that the simplest approximation for δ [cf. Eq. (6.71)] has maximum error, therefore this approximation will underestimate the depletion layer charge Q_b the most. However, the resulting error in I_{ds} calculations is not usually significant because Q_b is much smaller than Q_i. In fact, for I_{ds} calculations, any of the δ functions discussed above can be used depending upon the desired accuracy and speed of calculation. However, accuracy in δ approximations are important for

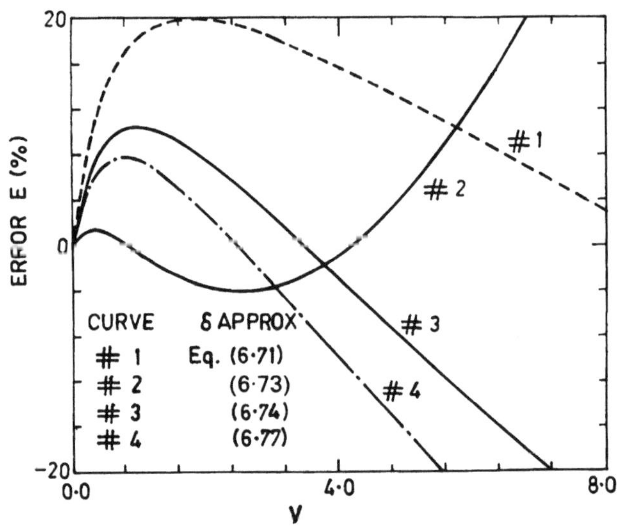

Fig. 6.12 Error between the exact and approximate square-root function $F(V, V_0)$ for different δ approximations

MOSFET capacitance calculations, where small error in Q_b can cause large errors in the capacitances. For this reason Eqs. (6.73) or (6.74) are most appropriate for circuit design, although these expressions can create problem in the capacitance calculations as we shall see in Chapter 7.

6.4.4 Drain Current Equation with Square-Root Approximation

With the square-root approximation (6.69), Eq. (6.63) for $Q_b(y)$ becomes

$$Q_b(y) = -C_{ox}\gamma[\delta V(y) + \sqrt{2\phi_f + V_{sb}}] \tag{6.78}$$

while Eq. (6.64) for $Q_i(y)$ reduces to

$$\boxed{Q_i(y) = -C_{ox}[V_{gs} - V_{th} - \alpha V(y)]} \tag{6.79}$$

where we have made use of Eq. (6.46) for V_{th} and α is defined as

$$\boxed{\alpha = 1 + \delta\gamma.} \tag{6.80}$$

Note the similarity of Eqs. (6.79) and (6.45); the only difference being the presence of the α term which takes into account variations in the bulk charge Q_b along the channel. Using the above value of $Q_i(y)$ in Eq. (6.41) and integrating we get the current in the linear region as

$$\boxed{I_{ds} = \beta[V_{gs} - V_{th} - 0.5\alpha V_{ds}]V_{ds}} \quad V_{gs} > V_{th}. \tag{6.81}$$

Comparing this equation with Eq. (6.65) we see that just by approximating the square root term in Q_b we could get a much simpler expression for I_{ds}. It is this current equation which is used in most of the newly developed MOSFET models for circuit simulation. For example, SPICE MOS Level 3 [23] and Level 4 [25] use Eq. (6.81) for I_{ds}; however, Level 3 uses δ given by Eq. (6.71), while Level 4 (BSIM model) uses δ given by Eq. (6.73). Differentiating Eq. (6.81) and equating the resulting expression to zero gives the following simple expression for V_{dsat}, namely

$$V_{dsat} = \frac{V_{gs} - V_{th}}{\alpha}. \tag{6.82}$$

Substituting this equation into Eq. (6.81) yields the saturation region current, without CLM, as

$$I_{ds} = \frac{\beta}{2\alpha}(V_{gs} - V_{th})^2 \quad V_{ds} > V_{dsat}. \tag{6.83}$$

To summarize, we now have a more accurate drain current model that takes into account the bulk charge variation along the channel region and is represented by the following set of equations:

$$I_{ds} = \begin{cases} 0 & V_{gs} \le V_{th} & \text{(cutoff region)} \\ \beta(V_{gs} - V_{th} - 0.5\alpha V_{ds})V_{ds} & V_{gs} > V_{th}, V_{ds} \le V_{dsat} & \text{(linear region)} \\ \dfrac{\beta}{2\alpha}(V_{gs} - V_{th})^2 & V_{gs} > V_{th}, V_{ds} > V_{dsat} & \text{(saturation region).} \end{cases} \tag{6.84}$$

It is worth pointing out that the charge-sheet model, discussed in section 6.3, can also be simplified using the square root-approximation. In this case, the final equation for I_{ds} in the linear region looks similar to Eq. (6.81). This can be seen as follows: replacing the square-root dependence of Q_b in Eq. (6.27) with a linear approximation [cf. Eq. (6.69)] we get

$$Q_b(y) \approx -\gamma C_{ox}[\sqrt{\phi_{s0}} + \delta(\phi_s(y) - \phi_{s0})] \tag{6.85}$$

where δ is given by any of the expressions discussed earlier in section 6.4.3. Using this value of $Q_b(y)$ in Eq. (6.28) yields

$$Q_i(y) = -C_{ox}[V_{gb} - V_n - \alpha\phi_s(y)] \tag{6.86}$$

where

$$V_n = V_{fb} + \gamma\delta\phi_{s0} - \gamma\sqrt{\phi_{s0}}$$

and α is given by Eq. (6.80). Substituting Q_i from Eq. (6.86) in Eq. (6.33) and carrying out the integration under the boundary conditions given in Eq. (6.34) yields [18]

$$I_{ds} = I_{ds1} + I_{ds2} = \beta[V_{gb} - V_n + \alpha V_t - 0.5\alpha(\phi_{sL} - \phi_{s0})](\phi_{sL} - \phi_{s0}). \tag{6.87}$$

Note that unlike Eq. (6.84), the above equation is continuous in all the regions of device operation (subthreshold, linear and saturation). Comparing Eq. (6.87) with Eq. (6.84) in the linear region we see that there is an extra term $\alpha V_t(\phi_{sL} - \phi_{s0})$ in Eq. (6.87). This is due to the diffusion component of the current that is neglected in Eq. (6.81).

Figure 6.13 shows comparison of the calculated $I_{ds} - V_{ds}$ characterstics using the charge-sheet model, the rudimentary (first order) model and the bulk-charge model. The model parameters used are those shown in Table 6.1. Note that the piece-wise models (rudimentary and bulk-charge models) overpredict the drain current compared to the charge-sheet model. This can be explained as follows. In deriving the piece-wise drain current models in the previous sections we assumed that in strong inversion the potential drop ϕ_s across the silicon was pinned at $2\phi_f + V_{cb}$. In reality this is not

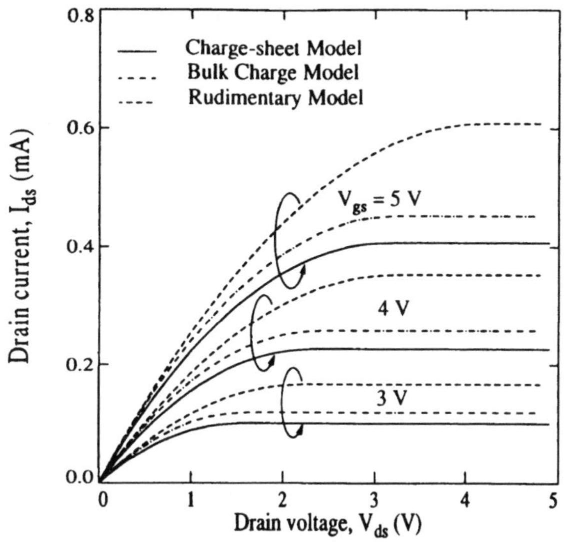

Fig. 6.13 Comparison of the MOSFET output characteristics using (a) charge-sheet model, (b) bulk-charge model and (c) rudimentary (classical model). The classical model overpredicts current

true and indeed the potential does increase by few times the thermal voltage ($\sim 4V_t$) as was discussed in chapter 5. This shows that piece-wise models underestimate ϕ_s and hence the bulk charge Q_b. For a given gate voltage, underestimating ϕ_s means overestimating V_{ox}, the voltage across the oxide [cf. Eq. (4.16)]. Overestimating V_{ox} leads to an overestimation of silicon charge Q_s, which in turn means overestimating Q_i because Q_b is being underestimated. The overestimation of inversion charge Q_i in the channel results in an overestimation of drain current. Indeed it has been found that the piece-wise multisection model overestimates the drain current by 15–20% [11]. In spite of its inaccuracy, the multisection (piece-wise) model [cf. Eq. (6.84)] is the one used in today's widely used circuit simulators because of its simplicity.

6.4.5 Subthreshold Region Model

While deriving I_{ds} Eqs. (6.62) and (6.84), it was assumed that the current flow is due to drift only (assumption 6). This resulted in $I_{ds} = 0$ for $V_{gs} < V_{th}$, that is, there is no current flow for V_{gs} below threshold. In reality this is not true and I_{ds} has small but finite values for $V_{gs} < V_{th}$. For the device shown in Figure 6.5 this current is of the order 10^{-7} A to 10^{-8} A when V_{gs} approaches V_{th} and then decreases exponentially below V_{th}. In fact the device behavior changes from square law to exponential when V_{gs} approaches V_{th}. This current below V_{th} is called the *subthreshold or weak inversion current* and occurs when $V_{gs} < V_{th}$, or $\phi_f \geq \phi_s \geq 2\phi_f$. *Unlike the strong inversion region where drift current dominates, the subthreshold region conduction is dominated by diffusion current.* It should be emphasized that the transition from weak to strong inversion is not well defined, as was discussed in chapters 4 and 5. This region of device (subthreshold) current is important in that it is a leakage current that affects dynamic circuits and determines CMOS standby power. In this region of operation, Eqs. (6.62) or (6.84) are no longer valid.

In the subthreshold region of operation, the surface potential ϕ_s, or the band bending, is nearly constant from the source to the drain end because the inversion charge density Q_i is several orders of magnitude smaller than the bulk charge density Q_b (cf. section 4.2). This means that we can replace $\phi_s(y)$ in subthreshold region by some constant value, say ϕ_{ss}. With this assumption, the bulk charge Q_b can be expressed as

$$Q_b = -C_{ox}\gamma\sqrt{\phi_s(y)} = -C_{ox}\gamma\sqrt{\phi_{ss}}. \tag{6.88}$$

Further, since $Q_i \ll Q_b$, we have $Q_s \approx Q_b$, so that Eq. (6.19) becomes

$$V_{gb} = V_{fb} + \phi_{ss} - \frac{Q_b}{C_{ox}}. \tag{6.89}$$

Solving Eqs. (6.88) and (6.89) for ϕ_{ss} we get

$$\phi_{ss} = V_{gb} - V_{fb} + \frac{\gamma^2}{2}\left[1 - \sqrt{1 + \frac{4}{\gamma^2}(V_{gb} - V_{fb})}\right]$$

or

$$\phi_{ss} = \left[-\frac{\gamma}{2} + \sqrt{\frac{\gamma^2}{4} + V_{gb} - V_{fb}}\right]^2. \tag{6.90}$$

This shows that ϕ_{ss} is nearly linearly dependent on V_{gs}. It should be emphasized that the surface potential ϕ_{ss} in the subthreshold region is constant from source to drain only for a long channel device. As the channel length become shorter, ϕ_{ss} no longer remains constant over the whole channel length.

Because ϕ_{ss} is constant, the electric field $\mathscr{E}_y$ is zero. Hence, the only current that can flow is diffusion current as can be seen from Eq. (6.2) and is given by

$$J_n(\text{diffusion}) = qD_n\frac{dn}{dy} \tag{6.91}$$

Integrating this equation across the channel of thickness t_{ch} and making use of Eq. (6.13) we can write the drain current I_{ds} (due to diffusion) in the subthreshold region as

$$I_{ds} = \mu_s W V_t \frac{dQ_i}{dy} \quad \text{(subthreshold region)} \tag{6.92}$$

where we have made use of the Einstein relation $D_n = \mu_s V_t$ [cf. Eq. (2.34)] and made the assumption that $d\mu_s/dx = 0$. Integrating the equation above from $y = 0$ to $y = L$ we get

$$I_{ds} = \frac{W}{L}\mu_s V_t \int_{Q_{is}}^{Q_{id}} dQ_i = \frac{W}{L}\mu_s V_t(Q_{id} - Q_{is}) \tag{6.93}$$

where Q_{is} and Q_{id} are the inversion charge densities at the source and the drain end respectively when the device is in the subthreshold or weak inversion region. Following the MOS capacitor case, the inversion charge density $Q_i(y)$ in weak inversion [cf. Eq. (4.43)] is given by

$$Q_i(y) = \sqrt{\frac{\epsilon_0 \epsilon_{si} q N_b}{2\phi_{ss}}}\, V_t e^{(\phi_{ss} - 2\phi_f - V_{cb}(y))/V_t} = \frac{\gamma C_{ox}}{2\sqrt{\phi_{ss}}} V_t e^{(\phi_{ss} - 2\phi_f - V_{cb}(y))/V_t} \tag{6.94}$$

where we have replaced ϕ_s by ϕ_{ss} and have made use of Eq. (6.23) for γ and Eq. (6.22) for V_{gb}. Remembering that $V_{cb}(y = 0) = V_{sb}$ and $V_{cb}(y = L) = V_{sb} + V_{ds}$, the inversion charge Q_{is} and Q_{id} at the source and drain ends, respectively, can be written as

$$Q_{is}(\text{source end}) = \frac{\gamma C_{ox}}{2\sqrt{\phi_{ss}}} V_t e^{(\phi_{ss} - 2\phi_f - V_{sb})/V_t} \tag{6.95a}$$

and

$$Q_{id}(\text{drain end}) = \frac{\gamma C_{ox}}{2\sqrt{\phi_{ss}}} V_t e^{(\phi_{ss} - 2\phi_f - V_{sb} - V_{ds})/V_t}. \tag{6.95b}$$

Using these values of Q_{is} and Q_{id} in Eq. (6.93) yields

$$I_{ds} = \frac{\mu_s W C_{ox} \gamma}{2L\sqrt{\phi_{ss}}} V_t^2 e^{(\phi_{ss} - 2\phi_f - V_{sb})/V_t}(1 - e^{-V_{ds}/V_t}). \tag{6.96a}$$

Above equation takes the following form, after eliminating ϕ_f using Eq. (2.15) and making use of Eq. (6.50) for β,

$$I_{ds} = \frac{\beta}{2\sqrt{\phi_{ss}}}\gamma\left(V_t \frac{n_i}{N_b}\right)^2 e^{(\phi_{ss} - V_{sb})/V_t}(1 - e^{-V_{ds}/V_t}). \tag{6.96b}$$

This is the current equation for the subthreshold region. For each V_{gs} we first calculate ϕ_{ss} from Eq. (6.90), which in turn is used to calculate I_{ds} from

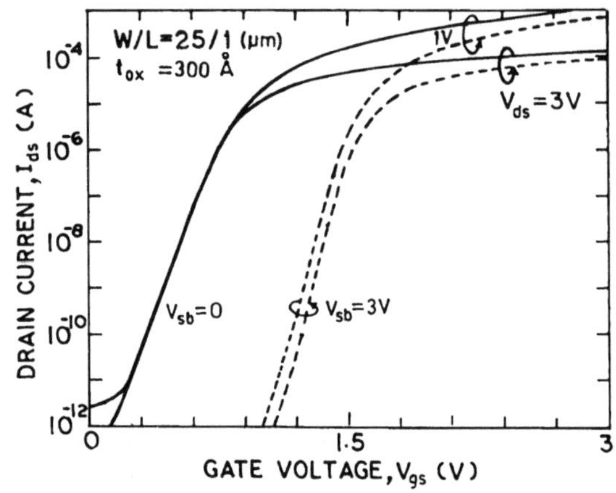

Fig. 6.14 Typical device $I_{ds} - V_{gs}$ characteristics in the subthreshold or weak inversion region for two different back bias

Eq. (6.96). The following conclusions about subthreshold conduction can be drawn from Eq. (6.96):

- The subthreshold current increases exponentially with the surface potential ϕ_{ss} and hence V_{gs} [cf. Eq. (6.90)]. This is evident from Figure 6.14 where measured I_{ds} is plotted against V_{gs} for different values of V_{sb} and V_{ds} for a nMOST fabricated using 1 μm CMOS technology.
- The current is dependent upon an exponentially decreasing term which for V_{ds} larger than $4V_t$ ($\sim 100\,\text{mV}$) is negligible, becoming independent of V_{ds}. It should be pointed out that this is true only for long channel devices. In fact for short channel devices, this region of drain current exhibits a significant dependence on the drain voltage as we will see in section 6.9.
- The subthreshold current is strongly dependent on temperature due to its dependence on the square of the intrinsic carrier concentration n_i, resulting in steeper slopes at low temperatures. The temperature dependence of subthreshold current is discussed in section 6.9.

Often Eq. (6.96) is written in terms of the surface concentration n_s as[6]

$$I_{ds} = \frac{WD_n t_{ch} q n_s}{L}(1 - e^{-V_{ds}/V_t}). \tag{6.97}$$

[6] Equation (6.97) can be derived as follows: The inversion channel is confined by the potential well created by the oxide to the silicon interface on one side and on the other side by the perpendicular electric field $\mathscr{E}_s$ at the surface in the substrate. Since $Q_i \ll Q_b$ in weak inversion, $\mathscr{E}_s$ is equal to the depletion field, that is

$$\mathscr{E}_s = \frac{Q_b}{\epsilon_0 \epsilon_{si}} = \frac{\gamma C_{ox}\sqrt{\phi_{ss}}}{\epsilon_0 \epsilon_{si}}.$$

(Continued next page)

Most of the expressions reported in the literature for I_{ds} in weak inversion region are variations of Eq. (6.96) [4], [29]–[32]. For circuit simulation models, often a simplified form of this equation is used. Since Q_b is a weak function of ϕ_{ss}, we can expand Q_b using Taylor series around ϕ_{s0} which lies between ϕ_f and $2\phi_f$. Retaining the first two terms of the Taylor series, we get

$$Q_b(\phi_{ss}) \cong Q_b(\phi_{s0}) + (\phi_{ss} - \phi_{s0})\frac{\partial Q_b}{\partial \phi_{ss}}. \tag{6.98}$$

From Eq. (6.88) we get

$$\frac{\partial Q_b}{\partial \phi_{ss}} \equiv C_d = \frac{\gamma C_{ox}}{2\sqrt{\phi_{ss}}} \tag{6.99}$$

where C_d is called the *depletion region capacitance.*[7] Combining Eqs. (6.98) and (6.99) with (6.89) yields

$$V_{gb} = V_{fb} + \phi_{s0} + \frac{Q_b(\phi_{s0})}{C_{ox}} + (\phi_{ss} - \phi_{s0})\frac{C_d}{C_{ox}}. \tag{6.100}$$

For calculating I_{ds}, it is more appropriate to take ϕ_{s0} in the middle of the subthreshold region (i.e., $\phi_{s0} = 1.5\phi_f + V_{sb}$) because ϕ_{ss} lies between $2\phi_f + V_{sb}$ and $\phi_f + V_{sb}$. However, by assuming $\phi_{s0} = 2\phi_f + V_{sb}$, the condition for the onset of strong inversion, we arrive at an expression for I_{ds} that is often used in circuit models. Thus, assuming $\phi_{s0} = 2\phi_f + V_{sb}$, Eq. (6.100) becomes

$$\phi_{ss} - 2\phi_f - V_{sb} = \frac{V_{gs} - V_{th}}{\eta}, \tag{6.101}$$

The average thermal energy of the carriers for motion perpendicular to the surface is kT. The average thickness t_{ch} of the weak inversion channel is, therefore, given by

$$q\mathscr{E}_s t_{ch} = kT.$$

Solving these two equations for $\sqrt{\phi_{ss}}$, by eliminating $\mathscr{E}_s$, and then combining Eqs. (6.96a) and Eq. (6.10) results in the desired equation.

[7] Rewriting Eq. (6.99) in the following form

$$C_d = \frac{\gamma C_{ox}}{2\sqrt{\phi_{ss}}} = \sqrt{\frac{\epsilon_0 \epsilon_{si} q N_b}{2\phi_{ss}}} = \frac{\epsilon_0 \epsilon_{ox}}{X_{dm}}$$

$$= \frac{\epsilon_0 \epsilon_{ox}}{\text{thickness of the depletion region under the channel}}$$

clearly shows C_d as the depletion layer capacitance.

where we have made use of Eq. (6.46) for V_{th}, and

$$\eta = 1 + \frac{C_d}{C_{ox}} = 1 + \frac{\gamma}{2\sqrt{2\phi_f + V_{sb}}}. \tag{6.102}$$

Typical values of η range from 1 to 3. Physically, η signifies the capacitive coupling between the gate and silicon surface. If there is a significant interface trap density, the capacitance C_{it} associated with this trap is in parallel with the depletion layer capacitance C_d, and therefore Eq. (6.102) becomes

$$\eta = 1 + \frac{C_{it}}{C_{ox}} + \frac{C_d}{C_{ox}} = \eta_0 + \frac{C_d}{C_{ox}}. \tag{6.103}$$

This is the equation for η used in SPICE model Levels 2 and 3. In this equation C_{it}, called the *surface state capacitance*, is normally regarded as an adjustable parameter through η_0 and is used to fit the value of η to measured characteristics. Combining Eqs. (6.96), (6.99) and (6.101) yields

$$I_{ds} = \beta \frac{C_d}{C_{ox}} V_t^2 \exp\left[\frac{V_{gs} - V_{th}}{\eta V_t}\right](1 - e^{-V_{ds}/V_t}), \quad V_{gs} < V_{th} \tag{6.104a}$$

or

$$\boxed{I_{ds} = I_{pf} \exp\left[\frac{V_{gs} - V_{th}}{\eta V_t}\right](1 - e^{-V_{ds}/V_t}), \quad V_{gs} < V_{th}} \tag{6.104b}$$

where $I_{pf} = \beta(C_d/C_{ox})V_t^2 = \beta(\eta - 1)V_t^2$, is a prefactor term. This is the most commonly used drain current equation for the subthreshold region of device operation. It clearly shows that the subthreshold current ($V_{gs} < V_{th}$) increases exponentially with V_{gs} and for V_{ds} larger than about $3V_t$, the current becomes independent of V_{ds}. Further, since the parameter η is inversely proportional to the square root of V_{sb}, the subthreshold slope becomes steeper at higher values of V_{sb}. This indeed is the case as can be seen from Figure 6.14 which is a plot of $\log(I_{ds})$ versus V_{gs} for an experimental device. Note that the curve is linear (on the log scale) until the device starts to turn on. When V_{gs} approaches V_{th}, Eq. (6.104) is no longer valid and the current will increase either linearly (linear region) or as the square of ($V_{gs} - V_{th}$) (saturation region) depending upon the value of V_{ds}.
Very often the following simpler version of Eq. (6.104) is used for circuit models [34]

$$I_{ds} = \beta m(\eta V_t)^2 \exp\left[\frac{V_{gs} - V_{th}}{\eta V_t}\right] \tag{6.105}$$

where V_{ds} dependence is ignored because its effects on I_{ds} is negligible for $V_{ds} > 3V_t$. The parameter m is inserted to correct for various approximations made in the derivation of Eq. (6.104) and is calculated in the same way as η_0, that is, by curve fitting the experimental data.

Subthreshold Slope. An important parameter characteristic of the subthreshold region is the *gate voltage swing* required to reduce the current from its 'on' value to an acceptable 'off' value. This gate voltage swing, also called the *subthreshold slope S*, is *defined as the change in the gate voltage V_{gs} required to reduce subthreshold current I_{ds} by one decade.* For a device to have good turn-on characteristics, S should be as small as possible. Clearly, S is a convenient measure of the turnoff characteristics of a MOSFET. By this definition

$$S = \frac{dV_{gb}}{d(\log I_{ds})} = 2.3\left[\frac{dV_{gb}}{d(\ln I_{ds})}\right] \quad \text{(V/decade)} \tag{6.106}$$

where the factor 2.3 accounts for the conversion from "log" (logarithm to the base 10) to "ln" (logarithm to the base e). Strictly speaking, S varies with the current level. However, this variation is small over one decade of current so that S can be taken as gate swing per decade [32]. We can rewrite Eq. (6.106) for S as

$$S = 2.3\left[\frac{dV_{gb}}{d\phi_{ss}} \bigg/ \frac{d(\ln I_{ds})}{d\phi_{ss}}\right]. \tag{6.107}$$

Differentiating Eq. (6.89) we get

$$\frac{dV_{gb}}{d\phi_{ss}} = 1 + \frac{\gamma}{2\sqrt{\phi_{ss}}} = 1 + \frac{C_d}{C_{ox}} \tag{6.108}$$

where we have made use of Eq. (6.99) for C_d. Assuming $V_{ds} > 3V_t$ and taking the logarithm of both sides of Eq. (6.96b), we get

$$\ln I_{ds} = \ln I_0 + \frac{(\phi_{ss} - V_{sb})}{V_t} - \frac{1}{2}\ln(\phi_{ss}) \tag{6.109}$$

where

$$I_0 = \frac{1}{2}\frac{W}{L}\mu_s C_{ox}\gamma\left(V_t\frac{n_i}{N_b}\right)^2. \tag{6.110}$$

Now differentiating Eq. (6.109), we get

$$\frac{d(\ln I_{ds})}{d\phi_{ss}} = \frac{1}{V_t} - \frac{1}{2\phi_{ss}} = \frac{1}{V_t}\left[1 - \frac{2V_t}{\gamma^2}\left(\frac{C_d}{C_{ox}}\right)^2\right] \tag{6.111}$$

where again we have made use of Eq. (6.99) for C_d. Substituting Eqs. (6.108) and (6.111) in Eq. (6.107), we get

$$S = 2.3V_t\left[\left(1+\frac{C_d}{C_{ox}}\right)\bigg/\left\{1-\frac{2V_t}{\gamma^2}\left(\frac{C_d}{C_{ox}}\right)^2\right\}\right] \quad \text{(V/decade)}. \tag{6.112}$$

For $\gamma \gg C_d\sqrt{V_t}/C_{ox}$, the subthreshold swing becomes

$$S \cong 2.3V_t\left(1+\frac{C_d}{C_{ox}}\right) = 2.3\cdot V_t\cdot\eta \tag{6.113}$$

where we have made use of Eq. (6.102) for η. This shows that the theoretical minimum swing $S_{\min}$ is given by

$$S_{\min} = 2.3\cdot V_t \cong 60 \quad \text{(mV/decade)} \tag{6.114}$$

that is, the *minimum attainable subthreshold slope for any device is approximately 60 mV per decade* at room temperature. Since η lies in the range 1–3, typical values of S lie in the range of 60 to 180 mV/decade. If there is a substantial interface trap density, then C_d in Eq. (6.113) should be replaced by $(C_d + C_{it})$. Thus, the *subthreshold slope is a convenient measure of the importance of the interface traps on device performance.*

Note that C_d is a function of ϕ_{ss} and the value of ϕ_{ss} chosen to calculate C_d affects S to some degree. For circuit models, we can assume that $\phi_{ss} = b\phi_f + V_{sb}$ where $2 > b > 1$. However, Brews [32] determined ϕ_{ss} in terms of current level. Therefore, to find ϕ_{ss} we first choose a certain current level, say $I_{ds} = 10^{-10}$ A. Rearranging Eq. (6.109), we get

$$\phi_{ss}^{i+1} = V_{sb} + V_t\ln\left(\frac{I_{ds}}{I_0}\right) + \frac{V_t}{2}\ln(\phi_{ss}^i). \tag{6.115}$$

Assuming a certain current level, Eq. (6.115) is used to calculate ϕ_{ss} by iteration. This iterative procedure converges very rapidly [32].

The plot of subthreshold swing S as a function of body factor γ for three different gate oxide thicknesses ($t_{ox} = 100$, 300 and 500 Å) is shown in Figure 6.15. In these curves C_d is calculated using Eq. (6.99) with $\phi_{ss} = 1.5\phi_f + V_{sb}$, although one can also use Eq. (6.115) for ϕ_{ss}. Note that even for $\gamma = 0$, there is a minimum swing of 60 mV/decade. The swing varies linearly[8] with γ and is substrate bias dependent. The higher the V_{sb}, the higher the ϕ_{ss}, and therefore the lower the depletion capacitance C_d which then results in S being lower. This shows that use of substrate bias can improve subthreshold turn off.

[8] Increased γ means higher substrate doping N_b. The higher the N_b, the lower the depletion width; this, in turn leads to a higher value for C_d which results in higher value for S.

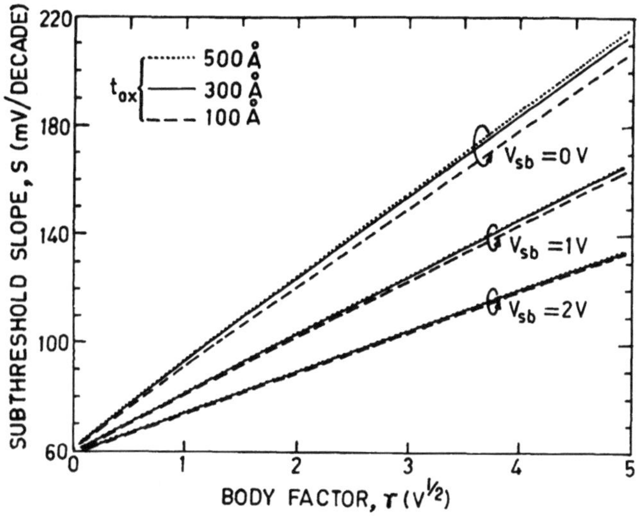

Fig. 6.15 Subthreshold slope S versus body factor γ for different substrate bias. Variation in S at contrast γ, when oxide thickness t_{ox} varies from 100 Å to 500 Å, is also shown

6.4.6 *Limitations of the Model*

In the multisection drain current model developed above we have assumed that I_{ds} in the subthreshold region (weak inversion) consists of a diffusion component only, whereas in the linear and saturation regions (strong inversion) it consists of a drift component only. Hence, one can not expect a smooth transition between the two regions. The non-continuous transition between these (weak and strong inversion) regions is a severe drawback of the simplified model discussed so far. For the model to be implemented in a circuit simulator it is necessary that there be a smooth transition between the two respective regions. The simplest way to achieve a continuous transition is to assume that the charge Q_i in the weak inversion region is a tangent to the strong inversion region charge. [30]. The point of tangency V_{on} is the dividing point above which strong inversion region equations will be valid and below which weak inversion region equations will be valid, as shown in Figure 6.16. Under the assumption of low $V_{ds}(\sim 4V_t)$, using Eqs. (6.95) and (6.101), the weak inversion charge Q_i becomes

$$Q_i = C_d V_t \exp\left(\frac{V_{gs} - V_{th}}{\eta V_t}\right) \quad \text{(weak inversion, } V_{ds} \sim 0.1\,\text{V).} \qquad (6.116)$$

Under the same conditions, that is low V_{ds}, the strong inversion charge, from Eq. (6.79), becomes

$$Q_i = C_{ox}(V_{gs} - V_{th}) \quad \text{(strong inversion, } V_{ds} \sim 0.1\,\text{V).} \qquad (6.117)$$

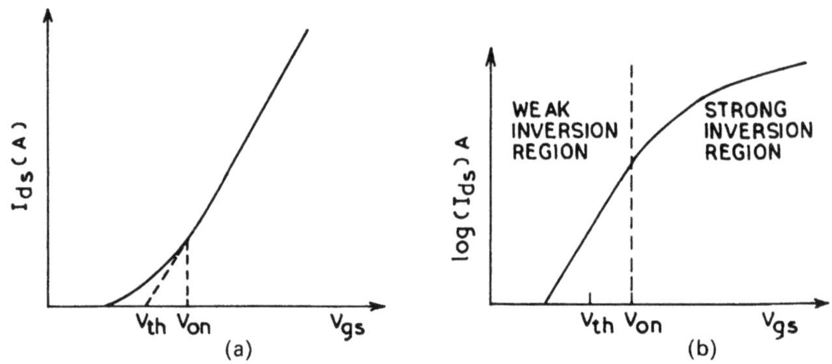

Fig. 6.16 Gate voltage V_{on} dividing the weak and strong inversion region model, (a) linear scale, and (b) log scale

Thus, equating Eqs. (6.116) and (6.117) and their derivatives with respect to V_{gs}, we get at $V_{gs} = V_{on}$ [30]

$$\boxed{V_{on} = V_{th} + \eta V_t.} \tag{6.118}$$

When $V_{gs} \geq V_{on}$, the drain current I_{ds} is given by (6.84) while, for $V_{gs} < V_{on}$, I_{ds} can be calculated from Eq. (6.105) by replacing V_{th} with V_{on}. Thus,

$$I_{ds}(\text{subthreshold}) = I_{on}\exp\left(\frac{V_{gs} - V_{on}}{\eta V_t}\right), \quad V_{gs} < V_{on} \tag{6.119}$$

where I_{on} is the current calculated from Eq. (6.84) using $V_{gs} = V_{on}$. *Thus, V_{on} acts as a point at which behaviors of strong and weak inversion are pieced together.* This is the approach used in SPICE Levels 2 and 3. Combining Eqs. (6.119) with (6.84) we now have a complete long-channel DC MOSFET model, which is continuous in all regions,

$$I_{ds} = \begin{cases} I_{on}\exp\left(\dfrac{V_{gs} - V_{on}}{\eta V_t}\right) & V_{gs} \leq V_{on} \quad \text{(cutoff region)} \\ \beta(V_{gs} - V_{th} - 0.5\,\alpha V_{ds})V_{ds} & V_{gs} > V_{on}, V_{ds} \leq V_{dsat} \quad \text{(linear region)} \\ \dfrac{\beta}{2\alpha}(V_{gs} - V_{th})^2 & V_{gs} > V_{on}, V_{ds} > V_{dsat} \quad \text{(saturation region).} \end{cases} \tag{6.120}$$

The transfer characteristics of a nMOST ($W_m/L_m = 3/1, t_{ox} = 150$ Å) is shown in Figure 6.17, where continuous line is calculated based on Eq. (6.120) while symbols are experimental data.

Although Eq. (6.118) results in a continuous transition from weak to strong inversion (see Figure 6.17), there are large errors in the I_{ds} calculations around the *transition region*, often called the *moderate inversion* region [15].

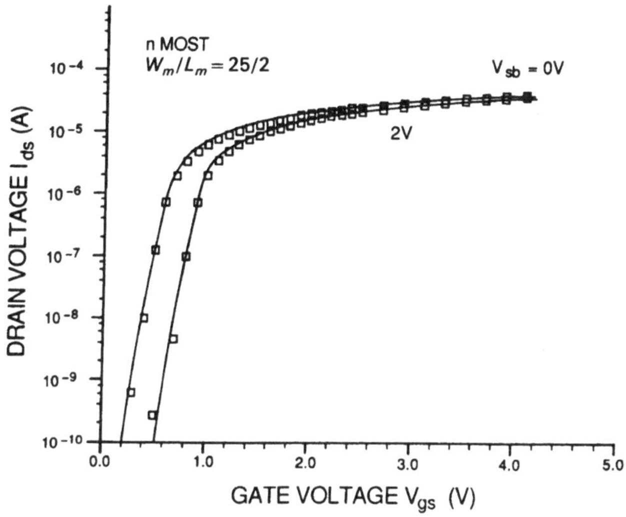

Fig. 6.17 Device $I_{ds} - V_{gs}$ characteristics in the subthreshold or weak inversion region. Squares are experimental points while lines are model based on Eq. (6.119)

However, for most of the digital applications this error is not significant due to the low magnitude of the current in this region.

A slightly different approach, that ensures continuity of the weak and strong inversion current and its derivative, is to replace V_{gs} in Eq. (6.120) by an effective gate voltage V_{gsx} defined as [34]

$$V_{gsx} = \eta V_t \ln\left[1 + \exp\left(\frac{V_{gs} - V_{th}}{\eta V_t}\right)\right] + V_{th}. \tag{6.121}$$

When the gate voltage is a few V_t above V_{th}, V_{gsx} reduces to V_{gs} as is required. When the gate voltage is a few V_t below V_{th}, the effective gate voltage becomes

$$V_{gsx} = \eta V_t \exp\left(\frac{V_{gs} - V_{th}}{\eta V_t}\right) + V_{th} \tag{6.122}$$

which indicates the exponential dependence of V_{gsx} on V_{gs} when $V_{gs} < V_{th}$. Thus, replacing V_{gs} in Eq. (6.84) by V_{gsx} ensures continuity of the current. In fact the two approaches are not very different. The large errors in the middle inversion region still exist. However, the advantage of using (6.121) is that we need only two equations instead of three in (6.120).

6.5 Drain Current Model for Depletion Devices

Strictly speaking, the drain current models developed in the previous sections are valid for enhancement devices only. However, in SPICE these models have been used for depletion type devices also simply by changing the sign of the threshold voltage as was pointed out in section 5.2.2. If the depletion device is used only as a load element (source and gate tied together) in a circuit, then this zero order model is quite satisfactory. However, for device to be used in a more general configuration requires a separate model. Although a general model, similar to the charge sheet model for the enhancement devices, has been developed [35] but it will not be discussed here due to its complexity. Moreover, such models are not very suitable for circuit simulators. Here we will discuss only piece-wise models that are normally used for circuit simulations [36]–[45].

Recall that depletion devices have a deep channel implant which is of opposite type to that of the substrate, thereby forming a *pn* junction underneath the gate. Unlike the enhancement devices, the depletion devices conduct even at zero V_{gs} and have many modes of operation depending upon the applied gate and drain voltages, channel doping concentration and implant depth [38]–[45]. These different modes of operation are named according to the condition at the silicon surface. Thus, if the entire surface is accumulated, depleted or inverted, the device is said to be operating in accumulation, depletion or inversion mode, respectively, as shown in Figure 6.18. In addition, there can be mixed mode of operation such as accumulation at the source end and depletion at the drain end, called the accumulation/depletion mode. In this section we will develop a drain current model for different modes of operation of the depletion devices.

Figure 6.19a shows the cross-section of an *n*-channel depletion MOSFET in the direction of the channel current flow. The implanted or buried *n*-type channel region is approximated by a step profile of depth X_i and uniform concentration N_s. This is the approximation we had used earlier to calculate the threshold voltage of depletion devices (cf. section 5.2.2). The boundaries

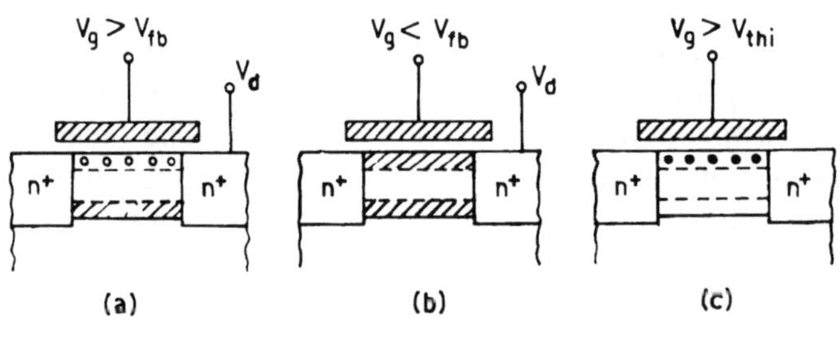

Fig. 6.18 Different modes of operation in depletion devices (a) accumulation (b) depletion and (c) inversion

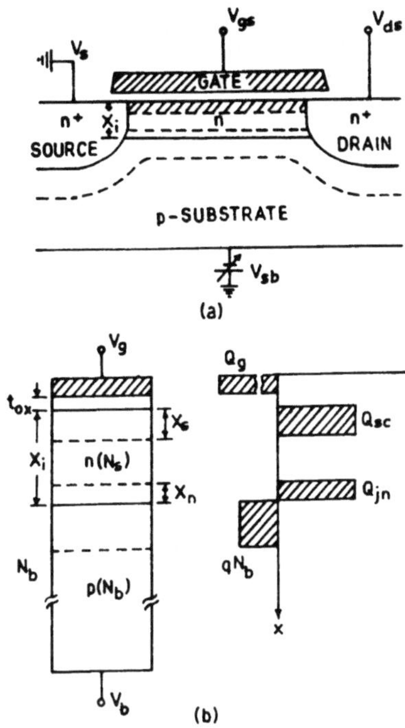

Fig. 6.19 (a) Cross-section of a *n*-channel depletion device, (b) depletion widths and charges for the device in (a)

of the two space charge regions, one at the surface and the other due to the *pn* junction formed by the channel and the substrate, are shown as dotted lines. The channel *pn* junction depletion width X_n is controlled by the channel voltage V_{cb}. The surface space-charge region X_s is due to the combined effect of the gate and channel voltage. An elemental section of the device at a position y together with the charge and potential distribution in the x direction is shown in Figure 6.19b. For the sake of algebraic manipulation it is convenient to define the following modified voltages

$$\begin{aligned} V_{mg} &= V_{gs} + V_{sb} - V_{fb} + \phi_j \\ V_{md} &= V_{ds} + V_{sb} + \phi_j \\ V_{ms} &= V_{sb} + \phi_j \\ V_y(y) &= V(y) + V_{sb} + \phi_j \end{aligned} \tag{6.123}$$

where $V(y)$ is the channel voltage which is zero at the source end and V_{ds} at the drain end, ϕ_j is the built-in potential of the channel *pn* junction. Using the GCA we can write the mobile charge density Q_m in the channel as [38]

$$Q_m = -Q_{im} + Q_{jn} + Q_{sc} \tag{6.124}$$

where Q_{im}, Q_{jn} and Q_{sc} are the implanted layer charge density, the channel *pn* junction space-charge density and the surface space-charge density, respectively. The implanted layer charge density Q_{im} is simply [cf. Eq. (5.50)]

$$Q_{im} = qN_sX_i. \tag{6.125}$$

The substrate *pn* junction space-charge density Q_{jn} was calculated as [cf. Eq. (5.55)]

$$Q_{jn}(y) = \gamma_e C_{ox}\sqrt{V_y(y)} \tag{6.126}$$

where γ_e is given by Eq. (5.54). The surface space-charge density Q_{sc} takes the following values depending upon the gate and drain voltages [38]

$$Q_{sc}(y) = \begin{cases} -C_{ox}(V_{mg} - V_y(y)) & \text{(surface accumulation)} \quad (6.127a) \\ \gamma_s C_{ox}\left[-\dfrac{\gamma_s}{2} + \left(\dfrac{\gamma_s^2}{4} + V_y(y) - V_{mg}\right)^{1/2}\right] & \text{(depletion at the surface)} \quad (6.127b) \\ \gamma_s C_{ox}\sqrt{V_y(y)} & \text{(surface inversion)} \quad (6.127c) \end{cases}$$

where γ_s is given by Eq. (5.57a).

The sets of equations (6.125)–(6.127) are valid at any point along the surface between the source and drain. Whether all the conditions mentioned in Eqs. (6.127) actually occur in a given device depends on the doping concentration in the implanted layer, the thickness of the layer and the channel voltage.

Knowing the mobile charge density Q_m, we can now calculate the drain current in the depletion devices. Neglecting the diffusion current, the drain current can be written as [cf. Eq. (6.14)]

$$I_{ds} = -\frac{\mu W}{L}\int_{V_{ms}}^{V_{md}} Q_m dV_y. \tag{6.128}$$

Note that here μ is not the surface mobility, but rather more closer to bulk mobility because in this case current is flowing away from the surface in the burried channel.

Substituting Q_m from Eq. (6.124) into (6.128) and integrating we obtain [38]

$$\boxed{I_{ds} = \frac{\mu W}{L}\left[Q_{im}V_{ds} - \frac{2}{3}\gamma_s C_{ox}(V_{md}^{3/2} - V_{ms}^{3/2}) - F_s\right]} \tag{6.129}$$

where F_s is the contribution of the surface space-charge region to the drain current, and is defined as

$$F_s = \int_{V_{ms}}^{V_{md}} Q_{sc} dV_y. \tag{6.130}$$

The function F_s takes different values depending upon the condition existing at the surface. We now evaluate the function F_s for different conditions at the surface ranging from inversion to accumulation.

1. *Inversion Along the Entire Surface.* This condition exists for the gate voltages satisfying $V_{mg} \le \gamma_s \sqrt{V_{mg}}$. In this case using Eqs. (6.127c) and (6.130) we get

$$F_s = \frac{2}{3}\gamma_s C_{ox}(V_{md}^{3/2} - V_{ms}^{3/2}). \tag{6.131}$$

2. *Inversion at the Source, Depletion at the Drain.* This condition exists for the gate voltages satisfying $-\gamma_s\sqrt{V_{md}} < V_{mg} < \gamma_s\sqrt{V_{ms}}$. The surface in this case is inverted at the source end and up to the point along the surface where $V_y = V_{id} = (V_{mg}/\gamma_s)^2$. Beyond this point and up to the drain, the surface is depleted. In this case, using Eqs. (6.127b and c) in (6.130) yields

$$F_s = \int_{V_{ms}}^{V_{id}} Q_{sc}(\text{inversion}) dV_y + \int_{V_{id}}^{V_{md}} Q_{sc}(\text{depletion}) dV_y$$

or

$$F_s = \frac{2}{3}\gamma_s C_{ox}\left[(V_{id}^{3/2} - V_{ms}^{3/2}) + \left(\frac{\gamma_s^2}{4} + V_{md} - V_{mg}\right)^{3/2} - \left(\frac{\gamma_s^2}{4} + V_{id} - V_{mg}\right)^{3/2} - \frac{3}{4}\gamma_s(V_{md} - V_{id})\right]. \tag{6.132}$$

3. *Depletion Along the Entire Surface.* This condition exists for the gate voltages satisfying $-\gamma_s\sqrt{V_{ms}} < V_{mg} \le \sqrt{V_{ms}}$. In this case using Eqs. (6.127b) in (6.130) yields

$$F_s = \frac{2}{3}\gamma_s C_{ox}\left[\left(\frac{\gamma_s^2}{4} + V_{md} - V_{mg}\right)^{3/2} - \left(\frac{\gamma_s^2}{4} + V_{ms} - V_{mg}\right)^{3/2} - \frac{3}{4}\gamma_s(V_{md} - V_{ms})\right]. \tag{6.133}$$

4. *Accumulation at the Source, Depletion at the Drain.* This condition exists for the gate voltages satisfying $V_{ms} < V_{mg} \le V_{md}$. In this case using Eqs.

(6.127a) and (6.127b) in Eq. (6.130) yields

$$F_s = \int_{V_{ms}}^{V_{mg}} Q_{sc}(\text{accumulation})dV_y + \int_{V_{mg}}^{V_{md}} Q_{sc}(\text{depletion})dV_y$$

or

$$F_s = \frac{1}{2}C_{ox}[V_{mg}^2 - (V_{ms}^2 - 2V_{mg}V_{ms})]$$

$$+\frac{2}{3}\gamma_s C_{ox}\left[\left(\frac{\gamma_s^2}{4} + V_{md} - V_{mg}\right)^{3/2} - \left(\frac{\gamma_s^2}{4}\right)^{3/2} - \frac{3}{4}\gamma_s(V_{md} - V_{ms})\right]. \tag{6.134}$$

5. *Accumulation Along the Entire Surface.* In this case, gate voltage is always greater than V_{md}. Using Eqs. (6.127a) in (6.130), we get

$$F_s = C_{ox}[\tfrac{1}{2}(V_{md}^2 - V_{ms}^2) - V_{mg}(V_{md} - V_{ms})]. \tag{6.135}$$

It should be noted that in the surface accumulation case the surface mobility μ_s should be used instead of the bulk mobility μ. It has been found that μ_s required to fit the data is about one half the bulk mobility value [38]. However, to better fit the data it is more appropriate to use the following empirical expression for the mobility at the surface (when surface is partly or fully accumulated) [42]

$$\mu' = \frac{\mu_s/\mu}{1 + \theta_1 V_{gs}}$$

which is gate bias dependent (see section 6.6.1).

It is important to note that depending upon the implanted region, some of the conditions at the surface may not be encountered. For example, for shallow implants, inversion may not occur at the surface, and in this case, the whole range of device currents are covered by conditions (3)–(5) above. Such would also be the case if the surface region is completely isolated from the substrate. Also note that unlike in enhancement devices, the threshold voltage V_{th} does not appear explicitly in the drain current equations for depletion devices. In spite of this drawback, the model fairly accurately represents the measured I–V characteristics for long channel depletion devices. Although this is the most commonly used drain current model based on a step profile in the channel; there are other more accurate but more complex models based on linearly graded profiles [27], [36].

Saturation Voltage. Similar to the case of enhancement devices, the saturation voltage V_{dsat} for depletion devices is also defined as the drain voltage at which the drain current reaches its maximum value for a fixed gate voltage. With no velocity saturation effect, it is equivalent to setting

$Q_m = 0$ in Eq. (6.124) and replacing V_y by $V_{msat} = V_{dsat} + V_{sb} + \phi_j$. If the surface at the drain end is depleted, then V_{dsat} is obtained by solving the following equation

$$\gamma_s C_{ox}\sqrt{V_{msat}} + \left[-\frac{\gamma_s}{2} + \left(\frac{\gamma_s^2}{4} + V_{msat} - V_{mg} \right)^{1/2} \right] = Q_{im}. \tag{6.136}$$

However, if the drain end is inverted, then one has to use the following equation for V_{dsat} calculation

$$C_{ox}(\gamma_s + \gamma_e)\sqrt{V_{msat}} = Q_{im}. \tag{6.137}$$

Note that in this case V_{dsat} is independent of the gate voltage, because the gate is screened from the channel by the inversion layer. However, if the region is accumulated the saturation of the drain current does not occur. Figure 6.20 shows a comparison of the calculated and measured drain current for an n-channel depletion device fabricated using an NMOS

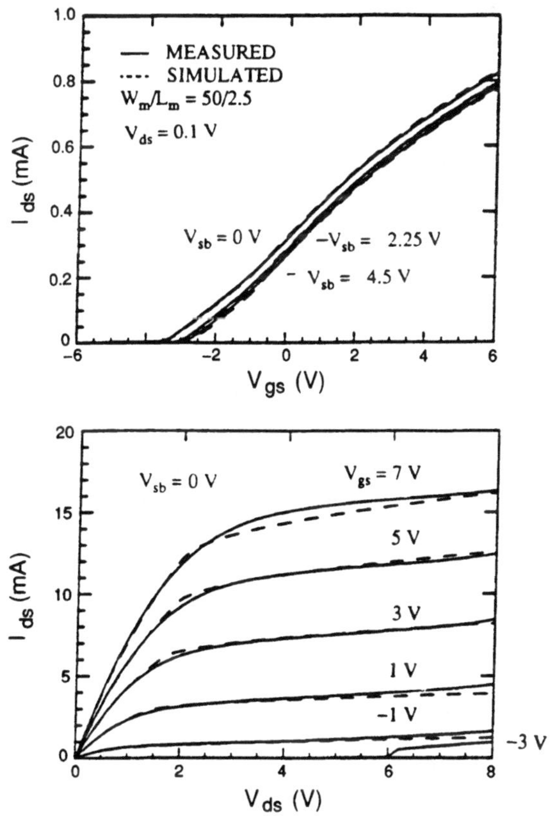

Fig. 6.20 Measured and calculated transfer and output characteristics at different back bias for an n-channel depletion device. (After Divekar and Dowell [42])

process with $W_m/L_m = 50/2.5$. Solid lines are measured data while dashed lines are calculated based on the model discussed above. The model fits the data fairly well [42].

6.6 Effective Mobility

The carrier inversion layer mobility μ_s for electrons varies in the range 400–700 cm^2/V·s while for the holes it is in the range 100–300 cm^2/V·s. These values are lower than the bulk mobility values (cf. section 2.4) because carriers in the channel undergo scattering by the charges at the surface boundary and by surface roughness, in addition to the scattering with the crystal lattice and ionized impurity atoms.[9] In fact, *carrier mobility of a MOSFET is a strong function of the Si–SiO$_2$ interface and is strongly influenced by processing techniques.*

While developing the drain current model, we assumed that the μ_s is constant, independent of the gate or the drain voltage (assumption 5). In reality this is not true because when carriers in the channel move under the influence of the normal electric field $\mathscr{E}_x$ and the lateral electric field $\mathscr{E}_y$ due to the gate voltage V_{gs} and drain voltage V_{ds}, respectively, they undergo increased scattering with increasing fields. The reason being that the normal field $\mathscr{E}_x$ acts in a direction so as to accelerate the charge carriers towards the surface causing carriers to scatter more frequently than in the absence of the gate field. On the other hand, the lateral field $\mathscr{E}_y$ causes charge carriers to move faster, so that at high enough V_{ds}, the carriers become velocity saturated. Clearly μ_s is not constant and depends on both $\mathscr{E}_x$ and $\mathscr{E}_y$, which is contrary to our earlier assumption of constant μ_s. It was on this assumption that μ_s was taken outside the integral in Eq. (6.14) and subsequent equations for I_{ds}. Strictly speaking, we must include μ_s inside the integral for calculating the current. However, in that case the resulting current equations become very complicated [46]. In order to keep the current equations manageable we normally define an *effective mobility* μ_{eff} as the average mobility of the carriers in a MOSFET, i.e.,

$$\mu_{\text{eff}} = \frac{\int_0^{t_{ch}} \mu_s(x, y) n(x, y) dx}{\int_0^{t_{ch}} n(x, y) dx} \tag{6.138}$$

[9] In general the different scattering mechanisms that affect the surface mobility behavior are (1) phonon scattering due to lattice vibrations, (2) Coulomb scattering due to charge centers such as ionized impurities, fixed oxide charges, etc., and (3) scattering due to roughness of the surface. The relative importance of these mechanisms depends on the magnitude of the electric field at the surface and the temperature.

such that when used in the following equation [cf. Eq. (6.41)]

$$I_{ds} = \frac{W}{L} \mu_{\text{eff}} \int_0^{V_{ds}} Q_i dV \tag{6.139}$$

the correct value of I_{ds} is predicted.

To develop a theoretical model for μ_{eff} is not easy because separation of the contributions of the various scattering mechanisms is difficult due to many parameters involved. Furthermore, theoretical analyses are complicated by the confinement of the channel region to a very small thickness in a potential well at the silicon surface. The theory is further complicated by quantum effects which play an important role, and because surface roughness at the Si-SiO_2 interface is poorly characterized. Recent theoretical models of carrier surface mobility, some of which are reviewed by Ando et al. [47], are insightful, but they are too complex to be useful even in device simulation, let alone circuit simulation [48]. So *to predict the effective mobility theoretically, we normally rely on experimental data and empirical equations* [49]–[66]. Although these empirical equations lack physical significance, they have worked fairly well in device modeling. The parameters in the empirical equations are adjusted until one obtains an acceptable fit to the experimental data.

6.6.1 Mobility Degradation Due to the Gate Voltage

Based on extensive measurements of surface mobility μ_s at low V_{ds}, Sabnis and Clemens [51] observed that μ_s, when plotted against the effective normal field $\mathscr{E}_{\text{eff}}$, show a "universal curve" independent of the substrate doping concentration N_b. The existence of a universal relationship between $\mathscr{E}_{\text{eff}}$ and μ_s was subsequently confirmed by many others [52]–[57]. For instance, Sun and Plummer [52] showed that for different doping concentrations ($N_b = 3.0 \times 10^{14}\,\text{cm}^{-3}$ to $1.4 \times 10^{17}\,\text{cm}^{-3}$) μ_s falls on the same curve except near the onset of inversion, as shown in Figure 6.21. They also showed that the resulting curve shifts downwards as the interface charge is increased and is unaffected by the substrate bias. This implies that the limiting Coulomb scattering mechanism due to the interface states and fixed oxide charges is the dominant scattering mechanism. It was observed that the dependence of mobility on the gate field can be described by an empirical relationship of the following form, [50], [52]

$$\mu_s = \mu_0 \left(\frac{\mathscr{E}_0}{\mathscr{E}_{\text{eff}}} \right)^{\nu} \tag{6.140}$$

where μ_0 is the *maximum extracted value of the mobility* at a given doping concentration, also some times called the *low field surface mobility*, whose

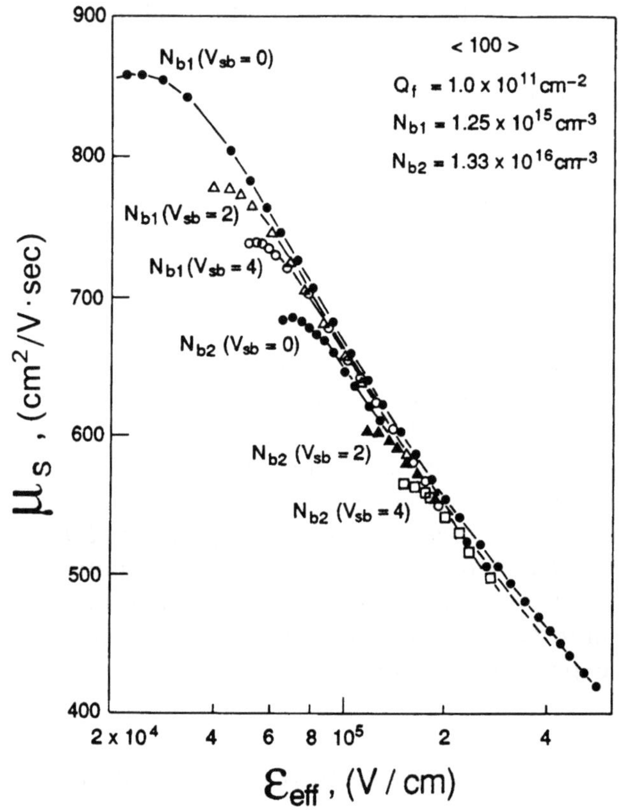

Fig. 6.21 Inversion layer electron mobility data for silicon at 300 K for two different substrate dopings $N_{b1} = 1.25 \times 10^{15}\,\text{cm}^{-3}$, $N_{b2} = 1.33 \times 10^{16}\,\text{cm}^{-3}$. (After Sun and Plummer [52])

value for electrons is 400–700 $\text{cm}^2/(\text{V}\cdot\text{s})$ and holes is 100–300 $\text{cm}^2/(\text{V}\cdot\text{s})$. $\mathscr{E}_0$ is the critical electric field below which $\mu_s = \mu_0$ and above which μ_s begins to decrease and ν is an empirical constant. An increase in $\mathscr{E}_{\text{eff}}$ causes carriers to be drawn closer to the interface, thus increasing surface scattering, and hence lower mobility.

Recently Liang et al. [57] studied the effective mobility at higher effective fields using various oxide thicknesses (53 Å, 88 Å, 169 Å and 418 Å) and a fixed substrate concentration of $5.0 \times 10^{16}\,\text{cm}^{-3}$ as shown in Figure 6.22. Their study showed that the μ_s versus $\mathscr{E}_{\text{eff}}$ relationship is independent of the oxide thickness down to 50 Å, and that the thin-oxide MOSFET transconductance is degraded by the finite inversion layer capacitance and not by decreased mobility for thin oxides. However, they suggested the following empirical formulation, which is slightly different from Eq. (6.140)

$$\mu_s = \frac{\mu_0}{1 + (\mathscr{E}_{\text{eff}}/\mathscr{E}_0)^{\nu}}. \tag{6.141}$$

The parameters $\mu_0, \mathscr{E}_0$ and ν are given in Table 6.3 [62].

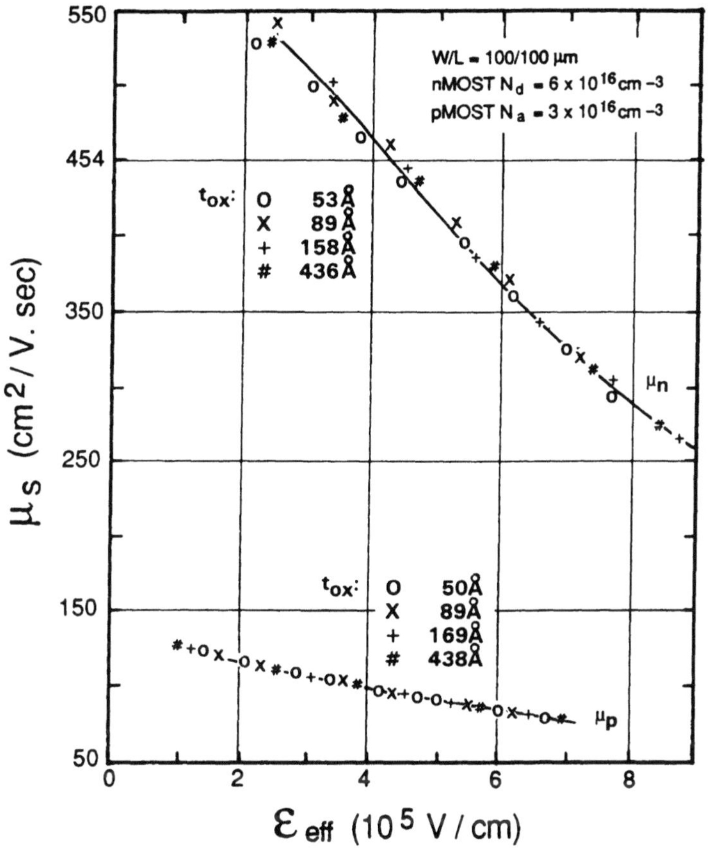

Fig. 6.22 Inversion layer electron and hole mobility for different oxide thicknesses. (After Liang et al. [57])

Table 6.3. *Parameters for the surface mobility model Eq. (6.141) for silicon at 300 K*

Parameter	μ_0(cm^2/V.sec)	$\mathscr{E}_0$(V/cm)	ν
Electrons (nMOST)	670	6.7×10^5	1.6
Holes (pMOST with p^+ Poly-Si)	160	7.0×10^5	1.0
Holes (pMOST with n^+ Poly-Si)	290	3.5×10^5	1.0

Note that hole mobility μ_0 for a pMOST with n-implant at the surface (p^+polysilicon gate) is much lower compared to the pMOST with a partially buried channel (n^+ polysilicon gate). This is because in the latter case current flows slightly below the silicon surface, thus resulting in less scattering and hence higher hole mobility. It has been observed that mobility is independent of the gate-oxide thickness, provided the Si–SiO_2 interface is of good quality (oxide charge Q_0 is less than 1.0×10^{10} c/cm^2 and thus negligible) and the channel inversion charge Q_i is properly calculated. Otherwise, a dependence of mobility on oxide thickness can be observed for thicknesses lower than 100 Å. These results lead to the conclusion that *mobility* μ_s *is more a function of the Si–SiO$_2$ interface than device parameters such as oxide thickness or doping concentration.*

A mobility model suitable for circuit simulation, which fits the observed experimental mobility data at low V_{ds}, for both p- and n-channel devices, is of the form [59], [63]

$$\boxed{\mu_s = \frac{\mu_0}{1 + \alpha_\theta \mathscr{E}_{\text{eff}}}} \tag{6.142}$$

where α_θ is called the scattering constant. Thus, to calculate μ_s we need to calculate $\mathscr{E}_{\text{eff}}$ to be discussed shortly.

It has been shown [56] that Eqs. (6.140)–(6.142) are valid for $\mathscr{E}_{\text{eff}} < 5.5 \times 10^5$ V/cm and that at higher fields there is a stronger dependence of mobility on $\mathscr{E}_{\text{eff}}$. The failure of the model at high fields is believed to be caused by the onset of quantum effects in the deep inversion channel potential well and the populating of the upper subbands in silicon. At higher fields Eq. (6.142) becomes invalid. A typical set of universal mobility field data, including high fields, is presented in Figure 6.23 [65]. Note that electron mobility falls off as $\mathscr{E}_{\text{eff}}^{-0.3}$ at intermediate fields with a transition to $\mathscr{E}_{\text{eff}}^{-2}$ at high fields for nMOST and $\mathscr{E}_{\text{eff}}^{-1}$ for pMOST. The $\mathscr{E}_{\text{eff}}^{-0.3}$ dependence is due to acoustical phonon scattering of the inversion layer carriers and at high fields, the $\mathscr{E}_{\text{eff}}^{-2}$ dependence is due to surface roughness scattering [28], [64]–[66]. The net mobility model is

$$\mu_s = \frac{\mu_0}{1 + \theta_1 \mathscr{E}_{\text{eff}}^{0.3} + \theta_2 \mathscr{E}_{\text{eff}}^{m}} \tag{6.143}$$

where θ_1 and θ_2 are fitting parameters, and $m = 2$ for electrons and $m = 1$ for holes. At lower fields where fixed oxide charge scattering and Coulomb scattering are important, this model becomes less accurate.

Calculation of the Effective Field $\mathscr{E}_{\text{eff}}$. It is easy to find an expression for $\mathscr{E}_{\text{eff}}$, if we interpret $\mathscr{E}_{\text{eff}}$ as the *average electric field* $\mathscr{E}_{\text{avg}}$ experienced by the carriers in the inversion layer, that is,

$$\mathscr{E}_{\text{eff}} \equiv \mathscr{E}_{\text{avg}} = \frac{\mathscr{E}_{x1} + \mathscr{E}_{x2}}{2} \tag{6.144}$$

where $\mathscr{E}_{x1}$ and $\mathscr{E}_{x2}$ are the electric field normal to the surface at the Si-SiO$_2$ interface and the channel-depletion layer interface respectively. Using

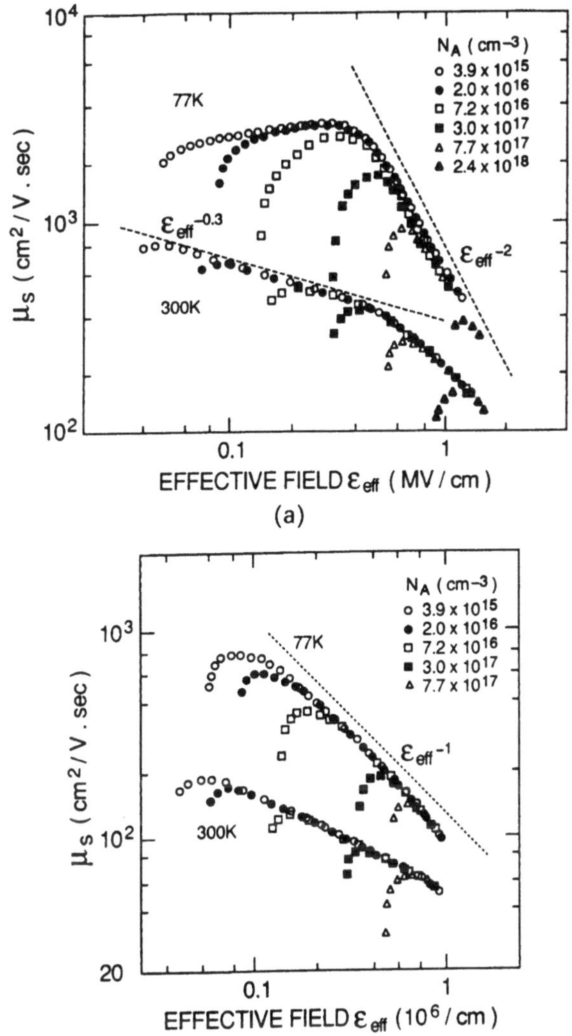

Fig. 6.23 The inversion layer mobility for (a) electrons and (b) holes at high fields. (After Takagi et al. [65])

Gauss' law it is easy to see that

$$\mathscr{E}_{x1} - \mathscr{E}_{x2} = \frac{Q_i}{\epsilon_0 \epsilon_{si}} \tag{6.145}$$

and

$$\mathscr{E}_{x2} = \frac{Q_b}{\epsilon_0 \epsilon_{si}}. \tag{6.146}$$

Solving Eqs. (6.145) and (6.146) for $\mathscr{E}_{x1}$ and $\mathscr{E}_{x2}$ and substituting in Eq. (6.144) yields

$$\mathscr{E}_{\text{eff}} = \frac{1}{\epsilon_0 \epsilon_{si}}(Q_b + 0.5Q_i). \tag{6.147}$$

Thus $\mathscr{E}_{\text{eff}}$ is related to the bulk depletion charge Q_b and to the inversion charge Q_i. This equation for $\mathscr{E}_{\text{eff}}$ when used in Eqs. (6.140)–(6.143) provides good fit to the experimental mobility data for *n*-channel devices. However, for *p*-channel devices it was found by Arora and Gildenblat [59] and later confirmed by other workers [60]–[61] that when the factor 0.5 in Eq. (6.147) is replaced by 0.3, the resulting $1/\mu_s$ vs. $\mathscr{E}_{\text{eff}}$ curve is independent of back bias. It is thus more appropriate to write $\mathscr{E}_{\text{eff}}$ as

$$\boxed{\mathscr{E}_{\text{eff}} = \frac{1}{\epsilon_0 \epsilon_{si}}(Q_b + \zeta Q_i)} \tag{6.148}$$

where $\zeta = 0.5$ for *n*-channel devices, and $\zeta = 0.25 - 0.30$ for *p*-channel devices.[10] Using Q_b from Eq. (6.78) and Q_i from Eq. (6.79) in Eq. (6.148) we get

$$\mathscr{E}_{\text{eff}} = \frac{\zeta C_{ox}}{\epsilon_0 \epsilon_{si}}\left[V_{gs} - V_{th} + \frac{1}{\zeta}\gamma\sqrt{2\phi_f + V_{sb}} - V\left(\alpha - \frac{1}{\zeta}\delta\gamma\right)\right].$$

In order to simplify this expression we assume a constant channel electric field in the lateral direction, so that the average value of $\mathscr{E}_{\text{eff}}$ becomes

$$\mathscr{E}_{\text{eff}} = \frac{1}{V_{ds}}\int_0^{V_{ds}} \frac{\zeta C_{ox}}{\epsilon_0 \epsilon_{si}}\left[V_{gs} - V_{th} + \frac{1}{\zeta}\gamma\sqrt{2\phi_f + V_{sb}} - V\left(\alpha - \frac{1}{\zeta}\delta\gamma\right)\right]dV$$

which on simplification, remembering $\delta\gamma = \alpha - 1$ [cf. Eq. (6.80)], yields

$$\mathscr{E}_{\text{eff}} = \frac{\zeta C_{ox}}{\epsilon_0 \epsilon_{si}}\left[V_{gs} - V_{th} + \frac{1}{\zeta}\gamma\sqrt{2\phi_f + V_{sb}} - 0.5\,V_{ds}\left(\frac{1}{\zeta} - \frac{1}{\zeta}\alpha + \alpha\right)\right]. \tag{6.149}$$

For $\zeta = 0.5$, the above equation reduces to

$$\mathscr{E} = \frac{C_{ox}}{2\epsilon_0 \epsilon_{si}}[V_{gs} - V_{th} + 2\gamma\sqrt{2\phi_f + V_{sb}} - 0.5\,V_{ds}(2 - \alpha)] \tag{6.150}$$

---

[10] Similar results were obtained by writing $\mathscr{E}_{\text{eff}}$ as [67]

$$\mathscr{E}_{\text{eff}} = \frac{1}{\epsilon_0 \epsilon_{si}}(\zeta Q_b + 0.5Q_i).$$

However, it has been shown that Eq. (6.148) is more appropriate to use [71].

which is the equation used in many circuit models [24], [68, 70]. Often, the V_{ds} dependence in Eq. (6.150) is ignored and the resulting value when used in Eq. (6.142) yields

$$\boxed{\mu_s = \frac{\mu_0}{1 + \theta(V_{gs} - V_{th} + 2\gamma\sqrt{2\phi_f + V_{sb}})}} \tag{6.151}$$

where

$$\theta = \frac{\alpha_\theta C_{ox}}{2\epsilon_0\epsilon_{si}} \quad (\mathrm{V}^{-1}) \tag{6.152}$$

is called the *mobility degradation coefficient*, whose value lies in the range 0.03–0.1 V^{-1}. A simplified version of Eq. (6.151) is also widely used in simulators [27]

$$\mu_s = \frac{\mu_0}{1 + \theta(V_{gs} - V_{th}) + \theta_b V_{sb}}. \tag{6.153}$$

For n-channel devices, the value of θ_b is usually small ($\sim 0.005\ V^{-1}$), and is often neglected. For p-channel devices θ_b improves the current-voltage data fit. The SPICE MOSFET Levels 3 and 4 models use $\theta_b = 0$ in Eq. (6.153) for both p- and n-channel devices resulting in the following equation for μ_s

$$\mu_s = \frac{\mu_0}{1 + \theta(V_{gs} - V_{th})}. \tag{6.154}$$

Physically speaking, this is an incorrect equation as it incorrectly predicts the V_{sb} dependence of μ_s [54]. This can be easily seen as follows: we know that an increase in V_{sb} causes an increase in V_{th}; therefore, from Eq. (6.154) μ_s should increase with increasing V_{sb}, while experimental data clearly shows that μ_s decreases as V_{sb} increases. In spite of this inaccuracy Eq. (6.154) has been used in circuit simulators because of its simplicity. Moreover, the results do not significantly differ from the observed mobility variation, particularly for n-channel devices, so that in this case θ becomes a completely empirical parameter.

The SPICE MOSFET Level 2 model uses the following equation for μ_s,

$$\mu_s = \mu_0\left[\frac{\epsilon_0\epsilon_{si}\mathscr{E}_0}{C_{ox}(V_{gs} - V_{th} - u_t V_{ds})}\right]^{\nu} \tag{6.155}$$

which is similar in form to Eq. (6.140). Here u_t is a fitting parameter whose value lies between 0 and 0.5 and represents the contribution to the field due to drain voltage. The exponent ν is approximately 0.25 for n-channel and 0.15 for p-channel devices [50]. For devices in strong inversion,

Eq. (6.155) yields good agreement between Eq. (6.155) and experimental data in the absence of short and narrow channel effects.
Another model for μ_s which includes L and W dependence has been suggested and has the following form [53]

$$\mu_s = \frac{\mu_0}{1 + A/L + B/W + \theta(V_{gs} - V_{th})} \tag{6.156}$$

which simplifies to Eq. (6.154) when $A = B = 0$. In one case it was found that $A = 0.34\,\mu\text{m}$, $B = 0.07\,\mu\text{m}$ and $\theta = 0.06\,\text{V}^{-1}$ [53].

6.6.2 *Mobility Degradation Due to the Drain Voltage*

The mobility degradation due the lateral field $\mathscr{E}_y$ (drain voltage) has a more significant effect on the device current equations than does the normal field $\mathscr{E}_x$ (gate voltage). This is because an increase in the lateral field eventually causes velocity saturation of the carriers. For a given normal field $\mathscr{E}_x (= \mathscr{E}_{\text{eff}})$, the velocity v of a carrier is proportional to $\mathscr{E}_y$ at low lateral fields, and the proportionality constant is the mobility μ_s discussed in the previous section and plotted in Figure 6.24. However, as $\mathscr{E}_y$ increases, the carrier velocity tends to saturate. Experimental measurements show that the velocity-field relationship for electrons and holes is well described by Eq. (2.24), repeated here for convenience [72]–[77].

$$v = \frac{\mu_s \mathscr{E}_y}{[1 + (\mathscr{E}_y/\mathscr{E}_c)^{\nu}]^{1/\nu}}, \qquad \mathscr{E} > \mathscr{E}_c \tag{6.157}$$

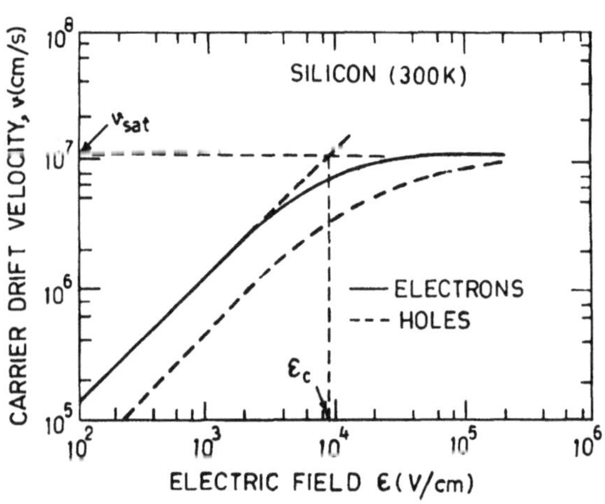

Fig. 6.24 Carrier (electrons and holes) velocity v versus lateral electric $\mathscr{E}$ in silicon at 300 K

where $\nu = 2$ for electrons and $\nu = 1$ for holes; $\mathscr{E}_c$ is the critical field at which carriers become velocity saturated and is related to the saturated carrier velocity v_{sat} by the following expression

$$\mathscr{E}_c = \frac{v_{sat}}{\mu_s}. \tag{6.158}$$

The reported value of v_{sat}, for the MOSFET inversion layer, varies over a wide range [72]–[79]. For electrons v_{sat} varies between 6–9 × 10^6 cm/s, while for holes it is between 4–8 × 10^6 cm/s.
The use of Eq. (6.157) in Eq. (6.42) leads to an expression for the current which cannot be solved in the general case. By making some approximations, Eq. (6.157) has been used for the current calculation of n-channel MOSFET's [34], [77]–[82], though the final I_{ds} expression is too complicated to be suitable for circuit models. Various expressions, which closely approximate Eq. (6.157) for electrons ($\nu = 2$), have been proposed. These are of the general form [28], [82]–[83]

$$v = \frac{\mu_s \mathscr{E}_y}{1 + \delta_0(\mathscr{E}_y/\mathscr{E}_c)} \tag{6.159}$$

where δ_0 is usually a function of the field.[11] However, the final expression for the current is still very complicated. The circuit models invariably use $\delta_0 = 1$ for electrons. In other words, Eq. (6.157) with $\nu = 1$ is used for both p- and n-channel velocity-field relationship, that is,

$$v = \frac{\mu_s \mathscr{E}_y}{1 + (\mathscr{E}_y/\mathscr{E}_c)}. \tag{6.160}$$

Because Eq. (6.160) is a crude approximation of (6.157) for $\nu = 2$, in the CAD models $\mathscr{E}_c$ is assumed to be a fitting parameter whose value is obtained by curve fitting the equation to the experimental data.
A two-region piece-wise model for velocity saturation has also been proposed and is defined as [89]

$$v = \begin{cases} \dfrac{\mu_s \mathscr{E}_y}{1 + (\mathscr{E}_y/\mathscr{E}_c)} & \mathscr{E} \le \mathscr{E}_c \\[2ex] \dfrac{\mu_s \mathscr{E}_c}{2} & \mathscr{E} > \mathscr{E}_c \end{cases}. \tag{6.161}$$

[11] For example, one model assumes δ_0 to be [82]

$$\delta_0 = \left(\frac{\mathscr{E}_y}{\mathscr{E}_c}\right)\left[1 + 1.5\left(\frac{\mathscr{E}_y}{\mathscr{E}_c}\right)^{1/2}\right]^{-1}$$

while, for another model $\delta_0 = 0.42(\mathscr{E}_y/\mathscr{E}_c)^{0.8}$ [28]. Still in other approach δ_0 is a fitting parameter [83].

The difference between the models represented by Eqs. (6.160) and (6.161) can be seen as follows. In Eq. (6.160), the electric field $\mathscr{E}_y$ must be much larger than the critical field $\mathscr{E}_c$ (in fact $\mathscr{E}_y \rightarrow \infty$) for the velocity to approach υ_{sat}. However, Eq. (6.161) assumes $\mathscr{E}_y = \mathscr{E}_c$ when the carrier velocity equals υ_{sat}. The result is that the piece-wise linear model has a steeper slope at high fields and thus models the experimental data more precisely than Eq. (6.160). However, note that for $\mathscr{E} > \mathscr{E}_c$ carriers move with a velocity that is only one half of the theoretical saturated velocity. As we shall see later, this results in a lower V_{dsat} (cf. section 6.7.2).

A comparison of the three velocity saturation models for electrons is shown in Figure 6.25. The more complicated Eq. (6.157) (model B in Figure 6.25) fits the experimental data quite well, while the simple Eq. (6.160) (model C) significantly underestimates the velocity at moderate field if the same saturation velocity is assumed. However, Eq. (6.161) (model A) overpredicts the velocity in the saturation region. Seting $\delta_0 = 0.4$ corrects this overestimation. An important point to remember is that, as the normal field is increased, it takes higher lateral field for the carriers to reach the same velocity.

An advantage of using Eqs. (6.160)–(6.161) is that by replacing $\mathscr{E}_y = |(dV/dy)|$, the current equation can easily be integrated in a closed form expression as we will see in the next section. It should be pointed out that although the $\upsilon - \mathscr{E}$ relations (6.160)–(6.161) are simple to use, strictly speaking they are inaccurate. A more correct expression is obtained by eliminating the critical field $\mathscr{E}_c$ using Eq. (6.158) and introducing instead

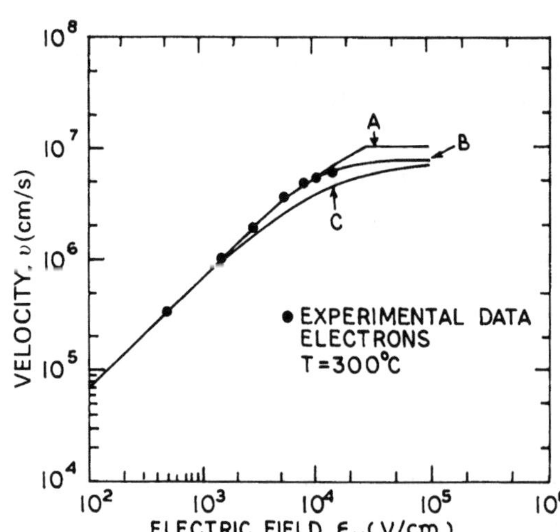

Fig. 6.25 Comparison of electron velocity versus electric field at 300 K for different models. Models A, B, and C are based on Eqs. (6.161), (6.157) and (6.160), respectively. (After Sodini et al. [89])

saturation velocity υ_{sat} as the parameter [73]. Thus, Eq. (6.160) would read as

$$\boxed{\upsilon = \frac{\mu_s \mathscr{E}_y}{1 + \mu_s \mathscr{E}_y/\upsilon_{sat}}.} \tag{6.162}$$

Remember that μ_s is not constant but is gate field dependent. Using Mathiessens's rule to combine the effect of mobility degradation due to the normal and lateral fields in the channel, we can write the effective mobility μ_{eff} as

$$\frac{1}{\mu_{eff}} = \frac{1}{\mu_s} + \frac{1}{\mu_\upsilon} \tag{6.163}$$

where μ_υ is the mobility due to lateral field. Thus, to a first approximation, we can write

$$\mu_{eff} = \frac{\mu_0}{(1 + \alpha_\theta \mathscr{E}_{eff})[1 + \delta_0(\mathscr{E}_y/\mathscr{E}_c)]}. \tag{6.164}$$

A slightly different form for μ_{eff} has also been used [24], [46], [68]

$$\mu_{eff} = \frac{\mu_0}{1 + \alpha_\theta \mathscr{E}_{eff} + \theta_l \mathscr{E}_y} \tag{6.165}$$

which has been generally expressed as [27], [81]

$$\mu_{eff} = \frac{\mu_0}{1 + \theta(V_{gs} - V_{th}) + \theta_b V_{sb} + \theta_c V_{ds}} \tag{6.166}$$

where $\theta_c = (L\mathscr{E}_c)^{-1}$. Note that Eqs. (6.165) or (6.166) follows from Eq. (6.164), if the products of mobility degradation factors α_θ and θ_l, due to the normal and longitudinal fields, respectively, are neglected. Thus, μ_{eff} is modeled by four parameters μ_0, θ, θ_b and υ_{sat} (or $\mathscr{E}_c$) which are generally obtained by curve fitting the experimental data to the model equation.

6.7 Short-Geometry Models

In the discussion so far we have assumed that device lengths and widths are large so that edge effects can be neglected. However, when the device physical dimensions are reduced, the following distinct features are observed in the device characteristics:

- The drain current I_{ds} increases with drain voltage V_{ds} beyond V_{dsat}. That is, the slope of $I_{ds} - V_{ds}$ curve increases from zero for long channel devices to some positive number for shorter devices. The shorter the device, the higher the slope. These devices also exhibit a softer breakdown that is not seen in long channel devices. Historically, this phenomena was the first short-channel effect to be studied.

- The threshold voltage V_{th} shifts from its long channel value. In other words, it becomes geometry dependent. In addition, the short-channel V_{th} becomes drain voltage dependent due to the drain induced barrier lowering (DIBL) effect. Short-channel and narrow-width threshold voltage behavior were discussed in Chapter 5.
- The subthreshold slope increases as the device becomes shorter (see Figure 6.26) [80]. In fact, when the device becomes very short, the gate no longer controls the drain current and the device can not be turned off. This is caused by the punch-through effect (cf. section 3.1). In addition, the subthreshold slope changes with the drain voltage.

These characteristics cannot be modeled using the one dimensional current flow equations developed in the previous sections. In fact, one needs 2- or 3-D analysis to account for short-channel and/or narrow-width behavior. However, for circuit models we invariably modify the 1-D equations developed earlier by including 2- or 3-D effects using some simple relations based on physical reasoning, or sometimes purely on empirical grounds.

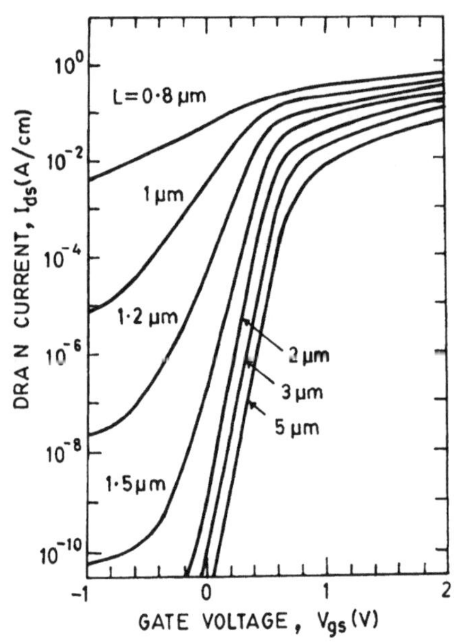

Fig. 6.26 Calculated subthreshold characteristics (using 2-D device simulator) of a MOSFET for various channel lengths. $V_{ds} = 2\,V$, $V_{sb} = 0\,V$, $X_j = 0.33\,\mu m$, $t_{ox} = 500\,Å$. (After Kotani and Kawazu [80])

For this reason, *the short-geometry circuit models always have some parameters whose values can best be determined by curve fitting the I–V data.* The following effects are generally considered necessary when modeling DC characteristics of small-geometry MOSFETs:

- surface mobility reduction due to the vertical and lateral field as discussed in section 6.6,
- carrier velocity saturation,
- channel length modulation,
- depletion charge sharing by the source/drain,
- source/drain series resistance.
- hot-carrier effect.

We will now discuss the small-geometry models, suitable for circuit simulators, which include these effects.

6.7.1 Linear Region Model

The linear region current equation for short-geometry devices is simply obtained by replacing the carrier velocity υ in Eq. (6.42) with Eq. (6.162), resulting in the following expression for I_{ds}

$$I_{ds} = -W\mu_s Q_i \frac{\mathscr{E}_y}{1 + (\mu_s/\upsilon_{sat})\mathscr{E}_y} \qquad (6.167)$$

where μ_s is given by Eq. (6.151) or any of the other expressions for μ_s discussed in section 6.6.1. For short-geometry devices, the inversion charge density Q_i can still be given by Eq. (6.79), although now the parameter α becomes length and width dependent. This can be seen as follows: Recall that for short-geometry devices, the bulk charge density Q_b is length and width dependent (cf. section 5.3). Suppose we use Eq. (5.68) for short-channel Q_b, then $Q_b(y)$ becomes [cf. Eq. (6.78)]

$$Q_b(y) = -F_l\gamma C_{ox}\sqrt{2\phi_f + V_{sb} + V(y)} \approx -F_l\gamma C_{ox}(\delta V + \sqrt{2\phi_f + V_{sb}})$$

where $F_l(<1)$ is a short-channel factor. This value of Q_b when used in Eq. (6.43) yields

$$Q_i(y) = -C_{ox}[V_{gs} - V_{fb} - 2\phi_f - F_l\gamma\sqrt{2\phi_f + V_{sb}} - (1 + F_l\gamma\delta)V(y)]$$

Replacing $(1 + F_l\gamma\delta)$ by α and recalling that the short-channel threshold voltage V_{th} is [cf. Eq. (5.69)]

$$V_{th} = V_{fb} + 2\phi_f + F_l\gamma\sqrt{2\phi_f + V_{sb}}$$

we see that $Q_i(y)$ for short-channel device becomes

$$Q_i(y) = -C_{ox}[V_{gs} - V_{th} - \alpha V(y)] \qquad (6.168)$$

which is of the same form as Eq. (6.79). However, note that now α and V_{th} are length dependent through the F_l term. In fact, whatever expression for the short-geometry Q_b is used (cf. section 5.3), the inversion charge Q_i can always be expressed by Eq. (6.168) where α and V_{th} become L and W dependent. Also recall that for short-channel devices V_{th} depends on V_{ds} through the DIBL effect [cf. Eq. (5.96)], i.e.,

$$V_{th}(V_{ds}) \equiv V_{thd} = V_{th} - \sigma V_{ds} \tag{6.168a}$$

where σ is the DIBL parameter which depends on the gate oxide thickness t_{ox}, the source/drain junction depth X_j, the doping concentration N_b, and the substrate bias V_{sb}. Therefore, for short-channel devices one must replace V_{th} in Eq. (6.168) with V_{thd}.

Substituting $Q_i(y)$ from Eq. (6.168) in (6.167) and replacing $\mathscr{E}_y$ by $|dV/dy|$ we get

$$I_{ds}\left(1 + \frac{\mu_s}{v_{sat}}\frac{dV}{dy}\right) = W\mu_s C_{ox}[V_{gs} - V_{th} - \alpha V(y)]\frac{dV}{dy}.$$

Where V_{th} is a function of V_{ds} as given by Eq. (6.18a).

Integrating this equation from $y = 0$ $(V = 0)$ to $y = L(V = V_{ds})$, we get

$$\boxed{I_{ds} = \frac{W\mu_s C_{ox}}{L\left(1 + \dfrac{\mu_s V_{ds}}{L v_{sat}}\right)}(V_{gs} - V_{th} - 0.5\alpha V_{ds})V_{ds}.} \tag{6.169a}$$

Remembering that $\beta = \mu_s C_{ox} W/L$, and for the sake of convenience replacing v_{vsat}/μ_s with $\mathscr{E}_c$, the above equation is often written as

$$I_{ds} = \frac{\beta}{(1 + V_{ds}/L\mathscr{E}_c)}(V_{gs} - V_{th} - 0.5\alpha V_{ds})V_{ds}. \tag{6.169b}$$

This is the most widely used MOSFET linear region drain current equation for short-channel I_{ds} calculations. It is used in the SPICE Levels 3 and 4 models. From this derivation for I_{ds}, it is evident that if we use Eq. (6.157) for the carrier velocity, then the integration of the resulting I_{ds} equation will not yield a simple closed form solution.

Comparing Eq. (6.169) with the corresponding Eq. (6.81) for the long channel device, we see that forms of the two equations are the same except for an additional factor related to $\mathscr{E}_c$. Since $\mathscr{E}_c$ and L appear as a product term in Eq. (6.169), it follows that a device behaves like a 'long-channel' device, if the product of L and $\mathscr{E}_c$ is large. For a given gate and drain voltage, a device with thinner gate oxide behaves more like a 'long-channel' device because the surface mobility μ_s is low (due to higher normal field) so that $\mathscr{E}_c$ is high, since v_{sat} is constant, physical property of the silicon.

Some what empirical expressions for short-channel I_{ds}, which are similar in form to Eq. (6.169), have also been reported and implemented in SPICE [84], [85]. In one model the linear region drain current is expressed as [84], [86]

$$I_{ds} = \beta(V_{gs} - V_{th} - 0.5\alpha_x V_{ds})V_{ds} \tag{6.170}$$

where

$$\alpha_x = \alpha_1 + \alpha_2(V_{gs} - V_{th}) \tag{6.171}$$

where α_1 and α_2 are the so called *short-channel parameters*. The expression for α_x is arrived at empirically based on short-channel effects, such as nonlinear velocity-field characterstics and modification of the bulk charge due to the presence of the mobile carriers. The parameter α_x effectively represents the ratio of mobility from linear to saturation region. The parameter α_2 which makes α_x dependent on the gate voltage, is called the velocity saturation factor.

6.7.2 Saturation Voltage

The saturation voltage for a long channel device was calculated assuming that the inversion charge density $Q_i(y)$ at the drain end was totally depleted, i.e., $Q_i(L)=0$, the so called *pinch-off* condition (cf. section 6.4.1). The condition $Q_i(L)=0$ is equivalent to equating the linear region current derivative, with respect to V_{ds}, to zero. However, this classical condition for pinch-off is not realistic for short-channel devices. This is because the carriers reach velocity saturation even before the pinch-off condition are fulfilled. Therefore, a more realistic way to calculate short-channel V_{dsat} is to define the saturated drain current I_{dsat} as the current when the carriers at the drain end are velocity saturated, that is,

$$I_{dsat} = -Wv_{sat}Q_{sat} = Wv_{sat}C_{ox}(V_{gs} - V_{th} - \alpha V_{dsat}) \tag{6.172}$$

where Q_{sat} is the value of Q_i at $V_{ds} = V_{dsat}$, and is obtained by replacing $V(y)$ with V_{dsat} in Eq. (6.168) for Q_i. Equation (6.172) assumes that the GCA underlying Eq. (6.168) still holds at point C in the channel where $V_{ds} = V_{dsat}$ (see Figure 6.27). We consider C as the point where carriers have reached velocity saturation. The I_{dsat} on the left hand side of the point C is still given by Eq. (6.169) with V_{ds} replaced by V_{dsat}, that is,

$$I_{dsat} = \frac{W\mu_s C_{ox}}{L(1 + V_{dsat}/L\mathscr{E}_c)}(V_{gs} - V_{th} - 0.5\alpha V_{dsat})V_{dsat}. \tag{6.173}$$

Therefore, by equating Eqs. (6.172) and (6.173) and solving the resulting quadratic equation in V_{dsat}, we get [34], [87]

$$V_{dsat} = L\mathscr{E}_c\left[\sqrt{1 + \frac{2(V_{gs} - V_{th})}{\alpha L\mathscr{E}_c}} - 1\right]. \tag{6.174}$$

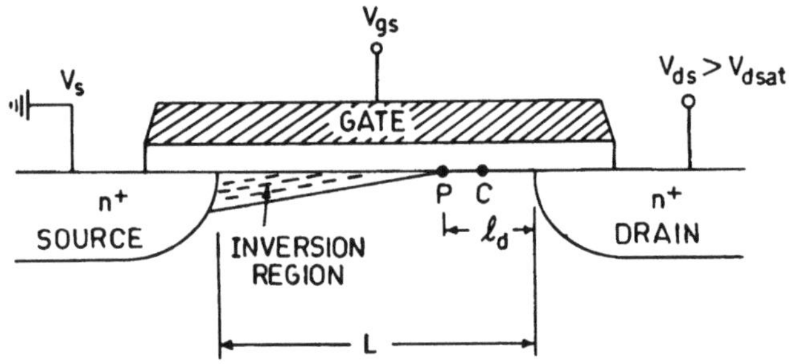

Fig. 6.27 Schematic diagram showing MOSFET operating in saturation region. P is the pinch-off point

For SPICE applications it is more appropriate to write this equation as

$$V_{dsat} = \frac{2(V_{gs} - V_{th})}{\alpha}\left[\sqrt{1 + \frac{2(V_{gs} - V_{th})}{\alpha L \mathscr{E}_c}} + 1\right]^{-1}.$$

This is the expression most widely used for short-channel V_{dsat} calculations and appears to fit the data well [33], [87], [88]. It is also used in the SPICE Level 4 model (BSIM)[12] [25]. Equation (6.174) shows that short-channel V_{dsat} depends upon channel length L and is reduced from its long channel value of $(V_{gs} - V_{th})/\alpha$ [cf. Eq. (6.82)]; the shorter the L, the lower the V_{dsat}. The measured and calculated V_{dsat} as a function of V_{gs} for different length devices is shown in Figure 6.28; continuous lines are based on Eq. (6.174), while symbols are measured V_{dsat} using methods discussed in section 9.11. Physically speaking, as one scales the channel length, a lower voltage at the drain is required to reach the critical field $\mathscr{E}_c$. In other words, V_{dsat} reduces when L becomes smaller. When L is large Eq. (6.174) approaches Eq. (6.82), the long channel V_{dsat}.
If we use $v_{sat} = \mathscr{E}_c \mu_s/2$ (piece-wise model) in Eq. (6.161), then equating the resulting equation to Eq. (6.173) yields the following expression for V_{dsat} [89]

$$V_{dsat} = \frac{(V_{gs} - V_{th})L\mathscr{E}_c}{\alpha L \mathscr{E}_c + (V_{gs} - V_{th})}. \tag{6.175}$$

The V_{dsat} predicted by this equation is lower compared to that of (6.174). This is because while deriving Eq. (6.175) we assumed that the carriers move through the saturation region at only one half of the theoretical saturated velocity ($v_{sat} = \mathscr{E}_c \mu_s/2$).

[12] The V_{dsat} expression in the BSIM model is exactly the same as Eq. (6.174), although it is written in a form that at first glance appears different.

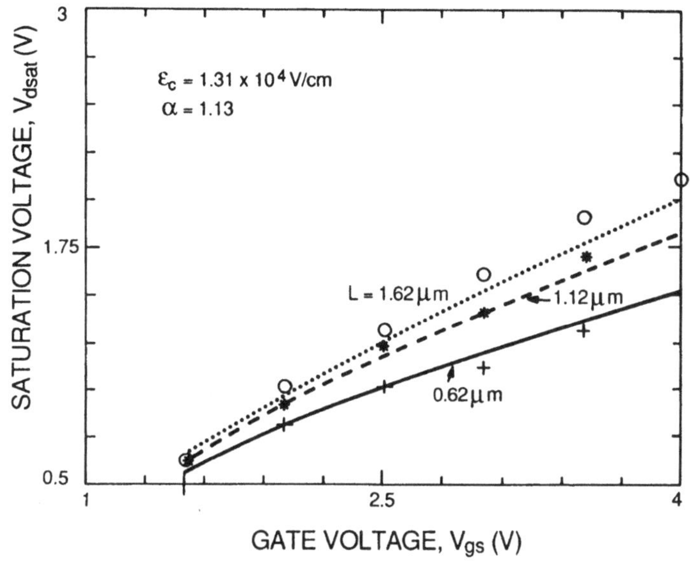

Fig. 6.28 Measured and calculated drain saturation voltage $V_{d\text{sat}}$ as a function of gate bias V_{gs} with channel length L as a parameter

A more appropriate value of $V_{d\text{sat}}$ is obtained by using the velocity field relationship (6.159). Following exactly the same procedure as was used in deriving Eq. (6.174), it is easy to show that in this case

$$V_{d\text{sat}} = \frac{(1-\delta_0)V_{ge} + L\mathscr{E}_c}{(1-2\delta_0)}\left[-1 + \sqrt{1 + \frac{2V_{ge}L\mathscr{E}_c(2\delta_0 - 1)}{((1-\delta_0)V_{ge} + L\mathscr{E}_c)^2}}\right] \tag{6.176}$$

where

$$V_{ge} = \frac{(V_{gs} - V_{th})}{\alpha}.$$

When $\delta_0 = 1$, Eq. (6.176) reduces to Eq. (6.174). However, if $\delta_0 = 0$ then

$$V_{d\text{sat}} = V_{ge} + L\mathscr{E}_c - \sqrt{V_{ge}^2 + (L\mathscr{E}_c)^2} \tag{6.177}$$

which is the $V_{d\text{sat}}$ expression used in the SPICE Level 3 model.
The $V_{d\text{sat}}$ expressions (6.174)–(6.177) assume that the pinch-off point is the carrier velocity saturation point. However, in reality the operating conditions that lead to pinch-off in short channel devices are not easy to define. A variety of suggestions to identify pinch-off have appeared in the literature. In one approach [92] pinch-off is defined as the condition *when dominant control of the channel charge Q_i by the gate field ceases.* This is also expressed as the failure of the GCA. This occurs when the field gradient along the channel $\partial\mathscr{E}_y/\partial y$ reaches a considerable fraction of the field gradient

perpendicular to the channel $\partial\mathscr{E}_x/\partial x$, that is

$$\frac{\partial\mathscr{E}_y}{\partial y} = \frac{1}{f}\frac{\partial\mathscr{E}_x}{\partial x} \tag{6.178}$$

where f is a constant factor. Two-dimensional numerical analysis shows that f depends on the bulk doping N_b, oxide thickness t_{ox}, drain junction depth X_j, and free carrier charge density Q_i. The point P in the channel (see Figure 6.27) is the point at which the pinch-off condition (6.178) is satisfied. Beyond point P, the gate no longer controls the current as the drain field takes over. Therefore, we consider P as the point where the normal field $\mathscr{E}_x$ at the interface changes sign (i.e., field is inverted) and beyond point P, free carriers are being pushed away from the surface into the bulk and move towards the drain with a saturated drift velocity. We define the value of the field and the potential at this point to be $\mathscr{E}_p$ and V_{dsat}, respectively. Assuming a constant f for a given transistor type, it has been shown [90] that the field $\mathscr{E}_p$ at point P depends upon the effective gate voltage $(V_{gs} - V_{th})$. It has been found in practise, however, that assuming constant $\mathscr{E}_p$ (taken at some average effective gate voltage) results in a model of adequate accuracy, and also yields a final expression for V_{dsat} in a closed form solution.

Assuming a bias independent $\mathscr{E}_p$ at the saturation point, the saturation current I_{dsat}, from Eq. (6.167), becomes

$$I_{dsat} = W\mu_s Q_{sat}\frac{\mathscr{E}_p}{(1 + \mathscr{E}_p/\mathscr{E}_c)} = W\mu_s C_{ox}(V_{gs} - V_{th} - \alpha V_{dsat})\frac{\mathscr{E}_p}{(1 + \mathscr{E}_p/\mathscr{E}_c)}. \tag{6.179}$$

Equating Eq. (6.169) and (6.172) and solving the resulting quadratic equation in V_{dsat} yields [90, 91]

$$V_{dsat} = \frac{V_{ge} + L\mathscr{E}_p}{(\mathscr{E}_p\mathscr{E}_c - 1)}\left[\sqrt{1 + \frac{2V_{ge}L\mathscr{E}_c(\mathscr{E}_p\mathscr{E}_c - 1)}{(V_{ge} + L\mathscr{E}_p)^2}} - 1\right]. \tag{6.180}$$

In the circuit models, $\mathscr{E}_p$ is often treated as a fitting parameter. When $\mathscr{E}_p \to \infty$, Eq. (6.180) reduces to Eq. (6.174) and when $\mathscr{E}_p = \mathscr{E}_c$, Eq. (6.175) results. Clearly Eqs. (6.174) and (6.175) are special cases of Eq. (6.180). In most of the circuit models $\mathscr{E}_p = \mathscr{E}_c$ is assumed.

It is interesting to note that Eq. (6.174) for short-channel V_{dsat} can also be derived by equating to zero the derivative with respect to V_{ds} of the linear region current equation (6.169). This is because, while arriving at Eq. (6.174), we had assumed v_{sat} equals $\mathscr{E}_c\mu_s$. However, this is true only when $\mathscr{E}_c$ is very large. Thus, in this case the velocity "pinch-down" condition is equivalent to the channel "pinch-off" condition [93].

6.7.3 Saturation Region—Channel Length Modulation

For a drain voltage V_{ds} larger than V_{dsat}, the drain current is in the saturation mode. When $V_{ds} > V_{dsat}$, the channel pinch-off point, or the velocity pinch-down point, starts to move towards the source. This movement is referred to as the *channel length modulation* (CLM). The point P in Figure 6.27 still has a voltage V_{dsat} but the portion of the drain voltage beyond V_{dsat} is now dropped across a depletion region of distance l_d from point P to the drain. The device under this condition behaves as if its channel length L was shortened by l_d so that I_{ds} becomes larger than its value at the onset of saturation when l_d was zero. Thus, in the saturation mode, the channel region splits into two different regions (Figure 6.27)—one on the source side where the GCA is valid and a second on the drain side where the GCA is violated. Since l_d increases with increasing drain voltage, the drain current continues to increase with drain voltage in the saturation region.

The approach normally used to calculate the drain current in the saturation region is to replace the normal channel length L with the reduced channel length $(L - l_d)$ and V_{ds} with V_{dsat} in the linear region current equation (6.169), where the GCA is valid. The drain current equation in the saturation region, therefore, becomes

$$I_{ds} = \frac{W\mu_s C_{ox}}{(L - l_d)[1 + V_{dsat}/(L - l_d)\mathscr{E}_c]}(V_{gs} - V_{th} - 0.5\alpha V_{dsat})V_{dsat}. \tag{6.181}$$

Combining this equation with Eq. (6.173), we get I_{ds} in the saturation region as

$$I_{ds} = I_{dsat}\left[\frac{L\mathscr{E}_c + V_{dsat}}{(L - l_d)\mathscr{E}_c + V_{dsat}}\right]. \tag{6.182}$$

Instead of Eq. (6.182) for I_{ds} in saturation, very often the following approximate expression is used [cf. Eq. (6.60)]

$$I_{ds} = I_{dsat}\frac{L}{L - l_d} \approx I_{dsat}\left(1 + \frac{l_d}{L}\right) \tag{6.183}$$

which can easily be derived from Eq. (6.182) assuming $L \gg V_{dsat}/\mathscr{E}_c$. It is the approximate Eq. (6.183) which is generaly used for CAD models, including the SPICE Levels 2 and 3 models. Irrespective of which equation is used, one needs to know l_d to calculate I_{ds}.

The most general approach to calculate l_d is to solve the 2-D Poisson equation near the drain

$$\frac{\partial^2 V}{\partial x^2} + \frac{\partial^2 V}{dy^2} = -\frac{\rho(x, y)}{\epsilon_0 \epsilon_{si}} \tag{6.184}$$

where ρ is the charge density in the pinch-off region. This equation can only be solved by using numerical techniques. In order to obtain an approximate analytical solution for $V(x, y)$ various simplifying assumptions are made. The method most widely used to solve Eq. (6.184) is to ignore the field gradient in the x direction so that Eq. (6.184) is reduced to

$$\frac{\partial^2 V}{\partial y^2} = -\frac{\rho(y)}{\epsilon_0 \epsilon_{si}}. \tag{6.185}$$

Assuming uniform doping in the substrate, the charge density ρ can be replaced by the sum of the depletion charge density (qN_b) and mobile charge density Q_i. Recall that while calculating V_{dsat} for long channel devices, we had assumed that Q_i was zero in the pinch-off region. Thus, following the long channel approximation, one assumes that no mobile carriers are present in the pinch-off region and only depletion charge exists; that is, $\rho = -qN_b$, so that Eq. (6.185) becomes

$$\frac{\partial^2 V}{\partial y^2} = \frac{qN_b}{\epsilon_{si}}. \tag{6.186}$$

Integrating this equation under the following boundary conditions

$$\begin{aligned} V(y) &= \begin{cases} V_{dsat} & \text{at } y = L - l_d \\ V_{ds} & \text{at } y = L \end{cases} \\ \mathscr{E}_y(y) &= -\frac{dV}{dy} = 0 \quad \text{at } y = L - l_d \end{aligned} \tag{6.187}$$

we get[13]

$$l_d = \sqrt{\frac{2\epsilon_0 \epsilon_{si}}{qN_b}(V_{ds} - V_{dsat})} = \sqrt{\frac{V_{ds} - V_{dsat}}{a}} \tag{6.188}$$

where

$$a = \frac{qN_b}{2\epsilon_0 \epsilon_{si}}.$$

Note that Eq. (6.188) is the same equation as that obtained for the depletion layer width in a step *pn* junction with a voltage $V_{ds} - V_{dsat}$ dropped across the junction. This is the model for the CLM effect, first proposed by Reddi and Sah [94], to account for non-zero output conductance. However, this

[13] Integration can easily be performed by redefining the coordinate system such that $y' = 0$ at $y = L - l_d$ and $y' = l_d$ at $y = L$, so that the limits of integration are from $y' = 0$ to $y' = l_d$.

formulation overestimates the output conductance [95]–[96]. This is because the approach completely ignores the presence of a gate electrode and treats the field problem along the channel the same as that of a *pn* junction between the substrate and drain regions. Further, this simple approach results in a discontinuity of the field at $y = L - l_d$. This is because while deriving Eq. (6.188) we assumed that $\mathscr{E}_y = 0$ at $y = L - l_d$; we also assumed that $Q_i = 0$ in the pinch-off region which means that at $y = L - l_d$ the field $\mathscr{E}_y$ is infinite (see Figure 6.29).

In the model proposed by Baum–Beneking [97] the discontinuity in the field at $y = L - l_d$ (or $y' = 0$) was removed by assuming that at $V = V_{dsat}$, the field $\mathscr{E} = \mathscr{E}_p$. Therefore, the boundary conditions given by Eq. (6.187) are now modified as follows

$$V(y) = \begin{cases} V_{dsat} & \text{at } y = L - l_d \quad (y' = 0) \\ V_{ds} & \text{at } y = L \quad (y' = l_d) \end{cases}$$

$$\mathscr{E}_y = -\frac{dV}{dy} = \mathscr{E}_p \quad \text{at } y = L - l_d \quad (y' = 0) \tag{6.189}$$

where $\mathscr{E}_p$ is the lateral field at the transition point. Assuming $\mathscr{E}_p$ at $V_{ds} = V_{dsat}$, the field becomes continuous from the linear to saturation region (see Figure 6.29). Again solving the Poisson Eq. (6.186), under the above boundary conditions, results in the following expression for l_d

$$l_d = \sqrt{\frac{V_{ds} - V_{dsat}}{a} + \left(\frac{\mathscr{E}_p}{2a}\right)^2} - \left(\frac{\mathscr{E}_p}{2a}\right). \tag{6.190}$$

This is the equation for l_d used in the SPICE Level 3 model. In a more elaborate formulation [98]–[100] mobile charges are included in Eq. (6.185),

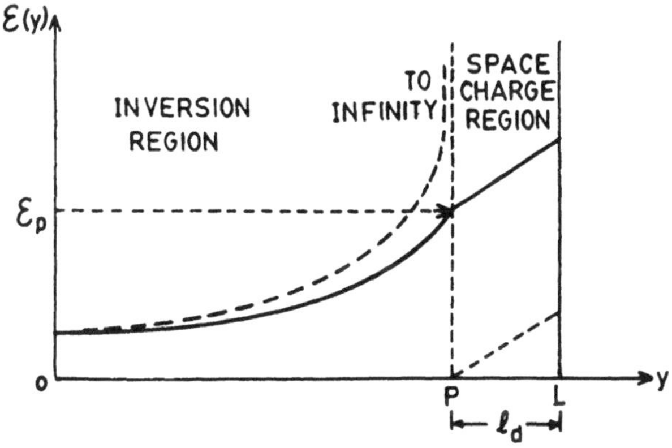

Fig. 6.29 Electric field along the channel of a MOSFET assuming field at a point *P* is (a) infinite (very large) (dotted line) and (b) finite value $\mathscr{E}_p$ (continuous line)

that is,

$$\frac{\partial^2 V}{\partial y^2} = \frac{qN_b}{\epsilon_{si}} + \frac{I_{dsat}}{W v_{sat} X_0} \tag{6.191}$$

where X_0 is the mean depth of current spreading near the drain end. This equation when solved under the boundary condition (6.189) yields

$$l_d = \sqrt{\frac{V_{ds} - V_{dsat}}{a(1 + bI_{dsat})} + \left\{\frac{\mathscr{E}_p}{a(1 + bI_{dsat})}\right\}^2} - \frac{\mathscr{E}_p}{a(1 + bI_{dsat})} \tag{6.192}$$

where $b = 1/qN_b W v_{sat} X_0$. Again when $b = 0$ (i.e., no mobile carriers in the depletion region), Eq. (6.192) reduces to Eq. (6.190). Note that using either Eq. (6.190) or (6.192), the slope at the transition point will be discontinuous.
The CLM models described above, though different in their exact formulations, all predict a constant field gradient in the CLM region due to the constant term in the right hand side of Eq. (6.185). However, using a 2-D device simulator it has been found that the channel field rises exponentially towards the drain. Thus, due to the incorrect channel field calculations, these models do not predict device output conductance accurately. This inaccuracy in the device output conductance is of more concern for analog circuit design than for digital design. For analog design, a more accurate expression for the channel field is thus desirable. An accurate knowledge of the field, particularly the maximum field, is also important for modeling substrate current, as we will see later in Chapter 8.
The inaccurate field calculation in the above models stems from the fact that we have ignored the oxide field in the analysis. To take the oxide field into account both empirical and pseudo-two dimensional analysis has been used.

Empirical Model. An empirical model that has been widely quoted to account for the CLM effect is the model proposed by Frohman-Bentchkowsky and Grove [95], according to which

$$l_d = \frac{V_{ds} - V_{dsat}}{\mathscr{E}_t} \tag{6.193}$$

where $\mathscr{E}_t$ is the average transverse electric field near the drain depletion region at the Si-SiO$_2$ interface and is the result of the following three fields (see Figure 6.30):

- The field $\mathscr{E}_1$ in the depletion region due to the *pn* junction comprised of the n^+ drain region and the *p*-substrate. Thus,

$$\mathscr{E}_1 = \sqrt{\frac{qN_b}{2\epsilon_0\epsilon_{si}}(V_{ds} - V_{dsat})}.$$

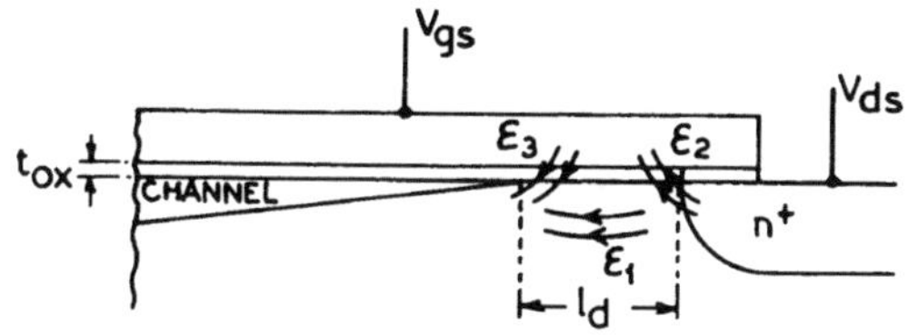

Fig. 6.30 Electric field distribution for a MOSFET operating in saturation showing components $\mathscr{E}_1, \mathscr{E}_2$ and $\mathscr{E}_3$ of the transverse field $\mathscr{E}_t$

- The fringing field $\mathscr{E}_2$ due to the potential difference $V_{ds} - V'_{gs}$, between the drain and the gate, where $V'_{gs} = V_{gs} - V_{fb}$. Thus,

$$\mathscr{E}_2 = A_2 \frac{\epsilon_{ox}}{\epsilon_{si}} \frac{V_{ds} - V'_{gs}}{t_{ox}}$$

where A_2 is the empirical fringing field factor associated with the field $\mathscr{E}_2$.

- The fringing field $\mathscr{E}_3$ due to the potential difference $V'_{gs} - V_{d\,sat}$ between the gate and the end of the inversion layer. Thus,

$$\mathscr{E}_3 = A_3 \frac{\epsilon_{ox}}{\epsilon_{si}} \frac{V'_{gs} - V_{\text{dsat}}}{t_{ox}}$$

where A_3 is the empirical fringing field factor associated with the field $\mathscr{E}_3$.

The total field $\mathscr{E}_t = \mathscr{E}_1 + \mathscr{E}_2 + \mathscr{E}_3$. Typical values for A_2 and A_3 are 0.2 and 0.6, respectively. This model incorporates a rather complete theory on the CLM supported by experimental data. Equation (6.193) for l_d has been used by many others [18], [101].

Pseudo-2D Model. The approach used by Frohman-Bentchkowsky and Grove is purely empirical. A more physical approach to calculate CLM factor l_d was proposed by El-Mansy and Boothroyd [102] and subsequently modified by others who also took into account the shape of the source/drain structures [62], [104]. A simplified form, that retains the essential features of these models, is summarized here.

The cross-section of the drain region, where the CLM effect is taking place, is shown schematically in Figure 6.31. To simplify the mathematics it is assumed that (1) the drain and source junctions are square in shape, (2) the drain current is confined to flow within the depth of the junction , and (3) the velocities of all carriers in the drain region are saturated. Assumption (2) limits the validity of the present analysis to conventional source/drain junctions, but the analysis can easily be extended to LDD junctions [103]–[109].

As shown in Figure 6.31, the drain region is bounded on one side by the line AB, which marks the beginning of the velocity saturation region, and

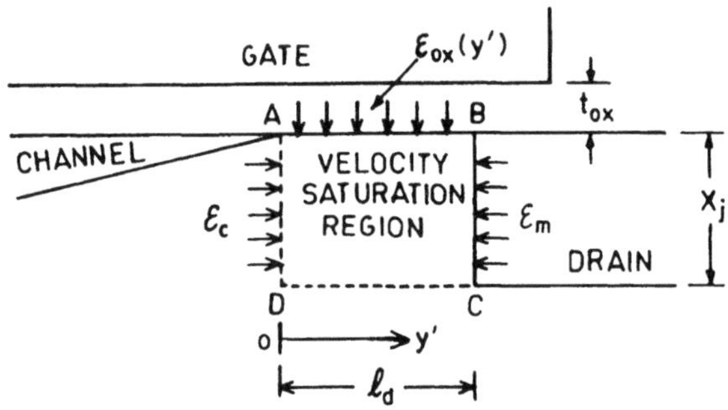

Fig. 6.31 Schematic diagram illustrating analysis of the velocity saturation region

on the other side at the drain junction edge by CD. Since there are no field lines crossing the line CD, the space charge is controlled only by the electric fields crossing the other 3 sides of the rectangle. Applying Gauss' law to the volume with sidewall ABCD and unit width W we get

$$-\int_0^{X_j} \mathcal{E}_y(0, y)dx + \int_0^{X_j} \mathcal{E}_y(x, y)dx + \frac{\epsilon_{ox}}{\epsilon_{si}}\int_0^{y} \mathcal{E}_{ox}(0, y)dy$$

$$= \frac{q}{\epsilon_0\epsilon_{si}} N_b X_j y + \frac{Q_m}{\epsilon_0\epsilon_{si}} y \tag{6.194}$$

where Q_m is the mobile charge density in the drain region and $\mathcal{E}_{ox}$ is the gate oxide field given by

$$\mathcal{E}_{ox} = \frac{V_{gs} - V_{fb} - 2\phi_f - V(y)}{t_{ox}}. \tag{6.195}$$

Differentiating Eq. (6.194) with respect to y we get

$$X_j \frac{d\mathcal{E}_y(y)}{dy} + \frac{\epsilon_{ox}}{c_{si}}\mathcal{E}_{ox}(0, y) = \frac{q}{\epsilon_0\epsilon_{si}} N_b X_j + \frac{Q_m}{\epsilon_0\epsilon_{si}}. \tag{6.196}$$

Since the velocity of the mobile carriers is assumed to be saturated in the drain region, the mobile charge density Q_m equals that at the point of saturation where $V(y) = V_{dsat}$. Therefore, from simple charge control analysis we get [cf. Eq. (6.43)]

$$Q_m = C_{ox}[V_{gs} - V_{fb} - 2\phi_f - V_{dsat}] - qN_b X_j. \tag{6.197}$$

Combining Eqs. (6.195)–(6.197) we get

$$\frac{d\mathcal{E}_y(y)}{dy} = \frac{V(y) - V_{dsat}}{l^2} \tag{6.198}$$

where

$$l^2 = \frac{\epsilon_{si}}{\epsilon_{ox}} t_{ox} X_j. \tag{6.199}$$

The right hand side of Eq. (6.198) is the amount of charge released by the oxide field as a results of a rise in the channel voltage equal to $(V(y) - V_{dsat})$, while the left hand side is the corresponding increase of the channel field gradient in order to support these charges. Redefining the coordinate system as $y' = 0$ at point D and $y' = l_d$ at point C, and solving Eq. (6.198) under the boundary condition

$$y' = 0 \begin{cases} \mathscr{E}_y(0) = \mathscr{E}_c \\ V(0) = V_{dsat} \end{cases}$$

we get

$$\mathscr{E}_y(y') = \mathscr{E}_c \cosh\left(\frac{y'}{l}\right) \tag{6.200a}$$

$$V(y') = V_{dsat} + l\mathscr{E}_c \sinh\left(\frac{y'}{l}\right). \tag{6.200b}$$

Equation (6.200) shows that the channel field increases exponentially towards the drain. Extensive 2-D numerical analysis confirms the basic form of Eq. (6.200). At the drain end of the channel where the field is maximum, denoted by $\mathscr{E}_m$, we have

$$\mathscr{E}_m - \mathscr{E}_y(y' = l_d) = \mathscr{E}_c \cosh\left(\frac{l_d}{l}\right) \tag{6.201a}$$

$$V_{ds} = V_{dsat} + l\mathscr{E}_c \sinh\left(\frac{l_d}{l}\right) \tag{6.201b}$$

Using the identity $\sinh^2 A + 1 = \cosh^2 A$, Eqs. (6.200) and (6.201) can be combined to give the following expression for l_d and $\mathscr{E}_m$ in the channel

$$l_d = l \ln\left[\frac{(V_{ds} - V_{dsat})}{l\mathscr{E}_c} + \frac{\mathscr{E}_m}{\mathscr{E}_c}\right] \tag{6.202a}$$

$$\mathscr{E}_m = \left[\left(\frac{V_{ds} - V_{dsat}}{l}\right)^2 + \mathscr{E}_c^2\right]^{1/2}. \tag{6.202b}$$

Further simplification for l_d can be obtained if we approximate $\mathscr{E}_m$ as

$$\mathscr{E}_m \approx \mathscr{E}_c + \delta'\left(\frac{V_{ds} - V_{dsat}}{l}\right) \tag{6.203}$$

where δ' allows $\mathscr{E}_m$ to fit more closely with Eq. (6.202b). With this approximation, Eq. (6.202) for l_d simplifies to

$$\boxed{l_d = l \ln\left[1 + \left(\frac{V_{ds} - V_{dsat}}{V_p}\right)\right]} \tag{6.204}$$

where $V_p = l\mathscr{E}_c/(1 + \delta'\mathscr{E}_c)$ and can be treated as a fitting parameter. This equation has been shown to fit the conductance very well [116].

Remarks on the Continuity of the Current and Conductance at the Transition Between the Linear and Saturation Regions. Any of the l_d expressions discussed here, when used in Eq. (6.182) or (6.183) result in a discontinuity of the slope at the transition point from linear to saturation region. This obviously is not desirable for circuit models. This discontinuity of the slope at the transition point can easily be removed by introducing an additional condition to be satisfied, that is

$$\left.\frac{dI_{ds}}{dV_{ds}}\right|_{V_{ds}=V_{dsat}} \quad \text{(linear region)}$$
$$= \left.\frac{dI_{ds}}{dV_{dsat}}\right|_{V_{ds}=V_{dsat}} \quad \text{(saturation region).} \tag{6.205}$$

With condition (6.205) satisfied, the parameter $\mathscr{E}_c$ or $\mathscr{E}_p$ (one of them if both are used) in l_d expressions can no longer be a fitting parameter, since its value will be dictated by the condition (6.205). Though this condition ensures that the first derivative will be continuous, it does not guarantee that the conductance g_{ds} will be smooth. *For g_{ds} to be smooth, the second derivative of I_{ds} must be continuous at $V_{ds} = V_{dsat}$.* Although a drain current model having a continuous conductance is not necessary for simulation of digital circuits, it is important for analog circuit simulation.
Another approach that ensures continuity of the drain current derivatives at the transition point from linear to saturation region is to introduce the following empirical function

$$F(V_{ds}, V_{dsat}) = 1 - \frac{1}{B}\ln\left[1 + e^{A(1 - V_{ds}/V_{dsat})}\right] \tag{6.206}$$

where $B = \ln(1 + e^A)$.
Figure 6.32 shows a plot of the function $F(V_{ds}, V_{dsat})$ versus (V_{ds}/V_{dsat}) for 3 different values of the parameter A. Large values of A yields steep transitions between the linear and saturation regions while small values result in smooth transitions. The value of $A = 10$ has been found to be a good choice [118]. The effective drain-source voltage V_{dsx}, which results in a smooth

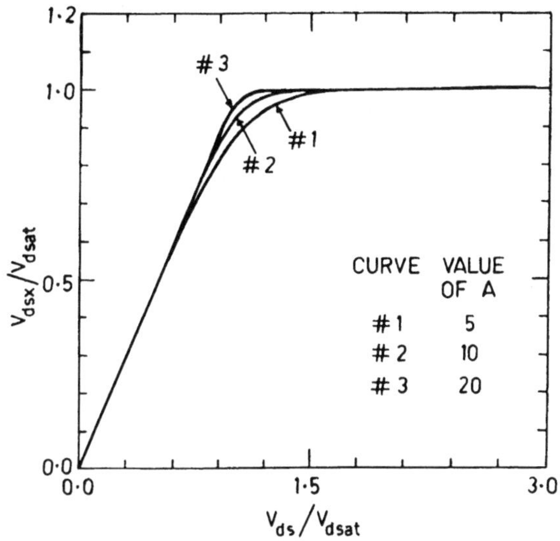

Fig. 6.32 Variation of the function $F(V_{ds}, V_{dsat})$ [Eq. (6.206)] as a function of V_{ds}/V_{dsat} for different values of A

transition from linear to saturation region, becomes

$$V_{dsx} = V_{dsat}\left\{1 - \frac{1}{B}\ln\left[1 + e^{A(1 - V_{ds}/V_{dsat})}\right]\right\}. \tag{6.207}$$

By replacing V_{ds} with V_{dsx} in the current Eqs. (6.169) and (6.182), a smooth transition is observed. This also ensures a smooth g_{ds}. The use of V_{dsx} for V_{ds} not only insures smooth current and conductances, but it also reduces two drain current equations in the linear and saturation regions of device operation to a single current equation as follows

$$I_{ds} = \frac{W\mu_s C_{ox}}{(L - l_d)(1 + V_{dsx}/(L - l_d)\mathscr{E}_c)}(V_{gs} - V_{th} - 0.5\alpha V_{dsx})V_{dsx}. \tag{6.208}$$

Note that in this approach l_d is used for both the linear and saturation regions and V_{ds} is replaced by V_{dsx} everywhere including l_d.
Equation (6.208) predicts that output resistance $R_o(= 1/g_{ds})$ of a short channel MOSFET in saturation increases with increasing V_{ds} due to increasing l_d. However, in real devices, particularly nMOST, R_o increases only up to moderate V_{ds} (beyond V_{dsat}), and at higher V_{ds} it starts to decrease (see Figure 7.21, which is a plot of g_{ds} vs. V_{ds}). This decrease in R_o is induced by the hot-carrier substrate current I_b (cf. section 3.4; also see chapter 8). The substrate current created near the drain flows towards the substrate contact and produces a voltage drop across the substrate resistance along its path as shown in Figure 6.33a. This voltage drop forward biases the

channel causing a reverse body-bias effect, which lowers the device threshold voltage V_{th} and thereby increases the drain current. DIBL also causes V_{th} to decrease, but the effect is much smaller and in general affects the drain current only near V_{th} (cf. section 5.3). In general, all three mechanisms – CLM, DIBL, and hot-carrier effect – affect the MOSFET output resistance, but their relative contributions strongly depend on the bias condition as shown in Figure 6.33b. To a first order the increase in the drain current, or decrease in the output resistance, can be modeled by including hot-electron induced substrate current I_b as [62]

$$I'_{ds} = I_{ds} + AI_b \text{ (for } I_{ds} > I_{dsat}) \tag{6.209}$$

where A is a fitting parameter. The expressions for I_b are discussed in Chapter 8.

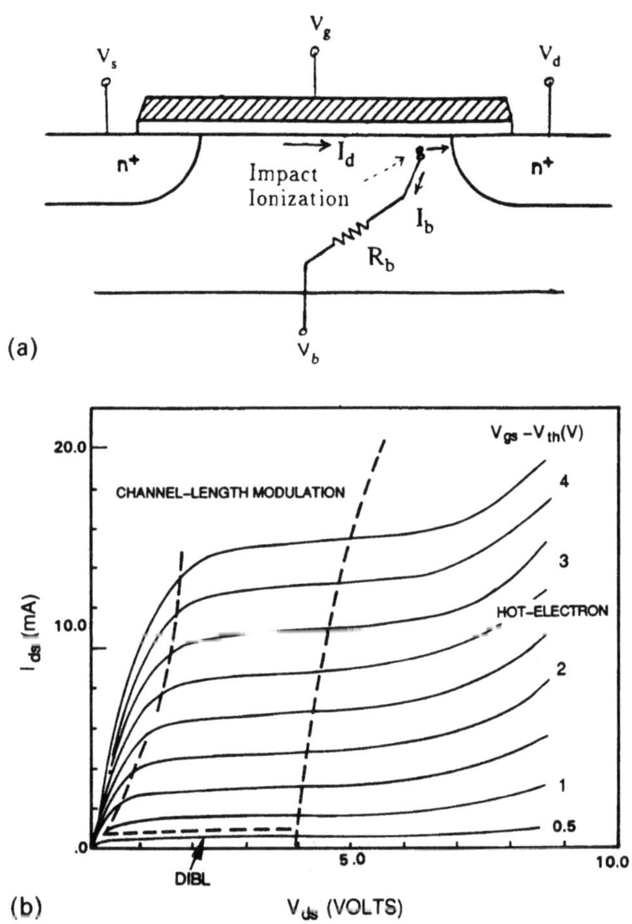

Fig. 6.33 (a) Schematic diagram to illustrate the effect of substrate current to MOSFET output resistance. (b) Drain current I_{ds} vs. drain voltage V_{ds} of a nMOST showing the dominant mechanisms affecting the current in different bias regions. (After Ko [62])

6.7.4 *Subthreshold Model*

For short channel devices, the surface potential ϕ_s is not constant along the length of the channel (see Fig. 5.25). Although the drain current remains exponentially dependent on the gate voltage, various physical arguments used in the derivation of (6.104) no longer apply (cf. section 5.3.3). Nevertheless, for short-channel subthreshold current calculations most of the CAD models use slightly modified form of Eq. (6.104) or (6.105). Since short-channel subthreshold currents show strong dependence on V_{ds}, it is normally included in the effective gate drive through the DIBL effect. Thus, V_{th} in Eq. (6.104) is replaced by V_{thd} [cf. Eq. (6.168a)]. Once V_{th} is replaced by V_{thd}, the different short-channel subthreshold current models differ only in the prefactor term I_{pf} [84], [110]–[115]. Starting from the diffusion current expression for n-channel [cf. Eq. (6.91)], it has been shown that I_{ds} for short channel devices can be approximated as [cf. Eq. (6.97)] [110]–[115].

$$I_{ds} = \frac{qWD_n\bar{t}_{ch}n_s}{L_{\text{eff}}} e^{(V_{gs}-V_{thd})/\zeta_1 V_t}(1 - e^{-V_{ds}/V_t}) \tag{6.210}$$

where D_n is the electron diffusion constant, n_s is the electron concentration in the channel at the source, and $\bar{t}_{ch}$ is the average channel thickness given by

$$\bar{t}_{ch} = \zeta' t_{ch} = \zeta' \sqrt{\frac{\epsilon_0 \epsilon_{si} V_t}{2qN_b[1 + (V_{sb} + \bar{\phi}_s)/V_t]}}$$

where ζ' is a fitting parameter which accounts for the fact that the real channel thickness is somewhat bigger than that derived from the square root term in the above equation. In the above equation $\bar{\phi}_s$ is the average surface potential and can be replaced by an average value of 0.5 [110], ζ_1 is given by

$$\zeta_1 = \frac{\eta_0}{1 + \zeta_0\sqrt{V_{sb} + \phi_s}}$$

where ζ_0 and η_0 are some fitting parameters. Finally, L_{eff} in Eq. (6.210) is given by

$$L_{\text{eff}} = L - X_{sd} - X_{dd} \tag{6.211a}$$

where X_{sd} and X_{dd} are the source and drain depletion widths, respectively, at the surface, given by

$$X_{sd} = \sqrt{\frac{2\epsilon_0\epsilon_{si}}{qN_b}(\phi_{bi} - \bar{\phi}_s)} \tag{6.211b}$$

and

$$X_{dd} = \sqrt{\frac{2\epsilon_0\epsilon_{si}}{qN_b}(\phi_{bi} - \bar{\phi}_s + V_{ds})} \tag{6.211c}$$

where ϕ_{bi} is the source/drain to bulk built-in potential, given by Eq. (3.3). Compare Eqs. (6.211a) and (6.211b) with Eq. (5.95), the depletion region expressions between the source/drain to substrate *pn* junctions.
In the BSIM model the prefactor term, $I_{pf} = \beta V_t^2 \exp(1.8)$, is an empirical factor that is based on matching the experimental data with the model [25]. Often in CAD models, Eq. (6.105) is also used for short-channel I_{ds} with m and η regarded as fitting parameters.

Transition from the Weak to Strong Inversion Region. Accurate modeling of the transition region can be achieved by including both drift and diffusion currents simultaneously without distinguishing between the weak and strong inversion regions. However, this leads to a more complicated equation for the current. Therefore, for circuit models, various simplified approaches have been suggested. In one approach, the following approximate formula is used to ensure continuity of the current from weak to strong inversion

$$I_{ds,t} = I_{pf} \ln\left[e^{I_{ds,s}/I_{pf}} + e^{(V_{gs}-V_{th})/\eta V_t}\right](1 - e^{-V_{ds}/V_t}) \tag{6.212}$$

where $I_{ds,s}$ is the strong inversion current due to the drift only such as given by Eq. (6.84) or (6.208). Equation (6.212) matches the following two extreme cases:

- When $V_{gs} \ll V_{th}$, $I_{ds,s}$ is zero so that Eq. (6.212) reduces to

  $$I_{ds,t} = I_{pf} \ln\left[1 + e^{(V_{gs}-V_{th})/\eta V_t}\right](1 - e^{-V_{ds}/V_t})$$

  or

  $$\approx I_{pf} e^{(V_{gs}-V_{th})/\eta V_t}(1 - e^{-V_{ds}/V_t})$$

  which is the same as Eq. (6.104).
- When $V_{gs} \gg V_{th}$, the drift current $I_{ds,s}$ is much larger than the diffusion current I_{ds} and therefore, Eq. (6.212) reduces to $I_{ds,t} \approx I_{ds,s}$.

Equation (6.212) models the transition region fairly accurately:
In the approach used in the SPICE Level 4 model (BSIM), the transition region is modeled based on the fact that when V_{gs} is increased above V_{th}, the subthreshold current approaches a constant value. This imposes an upper limit on the subthreshold current. This limiting current applies when $V_{gs} > 3V_t$ above V_{th} and is obtained from the current in the saturation region with $V_{gs} = V_{th} + 3V_t$, so that

$$I_{sub0} = \frac{\beta}{2}(3V_t)^2 \tag{6.213}$$

and the weak inversion, or subthreshold current, is modeled as

$$I_{ds,w} = \frac{I_{sub} I_{sub0}}{I_{sub} + I_{sub0}} \tag{6.214}$$

where I_{sub} is given by Eq. (6.104). The total drain current from weak to strong inversion now becomes [25]

$$I_{ds,t} = I_{ds,w} + I_{ds,s} \tag{6.215}$$

where $I_{ds,w}$ and $I_{ds,s}$ are given by Eqs. (6.214) and (6.84), respectively.

In still another approach the transition region, bounded by gate voltages V_{gx1} and V_{gx2} between weak and strong inversion, is modeled by a third order polynomial of the following form [112]

$$\log(I_{ds}) = aV_{gs}^3 + bV_{gs}^2 + cV_{gs} + d \tag{6.216}$$

where the coefficients a, b, c and d are calculated from $\log(I_{ds})$ and its derivative with respect to V_{gs} at two end points (V_{gx1} and V_{gx2}) of the transition region. Although this approach results in a continuous and smooth transition, the accuracy of the simulation in the transition region is totally dependent on the accuracy of the function values and their slopes at the two end points. Of the three approaches used to model the transition region, the one given by Eq. (6.212) is used by many others and seems to work well.

6.7.5 *Continuous Model*

The short channel models discussed so far are piece-wise model where different equations are used for different regions of device operation. In order to ensure that the current and its (at least) first derivatives are continuous at the transition points smoothing functions are often used. Thus, by using smoothing functions (6.122) and (6.207), one arrives at the following equation

$$I_{ds} = \frac{W\mu_s C_{ox}}{(L - l_d)(1 + V_{dsx}/(L - l_d)\mathscr{E}_c)}(V_{gsx} - V_{th} - 0.5\alpha V_{dsx})V_{dsx} \tag{6.217}$$

and is obtained by replacing V_{gs} in Eq. (6.208) with V_{gsx} given by Eq. (6.122). This is single equation which is valid in all region of device operation. Figure 6.34a shows measured and simulated $I_{ds} - V_{ds}$ characteristics for a p-channel device fabricated using submicron CMOS process with W_m/L_m (drawn dimensions in μm) = 10/0.5 and $t_{ox} = 105$ Å. Circles are experimental data points while continuous lines are based on Eqs. (6.142), (6.174) and (6.204) for μ_s, V_{dsat} and l_d, respectively. The corresponding $I_{ds} - V_{ds}$ and $I_{ds} - V_{gs}$ characteristics for n-channel device are shown in Figures 6.34b and 6.34c, respectively. In this case Eq. (176) was used for V_{dsat}. Best fit to the data was obtained using nonlinear optimization method as discussed later in section 10.6. The extracted value of v_{sat} is 8×10^6 cm/s for nMOST while the corresponding value for pMOST is 6×10^6 cm/s. These values are consistent with those measured experimentally as discussed in section 6.6.2. Note from Figures 6.34, while n-channel devices get velocity saturated

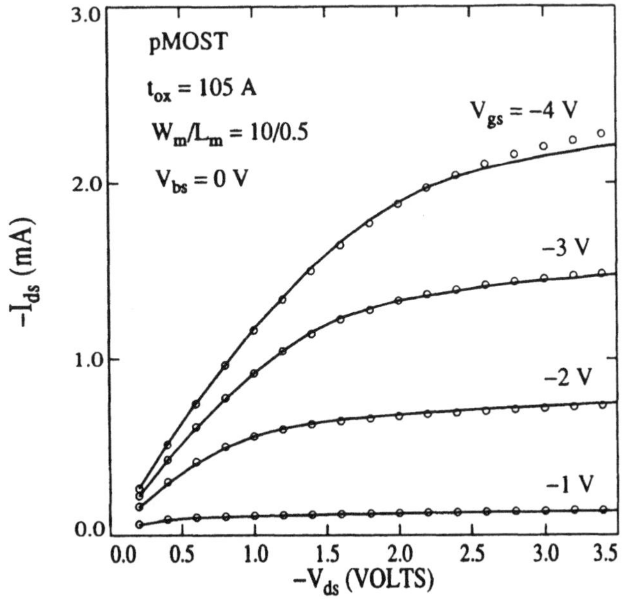

(a)

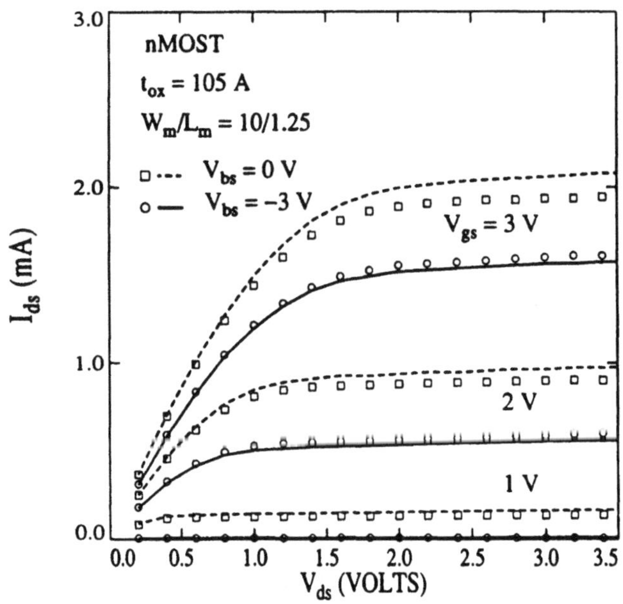

(b)

Fig. 6.34 Measured and calculated $I-V$ characteristics of a MOSFET fabricated using submicron CMOS *n*-well process ($t_{ox} = 150$ Å). (a) *p*-channel output characteristics at $V_{bs} = 0$V, (b) *n*-channel output characteristics at two V_{bs}, and (c) *n*-channel transfer characteristics for different channel lengths. All dimensions are drawn. Symbols are measured data, while lines are those calculated using Eq. (6.217)

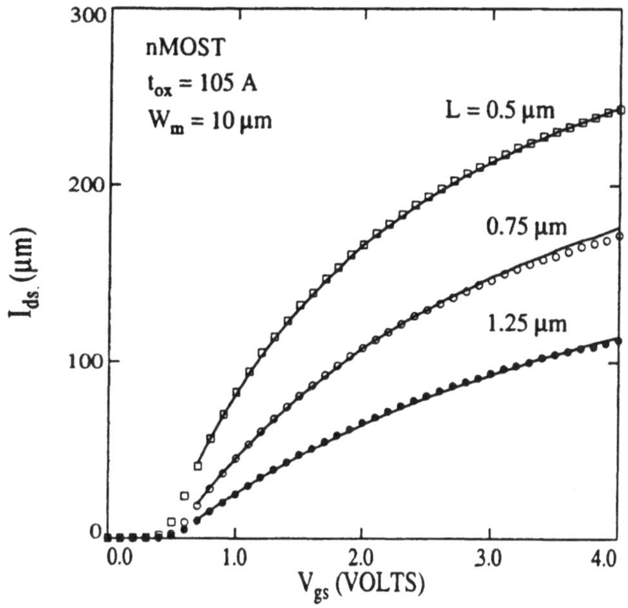

(c)

Fig. 6.34 (continued)

and have very little CLM effect, the corresponding *p*-channel devices have more CLM effect and less velocity saturation effect for the same applied field. The model fits the data fairly well with an average error of less than 3.0% for $V_{gs} > V_{th}$ over series of different length and width devices and back biases. However, for $V_{gs} < V_{th}$ the average error is over 15% due to large errors in the moderate inversion region[14] (near V_{th}). In fact all piece-wise models have generally high errors in this region.

The long channel charge-sheet model (cf. section 6.3), which is inherently continuous in all regions of device operation and is also accurate in the moderate inversion region, has been extended for short channel devices [17]–[19]. However, they are not generally used for VLSI simulations for reasons discussed in section 6.3, though they are being used for circuit simulations, particularly analog applications, when computation time is not of prime concern. The drain current models for narrow width devices are the same as for short channel devices, provided proper threshold voltage model for narrow widths (cf. section 5.3.2) is taken into account [115].

If the model is physically based, then a single set of model parameters should fit different geometry devices. However, often we introduce empirical

[14] It should be pointed out that the calculated current (continuous lines) shown in Figure 6.34b also takes into account the source/drain resistance, as discussed in section 6.8.

parameters in the circuit models in order to acheive good computation efficiency as well as accuracy. For this reason, many electrical parameters become geometry dependent. The most commonly used formulation for the geometry dependence of an electrical parameter P is [120]

$$P = P_0 + \frac{P_l}{L} + \frac{P_w}{W}$$

where P_0, P_l and P_w are fitting parameters. The BSIM model has 9 of its parameters given by the above equation (see section 11.5). While using such equations one must be careful in extracting the length and width dependent parameter values.

6.8 Impact of Source-Drain Resistance on Drain Current

In the discussion so far we had implicitly assumed that the voltage applied at the terminals of the device are the same as that across the channel region. In other words, the voltage drop across the source/drain region is negligible compared to the voltage applied at the terminals. As was discussed in section 3.6.1, this indeed is true only for long-channel devices. For short-channel devices, the impact of source/drain resistance is a reduction of the transconductance g_m and device current driving capability.
The effect of the source and drain resistance R_s and R_d, respectively, in calculating I_{ds} can be understood from the equivalent circuit shown in Figure 6.35. Clearly the effective drain and gate voltages V'_{ds} and V'_{gs}, respectively, are reduced below the voltages V_{ds} and V_{gs} applied at the external terminal of the device by the voltage drop across these resistors; that is,

$$V'_{gs} = V_{gs} - I_{ds}R_s \tag{6.218a}$$

$$V'_{ds} = V_{ds} - I_{ds}R_t \tag{6.218b}$$

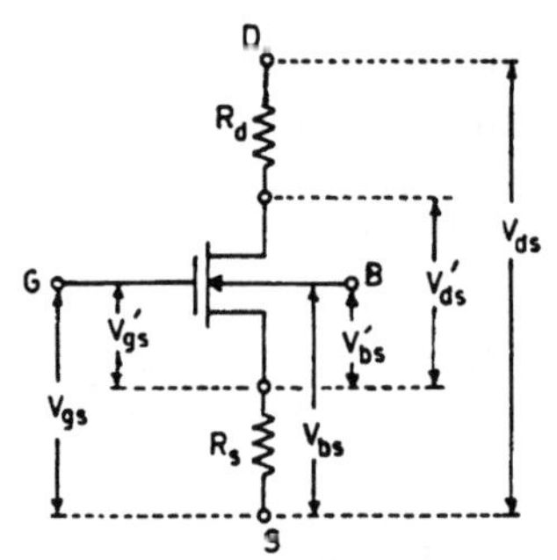

Fig. 6.35 MOSFET showing internal and external terminal voltages when source/drain resistance is taken into account

where $R_t = R_s + R_d$. It is generally assumed that $R_s = R_d = R_t/2$. Often circuit models (SPICE Levels 1–3) treat this effect of R_s and R_d as an external component of the device by including two additional nodes per transistor. However, this results in extra computational time. Rather than considering these resistors as external simulation elements, one can incorporate them into the device model explicitly, thereby reducing the computational time. This is the approach normally used in most of the recently developed device models, including SPICE Level 4 model. Although R_s and R_d are gate bias dependent, particularly for LDD devices (cf. sections 3.6.1), in what follows we will assume them to be bias independent. In spite of this assumption, the explicit inclusion of R_s and R_d in an analytical drain current model results in a complex equation for I_{ds} as seen below.
In terms of the intrinsic voltages, the linear region drain current model for short-channel devices is given by

$$I_{ds} = \beta_0 \frac{(V'_{gs} - V_{th} - 0.5\alpha V'_{ds})V'_{ds}}{\{1 + \theta(V'_{gs} - V_{th}) + \theta_b V_{sb}\}(1 + \theta_c V_{ds})} \tag{6.219}$$

where we have replaced V_{gs} and V_{ds} of Eq. (6.169) by the intrinsic voltages V'_{gs} and V'_{ds}, respectively, and $\theta_c = (L\mathscr{E}_c)^{-1}$. Using Eq. (6.218) in (6.219), it is easy to see that resulting equation in I_{ds} with explicit R_s and R_d will be difficult to solve. However, remembering that θ, θ_b, and θ_c are small so that the terms involving their products can be neglected, one can write the above equation as

$$I_{ds} = \beta_0 \frac{(V_{ds} - V_{th} - I_{ds}R_s)(V_{ds} - I_{ds}R_t) - 0.5\alpha(V_{ds} - I_{ds}R_t)^2}{1 + \theta(V_{gs} - V_{th} - I_{ds}R_s) + \theta_b V_{sb} + \theta_c V_{ds}}. \tag{6.220}$$

Equation (6.220) is quadratic in I_{ds} and can now be solved for I_{ds} giving

$$I_{ds} = \frac{a_2 - \sqrt{a_2^2 - 4a_1a_3}}{2a_1} \tag{6.221}$$

where

$$a_1 = 0.5\beta_0(1 - \alpha)R_t^2 + 0.5\theta R_t + \theta_c R_t \tag{6.222a}$$

$$a_2 = 1 + (\theta + \beta_0 R_t)V_{gt} + \theta_b V_{sb} + (\theta_c + 0.5R_t\beta_0 - \alpha\beta_0 R_t)V_{ds} \tag{6.222b}$$

$$a_3 = \beta_0(V_{gt} - 0.5\alpha V_{ds}^2) \tag{6.222c}$$

$$V_{gt} = V_{gs} - V_{th}. \tag{6.223}$$

In general, the term a_1 is much smaller than the other terms and in practice one often meets the condition [121]

$$\frac{a_1a_3}{a_2^2} < 0.1.$$

Using typical values for the parameters θ, θ_b and θ_c, the above condition, in a more practical form, becomes

$$I_{ds}R_t < 0.5\,\text{V}.$$

If the above condition is satisfied than the square root term in Eq. (6.221) can be expanded. Retaining the first two terms in the Binomial expansion of the expression under the square root, we get

$$I_{ds} \approx \frac{a_3}{a_2} = \beta_0 \frac{(V_{gt} - 0.5\alpha V_{ds})V_{ds}}{1 + (\theta + \beta_0 R_t)V_{gt} + \theta_b V_{sb} + \theta'_c V_{ds}} \tag{6.224}$$

where

$$\theta'_c = \theta_c - \beta_0 R_t(\alpha - 0.5).$$

The expression for I_{ds} in saturation is even more cumbersome. Since R_t degrades g_m in the linear region more severely than in the saturation region (cf. section 3.6.1), circuit models usually include R_t only in linear region current models.

When the effect of carrier velocity saturation is not important so that θ'_c can be ignored, then Eq. (6.224) can be approximated as

$$I_{ds} \approx \frac{\beta_0(V_{gs} - V_{th} - 0.5\alpha V_{ds})V_{ds}}{1 + (\theta + \beta_0 R_t)(V_{gs} - V_{th}) + \theta_b V_{sb}}. \tag{6.225}$$

The first order equations (6.224) and (6.225) clearly show that the effect of R_t is to reduce the drain current and hence the transconductance g_m. For long channel devices $\beta_0 R_t \approx 0$, which implies that the current is almost the same as if series resistance R_t is zero. However, for short-channel devices the $\beta_0 R_t$ term is not negligible, and therefore the effect of series resistance must be taken into account. From Eq. (6.225) we see that mathematically the effect of R_t is the same as of reducing the effective mobility μ_s due to the vertical field. Therefore, *in circuit models the effect of R_t is simply modeled by replacing mobility degradation factor θ with $(\theta + \beta_0 R_t)$ in the drain current equations.*

Note that replacing θ with $\theta + \beta_0 R_t$ not only affects the linear region current, but also reduces short channel current in the saturation region. This is because for short channel devices, V_{dsat} depends upon μ_s through $\mathscr{E}_c = v_{sat}/\mu_s$. Since R_t reduces μ_s which in turn increases $\mathscr{E}_c$, it results in an increase in the V_{dsat} [cf. Eq. (6.174)]. The modeled $I_{ds} - V_{ds}$ characteristics (continuous lines) shown in Figures 6.33 and 6.34 were based on inclusion of R_t through μ_s.

6.9 Temperature Dependence of the Drain Current

MOS transistor characteristics are strongly temperature dependent. Modeling the temperature dependence of the MOSFET characteristics is important in designing VLSI circuits since, in general, an IC is specified to be functional in a certain temperature range, for example $-55\,°C$ to $125\,°C$. In addition, operating the MOSFET below room temperature (low-temperature operation) results in improved device performance; a factor of two improvement in switching speed can be achieved by operating a 1 μm device at 77 K ($-196\,°C$) [122]–[124]. However, device degradation due to hot-carrier effects also increases with decreasing temperature (see section 8.6) [124].

The MOSFET drain current varies considerably with temperature. The change in drain current in the temperature range 0–100 °C for a typical *n*-channel device is over 20%, being slightly lower for the corresponding *p*-channel device. *The temperature coefficient of the drain current can be positive, negative, or zero depending upon the operating voltages.* This is shown in Figure 6.36 where measured $\sqrt{I_{ds}}$ in saturation is plotted against gate voltage for different temperatures for a nMOST with $W_m/L_m = 9.4/9.4$ (μm), $t_{ox} = 105$ Å, and $V_{th} = 0.56$ V. The Zero Temperature Coefficient (ZTC)

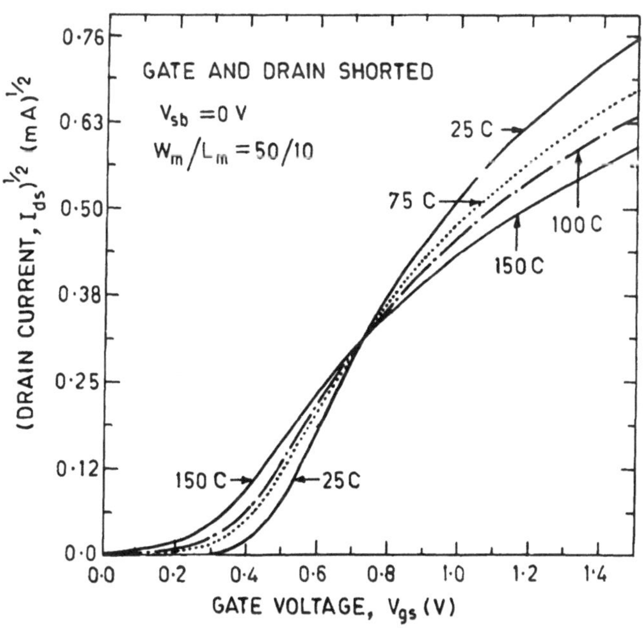

Fig. 6.36 Variation of I_{ds} versus V_{gs} in saturation region of device operation with temperature as a parameter. It shows temperature coefficient of I_{ds} is either positive, negative, or zero depending upon the operating bias

of the drain current could be either in the linear or saturation region of device operation. The gate voltage, which leads to ZTC is very close to the device threshold voltage, as we shall see shortly.
The negative temperature coefficient of I_{ds} (at higher temperatures) is primarily due to (1) decrease in the carrier mobility, (2) increase in the threshold voltage (cf. section 5.4), and (3) decrease in the carrier saturation velocity. There are many other parameters which are temperature dependent [126], but if the drain current model is physically based one can easily explain the temperature dependence of the current using the above three parameters. The positive temperature coefficient of I_{ds} occurs when the device is operating in the weak and moderate inversion region and is primarily due to an increase in the intrinsic carrier concentration n_i, in accordance with Eq. (6.96b) [127], [128].

6.9.1 Temperature Dependence of Mobility

Carrier mobility in the inversion layer is strongly temperature dependent. The temperature dependence of the mobility has been traditionally used to extract contributions from different scattering mechanisms. For high quality devices, electron surface mobility μ_s may range from 600 cm²/V.s at room temperature to 20,000 cm²/V.s at 4.2 K. Since different scattering mechanisms are effective in different ranges of temperature, circuit simulators normally use different mobility models for different temperature ranges. For $T >$ 200 K, the most commonly used temperature dependent mobility model is [59], [124]

$$\mu_s(T) = \frac{\mu_0(T)}{1 + \theta(T)\mathscr{E}_{\text{eff}}} \tag{6.226}$$

which is valid for both p- and n-channel devices, provided the field is not very high ($< 8 \times 10^4$ V/cm). In *general θ is a weak function of temperature; therefore, its temperature dependence is generally ignored.* A comparison of Eq. (6.226) with experimental data is shown in Figure 6.37. Note that for p-channel devices the linear relationship between $1/\mu_s$ and $\mathscr{E}_{\text{eff}}$ remains valid even at low temperatures, which indeed is not the case for n-channel devices. However, for thin gate oxides and higher gate voltages (high fields), it is more appropriate to use the following equation for mobility [28], [64]

$$\mu(T) = \frac{\mu_0(T)}{1 + \theta_1(T)\mathscr{E}_{\text{eff}}^2 + \theta_2(T)\mathscr{E}_{\text{eff}}^{-1/3}} \tag{6.227}$$

where θ_1 and θ_2 are functions of temperature and have been obtained by fitting the data to the model over a wide temperature range [28].
It has been observed that the functional form of the temperature dependence of low field mobility μ_0 is T^{-m}; that is, μ_0 at any temperature T, in the

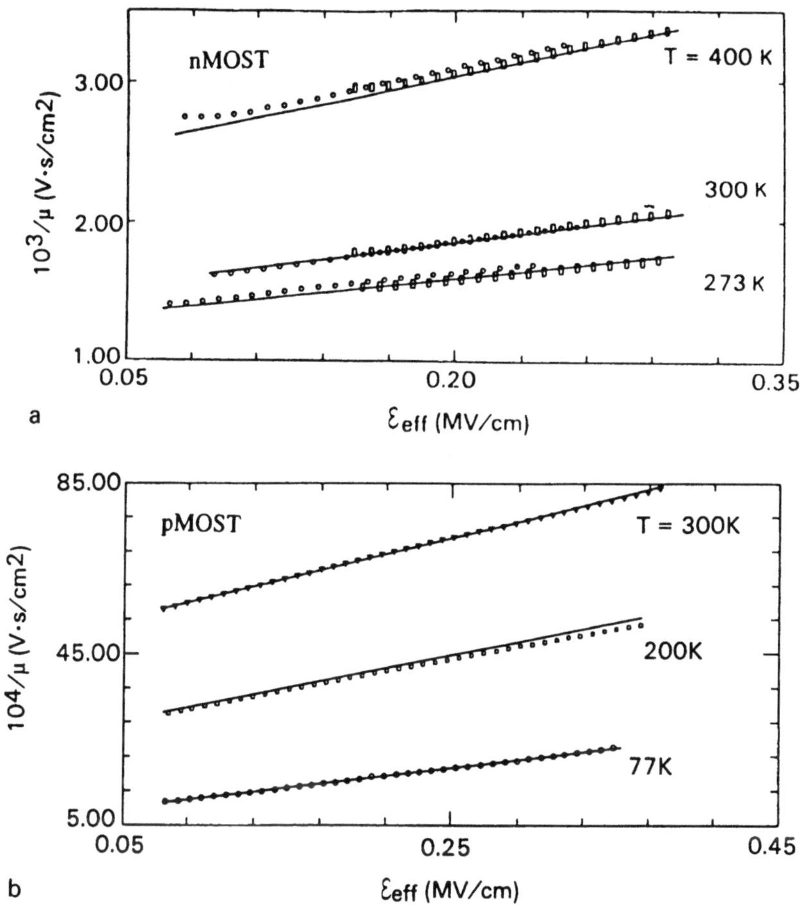

Fig. 6.37 Inversion layer mobility versus normal effective field at different temperatures (a) *n*-channel MOSFET $t_{ox} = 310$ Å and (b) (b) *p*-channel MOSFET $t_{ox} = 250$ Å. Solid lines represent Eq. (6.226) while symbols corresponds to experimental data. (After Arora and Gildenblat [59])

temperature range 200–400 K can be expressed as

$$\mu_0(T) = \mu_0(T_0)\left(\frac{T}{T_0}\right)^{-m} \tag{6.228}$$

where m is the slope of the line fitted to the low field mobility μ_0 versus temperature T curve plotted on log–log scale, and T_0 is the nominal or reference temperature. For *p*-channel devices, the value of m lies in the range 1.2–1.4 while for *n*-channel devices it ranges between 1.4 to 1.6 for the temperature range between 200 K–450 K. The value of m is a function of the gate field at which the mobility is measured and it tends to be higher at lower fields. The observed difference between the temperature dependence of the n- and p-channel mobilities is because electron and hole scattering processes are different. SPICE uses Eq. (6.228) with $m = 1.5$ for both p- and n-channel devices. Assuming $m = 1.5$, Eq. (6.228) yields the following expression for the temperature coefficient of mobility

$$\frac{1}{\mu}\frac{d\mu}{dT} = -\frac{1.5}{T}. \tag{6.229}$$

Temperature Dependence of Threshold Voltage. The temperature coefficient of V_{th} is approximately 1 mV/°C for modern CMOS devices as was discussed in section 5.4. Recall that the threshold voltage exhibits a linear dependence on temperature over a wide temperature range. Threshold voltage V_{th} increases with increasing temperature, and therefore, drain current decreases.

Zero Temperature Coefficient of Drain Current. Differentiating the classical saturation drain current equation (6.57) with respect to the temperature T, yields the following equation

$$\frac{1}{I_{ds}}\frac{dI_{ds}}{dT} = \frac{1}{\mu}\frac{d\mu}{dT} - \frac{dV_{th}}{dT}\frac{2}{(V_{gs} - V_{th})}. \tag{6.230}$$

The gate voltage which corresponds to zero temperature coefficient (ZTC) of I_{ds} is simply obtained by equating above equation to zero. Combining the resulting equation with (6.229) yields

$$V_{gs} = V_{th} - \frac{dV_{th}}{dT}\frac{T}{0.75}. \tag{6.231}$$

Assuming $dV_{th}/dT = 1\,\text{mV/°C}$, it is easy to see that the gate voltage corresponding to ZTC of I_{ds} is very close to V_{th}. Similar results are obtained if we use the linear region current equation.

Negative Temperature Coefficient of Drain Current. For long channel devices, temperature dependence of I_{ds} is accurately modeled using only the temperature dependence of μ_0 and V_{th}. However, for short channel devices one needs to take into account the temperature dependence of carrier saturation velocity v_{sat} or saturation field $\mathscr{E}_c$. The following linear relation for v_{sat} fits the data well [119]

$$v_{sat}(T) = v_{sat}(T_0) - \beta_v(T - T_0) \tag{6.232}$$

where $v_{sat}(T_0)$ is the value of v_{sat} at $T = T_0$ and β_v is a fitting constant. Figure 6.38 shows temperature dependence of I_{ds} for a p-channel device

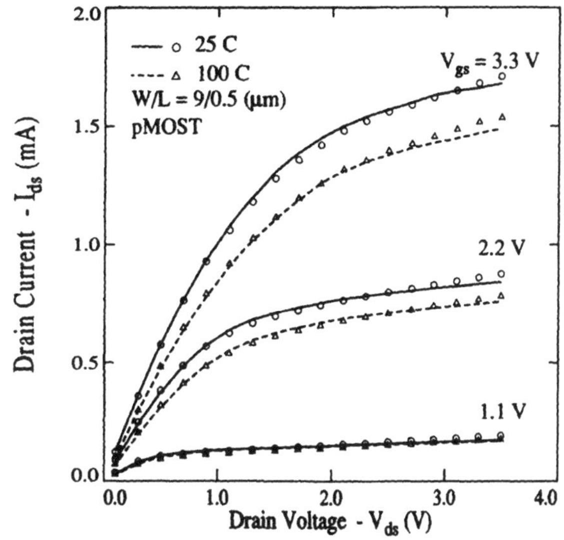

Fig. 6.38 Measured and calculated output characteristics at two temperatures

($W_m/L_m = 10/0.75$, $t_{ox} = 105$ Å) at 25 °C and 100 °C. Circles are experimental data while continuous lines are calculated current based on Eq. (6.208). The only parameters that have been changed from 25 to 100 °C are V_{th}, μ_0 and v_{sat}. It should be pointed out that if the drain current model is not physical, but more empirical in nature, then in order to fit the data well, one probably requires more temperature dependent parameters than discussed here.

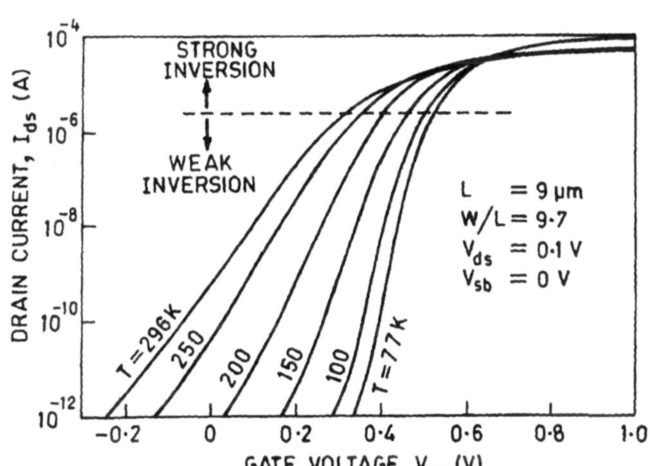

Fig. 6.39 Device $I_{ds} - V_{gs}$ characteristics in the subthreshold or weak inversion region at different temperatures. (After Gaensslen et al. [122])

Positive Temperature Coefficient of Drain Currents. As temperature increases, the subthreshold current increases [cf. Eq. (6.96b)] resulting in a decrease in the subthreshold slope. Figure 6.39 shows drain current as a function of gate voltage showing subthreshold behavior as a function of temperature [122]. Note that the subthreshold slope S has decreased from 86 mV/decade at 296 K to 22 mV/decade at 77 K. This shows that the device can be turned-off much more easily at low temperatures than at high temperatures. This is the so called positive temperature coefficient of I_{ds} and can be modeled fairly accurately using Eq. (6.104).

References

[1] S. Selberherr, A. Schutz, and H. W. Potzl, 'MINIMOS—A two-dimensional MOS transistor analyzer', IEEE Trans. Electron. Devices, ED-27, pp. 1540–1550 (1980).

[2] M. R. Pinto, C. S. Rafferty, and R. W. Dutton, 'PISCES-II: Poisson and continuity equation solver', Stanford Electronic Lab. Tech. Rep., Sept. 1984.

[3] C. L. Wilson and J. L. Blue, 'Two-dimensional finite element charge-sheet model of a short channel MOS transistor', Solid-State Electron., 25, pp. 461–477 (1982).

[4] H. C. Pao and C. T. Sah, 'Effects of diffusion current on characteristics of metal-oxide (insulator)-semiconductor transistors', Solid-State Electron., 9, pp. 927–937 (1966).

[5] R. F. Pierret and J. A. Shields, 'Simplified long-channel MOSFET theory', Solid-State Electron., 26, pp. 143–147 (1983).

[6] A. Nussbaum, R. Sinha, and D. Dokos, 'The theory of the long-channel MOSFET', Solid-State Electron., 27, pp. 97–107 (1984).

[7] J. R. Brews, 'A charge sheet model of the MOSFET', Solid-State Electron., 21, pp. 345–355 (1978).

[8] J. R. Brews, 'Physics of MOS transistor', in *Silicon Integrated Circuits*, Part A, Ed. D. Kahng, Applied Solid-State Science Series, Academic Press, New York, 1981.

[9] P. P. Guebels and F. Van de Wiele, 'A small geometry MOSFET models for CAD applications', Solid-State Electron., 26, pp. 267–263 (1983).

[10] G. Baccarani, M. Rudan, and G. Spadini, 'Analytical i.g.f.e.t model including drift and diffusion', IEE J. Solid-State and Electron Devices, 2, pp. 62–68 (1978).

[11] F. Van de Wiele, 'A long channel MOSFET model', Solid-State Electron., 22, pp. 991–997 (1979).

[12] C. Turchetti and G. Masetti, 'A CAD-oriented analytical MOSFET model for high-accuracy applications', IEEE Trans. Computer Aided Design, CAD-3, pp. 117 122 (1984).

[13] A. M. Ostrowsky, *Solutions of Equations and Systems of Equations*, Academic Press, New York, 1973.

[14] Y. Tsividis, 'Moderate inversion in MOS devices', Solid-State Electron., 25, pp. 1099–1104 (1982), also see "Erratum", Solid-State Electron., 26, p. 823 (1983).

[15] Y. P. Tsividis, *Operation and Modeling of the MOS Transistor*, McGraw-Hill Book Company, New York, 1987.

[16] P. P. Guebels and F. Van de Wiele, 'A small geometry MOSFET models for CAD applications', Solid-State Electron., 26, pp. 263–267 (1983).

[17] S. Yu, A. F. Franz, and T. G. Mihran, 'A physical parametric transistor model for CMOS circuit simulation', IEEE Trans. Computer-Aided Design, CAD-7, pp. 1038–1052 (1988).

[18] H. J. Park, P. K. Ko, and C. Hu, 'A charge sheet capacitance model of short channel

MOSFET's for SPICE', IEEE Trans. Computer-Aided Design, CAD-10, pp. 376–389 (1991).
[19] A. R. Boothryod, S. W. Tarasewicz, and C. Slaby, 'MISNAN—A physically based continuous MOSFET model for CAD applications', IEEE Trans. Computer-Aided Design, CAD-10, pp. 1512–1529 (1991).
[20] C. T. Sah, 'Characteristics of the metal–oxide–semiconductor transistors', IEEE Trans. Electron. Devices, ED-11, pp. 324–345 (1964).
[21] H. Schichman and D. A. Hodges, 'Modeling and simulation of insulated-gate field-effect transistor switching circuits', IEEE J. Solid-State Circuits, SC-3, pp. 285–289 (1968).
[22] H. K. J. Ihantola and J. L. Moll, 'Design theory of a surface field-effect transistor', Solid-State Electron., 7, pp. 423–430 (1964).
[23] A. Vladimirescu and S. Liu, 'The simulation of MOS integrated circuits using SPICE2', Memorandum No. UCB/ERL M80/7, Electronics Research Laboratory, University of California, Berkeley, October 1980.
[24] H. I. Hanafi, L. H. Camnitz, and A. J. Dally, 'An accurate and simple MOSFET model for computer-aided design', IEEE J. Solid-State Circuits, SC-17, pp. 882–891 (1982).
[25] B. J. Sheu, D. L. Scharfetter, P. K. Ko, and M. C. Jeng, 'BSIM: Berkeley short-channel IGFET model for MOS transistors', IEEE J. Solid-State Circuits, SC-22, pp. 558–565 (1987).
[26] N. D. Arora and L. M. Richardson, 'MOSFET modeling for circuit simulation, in *Advanced MOS Device Physics* (N. G. Einspruch and G. Gildenblat, Eds.), VLSI Electronics: Microstructure Science, Vol. 18, pp. 236–276, Academic Press Inc., New York, 1989.
[27] H. C. de Graaff and F. M. Klaassen, *Compact Transistor Modelling for Circuit Design*, Springer-Verlag, Wien-New York, 1990.
[28] C.-L. Huang and G. Sh. Gildenblat, 'Measurements and modeling of the n-channel MOSFET inversion layer mobility and device characteristics in the temperature range 60–300 K', IEEE Trans. Electron. Devices, ED-37, pp. 1289–1300 (1990).
[29] M. B. Barron, 'Low level currents in insulated gate field effect transistors', Solid-State Electron., 21, pp. 293–309 (1972).
[30] R. M. Swanson and J. D. Meindel, 'Ion-implanted complementary MOS transistors in low voltage circuits', IEEE J. Solid-State Circuits, SC-7, pp. 146–153 (1972).
[31] W. Fichtner and H. W. Poetzl, 'MOS modeling by analytical approximation. I. Subthreshold current and threshold voltage', Int. J. Electron., 46, pp. 33–55 (1979).
[32] J. B. Brews, 'Subthreshold behavior of uniformly and nonuniformly doped long-channel MOSFET', IEEE Trans. Electron Devices, ED-26, pp. 1282–1291 (1979).
[33] G. T. Wright, 'Simple and continuous MOSFET models for the computer-aided design of VLSI', IEE Proc. I, Solid-State and Electron Devices, 132, pp. 187–194 (1985).
[34] G. T. Wright, 'Physical and CAD model for the implanted-channel VLSI MOSFET', IEEE Trans. Electron Devices, ED-34, pp. 823–833 (1987).
[35] C. Turchetti and G. Masetti, 'Analysis of the depletion-mode MOSFET including diffusion and drift currents', IEEE Trans. Electron Devices, ED-32, pp. 773–782 (1985).
[36] G. R. Mohan Rao, 'An accurate model for a depletion mode IGFET used as a load device', Solid-State Electron., 21, pp. 711–714 (1978).
[37] G. Baccarani, F. Landini, and B. Ricco, 'Depletion-mode MOSFET model including field dependent surface mobility', Proc. IEE., 127, pt. I, pp. 230 (1980).
[38] Y. A. El-Mansy, 'Analysis and characterization of depletion mode MOSFET', IEEE J. Solid-State Circuits, SC-15, pp. 331–340 (1980).

[39] P. Ratnam and A. B. Bhattacharyya, 'Accumulation-punchthrough model of operation of buried-channel MOSFET's', IEEE Trans. Electron Device Lett., EDL-3, pp. 203–204 (1982).
[40] T. Yamaguchi and S. Morimoto, 'Analytical model and characteristics of small geometry buried-channel depletion MOSFETs', IEEE J. Solid-State Circuits, SC-18, pp. 784–793 (1983).
[41] S. H. Ahmed and C. A. T. Salama, 'Depletion mode MOSFET modeling for CAD', IEE Proc. Part I, Solid-State & Electron Dev., 130, pp. 281–286 (1983).
[42] D. A. Divekar and R. I. Dowell, 'A depletion-mode MOSFET model for circuit simulation', IEEE Trans. Computer-Aided Design, CAD-3, pp. 80–87 (1984).
[43] D. Ma, 'A physical and SPICE-compatible models for the MOS depletion device', IEEE Trans. Computer-Aided Design, CAD-4, pp. 349–356 (1985).
[44] C. Y. Yu and K. C. Hsu, 'Mobility models for the I–V characterstics of buried-channel MOSFETs', Solid-State Electron., 28, pp. 917–923 (1985).
[45] M. J. Van de Tol and S. G. Chamberlain, 'Buried-channel MOSFET model for SPICE', IEEE Trans. Computer-Aided Design, CAD-10, pp. 1015–1035 (1991).
[46] G. Merckel, J. Borel, and N. Z. Cupcea, 'An accurate large-signal MOS transistors model for computer-aided-design', IEEE Trans. Electron Devices, ED-19, pp. 681–690 (1972).
[47] T. Ando, A. B. Fowler, and F. Stern, 'Electronic properties of two-dimensional systems', Reviews of Modern Phys., 54, pp. 437–672, 1982.
[48] D. K. Ferry, K. Hess, and P. Vogl, 'VLSI Electronics: Microstructure Science, Vol. 2 (N. G. Einspruch, Ed.), p. 67. Academic Press, New York, 1981.
[49] F. N. Trofimenkoff, 'Field-dependent mobility analysis of the field effect transistor', Proc. IEEE, 53, pp. 1765–1766 (1965).
[50] D. Frohman-Bentchkowsky, 'On the effect of mobility variation on MOS device characteristics', IEEE Proc., 56, pp. 217–218 (1968).
[51] A. G. Sabnis and J. T. Clemens, 'Characterization of the electron mobility in the inverted ⟨100⟩ Si surface', IEEE IEDM, *Tech. Dig.*, pp. 18–21 (1979).
[52] S. C. Sun and J. D. Plummer, 'Electron mobility in inversion and accumulation layers on thermally oxidized silicon surface', IEEE Trans. Electron Devices, ED-27, pp. 1497–1508 (1980).
[53] P. P. Wang, 'Device characteristics of short-channel and narrow-width MOSFETs', IEEE Trans. Electron Devices, ED-25, pp. 779–786 (1978).
[54] M. H. White, F. van de Wiele, and J. P. Lambot, 'High-accuracy MOS models for computer aided design', IEEE Trans. Electron Devices, ED-27, pp. 899–906 (1980).
[55] K. Y. Fu, 'Mobility degradation due to the gate field in the inversion layer of MOSFETs', IEEE Trans. Electron Device Lett., EDL-3, pp. 292–293 (1982).
[56] M. S. Lin, 'The classical versus the quantum mechanical model of mobility degradation due to the gate field in MOSFET inversion layers', IEEE Trans. Electron Devices, ED-32, pp. 700–710 (1985).
[57] M. S. Liang, J. Y. Choi, P. K. Ko, and C. M. Hu, 'Inversion-layer capacitance and mobility of very thin gate-oxide MOSFETs', IEEE Trans. Electron Devices, ED-33, pp. 409–413 (1986).
[58] B. Majkusiak and A. Jakubowski, 'The dependence of MOSFET surface carrier mobility on gate oxide thickness', IEEE Trans. Electron Devices, ED-33, pp. 1717–1721 (1986).
[59] N. D. Arora and G. Sh. Gildenblat, 'A semi-empirical model of the MOSFET inversion layer mobility for low-temperature operation', IEEE Trans. Electron Devices, ED-34, pp. 89–93 (1987).
[60] J. W. Watt and J. D. Plummer, 'Universal mobility-field curves for electrons and holes in MOS inversion layers', Proc. Symp. VLSI Tech., pp. 81–82 (1987).

[61] D. T. Amm, H. Mingam, P. Delpech, and T. T. D'ouville, 'Surface mobility in n^+ and p^+ doped polysilicon gate PMOS transistors', IEEE Trans. Electron Devices, ED-36, pp. 963–967 (1989).
[62] P. K. Ko, 'Approaches to scaling' in *Advanced MOS Device Physics* (N. G. Einspruch and G. Gildenblat, Eds.), VLSI Electronics Vol. 18, Academic Press Inc., New York, 1989.
[63] S. W. Lee, 'Universality of mobility-gate field characteristics of electrons in the inversion charge layer and its application in MOSFET modeling', IEEE Trans. Computer-Aided Design, CAD-8, pp. 742–730 (1989).
[64] A. J. Walker and P. H. Woerlee, 'Mobility model for silicon inversion layers', EESDERC 1987, *Tech. Dig.*, pp. 667–670 (1987).
[65] S. Takagi, M. Iwase, and A. Toriumi, 'On universality of inversion-layer mobility in n- and p-channel MOSFETs', IEDM, *Tech. Dig.*, pp. 398–401 (1988).
[66] G. M. Yeric and Al F. Tasch, 'A universal MOSFET mobility degradation model for circuit simulation', IEEE Trans. Computer-Aided Design, CAD-9, pp. 1123–1126 (1990).
[67] T. J. Krutsick and M. H. White, 'Consideration of doping profiles in MOSFET mobility modeling', IEEE Trans. Electron Devices, ED-35, pp. 1153–1155 (1988).
[68] J. H. Satter, 'The S-model: A highly accurate MOST model for CAD', Solid State Electron., 29, pp. 990–997 (1986).
[69] B. J. Moon, C. K. Park, K. M. Rho, K. Lee, and M. Shur, 'New short-channel n-MOSFET current-voltage model in strong inversion and unified parameter extraction method', IEEE Trans. Electron Devices, ED-38, pp. 592–602 (1991).
[70] B. J. Moon, C. K. Park, K. M. Rho, K. Lee, M. Shur, and T. A. Fjeldly, 'Analytical model for p-channel MOSFET's', IEEE Trans. Electron Devices, ED-38, pp. 2632–2646 (1991).
[71] G. Sh. Gildenblat, C.-L. Huang, and N. D. Arora, 'Split C-V measurements of low temperature MOSFET inversion layer mobility,' Cryogenics, 29, pp. 1163–1166, (1989).
[72] T. Sato, Y. Takeishi, H. Tango, H. Ohnuma, and Y. Okamoto, 'Drift velocity saturation of charge carriers in Si inversion layers', J. Phys. Soc. Japan, 31, pp. 1846–1849 (1971).
[73] K. K. Throuber, 'Relation of drift velocity to low field mobility and high field saturation velocity', J. Appl. Phys., 51, pp. 2127–2136 (1980).
[74] R. Coen and R. S. Muller, 'Velocity of surface carriers in inversion layers of silicon', Solid-State Electron., 23, pp. 35–40 (1980).
[75] J. P. Leburton and G. E. Dorda, 'υ-E dependence in small-sized MOS transistors', IEEE Trans. Electron Devices, ED-29, pp. 1168–1171 (1981).
[76] S. A. Schwarz, 'Semi-empirical equations for electron velocity in silicon: Part II—MOS inversion layer', IEEE Trans. Electron Devices, ED-30, pp. 1634–1639 (1983).
[77] G. W. Taylor, 'Velocity-saturated characteristics of short-chabbel MOSFETs', AT&T Bell System Technical J., 63, pp. 1325–1381 (1984).
[78] J. A. Cooper, Jr. and D. F. Nelson, 'Measurement of the high-field drift velocity of electrons in inversion layers in silicon', IEEE Electron Devices Lett., EDL-2, pp. 171–173 (1981). See also J. A. Cooper, Jr., D. F. Nelson, S. A. Schwarz, and K. K. Thornber, 'VLSI Electronics: Microstructure Science' Vol. 10 (N. G. Einspruch and R. S. Bauer, Eds.), p. 323, Academic Press, New York, 1985.
[79] T. Y. Chan, S. W. Lee, and H. Gaw, 'Experimental characterization and modeling of electron saturation velocity in MOSFET's inversion layer from 90 to 350 K', IEEE Electron Devices Lett., EDL-2, pp. 466–468 (1990).
[80] N. Kotani and S. Kawazu, 'Computer analysis of punch-through in MOSFETs', Solid-State Electron., 22, pp. 63–70 (1979).

[81] L. Risch, 'Electron mobility in short-channel MOSFETs with series resistance', IEEE Trans. Electron Devices, ED-30, pp. 959–961 (1983).
[82] T. Grotjohn and B. Hoefflinger, 'A parametric short-channel MOS transistor model for subthreshold and strong inversion current', IEEE Trans. Electron Devices, ED-31, pp. 234–246 (1984).
[83] S. L. Wong and C. A. T. Salama, 'Improved simulation of *p*- and *n*-channel MOSFETs using an enhanced SPICE MOS3 model', IEEE Trans. Computer-Aided Design, CAD-6, pp. 586–591 (1987).
[84] P. Yang and P. K. Chatterjee, 'SPICE modeling for small geometry MOSFET circuits', IEEE Trans. Computer-Aided Design, Vol. CAD-1, pp. 169–182 (1982).
[85] B. J. Sheu, D. L. Scharfetter, P. K. Ko, and M. C. Jeng, 'BSIM: Berkeley short-channel IGFET model for MOS transistors', IEEE J. Solid-State Circuits, SC-22, pp. 558–565 (1987).
[86] C. Duvvury, 'Guide to short-channel effects in MOSFETs', IEEE Circuits and Device Mag., pp. 6–10 (1986).
[87] A. L. Silburt, R. C. Foss, and W. F. Petrie, 'An efficient MOS transistor model for computer-aided design', IEEE Trans. Computer-Aided Design, CAD-3, pp. 104–110 (1984).
[88] N. D. Arora and M. Sharma, 'MOSFET substrate current model for circuit simulation', IEEE Trans. Electron Devices, ED-38, pp. 1392–1398 (1991).
[89] C. G. Sodini, P. K. Ko, and J. L. Moll, 'The effect of high fields on MOS device and circuit performance', IEEE Trans. Electron Devices, ED-31, pp. 1386–1396 (1984).
[90] B. Hoefflinger, H. Sibbert, and G. Zimmer, 'Model and performance of hot electron MOS transistor for VLSI', IEEE J. Solid State Circuits, SC-14, pp. 435–442 (1979).
[91] F. M. Klaassen and W. C. J. de Groot, 'Modeling of scaled-down MOS transistors', Solid-State Electron., 23, pp. 237–242 (1980).
[92] B. Hoefflinger, 'Output characteristics of short-channel field-effect transistors', IEEE Trans. Electron Devices, ED-28, pp. 971–976 (1981).
[93] B. T. Murphy, 'Unified field-effect transitor theory including velocity saturation', IEEE J. Solid State Circuits, SC-15, pp. 325–328 (1980).
[94] V. G. K. Reddi and C. T. Sah, 'Source to drain resistance beyond pinch-off in metal-oxide-semiconductor transistors (MOST)', IEEE Trans. Electron Devices, ED-17, pp. 139–141 (1965).
[95] D. Frohman-Bentchkowsky and A. S. Grove, 'Conductance of MOS transistors in saturation', IEEE Trans. Electron Devices, ED-16, pp. 108–113 (1969).
[96] F. S. Jenkins, E. R. Lane, W. W. Lattin, and W. S. Richardson, 'MOS-device modeling for computer implementation', IEEE Trans. Circuit Theory, CT-20, pp. 649–658 (1973).
[97] G. Baum and H. Beneking, 'Drift velocity saturation in MOS transistors', IEEE Trans. Electron Devices, ED-17, pp. 481–482 (1970).
[98] A. Popa, 'An injection level dependent theory of the MOS transistor in saturation', IEEE Trans. Electron Devices, ED-19, pp. 774–781 (1972).
[99] P. Rossel, H. Martinot, and G. Vassilieff, 'Accurate two sections model for MOS transistor in saturation', Solid-State Electron., 19, pp. 51–56 (1976).
[100] T. Poorter and J. H. Satter, 'A DC model for an MOS-transistor in the saturation region', Solid-State Electron., 23, pp. 765–772 (1980).
[101] P. Ratnam and C. A. T. Salama, 'A new approach to the modeling of nonuniformly doped short-channel MOSFETs', IEEE Trans. Electron Devices, ED-31, pp. 1289–1298 (1984).
[102] Y. A. El-Mansy and A. R. Boothroyd, 'A simple two-dimensional model for IGFET', IEEE Trans. Electron Devices, ED-24, pp. 254–262 (1977).
[103] F. J. Lai and J. Y. Sun, 'An analytical one-dimensional model for lightly doped drain

(LDD) MOSFET devices', IEEE Trans. Electron Devices, ED-32, pp. 2803–2811 (1985).
[104] M. E. Banna and M. E. Nokali, 'A pseudo-two-dimensional analysis of short channel MOSFETs', Solid-State Electron., 31, pp. 269–274 (1988).
[105] G. S. Huang and C. Y. Wu, 'An analytic I–V model for lightly doped drain (LDD) MOSFET devices', IEEE Trans. Electron Devices, ED-34, pp. 1311–1321 (1987).
[106] C. Turchetti and G. Masetti, 'A charge-sheet analysis of short-channel enhancement-mode MOSFET's', IEEE J. Solid-State Circuits, SC-21, pp. 267–275 (1986).
[107] K. Y. Toh, P. K. Ko, and R. G. Meyer, 'An engineering model for short-channel MOS devices', IEEE J. Solid-State Circuits, SC-23, pp. 950–958 (1988).
[108] K. Mayaram, J. C. Lee, and C. Hu, 'A model for the electric field in lightly doped drain structures', IEEE Trans. Electron Devices, ED-34, pp. 1509–1518 (1987).
[109] Y. Hu, R. V. H. Booth, and M. H. White, 'An analytical model for the lateral channel electric field in LDD structures', IEEE Trans. Electron Devices, ED-37, pp. 2254–2263 (1990).
[110] G. W. Taylor, 'Subthreshold conduction in MOSFETs', IEEE Trans. Electron Devices, ED-25, pp. 337–350 (1978).
[111] P. Antognetti, D. D. Caviglia, and E. Profumo, 'CAD model for threshold and subthreshold conduction in MOSFETs', IEEE J. Solid-State Circuits, SC-17, pp. 454–458 (1982).
[112] P. C. Chan, R. Liu, S. K. Lau, and M. Pinto-Guedes, 'A subthreshold conduction model for circuit simulation of submicron MOSFET', IEEE Trans. Computer-Aided Design, CAD-6, pp. 574–581 (1987).
[113] S. S. Chung and C. T. Sah, 'A subthreshold model of the narrow-gate effects in MOSFET's', IEEE Trans. Electron Devices, ED-34, pp. 2521–2528 (1987).
[114] A. L. Silburt, A. R. Boothroyd, and M. Digiovanni, 'Automated parameter extraction and modeling of the MOSFET below threshold', IEEE Trans. Computer-Aided Design, CAD-7, pp. 484–488 (1988).
[115] S. S. Chung, 'A complete model of the I–V characteristics for narrow-gate MOSFETs', IEEE Trans. Electron Devices, ED-37, pp. 1020–1030 (1990).
[116] F. M. Klaassen and R. M. D Velghe, Proceedings ESSDERC 89, pp. 418–422, Springer-Verlag, Vienna (1989).
[117] H. Masuda, J. I. Mano, R. Ikematsu, H. Sugihara, and Y. Aoki, 'A submicrometer MOS transistor I–V model for circuit simulation', IEEE Trans. Computer-Aided Design, CAD-10, pp. 161–170 (1991).
[118] J. A. Power and W. A. Lane, 'Enhanced SPICE MOSFET model for analog applications including parameter extraction schemes', Proc. IEEE 1990 Int. Conf. Microelectronics Test Structures, 3, pp. 129–134 (1990).
[119] N. D. Arora, 'A continuous MOSFET model for VLSI simulation' (to be published).
[120] M. G. Hsu and B. J. Sheu, 'Inverse-geometry dependence of MOS transistor electrical parameters', IEEE Trans. Computer-Aided Design, CAD-6, pp. 582–585 (1987).
[121] F. M. Klaassen, P. T. J. Biermans, and R. M. D. Velghe, 'The series resistance of submicron MOSFETs and its effect on their characteristics', Proc. ESSDERC 1988, J. De Physique, pp. 257–260 (1988).
[122] F. H. Gaensslen, V. L. Rideout, E. J. Walker, and J. J. Walker, 'Very small MOSFETs for low temperature operation', IEEE Trans. Electron Devices, ED-24, pp. 218–219 (1977).
[123] S. K. Tewksbury, 'N-channel enhancement-mode MOSFET characteristics from 10–300 K', IEEE Trans. Electron Devices, ED-28, pp. 1519–1529 (1981).
[124] G. Gildenblat, 'Low-temperature operation' in *Advanced MOS Device Physics* (N. G. Einspruch and G. Gildenblat, Eds.), VLSI Electronics Vol. 18, pp. 191–232, Academic Press Inc., New York, 1989.

[125] F. Shoucair, W. Hwang, and P. Jain, 'Electrical characteristics of large scale integration (LSI) MOSFETs at very high temperatures', Microelect. Reliab., 24, part I, pp. 465–485 and part II, pp. 487–510 (1984).
[126] C. P. Wan and B. J. Sheu, 'Temperature dependence modeling for MOS VLSI circuit simulation', IEEE Trans. Computer-Aided Design, CAD-8, pp. 1065–1073 (1989).
[127] M. Nishida and H. Ohyabu, 'Temperature dependence of MOSFET characteristics in weak inversion', IEEE Trans. Electron Devices, ED-24, pp. 1245–1248 (1977).
[128] H. C. Card and R. W. Ulmer, 'On the temperature dependence of subthreshold currents in MOS electron inversion layers', Solid-State Electron., 22, pp. 463–465 (1979).

7 Dynamic Model

The MOS transistor DC models developed in the last chapter are applicable when applied voltages do not vary with time. In this chapter we will develop transistor dynamic models which are applicable when the device terminal voltages are varying with time. The variation in the applied voltages, if sufficiently small, results in the *small signal* model. However, if the variation in the voltages is large, the *large signal model* results. Both types of models are required for a circuit simulator, as was discussed in Chapter 1.
The dynamic behavior of a MOSFET is due to the device capacitive effects, which in turn are the results of the charges stored in the device. This is in addition to steady-state current (DC) as discussed in Chapter 6. The capacitive characteristics are in fact the sum of the intrinsic (channel region) and extrinsic (source/drain junction region) capacitances as discussed in section 3.2. Of key importance in calculating the MOSFET capacitances is an accurate description of the various charges in the device and how they depend on externally applied voltages. These capacitances are an essential part of the large signal as well as small signal model for frequencies of operation greater than about 1 KHz.
In this chapter we will first develop models for the intrinsic charges and capacitances of a large and wide MOSFET and then discuss models for short channel devices. This will give us the so called large signal model. This will be followed by small signal linear model parameters required for small signal analysis.

7.1 Intrinsic Charges and Capacitances

Under steady-state conditions, the only current in a MOSFET results from mobile carriers (electrons in nMOST and holes in pMOST) flow from source to drain.[1] In transient analysis this current is referred to as the *trans-*

[1] This of course assumes that substrate and gate current is zero (or negligible), which indeed is true for normal device operation.

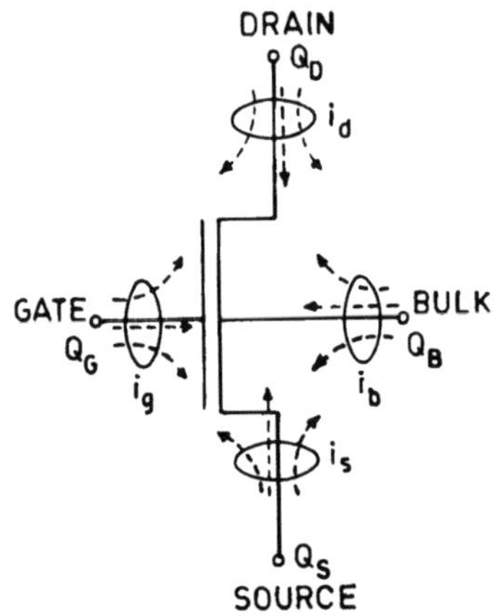

Fig. 7.1 Schematic of the transient current flowing through a MOSFET

port current. In dynamic situations, additional currents that are associated with the stored charges at the device terminals, called the *charging current*, also exist. Figure 7.1 shows transient (dynamic) currents i_g, i_s, i_d and i_b flowing through the gate(g), source(s), drain(d) and bulk(b) terminals, respectively, of a MOSFET. Here Q_G, Q_S, Q_D and Q_B are the total source, drain, gate and bulk charges, respectively[2] corresponding to the four terminals of the MOSFET. It is important to note that the terminal charges are not independent, but are functions of the four different terminal voltages V_g, V_s, V_d and V_b. Thus, in general

$$Q_j = f(V_g, V_s, V_d, V_b), \quad j = \mathrm{G, S, D, B}. \tag{7.1}$$

Since Kirchhoff's current law (KCL) holds for the total current, we have

$$i_g + i_s + i_d + i_b = 0 \tag{7.2}$$

and from the charge conservation law we have

$$Q_G + Q_S + Q_D + Q_B = 0. \tag{7.3}$$

While calculating various charges of a MOSFET we will assume that the terminal voltages vary sufficiently slowly so that the distribution in the stored charges Q_G, Q_B, Q_S and Q_D can follow the voltage variations.[3] Stated

---

2 So for we have talked about the charges per unit area denoted by Q with the lower case subscript (e.g., Q_g). In this chapter we will also be dealing with the total charge, which will be denoted by Q with the upper case subscript (e.g., Q_G).

3 Strictly speaking, Q_S and Q_D are not the stored charges in the same sense as Q_G and Q_B. The inversion charge Q_I (from which Q_S and Q_D are derived) is the result of inversion carriers entering the source and leaving the drain and being continuously replaced by new carriers from the source. Nonetheless, Q_I and hence Q_S and Q_D, can loosely be called stored charges.

another way, the terminal currents vary instantaneously with the terminal voltages. In other words, the charge per unit area at any time t is the same as obtained by using the DC voltage at that time. This is called the *quasi-static operation* of the device and the resulting dynamic model is called the quasi-static model. In practice, the quasi-static model works quite well for much of circuit work. However, *it should be kept in mind that this approach may fail, especially with long channel devices operating at high switching speeds, or when the load capacitance is very small.*

Assuming quasi-static operation, we can write the transient currents as the sum of the time dependent transport current and a charging current as

$$i_s(t) = -I_s(V(t)) + \frac{dQ_S}{dt} \tag{7.4a}$$

$$i_d(t) = I_d(V(t)) + \frac{dQ_D}{dt} \tag{7.4b}$$

$$i_g(t) = \frac{dQ_G}{dt} \tag{7.4c}$$

$$i_b(t) = \frac{dQ_B}{dt} \tag{7.4d}$$

remembering that no transport current is flowing to the gate ($I_g = 0$) and substrate ($I_b = 0$). While writing Eqs. (7.4) we have tactically assumed that charges at the source and drain are known explicitly, while in reality we only know total inversion or channel charge Q_I, that is,

$$i_s + i_d = I_{ds}(V(t)) + \frac{dQ_I}{dt}. \tag{7.5}$$

This equation presents some problem because, from the circuit simulation point of view, we need i_s and i_d individually, not just their sum. We will come to this point a bit later.

Thus, to develop a dynamic model, it is necessary to obtain expressions for the total gate, bulk and inversion charges, Q_G, Q_B and Q_I, respectively, as functions of terminal voltages. From the steady-state (DC) analysis we already know Q_g, Q_b and Q_i, the charges per unit area, which in general depend on the position y along the length of the channel. By integrating these charges over the area of the active gate region we can obtain the corresponding total charge Q_G, Q_B and Q_I. For example, the gate charge contained in a small area of width W (device width) and length dy is $Q_g \cdot W \cdot dy$. Integrating this charge over the channel length L gives the total gate charge Q_G as

$$Q_G = W \int_0^L Q_g(y)dy \quad \text{(C)}. \tag{7.6}$$

Similarly, we have

$$Q_I = W\int_0^L Q_i(y)dy \quad \text{(C)} \tag{7.7a}$$

$$Q_B = W\int_0^L Q_b(y)dy \quad \text{(C)} \tag{7.7b}$$

and from the charge conservation principle

$$Q_G + Q_I + Q_B = 0. \tag{7.8}$$

Since these are distributed charges, the corresponding intrinsic capacitances should be modeled as a distributed capacitances. However, such a model would be too complex for use in circuit simulators. Instead, these distributed capacitances are usually modeled as lumped two terminal capacitances appearing between the gate, source, drain, and bulk or substrate terminals of the MOSFET.

Several models for the intrinsic capacitances of a MOSFET have been proposed. The model which has been almost universally used for many circuit simulators, until very recently, is the Meyer model [2] that was derived for long-channel devices. The most serious error in this model appears as charge nonconservation (see section 7.1.2). Charge conservation is very important in such circuits as dynamic RAMs and switched capacitor filters. Drawbacks in the Meyer model were overcome by using charge as a state variable. This resulted in the development of new models, normally known as charge based capacitance models [3]–[8]. However, due to the intrinsic simplicity of the Meyer model, it has been extensively used in simulating circuits that do not have charge-conservation problems. The Meyer model is the default capacitance model for SPICE Levels 1–4. However, for Levels 2 and 4 a charge based model is also available. We will first discuss the Meyer model and then develop a more accurate charge based capacitance model.

7.1.1 Meyer Model

In the Meyer model, the distributed gate-channel capacitances are split into three lumped capacitances; gate-to-source (C_{GS}), gate-to-drain (C_{GD}), and gate-to-bulk[4] (C_{GB}). They are defined as the derivative of the total gate

[4] In the original work by Meyer [2], the capacitance C_{GB} formed by the depletion charge within the bulk of the device is ignored. However, most implementations of the Meyer model in use include this capacitance.

charge Q_G with respect to the source, drain and bulk respectively, that is

$$\begin{aligned} C_{GS} &= \left.\frac{\partial Q_G}{\partial V_{gs}}\right|_{V_{gd},V_{gb}} \quad \text{(F)} \\ C_{GD} &= \left.\frac{\partial Q_G}{\partial V_{gd}}\right|_{V_{gs},V_{gb}} \quad \text{(F)} \\ C_{GB} &= \left.\frac{\partial Q_G}{\partial V_{gb}}\right|_{V_{gs},V_{gd}} \quad \text{(F)} \end{aligned} \tag{7.9}$$

where $V_{gd} = (V_{gs} - V_{ds})$ and $V_{gb} = (V_{gs} - V_{bs})$. Notice that the capacitance definitions (7.9) imply that these capacitances are *reciprocal*; i.e., both terminals of a capacitor are equivalent and the capacitance is symmetric. For example, C_{GD} is the same as C_{DG}; in other words, change in the charge Q_G due to V_{gd} may be due to the change either in the gate voltage V_g or drain voltage V_d. The Meyer model derives its capacitances on the following assumptions:

1. MOSFET capacitances are *reciprocal.*
2. The bulk charge density Q_b is constant along the length of the channel depending only on the applied gate to bulk voltage V_{gb} but independent of the applied source to drain voltage V_{ds}. This implies that *bulk-to-source(drain) capacitance $C_{BS}(C_{BD})$ is zero.*[5]

Assumption 2, though physically incorrect, is not the cause of great error because when the device is conducting the channel acts as a screen for the bulk charge Q_B, and therefore the change in Q_B is very small. However, assumption 1 has profound implications as we will see later. For a piecewise multisection model, the charge Q_g (and hence Q_G) is different in different regions of device operation; therefore, the corresponding capacitance values will also be different for different regions.
Using the charge conservation rule [cf. Eq. (7.8)] the total gate charge Q_G can be expressed as

$$Q_G = -(Q_I + Q_B) = -W\int_0^L Q_i(y)dy - W\int_0^L Q_b(y)dy \tag{7.10}$$

where we have made use of Eq. (7.7). Since bulk charge density Q_b is constant along the length of the channel (assumption 2), it can be taken out of the

[5] These capacitances should not be confused with the source(drain)-to-bulk *pn* junction capacitances, which are extrinsic to the device.

integral. Thus, the above equation becomes

$$Q_G = -W\int_0^L Q_i(y)dy - Q_B \tag{7.11}$$

where

$$Q_B = WLQ_b.$$

Strong Inversion. In principle, to calculate the gate charge Q_G, any expression for Q_i which is used to calculate I_{ds} can be used. However, in the Meyer model the following long channel expressions for Q_i and Q_b are used [cf. Eqs. (6.44) and (6.45)]

$$Q_i(y) = -C_{ox}(V_{gs} - V_{th} - V(y)) \quad (\text{C/cm}^2) \tag{7.12a}$$

$$Q_b = -C_{ox}\gamma\sqrt{2\phi_f + V_{sb}} \quad (\text{C/cm}^2) \tag{7.12b}$$

where C_{ox} is the gate oxide capacitance per unit area, V_{th} is the threshold voltage and $V(y)$ is the voltage at any point y along the length of the channel from the source to drain. Since we know Q_i as a function of V, to integrate Eq. (7.11) we first change the variable of integration from 'dy' to 'dV' by making use of the following equation [cf. Eq. (6.13)]

$$dy = \frac{\mu_s W}{I_{ds}} Q_i(y)dV. \tag{7.13}$$

Combining Eqs. (7.11)–(7.13) yields

$$Q_G = \frac{\mu_s W^2 C_{ox}}{I_{ds}} \int_0^{V_{ds}} (V_{gs} - V_{th} - V)^2 dV - Q_B. \tag{7.14}$$

Using Eq. (7.12) in (7.13) and integrating the resulting equation from source to drain, we get the following equation for I_{ds} [cf. Eq. (6.49)]

$$I_{ds} = \frac{W\mu_s C_{ox}}{L}(V_{gs} - V_{th} - 0.5V_{ds})V_{ds}, \quad (\text{A})$$

which is algebraically equivalent to

$$I_{ds} = \frac{W\mu_s C_{ox}}{2L}[(V_{gs} - V_{th})^2 - (V_{gd} - V_{th})^2]. \tag{7.15}$$

Now combining Eq. (7.15) with (7.14), and carrying out the integration results in the following expression for the gate charge Q_G in the linear region

$$Q_G = \frac{2}{3}WLC_{ox}\left[\frac{(V_{gd} - V_{th})^3 - (V_{gs} - V_{th})^3}{(V_{gd} - V_{th})^2 - (V_{gs} - V_{th})^2}\right] - Q_B. \tag{7.16}$$

Differentiating the above equation with respect to V_{gs}, V_{gd} and V_{gb}, we get the gate capacitances C_{GS}, C_{GD} and C_{GB}, respectively, in the linear region as

$$C_{GS} = \frac{\partial Q_G}{\partial V_{gs}} = \frac{2}{3} WLC_{ox}\left[1 - \frac{(V_{gd} - V_{th})^2}{(V_{gd} + V_{gs} - 2V_{th})^2}\right] \tag{7.17a}$$

$$C_{GD} = \frac{\partial Q_G}{\partial V_{gd}} = \frac{2}{3} WLC_{ox}\left[1 - \frac{(V_{gs} - V_{th})^2}{(V_{gd} + V_{gs} - 2V_{th})^2}\right] \tag{7.17b}$$

$$C_{GB} = \frac{\partial Q_G}{\partial V_{gb}} = 0. \tag{7.17c}$$

Note that in strong inversion C_{GB} is zero. Physically speaking, this makes sense because the formation of an inversion layer in the channel region provides an electrostatic shield between the gate and the substrate, so that Q_G ceases to respond to changes in the substrate bias V_{sb}.
In the saturation region, the gate charge Q_G can be obtained by replacing V_{ds} (through $V_{gd} = V_{gs} - V_{ds}$) in Eq. (7.16) with $V_{dsat}(= V_{gs} - V_{th})$, the long-channel saturation voltage. This is algebraically and intuitively equivalent to replacing V_{gd} in Eq. (7.16) with V_{th}. Thus, in saturation we have

$$Q_G = \frac{2}{3} WLC_{ox}(V_{gs} - V_{th}) - Q_B \text{ (saturation region)} \tag{7.18}$$

which results in the following capacitances in saturation

$$C_{GS} = \frac{\partial Q_G}{\partial V_{gs}} = \frac{2}{3} WLC_{ox} \tag{7.19a}$$

$$C_{GD} = \frac{\partial Q_G}{\partial V_{gd}} = 0 \tag{7.19b}$$

$$C_{GB} = \frac{\partial Q_G}{\partial V_{gb}} = 0. \tag{7.19c}$$

Note that the capacitances in saturation are independent of the drain voltage. Intuitively this makes sense, because in saturation the drain is cutoff from the channel due to channel pinch-off. Therefore, when V_{ds} is varied in saturation, the intrinsic device is not affected and hence the capacitances remain the same.

Weak Inversion or Subthreshold Region. When a MOSFET is in the weak inversion or subthreshold region ($V_{gs} < V_{th}$), the inversion layer charge Q_I is negligible through out the channel, and therefore the gate charge Q_G, to a first approximation, can be written as

$$Q_G \approx -Q_B = W\int_0^L Q_b dy = WLQ_b. \tag{7.20}$$

Under the depletion approximation, the bulk charge Q_b for a long channel device is given by [cf. Eq. (6.88)]

$$Q_b = -\gamma C_{ox}\sqrt{\phi_{ss}} \tag{7.21}$$

where the surface potential ϕ_{ss} in weak inversion is given by [cf. Eq. (6.90)]

$$\phi_{ss} = \left[-\frac{\gamma}{2} + \sqrt{\frac{\gamma^2}{4} + V_{gb} - V_{fb}}\right]^2 \tag{7.22}$$

which is practically independent of the position y along the channel. This means that Q_b is independent of the position, and therefore from Eq. (7.20) the gate charge in weak inversion becomes

$$Q_G = -Q_B = -\frac{1}{2}WLC_{ox}\gamma^2\left[1 - \sqrt{1 + \frac{4}{\gamma^2}(V_{gb} - V_{fb})}\right]. \tag{7.23}$$

Differentiating Eq. (7.23) with respect to V_{gb} gives the gate to bulk capacitance C_{GB} in the subthreshold or weak inversion region as

$$C_{GB} = \frac{\partial Q_G}{\partial V_{gb}} = WLC_{ox}\left[1 + \frac{4}{\gamma^2}(V_{gb} - V_{fb})\right]^{-1/2} \tag{7.24}$$

where we have assumed that γ is constant independent of V_{bs}. Note that this is true only for a uniformly doped substrate. For practical MOSFET's which are nonuniformly doped, γ is bias dependent as was discussed in Chapter 5. Therefore, proper γ value and derivative need to be used.

Since in weak inversion Q_G does not depend on V_{ds},

$$C_{GS} = C_{GD} = 0. \tag{7.25}$$

Although $C_{GB}(=\partial Q_G/\partial V_{gb})$ calculated from Eq. (7.24) is inaccurate at $V_{gs} = V_{fb}$ owing to the failure of the depletion approximation used in arriving at equation (7.24), it is still used because of its simplicity.

Figure 7.2 shows a plot of three capacitances as a function of V_{gs} for two different V_{ds} of 1 and 3V. Note that the capacitance is normalized with respect to the total gate oxide capacitance C_{oxt} given by

$$\boxed{C_{oxt} = WLC_{ox} \quad \text{(F).}} \tag{7.26}$$

Maximum capacitance of a device is the gate oxide capacitance C_{oxt} which occurs in accumulation. In the inversion region, where the transistor action

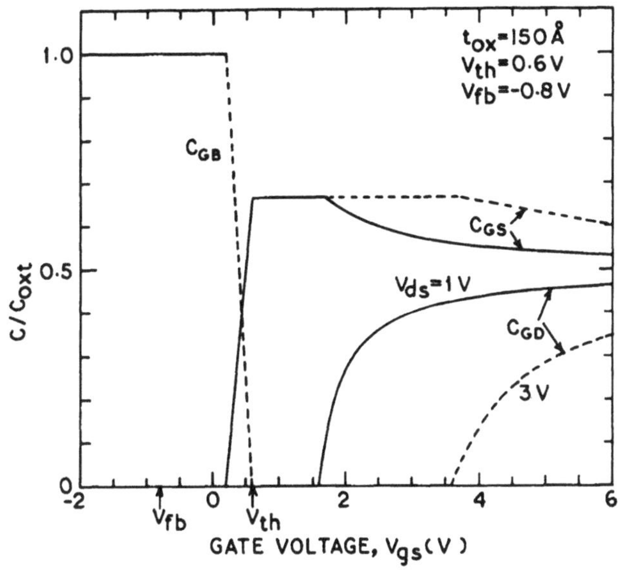

Fig. 7.2 The capacitances C_{GS}, C_{GD} and C_{GB} associated with the gate terminal of a MOSFET as a function of gate voltage V_{gs} for two values of drain voltage V_{ds}(1 V and 3 V). Simulated results are based on Eqs. (7.17), (7.19) and (7.24)

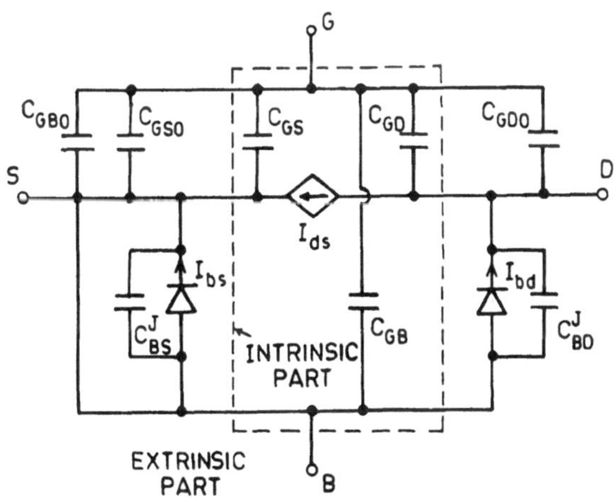

Fig. 7.3 Complete equivalent circuit of a MOSFET showing both extrinsic and intrinsic capacitances (Meyer model)

takes place, the maximum capacitance occurs in saturation and is equal to $2/3C_{oxt}$ (cf. Eq. (7.19a)].
The Meyer model can be represented by a simple equivalent circuit shown in Figure 7.3. The capacitance equations representing the Meyer model, as implemented in SPICE, are discussed in section 11.2.2.

7.1.2 Drawbacks of the Meyer Model

The Meyer model, though simple, becomes inadequate for simulating circuits like dynamic RAMs and switched capacitor circuits, which are sensitive to the capacitive component of the MOSFET currents. This can be understood as follows: Suppose the stored charge Q is a function of a single voltage V as

$$Q = f(V)$$

so that

$$i = \frac{dQ}{dt} = \frac{\partial Q}{\partial V}\cdot\frac{\partial V}{\partial t} = C(V)\frac{\partial V}{\partial t}. \tag{7.27}$$

By integrating from the present time point t_1 to the next time point t_2 we get

$$Q = \int_{t_1}^{t_2} i dt = \int_{t_1}^{t_2} C(V) dV. \tag{7.28}$$

Here comes the problem. The function $C(V)$ is not known over the time interval $\Delta t = t_1 - t_2$; what is known is its value at times t_1 and t_2. Assuming $\bar{C}(V)$ to be the average value of $C(V)$ in the time interval Δt, then we can integrate Eq. (7.28) resulting in

$$Q = \bar{C}[V(t_2) - V(t_1)]. \tag{7.29}$$

Thus, the Meyer model, as implemented in SPICE, calculates the terminal charges required for the transient analysis from the corresponding *average capacitance* whose value depends only on the value of the terminal voltages at the beginning and at the end of the time interval Δt. Thus, for example, the source charge Q_S is calculated as

$$Q_S = \bar{C}_{GS}\Delta V_{gs}, \tag{7.30}$$

where $\bar{C}_{GS}$ is the average value of C_{GS} given by

$$\bar{C}_{GS} = \tfrac{1}{2}(C_{GS}^{k+1} + C_{GS}^{k}),$$

and

$$\Delta V_{gs} = (V_{gs}^{k+1} - V_{gs}^{k}).$$

Here k and $(k + 1)$ are the two consecutive time points t_1 and t_2, respectively, having time interval Δt. Similarly,

$$Q_D = \bar{C}_{GD}\Delta V_{gd},$$

$$Q_B = \bar{C}_{GB}\Delta V_{gb}$$

so that

$$\int dQ_G = \bar{C}_{GS}\Delta V_{gs} + \bar{C}_{GD}\Delta V_{gd} + \bar{C}_{GB}\Delta V_{gb}. \tag{7.31}$$

Clearly the above is not the same as the exact expression

$$\int dQ_G = \int C_{GS}\cdot dV_{gs} + \int C_{GD}\cdot dV_{gd} + \int C_{GB}\cdot dV_{gb}. \tag{7.32}$$

Thus, in the Meyer's model, *the gate charge is obtained from a nonexact differential that over a period of time could give rise to a non-zero value of the total charge, i.e., to a charge conservation problem.* This clearly shows that charge nonconservation in the Meyer model is the result of incorrect numerical integration and not due to the inaccuracy in the model. In fact, inaccuracy of the model and charge nonconservation are two independent things.

The effects of charge nonconservation in a circuit appear as unlimited increasing or deceasing of terminal voltages [7], [16]. Note that if the time step is so small that the integration error becomes smaller than the convergence criteria, though an impractical constraint, then the non-conservation situation will not be noticed. Recently, it has been shown that the Meyer model, if properly implemented, will result in charge conservation [16, 17].

In addition to the charge nonconservation problem, the assumption of capacitance reciprocity in the Meyer model is more critical. In fact it is easy to see that the assumption of reciprocity is inconsistent with the charge conservation law [18, 19]. This can be seen as follows: Following Smedes [18], we start with the definition of gate capacitance (7.9) and use the reciprocity principle ($C_{ij} = C_{ji}$) to get

$$C_{GS} = \frac{\partial Q_G}{\partial V_{gs}} \equiv C_{SG} = \frac{\partial Q_S}{\partial V_{sg}} = -\frac{\partial Q_S}{\partial V_{gs}} \tag{7.33a}$$

$$C_{GD} = \frac{\partial Q_G}{\partial V_{gd}} \equiv C_{DG} = \frac{\partial Q_D}{\partial V_{dg}} = -\frac{\partial Q_D}{\partial V_{gd}} \tag{7.33b}$$

$$C_{GB} = \frac{\partial Q_G}{\partial V_{gb}} \equiv C_{BG} = \frac{\partial Q_B}{\partial V_{gb}} = -\frac{\partial Q_B}{\partial V_{gb}}. \tag{7.33c}$$

By differentiating the charge conservation Eq. (7.3) with respect to V_{gs}, V_{gd} and V_{gb}, respectively, and using Eq. (7.33) we find

$$\frac{\partial Q_D}{\partial V_{gs}} + \frac{\partial Q_B}{\partial V_{gs}} = 0 \tag{7.34a}$$

$$\frac{\partial Q_S}{\partial V_{gd}} + \frac{\partial Q_B}{\partial V_{gd}} = 0 \tag{7.34b}$$

$$\frac{\partial Q_D}{\partial V_{gb}} + \frac{\partial Q_S}{\partial V_{gb}} = 0. \tag{7.34c}$$

It can easily be verified that the above equations can be rewritten as

$$\begin{aligned} C_{DS} + C_{BS} &= 0 \\ C_{SD} + C_{BD} &= 0 \\ C_{DB} + C_{SB} &= 0. \end{aligned} \tag{7.35}$$

Together with the reciprocity law ($C_{ij} = C_{ji}$), this leads us to

$$C_{DS} = -C_{BS} = C_{SB} = C_{DB} = C_{BD} = C_{SD} = -C_{DS}. \tag{7.36}$$

This is only possible if all derivatives in Eq. (7.34) are zero, which indeed is in contradiction with experiments and physical intuition. For example, bulk charge is a function of the source voltage and therefore C_{BS} can not be zero. Furthermore, Eq. (7.36) implies that the channel charge must be separated in a part $Q_S(V_{gs})$ and $Q_D(V_{gd})$. Since the channel charge depends non-linearly upon both voltages, this separation is not possible. *Thus, charge nonconservation and reciprocity are mutually exclusive properties of a MOSFET charge model.*

Now if the expression for the charges as a function of terminal voltages are available, the integration of Eq. (7.27) can be carried out in the following way, which avoids all problems of charge nonconservation. Note that in general[6]

$$i_j(t) = \frac{dQ_j(t)}{dt}. \tag{7.37}$$

By integrating from t_1 to t_2 we get

$$\int_{t_1}^{t_2} i_j dt = Q_j(t_2) - Q_j(t_1) = f(v(t_2)) - f(v(t_1)). \tag{7.38}$$

Since $f(v(t_2))$ will be evaluated at the new time point t_2, one can approximate it by performing a Taylor series expansion about the voltage at the last iteration to obtain the companion model used in the Newton–Raphson iteration. The integration on the left hand side of Eq. (7.38) can easily be carried out using either trapezoidal or the Gear integration formula. Note

---

[6] The subscript j stands for G, S, D or B for the charge Q and capacitance C, as we are now dealing with the total charge or total capacitance. However, for current and voltage, the subscript j represents g, s, d and b.

that changing variables of integration from $C(V)$ to $Q(V)$ reduces numerical errors (not eliminate them), although mathematically they appear to be the same.

7.2 Charge-Based Capacitance Model

Since the terminal charge Q_j ($j = \mathrm{G, S, D, B}$) in general is a function of terminal voltages V_g, V_s, V_d and V_b, we can write the terminal current i_j as

$$i_j = \frac{dQ_j}{dt} = \frac{\partial Q_j}{\partial V_g}\frac{\partial V_g}{\partial t} + \frac{\partial Q_j}{\partial V_s}\frac{\partial V_s}{\partial t} + \frac{\partial Q_j}{\partial V_d}\frac{\partial V_d}{\partial t} + \frac{\partial Q_j}{\partial V_b}\frac{\partial V_b}{\partial t}. \tag{7.39}$$

From this equation it is evident that each terminal has a capacitance with respect to the remaining three terminals. Thus, a four terminal device will have 16 capacitances, including 4 self capacitances corresponding to its 4 terminals. Excluding the self capacitances, there will be 12 intrinsic capacitances which in general are *nonreciprocal.* The 16 capacitances form the so called *indefinite admittance matrix* (IAM). *Each element C_{ij} of this capacitance matrix describes the dependence of the charge at the terminal i with respect to the voltage applied at the terminal j with all other voltages held constant.* For example, C_{GS} specifices the rate of change of Q_G with respect to the source voltage V_s with voltages at the other terminals (V_g, V_d and V_b) held constant. Thus, in general,

$$C_{ij} = \begin{cases} -\dfrac{\partial Q_i}{\partial V_j}, & i \neq j \quad i, j = \mathrm{G, S, D, B} \\[2ex] \dfrac{\partial Q_i}{\partial V_j}, & i = j \end{cases} \tag{7.40}$$

where the signs of the C_{ij}'s are chosen to keep all of the capacitance terms positive for well-behaved devices, i.e., devices for which the charge at a node increases with an increase in the voltage at that node and decreases with an increase in the voltage at any other node. All 16 capacitances of the matrix C_{ij}, shown below, are not independent

$$C_{ij} \equiv \begin{bmatrix} C_{GG} & -C_{GD} & -C_{GS} & -C_{GB} \\ -C_{DG} & C_{DD} & -C_{DS} & -C_{DB} \\ -C_{SG} & -C_{SD} & C_{SS} & -C_{SB} \\ -C_{BG} & -C_{BD} & -C_{BS} & C_{BB} \end{bmatrix}. \tag{7.41}$$

Each row must sum to zero for the matrix to be reference-independent, and each column must sum to zero for the device description to be charge-conservative, which is equivalent to obeying KCL. One of these four capacitances, corresponding to each terminal of the device, is the self capacitance which is the sum of the remaining three capacitances. Thus, for example, the gate capacitance C_{GG} is

$$C_{GG} = C_{GS} + C_{GD} + C_{GB}. \tag{7.42}$$

The twelve *internodal* or *intrinsic capacitances* (excluding self capacitances C_{GG}, C_{DD}, C_{SS} and C_{BB}) of a MOSFET are also called the *transcapacitances*. Further, these capacitances are non-reciprocal. Thus, for example, C_{DG} and C_{GD} differ both in value and physical interpretation. Note that of the 12 transcapacitances only 9 are independent. Therefore, if we choose to evaluate $C_{GB}, C_{GS}, C_{GD}, C_{BG}, C_{BS}, C_{BD}, C_{DG}, C_{DS}, C_{DB}$ then the other three capacitances C_{SG}, C_{SD}, C_{SB} can be determined from the following relations

$$C_{SG} = C_{GB} + C_{GD} + C_{GS} - C_{BG} - C_{DG}$$

$$C_{SD} = C_{BG} + C_{BD} + C_{BS} - C_{GB} - C_{DB}$$

$$C_{SB} = C_{DG} + C_{DB} + C_{DS} - C_{GD} - C_{BD}.$$

For the sake of comparison, the corresponding C_{ij} matrix for the Meyer model is shown below.

$$\begin{bmatrix} C_{GD} + C_{GS} + C_{GB} & -C_{GD} & -C_{GS} & -C_{GB} \\ -C_{GD} & C_{GD} & 0 & 0 \\ -C_{GS} & 0 & C_{GS} & 0 \\ -C_{GB} & 0 & 0 & C_{GB} \end{bmatrix}. \tag{7.43}$$

Thus, we see that a MOSFET has capacitances that are much more complex than the Meyer model assumes. It is thus evident from Eq. (7.40) that to calculate MOSFET intrinsic capacitances we need to calculate the charges Q_G, Q_D, Q_S and Q_B as a function of node voltages, and if we take these charges as independent variables then charge conservation will be guaranteed. It should be pointed out that though the Meyer model represents an inaccurate approximation of MOSFET capacitances, it is reported to predict the high frequency capacitances more accurately than the charge based reciprocal capacitance model to be discussed in sections 7.3 and 7.4. This is because a network with non-reciprocal capacitances based on quasi-static operation can generate infinite power at infinite frequency [20]. For this reason models based on quasi-static approximation fail at very high frequencies (see section 7.5).

Channel Charge Partition. The gate and bulk charges, Q_G and Q_B respectively, can easily be obtained by integrating the corresponding charge per unit area over the area of the active gate region as is given by Eqs. (7.6) and (7.7). However, calculation of the source and drain charges Q_S and Q_D, respectively, can only be determined from the channel charge Q_I, because both source and drain terminals are in intimate contact with the channel region. It is thus necessary to partition the channel charge into a charge Q_D associated with the drain terminal and a charge Q_S associated with the source terminal, such that

$$Q_I = Q_S + Q_D. \tag{7.44}$$

Although this partition of Q_I into $Q_S + Q_D$ is not accurate physically [1], nonetheless it does leads to MOSFET capacitance model which agrees with the experimental results.

Various approaches have been used in the literature to partition Q_I into Q_S and Q_D [3]–[12], some of these are discussed by Yang [11]. These different approaches vary from an equal division of Q_I across both terminals ($Q_S = Q_D = 0.5Q_I$) [6] to a Q_I multiplied by a 'linear partioning' or 'weighted function' [3]. The approach which can rigorously be shown to be correct and which agrees with the experimental results is that proposed by Ward [3] and is based on the 1-D continuity equation.

Neglecting recombination in the channel region, the 1-D continuity equation is given by

$$\frac{\partial I(y,t)}{\partial y} = -W\frac{\partial Q_i(y,t)}{\partial t}. \tag{7.45}$$

Integrating the above equation along the channel from the source ($y = 0$) to an arbitrary point y along the channel yields:

$$\int_0^y \frac{\partial I(y',t)}{\partial y'}dy' = -W\int_0^y \frac{\partial Q_i(y',t)}{\partial t}dy'$$

or

$$I(y,t) - I(0,t) = -W\int_0^y \frac{\partial Q_i(y',t)}{\partial t}dy'. \tag{7.46}$$

Integrating again Eq. (7.46) along the whole length of the channel results in:

$$\int_0^L I(y,t)dy - \int_0^L I(0,t)dy = -W\int_0^L\int_0^y \frac{\partial Q_i(y',t)}{\partial t}dy'dy. \tag{7.47}$$

The right hand side of the above equation can be rewritten by taking the time derivative outside the integral and integrating by parts. We finally obtain

$$I(0,t) = \frac{1}{L}\int_0^L I(y,t)dy + \frac{W}{L}\frac{\partial}{\partial t}\int_0^L\left(1 - \frac{y}{L}\right)Q_i dy. \tag{7.48}$$

We now have an expression for the current at the position $y = 0$ in the channel for any time t, that is, the total current flowing through the source contact. The first term on the right hand side is the average transport current in the channel at time t; this is the DC current under quasi-static operation. If we compare Eq. (7.48) with (7.4a), it is easy to see that the charge Q_S associated with the source is

$$Q_S = -W\int_0^L \left(1 - \frac{y}{L}\right) Q_i dy. \tag{7.49a}$$

A similar expression can be derived for the drain current, where the charge Q_D associated with the drain is given by

$$Q_D = -W\int_0^L \frac{y}{L} Q_i dy. \tag{7.49b}$$

Note that Q_S and Q_D sum up to the total inversion charge Q_I in the channel. It is this charge partioning scheme represented by Eq. (7.49) which is commonly used. This approach has been criticized on the ground that it predicts non-zero drain charge in the saturation region [7]. It is argued that since the drain is insulated from rest of the device, it should have zero charge in saturation. However, this is inconsistent because in saturation it is still possible for a charging current to flow through the channel via the drain.

We will now derive the charge expressions first for the long channel devices, and then modify those charge expressions for short-channel devices. While deriving the charge expressions, both assumptions of the Meyer model are removed. The information required for calculating the charge expressions is normally available from any model used to calculate the steady-state (DC) current in a MOSFET. Thus, we can use Q_i and Q_b from the charge-sheet model [22, 23]. However, we will compute the terminal charges using the piece-wise DC current model because that is the model commonly used in SPICE. This is discussed in the next section.

7.3 Long-Channel Charge Model

In this section we will compute the terminal charges using the piece-wise DC current model discussed in section 6.4.4. The charge model, similar to the DC model, will thus have different charge equations for different regions of device operation.

Strong Inversion. The channel charge density Q_i for a long-channel device was derived as [cf. Eq. (6.79)]

$$Q_i(y) = -C_{ox}[V_{gs} - V_{th} - \alpha V(y)] \tag{7.50}$$

while the bulk charge density is given by [cf. Eq. (6.78)]

$$Q_b(y) = -C_{ox}\gamma[\delta V(y) + \sqrt{2\phi_f + V_{sb}}]. \tag{7.51}$$

Since the total charge in the system must be zero, i.e., $Q_g + Q_i + Q_b = 0$, the gate charge density Q_g becomes

$$Q_g(y) = C_{ox}[V_{gs} - V_{fb} - 2\phi_f - V(y)] \tag{7.52}$$

where V_{th} is given by Eq. (6.45), and $\alpha = (1 + \gamma\delta)$ [cf. Eq. (6.80)]. Equations (7.50)–(7.52) can be used to calculate the terminal charges using Eqs. (7.6)–(7.7) and (7.49). Let us first calculate Q_S and Q_D using Eq. (7.49). Since $Q_i(y)$ is known as a function of V, we first change the variable of integration 'dy' in Eq. (7.49) to 'dV' using Eq. (7.13). This yields

$$Q_S = -\frac{\mu_s W^2 C_{ox}}{I_{ds}} \int_{V_s}^{V_d} \left(1 - \frac{y}{L}\right) Q_i \cdot Q_i dV \tag{7.53a}$$

$$Q_D = -\frac{\mu_s W^2 C_{ox}}{I_{ds}} \int_{V_s}^{V_d} \frac{y}{L} Q_i \cdot Q_i dV. \tag{7.53b}$$

To express y in the above equations in terms of V_{ds}, we integrate Eq. (7.13) from $y = 0$ to an arbitrary point in the channel. This yields

$$y = -\frac{\mu_s W}{I_{ds}} \int_0^V Q_i dV = \frac{\mu_s W C_{ox}}{I_{ds}} (V_{gs} - V_{th} - 0.5\alpha V)V. \tag{7.54}$$

At the drain end $y = L$, and $V = V_{ds}$, so that we have

$$I_{ds} = \frac{\mu_s C_{ox} W}{L} (V_{gs} - V_{th} - 0.5\alpha V_{ds}) V_{ds}.$$

Now combining Eq. (7.53) with Eqs. (7.50) and (7.54) and carrying out the integration, we get after lengthy algebra the following expression for Q_D and Q_S in the linear region of device operation

$$Q_D = -C_{oxt}[\tfrac{1}{2}V_{gt} - \tfrac{1}{3}\alpha V_{ds} + \mathscr{A}\mathscr{B}] \tag{7.55a}$$

$$Q_S = -C_{oxt}[\tfrac{1}{2}V_{gt} - \tfrac{1}{6}\alpha V_{ds} + \mathscr{A}(1 - \mathscr{B})] \tag{7.55b}$$

where

$$\mathscr{A} = \frac{\alpha^2 V_{ds}^2}{12(V_{gt} - 0.5\alpha V_{ds})} \tag{7.56a}$$

$$\mathscr{B} = \frac{5V_{gt} - 2\alpha V_{ds}}{10(V_{gt} - 0.5\alpha V_{ds})} \tag{7.56b}$$

and

$$V_{gt} = V_{gs} - V_{th} \qquad \text{and} \qquad C_{oxt} = WLC_{ox}.$$

When $V_{ds} = 0$, we find that $Q_S = Q_D = 0.5C_{oxt}V_{gt}$ as is expected from symmetry.

The total gate charge Q_G can be obtained by integrating the gate charge density Q_g over the area of the active gate region as

$$Q_G = W\int_0^L Q_g(y)dy = \frac{\mu_s W^2}{I_{ds}}\int_{V_s}^{V_d} Q_g \cdot Q_i dV \tag{7.57}$$

where we have replaced the differential channel length 'dy' with the corresponding differential potential drop 'dV' using Eq. (7.13). Substituting Q_i and Q_g from Eqs. (7.50) and (7.52), respectively, and carrying out the integration results in the following expression for the charge Q_G

$$Q_G = C_{oxt}\left[V_{gs} - V_{fb} - 2\phi_f - 0.5V_{ds} + \frac{1}{\alpha}\mathscr{A}\right]. \tag{7.58}$$

Similarly, the total bulk charge Q_B can be written as

$$Q_B = -W\int_0^L Q_b(y)dy = -\frac{\mu_s W^2}{I_{ds}}\int_{V_s}^{V_d} Q_b \cdot Q_i dV. \tag{7.59}$$

Substituting Q_i and Q_b from Eqs. (7.50) and (6.78), respectively, and carrying out the integration yields

$$Q_B = -C_{oxt}[\gamma\sqrt{2\phi_f + V_{sb}} + (\alpha - 1)V_{ds}\mathscr{D}] \tag{7.60}$$

where

$$\mathscr{D} = \frac{3V_{gt} - 2\alpha V_{ds}}{6(V_{gt} - 0.5\alpha V_{ds})}. \tag{7.60a}$$

Note that the bulk charge consists of two terms. The first term gives the total bulk charge due to the back bias V_{sb} and is related to the threshold voltage. The second term describes additional charge induced by the drain bias. As expected, it reduces to zero when $V_{ds} = 0$. In terms of V_{th}, one can write Q_B as

$$Q_B = -C_{oxt}[V_{th} - V_{fb} - 2\phi_f + (\alpha - 1)V_{ds}\mathscr{D}].$$

It is easy to verify that the sum of Q_G, Q_S, Q_D and Q_B is zero.

Equations (7.55), (7.58) and (7.60) are charges for the linear region of the device operation. The corresponding charges in the saturation region are obtained by replacing V_{ds} in these equations with $V_{dsat}(= V_{gt}/\alpha)$ [cf. Eq. (6.82)], resulting in the following expressions for Q_S, Q_D, Q_G and Q_B in the *saturation region*

$$Q_D = \tfrac{4}{15} C_{oxt} V_{gt} \tag{7.61a}$$

$$Q_S = -\tfrac{2}{5} C_{oxt} V_{gt} \tag{7.61b}$$

$$Q_G = C_{oxt}\left[V_{gs} - V_{fb} - 2\phi_f - \frac{V_{gt}}{3\alpha}\right] \tag{7.61c}$$

$$Q_B = -C_{oxt}\left[V_{th} - V_{fb} - 2\phi_f + (\alpha - 1)\frac{V_{gt}}{3\alpha}\right]. \tag{7.61d}$$

Adding Eqs. (7.61a) and (7.61b) we find inversion charge in saturation region as

$$Q_I = Q_S + Q_D = -\tfrac{2}{3} C_{oxt} V_{gt} \tag{7.62}$$

which is the same result as obtained in the Meyer model [cf. Eq. (7.18)] assuming $Q_B = 0$. Note from Eqs. (7.61) that none of the charges in saturation depends upon V_{ds}. This is because in saturation, due to the pinch-off, the drain has no influence on the behavior of the device. Also note that the mobility degradation factor θ due to the gate field does not appear in the charge expressions. This is because of the global way of modeling the mobility, which cancels out while deriving the charges. In fact 2-D device simulators confirm the analytical results that mobility degradation has little effect on the charges [18].

The model proposed by Yang et al. [7] and Sheu et al. [12] uses the same charge expressions as discussed above; except that in their model α, is replaced by α_x which is not a simple body factor term, but is rather effective gate voltage dependent [cf. Eq. (6.171)]. Figure 7.4 shows Q_S and Q_D, as a function of V_{ds} for different $V_{gs}(> V_{th})$, for a MOSFET with parameters shown in Table 7.1. It is clear that drain and source charges generally behave the same, except that the drain charge saturates to a smaller absolute value than the source charge. This is because the potential difference between the gate and channel decreases when going from source to drain. The bulk charge as a function of V_{ds} for different $V_{gs}(> V_{th})$ are shown in Figure 7.5a while the gate charge as a function of V_{gs} is shown in Figure 7.5b.

Weak Inversion Region. Although mobile charge at the interface is small when the device is in weak inversion, still these charges are important for the simulation of switching behavior of a MOSFET. Further, in this region bulk charge behaves differently as compared to the strong inversion condition because it is now not screened from the channel.

In order to arrive at the expression for the terminal charges in the weak inversion, we will assume that current transport occurs by diffusion only as was the case while deriving the subthreshold drain current expression [21]. Indeed this is a good approximation for low gate voltages. For higher gate voltages ($> V_{th}$), the diffusion current saturates and drift transport becomes more and more important, as discussed in Chapter 6. From

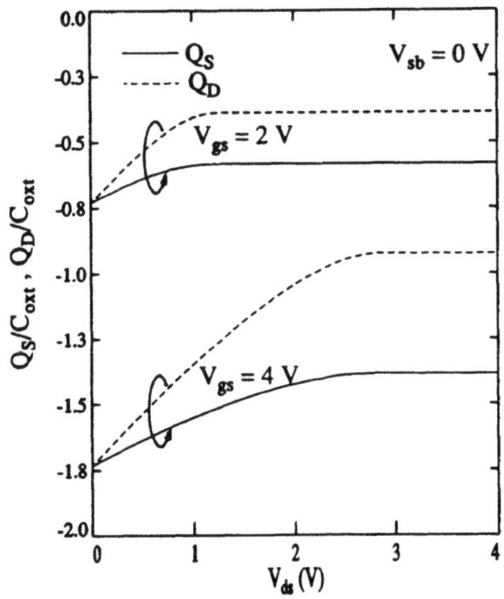

Fig. 7.4 The normalized source and drain charges Q_S and Q_D, respectively, as a function of V_{ds} for different V_{gs} in strong inversion. The normalization factor is total gate oxide capacitance $C_{oxt} = C_{ox}WL$

Table 7.1. *nMOST parameter values used for Figures 7.6–7.9*

Parameter symbol	Parameter description	Parameter value	Units
L	Effective channel length	50	μm
W	Effective channel width	50	μm
t_{ox}	Gate oxide thickness	150	Å
μ_s	Channel mobility	600	cm^2/V.s
V_{fb}	Flat band voltage	−0.8	V
V_{th}	Threshold voltage	0.6	V
N_b	Substrate concentration	3×10^{16}	cm^{-3}

Eq. (6.92) the drain current (due to diffusion) at any point y along the surface is given by

$$I_{ds} = \mu_s W V_t \frac{dQ_i}{dy} \tag{7.63}$$

which on integration yields

$$y = \frac{\mu_s W}{I_{ds}} V_t(Q_i - Q_{is}) \tag{7.64}$$

where $V_t = kT/q$ is the thermal voltage and Q_{is} is the mobile charge density at the source end [cf. Eq. (6.95)]. At the drain end $Q_i = Q_{id}$.

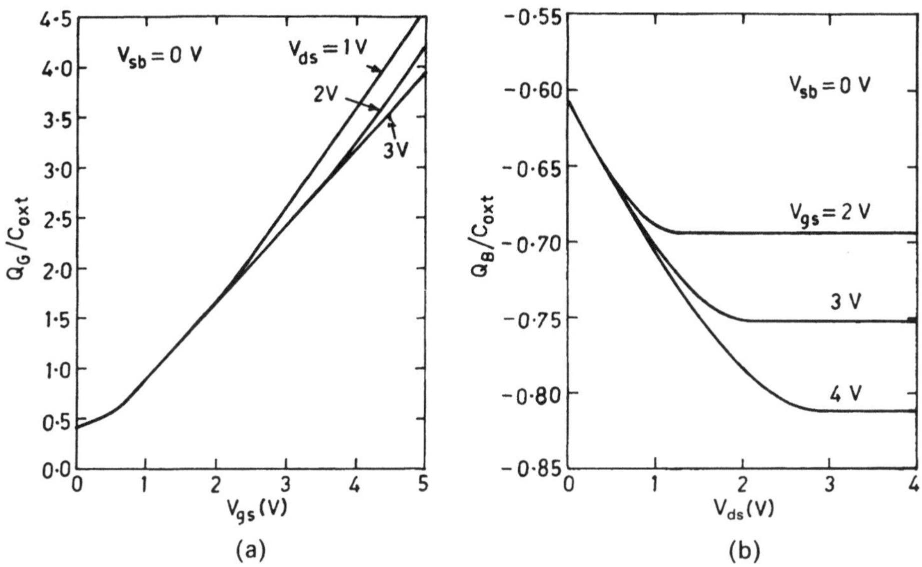

Fig. 7.5 The normalized (a) gate charge Q_G as a function of V_{gs} for different V_{ds}, (b) bulk charge Q_B as a function of V_{ds} for different V_{gs} in strong inversion

Let us first calculate the source and drain charge Q_D and Q_S, respectively. Application of Eqs. (7.63) and (7.64) with Eq. (7.49) results in

$$Q_D = \frac{W}{L}\left(\frac{\mu_s W}{I_{ds}}\right)^2 V_t^2 \int_{Q_{is}}^{Q_{id}} Q_i(Q_i - Q_{is})dQ_i \tag{7.65}$$

which on integration yields, after using Eq. (6.93) for I_{ds},

$$Q_D = \tfrac{1}{6}WL(2Q_{id} + Q_{is}). \tag{7.66}$$

We can now relate charge densities Q_{is} and Q_{id} using Eq. (6.95), resulting in the following equation for Q_D

$$Q_D = -\tfrac{1}{6}WLC_{ox}(\eta - 1)V_t \exp\left(\frac{V_{gs} - V_{th}}{nV_t}\right)(2e^{-V_{ds}/V_t} + 1) \tag{7.67}$$

where $\eta = (1 + C_d/C_{ox})$ [cf. Eq. (6.103)]. Similar procedures can be used for calculating the source charge Q_S and is found to be

$$Q_S = -\tfrac{1}{6}WLC_{ox}(\eta - 1)V_t \exp\left(\frac{V_{gs} - V_{th}}{\eta V_t}\right)(e^{-V_{ds}/V_t} + 2).$$

Note that when $V_{ds} = 0$, and $V_{gs} = V_{th}$, we have $Q_D = Q_S = -0.5C_{oxt}(\eta - 1)V_t$. From Eq. (7.67) it is evident that V_{ds} dependence on Q_S and Q_D is rather weak because for V_{ds} greater than a few V_t, the terms involving V_{ds} become negligible and we find $Q_S = 2Q_D$. Figure 7.6 shows drain and source charges

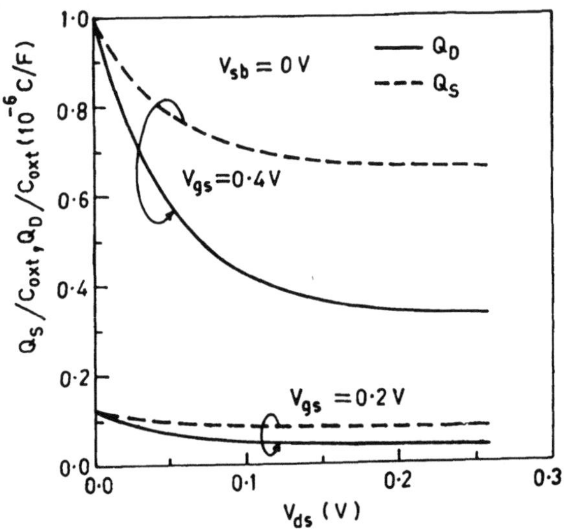

Fig. 7.6 The normalized source and drain charges Q_S and Q_D, respectively, as a function of V_{ds} for different V_{gs}, in weak inversion

in weak inversion as a function of V_{ds} for two $V_{gs}(< V_{th})$. The exponential behavior is clearly visible as well as a weak drain bias dependence. Note that the magnitudes of these charges are six orders of magnitude smaller than those in strong inversion.

From the strong inversion Eq. (7.55) note that at $V_{gs} = V_{th}, Q_D = Q_S = 0$, while from weak inversion Eq. (7.67) we get small but finite values of Q_S and Q_D. This results in a discontinuity of these charges at the transition from weak to the strong inversion. To avoid this discontinuity, the weak inversion charge must be added to the strong inversion charge. However, this does complicates the charge equations. Although it results in a continuous Q_S and Q_D, the corresponding capacitances at the transition point will still be discontinuous (see Figures 7.8–7.10). In order to avoid the discontinuity in the capacitance a smoothing function, such as Eq. (6.121) used in the drain current modeling, can be used. Because *these charges make only minor contributions to the total charges and they decrease exponentially with decreasing V_{gs}, we often assume Q_S and Q_D to be zero in weak inversion.*

Since in weak inversion the bulk charge Q_B is virtually independent of the source/drain voltage V_{ds}, we can use Eq. (7.23) for Q_B, which at the boundary of the strong inversion can be rewritten as

$$Q_B = -C_{oxt}\gamma\sqrt{2\phi_f + V_{sb}}.$$

This equation is the same as to the first term in Eq. (7.60).

If the channel charge is assumed zero ($Q_I = 0$) in the subthreshold region, the gate charge becomes equal to the bulk charge. Thus,

$$Q_G = -Q_B.$$

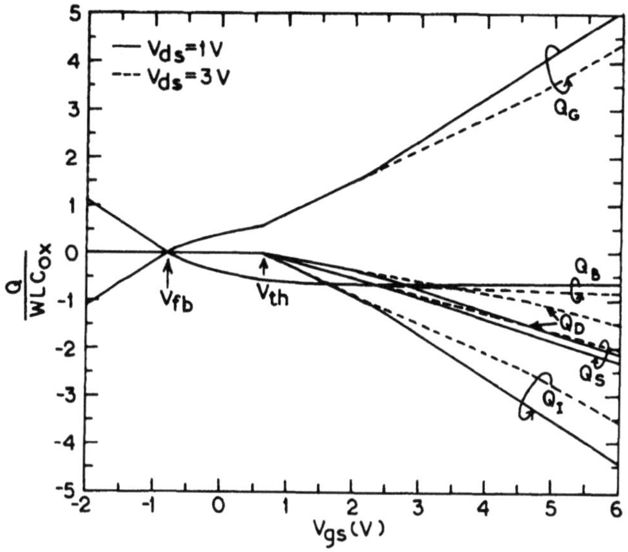

Fig. 7.7 The normalized plot of the charges Q_G, Q_B, Q_S and Q_D associated with the gate, bulk, drain and source terminals, respectively

Accumulation Region. For the sake of completeness, we discuss charges in the accumulation region of operation where $V_{gb} < V_{fb}$. In accumulation, a thin layer of majority carriers are formed at the interface, thus forming a parallel plate capacitor with the gate. In this case, the bulk charge Q_B is simply written as

$$Q_B = -C_{oxt}(V_{gs} + V_{sb} - V_{fb}). \tag{7.68}$$

Since there is no current flow, the gate charge is given by

$$Q_G = -Q_B = C_{oxt}(V_{gs} + V_{sb} - V_{fb}). \tag{7.69}$$

Figure 7.7 shows charges Q_G, Q_B, Q_S and Q_D associated with the gate, bulk, source, and drain terminals, respectively, as a function of gate voltage V_{gs} for 2 different drain voltages V_{ds} and fixed substrate bias $V_{bs} = 0\,V$. The parameters used for simulations are shown in Table 7.1. They are based on the assumption that $Q_S = Q_D = 0$ in inversion.

7.3.1 Capacitances

Using the expressions derived for various charges in different regions of device operation and the definition (7.40) we can now find the capacitances associated with a MOSFET. The mathematics, though quite basic, is however some times very lengthy. The final expression for 12 capacitances are given in Appendix F using charges given in section 7.3. Figure 7.8 shows

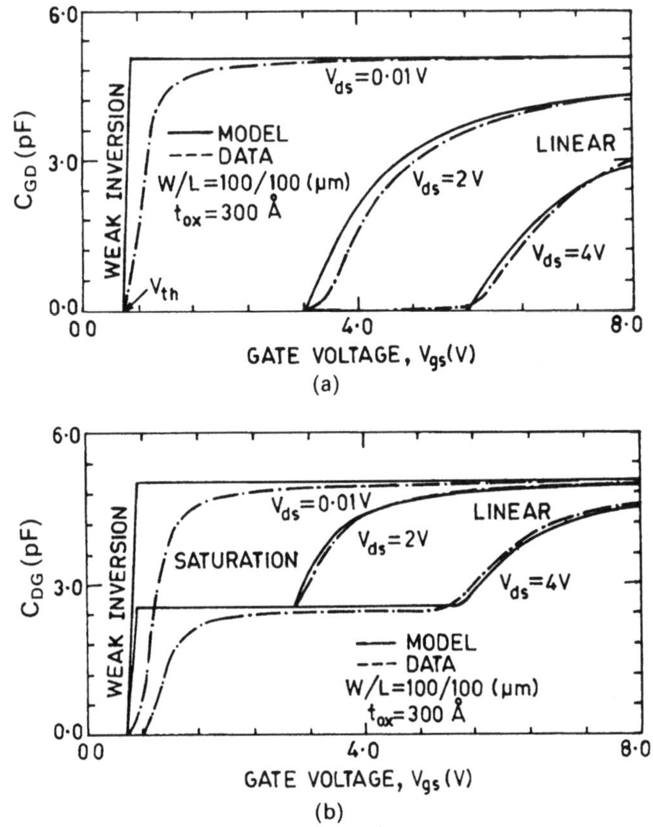

Fig. 7.8 Measured and calculated capacitance (a) gate-to-drain C_{GD} and (b) drain-to-gate C_{DG} as a function of V_{gs} with V_{ds} as a parameter

C_{GD} and C_{DG} as a function of V_{gs} for different V_{ds}. Continuous lines are from the model [cf. Eqs. (7.70)], while dashed lines are measured data for a long channel device ($W/L = 100/100\ \mu$m, $V_{th} = 0.8$ V, $t_{ox} = 305$ Å). Remember that measured capacitances also include gate overlap capacitances which have been subtracted out in the data shown in this figure. The equations for C_{GD} and C_{DG} are obtained by differentiating Q_D [Eq. (7.55a)] with respect to V_g (or V_{gs}) and Q_G [Eq. (7.58)] with respect to V_d (or V_{ds}), respectively, and using $\mathscr{A}$ and $\mathscr{B}$ defined in Eq. (7.56), that is,

$$C_{GD} = \frac{\partial Q_G}{\partial V_d} = 0.5C_{oxt}\left[-1 + \frac{1}{V_{gt} - 0.5\alpha V_{ds}}\left(\mathscr{A} + \frac{1}{3}\alpha V_{ds}\right)\right] \quad (7.70a)$$

$$C_{DG} = \frac{\partial Q_D}{\partial V_g} = 0.5C_{oxt}\left[1 + \frac{\mathscr{A}}{V_{gt} - 0.5\alpha V_{ds}}(1 - 4\mathscr{B})\right]. \quad (7.70b)$$

These are the capacitances in the linear region. The corresponding capacitances in the saturation region are obtained, either differentiating the saturation region charge [cf. Eq. (7.61)] or replacing V_{ds} with $V_{dsat} = (V_{gt}/\alpha)$ in Eq. (7.70), resulting in the following expressions

$$C_{GD} = 0 \tag{7.71a}$$

$$C_{DG} = -\tfrac{4}{15}C_{oxt}. \tag{7.71b}$$

Figure 7.8 clearly shows the non-reciprocal nature of the capacitances. It also shows that the model fits the data fairly well. Note that though the transition from linear to saturation regions is smooth, the same is not the case for transition from saturation to subthreshold regions due to our assumption of $Q_I = 0$ in the subthreshold region. Although continuity of the capacitances is desirable, particularly in small signal analysis, the discontinuity does not pose any convergence problem in SPICE. This is because the capacitance value is multiplied by the voltage difference term which vanishes as convergence is reached. Also note that $C_{DG} = 0$ in the saturation region. This is because of our assumption of the pinch-off condition ($Q_I = 0$ at the drain end, which has resulted in $V_{dsat} = V_{gt}/\alpha$) in the charge expressions. For long channel devices, this indeed is observed experimentally because pinch-off shields the channel from any further drain voltage increase. It should be pointed out that C_{GD} is most important among the gate capacitances because its effect is multiplied by the voltage gain between the drain and gate nodes due to the Miller effect.

Figure 7.9 shows C_{GS} and C_{SG} as a function of V_{gs} for different V_{ds}. Again, continuous lines are from the model [cf. Eq. (7.72)], while dashed lines are measured data for a long channel device ($W/L = 100/100\,\mu$m, $V_{th} = 0.8$ V, $t_{ox} = 305$ Å). The C_{GS} and C_{SG} are obtained by differentiating Q_G [Eq. (7.58)] with respect to V_s and Q_s [Eq. (7.55b)] with respect to V_g (or V_{gs}), respectively, and using $\mathscr{A}$ and $\mathscr{B}$ defined in Eq. (7.56), that is,

$$C_{GS} = \frac{\partial Q_G}{\partial V_s} = C_{oxt}\Bigg[-0.5 - \mathscr{A}\left(\frac{\partial \alpha}{\partial V_{bs}}\frac{1}{\alpha^2} + \frac{2}{\alpha V_{ds}}\right) - \frac{\mathscr{A}}{\alpha(V_{gt} - 0.5\alpha V_{ds})}\left(-1 + \frac{\partial V_{th}}{\partial V_{bs}} + 0.5\alpha + 0.5\frac{\partial \alpha}{\partial V_{bs}}V_{ds}\right)\Bigg] \tag{7.72a}$$

$$C_{SG} = \frac{\partial Q_S}{\partial V_g} = 0.5C_{oxt}\left[1 - \frac{\mathscr{A}}{V_{gt} - 0.5\alpha V_{ds}}(3 - 4\mathscr{B})\right]. \tag{7.72b}$$

In the saturation region we have

$$C_{GS} = C_{oxt}\left[-1 - \frac{1}{3\alpha}\left(-1 + \frac{\partial V_{th}}{\partial V_{bs}}\right) - \frac{V_{gt}}{3\alpha^2}\frac{\partial \alpha}{\partial V_{bs}}\right] \tag{7.73a}$$

$$C_{SG} = \tfrac{1}{5}C_{oxt}. \tag{7.73b}$$

Again the non-reciprocal nature of the capacitance is self evident.

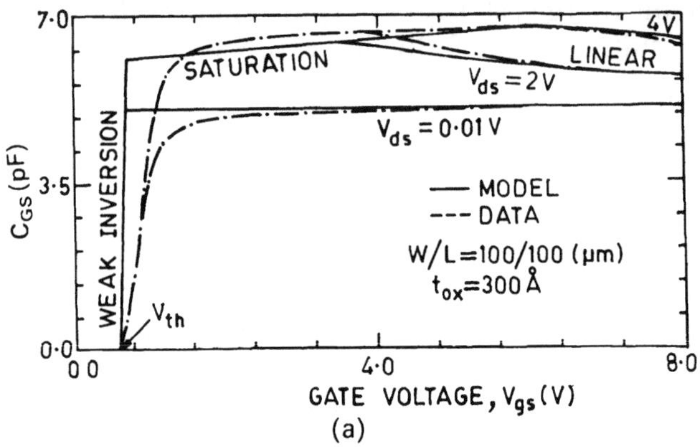

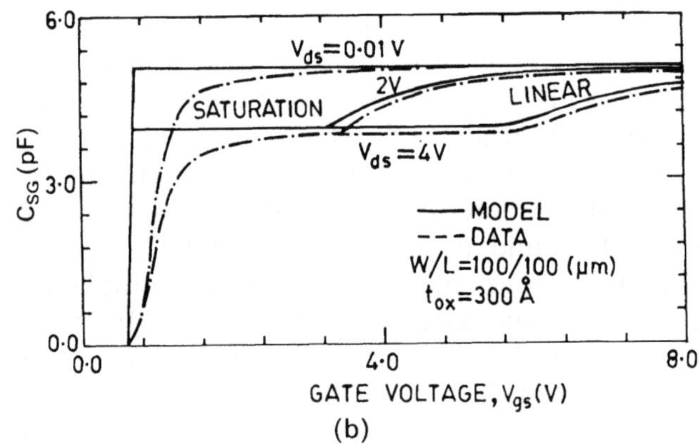

Fig. 7.9 Measured and calculated capacitance (a) gate-to-source C_{GS} and (b) source-to-gate C_{SG} as a function of V_{gs} with V_{ds} as a parameter

The gate-to-bulk capacitance C_{GB} is shown in Figure 7.10 as a function of V_{gs} for different V_{ds}. The model equation (continuous line) for C_{GB} is given in Appendix F. Although this capacitance is much smaller in strong inversion, it is the main capacitance in weak inversion and accumulation.

Figure 7.11 shows plots of nine internodal capacitances as a function of V_{ds}. The capacitances are normalized to the total gate capacitance $C_{oxt}(= WLC_{ox})$. For the sake of clarity, these capacitances are plotted at one bias, $V_{ds} = 3$ V and $V_{bs} = 0$ V. Note from this figure that the capacitances C_{DS} and C_{SD} are negative. This shows that MOS capacitors are not only non-reciprocal but are negative too. This negative capacitance could be explained as follows. Consider C_{DS} when the device is biased with say $V_{ds} = 1$ V. This capacitance is the result of a small change in the drain charge due to change in the source voltage keeping all other voltages constants. From Eq. (7.50) it is evident that a small increase in the source voltage will result in an increase in the inversion charge Q_I, i.e., the total number of mobile electrons in the channel will increase. Since the device is biased

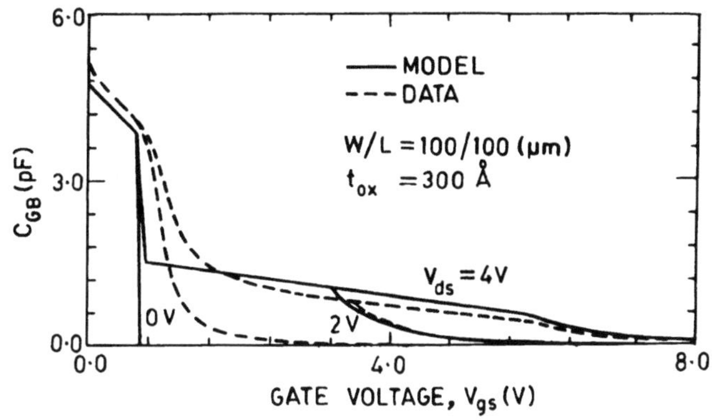

Fig. 7.10 Measured and calculated gate-to-bulk capacitance C_{GB} as a function of V_{gs} with V_{ds} as parameter

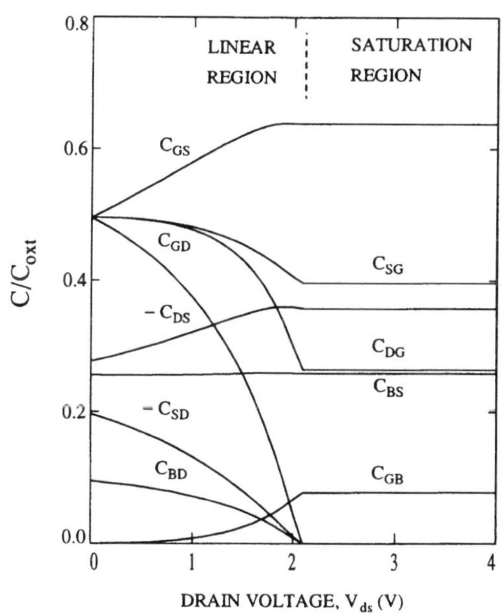

Fig. 7.11 Normalized plots of 9 internodal capacitances versus drain voltage at $V_{gs} = 3.0$ V and $V_{sb} = 0$ V

symmetrically, some of this increase in charge will be supplied by the drain, and if the drain supplies positive charge when the source voltage increases, a negative capacitance is observed by definition [cf. Eq. (7.40)].

Also note from Figure 7.11 that $C_{BS} \neq C_{BD}$ at $V_{ds} = 0$ V, although by symmetry they should be equal. The reason for this discrepancy is the value of δ (in α) used for the square root approximation (cf. section 6.4.3). By substituting $V_{ds} = 0$ in Eqs. (F.3a) and (F.3b) (Appendix F) for C_{BS} and

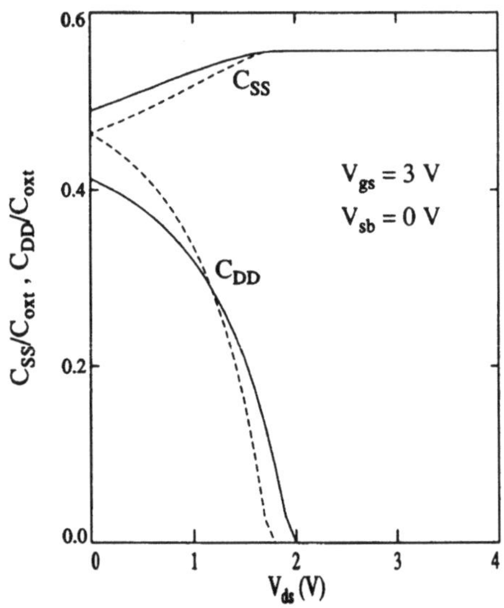

Fig. 7.12 Normalized plot of the drain-to-source and source-to-drain capacitance C_{SD} and C_{DS}, respectively, for two different expressions for δ function. Solid lines are based on δ value from Eq. (6.73), while dashed lines correspond to δ given by Eq. (6.70)

C_{BD}, respectively, we get

$$C_{BS} = C_{oxt}\left[\frac{\partial V_{th}}{\partial V_{bs}} - 0.5(\alpha - 1)\right]$$
$$C_{BD} = 0.5C_{oxt}(\alpha - 1). \tag{7.74}$$

At $V_{ds} = 0$ V, we get $C_{BD} = C_{BS} = 0.5\gamma\delta$ provided we assume $\delta = 0.5/\sqrt{2\phi_f + V_{sb}}$ [cf. Eq. (6.71)] in the bulk charge approximation. For the drain current modeling it is common practise to slightly modify the value for δ to obtain better fits in the drain current versus drain voltage plot (cf. section 6.5). However, this will lead to a small discontinuity in the capacitance. This difference is more evident when we plot drain and source capacitances C_{DD} and C_{SS}, respectively, as a function of V_{ds}. This is shown in Figure 7.12, where dashed lines assume Eq. (6.71) for δ, while continuous lines assume Eq. (6.73) for δ. The situations in which these discrepancies arise have comparatively small capacitances, therefore, it is not the cause of any significant error in circuit simulation when all capacitances at a node point are added together.

7.4 Short-Channel Charge Model

In the long channel model discussed in the previous section we have neglected velocity saturation, channel length modulation and series resistance, as these effects are important only for short-channel devices (cf.

section 6.7). As in the case of drain current calculations, we need to take these effects into account while calculating charges for short channel devices [14, 15], [18], [25]. Indeed the final charge equations become more complex.

Often for simulating short-channel capacitances, the long channel charge model has been used by modifying the body factor α [7], [12]. Thus, in the model proposed by Yang et al. [7], the α term in the long channel charge expressions (cf. section 7.3) is replaced by $\alpha_x = \alpha_1 + \alpha_2(V_{gs} - V_{th})$ where α_1 and α_2 are short-channel fitting parameters [7]. They also assume $Q_D = 0$ in the saturation region due to the fact that the channel is isolated from the drain. In the BSIM model (SPICE Level 4 model) [12], the α term in the long channel charge expressions is replaced by α_x such that $\alpha_x = \alpha(1 + \theta(V_{gs} - V_{th}))$. In this case α_x is no longer a simple body factor term, but is now effective gate voltage dependent, similar to the Yang et al. [7] model. However, to arrive at more accurate charge and capacitance expressions for short-channel devices, one must take into account short-channel effects such as carrier velocity saturation, channel length modulation and source/drain series resistance. We will now show how to include these effects in the charge equations, which in turn will be used for the derivation of short-channel capacitances.

Recall that I_{ds} for short-channel devices in the linear region is given by (cf. section 6.7.1)

$$I_{ds} = WC_{ox}(V_{gt} - \alpha V)\frac{\mu_s \mathscr{E}_y}{1 + \mathscr{E}_y/\mathscr{E}_c}. \tag{7.75}$$

Replacing $\mathscr{E}_y$ by $|dV/dy|$ and rearranging we get

$$dy = \left[\frac{\mu_s W C_{ox}}{I_{ds}}(V_{gt} - \alpha V) - \frac{1}{\mathscr{E}_c}\right] dV. \tag{7.76}$$

Integrating this equation yields

$$y = \left[\frac{\mu_s W C_{ox}}{I_{ds}}(V_{gt} - 0.5\alpha V) - \frac{1}{\mathscr{E}_c}\right] V. \tag{7.77}$$

Substituting $y = L$ and $V = V_{ds}$ (at the drain end) in the above equation permits solution for I_{ds}. Remember that $\mathscr{E}_c = v_{sat}/\mu_s$ [cf. Eq. (6.158)], where μ_s depends upon S/D resistance.

Let us first calculate Q_D and Q_S. Following the same procedure as was used for long channel devices (cf. section 7.3), we get the following expressions for the source and drain charges in the linear region of device operation

$$Q_D = -C_{oxt}[\tfrac{1}{2}V_{gt} - \tfrac{1}{3}\alpha V_{ds} - \mathscr{A}'\mathscr{B}'] \tag{7.78a}$$

$$Q_S = -C_{oxt}[\tfrac{1}{2}V_{gt} - \tfrac{1}{6}\alpha V_{ds} + \mathscr{A}'(1 + \mathscr{B}')] \tag{7.78b}$$

where

$$\mathscr{A}' = \mathscr{A}\left(1 + \frac{V_{ds}}{L\mathscr{E}_c}\right) \tag{7.79a}$$

$$\mathscr{B}' = \frac{1}{2L\mathscr{E}_c}\mathscr{B}\left(1 + \frac{V_{ds}}{L\mathscr{E}_c}\right) \tag{7.79b}$$

and $\mathscr{A}$ and $\mathscr{B}$ itself are given by Eqs. (7.56a) and (7.56b), respectively. Comparing the above equations with long channel Q_S and Q_D equations (7.55) we see that the two equations have the same form, except that the auxiliary functions $\mathscr{A}'$ and $\mathscr{B}'$ now contain a velocity saturation factor. For the long-channel case, when the product $L\mathscr{E}_c$ is very large, Eq. (7.78) reduces to Eq. (7.55) as is expected.
The remaining charges can also be derived in a similar way as for long channel devices. Thus, the gate charge for short-channel devices can be derived as

$$Q_G = \frac{\mu_s W^2}{I_{ds}}\int_{v_s}^{V_d} Q_g \cdot Q_i dV - \frac{W}{\mathscr{E}_c}\int_{V_s}^{V_d} Q_g dV. \tag{7.80}$$

Substituting Q_i and Q_g from Eq. (7.50) and (7.52) and carrying out the integration we get, after lengthy algebra, the following equation for Q_G in the linear region

$$Q_G = C_{oxt}\left[V_{gs} - V_{fb} - 2\phi_f - 0.5V_{ds} + \frac{1}{\alpha}\mathscr{A}'\right]. \tag{7.81}$$

Here again, for long channel devices the above equation reduces to Eq. (7.58). Similarly one can derive the bulk charge expression as

$$Q_B = -C_{oxt}[\gamma\sqrt{2\phi_f + V_{sb}} + (\alpha - 1)V_{ds}\mathscr{D}'] \tag{7.82}$$

where

$$\mathscr{D}' = \mathscr{D} - \frac{1}{12(V_{gt} - 0.5\alpha V_{ds})}\frac{\alpha V_{ds}}{L\mathscr{E}_c} \tag{7.82a}$$

and $\mathscr{D}$ is given by Eq. (7.60a).
Recall that while deriving the long-channel charges in saturation, we simply replaced the drain voltage V_{ds} in the linear region charge expressions by the drain saturation voltage V_{dsat}. However, for short-channel devices, where velocity saturation and channel length modulation (CLM) become important, the charges in saturation consists of two components. One is the charge near the source region (region I in Figure 7.13) where the gradual channel approximation (GCA) can be applied and the other is charge near the drain end (region II in Figure 7.13) where carrier velocity saturates.

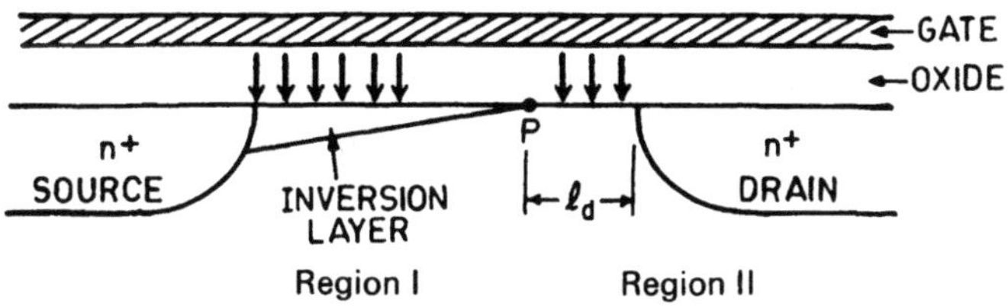

Fig. 7.13 Two-section model for calculating short-channel capacitance in saturation

Thus, in general

$$Q_j(\text{saturation}) = Q_{j1}(\text{linear})|_{V_{ds} \to V_{dsat}} + Q_{j2}(\text{over the distance } l_d)$$

where l_d is the CLM region near the drain end (cf. section 6.7.3). Assuming that over the distance l_d carriers travel with saturated velocity, we can write Q_{j2} as

$$Q_{j2} = \frac{I_{ds}}{v_{sat}} l_d.$$

This two section model does create a discontinuity in the capacitances from linear to saturation regions, similar to the case of drain current modeling. Therefore, often Q_{j2} is ignored for short-channel modeling, unless one can use smoothing functions such as discussed in section 6.7.4.
The effect of including velocity saturation in the charge expressions is a reduction in the amount of charge from its long channel value, which intuitively makes sense, because carriers are velocity saturated. This is shown in Figure 7.14 for Q_S and Q_D with and without velocity saturation,

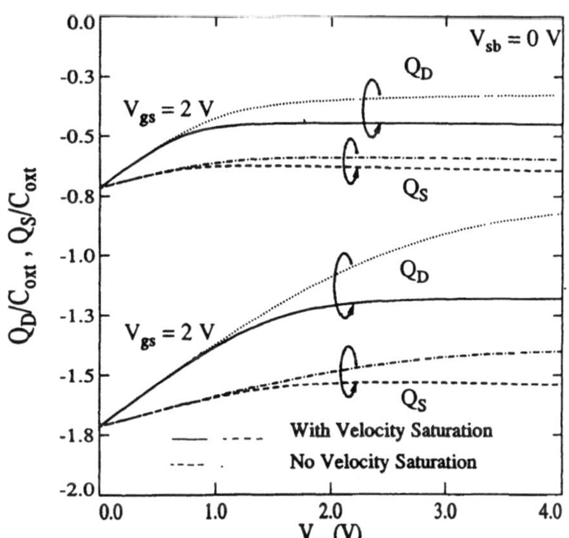

Fig. 7.14 The normalized source and drain charges, Q_S and Q_D, respectively, as a function of V_{ds} for different V_{gs}, with and without velocity saturation

respectively, and assuming $l_d = 0$. Although the effect of S/D resistance is taken into account it is possible to include its effect externally [18].
In the weak inversion, Q_I, hence Q_S and Q_D, is assumed zero, similar to the long channel case. This means that $Q_G = -Q_B$ in the weak inversion. The bulk charge Q_B for short channel devices is still given by Eq. (7.23), but with the long channel body factor γ replaced by the effective γ, which takes into account the reduction in the bulk charge density due to short-channel and narrow-width effects as discussed in Chapter 5.

7.4.1 Capacitances

Once charges are known, the corresponding capacitances can easily be calculated using the same procedure as discussed earlier for the long channel case. The mathematics is basic, but lengthy. We will not derive the final expressions for the capacitances. Instead, here we will show some experimental data for short channel devices and compare their behavior with long channel devices.
It should be pointed out that unlike the long channel capacitances, the short channel capacitance measurement is not a trivial task. This is because of very small values of the capacitances involved (in the aF range); the details of measurements are discussed in section 9.7. Moreover, for short channel devices, due to the large steady-state current (I_{ds}) it is very difficult to separate out small transient currents due to the capacitances associated with the source and drain terminals. For this reason, only short channel capacitances that have been measured and reported todate are the gate capacitances C_{GS}, C_{GD} and C_{GB}. Figure 7.15 shows measured C_{GS} (normalized to C_{oxt}) for an n-channel LDD MOSFET with $W/L = 50/0.65$ and $t_{ox} =$ 105 Å. The measured data for long channel device $W/L = 50/50$ is also shown for comparison. These are devices fabricated using 0.75 μm CMOS technology with $\Delta L = 0.25\,\mu$m. Note that in the linear region the short channel C_{GS} is larger than the long channel C_{GS}, which is more evident at higher V_{ds}. This is due to the velocity saturation effect, which causes Q_I to be proportional to I_{ds}, and hence modulating V_s has an additional effect on Q_I through change in I_{ds}. In the saturation region, the short channel C_{GS} decreases with increasing V_{ds} due to the CLM effect, while the long channel C_{GS} is independent of V_{ds} in saturation. Unlike constant C_{GS} in the cut-off region ($V_{gs} < V_{th}$) for long channel devices, the short channel C_{GS} increases due to channel side fringing field effect at the source end.
The short channel C_{GD} is shown in Figure 7.16. For comparison, long channel C_{GD} is also shown. Here again changes in short channel C_{GD} behavior can be explained by velocity saturation, channel length modulation and channel fringing field effect. Figure 7.17 shows C_{GB} as a function of

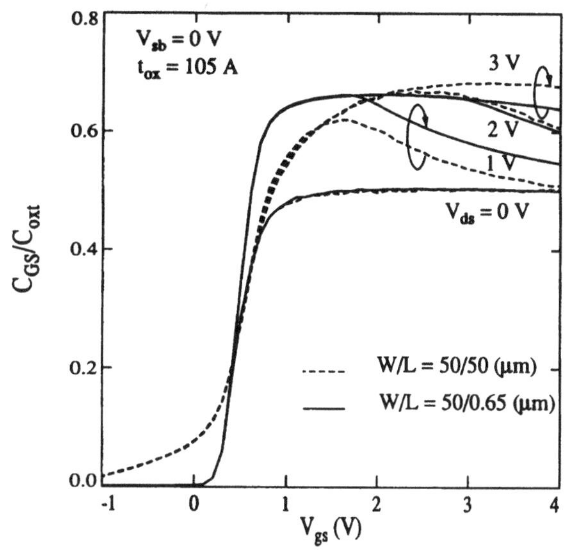

Fig. 7.15 Measured short and long channel gate-to-source capacitance C_{GS} as a function of V_{gs} with V_{ds} as a parameter

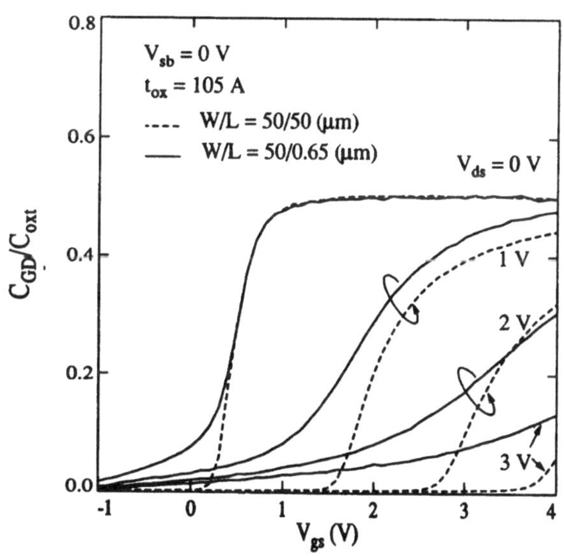

Fig. 7.16 Measured short and long chemical gate-to-drain capacitance C_{GD} as a function of V_{gs} with V_{ds} as a parameter

gate voltage. Note that higher V_{ds} results in smaller C_{GB} in the cut-off region. As V_{ds} increases, more bulk charge will be associated with the drain junction which results in less bulk charge available to modulate the gate charge. In the strong inversion region, the short channel C_{GB} is much smaller than the long channel C_{GB} for the same reason. In all cases the

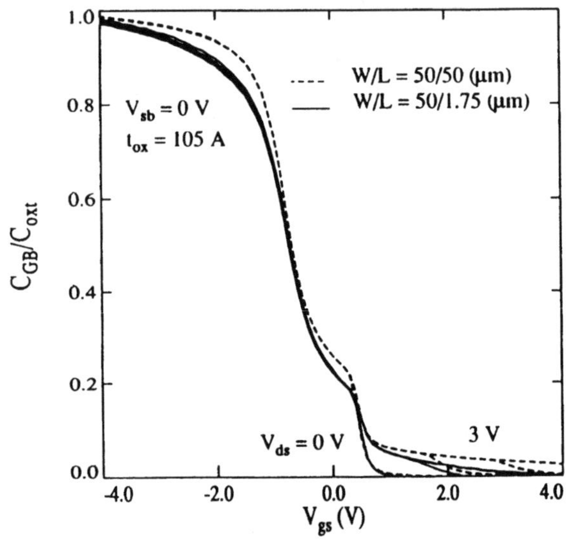

Fig. 7.17 Measured short and long channel gate-to-source capacitance C_{GB} as a function of V_{gs} with V_{ds} as a parameter

short channel gate capacitances varies more gradually from one region to the other as compared to the long channel device.

The measured capacitances shown in Figures 7.15–7.17 includes the overlap capacitances and as such, they are not intrinsic capacitances. Note from Figure 7.16, the overlap capacitance (measured capacitance in accumulation) is drain and gate bias dependent. This is true particularly for short channel LDD devices [18]. However, no such bias dependent overlap is generally observed in short-channel conventional source/drain junctions. The bias dependence of the overlap capacitance is due to the modulation of the

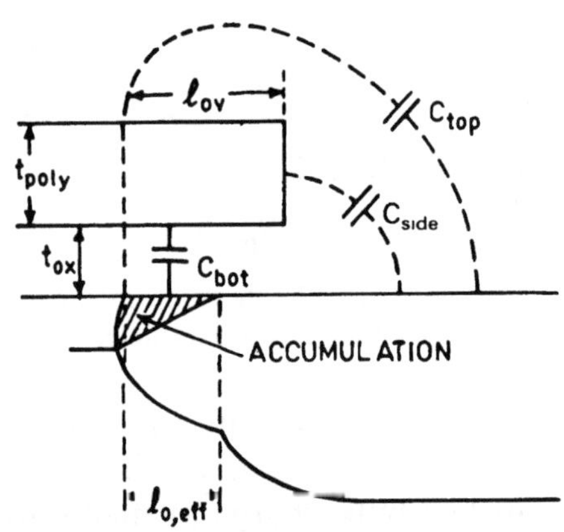

Fig. 7.18 Different components constituting MOSFET overlap capacitance in a LDD device. (After Smedes [18])

lightly doped n-region. It has been modeled using parallel combination of three components associated with the bottom, the sidewall and the top of the gate $C_{\text{bot}}, C_{\text{side}}$ and C_{top}, respectively, and is given by (see Figure 7.18) [18]

$$C_{ovl} = WC_{ox}\left[l_{o,\text{eff}} + t_{ox}\ln\left(1 + \frac{t_{\text{poly}} + l_{ov}}{t_{ox}}\right)\right] \tag{7.83}$$

with

$$l_{o,\text{eff}} = \min\left\{\frac{V_{gs} - V_{fb} - V_{\text{dsat}}}{V_{ds} - V_{\text{dsat}}}\cdot l_{ov}, l_{ov}\right\} \tag{7.84}$$

where l_{ov} is the geometrical overlap.

7.5 Limitations of the Quasi-Static Model

The MOSFET charge and capacitance models discussed so far are based on the quasi-static assumption; that is, terminal voltages vary sufficiently slowly so that the stored charge (Q_G, Q_B, Q_S and Q_D) can follow voltage variations. It has been found that for much of the digital circuit work the quasi-static model gives acceptable accuracy if the rise time t_r of the waveforms involved is such that [1]

$$t_r < 15\tau_t \tag{7.85}$$

where τ_t is the transit time associated with the DC operation of the device. It is defined as the average time the inversion carriers take to travel the length of the channel, that is,

$$\tau_t = \frac{|Q_I|}{I_{ds}}. \tag{7.86}$$

Using Eq. (7.62) for Q_I and Eq. (6.84) for I_{ds} in saturation, we get[7]

$$\tau_t = \frac{4}{3}\alpha\frac{L^2}{\mu_s(V_{gs} - V_{th})} = \frac{4}{3}\alpha\frac{L^2}{V_{dsat}}. \tag{7.87}$$

This shows that transit time is proportional to L^2. The shorter the L, the smaller the transit time, and thus the higher the speed. If the carriers are velocity saturated then Eq. (7.87) becomes invalid and one needs to use Q_I and I_{ds} equations discussed in section 6.7. However, a simple estimate for τ can still be made assuming carriers are moving from source to drain with

---

[7] In the linear region, under the condition of $V_{ds} = 0$, we have $Q_I = WLC_{ox}(V_{gs} - V_{th})$ and therefore $\tau \approx L^2/V_{ds}$.

their scatter limited saturation velocity v_{sat} for the whole length of the channel rather than only part of the channel. Since carriers cannot move faster than v_{sat}, the time required for the drain current to respond to the changes in the gate voltage is simply v_{sat}/L. Thus, in general

$$\tau_t > \frac{L}{v_{sat}}. \tag{7.88}$$

Assuming $L = 1\,\mu m$ and $v_{sat} = 10^7$ cm/s, the transit time is around 10 ps. For a typical ring oscillator circuit with 1 μm channel length MOSFETs, the measured delay is of the order of 10 ns. This shows that switching is limited by the parasitic capacitances rather than the time required for the charge redistribution within the transistor itself. Thus, quasi-static operation is good enough for most of the cases.

It should be pointed out that Eq. (7.85) is only a rough rule of thumb and often, due to the significant extrinsic parasitic capacitances, this rule is not restrictive. In fact the parasitic capacitances can mask the error due to the quasi-static assumption. However, *if parasitic capacitances are indeed low and input changes in the waveforms are too fast then the quasi-static model will break down.* In such situations, one way to extend the quasi-static model is to consider the device as a connection of several sections, each section being short enough to be modeled quasi-statistically [1]. However, more correctly, one needs to include time dependence in the basic charge equations. The resulting analysis is called non-quasistatic (NQS) analysis. The NQS is not covered here and interested readers are referred to the references cited [1], [20], [30]–[34].

7.6 Small-Signal Model Parameters

In this section we will discuss MOSFET small-signal parameters discussed in section 3.2.1, namely g_m, g_{ds} and g_{mbs}. These parameters are required for the small-signal analysis. In addition they are also required for linearizing nonlinear drain current models. The output conductance g_{ds} and transconductance g_m are important parameters in analog circuit design. As was pointed out earlier, these parameters can easily be derived from the device drain current model discussed in Chapter 6. This means that I_{ds} equations must be differentiable with respect to all terminal voltages. This is also important for SPICE convergence process since discontinuous derivatives can result in nonconvergence of the solution.

For the sake of simplicity, let us consider the I_{ds} equation discussed in section 6.4.4. Application of definition (3.11) to the drain current Eq. (6.47) and assuming μ_s is constant independent of V_{gs} (to first order), yields

$$g_{ds} = \begin{cases} \beta(V_{gt} - \alpha V_{ds}) & \text{(linear region, } V_{ds} \le V_{dsat}) \\ 0 & \text{(saturation region, } V_{ds} > V_{dsat}). \end{cases} \tag{7.89}$$

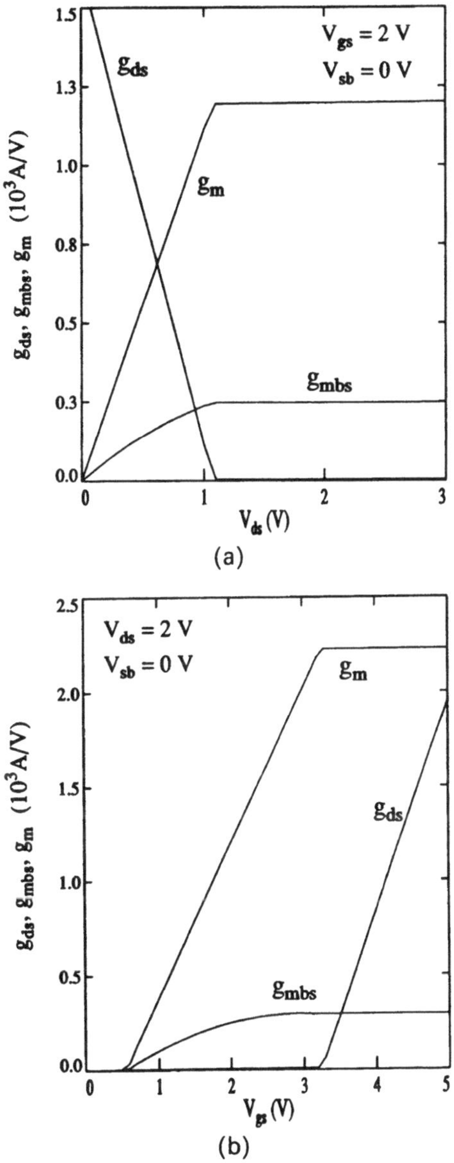

Fig. 7.19 MOSFET small-signal parameters (conductances) as a function of (a) drain voltage V_{ds} (b) gate voltage V_{gs} based on Eqs. (7.89), (7.90) and (7.91)

Note that in saturation $g_{ds} = 0$, while in practise g_{ds} has nonzero values in saturation and depends upon both V_{gs} and V_{ds}. This can be seen from Figure 7.19 where g_{ds} versus V_{ds} is plotted for $V_{gs} = 2$V for the parameter shown in Table 7.1 Nonzero g_{ds} in saturation is generally obtained from an I_{ds} equation containing CLM factor l_d, which is gate and drain bias dependent as discussed in section 6.7.3. However, as discussed earlier, this does not insure smooth g_{ds}. Since the output conductance depends upon the slope of $I_{ds} - V_{ds}$, an I_{ds} model that fits the data well in the linear and saturation regions will not necessarily give a good fit to the g_{ds} data (see section 10.3.2; Figures 10.3 and 10.4). The Eq. (6.208), which insures continuity of the current in all regions of operation through smoothing functions in the transition regions, insures smooth conductance and if proper models are chosen for V_{dsat} and l_d one can achieve a good fit of the g_{ds} model to the data as shown in Figure 7.20. Note the increase in the measured conductance (shown as circles) at higher V_{ds}. This increase is due to hot-electron effect as discussed in section 6.7.3 [cf. Eq. (6.209)] and is not taken into account in the model (continuous line).

If we use definition (3.8) and Eq. (6.57) and assume μ_s is constant independent of V_{gs} (to first order), we obtain

$$g_m = \begin{cases} \beta V_{ds} & \text{(linear region, } V_{ds} \leq V_{dsat}) \\ \dfrac{\beta}{2\alpha} & \text{(saturation region, } V_{ds} > V_{dsat}). \end{cases} \tag{7.90}$$

This shows that g_m in the linear region is independent of V_{gs}, while in reality it does depends on V_{gs} due to the mobility degradation factor.

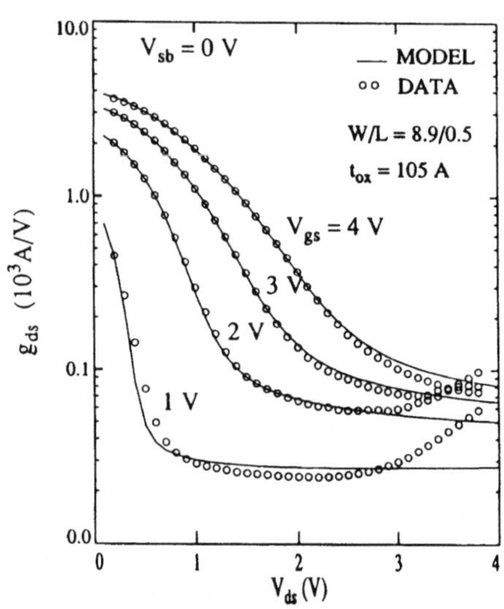

Fig. 7.20 Measured (circles) and calculated (continuous lines) output conductance g_{ds} of a short-channel nMOST as a function of V_{ds} for different V_{gs}. Calculated results are based on model Eq. (6.208)

The measured and calculated g_m shown in Figures 7.21 where circles are measured conductance while solid lines are those obtained from drain current model discussed in section 6.7.5. Note that both the measured and calculated conductances are continuous at the transition point because the model is inherently continuous in all regions of device operation, as discussed earlier. If the second derivative of I_{ds} is not continuous at the transition point then one needs to use model parameter extraction procedure discussed in section 10.3.2, although it still does not insure continuous and smooth conductances.

Similarly, one can obtain g_{mbs} by differentiating Eq. (6.47) with respect to V_{bs} giving

$$g_{mbs} = \begin{cases} \beta V_{ds}\left(-\dfrac{\partial V_{th}}{\partial V_{sb}} + 0.5\dfrac{\partial \alpha}{\partial V_{sb}}\right) & \text{(linear region, } V_{ds} \leq V_{dsat}) \\ \dfrac{\beta}{2\alpha}\left(-\dfrac{\partial V_{th}}{\partial V_{sb}} + 0.5\left(\dfrac{V_{gs} - V_{th}}{\alpha}\right)\dfrac{\partial \alpha}{\partial V_{sb}}\right) & \text{(saturation region, } V_{ds} > V_{dsat}). \end{cases} \tag{7.91}$$

Combining Eqs. (7.90) and (7.91) it is easy to see that at low V_{ds} we get

$$\frac{g_{mbs}}{g_m} = b = \frac{\partial V_{th}}{\partial V_{sb}}. \tag{7.92}$$

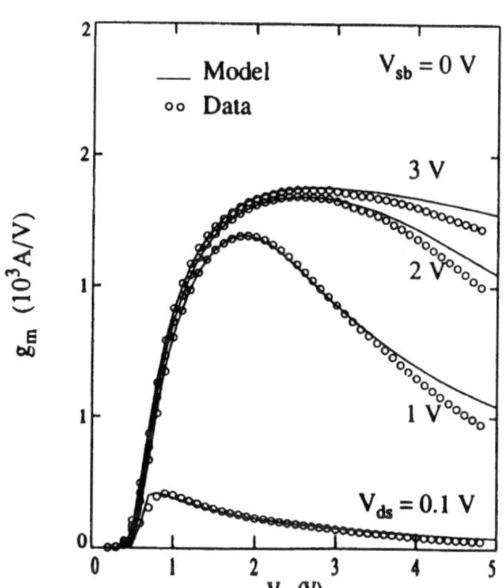

Fig. 7.21 Measured (circles) and calculated (continuous lines) transconductance g_m as a function of V_{gs} for different V_{ds} for the device whose g_{ds} is shown in Figure 7.20

The conductance g_{ds} and the transconductance g_m and g_{mbs} based on Eqs. (7.89), (7.90) and (7.91), respectively, as a function of V_{ds} and V_{gs} are shown in Figure 7.19. The sharp corners are artifact of the model. In real devices there will be smooth transitions for the conductance values.

References

[1] Y. P. Tsividis, *Operation and Modeling of the MOS Transistor*, McGraw-Hill Book Company, New York, 1987.

[2] J. Meyer, 'MOS models and circuit simulation', RCA Review, 32, pp. 42–63 (1971).

[3] D. Ward and R. W. Dutton, 'A charge-oriented model for MOS transistor capacitances', IEEE J. Solid-State Circuits, SC-13, pp. 703–707 (1978). Also see D. Ward, 'Charge-based modeling of capacitances in MOS transistors', Stanford University Tech. Rep., G201-11, 1982.

[4] S. Y Oh, D. E. Ward, and R. W. Dutton, 'Transient analysis of MOS transistors', IEEE Trans. Electron Devices, ED-27, pp. 1571–1578 (1980).

[5] M. F. Sevat, 'On the channel charge division in MOSFET modeling', in: *Digest of IEEE Int. Conf. on Computer-Aided Design*, ICCAD-87, pp. 204–207 (1987).

[6] G. W. Taylor, W. Fichtner, and J. G. Simmons, 'A description of MOS internodal capacitances for transient simulation', IEEE Trans. Computer-Aided Design, CAD-1, pp. 150–156 (1982).

[7] P. Yang, B. D. Epler, and P. Chatterjee, 'An investigation of the charge conservation problem for MOSFET circuit simulation', IEEE J. Solid-State Circuits, SC-18, pp. 128–138 (1983).

[8] K. Y. Tong, 'AC model for MOS transistors from transient-current computation', IEE Proc., Vol. 130, Pt I, pp. 33–36 (1983).

[9] J. G. Fossum, H. Jeong, and S. Veeraraghavan, 'Significance of the channel-charge portion in the transient MOSFET model', IEEE Trans. Electron Devices, ED-33, pp. 1621–1623 (1986).

[10] B. J. Sheu, D. L. Scharfetter, C. M. Hu, and D. O. Pederson, 'A compact IGFET charge model', IEEE Trans. on Circuits and Systems, CAS-31, pp. 745–748 (1984).

[11] P. Yang, 'Capacitance modeling for MOSFET', in: *Circuit Analysis, Simulation and Design* (A. E. Ruehli, ed.), Elsevier, Amsterdam, 1986.

[12] B. J. Sheu, W. J. Hsu, and P. K. Ko, 'An MOS transistor charge model for VLSI design', IEEE Trans. Computer-Aided Design, CAD-7, pp. 520–527 (1988).

[13] H. Masuda, Y. Aoki, J. Mano, and O. Yamashiro, 'MOSTSM: A physically based charge conservative MOSFET model', IEEE Trans. Computer-Aided Design, CAD-7, pp. 1229–1235 (1988).

[14] R. Gharabagi and A. El-Nokali, 'A model for the intrinsic gate capacitances of short-channel MOSFETYs', Solid-State Electron., 32, pp. 57–63 (1989).

[15] R. Gharabagi and A. El-Nokali, 'A charge-based model for short-channel MOS transistor capacitances', IEEE Trans. Electron Devices, ED-37, pp. 1064–1072 (1990).

[16] C. Turchetti, P. Prioretti, G. Masetti, E. Profumo, and M. Vanzi, 'A Meyer-like approach for the transient analysis of digital MOS ICs', IEEE Trans. Computer-Aided Design, CAD-5, pp. 499–506 (1986).

[17] M. A. Cirit, 'The Meyer model revisited: Why is charge not conserved?', IEEE Trans. Computer-Aided Design, CAD-8, pp. 1033–1037 (1989).

[18] T. Smedes and F. M. Klaassen, Effects of the lightly doped drain configuration on capacitance characteristics of submicron MOSFETs', IEEE IEDM-90, *Technical Digest*, pp. 197–200 (1990). Also see T. Smedes, 'Compact modeling of the dynamic

behavior of MOSFETs', *Ph.D. thesis*, 1991, Technical University of Eindhoven, Eindhoven, The Netherlands.

[19] K. A. Sakallah, Y. T. Yen, and S. S. Greenberg, 'A first order charge conserving MOS capacitor model', IEEE Trans. Computer-Aided Design, CAD-9, pp. 99–108 (1990).

[20] J. J. Paulos and D. A. Antoniadis, 'Limitations of quasi-static capacitance models for the MOS transistor', IEEE Electron Device Lett., EDL-4, 221–224 (1983).

[21] A. Afzali-Kushaa and A. El-Nokali, 'Modeling subthreshold capacitances of MOS transistors', Solid-State Electron., 35, pp. 45–49 (1992).

[22] W. Budde and W. H. Lamfried, 'A charge-sheet capacitance model based on drain current modeling', IEEE Trans. Electron Devices, ED-37, pp. 1678–1687 (1990).

[23] H. J. Park, P. K. Ko, and C. Hu, 'A charge sheet capacitance model of short channel MOSFET's for SPICE', IEEE Trans. Computer-Aided Design, CAD-10, pp. 376–389 (1991).

[24] C. T. Yao, I. A. Mack, and H. C. Lin, 'Short-channel effects on MOSFET terminal capacitances', *Tech. Digest*, IEEE Custom Integrated Circuits Conference, CICC-87, pp. 400–404 (1987).

[25] B. J. Sheu and P. K. Ko, 'Measurement and modeling of short-channel MOS transistor gate capacitances', IEEE J. Solid-State Circuits, SC-22, pp. 464–472 (1987).

[26] Y. T. Yeow, 'Measurement and numerical modeling of short channel MOSFET gate capacitances', IEEE Trans. Electron Devices ED-35, pp. 2510–2519 (1987).

[27] Y. Ohkura, T. Toyabe, and H. Masuda, 'Analysis of MOSFET capacitances and their behavior at short-channel lengths using a AC device simulator', IEEE Trans. Computer-Aided Design, CAD-6, 423–429 (1987).

[28] H. Iwai, M. R. Pinto, S. C, Rafferty, J. E. Oristian, and R. W. Dutton, 'Velocity saturation effect on short-channel MOS transistor capacitance,' IEEE Electron Device Lett., EDL-8, pp. 120–122 (1985).

[29] K. W. Chai and J. J. Paulos, 'Comparison of quasi-static and non-quasi-static capacitance models for the four terminal MOSFET,' IEEE Electron Device Lett., EDL-8, pp. 377–379 (1987).

[30] M. H. Bagheri and Y. Tsividis, 'A small signal dc-to-high frequency nonquasistatic model for the four-terminal MOSFET valid in all regions of operation', IEEE Trans. Electron Devices, ED-32, pp. 2383–2391 (1985).

[31] P. Mancini, C. Turchetti, and G. Masetti, 'A non-quasi-static analysis of the transient behavior of the long-channel MOST valid in all regions of operation,' IEEE Trans. Electron Devices, ED-34, pp. 325–335 (1987).

[32] P. J. V. Vandeloo and W. M. C. Sansen, 'Modeling of the MOS transistor for high frequency analog design', IEEE Trans. Computer-Aided Design, CAD-8, pp. 713–723 (1989).

[33] L. J. Pu and Y. Tsividis, 'Small-signal parameters and thermal noise of the four-terminal MOSFET in non-quasistatic operation,' Solid-State Electron., 33, pp. 513–521 (1990).

[34] H. J. Park, P. K. Ko, and C. Hu, 'A charge conserving non-quasistatic (NQS) MOSFET model for SPICE transient analysis,' IEEE Trans. Computer-Aided Design, CAD-10, pp. 629–642 (1991).

8 Modeling Hot-Carrier Effects

Over the past decade, the downward scaling of device dimensions has resulted in a reduction in gate-oxide thickness by a factor of four. While scaling continued, the supply voltage remained constant (normally 5 V) due to the constraints of retaining compatibility with existing systems. This has resulted in increased vertical electric fields in the oxide which have already reached above 1 MV/cm in thin oxides. The scaling of channel length, meanwhile, has lead to large lateral electric field in the channel. In spite of reducing the supply voltage to 3.3 V, a strong push still remains towards higher channel electric field as scaling continues. The increased channel electric field has caused hot-carrier effects that are becoming a limiting factor in realizing submicron level VLSI. This is because hot-carrier effects impose more severe constraints on VLSI device design as device dimensions are reduced.

The hot-carrier effect is a reliability problem which occurs when hot (energetic) carriers cause $Si–SiO_2$ interface damage and/or oxide trapping. This leads to the degradation of the current drive capability of the transistor, thus eventually causing circuit failure. The origin of this degradation is the high electric field near the drain end, as was discussed in section 3.4. One of the most effective ways to control the hot-carrier effect is to include a field reducing region in the transistor structure. These regions, called the LDD (lightly doped drain) or MDD (moderately doped drain), reduce the amount of damage a device suffers, and consequently increase its operational lifetime.

The two basic monitors that are important in assessing the overall effect of hot-carriers on device performance are the substrate current and the gate current. In this chapter we will first develop models for the substrate and gate currents and then discuss the measurement of the hot-carrier degradation, i.e., device lifetime models. With these models implemented in circuit simulators, one can determine the effect of hot-carrier induced device degradation on circuit level performance.

8.1 Substrate Current Model

The substrate current I_b in an n-channel MOSFET is due to holes which are generated by the impact ionization induced by the channel hot-electrons as they travel from the source to the drain. Mathematically, I_b can be expressed as [1,2]

$$I_b = (M-1)I_d \tag{8.1}$$

where I_d is the drain current[1] and M is the avalanche multiplication factor due to impact ionization and is given by

$$M = \frac{1}{1 - \int \alpha_n dy} \tag{8.2}$$

where α_n is the electron impact ionization coefficient per unit length and is a strong function of the channel electric field $\mathscr{E}$. Since the substrate current I_b, resulting from the channel hot electrons impact ionization process, is 3–5 orders of magnitude smaller than the drain current I_d, it can be considered a low-level avalanche current. For low-level multiplication $M \approx 1$, and therefore Eq. (8.1) becomes

$$I_b = I_d \int_{y=0}^{l_i} \alpha_n dy \tag{8.3}$$

where y is the distance along the channel with $y = 0$ representing the start of the impact ionization region, and l_i is the length of the drain section where impact ionization takes place. Several forms for α_n have been proposed but the most commonly used form is [3]

$$\alpha_n = A_i \exp\left(-\frac{B_i}{\mathscr{E}}\right) \quad (\text{cm})^{-1} \tag{8.4}$$

where A_i and B_i are called the impact ionization constants. Most of the reported data on α_n were measured in bulk silicon and the constants A_i and B_i show a wide range of values. It is only recently that Slotboom et al. [4] have measured α_n at the surface and in bulk silicon and found the following values for the constants:

---

[1] In this chapter we will represent drain current by I_d rather than I_{ds} in order to be consistent with the representation of substrate and gate current by I_b and I_g respectively.

Table 8.1. *Values of the parameters A_i and B_i for electrons in silicon*

α_n	A_i cm^{-1}	B_i V/cm
Surface	2.45×10^6	1.92×10^6
Bulk	0.703×10^6	1.23×10^6

Since the impact ionization in a MOSFET can occur at the surface and/or in the bulk and is important even at lower fields, it is more appropriate to leave A_i and B_i as adjustable parameters for the substrate current model. Due to the exponential dependence of α_n on electric field, it is easy to see that impact-ionization will dominate at the position where the electric field is maximum. In a MOSFET the maximum electric field is present at the drain end, therefore, we expect the ionization integral in Eq. (8.3) to be dominated by the maximum electric field $\mathscr{E}_m$ at the drain end. Substituting Eq. (8.4) in Eq. (8.3) we get

$$I_b = I_d A_i \int_0^{l_i} \exp\left(-\frac{B_i}{\mathscr{E}}\right) dy. \tag{8.5}$$

To solve the above integral we need first to calculate the electric field in the channel. Based on a quasi two-dimensional model it was shown in section 6.7.3 that the channel electric field $\mathscr{E}$ can be expressed as [cf. Eq. (6.201)]

$$\mathscr{E}(y) = -\frac{dV}{dy} = \sqrt{\frac{(V(y) - V_{dsat})^2}{l^2} + \mathscr{E}_c^2} \tag{8.6}$$

where $\mathscr{E}_c$ represents the channel field at which the carriers reach velocity saturation (at $y = 0$, $\mathscr{E} = \mathscr{E}_c$) and the corresponding voltage at that point is the saturation voltage V_{dsat}. $\mathscr{E}_c$ is about 4×10^4 V/cm for electrons. The term l can be treated as an effective ionization length and is given by [cf. Eq. (6.200)]

$$l^2 = \frac{\varepsilon_{si}}{\varepsilon_{ox}} t_{ox} X_j \tag{8.7}$$

where t_{ox} is the gate oxide thickness and X_j is the junction depth. Although Eqs. (8.6) and (8.7) were derived for conventional source/drain junctions, they are still used for LDD source/drains. For LDD devices, X_j is the junction depth of the LDD (n^-) region. The maximum field $\mathscr{E}_m$, which occurs at the drain end, can easily be obtained replacing V by V_{ds} in Eq. (8.6).

Since, in general, $\mathscr{E}_c$ is small compared to other terms in Eq. (8.6), neglecting $\mathscr{E}_c$ results in the following approximate expression for $\mathscr{E}_m$

$$\mathscr{E}_m \approx \frac{(V_{ds} - V_{dsat})}{l}. \tag{8.8}$$

Changing the variable of integration from dy to $(dy/d\mathscr{E})d\mathscr{E}$ in Eq. (8.5) and integrating from $\mathscr{E}_c$ to $\mathscr{E}_m$, we get

$$I_b = I_d A_i l \int_{\mathscr{E}_c}^{\mathscr{E}_m} \frac{1}{\sqrt{\mathscr{E}^2 - \mathscr{E}_c^2}} \exp\left(-\frac{B_i}{\mathscr{E}}\right) d\mathscr{E}. \tag{8.9}$$

The above integral has no closed form solution. However, it can be approximated[2] fairly accurately as

$$I_b \approx I_d A_i l \frac{1}{\sqrt{(\mathscr{E}_m^2 - \mathscr{E}_c^2)}} \frac{\mathscr{E}_m^2}{B_i} \exp\left(-\frac{B_i}{\mathscr{E}_m}\right) \tag{8.10}$$

which under the assumption of $\mathscr{E}_c < \mathscr{E}_m$ can further be approximated as

$$\boxed{I_b \approx C_1 I_d \exp\left(-\frac{B_i}{\mathscr{E}_m}\right)} \tag{8.11}$$

where $C_1 = A_i l \mathscr{E}_m / B_i$ is assumed constant. This is the most widely quoted approximate expression for the substrate current calculations [6]–[11] and is supported by the results of numerical simulation based on the 2-D device simulator MINIMOS [12]. Again, assuming $\mathscr{E}_c < \mathscr{E}_m$ and using Eq. (8.8) for $\mathscr{E}_m$ in Eq. (8.10), results in the following expression for I_b

$$\boxed{I_b = I_d \frac{A_i}{B_i}(V_{ds} - V_{dsat}) \exp\left(-\frac{l B_i}{V_{ds} - V_{dsat}}\right).} \tag{8.12}$$

---

[2] The integral in Eq. (8.9) is of the form

$$f(x) = \int_\alpha^\beta g(t) e^{x h(t)} dt$$

If x is positive, large such that $h(\beta) > h(t)$ and $\alpha \le t < \beta$, then $f(x)$ can be approximated as [5]

$$f(x) \approx e^{x h(\beta)} g(\beta) \frac{1}{x |h'(\beta)|}$$

This equation for I_b is often used for substrate current modeling [6]–[10]. It differs from Eq. (8.11) only in not lumping $\mathscr{E}_m$ in the constant C_1. The model proposed by Mar et al. [13] has a functional form somewhat similar to Eq. (8.12), but is more complicated as it involves iterative solution to get l and V_{dsat}. It is interesting to point out that the term $\exp(-B_i/\mathscr{E}_m)$ in Eq. (8.10) can also be fitted to the form $\mathscr{E}_m^n$. In that case, again under the assumption $\mathscr{E}_c < \mathscr{E}_m$, Eq. (8.10) becomes

$$I_b = I_d \frac{A_i}{B_i} l \mathscr{E}_m^{n+1}. \tag{8.13}$$

Using Eq. (8.8) for $\mathscr{E}_m$, the above equation becomes

$$I_b = aI_d(V_{ds} - V_{dsat})^b \tag{8.14}$$

where a and b are constants. This was the I_b model proposed by Sing and Sudlow [14] and later modified by Sakurai et al. [15] who proposed the following values for the constant a:

$$a = 2.24 \times 10^{-5} - 0.10 \times 10^{-5}\, V_{ds}$$

while $b = 6.4$.

It is customary to present substrate current data by plotting I_b as a function of gate voltage V_{gs} with the drain voltage V_{ds} as a parameter as is shown in Figure 8.1. The experimental data points (circles) are for a n-channel LDD device with $L = 0.78\,\mu$m, $t_{ox} = 150$ Å, $X_j = 0.2\,\mu$m at $V_{ds} = 4.6$ V and 3.8 V, $V_{sb} = 0$ V. The dotted lines were generated using model Eq. (8.12), while dashed lines are based on Eq. (8.14). Note that for a given V_{ds}, initially substrate current I_b increases with increasing V_{gs} due to an increase in the drain current I_d. Further increase in V_{gs} eventually results in a decrease in I_b due to the increase in V_{dsat}, which in turn reduces the channel field $\mathscr{E}$. Thus, I_b increases first, reach its peak value and then decreases resulting in a bell shape curve with its maximum occurring at a gate voltage $V_{gs} \approx (0.3 - 0.5)\, V_{ds}$. From Eq. (8.12) it is evident that the plot of $\ln(I_b/I_d(V_{ds} - V_{dsat}))$ versus $1/(V_{ds} - V_{dsat})$ will be a straight line. This indeed is found experimentally for both p- and n-channel devices as shown in Figure 8.2 [9], [16]. The slope of this line is lB_i. By making such plots for nMOST having different oxide thicknesses, junction depths and substrate doping concentrations, the following empirical relation for l was observed for long channel and thick gate oxide devices [17]

$$l = 0.22 t_{ox}^{1/3} X_j^{1/2}. \tag{8.15}$$

For $t_{ox} < 150$ Å and $L < 0.5\,\mu$m, Eq. (8.15) for l has been modified as [10]

$$l = 0.017 t_{ox}^{1/8} X_j^{1/3} L^{1/5} \tag{8.16}$$

with all quantities having units of cm. Thus, it is clear that the process

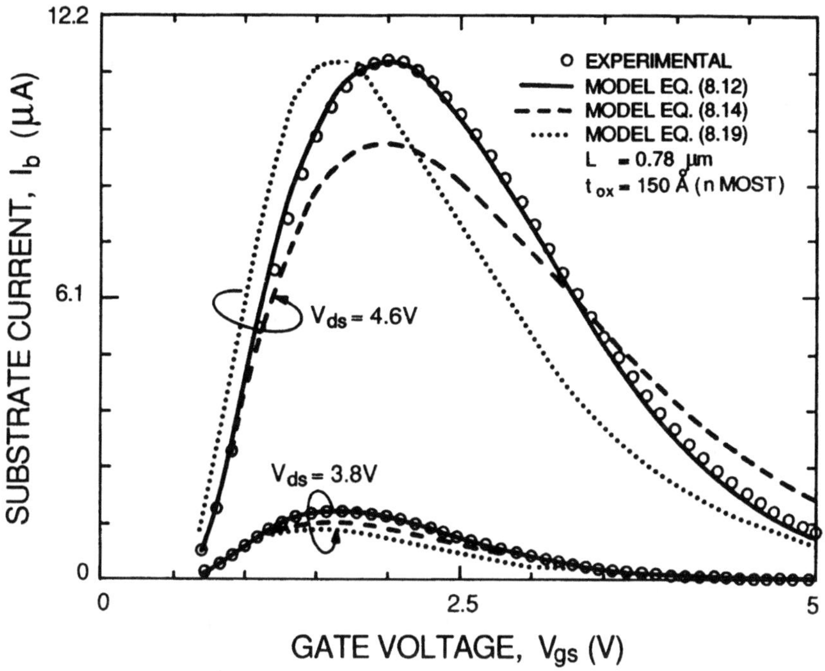

Fig. 8.1 Substrate current I_b as a function of V_{gs} at $V_{ds} = 4.6$ and 3.8 V and $V_{sb} = 0$ V for a typical MOSFET. Circles are experimental points for an n-channel LDD device ($L = 0.77\,\mu$m and $t_{ox} = 150$ Å). Solid, dashed and dotted lines are 3 different model equations (see text). (After Arora and Sharma [18])

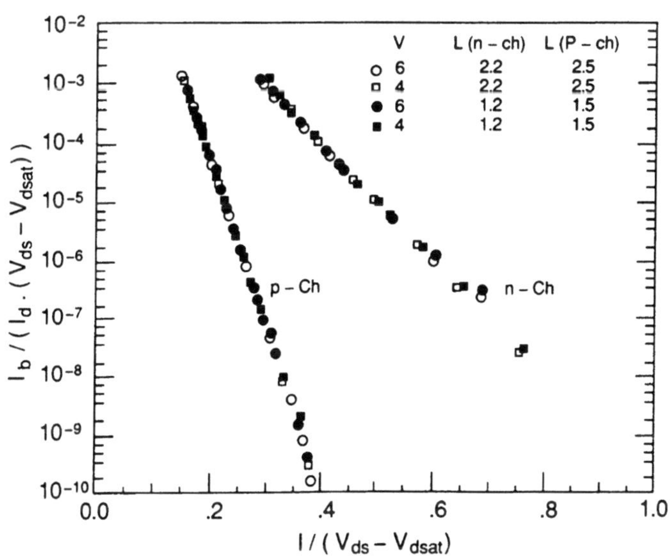

Fig. 8.2 Plot of $\log[I_b/I_d(V_{ds} - V_{dsat})]$ versus $1/(V_{ds} - V_{dsat})$ for different effective channel lengths L and gate voltage V_{gs} for both p- and n-channel devices with $t_{ox} = 152$ Å. (After Ong et al. [24])

parameters on which I_b depends are

- Gate oxide thickness t_{ox}; the thinner the t_{ox}, the higher the I_b.
- Source/drain junction depth X_j; the shallower the junctions, the higher the I_b.
- Effective channel length L; the smaller the L, the higher the I_b.
- Substrate concentration N_b; the higher the N_b, the higher the I_b.

As can be seen from Figures 8.1 and 8.2, Eq. (8.12) for I_b predicts results which are in general agreement with the experimental data, although the exact amount of I_b may not be represented by Eq. (8.12). However, for circuit models to predict device degradation, one needs to have more accurate model for I_b. It has been shown that for a given device geometry (i.e., given channel length, doping profile and oxide thickness) and bias conditions, the channel field $\mathscr{E}_m$ predicted by Eq. (8.8) is significantly lower compared to that predicted by MINIMOS (2-D device simulator), especially at higher drain voltages [18]. Since the substrate current depends exponentially on the peak field $\mathscr{E}_m$, even small errors in $\mathscr{E}_m$ can lead to substantial errors in I_b. Equation (8.8) used in the I_b model is somewhat oversimplified. There are other analytical equations proposed for $\mathscr{E}_m$ based on 2-D simulations [19], but for circuit modeling work it is more appropriate to use the following empirical expression for the peak field:

$$\mathscr{E}_m \approx \frac{(V_{ds} - \eta V_{dsat})}{l} \tag{8.17}$$

where $0 < \eta \leq 1$ is a technology dependent fitting parameter and is different for standard and LDD devices. Further, results from MINIMOS simulations also indicate that for a given t_{ox} and X_j, the effective ionization length l is a function of the gate and drain voltages. Based on 2-D device simulations, the following bias dependent equation for l has been suggested [18]

$$l = l_0 + l_1(V_{ds} - V_{geff}) + l_2(V_{ds} - V_{geff})^2 \tag{8.18}$$

where

$$V_{geff} = V_{gs} - V_{th0}$$

is the effective gate drive, V_{th0} is the threshold voltage at $V_{bs} = 0$ and l_0, l_1 and l_2 are fitting parameters. With these changes, the final equation for I_b becomes

$$I_b = I_d \frac{A_i}{B_i}(V_{ds} - \eta V_{dsat}) \times \exp\left[-\frac{B_i}{(V_{ds} - \eta V_{dsat})}\{l_0 + l_1(V_{ds} - V_{geff}) + l_2(V_{ds} - V_{geff})^2\}\right]. \tag{8.19}$$

This new model equation for I_b seems to fit the experimental data very well as can be seen from Figure 8.1 where the continuous line is based on Eq. (8.19). The model parameter values were obtained by fitting the data to Eq. (8.19) using a nonlinear optimization technique. The best fit values for the data shown in Figure 8.1 are $A_i = 0.536 \times 10^6\,\mathrm{cm}^{-1}$, $B_i = 1.92 \times 10^6\,\mathrm{V/cm}^{-1}$, $\eta = 0.57$, $l_0 = 16 \times 10^{-6}$, $l_1 = -2.34 \times 10^{-6}$ and $l_2 = 0.165 \times 10^{-6}$ [18]. Note that B_i is not optimized and is fixed to $1.92 \times 10^6\,\mathrm{V/cm}$ (see Table 8.1). This is because optimization results using a confidence region algorithm (see section 10.4) show that parameter B_i is redundant, and therefore can be set to its physical value while extracting other parameters. The model also fits well the back bias dependence of I_b as shown in Figure 8.3 where I_b versus V_{gs} data is plotted for $V_{sb} = 0$ and 3 V.
Although Eq. (8.12) was derived for nMOST's, it is also valid for pMOST's [24], [11]. This is evident from the plots of Figure (8.2). From these plots it is found that $B_i = 1.7 \times 10^6\,\mathrm{V/cm}$ and $3.7 \times 10^6\,\mathrm{V/cm}$ for nMOST and pMOST, respectively. This shows that value of B_i for pMOST's is 2.2 times that for nMOST's implying that a pMOST can take about twice the $(V_{ds} - V_{dsat})$ to generate the same I_b as an nMOST [9]. This means that for a given bias conditions I_b in pMOST will be smaller than in nMOST

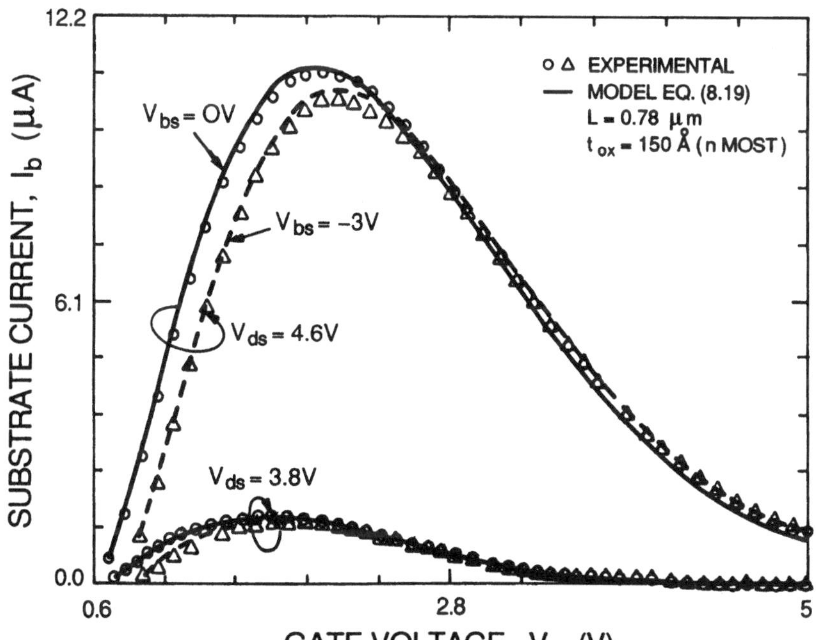

Fig. 8.3 Substrate current I_b as a function of V_{gs} at $V_{ds} = 4.6$ and 3.8 V, and $V_{sb} = 0$ V and -3 V, for a typical n-channel LDD device ($L = 0.77\,\mu\mathrm{m}$ and $t_{ox} = 150$ Å). Circles are experimental points while continuous lines are based on Eq. (8.19)

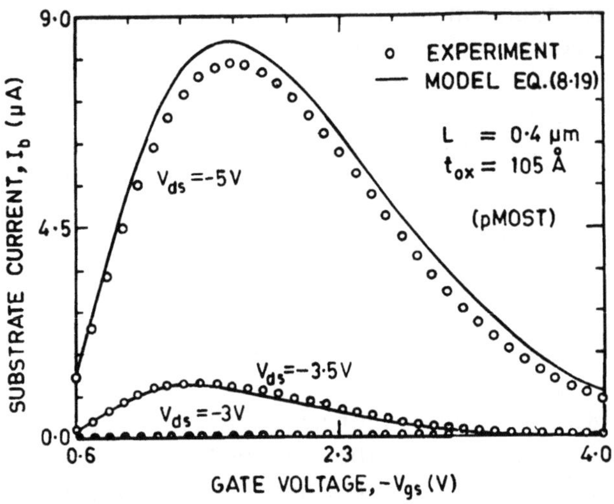

Fig. 8.4 Measured and calculated substrate current I_b as a function of V_{gs} at $V_{ds} = 4.6$ V for a *p*-channel device

fabricated using same the technology. This is consistent with the fact that the impact ionization of holes is 2–3 order of magnitude smaller than for electrons. Remember that substrate current I_b in *p*-channel devices is the result of electrons generated by the impact ionization of channel hot-holes.

It has been found that for a pMOST, the impact ionization length l is bias dependent, similar to Eq. (8.18), and therefore Eq. (8.19) can still be used for modeling pMOST substrate current. Figure 8.4 shows pMOST ($L = 0.4\,\mu$m and $t_{ox} = 105$ Å) substrate current[3] plotted against V_{gs} for different V_{ds}; circles are experimental data points and continuous lines are simulated using Eq. (8.19).

8.2 Gate Current Model

The gate current I_g is the result of channel hot electron (CHE) injection into the gate oxide, although gate current due to hot-hole injection has also been observed particularly in thin gate oxides ($t_{ox} < 150$ Å) *n*-channel MOSFETs. Since the energy required to surmount the Si–SiO_2 potential barrier is ~3.2 eV for electrons and ~4.9 eV for holes, the gate current due to hot holes is extremely small compared to hot electrons. Two different approaches have been used to model the gate current that lead to results

[3] Strictly speaking, the substrate current of a pMOST fabricated using n-well CMOS process should be called well current rather than substrate current.

in reasonable agreement. The first approach is the so called *lucky electron model* originally proposed by Shockley [20] for the study of transport phenomena in large electric fields. It was later followed by Verwey et al. [21] and Ning et al. [22] for the analysis of substrate hot-electron injection, and Hu and coworkers [6, 7], [23, 24] for the study of channel hot-electron injection.

A second approach to the problem is the so called *equivalent temperature* approach to the carrier energy [25]–[28]. This approach assumes that the electron heated by the channel electric field leads to a form of thermionic emission of hot electrons into the oxide. It is further assumed that the carrier energy distribution function is Maxwellian at the equivalent temperature $T_e = T_e(y)$, where y is the distance along the channel. The injected current density J_g is then calculated using the following Richardson formula for thermionic emission [26]

$$J_g = qn_s\left(\frac{kT_e}{2\pi m^*}\right)^{1/2} \exp\left(-\frac{\Phi_b}{kT_e}\right) \tag{8.20}$$

where n_s represents the electron concentration at the Si–SiO_2 interface, m^* is effective mass of the electron and Φ_b is electron potential barrier height at the Si–SiO_2 interface. Neglecting the effect of charge trapping in the oxide, the gate current I_g in the equivalent temperature model is expressed as

$$I_g = W\int_0^L J_g(y)dy. \tag{8.21}$$

The main difficulty with this approach is of determining the functional dependence of $T_e(y)$. No closed form solution is possible and only 2-D/3-D device simulators have been used for calculating I_g using this approach. On the other hand, the lucky electron model results in a closed form expression for I_g and hence is discussed next in some detail.

Following Tam et al. [23], the lucky electron model for CHE injection into the gate oxide of an n-channel MOSFET can be described as follows. In order for channel hot-electrons to reach the gate (1) the hot electrons must gain sufficient kinetic energy (in excess of the potential barrier at the Si–SiO_2 interface) from the channel field, (2) it must undergo an elastic collision, redirecting its momentum normal to the barrier, and (3) it must not experience any inelastic collision before reaching the interface. The different scattering events are illustrated in Figure 8.5. From point A to B a channel electron gains energy from the channel field and becomes 'hot'. At B, re-direction of the hot electron takes place. From point B to C (C is situated at the interface), the hot electron must not suffer any energy-robbing collision so that it will retain the energy required to surmount the Si–SiO_2 potential barrier. The hot electron must also not suffer collision in the oxide image-potential well located between C and D. Once the hot-

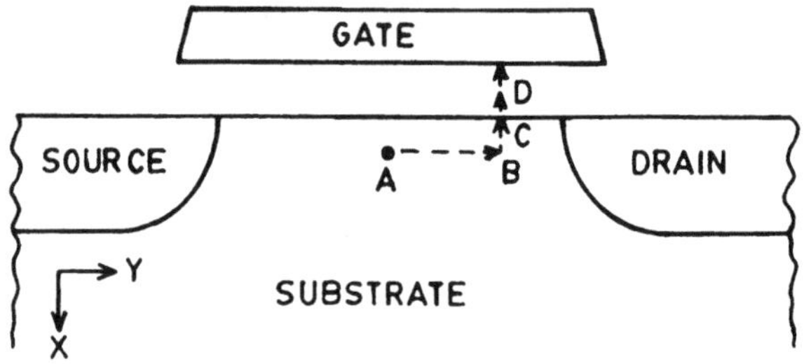

Fig. 8.5 Electron injection in gate oxide showing lucky electron model

electron arrives at location D, it will be swept toward the gate electrode by the aiding field. Since the processes are statistically independent, the resultant probability is the product of the probability for each individual event, i.e. [23]

$$I_g = I_d \int_0^L P_1 P_2 P_3 \left(\frac{dy}{\lambda_r} \right) \tag{8.22}$$

where λ_r is the redirectional scattering mean free path. The factor (dy/λ_r) can be interpreted as the probability of redirection over dy. P_1 is the probability for acquiring sufficient kinetic energy and normal momentum, P_2 is the probability that a hot electron travels to the Si–SiO_2 interface without suffering any inelastic collision, and P_3 is the probability to suffer no collision in the oxide image-potential well. Thus, to calculate I_g, we need to calculate the three probabilities P_1, P_2 and P_3. The essential processes involved for modeling channel hot-electron injection into the gate oxide is illustrated in Figure 8.6.

In order for the hot electron to surmount the Si–SiO_2 potential barrier Φ_b, its kinetic energy must be greater than $q\Phi_b$. To acquire kinetic energy $q\Phi_b$, the hot electron will have to travel a distance $d = \Phi_b/\mathscr{E}$, assuming the electric field $\mathscr{E}$ along the channel to be constant. The probability of a channel electron to travel a distance d or more without suffering collision can be written as $e^{-d/\lambda}$, where λ is the scattering mean free path of the hot electron [25]. Hence we can write $e^{-\Phi_b/\mathscr{E}\lambda}$ as the probability that an electron will acquire a kinetic energy greater than the potential barrier Φ_b. Now if the electron is to move into the oxide, its momentum must be redirected towards the Si–SiO_2 interface by elastic scattering so as to have sufficiently large momentum component perpendicular to the interface. It has been shown that the probability of an electron acquiring the required kinetic energy and retaining the appropriate momentum after redirection is [23]

$$P_1 = 0.25 \left(\frac{\mathscr{E}\lambda}{\Phi_b} \right) \exp\left(-\frac{\Phi_b}{\mathscr{E}\lambda} \right). \tag{8.23}$$

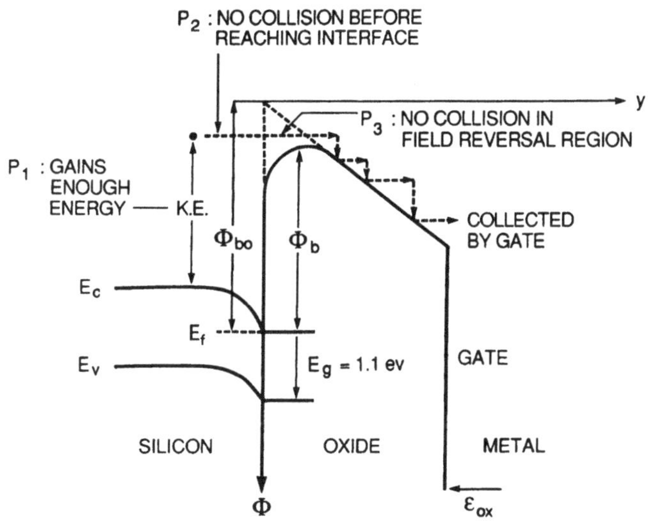

Fig. 8.6 The energy system for the MOS structure showing essential processes in the channel hot electron injection model. (After Tam et al. [23])

Since the potential barrier Φ_b is lowered by the image force effect, the net barrier height is generally expressed as [22, 23]

$$\Phi_b = \Phi_{bo} - 2.59 \times 10^{-4} \mathscr{E}_{ox}^{1/2} - a_0 \mathscr{E}_{ox}^{2/3} \tag{8.24}$$

where $\Phi_{bo} = 3.2\,\mathrm{eV}$ is the Si–SiO_2 interface barrier for the electrons, $\mathscr{E}_{ox}$ is the oxide field given by [cf. Eq. (6.195)]

$$\mathscr{E}_{ox} = \frac{(V_{gs} - V_{fb} - 2\phi_f - V_{ds})}{t_{ox}} \tag{8.25}$$

and a_0 is a constant whose value is obtained by fitting the experimental data; Ning et al. [22] have assumed $a_0 = 1 \times 10^{-5}\,(\mathrm{cm})$, while Tam et al. [23] find $a_0 = 4 \times 10^{-5}\,(\mathrm{cm})$ as a more appropriate value for their data. The second term in Eq. (8.24) represents the barrier lowering effect due to the image field, while the third term accounts phenomenologically for the finite probability of tunneling between the Si and SiO_2.
According to Tam et al. [23], the probability P_2 is given by

$$P_2 \approx \frac{5.66 \times 10^{-6} \mathscr{E}_{ox}}{(1 + \mathscr{E}_{ox}/1.45 \times 10^5)} \times \frac{1}{\{1 + 2 \times 10^{-3} L^{-1} \exp(-\frac{2}{3} \mathscr{E}_{ox} t_{ox})\}} + 2.5 \times 10^{-2} \tag{8.26}$$

while the probability P_3 of colision-free travel in the oxide-image potential well is given by

$$P_3 = \exp\left(-\frac{1}{\lambda_{ox}}\sqrt{\frac{q}{16\pi\mathscr{E}_{ox}\epsilon_0\epsilon_{ox}}}\right) \approx \exp\left(-\frac{300}{\sqrt{\mathscr{E}_{ox}}}\right) \tag{8.27}$$

where $\lambda_{ox} = 3.2\,\text{nm}$ is the electron mean free path in the oxide. Note that the product of P_2 and P_3 is essentially only a function of the gate oxide field $\mathscr{E}_{ox}$, therefore, it can be combined as $P_2P_3 = P(\mathscr{E}_{ox})$. It is found that $P(\mathscr{E}_{ox})$ is a weak function of $\mathscr{E}_{ox}$; its value is maximum at the drain end corresponding to the oxide field given by Eq. (8.25).

Since the probability P_1 depends exponentially on $\mathscr{E}$, which in turn varies exponentially with y [cf. Eq. (6.201)a], the integrant in Eq. (8.22) is a sharply peaking function. Combining Eqs. (8.22)–(8.27) gives an approximate expression for the gate current as

$$I_g \approx I_d \frac{P(\mathscr{E}_{ox})}{\lambda_r}\left(\frac{\mathscr{E}_m\lambda}{2\Phi_b}\right)\frac{1}{d\mathscr{E}/dx}\int_{\mathscr{E}_c}^{\mathscr{E}_m} \exp\left(-\frac{\Phi_b}{\mathscr{E}\lambda}\right)d\mathscr{E} \tag{8.28}$$

where $\mathscr{E}_m$ is the maximum channel field and $d\mathscr{E}/dx \approx \mathscr{E}_m/l_{che}$ is assumed to be constant over the length l_{che} where CHE injection is significant. Since value for l_{che} is not known, it can be treated as a fitting parameter; however, it can be replaced by t_{ox} without any loss of accuracy in the equation above [23]. The Eq. (8.28) can now be integrated to give a closed form expression for the gate current as

$$\boxed{I_g = 0.5\frac{I_d t_{ox}}{\lambda_r}\left(\frac{\lambda\mathscr{E}_m}{\Phi_b}\right)^2 P(\mathscr{E}_{ox})\exp\left(-\frac{\Phi_b}{\lambda\mathscr{E}_m}\right).} \tag{8.29}$$

To a first order above equation can be written as [6]

$$I_g \approx C_2 I_d \exp\left(-\frac{\phi_b}{\lambda\varepsilon_m}\right) \tag{8.29a}$$

where C_2 is about 2×10^{-3} for $V_{gs} > V_{ds}$. Note that the only fitting parameters in Eq. (8.29) are λ and λ_r. It was found that the gate current data is insensitive to the value of λ_r and has been chosen to be 61.6 nm based on theoretical considerations [23]. The value of λ which fits the data well is found to be 9.2 nm. It is worth noting that *while the substrate current I_b depends only on the channel electric field $\mathscr{E}_m$, the gate current I_g is a function of both the channel field $\mathscr{E}_m$ and the normal oxide field $\mathscr{E}_{ox}$.*

The gate current resulting from the channel hot-electrons in a nMOST is shown in Figure 8.7 where circles are experimental data points while continuous lines are calculated based on Eq. (8.29). Although the model is not very accurate near the peak current, it nonetheless does model the

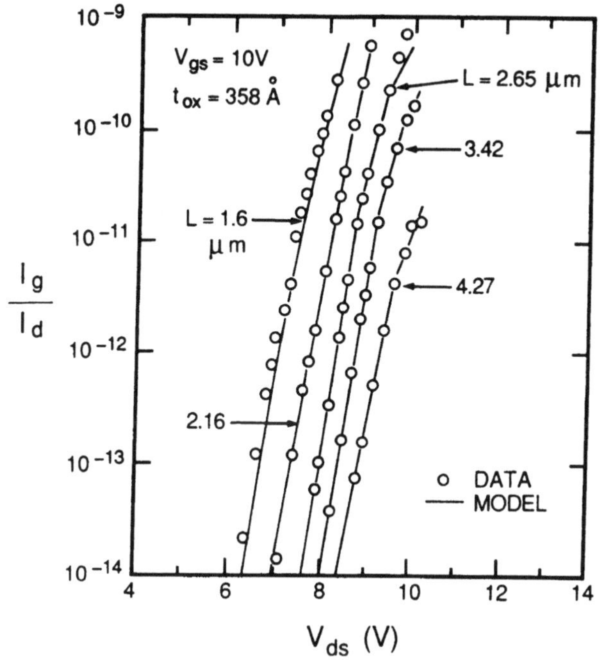

Fig. 8.7 Gate current I_g in an nMOST as a function of V_{ds} at $V_{gs} = 10$ V. (After Tam et al. [23])

general gate current behavior. The dependence of the gate current on the channel length is apparent. Reduction of the channel length reduces V_{dsat}. Therefore, for the same V_{ds} the channel electric field $\mathscr{E}_m$, and hence I_g, is higher in shorter channel devices. The devices with thinner gate oxides have higher gate current because of higher $\mathscr{E}_m$ and $\mathscr{E}_{ox}$.

Figure 8.8 shows both gate and substrate current for an nMOST with $t_{ox} = 200$ Å and $L = 1.1\ \mu$m. Note that peak gate current occurs at $V_{gs} \approx V_{ds}$ which is different from the peak of substrate current that occurs around $V_{gs} \approx V_{ds}/2$. For a given V_{gs}, the gate current I_g increases with increasing V_{ds} due to increasing $\mathscr{E}_m$ until $V_{gs} = V_{ds}$. For $V_{gs} > V_{ds}$, MOSFET is driven into the linear region of operation resulting in a reduction in $\mathscr{E}_m$ and hence I_g.

The gate current shown in Figures 8.7 and 8.8 is due to CHE injection into the gate oxide. However, it has been observed experimentally that gate current in nMOST can also be generated by injection of hot holes into the oxide (particularly thin gate oxide, $t_{ox} < 150$ Å) (see section 8.4) [27]–[30]. These holes are produced by impact ionization of the channel hot-electrons and are accelerated by the channel field. In order to evaluate this gate current component, the hole generation due to impact ionization and lucky electron probabilities for hole injection into the oxide must be modeled.

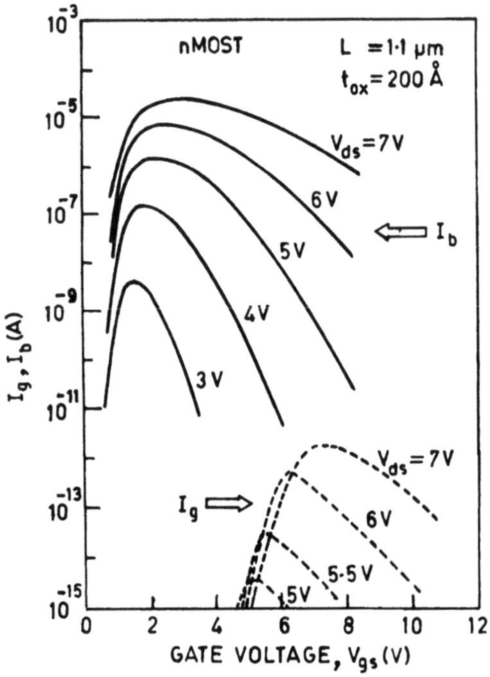

Fig. 8.8 Gate and substrate currents I_g and I_b, respectively, as a function of gate voltage V_{ds} for different drain voltage V_{ds} for a nMOST. (After Takeda et al. [26])

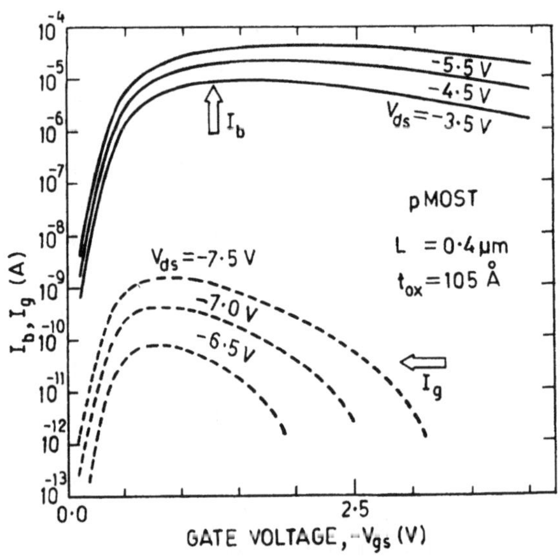

Fig. 8.9 Gate and substrate currents I_g and I_b, respectively, as a function of gate voltage V_{ds} for different drain voltage V_{ds} for a pMOST

The equivalent temperature model has also been used to model such hot-hole injection [28].

The gate current in a typical pMOST as a function of V_{gs} and V_{ds} is shown in Figure 8.9; for the sake of comparison the substrate current is also shown. Note that unlike in an nMOST, the peak of the gate current in a pMOST occurs at much lower gate voltage, similar to that for the substrate current. *From the direction of the gate current measured at low and mid V_{gs}, it is found that pMOST gate current is due to the avalanche hot-electrons (created by impact ionization of holes) rather than the channel hot-holes* [24], [31]–[33]. At higher $|V_{gs}|$ one expects the pMOST gate current to be composed of hot holes, but measurable channel hot-hole injection current in pMOST has not been reported. This is probably because of the large hole barrier height and much shorter mean free path for holes in the oxide. *The electron gate current in pMOST is often larger than the corresponding nMOST gate current,* despite the fact that the number of available avalanche hot-electrons in pMOST's is several orders of magnitude smaller than in nMOST's. This happens because the direction of $\mathscr{E}_{ox}$ is such that it aids electron injection in pMOST while it opposes electron injection in nMOST for $V_{gs} \ll V_{ds}$. For $V_{gs} > V_{ds}$, $\mathscr{E}_{ox}$ is favorable but then its value is too small. Furthermore, pMOST can take twice as large channel field as nMOST before breakdown. The lucky electron model discussed earlier for the nMOST has also been used to model the gate current in pMOST's [24]. Since the source of hot electrons resulting in the gate current in pMOST is from impact ionization process which also produces substrate current I_b, the pMOST gate current

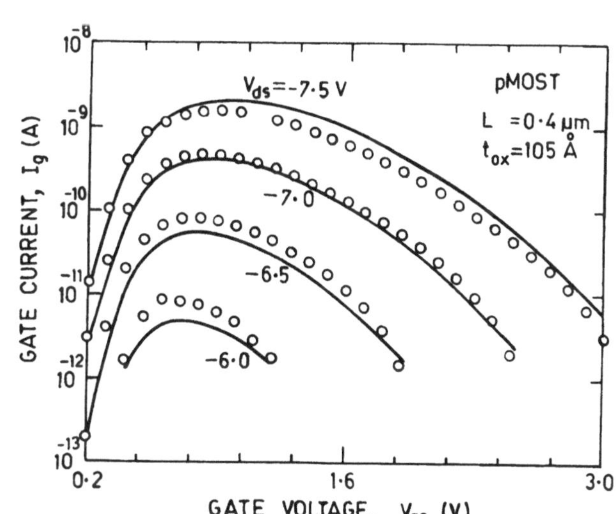

Fig. 8.10 Gate current I_g as a function of V_{gs} at different V_{ds} for pMOST. Circles are experimental points

can be expressed as

$$I_g \approx 0.5 \frac{I_b t_{ox}}{\lambda_r} \left(\frac{\lambda \mathscr{E}_m}{\Phi_b} \right)^2 P(\mathscr{E}_{ox}) \exp\left(-\frac{\Phi_b}{\lambda \mathscr{E}_m} \right) \quad \text{(pMOST)} \tag{8.30}$$

and is obtained by replacing I_d in Eq. (8.29) by I_b. The pMOST gate current calculated using Eq. (8.30) is shown in Figure 8.10 as continuous lines, circules are measured data. The reasonable agreement between the model and data validates Eq. (8.30).

8.3 Correlation of Gate and Substrate Current

Since the hot electrons responsible for the gate current and those responsible for the substrate current are heated by the same field, it is expected that the two currents will be correlated [34, 35]. We can write Eq. (8.11) as

$$I_b \approx C_1 I_d \exp\left(-\frac{B_i}{\mathscr{E}_m} \right) = C_1 I_d \exp\left(-\frac{\Phi_i}{\lambda \mathscr{E}_m} \right). \tag{8.31}$$

The above equation simply rewrites $B_i = \Phi_i/\lambda$, where λ is the hot-electron mean free path. In analogy with Φ_b, Φ_i can be interpreted as the energy that an hot electron must have in order to create an electron-hole pair through impact ionization, and $\exp(-\Phi_i/\lambda\mathscr{E}_m)$ is the probability that an electron travel a distance $d = \Phi_i/\mathscr{E}_m$ to gain energy $q\Phi_i$ or more without

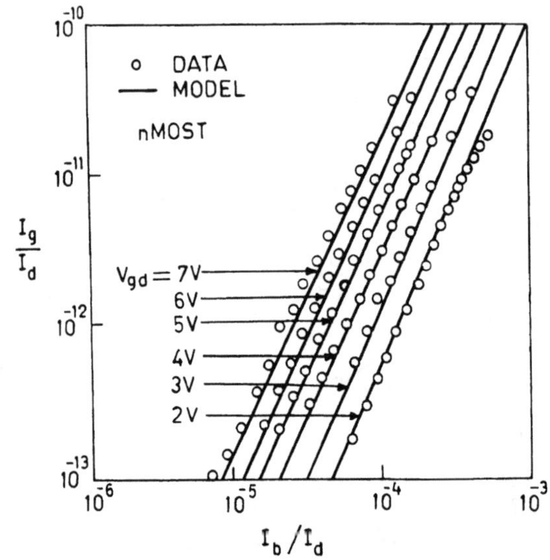

Fig. 8.11 Gate current I_g against substrate current I_b (both normalized to source current) for constant values of $V_{gs} - V_{ds}$, and therefore of $\mathscr{E}_{ox}$. (After Tam et al. [23])

suffering collision. Eliminating $\mathscr{E}_m$ from the exponential term in Eq. (8.29a) and (8.31) we get

$$\frac{I_g}{I_d} = C_3\left(\frac{I_b}{I_d}\right)^{\Phi_b/B_i\lambda}. \tag{8.32}$$

Such a power law relationship is indeed observed as shown in Figure 8.11. The slope of $\ln(I_g/I_d)$ versus $\ln(I_b/I_d)$ gives the quantity $\Phi_b/B_i\lambda$. Since B_i and λ are independent of oxide field $\mathscr{E}_{ox}$, the slope can be used to find Φ_b as a function of $\mathscr{E}_{ox}$.

8.4 Mechanism of MOSFET Degradation

The hot-carrier effects result from large electric field in the channel (particularly near the drain end), which causes damage to the gate oxide (by charge trapping in the oxide) and/or to the Si–SiO_2 interface (by generating interface states). This leads to degradation of the n-channel MOSFET current drive capability and affects parameters such as the threshold voltage V_{th}, the linear region transconductance g_m, the subthreshold slope S, and the saturation region drive current I_{dsat}. Whether carrier (electron/hole) trapping or interface generation is primarily responsible for the degradation is still debated. But usually a net negative charge density is observed after long time stressing as is evidenced by a threshold voltage (V_{th}) increase in nMOST's.

Figure 8.12 shows typical linear region $I_{ds} - V_{gs}$ characteristics, before and after stressing, which results in changes in V_{th} and the peak transconductance g_m (slope of the linear portion of the curve) [6]. The device was a 100/2 nMOST with gate oxide thickness $t_{ox} = 358$ Å; and was stressed at $V_{gs} = 6$ V, $V_{ds} = 7.5$ V for 90 minutes.[4] Notice that the drain current reduces after stressing and that the post-stress I–V characteristics are not symmetrical with respect to the source/drain terminal because the damage is localized at the drain end. This asymmetry is small in the linear region and is much larger in the saturation region. This can be seen from Figure 8.13 which shows typical $I_{ds} - V_{ds}$ characteristics for a nMOST ($L = 1.2\,\mu$m, $t_{ox} = 200$ Å) before and after stress [24]. From this figure it is evident that *the drain current reduction in saturation is much more severe in the reverse mode compared to the forward mode.* Thus, device parameters change if the roles

[4] Note that device stressing is done at accelerated voltages rather than at the normal operating voltages. The underlying philosophy is that a phenomenon which occurs over a short period under the action of accelerating stresses is indicative of a similar phenomenon which will occur over a much longer period when the device is operating normally. Accelerated stressing is necessary to study degradation in a reasonably short time.

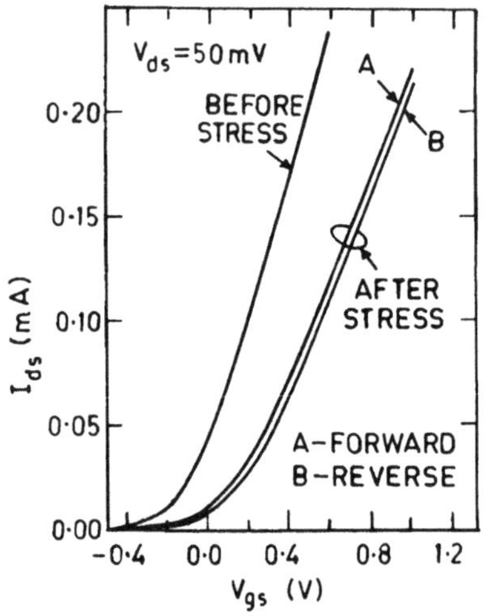

Fig. 8.12 Degradation of nMOST linear region characteristics due to hot carrier injection before and after stress. (After Hu et al. [6])

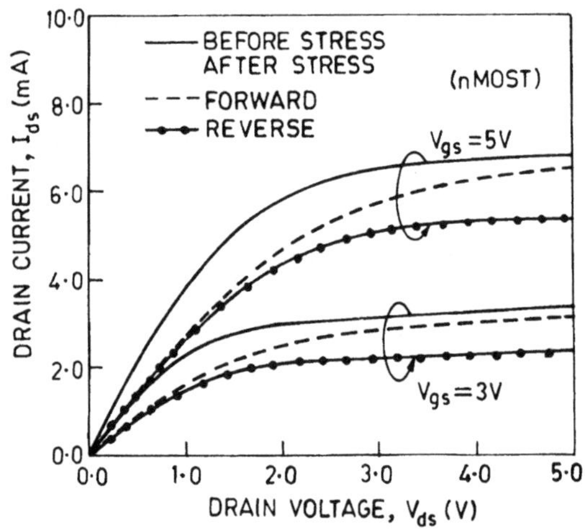

Fig. 8.13 $I_{ds} - V_{ds}$ characteristics of a nMOST ($L = 1.2\,\mu$m and $t_{ox} = 200$ Å) before and after stress. Stress voltages $V_{ds} = 7.5$ V and $V_{gs} = 3$ V. Stress time 5 min. (After Ong et al. [24])

of source and drain are reversed after stressing, a condition that occurs in transfer gates. An example of the degradation of a nMOST ($L = 0.77\,\mu$m) on a log–log scale is shown in Figure 8.14 [40]. Here $\Delta g_m = g_m(0) - g_m(t)$ is the difference between the device transconductance at times 0 and t. The devices are stressed at $V_{gs} = 3$ V and $V_{ds} = 7$ V that corresponds to stressing under peak substrate current condition.

The classical interpretation of the device degradation in n-channel devices has been that only hot electrons can be injected into the gate oxide. How-

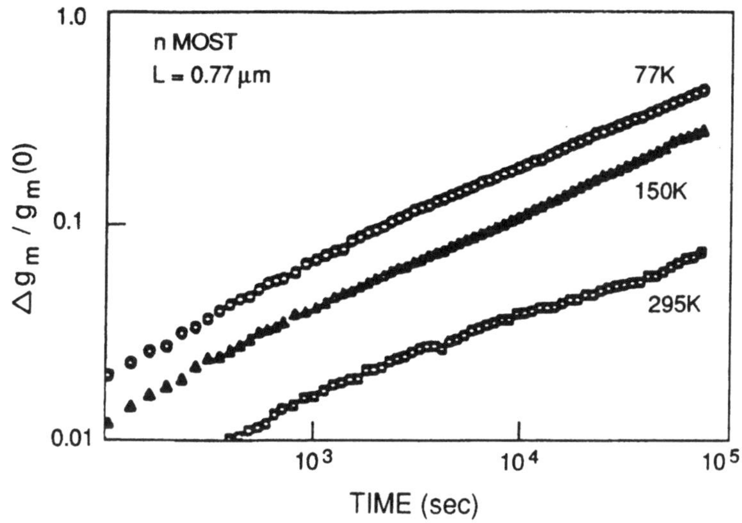

Fig. 8.14 The degradation of n-channel g_m at different temperatures. (After Yao et al. [40])

ever, recent studies show that hot hole injection is also possible [29]–[30]. These holes are produced by impact ionization and accelerated by the channel field. This hole injection into the oxide is referred to as *hole current* and is usually very small, but it may have significant role in the degradation of the device characteristics especially when $V_{gs} \leq V_{ds}/2$ [41]. In fact, holes need not even overcome the barrier but their field assisted tunneling is adequate to cause serious damage to Si–SiO_2 interface. This is because once holes are injected into the oxide, they are more likely to get trapped than the electrons; the trapping efficiency of holes being close to 1, while for electrons it is less than 10^{-5}.

The hot-carrier effect involves the generation, injection and trapping of carriers in the gate oxide. *Carrier injection is a localized phenomenon; it takes place over only a fraction of the total length of the channel.* Four kinds of hot-carrier generation/injection mechanism have been reported for nMOST [25], [29], [37]. These are

(a) *Channel Hot Electrons* (CHE) which are heated up in the channel particularly near the drain end with the MOSFET operated at $V_{gs} = V_{ds}$, called the *lucky electrons.* As shown in Figure 8.15a, lucky electrons are those flowing from source to drain gaining sufficient energy to surmount the Si–SiO_2 barrier without suffering an energy loosing collision in the channel, and thus move into the gate oxide resulting in the so called *gate current* I_g. This injection of hot electrons into the oxide is referred to as channel hot electron (CHE) injection [37]. The gate currents shown in Figures 8.5–8.6 are due to CHE injection.

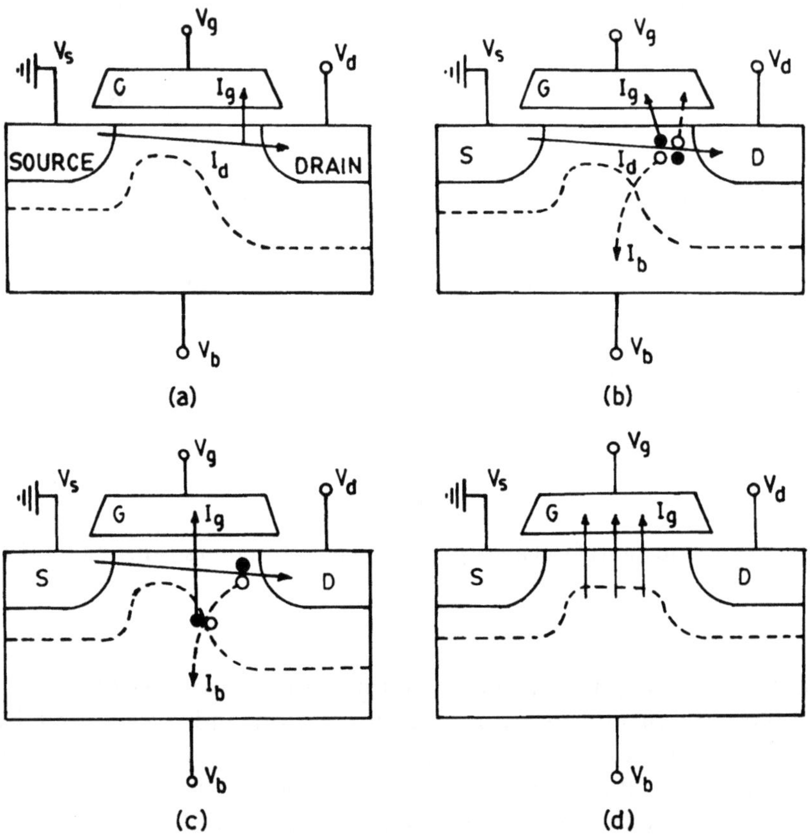

Fig. 8.15 Four different injection mechanisms. (a) Channel Hot Electrons (CHE) (b) Drain Avalanche Hot Carriers (DAHC), (c) Substrate Hot Electrons (SHE), and (d) Secondarily Generated Hot Electron (SGHE)

(b) *Drain Avalanche Hot Carriers* (DAHC) which are due to the high electric field near the drain region and promotes avalanche multiplication. The electrons from the channel gain enough energy so that they produce electron-hole pair by impact ionization which in turn produce further electron-hole pairs resulting in an avalanche process. It is these avalanche hot electrons and hot holes that are injected into the gate oxide, resulting in a gate current with two peaks in the gate current versus gate voltage curves, in addition to the CHE injection peak, as shown in Figure 8.16. It is mostly observed at the bias condition $V_{ds} > V_{gs} > V_{th}$ in nMOST with $t_{ox} < 150$ Å. Figure 8.15b schematically illustrates the DAHC mechanism [29]. *The DAHC injection mechanism causes the most severe device degradation as both holes and electrons are injected into the gate oxide.*

(c) *Substrate Hot Electrons* (SHE), which is due to the injection of thermally generated or injected electrons from the substrate near the surface into the

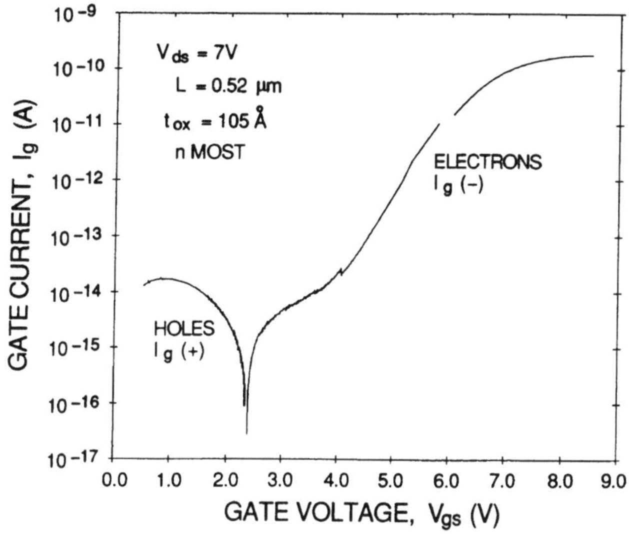

Fig. 8.16 Measured gate current showing both electron and hole injection in *n*-channel gate oxide

SiO_2. It occurs when $V_{ds} = 0$, $V_{gs} > 0$ and large back bias V_{bs}, such as arises in bootstrap circuits (Figure 8.15c). Electrons generated in the depletion region, or diffusing from the bulk neutral region of the substrate, drift towards the Si–SiO_2 interface. These electrons gain energy from the high field in the surface depletion region, some of them having gained enough energy to surmount the barrier. SHE injection, although less important from a practical view point, due to the small number of thermally generated electron-hole pairs, nevertheless has been thoroughly investigated in the past [37].
(d) *Secondarily Generated Hot electron* (SGHE), which is that of secondary minority carriers originated from secondary impact ionization of the substrate current (Figure 8.15d). It occurs when substrate hole current, produced by avalanche effect near the drain, generates further electron-hole pairs. These secondary electrons are then injected into the oxide, as in the case of SHE injection. This type of injection becomes particularly pronounced for large back bias V_{sb} and thin gate oxides ($t_{ox} < 100$ Å). In fact, interface generation due to hot holes and hot electrons has been reported for 0.25 μm pMOST leading to a reduction in g_m and I_d with time [38].
The hot-carrier effects in pMOST have been studied to a lesser extent than nMOST. This is because degradation in pMOST for $L > 0.5\,\mu$m is considered a minor problem, due to the fact that the change in pMOST characteristics after stress tends to saturate within an acceptable percentage. One reason is higher barrier heights for holes (compared to electrons) at the Si–SiO_2 interface. A further reason is the lower effectiveness of holes

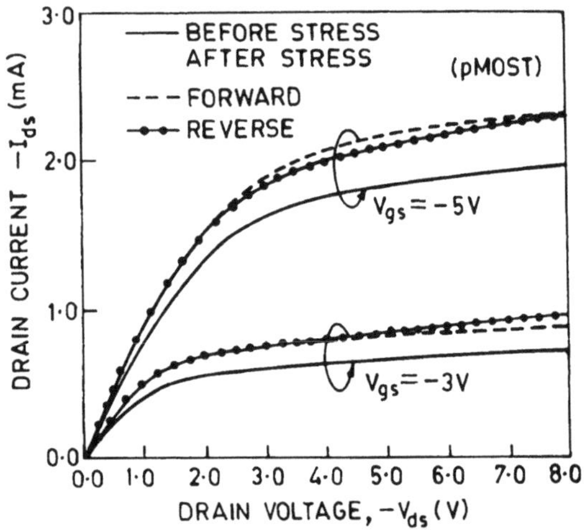

Fig. 8.17 $I_{ds} - V_{ds}$ characteristics of a pMOST ($L = 1.2\,\mu$m and $t_{ox} = 200$ Å) before and after stress. Stress voltages $V_{ds} = 7.5$ V and $V_{gs} = 3$ V. Stress time 5 min. (After Ong et al. [24])

in generating electron-hole pairs (i.e., smaller hole ionization coefficient). This situation may change for deep submicron ($L < 0.5\,\mu$m) devices with pMOST becoming of concern.

Figure 8.17 shows typical $I_{ds} - V_{ds}$ characteristics for pMOST before and after stress [24]. Note that while the drain current I_d reduces after stress in nMOST (see Figure 8.13), it increases in pMOST and is generally considered to be unharmful. In fact, after stress pMOST $|V_{th}|$ decreases (except at very high $|V_{gs}|$), g_m increases, and subthreshold leakage current increases (i.e., punchthrough voltage decreases) [33]. This is in contrast with increase in V_{th} and decrease in g_m in nMOST. It is generally believed that after stressing of pMOST, avalanche hot electrons are trapped in the gate oxide resulting in a negative charge near the drain. This leads to effective shortening of the channel length and thus in an increase in the drain current. Channel hot holes in pMOST do not play any significant role. However, in nMOST both channel hot electrons and avalanche hot holes are important in hot carrier induced degradation.

8.5 Measure of Degradation—Device Lifetime

It is common to characterize the device degradation by measuring shifts in the threshold voltage ΔV_{th}, change in the transconductance degradation $\Delta g_m/g_m$, or change in the drain current $\Delta I_d/I_d$ before and after the device is stressed. It has been observed that V_{th} shift, or g_m degradation, can well be expressed as [7], [39]

$$\Delta I_d/I_d \quad (\text{or} \quad \Delta V_{th}, \quad \text{or} \quad \Delta g_m/g_m) = A \cdot t^n \tag{8.33}$$

where t is the stress time. Equation (8.33) is valid for almost all MOS devices, in particular, at short stress time; at long stress time V_{th} shift and/or g_m degradation rather saturates. The slope n in a log–log plot of t versus ΔV_{th} is strongly dependent on V_{gs} but has a week dependence on V_{ds}. This suggests that n changes according to hot-carrier injection mechanism. In case of DAHC mechanism $n \approx 0.5$–0.7 for devices with $t_{ox} = 68$–200 Å and $L = 0.35$–$2\ \mu$m. On the other hand A, which represents the magnitude of degradation, is strongly dependent on V_{ds} [$A \propto \exp(-1/V_{ds})$]. Figure 8.18a is a plot of Δg_m versus stress time on a log-log scale for nMOST ($L = 0.48\ \mu$m and $t_{ox} = 105$ Å). All devices are stressed under peak substrate current conditions. For pMOST $n \approx 0.15$–0.25 [40], which is much smaller than for nMOST, showing smaller degradation for pMOST. The g_m degradation in pMOST is shown in Figure 8.18b. Note that pMOST do not obey the power law equation (8.33) but rather has been observed to obey a logarithmic time dependence [43–44]. This has been interpreted as being due either to a reduction in the lateral electric field with stressing time [43], or due to a shifting point of carrier injection.

Figure 8.19 shows the relationship between g_m degradation, generated surface states N_{it} and substrate current I_b in an nMOST with $L = 0.8\ \mu$m and $t_{ox} = 200$ Å. The stress conditions were $V_{ds} = 6.6$ V, $V_{sb} = 3$ V and stress time $= 10^4$ sec. A remarkable correlation between the peak of the substrate current I_b, g_m degradation and N_{it} generation leads one to conclude that the device degradation can be monitored using the substrate current. In contrast, in this bias range the gate current I_g increases exponentially suggesting that degradation may not be correlated to the gate current (for nMOST).

If we define lifetime τ as the stress time at which the change Δ in V_{th}, g_m or I_{dsat} reaches a certain failure criterion such that $\Delta V_{th} = 10$ mV, $\Delta g_m/g_m = 10\%$ or $\Delta I_d/I_d = 10\%$, then under conditions of DC stress we find [39]

$$\tau = C \cdot I_b^{-m} \tag{8.34}$$

where C is a process dependent constant, while $m \approx 3$ is constant for a large number of NMOS/CMOS technologies with different t_{ox}, S/D structure and channel length [6, 7]. *To determine τ from Eq. (8.34), devices are generally stressed at various values of V_{ds} with V_{gs} adjusted for maximum substrate current (which is found to correspond to maximum degradation).*

It should be pointed out that Eq. (8.34) is valid so long as V_{gs} is not varied too extensively as degradation and substrate current do not correlate perfectly; i.e., the peak of degradation does not exactly coincide with the peak of I_b [41]. In such situations it is more appropriate to use the following expression for lifetime due to DC stress conditions [7], [41]

$$\tau = C_1 (I_b/I_d)^{-m}/I_d \tag{8.35}$$

where m varies from 3–5. The plot of $\tau I_d/W$ versus I_b/I_d on a log–log scale will be a straight line, the slope and intercept of which gives the degradation

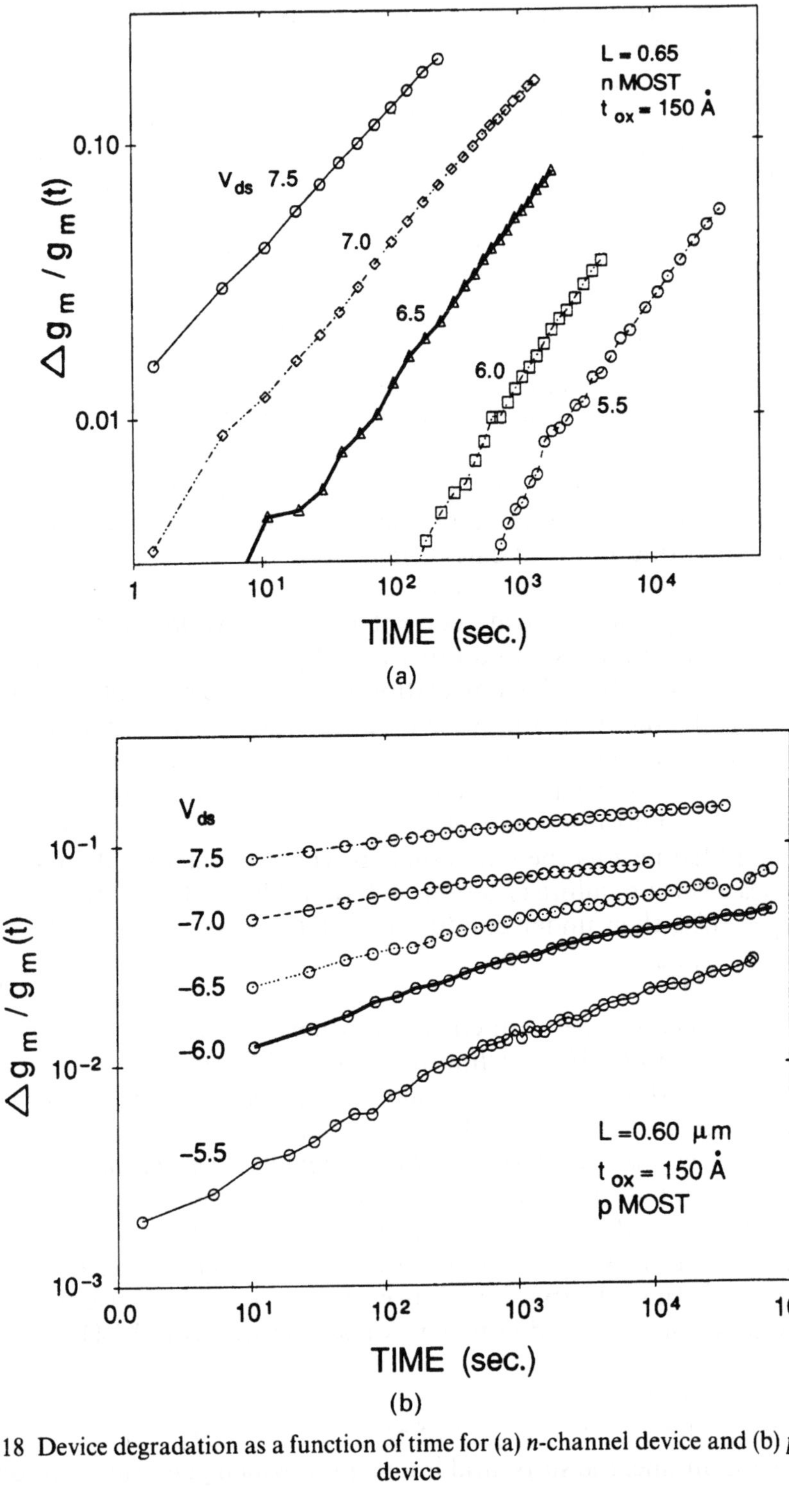

Fig. 8.18 Device degradation as a function of time for (a) *n*-channel device and (b) *p*-channel device

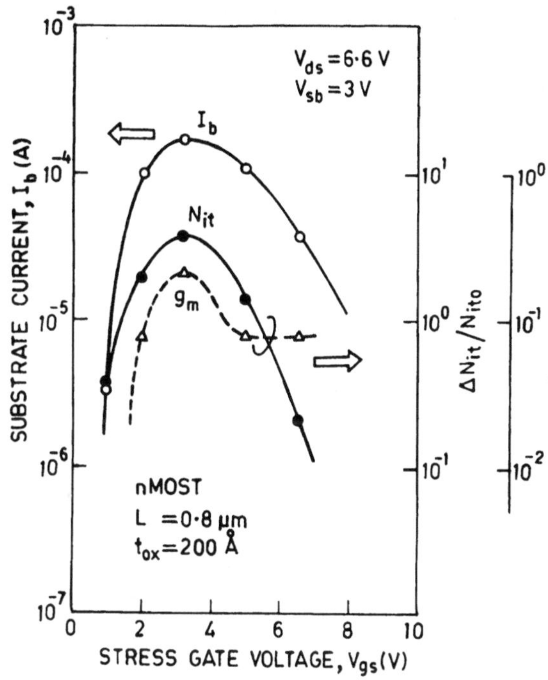

Fig. 8.19 Correlation between transconductance degradation g_m, substrate current I_b and density of interface states N_{it}, exhibiting similar variation with V_{gs}. $L = 0.8\,\mu$m, $t_{ox} = 200$ Å, $V_{ds} = 6.6$ V and $V_{sb} = 3$ V. (After Takeda et al. [36])

parameters m and C_1. Note that the drain current I_d is per unit width W (I_d/W).

Previous studies on near micron devices showed that nMOST degradation is technology dependent and is relatively independent of the channel length for stress at the same I_b [6]. However, recent studies have shown that the effect of device degradation on device performance is more prominent in short-channel submicron regime nMOST [10]. This is because device degradation is a localized phenomena, therefore, it is expected that hot-carrier created damage near the drain end will be independent of the channel length for the same amount of stress ($I_b^m \cdot t = \text{const}$). In other words, the ratio of the damaged interface area to the total channel area increases as the channel length decreases, and thus device lifetime decreases because the relative amount of degradation increases. Equations (8.34) and (8.35) have been slightly modified to take account for the channel length dependence on device degradation [42]. Thus, Eq. (8.34) is modified as

$$\tau = C_2 L^{n_2} \cdot I_b^{-m} \tag{8.36}$$

where $n_2 \approx 2$–3. Equation (8.35) can be modified in a similar way to take into account the dependence of τ on L.

For n-channel MOSFETs, I_b or (I_b/I_d) is a well accepted monitor for hot-carrier induced degradation. However, for p-channel MOSFETs both I_b [45] and I_g [46] have been used as monitors, although degradation follows I_g better than I_b [43]. It has been suggested that for electron trapping damage in pMOST, where g_m and I_d increase, I_g should be used; whereas for interface state generation in very short channel pMOST ($L < 0.5\,\mu$m), where g_m and I_d decrease, I_b should be used as the monitor for τ measurement [38]. If I_g is taken as the monitor, then pMOST lifetime can be expressed as

$$\tau = C_3 I_g^{-m} \quad \text{(pMOST)} \tag{8.37}$$

where constant $m = 1.5$ [24] as against 3 for nMOST.

Dynamic Stressing. Although MOSFETs in circuits are subjected to transient gate and drain voltage conditions, their hot carrier reliability has often been evaluated based upon the model for static or DC stress, as given by Eqs. (8.34)–(8.35). In many of these studies AC stress life time τ_{AC} has been compared to the lifetime predicted by quasi-static application of Eq. (8.35) for nMOST [8], [41] and Eq. (8.37) for pMOST [47]. Thus, for example, τ_{AC} for nMOST is given by

$$\frac{1}{\tau_{AC}} = \frac{1}{H \cdot T} \int_0^T (I_b^m / I_d^{m-1})\,dt \tag{8.38}$$

where T is the full cycle time, I_b and I_d are the currents at time $t(\leq T)$. The degradation parameters m and H are in general gate and drain bias dependent [8]. However, it has been observed that stress under AC, or dynamic conditions, can be significantly worse than might be expected from the quasi-static sum of DC stresses given by Eq. (8.38). Recently much attention has been focused on this enhanced AC stress effect [48]–[55]. Due to severe degradation in nMOST, dynamic or AC stress analysis has been studied mainly in nMOST. What follows is for n-channel devices.

Early reports showed that enhanced AC degradation was the result of enhanced substrate currents during falling gate voltage edges and shorter transition times [48]–[52]. The phenomenological link between substrate current and hot-carrier degradation [see Eq. (8.34)] then explained the enhanced AC degradation. However, later reports failed to confirm any substrate current enhancement, at least for rise/fall times as low as 3 ns, and the apparent increase in the substrate current was linked to the measurement difficulties [53]–[55]. It was also pointed out that the discrepancy between DC and AC stress could be due to the fact that Eqs. (8.34) or (8.35) do not adequately model all aspect of hot-carrier damage. Indeed in the absence of any 'transient effect', this is likely the case as has been pointed out by Mistry and coworkers [55]–[58]. They have shown that enhanced AC degradation is due to the presence of three different damage modes, rather than the one mode which traditionally has been associated with peak substrate current region, and is thought to be due to interface state generation. The three modes of degradations are (1) electron trap creation and interface state generation by hot holes ($N_{ox,h}$) taking place at *low gate voltages*, (2) electron trapping by hot electrons ($N_{ox,e}$) occurring at *high gate voltages*, and (3) interface state creation (N_{ss}) which occurs at *intermediate gate voltages*, around the peak of the substrate current. All the three types of damage contribute to device degradation during AC stress. The lifetime due to these damages are empirically modeled as [58]

$$1/\tau_{N_{ss}} = A_1 \cdot I_b^{-m_1} \tag{8.39a}$$

$$1/\tau_{N_{ox,h}} = A_2 \cdot (I_b/I_d)^{-m_2} \cdot I_d \tag{8.39b}$$

$$1/\tau_{N_{ox,e}} = A_3 \cdot (I_g/I_d)^{-m_3} \cdot I_d \tag{8.39c}$$

where A_1, A_2, A_3 and m_1, m_2, m_3 are empirical constants. Note that Eq. (8.39c) is valid only for V_{gs} such that the gate current is negative (i.e., consists primarily of electrons). As an approximation, it is valid for $V_{gs} > V_{ds}/2$.
In order to estimate AC stress lifetimes, we must first calculate the quasi-static contributions for the three damage modes by integrating Eqs. (8.39) over the time period T of the AC stress waveform. For example, the value of $\tau_{N_{ox,h}}$ is calculated as

$$\frac{1}{\tau_{N_{ox,h}}} = A_2 \cdot \frac{1}{T} \int_0^T (I_b/I_d)^{m_2} \cdot I_d dt \tag{8.40}$$

where quasi-static values are used for all currents. The values of $\tau_{N_{ss}}$ and $\tau_{N_{ox,e}}$ are similarly calculated. In this integration procedure, $1/\tau$ is treated as a damage function which is integrated over the time period of the AC stress waveform for each of the three damage modes. The following Matthiessen-like rule is then used to calculate the lifetime taking all three damage modes into account [58]

$$1/\tau_{AC} = 1/\tau_{N_{ss}} + 1/\tau_{N_{ox,h}} + 1/\tau_{N_{ox,e}}. \tag{8.41}$$

The damage functions for the three damage modes are added together in order to calculate the total damage. Figure 8.20a shows the measured AC stress lifetime (dotted lines) compared to that calculated (continuous lines) using the above model for a stress waveform resembling inverter-like AC stress. Figure 8.20b shows the damage contributions of the three damage modes.
Instead of using three damage mode equations as discussed above, Hu and coworkers have used Eq. (8.38) with H and m as bias dependent parameters to account for higher degradation under dynamic stressing [8], [59]–[61]. Phenomenologically bias dependent of H and m accounts for different damage mechanism under different bias conditions.

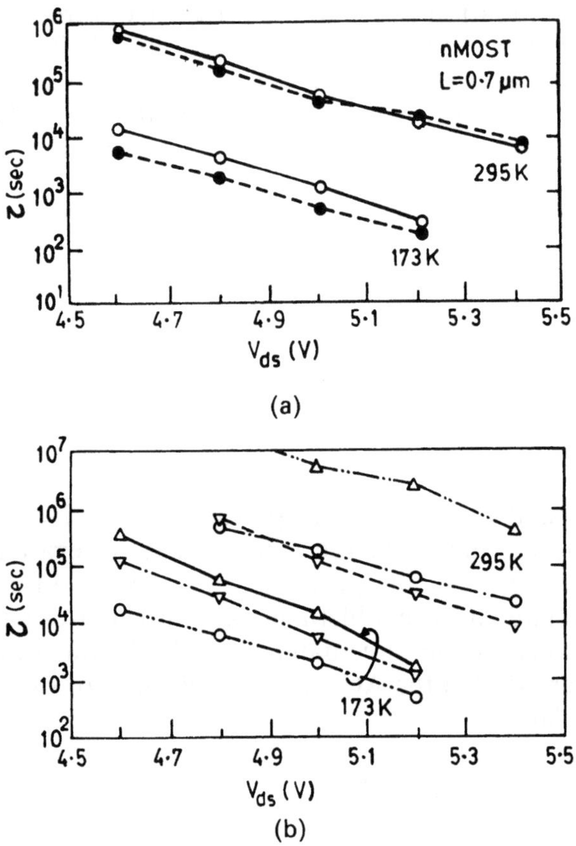

Fig. 8.20 (a) Measured (●) and calculated (○) AC stress lifetimes for inverter-like stress versus $V_{ds} = 4.3$ V. (b) Calculated contributions of the three damage modes to the AC lifetimes for N_{oxh}(○), N_{ss}(▽), and N_{oxe}(△). (After Mistry et al. [58])

8.6 Impact of Degradation on Circuit Performance

In the previous sections we have discussed models for MOSFET substrate and gate currents that are related to the device lifetime models based on device-level degradation parameters ΔV_{th}, $\Delta g_m/g_m$, etc. By combining these models in a pre- and post-processor configuration to a circuit simulator such as SPICE, one can calculate lifetime of each device in a circuit under operating conditions. Thus, the device lifetime can be estimated in a circuit environment. This is the approach used in most of the circuit reliability simulators to assess the circuit level performance as a function of hot-carrier stress [8], [18], [60]–[64]. One such simulator called SCALE (Substrate Current And Lifetime Evaluator) was developed at the University of California, Berkeley [8]. In a pre-processor configuration SCALE calls SPICE to calculate the transient voltage waveforms at the drain, gate,

source and substrate of the user selected devices. The post-processor then calculates the transient substrate current based on transient terminal voltages. The substrate current in turn is used to calculate device lifetime. In the Berkeley version of SCALE, drain currents are obtained from the BSIM model (Level = 4) and the device lifetime is calculated using Eq. (8.38). However, one can implement substrate current and device lifetime models in SCALE that are more appropriate for a particular technology [18].
Although using SCALE one can flag devices that have high substrate current and hence low lifetime, the relationship between individual device degradation and circuit degradation as a whole remains ambiguous. This is because not all transistors affect circuit behavior in the same way [60]–[64]. For example, in a circuit one transistor M_1 may degrade much more severely than other transistor M_2, but circuit performance may depend more on M_2 than M_1. The sensitivity of this dependence may also change depending on what characteristic of the circuit is studied. Simple device failure criterion such as setting device lifetime at $\Delta I_{ds}/I_{ds} = 10\%$ may often be misleading when applied generally. It is, therefore, imperative that a simulator be able (1) to predict the degradation of each transistor while operating in a circuit environment for user-definable length of time and, (2) to directly simulate the entire circuit using degraded device parameters obtained from the information in step 1. The simulator CAS (Circuit Aging Simulator)[5] simulates circuits undergoing dynamic degradation for a user defined length of time [59]–[60]. CAS incorporates the structure and model of SCALE; in fact SCALE is a subset of CAS. A new parameter *Age*, is introduced to quantify the amount of degradation each device experiences during circuit operation and is defined as

$$Age = \int \frac{1}{WH}\left(\frac{I_b}{I_d}\right)^m \cdot I_d \cdot dt \tag{8.42}$$

for nMOST, while for pMOST the ratio I_b^m/I_d^{m-1} is replaced by I_g^m or sum of the two with weighting factors [61]. In Eq. (8.42) H and m are gate and drain bias dependent degradation parameters, t is the circuit operating time, and W is device width. During circuit simulation, the *Age* is calculated for each device at each time-step, then integrated to obtain the total Age for the SPICE analysis. After the *Age* of each transistor in the circuit is calculated by this quasi-static method, the aged process files corresponding to the individual transistors is then used to simulate the actual circuit degradation for a user specified period of time.
Both SCALE and CAS are based on the assumptions that (1) SPICE analysis must be transient analysis since aging is based on time; and (2)

[5] CAS is now replaced by BErkeley Reliability Tool called BERT [60].

circuit behavior is assumed to be periodic with the period equal to the length of the SPICE analysis.

8.7 Temperature Dependence of Device Degradation

The device degradation depends upon gate and substrate current, which in turn depends upon drain current. Since I_d is temperature dependent (cf. section 6.9), it is expected that I_g and I_b will be temperature dependent. Experimentally it is found that the device degradation increases as temperature is lowered (see Figure 8.14) and hence device life-time becomes shorter at lower temperature (see Figure 8.20) [65]–[68]. Intuitively this could be understood as follows. Lowering the temperature results in an increased injection of hot electrons into the gate oxide and hence gate current increases [69]. Similarly lowering the temperature increases the optical-phonon mean free path λ and thus the energy ($\lambda\mathscr{E}_m$) acquired by hot carriers by the field, thereby increasing substrate current [cf. Eq. (8.31)].
The temperature coefficient of the gate current ($d(\ln(I_g/I_d))/dT$) is higher (-0.0256/°C) compared to the substrate current (-0.0132/°C) [23]. From Eq. (8.32) one expects the slopes of I_g/I_d and I_b/I_d versus temperature to differ by a factor of $\Phi_b/(B_i\lambda) \approx 2.1$. Increased device degradation at lower temperature can also be understood from the fact that lowering the temperature increases rate of hot-electron trapping. Higher density of traps and/or oxide charge will result in higher mobility degradation and threshold voltage shift.

Modeling Temperature Dependence of Substrate Current. Experimentally it is observed that I_b exhibits higher temperature dependence compared to I_d. Figure 8.21 shows a plot of I_b versus V_{gs} for an n-channel device ($L = 0.78\ \mu$m) at three temperatures 0, 25 and 100 °C. As can be seen from this figure I_b decreases as temperature increases, consistent with the I_d temperature dependence. The normalized temperature coefficient of I_b in the temperature range 0–120°C is approximately given by [18]

$$\frac{1}{I_b}\frac{\partial I_b}{\partial T} \approx 5 \times 10^{-3} \quad (/°\text{C}) \tag{8.43}$$

while the normalized temperature coefficient of I_d, in saturation, for the same device as shown in Figure (8.21) is

$$\frac{1}{I_d}\frac{\partial I_d}{\partial T} \approx 3 \times 10^{-3} \quad (/°\text{C}) \tag{8.44}$$

which is very close to the temperature coefficient of I_b. This implies that the rest of the parameters in Eq. (8.12) account for the difference.

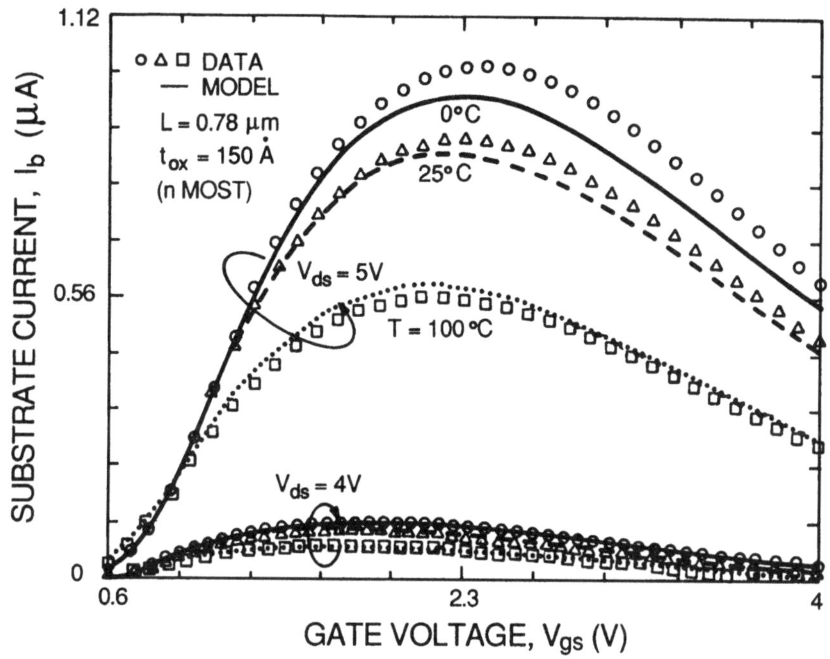

Fig. 8.21 Substrate current I_b as a function of V_{gs} at $V_{ds} = 4.6$ V, $V_{sb} = 0$ V for a typical nMOST at three temperatures 0, 25 and 100 C

It has been shown [18], [66] that Eq. (8.12) agrees with the experimental data in the temperature range 400 K–77 K provided temperature dependence of (1) the drain current, I_d, (2) the ionization coefficients A_i and B_i, and (3) the effective ionization length, l are taken into account. It has been observed that the ionization coefficient A_i is almost independent of temperature and it is the temperature dependence of B_i which accounts for the temperature dependence of the ionization rate α_n [18]. The following linear relation of the coefficient B_i with temperature is assumed

$$B_i(T) = B_{io}[1 + \beta_{B_i}(T - T_0)] \tag{8.45}$$

where B_{io} represent the values of the ionization constants B_i at the reference temperature $T = T_0$ (say, 300 K), and β_{B_i} is the temperature coefficients of B_i whose value is found to be [18]

$$\beta_{B_i} \equiv \frac{1}{B_i}\frac{dB_i}{dT} = 9.28 \times 10^{-4} \quad \mathrm{K}^{-1} \quad \text{(electrons)}. \tag{8.46}$$

This value is consistent with that reported by Grant [70].
The following linear relation for the temperature dependence of l is generally used

$$l(T) = l(T_0)[1 + \beta_l(T - T_0)] \tag{8.47}$$

where the parameter, β_l, is the temperature coefficient of l and is obtained by curve fitting the experimental I_b data to Eq. (8.12) with B_i and l given by Eqs. (8.45) and (8.47), respectively. This approach is adequate to model the temperature dependence of I_{sub}, which has been tested in the temperature range 273 K–400 K, as shown by the continuous line in Figure 8.21.

References

[1] R. R. Troutman, 'Low-level avalanche multiplication in IGFETs', IEEE Trans. Electron Devices, ED-23, pp. 419–425 (1976).

[2] Y. A. El-Mansy and D. M. Caughey, 'Modeling weak avalanche multiplication currents in IGFETS and SOS transistors for CAD', IEEE IEDM-75, *Dig. Tech. Papers*, pp. 31–34 (1975).

[3] S. Selberherr, *Analysis and Simulation of Semiconductor Devices*, Springer-Verlag, Wien, New-York, 1984.

[4] J. W. Slotboom, G. Streutker, G. J. T. Davids, and P. B. Hartog, 'Surface impact ionization in silicon devices', IEEE IEDM-87, *Dig. Tech. Papers*, pp. 494–497 (1987).

[5] A. Erdelyi, *Asymptotic Expansions*, Dover Publications Inc, New York, 1956.

[6] C. Hu, S. C. Tam, F. C. Hsu, P. K. Ko, T. Y. Chan, and K. W. Terrill, 'Hot-electron induced MOSFET degradation—model, monitor, and improvement', IEEE Trans. Electron Devices, ED-32, pp. 375–385 (1985).

[7] C. Hu, 'Hot carrier effects', in *Advanced MOS Device Physics* (N. G. Einspruch and G. Gildenblat, eds.), VLSI Electronics Vol. 18, pp. 119–139, Academic Press Inc., New York, 1989.

[8] M. M. Kuo, K. Seki, P. M. Lee, J. Y. Choi, P. K. Ko, and C. Hu, 'Simulation of MOSFET lifetime under AC hot-electron stress', IEEE Trans. Electron Devices, ED-35, pp. 1004–1010 (1988).

[9] T.-C. Ong, P. K. Ko, and C. Hu, 'Modeling of substrate current in p-MOSFET's', IEEE Trans. Electron Device Lett., EDL-8, pp. 413–416 (1987).

[10] J. Chung, M. C. Jeng, G. May, P. K. Ko, and C. Hu, 'Hot-electron currents in deep-submicrometer MOSFETs', IEEE IEDM-88, *Dig. Tech. Papers*, pp. 200–203 (1988).

[11] Y. Tang, D. M. Kim, Y.-H. Lee, and B. Sabi, 'Unified characterization of two-region gate bias stress in submicrometer p-channel MOSFETs', IEEE Electron Device Lett., EDL-11, pp. 203–205 (1990).

[12] J. Faricelli and G. Gildenblat, 'Numerical verification of substrate current model in silicon IGFET's', Solid-State Electron., 30, pp. 655–660 (1987).

[13] J. Mar, S. S. Li, and S. Y. Yu, 'Substrate current modeling for circuit simulation', IEEE Trans. Computer-Aided Design, CAD-1, pp. 183–186 (1982).

[14] Y. W. Sing and B. Sudlow, 'Modeling and VLSI design constraints of substrate current', IEEE IEDM-80, *Dig. Tech. Papers*, pp. 732–735 (1980).

[15] T. Sakurai, K. Nogami, M. Kakumu, and T. Iizuka, 'Hot-carrier generation in submicrometer VLSI environment', IEEE J. Solid-State Circuits, SC-22, pp. 256–259 (1987).

[16] T. Y. Chan, P. K. Ko, and C. Hu, 'A simple method to characterize substrate current in MOSFETs', IEEE Trans. Electron Device Lett., EDL-5, pp. 505–507 (1984).

[17] T. Y. Chan, P. K. Ko, and C. Hu, 'Dependence of channel electric field in device scaling', IEEE Electron Device Lett., EDL-6, pp. 551–553 (1985).

[18] N. D. Arora and M. Sharma, 'MOSFET substrate current model for circuit simulation', IEEE Trans. Electron Devices, ED-38, pp. 1392–1398 (1991).

[19] Y. Tang, D. M. Kim, 'Modeling of on-state MOSFET operation and derivation of maximum channel field', IEEE Trans. Electron Devices, ED-38, pp. 2472–2480 (1991).

[20] W. Shockley, 'Problems related to *pn* junction in silicon', Solid-State Electron., 2, pp. 35–67 (1961).

[21] J. F. Verwey, R. P. Kramer, and B. J. de Maagt, 'Mean free path of hot electrons at the surface of boron doped silicon', J. Appl. Phys., 46, pp. 2612–2619 (1975).

[22] T. H. Ning, C. M. Osburn, and H. N. Yu, 'Emission probability of hot electrons from silicon into silicon dioxide', J. Appl. Phys., 48, pp. 286–293 (1977).

[23] S. Tam, P. K. Ko, and C. Hu, 'Lucky-electron model of channel hot electron injection in MOSFETs', IEEE Trans. Electron Devices, ED-31, pp. 1116–1125 (1984).

[24] T.-C. Ong, P. K. Ko, and C. Hu, 'Hot-carrier current modeling and device degradation in surface-channel p-MOSFETs', IEEE Trans. Electron Devices, ED-37, pp. 1658–1666 (1990).

[25] P. E. Cottrell, R. R. Troutman, and T. H. Ning, 'Hot-electron emission in *n*-channel IGFETs', IEEE Trans. Electron Devices, ED-26, pp. 520–533 (1979).

[26] E. Takeda, H. Kume, T. Toyabe, and S. Asai, 'Submicrometer MOSFET structure for minimizing hot-carrier generation', IEEE Trans. Electron Devices, ED-29, pp. 611–618 (1982).

[27] K. R. Hofmann, C. Werner, W. Weber, and G. Dorda, 'Hot-electron and hole-emission effects in short *n*-channel MOSFETs', IEEE Trans. Electron Devices, ED-32, pp. 691–699 (1985).

[28] M. Miura-Mattausch, A. V. Schweri, W. Weber, C. Werner, and G. Dorda, 'Gate currents in thin oxide MOSFETs', IEE Proceedings, 134, Pt. I, pp. 111–115 (1987).

[29] E. Takeda, 'Hot-carrier effects in submicrometer MOS VLSI', IEE Proceedings, 131, Pt. I, pp. 153–164 (1984).

[30] N. S. Saks, P. L. Heremans, L. Van den Hove, H. E. Maes, R. F. De Keersmaecker, and G. J. Declerck, 'Observation of hot-hole injection in NMOS transistors using a modified floating gate technique', IEEE Trans. Electron Devices, ED-33, pp. 1529–1534 (1986).

[31] K. K. Ng and G. W. Taylor, 'Effects of hot-carrier trapping in *n*- and *p*-channel MOSFETs', IEEE Trans. Electron Devices, ED-30, pp. 871–876 (1983).

[32] T. Tsuchiya and J. Frey, 'Relationship between hot-electrons/holes and degradation for *p*- and *n*-channel MOSFETS', IEEE Electron Device Lett., EDL-6, pp. 8–11 (1985).

[33] M. Koyanagi, A. G. Lewis, J. Zhu, R. A. Martin, T. Y. Huang, and J. Y. Chen, 'Hot electron induced punchthrough (HEIP) effect in submicron PMOSFETs', IEEE Trans. Electron Devices, ED-34, pp. 839–844 (1987).

[34] S. Tam, P. K. Ko, C. Hu, and R. S. Muller, 'Correlation between substrate and gate currents in MOSFETs', IEEE Trans. Electron Devices, ED-29, pp. 1740–1744 (1982)

[35] S. Tanaka and S. Watanabe, 'A model for the relation between substrate and gate currents in n-channel MOSFETs', IEEE Trans. Electron Devices, ED-30, pp. 668–675 (1983).

[36] E. Takeda, A. Shimizu, and T. Hagiwara, 'Role of hot-hole injection in hot-carrier effects and the small degraded channel region in MOSFETs', IEEE Electron Device Lett., EDL-4, pp. 329–331 (1983).

[37] T. H. Ning, P. W. Cook, R. H. Dennard, C. M. Osburn, S. E. Schuster, and H.-N. Yu, '1 μm MOSFET VLSI technology: Part IV: Hot-electron design constraints, IEEE Trans. Electron Devices, ED-26, pp. 346–353 (1979).

[38] T. Tsuchiya, Y. Okazaki, M. Miyaka, and T. Kobayashi, 'New hot-carrier degradation mode in lifetime prediction method in quarter-micrometer PMOSFET', IEEE Trans. Electron Devices, ED-39, pp. 404–408 (1992).

[39] E. Takeda and N. Suzuki, 'An empirical model for device degradation due to hot-carrier injection', IEEE Electron Device Lett., EDL-4, pp. 111–113 (1983).

[40] C. Yao, J. Tzou, and R. Cheung, 'Temperature dependence of CMOS device reliability', IEEE IRPS-86, *Tech. Dig.*, pp. 175–182 (1986).

[41] W. Weber, C. Werner, and A. Schwerin, 'Lifetimes and substrate currents in static and dynamic hot carrier degradation', IEE-IEDM86, *Tech. Dig.*, pp. 390–393 (1986).
[42] K. R. Mistry and B. S. Doyle, 'An empirical model for the L_{eff} dependence of hot-carrier lifetimes of *n*-channel MOSFETs', Electron Device Letters, EDL-10, pp. 500–502 (1989).
[43] B. S. Doyle and K. R. Mistry, 'A lifetime prediction method for hot-carrier degradation in surface-channel p-MOS devices', IEEE Trans. Electron Devices, ED-37, pp. 1301–1307 (1990).
[44] M. Brox, E. Wohlrab, and W. Weber, 'A physical lifetime prediction method for hot-carrier-stressed p-MOS transistors', IEEE IEDM-91, *Tech. Dig.*, pp. 525–528 (1988).
[45] W. Weber and F. Lau, 'Hot-carrier drifts in submicrometer p-channel MOSFETs', IEEE Electron Device Lett., EDL-8, pp. 208–211 (1987).
[46] M. P. Brassington, M. W. Poulter, and M. El-Diwanay, 'Suppression of hot-carrier effects in submicrometer surface-channel PMOSFETs', IEEE Trans. Electron Devices, ED-35, p. 1149 (1988).
[47] T.-C. Ong, K. Seki, P. K. Ko, and C. Hu, 'Hot-carrier-induced degradation in *p*-MOSFET's under AC stress', IEEE Trans. Electron Device Lett., EDL-9, pp. 211–213 (1988).
[48] K. L. Chen, S. Saller, and R. Shah, 'The case of AC stress in the hot carrier effect', IEEE Trans. Electron Devices, ED-33, pp. 424–426 (1986).
[49] J. Y. Choi, P. K. Ko, and C. Hu, 'Hot-carrier-induced MOSFET degradation under AC stress', IEEE Electron Device Letters, EDL-8, pp. 333–335 (1987).
[50] H. Wang, M. Davis, and R. Lahri, 'Transient substrate current effects on *n*-channel MOSFET device lifetime', IEEE IEDM-88, *Tech. Dig.*, pp. 216–219 (1988).
[51] H. Wang, S. Bibyk, M. Davis, H. De, and Y. Nissan-Cohen, 'Transient hot-electron effect on *n*-channel device degradation', IEEE IEDM-89, *Tech. Dig.*, pp. 79–83 (1989).
[52] R. Bellens, P. Heremans, G. Groenseneken, and H. E. Maes, 'Analysis of the mechanisms for the enhanced degradation during AC hot carrier stress of MOSFETs', IEEE IEDM-88, *Tech. Dig.*, pp. 212–215 (1988).
[53] W. Hansch and W. Weber, 'The effect of transients on hot carriers', IEEE Electron Device Letters, EDL-10, pp. 252–255 (1989).
[54] R. Bellens, P. Heremans, G. Groeseneken, H. E. Maes, and W. Weber, 'The influence of measurement setup on enhanced AC hot carrier degradation of MOSFETs', IEEE Trans. Electron Devices, ED-37, pp. 310–313 (1990).
[55] K. R. Mistry and B. S. Doyle, 'The role of electron trap creation in enhanced hot-carrier degradation during AC stress', IEEE Electron Device Letters, EDL-11, pp. 267–269 (1990).
[56] B. S. Doyle, M. Bourecerie, C. Bergonzoni, R. Benecchi, A. Bravis, K. R. Mistry, and A. Boudou, 'The generation and characterization of electron and hole traps created by hole injection during low gate voltage hot-carrier stressing of n-MOS transistors', IEEE Trans. Electron Devices, ED-37, pp. 1869–1876 (1990).
[57] K. R. Mistry and B. S. Doyle, 'A model for AC hot-carrier degradation in *n*-channel MOSFETs', Electron Device Letters, EDL-12, pp. 492–494 (1991).
[58] K. R. Mistry, B. S. Doyle, A. Philipossian, and D. B. Jackson, 'AC hot carrier lifetimes in oxide and ROXNOX *n*-channel MOSFETs', IEEE-IEDM91, *Tech. Dig.*, pp. 727–730 (1991).
[59] P. M. Lee, M. M. Kuo, K. Seki, P. K. Ko, and C. Hu, 'Circuit aging simulator (CAS)', IEEE-IEDM88, *Tech. Dig.*, pp. 134–138 (1988).
[60] P. M. Lee, M. M. Kuo, P. K. Ko, and C. Hu, 'BERT—A circuit aging simulator', Memo. No. UCB/ERL M90/2, Electronics Research Lab., University of California, Berkeley, 1990.

[61] K. N. Quader, P. K. Ko, and C. Hu, 'Simulation of CMOS circuit degradation due to hot-carrier effects', IEEE IRPS-92, *Tech. Dig.*, pp. 16–23 (1992).
[62] T. S. Hobol and L. A. Glasser, 'Relic: A reliability simulator for integrated circuits', IEEE Proc. Int. Conf. Computer-Aided Design (Santa Clara CA), pp. 517–520 (1986).
[63] S. Aur, D. E. Hocevar, and P. Yang, 'Hotron: A circuit hot electron effect simulator', IEEE-IEDM87, *Tech. Dig.*, pp. 498–501 (1987).
[64] B. J. Sheu, W.-J. Hsu, and B. W. Lee, 'An integrated-circuit reliability simulator—RELY', IEEE J. Solid-State Circuits, 2, pp. 473–477 (1989).
[65] F. C. Hsu and K. Y. Chiu, 'Temperature dependence of hot-electron-induced degradation in MOSFETs', IEEE Electron Device Lett., EDL-5, pp. 148–150 (1984).
[66] D. Lau, G. Gildenblat, C. G. Sodini, and D. E. Nelsen, 'Low temperature substrate current characterization of *n*-channel MOSFETs', IEEE IEDM-85, *Tech. Dig.*, pp. 565–568 (1985).
[67] G. Gildenblat, 'Low-temperature operation' in: *Advanced MOS Device Physics* (N. G. Einspruch and G. Gildenblat, eds.), VLSI Electronics Vol. 18, pp. 191–232, Academic Press Inc., New York, 1989.
[68] P. Heremans, G. V. Den Bosch, R. Bellens, G. Groeseneken, and H. E. Maes, 'Temperature dependence of the channel hot-carrier degradation of *n*-channel MOSFETs', IEEE Trans. Electron Devices, ED-37, pp. 980–992 (1990).
[69] I. Kato, H. Oka, H. Hijiya, and T. Nakamura, IEEE IEDM-84, *Tech. Dig.*, pp. 601–604 (1984).
[70] W. N. Grant, 'Electron and hole ionization rates in epitaxial silicon at high electric fields', Solid-State Electron., 16, pp. 1189–1203 (1973).
[71] J. J. Sanchez, K. K. Hsueh, and T. A. DeMassa, 'Drain-engineered hot-electron-resistant device structures—A review', IEEE Trans. Electron Devices, ED-36, pp. 1125–1131 (1989)
[72] S. Tam, F.-C. Hsu, C. Hu, R. S. Muller, and P. K. Ko, 'Hot-electron currents in very short channel MOSFETs', IEEE Electron Device Letters, EDL-4, pp. 249–252 (1983).

9 Data Acquisition and Model Parameter Measurements

The accuracy of the device model predictions of the device characteristics are fully dependent on the model parameter values being used. Most of the circuit models discussed in the previous chapters are semi-empirical analytical models. These models always contain some fitting parameters that do not have physically well defined values, and very often physical values of model parameters do not always give the best fit to the actual device characteristics. For this reason, device model parameters are determined from the device data obtained from electrical measurement on different length and width devices and under different bias conditions. Collecting measured data and processing these data to accurately determine model parameter values is an essential task for the complete characterization of a transistor model for use in the circuit simulator.

In this chapter we will first discuss experimental setups for measuring the device data that is required for extracting DC and AC model parameters. We then discuss different methods of determining basic MOSFET parameters such as substrate doping concentration N_b (or doping profile), threshold voltage V_{th}, carrier mobility μ, device effective or electrical channel length L and width W, gate oxide capacitance C_{ox}, etc., which are essential for the characterization of any MOSFET model. These basic parameters are also used routinely for electrical (E)-test measurements. The general model parameter extraction using a nonlinear optimization technique will be the topic of discussion for the next chapter.

Before discussing data acquisition and the measurement setup, the following points should be noted. These are also applicable for all the device parameter measurements described in this book:

- The use of brand name equipment simply indicates the type of the equipment needed for the measurement and by no means implies that only the specified brands should be used.
- The measurement methods described have been tested and found suitable for the purpose of circuit simulation. However, this does not preclude

other measurements methods that could also be used but are not covered here.

9.1 Data Acquisition

For VLSI device characterization and model parameter extraction, measurements (such as device current, conductance, transconductance, and capacitance, etc.) are performed on the device at wafer level on test structures. This is because for device characterization we need large data sets, and therefore, it is almost always more convenient to do the measurements using a wafer *probe station* rather than encapsulating each individual device and then doing measurements. The probe station is an important piece of equipment which holds the wafer in place (see Figure 9.1). It consists of a platinum/gold plated *wafer chuck* that holds the wafer and a microscope with magnification to 100X or more. The microscope and/or the chuck have $X - Y$ movement. Often the chuck can be heated to vary the temperature of the wafer. Contacts to the device pads are made through standard (coaxial) probes mounted on magnetic (or vacuum) bases for ease of movement. The entire unit is generally enclosed in a metal box to eliminate light that might cause excess carriers to be generated at the device surface, and also provide efficient electromagnetic shielding for low current and capacitance measurements. The metal enclosure is known as a Faraday box and is normally grounded. Any leakage current between the terminals of the probe station and ground may be avoided by maintaining a dry ambient in the box. The

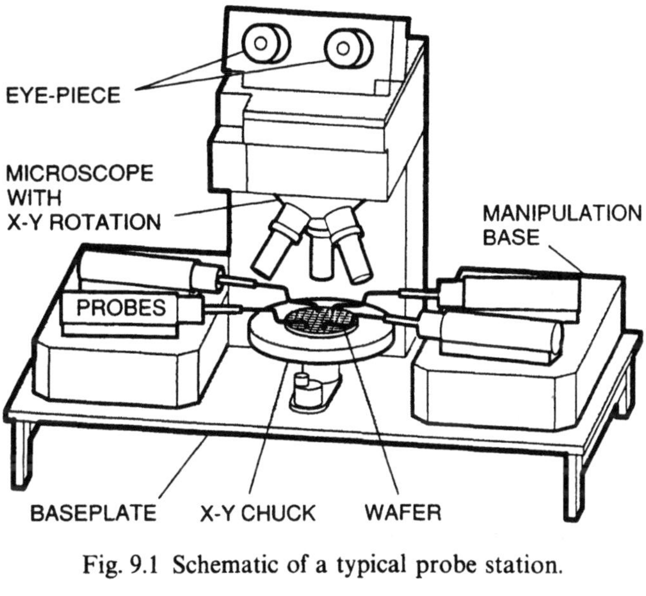

Fig. 9.1 Schematic of a typical probe station.

probe station may be vibrationally isolated by the use of an air suspension table.

One can construct one's own probe station as described in Chapter 12 (pp. 618–622) of reference [1]. Commercially available wafer probers are either manually operated or could be semi or fully automated. In the latter case the prober has internally programmed wafer stepping and probing capabilities and is controlled by a desk top computer such as a DEC μVAX, HP 9836, IBM PC or any other equivalent. Setting up an automatic prober involves turning on the prober, initializing the prober I/O system, and aligning a wafer prior to external control by the computer. Semiautomatic or fully automatic operation requires that a prober file containing such information as die size, device location, and device sizes be stored on the computer system. As opposed to semi or fully automatic operation, in manual mode the prober is used in a single device prober operation with the sole purpose of contacting to the pads, and does not receive any external commands.

Since MOS transistor behavior depends on both device dimensions and bias voltage, a very large number of measurements are required to generate a single set of model parameters that will be valid for all the transistors on a VLSI chip. Typically 3 to 4 transistors of varying length and constant width, 3 to 4 transistors of constant length and varying width, and 2 to 3 minimum geometry and square geometry transistors are used for each transistor type (n-channel/p-channel enhancement or depletion type, etc). For example, a CMOS technology will require 16 to 20 transistors (8 to 10 each for nMOST and pMOST) on which $I-V$ characteristics are measured with different gate, drain and bulk voltages in order to obtain model parameters for a particular test die. For example, a 2 μm CMOS technology may have drawn (or mask) channel width to length ratios (W_m/L_m) of 20/2, 20/3, 20/4, 20/6, 4/20, 6/20, 8/20, 4/2, 4/4, 20/20, or similar structures (see Figure 9.2). Different length and width devices are needed to extract the

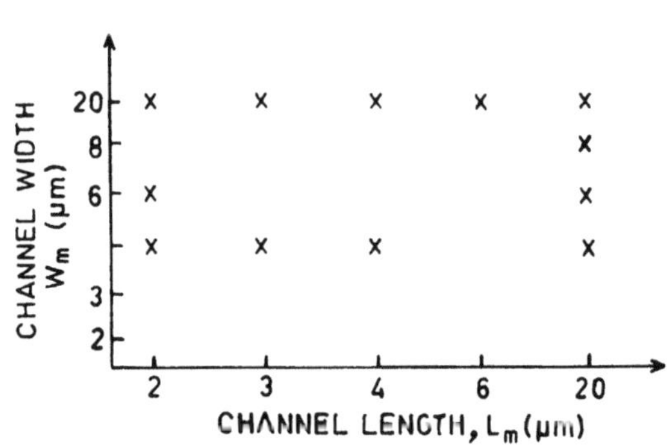

Fig. 9.2 A typical test transistor matrix (2 μm CMOS process) for model parameter extraction

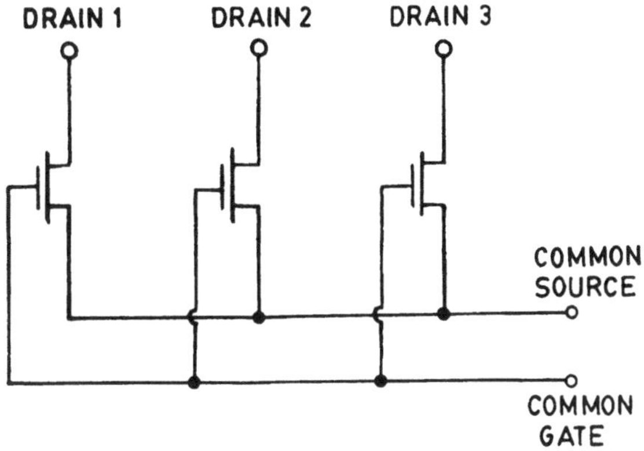

Fig. 9.3 Test transistor configuration in a test chip; transistors are laid down with common source and gate

geometry dependence of the parameters. The long and wide channel device is normally used as a reference device. Sometimes it is better to choose the most commonly used device as the reference device so that it could be modeled with the maximum accuracy.

These so called *test transistors* are normally included in scribe lanes in regular production wafers, or they could be part of a stand alone test structure. In the test structure one device type (nMOST, pMOST, etc) is normally laid out with common gate and source for maximum utilization of the connecting pads, as shown in Figure 9.3.

In order to measure statistically significant amounts of device data in a reasonable time, it is important to have a fully automated measurement system, called the *parametric test system*[1] [2]–[7]. For statistical analysis of the data, device measurements are obtained not only at the individual locations on a wafer (die-level), but also on individual wafers within a lot (wafer-level). The parametric test measurement data may be collected at any point where electrical measurements of the wafers are practical, but is commonly performed on wafers which have a protective oxide or nitride layer and have metallization to provide good electrical contacts. This data is important not only for model parameter extraction, but is often used to monitor the effect of many factors in the fabrication environment for process control and to improve device characteristics.

Figure 9.4 shows the block diagram of a device characterization system presently common throughout the industry. It consists of the following units:

---

[1] There are various commercially available Parameteric Test Systems and companies such as Hewlett-Packard [5] and Keithley [7] offer not only hardware but also parameter extraction software that runs on their tester.

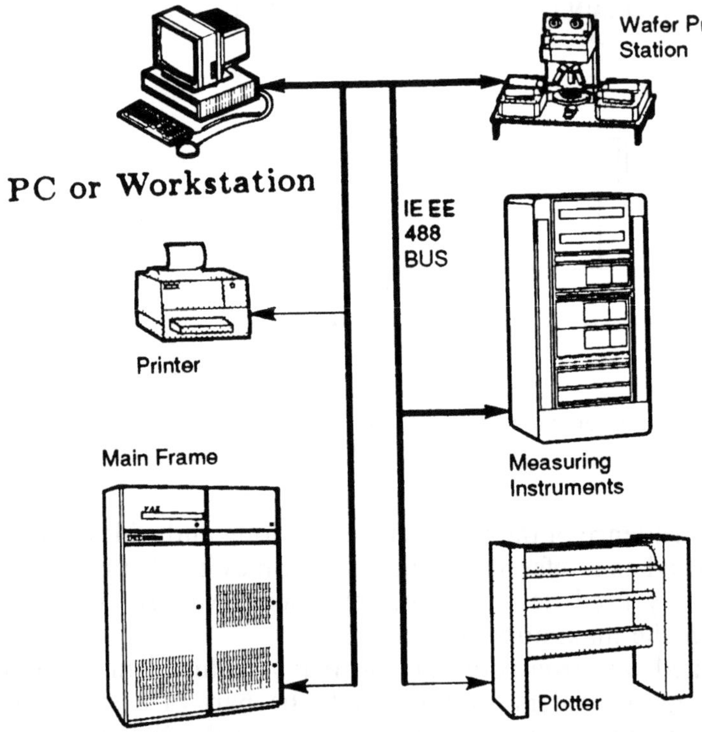

Fig. 9.4 Block diagram of the device characterization system

1. An automatic wafer prober unit which connects test structures, by means of a *probe card*, with the switching matrix of the measuring unit. Wafer adjustment and test structure selection are done automatically.
2. A measuring unit which can perform DC and AC measurements, like the HP 4145A semiconductor parameter analyzer [8] for DC measurements[2], and the HP 4275A multiple frequency LCR meter [9] for C–V measurement, or any other equivalent system such as a Keithley instrumentation package containing various instruments for voltage and current forcing and sensing.
3. A switching matrix box containing a fixed relay matrix that can be directed over the bus for interconnection of instruments to the device under test.
4. A DEC μ-VAX computer, or any other equivalent, which acts as a system

[2] HP4145A Semiconductor Parameter Analyzer has some MOS parameter extraction capabilities, and internal graphics that permit it to be used as stand alone system for measuring a single transistor. It has four stimulus measurement units capable of being setup as constant or stepped voltages up to 100 V or current sources from 1 pA to 100 mA [8].

controller. The computer controls the wafer prober and the measurement hardware and also stores the data.
5. A printer/plotter to produce hard copy of program and test results.
6. An ethernet link between the test controller and the network of a DEC VAX main frame computer for allowing data transfer, giving access to further data storage and analysis and providing the user with remote access.

The elements of this system, with the exception of the VAX mainframes, are linked together with an IEEE-488 bus. Another link connects the test controller to a plotter or printer. Normally a Shared Resource Management (SRM) software system enables connection of the test system to printer or plotter. It also enables connection to different test equipments, for example a C–V measuring system.

Conventionally, the device characterization and model parameter extraction require separate programs for different test structures, and therefore, a great deal of time and resources are needed to maintain the different softwares. Recently it has become more common to develop a single program which standardizes testing for different test structures. These are menu driven programs which generate test procedures to produce the database required to obtain DC and AC model parameters in support of circuit design and process monitoring.

Programs are written to automatically set up and control the acquisition of the device data and perform analysis [2]–[7]. A high level language, like Fortran or C, is used for writing the tester software in order to permit large program size, high degree of modularization and rapid execution. These test programs are normally divided into different segments to perform various tasks. The hardware segment automatically examines the device under test (DUT). The measurement hardware is interfaced to the DUT via a probe card. A special probe card simulating the device set can be used for the debug of this test software without requiring an actual DUT. Also, under software control, the user can interact with the hardware of the test system, thereby insuring the validity of the data obtained from wafer testing. The user also controls the stepping of the prober through the software. The wafer stepping software is independent of the hardware test segment. Once testing of the hardware test and wafer steeping segments are completed, the user can begin specifying particular electrical tests (E-tests) to be performed in the wafer test segment. It is in this segment of the program where wafer testing is done and data files are created.

While there are different ways to store data in files, these generally consist of sequential lists of records. Each record contains all the data collected from a device during a test session. For example, a device record might consist of three blocks: (1) the *header block* containg a description of the processing parameters, test operating environment data, and devices. It has the information on device geometry and type, implant data, die location, die identification, and test temperature. The header block primarily helps when retrieving data for analysis. (2) The *data block* consists of measured characteristics like $I-V$ data, threshold voltage, breakdown voltage etc. (3) The *calculation block* where the calculation results for the parameter values are stored. Retrieval programs are generally written to allow flexibility and simplicity in selecting the data.

Low-Current Measurement Considerations. At wafer level stable low current measurements, specially below $1\,pA$, are normally difficult to perform because the probe needles and wafer chuck act as antennas and pick up external noise. Therefore, to measure low currents the following precautions must be taken:

- In order to reduce noise due to external fields, the entire wafer probe and test leads should be enclosed in a metal box, called the *electrostatic shield box* or *Faraday cage*.
- The end of the measurement cables from the instruments are attached directly to the shielding box, which itself is connected to the system ground. Within the shielding box, connection to the probe is made by coaxial test leads. The cable must be low-noise cable[3], such that it is virtually free of noise due to the friction or twists in the cables.
- *Coaxial probe needles, rather than ordinary probe needles, must be used.* This is necessary in order to reduce stray capacitance between the probes. This will also reduce noise and inter-terminal signal crossover and suppress unwanted oscillations caused by electromagnetic fields.
- The AC power source of the prober stepper motor, light source, and thermal chuck are prime sources of noise. Therefore, it is important to use DC power supplies for these purposes.
- The wafer should not be exposed to light; otherwise, incident light will cause photogeneration current to flow in the device making reproducible measurements impossible.
- The switching matrix, if used, must have high-insulation relays for switching connections.
- To eliminate leakage currents (< fA) from the measurement system, measurements must be made in a dry ambient by flowing nitrogen gas over the wafer.

Device Functionality Test. For the purpose of taking device data for automatic parameter extraction, it is important that data be collected only

[3] For low level of current measurements in the pA range, a *triaxial cable* is often used. It is a double shielded coaxial cable in which the center conductor is surrounded by two concentric, independent shield conductors. It helps in maximizing the attenuation of radiated signals and minimizes the pickup of external interference.

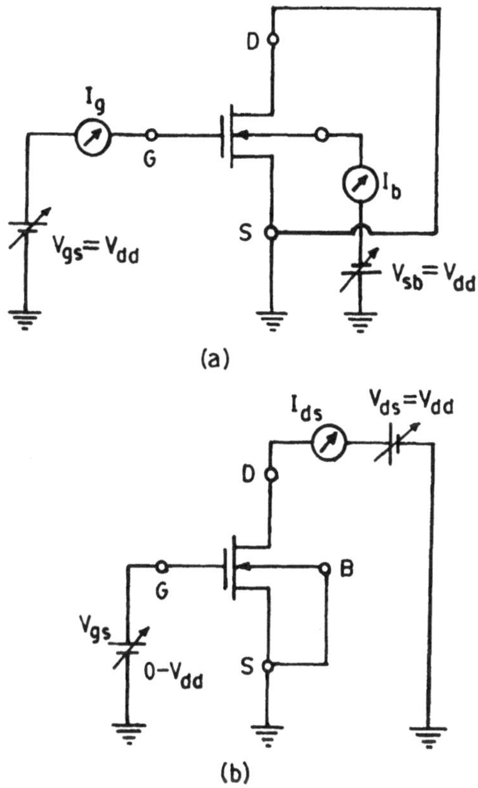

Fig. 9.5 Circuit schematic for the MOSFET functionality test; (a) indicates shorted gate or source/drain junctions and polarity of the device (*n*- or *p*-channel), (b) indicates shorted source and drain (see text for details)

for normal device behavior. Therefore, *the first step is to check for device functionality*. This is done by checking the source-drain junction leakage current, gate oxide leakage and shorted devices. These tests are done by biasing the transistor as shown in Figure 9.5a. Note that gate potential is at $V_{dd}(= 5\,\mathrm{V})$ and source and drain are ground potential, while the body is at $-V_{dd}$ or V_{dd}. Under these conditions, measuring the gate current I_g and substrate current I_b results in the detection of a short to one or the other terminal. These currents can easily be measured by direct methods using picoAmmeter such as HP4140B [10] or HP4145B parameter analyzer. A gate current $I_g > 0.1\,\mu\mathrm{A}$ normally indicates a shorted gate. If the substrate current $|I_b| > 10\,\mu\mathrm{A}$ for both $V_{sb} = \pm V_{dd}$, it indicates a shorted source/drain junction. The bias setup shown in Figure 9.5a can also be used for testing type of the transistor as to whether it is *n*-type or *p*-type. If $V_{sb} = V_{dd}$ and $I_b > 10\,\mu\mathrm{A}$, then the device is n-channel (nMOST), but if $V_{sb} = -V_{dd}$ and $I_b < -10\,\mu\mathrm{A}$ then the device is p-channel (pMOST).

If the drain is biased at V_{dd} and I_{ds} is measured as V_{gs} is changed from 0 to V_{dd}, as shown in Figure 9.5b, and if I_{ds} remains the same in the two cases ($V_{gs} = 0$ and $V_{gs} = 5$ V) it indicates the source and drain are shorted together.

9.1.1 Data for DC Models

The DC model parameters are obtained from the device $I - V$ measurements collected across the complete range of channel widths (W_m), lengths (L_m) and bias voltages appropriate for the process. Normally three types of $I - V$ data are required, corresponding to three different regions of device operation—linear, saturation and subthreshold region. These are:

- A family of I_{ds} curve as a function of V_{gs} for different values of V_{sb} and constant V_{ds}. The drain voltage V_{ds} is fixed, typically at 50 mV to make sure that the device is operating in the linear region. For the sake of reference we call this linear region data as data set A.
- The current I_{ds} is measured as a function of V_{ds} for varying values of V_{gs}, keeping constant V_{sb}. The whole set is then repeated for a different V_{sb}. This data covers the linear and saturation regions of the device characteristics. This data will be referred to as data set B.
- The current I_{ds} is measured as a function of V_{gs} for varying values of V_{ds} keeping constant V_{sb}. The $I_{ds} - V_{gs}$ data at different V_{ds} and V_{sb} covers subthreshold and saturation regions of device operation. We call this as data set C.

The data set A is used to calculate device threshold voltage V_{th} as a function of V_{sb}, as will be discussed in section 9.4. This V_{th} versus V_{sb} data, in turn, is used to calculate threshold voltage model parameters. In addition data set A is also used to determine linear region parameters, such as effective channel length and width parameters ΔL and ΔW, respectively, and mobility parameters, μ_0, θ, etc. The data set B is normally used to calculate saturation region parameters; however, often this data set is also used to extract both linear and saturation regions parameters. Finally, the data set C is used to calculate subthreshold slope and subthreshold region parameters including DIBL parameters. However, more often DIBL parameters are determined along with saturation region parameters using data set B.
The device $I - V$ data is easily obtained using either the Parametric Test System (PTS) described earlier or using an HP4145A parameter analyzer. Normally, for statistical analysis, the automated PTS system is used. However, for measurements on a few dice the HP4145 is more useful.

Substrate and Gate Current Measurements. For normal device characterization it is the drain current I_d which is routinely measured, while gate and substrate current I_g and I_b, respectively, are assumed zero. However,

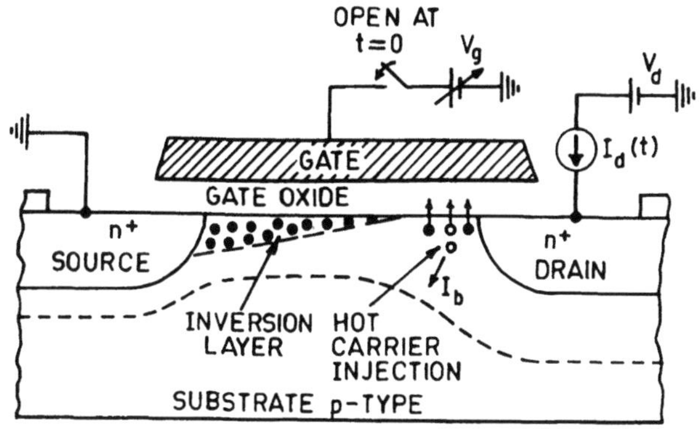

Fig. 9.6 Experimental set up for floating-gate measurements for small gate oxide currents ($\sim fA$ range)

with the scaling of the device dimensions, it has become important to characterize both I_b and I_g (particularly I_b) along with I_d because of device reliability considerations. The substrate current I_b, being of the order of 10^{-9} A and higher, is relatively easy to measure. On the other hand the gate current I_g normally being less than 10^{-12} A (particularly for n-channel devices) is difficult to measure accurately by the direct method as used for I_d and I_b measurements. Since I_g and I_b are correlated and since it is easier to measure I_b, the latter is often used to determine device lifetime, particularly for nMOST (see Chapter 8). However, there are situations, such as studying device degradation mechanisms due to hot-carrier injection into the gate oxide, when measurements of gate current are important. In such cases, I_g is measured using the so called *floating gate technique* [11]–[12]. With this technique gate current as low as 10^{-18} A can be measured over a wide range of gate voltages. Since measurements are done under conditions of hot-carrier injection, it allows the measurement of both electron and hole currents.

In the floating gate technique, the MOSFET is biased appropriately so as to generate gate current. At time $t = 0$ the probe connected to the gate is lifted (floating the gate) and the drain current I_d is measured as a function of time (see Figure 9.6). If electrons (holes) are injected from the substrate into the gate oxide, they will be collected by the gate causing the gate to discharge(charge up) and I_d will decrease(increase) with time. Figure 9.7 shows experimental data showing drain current I_d versus time for a MOSFET with $L = 0.52\,\mu$m and $t_{ox} = 105$ Å. The drain is biased at 7V; the continuous line is with an initial gate voltage $V_g = 8$V while the dotted line is with an initial $V_g = 0.5$V. If the $I_d - V_g$ relationship is known for the device, then the V_g versus time curve can be found from which dV_g/dt can be computed. The gate current as a function of V_g is then calculated using

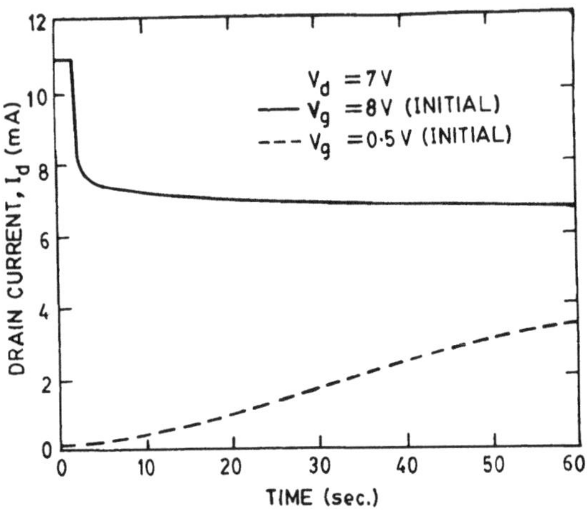

Fig. 9.7 Experimental floating-data showing drain current I_d versus time for nMOST with $L=0.52\,\mu$m and $t_{ox}=105$ Å. The drain is biased at 7.0 V with different initial gate voltages. (a) With initial gate voltage set to 8.0 V, injected electrons act to discharge the gate and cause a reduction in I_d (continuous line). (b) With initial gate voltage set to 0.5 V, injected holes act to charge the gate and cause an increase in I_d (dashed line)

the following relationship [12],

$$I_g = C_G \frac{dV_g}{dt} \tag{9.1}$$

where C_G is the total gate capacitance or measurement system capacitance. Thus, to measure I_g using the floating gate technique the following procedure is used:

1. Measure $I_d - V_g$ of an MOSFET at the high V_{ds} value of interest.
2. Bias the MOSFET terminals and then lift the gate probe and measure the $I_d - t$ curve. To lift the gate probe off the gate pad, a solenoid powered by DC voltage can be used.
3. For each drain current data point in the $I_d - t$ plane, calculate the corresponding gate voltage by interpolating the drain current from the $I_d - V_g$ curve obtained in step 1. This results in a $V_g - t$ curve.
4. Determine dV_g/dt by calculating the slope at each point along the $V_g - t$ curve from step 3. The number of points used in calculating the slope is important. For small gate currents, V_g will vary slowly with time, therefore, more points should be used in the linear regression to determine the slope.
5. Measure the gate system capacitance C_G using conventional high frequency C–V technique [12]. However, parasitics related to the gate probe may affect C_G. Since accuracy of the measured I_g relies directly on the accuracy of the measured value of C_G, care must be exercised in

measuring C_G. Alternatively, C_G can be extracted as a normalization factor between the $I_g - V_g$ data as measured with the conventional technique and with floating gate technique.[4]

6. Calculate I_g using Eq. (9.1).

In general, the implementation of this procedure is relatively simple, requiring only basic measurement instruments. However, the following points must be taken into consideration:

1. The gate probe must be disconnected in such a way so as to prevent noise and leakage from affecting the measurements.
2. If the bias conditions are such that the device under test is degraded, then $I_d - V_g$ data used as a reference for computing $V_g - t$ may not be valid resulting in an inaccurate value of I_g. The solution to the problem is the use of a companion device as a monitor. The two MOSFETs share common gate, source and substrate but have separate drains. One device may then be biased under conditions of carrier injection, while the companion device may be biased under conditions that ensure no device degradation. $I_g - V_g$ can therefore be computed from the $I_d - V_g$ and $I_d - t$ data from the companion device.
3. To reduce the gate leakage current to typically a few tens of 10^{-18} A or less, measurements must be made in a dry nitrogen gas ambient.

It should be pointed out that when C_G is small it is difficult to measure high gate currents ($> 10^{-12}$ A) using this technique. In such cases I_g is measured by direct methods using a picoAmmeter such as the HP4140 or HP4145B parameter analyzer. Figure 9.8 shows gate current (magnitude) measured using the floating gate technique for a MOSFET whose $I_d - t$ characteristics are shown in Figure 9.7; V_d is set to 7 V. Both electron injection and hole injection currents are shown. The gate current higher than 10 pA was measured using an HP4145B. For comparison the substrate current measured for the same device at $V_d = 7$ V is also shown.

Source/Drain Current Data. In addition to the $I - V$ data, one also needs to characterize the parasitic source/drain (S/D) *pn* junction diodes. For normal MOSFET operation, the parasitic S/D diodes are reversed biased, therefore, in terms of diode DC parameters the leakage current associated with the diode is the only parameter which needs to be known. Since these parasitic diodes have very small geometries, it is difficult to perform accurate measurements of the diode characteristics. Furthermore, for the MOSFET S/D diodes it is customary to distinguish between the area (bottom-wall)

---

[4] In order to measure any gate current with a conventional method, the bias conditions will be somewhat excessive but should be chosen so as not to damage the device; i.e., gate current measurements should be repeatable.

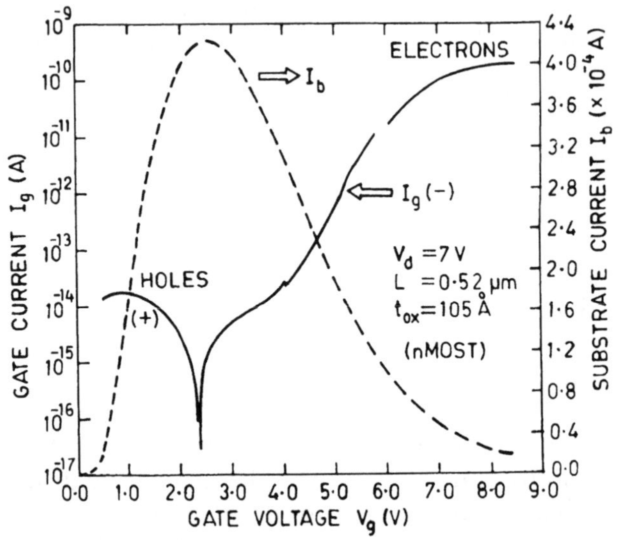

Fig. 9.8 Hot-carrier induced gate current I_g measured using floating gate technique for a nMOST whose $I_d - t$ characteristics are shown in Figure 9.7. The drain is biased at 7.0 V. For comparison, the substrate current I_b, measured on the same device, using direct method is also shown. (After Mistry and Doyle [13a])

and periphery (side-wall) of the diode. Therefore, special structures are normally fabricated to calculate the area and periphery diode currents, as was discussed earlier in section 2.8.

9.1.2 Data for AC Models

The characterization of the device AC models requires measurements of various device capacitances, including MOSFET intrinsic capacitances, source/drain junction capacitances, overlap capacitances and gate oxide capacitance. For CMOS technologies the well capacitance, which is a *pn* junction capacitance, also needs to be characterized.

The principle behind the capacitance measurement is simple. A DC voltage is applied between any two device terminals whose capacitance is to be determined. A small AC signal (typically 10–50 mV peak-to-peak with frequency 10 KHz–1 MHz) is then superimposed on the applied DC voltage and the capacitance is measured by the impedance or capacitance meter. If i_k is the small signal current flowing through the terminal k when terminal l is excited by a small signal potential v_l of frequency f, then in general

$$i_k = Y_{kl}v_l = (G_{kl} + j\omega C_{kl})v_l \qquad (9.2)$$

where $\omega = 2\pi f$; a small signal impedance Y_{kl} is measured and subsequently separated into a conductance G_{kl} and a capacitance C_{kl} according to Eq. (9.2). Thus, the capacitance C_{kl} between the terminals k and l is found from the quadrature current flowing through the terminal k when terminal l is excited. From Eq. (9.2) it is evident that for accurate capacitance measurement the imaginary part of the current must not be much smaller than the real part. This poses a constraint on the measurement. Since generally the capacitance scales with the area of the channel, and conductances with the width to length ratio, a first order estimation of the current ratio is given by

$$\frac{j(2\pi f)C_{kl}}{G_{kl}} \sim 2\pi f L^2.$$

This implies that for small MOSFETs a high measurement frequency is necessary. Since generally the gate conductances are negligible compared to the channel conductances, it is much easier to measure the corresponding gate capacitances (C_{GS}, C_{GD} and C_{GB}) compared to the channel capacitances (C_{SD}, C_{DS}, etc.)

Commercially available capacitance meters generally used for semiconductor measurements are either of three terminal types (like Boonton Model 75-C, General Radio model 1615-A, etc.), or four terminal types (like HP model 4275A, etc.) [1], [13]. These meters measure capacitance independent of the cable and stray capacitance to ground. Various other systems for MOS capacitance measurements such as dedicated C–V measuring instruments, electrometers, and lock-in-amplifiers are also used. Some of these are described in Chapter 12 of reference [1] and elsewhere [16].

Let us see what are the typical values of MOS capacitances. For a MOSFET with gate oxide thickness $t_{ox} = 200\,\text{Å}$, the gate capacitance is about $1.7\,\text{fF}/\mu\text{m}^2$ as given by Eq. (4.1). A typical transistor having this oxide thickness may have a channel length of $2\,\mu$m and channel width of $12\,\mu$m, resulting in a total gate capacitance of about 20 fF. The terminal capacitances will be only a fraction of this total gate capacitance. Therefore, direct measurement of these capacitances require capacitance meters with resolution of 1 fF or better. Commercially available capacitance meters can normally measure capacitance of about 1 pF, although some meters like HP 4275A can measure down to 0.1 pF with a resolution of 0.1 fF. However, in reality it is not easy to achieve this low capacitance measurements because of the large stray capacitance associated with the probe, and the noise generated at the measuring node. Therefore, *in order to reduce the requirements on the measurement setup, often wide transistors are used for the capacitance measurements.*

Most of the capacitance measurements shown in this text are based on the HP4275A LCR meter [9], [14]. The meter has four measurement ports, low current (L_c), low potential (L_p), high current (H_c), and high potential

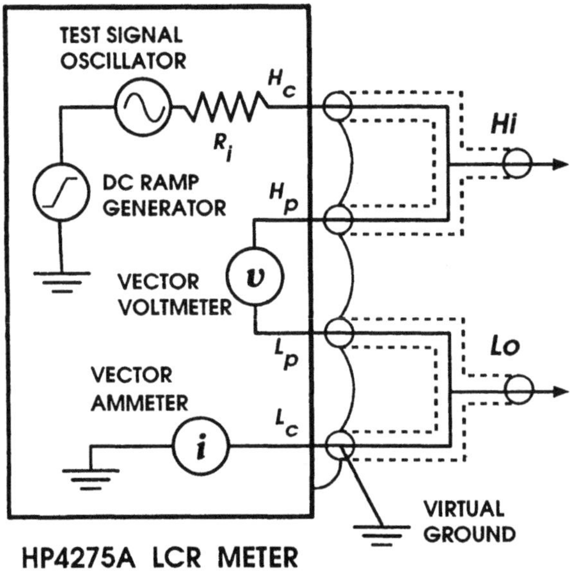

Fig. 9.9 Block diagram of an HP4275A LCR meter. The 'Lo' and 'Hi' terminals are connected to device under test whose capacitance is to be measured

(H_p) as shown in Figure 9.9. It has an internal DC bias voltage which is supplied from the H_c terminal; H_p is only for monitoring voltages and is kept in a high impedance state. The LCR meter monitors L_p and keeps its potential equal to the ground level by controlling the current into the L_c terminal. Since L_p is always at ground level, there is no charge-discharge current due to the wiring capacitance or any other floating capacitances. The meter has a resolution of 0.1 fF [9].

While measuring capacitance it may become necessary to correct the capacitance meter reading for the parasitic capacitance and inductance of the test fixture. Electrically, the test fixture (chuck, probes, and cables) can be represented by three parasitic capacitances as shown in Figure 9.10a. C_1 and C_2 are the stray and wiring capacitances to ground associated with the probe and the chuck, respectively. As mentioned earlier these capacitances are ignored by three and four terminal capacitance meters. The capacitance C_p between the chuck and the probe, for two terminal devices such as MOS capacitor, can be reduced to less than 0.01 pF by proper grounding and shielding (cf. section 9.1, low current measurement considerations). However, for three or four terminal devices such as MOSFET, C_p is the sum of (1) the capacitance C_{mp} between the polysilicon gate line and the metal lines of the source and the drain contacts (pads), and (2) the interprobe capacitance C_{ip}, so that $C_p = C_{mp} + C_{ip}$ (see Figure 9.10b). The gate-source or gate-drain line parasitic capacitance C_{mp} is fixed and

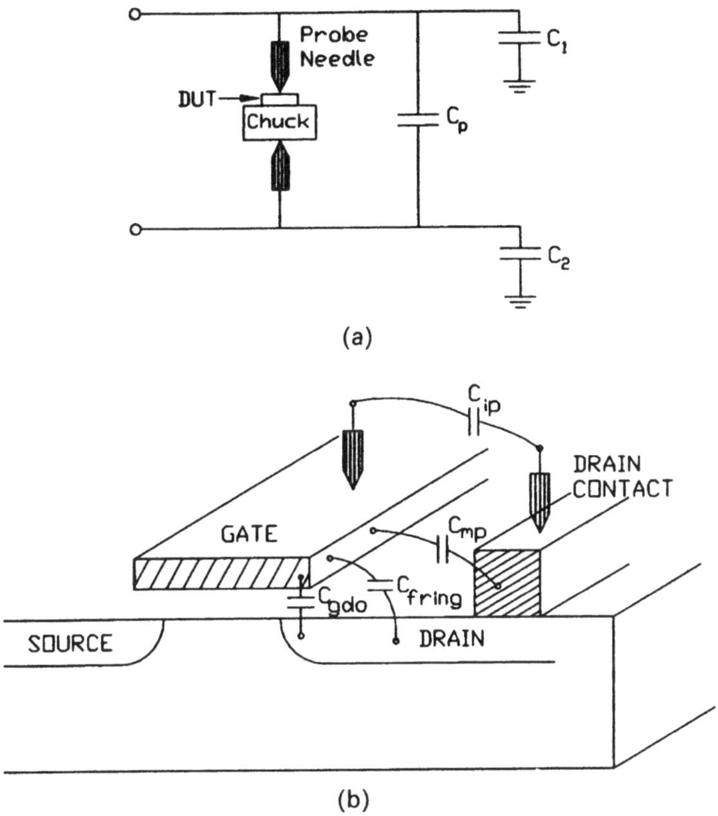

Fig. 9.10 Schematic showing capacitances associated with the test fixture and device under test (DUT) for capacitance measurements of (a) an MOS capacitor, (b) an MOSFET.

independent of the terminal voltages. Generally, test structures are used that minimize C_{mp}. The interprobe capacitance C_{ip} is minimized using coaxial probes so that the gate electrode is shielded from the drain/source probes. Though C_{ip} *can be reduced to 4–5 fF using coaxial probes, compared to a few pF for ordinary probes*, it cannot be made zero. This is because the tip of the probe, where most stray capacitance is picked up, is not shielded; the smaller the unshielded portion of the tip, the smaller the interprobe capacitances. In practice it is possible to effectively zero out the interprobe capacitance C_{ip} by subtracting the open circuit capacitance.

The series inductance is due to the loop of the feed wire that connects the probe and the chuck to the inner conductors of the cables. Typically, its value lies between 20 nH and 200 nH. This inductance is important only if one needs accurately the impedance of the capacitance structure. *The effect of stray capacitances and inductances is minimized by keeping the connecting cables short, no more than a foot, as it adds up capacitance to ground.*

During MOSFET capacitance measurements, often several DC bias voltages are applied to the device terminals. Since these voltages are provided by separate power supplies, it is essential that *ground terminals of all equipments be connected together.* Otherwise, any difference (typically a few mV) in the ground potentials of different equipment will appear on the device terminal which is unaccounted for. Further, in order to avoid any AC signal flowing into the power supplies, large capacitors (typically 1 μF) are connected across the terminals of the power supplies to bypass the AC signals to ground.

9.1.3 MOS Capacitor C–V Measurement

As was pointed out in section 4.6, the MOS capacitor as a test device is routinely used to measure low frequency (LF) and high frequency (HF) C–V curves for MOS process and device characterization. We will now discuss the experimental setup for measuring these curves [1], [13], [15].

Low Frequency C–V Measurement. The low frequency C–V curve requires a frequency as low as 1 Hz, which is difficult to achieve experimentally. Therefore, in practice an alternative method, called the *quasi-static* technique, is often used [17], [1, pp. 383–389]. In this technique, the displacement current through the MOS capacitor is measured. In this respect this method is slightly different from normal AC capacitance measurements. Figure 9.11 shows a block diagram of the setup for an LF C–V measurement. The MOS capacitor to be measured is connected to a linear voltage ramp generator and a sensitive current meter (an electrometer or a picoAmmeter

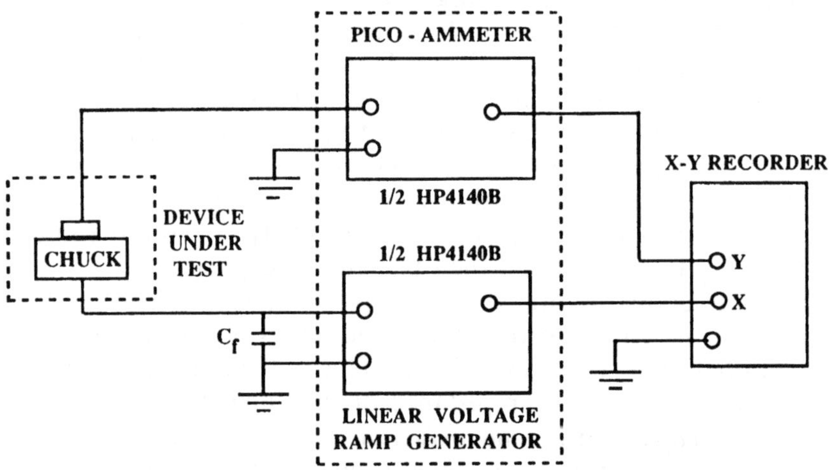

Fig. 9.11 Block diagram of the quasi-static C–V measurement setup

such as the HP4140B). The voltage output of the ramp generator can be expressed as

$$V = V_0 + rt \tag{9.3}$$

where V_0 is the voltage at $t = 0$ and $r = dV/dt$ is the ramp rate. Due to the varying ramp voltage, a displacement current i flows in the capacitor

$$i = \frac{dQ}{dt} = \frac{dQ}{dV} \cdot \frac{dV}{dt} = C_G r \quad \text{(A)} \tag{9.4}$$

measured by the current meter, and is directly proportional to the capacitance C_G of the MOS capacitor. Since there is no AC signal involved, the capacitance corresponds to the limiting case of zero frequency. The ramp rate r should be sufficiently slow to maintain equilibrium conditions in the interface state charge and minority carrier distribution. However, it must be fast enough so that the signal-to-noise ratio is as high as possible. Experimentation is generally necessary to obtain the optimum ramp rate. Typical values of the ramp rate used are 10–100 mV/s. This implies that for a capacitor of 100 pF, the (displacement) current is somewhere between 10 pA to 0.1 pA. The ramp generator should exhibit linearity better than 1% in order to reach an accuracy of 1% in the measurement.

To measure these low currents, the method of interconnection, grounding and shielding are of great importance (cf. section 9.1). The HP 4140B serves a dual purpose as it contains both an ultralinear ramp voltage generator and a picoAmmeter with resolution down to 0.01 pA. The filter capacitor C_f suppresses the noise (voltage spike) of the ramp generator; it is composed of a non-polar electrolytic capacitor and a small ceramic disc capacitor for shunting higher frequencies.

The quasi-static method gives an accurate LF C–V curve provided there is no leakage through the oxide or through the bulk silicon. *If leakage is present, the measured capacitance, in general, will be higher than the actual value.* Figure 9.12a shows a quasi-static curve at a slow ramp rate of 50 mV/s for a slightly leaky MOS capacitor; the inset shows current that leaks through the sample. If the measured capacitance is now converted to the current using Eq. (9.4) and then the leakage current is subtracted from it and converted back into capacitance, we get the actual capacitance as shown in Figure 9.12b. In this case the real oxide capacitance is 2×10^{-10} F, as against 2.69×10^{-10} F measured from the first curve. Thus, care must be taken with the leakage currents while measuring LF C–V curves.

High Frequency C–V Measurement. The commonly used frequency for HF C–V plots is 100 KHz–1 MHz, although this frequency is not necessarily high enough to exclude the response of the interface states close to the majority carrier band edges and may lead to a measurement error between

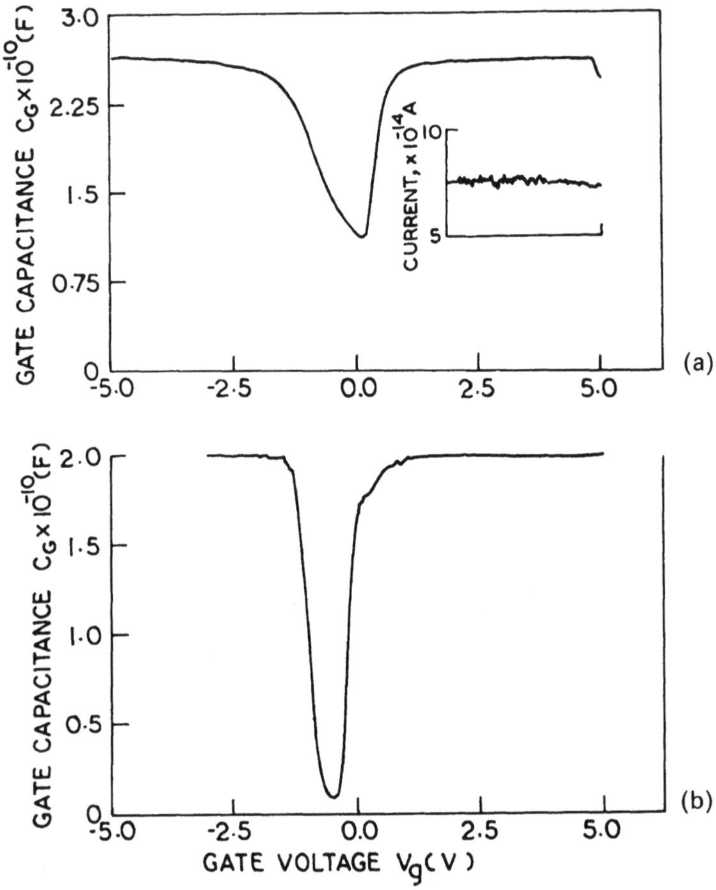

Fig. 9.12 (a) Experimental quasi-static C–V curve of an MOS capacitor; the inset shows DC leakage current. (b) C–V curve after subtracting the leakage current

the flat band voltage and accumulation. The magnitude of the AC signal applied to the sample is typically 10–20 mV, so that nonlinear effects are negligible. The DC ramp voltage sweep may be started at any point, but it is recommended to sweep the voltage from $+V_g$ to $-V_g$ for *p*-type substrates ($-V_g$ to $+V_g$ for *n*-type substrates) in order to avoid problems with minority carrier buildup in the inversion. If the bias is swept from $-V_g$ to $+V_g$, there is a tendency for the C–V curve to go into partial deep depletion and the resulting capacitance is smaller than the true value. However, if the sweep is from $+V_g$ to $-V_g$, the resulting capacitance is above the true value but the deviation in this case is small. The true capacitance can be obtained by setting the bias voltage to some value and then waiting for the device to come to equilibrium before applying the next voltage step [15].

The high frequency (HF) C–V curve is the most commonly measured C–V curve. It is usually not difficult to obtain a valid HF curve, as long as the sample does not have a large *series resistance.* A series resistance can exist if the back contact is not ohmic, or there is an oxide or insulating layer on top of the gate material, or the polysilicon gate is inadequately doped. *If this series resistance is high, the measured capacitance will be lower than the actual value.* As a consequence, an error will be introduced in the parameters to be determined from the HF C–V curve [18]. The presence of series resistance effects can easily be checked by measuring the MOS capacitance in both the series and parallel mode at a particular bias and frequency.[5] *If a significant difference exists between the capacitance measured in series and parallel mode then a series resistance is indicated.* The value of this resistance can be obtained by biasing the MOS capacitor in heavy accumulation, and reading the series resistance directly from the LCR meter. Once R_s is known the actual capacitance can be calculated; for the details of which the reader is referred to [1, p. 222].

9.2 Gate-Oxide Capacitance Measurement

The gate-oxide capacitance per unit area, C_{ox}, is a very important parameter, since knowledge of its value is essential for accurate modeling of the MOS transistor. C_{ox} can be determined from Eq. (4.1) if t_{ox}, the gate oxide thickness, is known. t_{ox} can be determined by several methods. The most prevalent methods used during the manufacturing process are optical methods such as ellipsometery [15], [19], and electrical methods such as the capacitance–voltage (C–V) method [1], [21]–[24].

9.2.1 Optical Method—Ellipsometry

Ellipsometry is the most commonly used optical method for gate oxide thickness (in fact any film thickness) measurement. It involves irradiating the surface of the film with a collimated beam of polarized,[6] monochromatic light and analyzing the difference in the polarization between the incident and reflected beams. The measured difference in the state of polarization, in combination with a physical model of the films covering the surface, permit various properties of the films to be computed. The physical model uses the Fresnel equations that describe the reflection of light from the films substrate structure [19].

---

[5] Most capacitance meters, including the HP 4572A LCR meter, are capable of measuring the capacitance either in series or parallel mode.

[6] A plane electromagnetic wave consists of mutually orthogonal electric and magnetic fields. If the amplitude of these fields is a real constant independent of time, the electromagnetic wave is said to be *linearly or plane polarized.* The direction of the polarization, by convention, is the direction of electric field. An *elliptically polarized* wave corresponds to two linearly polarized waves of unequal amplitudes at right angles to each other.

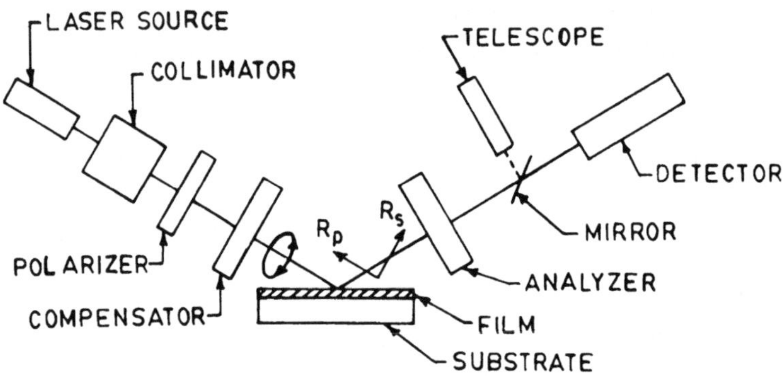

Fig. 9.13 Schematic diagram of a typical ellipsometer

Figure 9.13 shows a schematic diagram of a typical ellipsometer, known as the null ellipsometer configuration.[7] A known controllable state of polarization is obtained using a helium-neon laser, a polarizer, and a quarter-wave compensator. The elliptically polarized light incident upon the sample surface is converted upon reflection into linearly polarized light and passes through an analyzer to the photodetector. If R_p and R_s are the complex reflection coefficients for the parallel and perpendicular electric field vectors with respect to the plane of incidence, then their ratio gives the following basic equation of ellipsometry:

$$\frac{R_p}{R_s} = (e^{j\Delta} \cdot \tan \psi) \tag{9.5}$$

where Δ ($0° \leq \Delta \leq 360°$) represents the change in the phase difference between the parallel and perpendicular components upon reflection and $\tan \psi$ ($0° \leq \psi \leq 90°$) represents the change in the amplitude ratio upon reflection. Both angles ψ and Δ are complicated functions of the index of refraction and film thickness. However, each point in the (ψ, Δ) plane corresponds to a unique pair of film index of refraction and film thickness values. Considerable computation is required to obtain a solution and unambiguous results are obtained only if the range of thickness is known. For a detailed mathematical description of the technique and particulars of the computer program the reader is referred to the literature [19].

9.2.2 Electrical Method

The most commonly used electrical method of determining C_{ox} is by measuring the capacitance of an MOS capacitor in accumulation. In fact,

[7] There are other types of ellipsometers, such as rotating analyzer ellipsometer, but most commercial ellipsometers are based on null ellipsometer configuration.

the MOS capacitor as a test structure is routinely used to determine C_{ox}. Recall from the discussion in section 4.3, the capacitance of the MOS capacitor measured in accumulation is the gate oxide capacitance [cf. (Eq. 4.62)]. Typically, the capacitance is measured at -3 V to -5 V (for p-type substrate) to insure that the device is in accumulation. Since the measured capacitance in accumulation will be the total gate oxide capacitance C_{OX}, dividing C_{OX} by the gate area A_g will give $C_{ox}(= C_{OX}/A_g)$. In principle, either the quasi-static or the HF C–V methods, discussed earlier, could be used. However, the HF C–V method is most commonly used to determine C_{ox}, since it is comparatively easy to measure, although care must be taken with series resistance of the MOS capacitor as discussed in section 9.1.3. Note that the measured C_{OX} includes the parasitic capacitance C_p of the measuring system which must be subtracted from the measured value before calculating C_{ox}. However, by using a very large area MOS capacitor, C_p can often be ignored.

Sometimes the measured HF C–V curve does not saturate even in deep accumulation as shown in Figure 9.14. This often is the case for thin insulators ($t_{ox} < 100$ Å). In such cases the measured capacitance using the HF C–V method in strong accumulation may not coincide with actual C_{ox} due to the measured C_{ox} being bias dependent as is evident from Figure 9.14. In such situations the correct C_{ox} (corresponding to actual t_{ox}) can be determined using derivatives of C–V curve in accumulation [20], [22]. In a method proposed by McNutt and Sah [20], C_{ox} is determined graphically

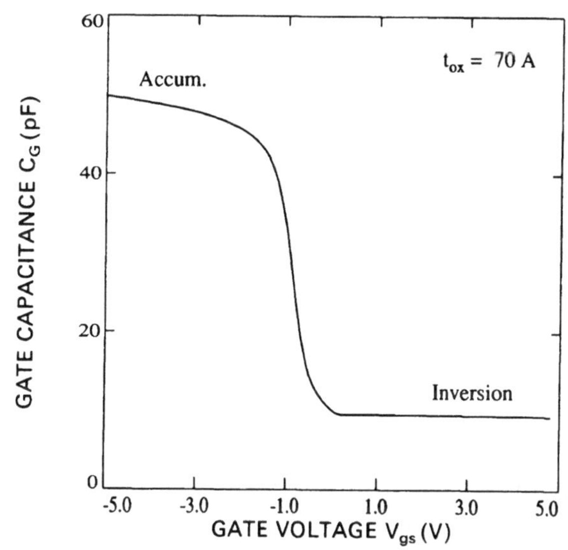

Fig. 9.14 Typical high-frequency C–V plot of an MOS capacitor (p-substrate) with an oxide thickness of 70 Å

using the following formulation, which is based on Eq. (4.67),

$$\frac{dC_{hf}}{dV_g} = \frac{1}{V_t} \cdot \frac{(C_{ox} - C_{hf})^2}{C_{ox}} \tag{9.6}$$

where $V_t (= kT/q)$ is the thermal voltage and C_{hf} is the capacitance of the MOS capacitor in F/cm^2. The latter is obtained by dividing the measured capacitance by the gate area A_g of the MOS capacitor. To find the derivative in Eq. (9.6), we normally use the following approximation

$$\frac{dC_{hf}}{dV_g} \approx \frac{C_{hf1} - C_{hf2}}{V_{g1} - V_{g2}} \tag{9.7}$$

although the three and five points weighted average technique can also be used [25]. Here C_{hf1} and C_{hf2} are HF capacitance values per unit area at gate voltages V_{g1} and V_{g2}, respectively, in accumulation.
It has been recently suggested that C_{ox} can be directly obtained by transforming Eq. (9.6) in the following form [24]

$$C_{ox} = b_1 + \sqrt{b_1^2 - C_{hf1} C_{hf2}} \tag{9.8}$$

where

$$b_1 = \frac{C_{hf1} + C_{hf2}}{2} + V_t \frac{dC_{hf}}{dV_g}.$$

The advantage of Eq. (9.8) over (9.6) is that the graphical extrapolation is replaced by a simple calculation.
A more sophisticated method, which involves the second derivative of the capacitance and which determines C_{ox} at flat band condition, has also been proposed [22]. It has been shown that using Eq. (4.67) one can arrive at a function $F(V_g)$ such that[8]

$$F(V_g) = \frac{C''_{hf} C_{hf}}{C'^2_{hf}} + \frac{3}{2} \left(\frac{C_{hf}}{3V_t |C'_{hf}|} \right)^{1/2} - 3 \tag{9.9}$$

becomes zero at the flat band voltage [22]. Here C'_{hf} and C''_{hf} indicate the first and second derivatives, respectively, of the measured HF capacitance C_{hf} with respect to the gate voltage V_g. Remember that C_{hf} is in F/cm^2 and is obtained by dividing the measured HF capacitance C_{HF} (Farads) by the gate area A_g of the MOS capacitor. The value of V_g that makes F zero could then be used to calculate C_{ox} as follows:

1. Calculate the function F from the measured C–V curve using Eq. (9.9)
2. Determine the gate voltage $V_g = V_{g0}$ at which $F(V_{g0}) = 0$

[8] For details of the derivation of Eq. (9.9) the reader is referred to the original paper [22].

3. Calculate C_s at the flat band condition from the following equation

$$C_s^2 = \pm \frac{C_{hf0}^3}{3V_t C'_{hf0}} \tag{9.10}$$

where the + and − signs are for n- and p-substrate, respectively, and C_{hf0} and C'_{hf0} are values of C_{hf} and C'_{hf}, respectively, at $V_g = V_{g0}$.

4. Extract C_{ox} using Eq. (4.61) as

$$\frac{1}{C_{ox}} = \frac{1}{C_{hf0}} - \frac{1}{C_s}. \tag{9.11}$$

This method of determining C_{ox}, and hence t_{ox}, has been found to agree within ± 2 Å with t_{ox} measured using Transmission Electron Microscopy (TEM) [22]. Note that the voltage V_{g0} at which the function F goes to zero is the flat band voltage V_{fb}, provided the substrate has uniform doping or is almost constant within a Debye length from the Si–SiO$_2$ interface. Further, since C_s is known at flat band from step 3 above, from Eq. (4.68) for C_s and Eq. (4.50) for L_d, we can calculate the substrate concentration N_b.

The only disadvantage of this method of determining C_{ox} is that we need to calculate first and second derivatives of C_{hf}. It is well known that numerical differentiation in general is less accurate than the parent data. The voltage step used in the C–V measurement is therefore very important. *In order to improve results, very small bias steps (from 10 to 50 mV) are normally used near V_{g0} and the capacitance is also calculated as the average of several values (up to 20) measured at the same voltage.* This averaging technique is important to reduce the noise in C–V data. The remaining noise in the data is further eliminated by using numerical methods, such as the one proposed by Savitsky and Golay [25]. Their algorithm does not distinguish between smoothing and differentiation, both being handled by a weighted moving average. The weights are selected to give either a smoothing or a derivative of any desired order. The weights are determined by performing a least squares procedure on an assumed cubic, quintic or sexie. The method has proven to be very successful and simple to implement.

Gate Oxide Capacitance Measurement Using MOSFET. The MOS capacitor method can easily be extended to measure C_{ox} using a large MOSFET. The experimental setup is shown in Figure 9.15a. The shorted source and drain are connected to the substrate. The gate of the MOSFET is connected to the 'Hi' terminal (H_p, H_c) of the HP4275A LCR meter, whose 'Lo' terminal (L_p, L_c) is connected to the substrate. A small AC voltage (20–30 mV peak-to-peak) of frequency 100 KHz is superimposed on the gate voltage which is ramped from $-V_g$(max) (accumulation) to $+V_g$(max) (inversion) and the corresponding gate-to-substrate capacitance C_{GB} measured.

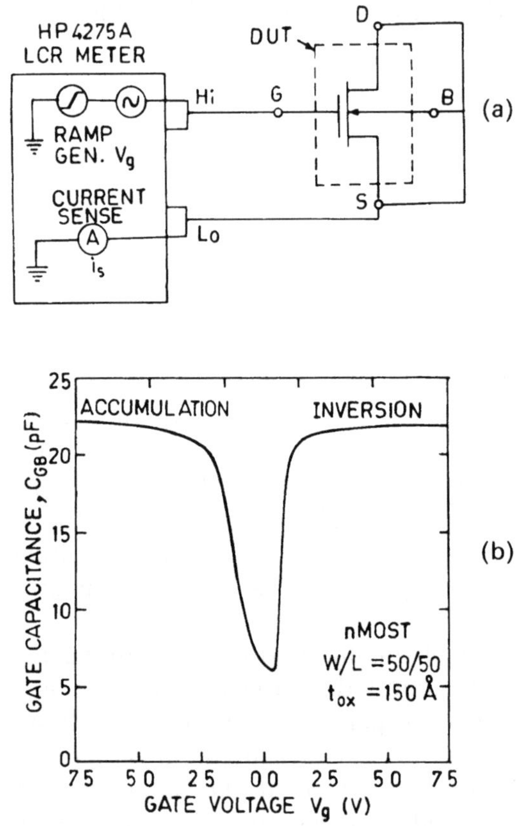

Fig. 9.15 Measurement of an MOSFET gate oxide capacitance using the C–V method; (a) experimental setup using LCR meter, and (b) gate-to-substrate capacitance C_{GB} as a function of gate voltage V_g measured using set up shown in (a); nMOST W/L = 50/50 (μm), $t_{ox} = 150$ Å

In accumulation the measured capacitance is

$$C_{GB} = C_{ox}WL + C_p \tag{9.12}$$

where W and L are the effective width and length, respectively, of the MOSFET. Figure 9.15b shows measured high frequency gate-to-substrate capacitance C_{GB} as a function of gate voltage V_g for a nMOST with W/L = 50/50 and $t_{ox} = 150$ Å. Knowing W and L, one can easily calculate C_{ox} (and hence t_{ox}) from Eq. (9.12).

It is important to note that for C_{GB} measurement, the parasitic capacitance C_p also includes the capacitance C_m of the interconnect lines to ground connecting the source/drain and gate, in addition to C_{mp} and C_{ip} (see Figure 9.10b). The effect of C_m is to shift the C–V curve upward resulting in an incorrect C_{ox}, if not taken into account.

Table 9.1. *Gate oxide thickness, t_{ox}, calculated using ellipsometery and HF C–V methods*

Electrical (C–V) Å	Optical (Ellipsometery) Å
214.0	209.8
146.0	143.5
99.2	96.9
89.0	85.9

We have discussed the two most commonly used methods of calculating t_{ox} (or, C_{ox}). Table 9.1 shows values of measured t_{ox} for MOS capacitors with different gate oxide thickness, using optical and electrical methods. It should be pointed out that t_{ox} values shown in Table 9.1 for optical method were obtained using a commercially available ellipsometer, Rudolph Auto EL IV, which is a multiwavelength, autonulling ellipsometer specifically suited for measuring thicknesses less than 250 Å. The t_{ox} from the C–V method was obtained in accumulation at $V_g = 5$ V.
Clearly, t_{ox} obtained by the two methods correlate well with each other. In general, the electrical method shows slightly higher t_{ox}; this difference is mainly caused by quantum effects at the semiconductor surface [21].

9.3 Measurement of Doping Profile in Silicon

The doping concentration, including the spatial variation (doping profile), in silicon is an important device parameter that has direct bearing on the device performance. Non-uniform doping occurs due to ion implantation of impurities into the silicon during MOSFET fabrication. It may also occur during oxidation due to dopant pile up at the interface or dopant gettering by the oxide. Many techniques have been developed for measuring doping profiles. These techniques could broadly be classified as destructive and non-destructive techniques. Here we will discuss in detail the most commonly used non-destructive electrical methods, while other methods will be briefly discussed. Some of the well known destructive techniques are [15], [26]:

Four-Point Probe Method. In this method the profile is obtained by measuring the sheet resistance ρ_s of the layer using the four-point probe method; then stripping a thin layer, for example, using anodic oxidation and measuring ρ_s again. The procedure is continued until the layers have been completely removed. Using the resistivity versus doping conversion curve, the impurity profile can be determined using Eq. (2.29) [27].

Hall-Effect Method. In this method Hall-effect measurements are added to the sheet resistivity measurements combined with layer removal technique to get the doping profile [15].

Spreading Resistance Method (SRP). In this technique, two probes are positioned on the surface of the sample to be profiled. The sample is first bevelled and polished to give vertical magnification. A small voltage (mV range) is then applied between the probes and spreading resistance measured. The probes are lifted, the stage moves over the adjusted distance Δx and the procedure is repeated. The spreading resistance values are then converted into a doping profile using a somewhat complicated conversion procedure [28]. This is a powerful technique due, firstly, to the speed at which profiles can be obtained and secondly, to the possibility of measuring profiles with lesser restriction as to the doping level.

Secondary Ion Mass Spectrometry (SIMS). In this technique, the sample to be profiled is mounted in a target chamber where high vacuum conditions ($\ll 10^{-6}$ torr) are maintained. The sample is then bombarded with a beam of fast ions, like Ar^+, Cs^+, etc. (typically 10 KeV energy) resulting in the ejection of atoms and molecules, in both neutral and charged state, from the substrate. The production of charged particles (secondary ions) coupled with high sensitivity mass spectrometry form the basis of the SIMS technique. The procedure of depth profiling is then to monitor the secondary ions signal of an element as a function of sputter time. The former can be translated to concentration, while the latter can be translated to depth through suitable calibration. It is a very attractive and sensitive technique and has been extensively used for the measurement of boron profiles in silicon. It has the advantage that layer removal (by ion sputtering) and measurement of ions removed are performed simultaneously.

9.3.1 Capacitance–Voltage Method

The most commonly used non-destructive method of profiling is the C–V method, wherein the sample to be profiled is fabricated as a MOS capacitor [1], [13, 15]. Since the methods explained next do not depend on the type of the doping (*n*- or *p*-type), the doping concentration will be designated by N only, without subscript.

Uniform Doping Concentration. For an MOS capacitor with uniformly doped substrate, the doping concentration can easily be computed by measuring the minimum and maximum capacitance C_{min} and C_{max}, respectively, of the HF C–V curve. This, so called the $C_{min} - C_{max}$ method, is widely used in the industry. Remember that the maximum capacitance $C_{max}(=C_{ox})$ occurs in accumulation, while C_{min} occurs in inversion. Combining Eqs. (2.15) and (4.74) and solving for the substrate concentration N yields

$$N = \left[\frac{2V_t}{\epsilon_0\epsilon_{si}qA_g^2}\right]\left[\frac{C_{max}}{C_{max}/C_{min}-1}\right]^2 \ln\left(\frac{N}{n_i}\right) \tag{9.13}$$

where A_g is the MOS capacitor gate area. This is a transcendental equation for N and therefore must be solved iteratively. To start the iterations, use a low level of doping, say $10^{13}\,cm^{-3}$, enter into the right hand side and calculate N. Next reenter the new value of N into the right hand side and recalculate N. This is repeated until a steady value is reached, e.g. one that does not change from one iteration to the next by $>10^{10}$. It is easily coded for an automatic measurement. For occasional bench testing and analysis, a graph of N as a function of C_{min} and t_{ox} is often used in place of Eq. (9.13). Based on a more accurate expression for C_{min} [cf. Eq. (4.75)], the following non-iterative expression for the doping concentration N has been suggested [29]

$$N = 1.786\times 10^7 T^{0.956}\exp\left(\frac{610}{T}\right)\cdot\left(\frac{C_{max}C_{min}}{C_{max}-C_{max}}\cdot\frac{1}{A_g}\right)^{2.174} \tag{9.14}$$

where A_g is the area of the capacitor, T is the temperature in Kelvin. At 296 K, the above equation reduces to

$$N = 3.23\times 10^{10}\left(\frac{C_{max}C_{min}}{C_{max}-C_{max}}\cdot\frac{1}{A_g}\right)^{2.174}. \tag{9.15}$$

Equations (9.13) and (9.15) can be used for non-uniform doping, but the value of N obtained will be the *average concentration* over the depletion width X_{dm}; i.e., it gives an effective doping density.

Nonuniform Doping. The high frequency C–V plot can be used to determine the dopant impurity profile of the substrate [30]–[37]. Replacing C_g in Eq. (4.66) with C_{hf} (to stress that we are dealing with HF capacitance) and differentiating with respect to V_g yields

$$\frac{dC_{hf}}{dV_g} = -\frac{C_{hf}^3}{q\epsilon_0\epsilon_{si}}\cdot\frac{1}{N}. \tag{9.16}$$

This equation states that the slope of the C_{hf} versus V_g plot is inversely proportional to the doping concentration N. Recall that Eq. (4.66) was derived on the assumption that the silicon is depleted up to a distance X_d (depletion width), beyond which it is neutral. Therefore, the value of N calculated from Eq. (9.16) will be at a distance X_d from the silicon surface.

Thus,

$$N(X_d) = -\frac{C_{hf}^3}{q\epsilon_0\epsilon_{si}}\left[\frac{dC_{hf}}{dV_g}\right]^{-1} \quad (\text{cm}^{-3}). \tag{9.17}$$

Remember that C_{hf} is in F/cm^2 and is obtained by dividing the measured HF capacitance C_{HF} by the area A_g of the MOS capacitor. For actual profiling work, it is more appropriate to write Eq. (9.17) in the following form

$$\boxed{N(X_d) = \frac{2}{q\epsilon_0\epsilon_{si}}\left[\frac{d}{dV_g}\left(\frac{1}{C_{hf}^2}\right)\right]^{-1} \quad (\text{cm}^{-3}).} \tag{9.18}$$

Thus, to obtain the ionized dopant impurity profile as a function of depth X_d, we calculate N at each value of V_g using Eq. (9.18). The corresponding value of X_d is obtained from Eqs. (4.63) and (4.64) as

$$\boxed{X_d = \epsilon_0\epsilon_{si}\left[\frac{1}{C_{hf}} - \frac{1}{C_{ox}}\right] \quad (\text{cm}).} \tag{9.19}$$

Although Eqs. (9.18) and (9.19) are derived for uniform substrate doping, they remain valid for the case of nonuniform doping. These are the basic equations for measuring doping profiles using the C–V method. With the high frequency C–V curve the maximum depth to which doping profile can be measured is limited by the gate voltage at which the surface inverts. In order to extend this bias range beyond the inverting gate voltage, one will measure the deep depletion capacitance rather than high frequency capacitance. In other words, *profiling requires a fast sweep voltage, fast enough to prevent the buildup of minority carriers (formation of an inversion layer) at the silicon surface.* Both rapidly varying ramp voltages and voltage pulses have been used. In this case doping profile will be limited by the onset of avalanche breakdown. The deep depletion C–V curve for an MOS capacitor with a Boron implanted substrate is shown in Figure 9.16. Also shown in this figure is the LF C–V curve (dotted line) for the same device. The resulting doping profile using Eqs. (9.18) and (9.19) is shown in Figure 9.17 as a continuous line.

The major source of error in C–V profiling is the presence of the derivative in Eq. (9.18). As was stated earlier in section 9.2.2 numerical differentiation results in a derivative which will always be less accurate than the parent data. In profiling, the voltage step used in the capacitance measurement is very important. If the data points are very closely spaced, differentiation of the experimental data results in very poor accuracy as the derivatives are dominated by rounding and random errors. On the other hand, if the

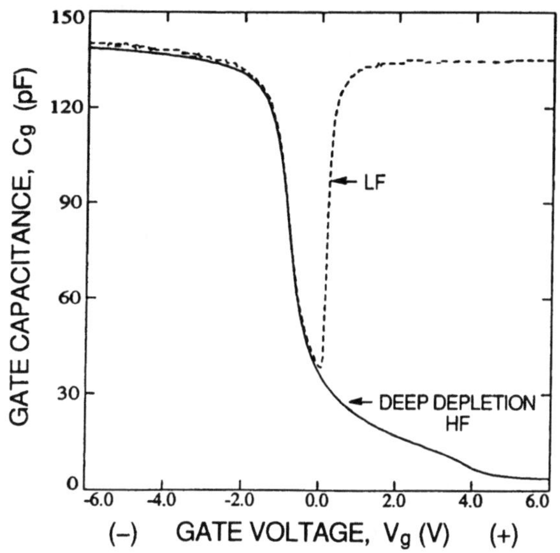

Fig. 9.16 Measurement of doping profile of a double boron implanted MOS capacitor; (i) deep depletion high frequency C–V curve (continuous line) (ii) quasi-static C–V curve (dotted line)

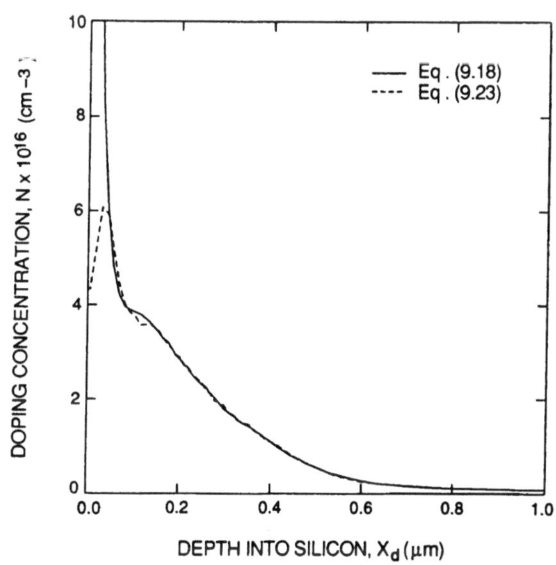

Fig. 9.17 Doping profile determined using the deep depletion high frequency C–V method (i) using Eqs. (9.18) and (9.19)—continuous line (ii) using Eqs. (9.23) and (9.19)—dotted line

profile varies rapidly then one needs to take closely spaced points in order to avoid any loss of the profile details. McGillivray et al. [37] have pointed out that an optimum voltage step ΔV can be taken as

$$\Delta V = \frac{q\epsilon_0\epsilon_{si}N(X_d)(200R)}{C_{hf}^3} \tag{9.20}$$

where R is a characteristic sum of the measurement system rounding and random errors and depends on the capacitance meter being used. For an HP4275A capacitance meter, R is approximately 1–2 times the least significant number for capacitances larger than 1 pF. To find $N(X_d)$ we normally approximate

$$\frac{d}{dV_g}\left(\frac{1}{C_{hf}^2}\right) \approx \frac{C_{hf1}^{-2} - C_{hf2}^{-2}}{V_{g1} - V_{g2}} \tag{9.21}$$

where C_{hf1} and C_{hf2} are HF capacitance values per unit area at gate voltages V_{g1} and V_{g2}, respectively. However, this simple procedure may result in a noisy doping profile unless the original C–V data is smoothed out.

Two remarks need to be made concerning the applicability of Eqs. (9.18) and (9.19). Recall that Eq. (4.66), from which (9.18) is derived, is based on the depletion approximation with abrupt space charge edges which implies that beyond the depth $x = X_d$, the profiling region is quasi-neutral. Since *in reality the space charge regions are not abrupt, Eq. (9.18) cannot be used for doping profile closer than within about three Debye length (L_d) of the surface*[9] [1]. This explains why the doping concentration diverges near the surface (see continuous line in Figure 9.17). In addition, the doping profile extracted using the depletion approximation represents the majority carrier concentration, rather than the doping concentration. Of course, the two are the same for a uniformly doped substrate [15].

Additionally Eq. (9.18) is valid so long as there are no interface traps. When such traps are present, the C–V curve is smeared out (cf. section 4.4), and therefore the extracted doping profile will be incorrect.[10] To correct this effect, a low frequency or quasi-static C–V curve must also be obtained. This is because at low frequency the interface traps get sufficient time to charge and discharge during the measurement and therefore it includes the capacitances related to the traps. Thus, in the presence of traps, both high frequency C_{hf} and low frequency C_{lf} capacitance curves are required and

---

[9] An abrupt space charge edge to apply under all conditions would imply a Debye length equal to zero.

[10] In the example of Figure 9.16, the LF and HF capacitances are almost the same in the depletion region, therefore, interface traps are negligible in this case.

in this case it has been shown [1] that

$$N(X_d) = 2\left(\frac{1 - C_{lf}/C_{ox}}{1 - C_{hf}/C_{ox}}\right)\left[q\epsilon_0\epsilon_{si}\frac{d}{dV_g}\left(\frac{1}{C_{hf}^2}\right)\right]^{-1}. \tag{9.22}$$

The width X_d is still given by Eq. (9.19).

Ziegler et al. [31] proposed a method, later modified by Lin and Reuter [32], by which the profile can be determined right up to the surface. In the Ziegler formulation, the doping profile may be computed from the following expression

$$N(X_d) = -\frac{C_{hf}^3}{q\epsilon_0\epsilon_{si}}\left[\frac{dC_{hf}}{dV_g}\right]^{-1} \cdot g_2 \quad (\text{cm}^{-3}) \tag{9.23}$$

and

$$X_d/L_d = \sqrt{2}L_d(g - \ln g - 1)^{1/2} \tag{9.24}$$

where L_d is the Debye length [cf. Eq. 4.50]. The doping concentration factor g_2 is defined as

$$g_2 = \frac{1}{1-g} - \left(\frac{X_d}{L_d}\right)^2 \frac{g}{(1-g)^3}. \tag{9.25}$$

It is convenient to define a function g_1 as

$$g_1 = \frac{kT}{q} C_{hf}^2 \left[\frac{d}{dV_g}\left(\frac{1}{C_{hf}^2}\right)\right]^{-1}. \tag{9.26}$$

This function is calculated directly from the measured data and is related to g as

$$\frac{1-g}{g - \ln g - 1} - \frac{2g}{1-g} - g_1 = 0. \tag{9.27}$$

The variable g_2 represents the ratio between the doping calculated with and without the depletion approximation. For depletion widths greater than $3L_d$, the value of g_2 reduces to unity making Eq. (9.23) and (9.17) identical. This will be the case when $g_1 \sim 0.1$ and corresponds to the depletion approximation. When depletion widths are less than 3 Debye lengths, which means that $g_1 > 0.1$, the value of g_2 begins to drop from its value at 1 pointing to a divergence between the exact [cf. Eq. (9.23)] and approximate profile [cf. Eq. (9.17)]. The difference is greatest at $X_d = 0$, which corresponds to the flat band condition, and occurs when $g_2 > 2/3$. In practice, the following steps are used to calculate doping profile using Eq. (9.23):

1. Obtain deep depletion HF C–V curve from an MOS capacitor.

2. Use Eq. (9.26) to calculate g_1.
3. Use Eq. (9.27) to calculate g by iterative Newton–Raphson technique.
4. Use of Eqs. (9.24) and (9.25) enables the calculation of g_2.
5. Finally from Eq. (9.23) calculate $N(X_d)$. Once $N(X_d)$ is known, the Debye length L_d may be computed using Eq. (4.50) and X_d found using Eq. (9.24).

The doping profile using this method is shown as a dotted line in Figure 9.17. Thus, one can see the difference between a more elaborate model which includes majority carriers in the depletion region and a simpler model where the depletion approximation is made.
When the actual doping profile is steep, the C–V extracted profile is often broader than the actual profile. This so called 'spillover' effect limits the usefulness of the C–V technique. However, gradient correction methods have been proposed which can correct this drawback, but this is outside the scope of this book and the interested reader is referred to references [33]–[36].
It can be concluded that the *deep depletion C–V technique offers a simple and fast method of measuring doping profile when doping levels are not too high (concentration below* $10^{18}\,\text{cm}^{-3}$*) and it does not vary very fast.* It offers sub-Debye length resolutions and the ability to determine the doping profile right up to the surface. The measurement, however, requires a precise value of the gate area, as capacitance varies as the square of the area. Since parasitic capacitance can distort the profile, large area devices are normally used so that parasitics can be neglected. However, this method is not suitable for high doping concentration such as found in the source/drain regions of MOSFET or heavily doped emitters. In these regions, one uses other techniques such as spreading resistance and secondary ion mass spectroscopy as discussed earlier.
There are other methods of determining the doping profile from MOS C–V curve that are based on a numerical solution of the Poisson's equation. They assume *a priori* a certain function for the doping profile and then match the resulting profile with measured CV curve [38].

Doping Profile Determination Using MOSFET's. The C–V method of profiling discussed above is not limited only to the MOS capacitor, but can easily be extended to determine the doping profile in an MOSFET channel region [39], [40]. The experimental setup is shown in Figure 9.18a. The gate of an nMOST is connected to the 'Hi' terminal of the HP4275A LCR meter, whose 'Lo' terminal is connected to the substrate. The source and drain are connected together and reverse biased relative to the substrate via the VA (voltage A) terminal of the HP4140B picoAmmeter. A small AC voltage (20–30 mV peak-to-peak) of frequency 1 MHz is superimposed on the gate voltage, which is stepped from $-V_g(\text{max})$ (accumulation) to

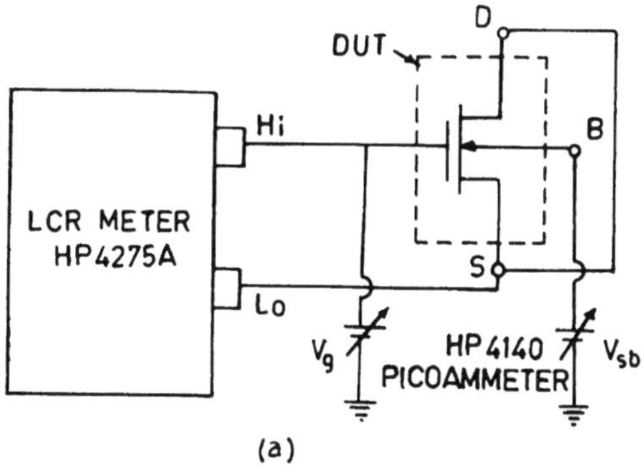

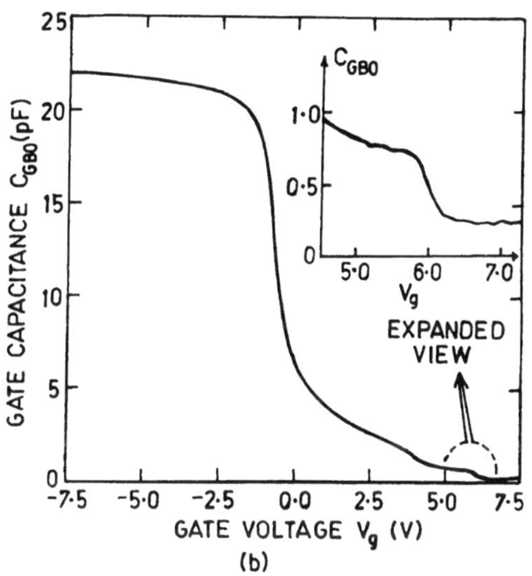

Fig. 9.18 Measurement of MOSFET channel doping profile using the C–V method (a) experimental setup using LCR meter, and (b) gate-to-substrate capacitance C_{GBO} as a function of gate voltage V_g at $V_d = 5\,\mathrm{V}$ measured using set up shown in (a); nMOST W/L = 50/50, $t_{ox} = 150\,\text{Å}$

$+V_g(\max)$ (inversion) and the corresponding gate-to-substrate capacitance C_{GBO} measured. Figure 9.18b shows C_{GBO} as a function of V_g for an nMOST with $W/L = 50/50$ and $t_{ox} = 150\,\text{Å}$ for a drain/source voltage $V_d = 5\,\mathrm{V}$. The magnitude of the drain voltage V_d determines the extent to which the depletion region extends into the semiconductor, i.e., more positive the V_d (for nMOST), the farther the depletion region extends into the silicon. For a given V_d as V_g is made more positive, starting from zero, the capacitance C_{GBO} decreases smoothly until a gate voltage is reached when capacitance rapidly decreases and finally vanishes (see inset in Figure 9.18b). This sudden decrease in the capacitance C_{GBO} and eventual vanishing of the capacitance for a given V_d is caused by the formation of an inversion channel under the

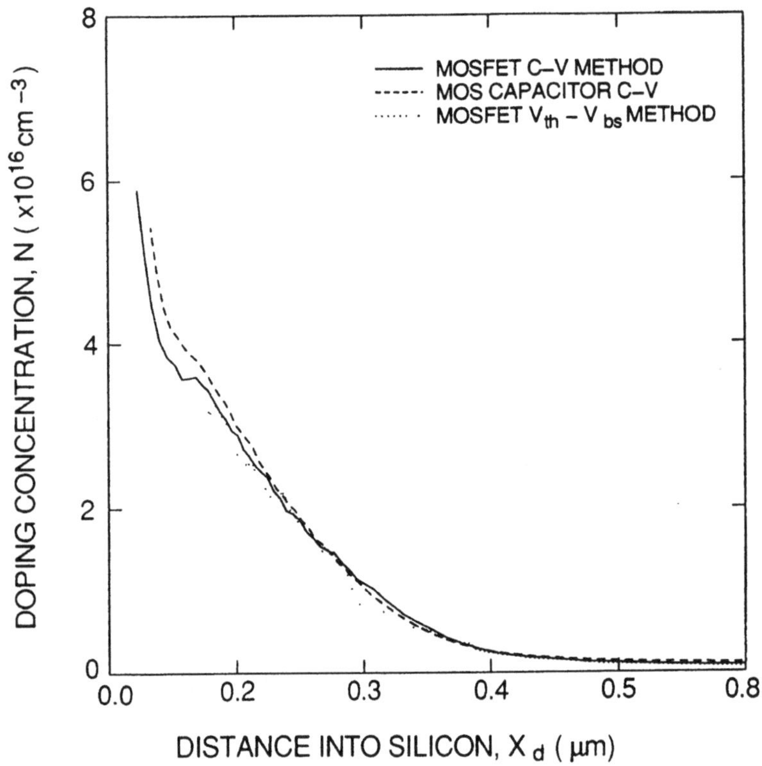

Fig. 9.19 The MOSFET channel doping profile determination using (i) the C–V method (continuous line) (ii) the threshold voltage method (dotted line). For comparison doping profile obtained using the MOS capacitor fabricated alongside the MOSFET is shown as dashed line

gate. The doping profile is then determined using Eqs. (9.18) and (9.19), the same equations as for the MOS capacitor, with $C_{hf} = C_{GBO}/WL$; the product WL is the area of the MOSFET. The parasitic capacitance C_p must be subtracted from C_{GBO} before calculating C_{hf} and hence the doping profile. The channel doping profile of a MOSFET obtained by measuring C_{GBO} is shown in Figure 9.19 as a continuous line. For comparison the doping profile obtained using an MOS capacitor, fabricated alongside the MOSFET, is also shown as dotted line. As expected the two profiles are almost the same. It should be pointed out that C_{GB} (see Fig. 9.15) and C_{GBO} are the same in accumulation, and therefore C_{GBO} can also be used to calculate C_{ox} using Eq. (9.12).

9.3.2 DC Method

In the C–V method, large area MOSFET's are generally required in order to reduce the stray capacitances and increase signal-to-noise ratio. However,

such measurements are difficult to make on small geometry MOSFETs due to the difficulties in determining accurate capacitances. This limitation is overcome in a method which allows the doping profile to be extracted from MOSFET threshold voltage measurements as a function of substrate bias [41]–[47]. It is a DC technique requiring only drain current-gate voltage measurements at low $V_{ds}(< 0.1\,\text{V})$. The method lends itself to measurements on small geometry MOSFETs [44].
The threshold voltage of a MOSFET can be expressed as

$$V_{th} = V_{fb} + 2\phi_f + \frac{1}{C_{ox}} qNX_{dm} \tag{9.28}$$

and is obtained by combining Eqs. (5.7), (5.9), and (5.12). Here V_{fb} is the flat band voltage, ϕ_f is the bulk Fermi potential, N is the substrate concentration, and X_{dm} is the maximum depletion width under the gate. The latter is given by Eq. (5.8), repeated here for convenience,

$$X_{dm} = \sqrt{\frac{2\epsilon_0\epsilon_{si}}{qN}(2\phi_f + V_{sb})} \quad \text{(strong inversion).} \tag{9.29}$$

Differentiating Eq. (9.28) we get

$$\frac{dV_{th}}{dV_{sb}} = \frac{q}{C_{ox}} \cdot N(X_{dm}) \cdot \frac{dX_{dm}}{dV_{sb}} \tag{9.30}$$

where it is assumed that $2\phi_f$ is constant for slowly varying dopant distribution under the channel. A differential increment of the body bias (dV_{sb}) accompanies a differential increment of the depth of the depletion region (dX_{dm}) and can be obtained by differentiating Eq. (9.29), that is,

$$X_{dm} \cdot \frac{dX_{dm}}{dV_{sb}} = \frac{2\epsilon_0\epsilon_{si}}{qN(X_{dm})}. \tag{9.31}$$

Solving Eqs. (9.30) and (9.31) we get

$$N(X_{dm}) = \frac{C_{ox}^2}{q\epsilon_0\epsilon_{si}} \left[\frac{d^2V_{sb}}{dV_{th}^2}\right]^{-1} \tag{9.32a}$$

and

$$X_{dm} = \frac{\epsilon_0\epsilon_{si}}{C_{ox}} \left(\frac{dV_{sb}}{dV_{th}}\right). \tag{9.32b}$$

Equations (9.32a, b) are the basis for dopant profile determination under the channel using the DC method. These equations were first derived by Shanon [41] and Buchler [42]. For actual profiling work it is more appropriate to write Eq. (9.32a) as [45]

$$N(X_{dm}) = \frac{2C_{ox}^2}{q\epsilon_0\epsilon_{si}}\left[\frac{dV_{sb}}{d(V_{th}/V_{sb})^{-2}}\right].$$

The relationship between $(V_{th}/V_{sb})^{-2}$ vs. V_{sb} becomes linear for the case of a uniform doped substrate, while it becomes nonlinear for nonuniform concentration. Therefore, the doping concentration N at a given X_{dm} can be obtained from the slope of $(V_{th}/V_{sb})^{-2}$ vs. V_{sb} curve. In order to obtain the impurity profile close to the surface, V_{sb} close to $2\phi_f$ should be applied. However, in this region the threshold voltage change is so small that high sensitivity measurements are required. The dotted line in Figure 9.19 shows the channel doping profile extracted using the threshold voltage method for the same MOSFET whose doping profile using the C–V method is shown as a continuous line. Note that the DC method results in a somewhat more noisy profile compared to the C–V method. This is because the C–V method requires only one derivative, while the DC method requires two derivatives; in general, the higher the derivatives, the more noisy the data tend to be.

The advantages of the DC method are that (1) $V_{th} - V_{sb}$ characteristics can be measured on relatively small MOSFET, and (2) no special test patterns are needed to fabricate the device as in the case of an MOS capacitor. In addition, the effect of surface states is completely eliminated since the surface is always in inversion. However, the method is subjected to the same Debye length limitations as the C–V method, that is, the profile cannot be reliably obtained closer than about 3 Debye lengths from the surface. The method allows for fast evaluation and therefore can be useful for process control. Errors in measuring V_{th} do affect the doping profile. It is recommended that the constant current method be used for determining V_{th} rather than the usual linear extrapolation method [43].

9.4 Measurement of Threshold Voltage

Experimentally, the threshold voltage V_{th} is determined by measuring the drain current I_{ds} for various values of gate voltage V_{gs}, at low V_{ds}(<0.1V, typically 50 mV). There are different ways in which $I_{ds} - V_{gs}$ data could be used to calculate V_{th}; however, the three most common methods are [48]–[52]:

Constant Current Method. In this method the gate voltage V_{gs} at low drain voltage V_{ds}(<0.1 V) and at a specified drain current I_{ds} is taken to be the threshold voltage. Normally

$$I_{ds} = I_{th}\frac{W_m}{L_m} = 10^{-7}\frac{W_m}{L_m} \quad \text{(A)} \tag{9.34}$$

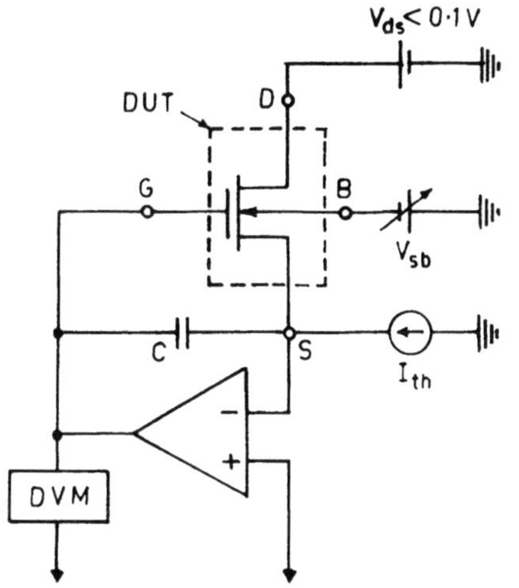

Fig. 9.20 Experimental setup for measuring threshold voltage using the constant current method. (After Lee et al. [46])

is assumed, where I_{th} is the threshold current, W_m and L_m are drawn (mask dimensions) device width and length, respectively. Since only one voltage measurement is required, this method is very fast and is often used for the purpose of process monitoring or to calculate V_{th} from 2-D device simulators such as MINIMOS. The value of $I_{th} = 10^{-7}$ A is chosen in a somewhat arbitrary manner and has no physical reasoning. Typical values for I_{th} are usually in the range of 10^{-6} to 10^{-9} A. This method can easily be implemented with a circuit shown in Figure 9.20 [48]. The threshold current I_{th} is forced at the MOSFET source terminal and the gate adjusts itself to V_{th} with the help of an operational amplifier (op-amp) and is read directly using DVM.

Linear Extrapolation Method. In this method, *threshold voltage is defined as that gate voltage which is obtained by extrapolating the linear portion of the I_{ds} versus V_{gs} characteristics, from the point of maximum slope, to zero drain current I_{ds}.* Note that the point of maximum slope is that point where the transconductance $g_m = \partial I_{ds}/\partial V_{gs}$ is maximum. This threshold voltage is often called the *extrapolated* V_{th}. *This is the most common method of determining V_{th} and is the defacto industry standard.*
To calculate V_{th} using the extrapolation method, the gate voltage V_{gs} is normally swept from $V_{gs} = 0$ to V_{gs}(max), say 5 V, in steps of say 50 mV, while simultaneously monitoring the corresponding drain current I_{ds}. To ensure operation in the linear region, drain voltage V_{ds} is fixed at ($\leqslant 0.1$ V), while the back bias V_{sb} is fixed at the desired value. Figure 9.21 shows a typical $I_{ds} - V_{gs}$ curve for a *n*-channel MOSFET obtained using the circuit

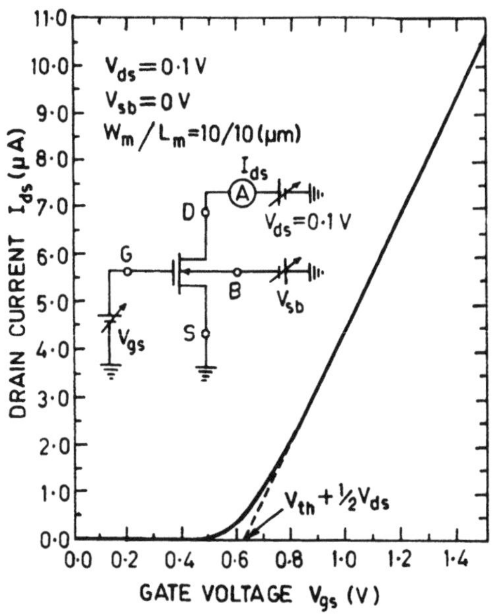

Fig. 9.21 Measurement of threshold voltage V_{th} from $I_{ds} - V_{gs}$ data in the linear region at low V_{ds} (≤ 0.1 V). Inset shows circuit schematic for $I_{ds} - V_{gs}$ measurement

schematic shown in the inset. Recall that the drain current I_{ds} at low V_{ds} (linear region) is given by Eq. (6.81), repeated here for convenience,

$$I_{ds} = \mu_s C_{ox}\left(\frac{W}{L}\right)[V_{gs} - V_{th} - \tfrac{1}{2}\alpha V_{ds}]V_{ds} \quad \text{(linear region).} \qquad (9.35)$$

The above equation shows that the plot of I_{ds} versus V_{gs} will be a straight line which crosses the V_{gs} axis at $V_{th} + 0.5\alpha V_{ds}$ as shown in Figure 9.21. The parameter α is back bias dependent; its value generally varies between 1.1–1.5 depending upon the value of V_{sb}. For V_{th} calculations, it is a good approximation to assume $\alpha = 1$ without introducing any significant error ($< 2\%$) because V_{ds} is generally small (< 0.1 V). Thus, *to calculate V_{th}, one needs to subtract half of the applied V_{ds} from the extrapolated V_{gs} value.* One can ask why extrapolation from the maximum slope? The reason being that I_{ds} does not vary precisely linearly with V_{gs} because carrier mobility μ_s begins to fall as V_{gs} increases. Therefore, extrapolation from the maximum slope portion of the curve insures that the degraded mobility is not inadvertently included in V_{th}.

Quadratic Extrapolation Method. The threshold voltage is also obtained by extrapolating the $\sqrt{I_{ds}}$ versus V_{gs} curve to $I_{ds} = 0$ with the MOSFET operated in the saturation region. The simplest way to place the transistor in the saturation is to connect the gate with the drain which makes $V_{gs} = V_{ds}$

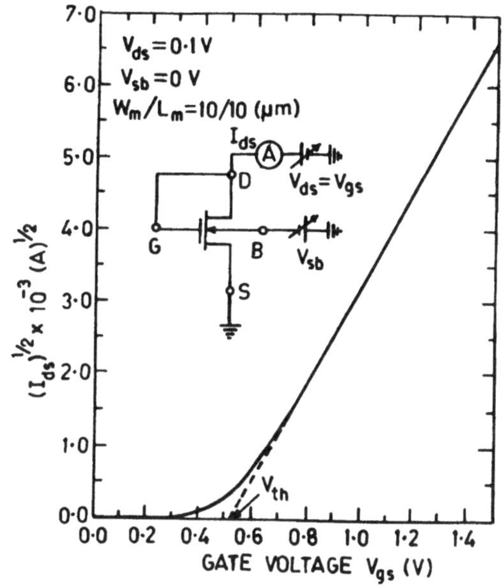

Fig. 9.22 Measurement of the threshold voltage from $I_{ds} - V_{gs}$ data in saturation region. Inset shows circuit schematic for $I_{ds} - V_{gs}$ measurement in saturation

(see inset in Figure 9.22). Recall that the drain current in saturation region is given by [cf. Eq. (6.84)]

$$I_{ds} = \frac{\mu C_{ox} W}{2L_\alpha}(V_{gs} - V_{th})^2 \quad \text{(saturation region).} \tag{9.36}$$

It is clear that a plot of $\sqrt{I_{ds}}$ versus V_{gs} gives a straight line which crosses the V_{gs} axis at V_{th} as shown in Figure 9.22. This method is not normally used for short channel devices. It is because V_{th} changes with V_{ds} for short channel devices due to the so called drain induced barrier lowering (DIBL) effect as discussed in section 5.3.3.

Obviously these different methods are not equivalent and therefore the V_{th} values obtained from the above methods are not always identical. It is not clear which threshold value is better, as they are each valid for a particular definition of the threshold condition. Although the constant current method is the simplest way to measure threshold voltage with reasonable accuracy, the V_{th} obtained depends upon the chosen value of the current and on the subthreshold slope. Since the devices used in circuit design are normally operated in the strong inversion regime, it is more appropriate to measure V_{th} in strong inversion rather than in weak inversion. *For this reason, the extrapolated methods are more appropriate.* The difference in the V_{th} values between the two extrapolation methods is due to the fact that in saturation I_{ds} varies slightly with V_{ds}, which is not taken into account for the first

Table 9.2. *Threshold voltage for a experimental n-channel enhancement MOSFET with $t_{ox} = 110$ Å and $W_m/L_m = 9.4/9.4$*

Method	Threshold voltage, V_{th}, (V)
Linear extrapolation	0.580
Saturation method	0.547
Constant current ($I_{th} = 10^{-7}$)	0.543

order equation above. This effect is more pronounced in short channel devices and therefore the difference in the measured V_{th} (from the two extrapolation methods) increases with decreasing channel length.

It is worth noting that the V_{th} value obtained by the three methods may differ by as much as 100 mV. Table 9.2 shows threshold voltage obtained from the different methods for an experimental *n*-channel device $W_m/L_m = 9.4/9.4$ fabricated using CMOS technology with $t_{ox} = 110$ Å. Note that, of the three methods discussed above, linear extrapolation gives the highest value of V_{th}.

It is interesting to note that the most commonly used linear extrapolation method is sensitive to the source/drain resistance R_t and mobility degradation factor θ. This is because devices with higher R_t and/or θ will have lower peak transconductance. Therefore the linear extrapolation line will have a smaller slope, leading to a lower extrapolated gate voltage. Thus, both R_t and θ influence the linearly extrapolated threshold voltage [49]–[51].

Other Methods. The linearly extrapolated V_{th} is the most common method of measuring V_{th}. However, this extrapolated V_{th} (shown as, V_{th}(E) in Fig. 9.23) does not match with the theoretically defined threshold voltage V_{th}(S) at strong inversion condition ($\phi_{si} = 2\phi_f + V_{sb}$). Various methods have been reported to determine a value of V_{th} which agrees with the classical threshold voltage criterion. In one method, the *threshold voltage is defined as that gate voltage at which the derivative of the low drain voltage transconductance dg_m/dV_{gs} is maximum.* This is called the *transconductance change* method [49]; the corresponding V_{th} is shown in Figure 9.23 as V_{th}(TC). This method gives threshold voltage V_{th}(TC) which is very close to the V_{th}(S) at $\phi_{si} = 2\phi_f + V_{sb}$ condition. It has the advantage that it is not affected by the device mobility degradation θ and series resistance R_t [49]. However, since in this method two derivatives of I_{ds} are required, this method tends to be very noisy.

Another method, called the *split C–V*, which normally is used in the inversion layer mobility measurements (see section 9.9.1), has also been used to give the classical threshold voltage [53] and has been taken as the true threshold voltage [50]. In this method gate and bulk currents are measured during quasi-static or high frequency C–V measurement with the transistor in the

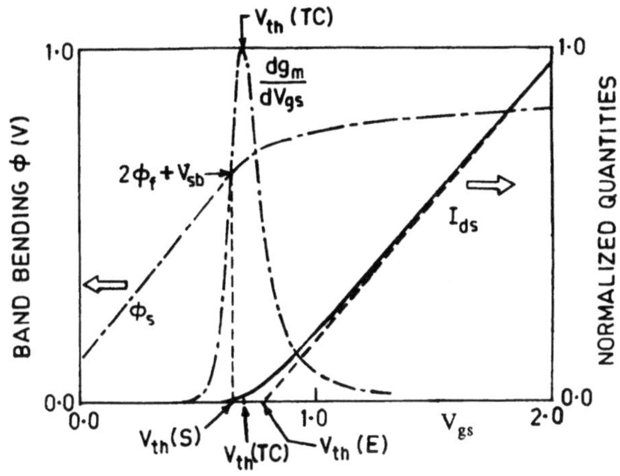

Fig. 9.23 Schematic illustration for determining V_{th} using transconductance change (TC) method. The threshold voltage corresponding to TC method is shown as V_{th}(TC). Also shown is V_{th} corresponding to classical '$2\phi_f$' point, denoted by V_{th}(S). The linearly extrapolated V_{th} is labeled as $V_{th}(E)$. (After Wong et al. [47])

gate controlled diode configuration. The threshold voltage is the gate voltage where the two currents are equal ($dQ_i/dV_{gs} = dQ_b/dV_{gs}$).
It is important to note that V_{th} at the $2\phi_f$ condition is only suitable for sub-threshold or weak inversion conduction as compared to the linear extrapolated V_{th}, which is obtained from the strong inversion condition. The difference between the classical (V_{th} at $2\phi_f$) and the extrapolated V_{th} is around $4\,V_t$ (~ 0.1 V) [50]–[52]. In practice, the extrapolated V_{th} is then matched with the model equation based on the strong inversion condition.

9.5 Determination of Body Factor γ

The body factor γ [cf. Eq. (5.11)] determines sensitivity of the threshold voltage V_{th} to the back bias V_{sb}; the higher the γ, the higher the body effect. For a uniformly doped substrate, the body factor γ is a constant number and is obtained from the device (long channel) V_{th} versus V_{sb} characteristics. If V_{TO} is the value of V_{th} measured at $V_{sb} = 0$, then rearranging Eq. (5.15) yields

$$V_{th} - V_{TO} = \gamma\sqrt{2\phi_f + V_{sb}} - \gamma\sqrt{2\phi_f} \quad \text{(V)} \tag{9.37}$$

where V_{fb} is the flat band voltage, and ϕ_f is the bulk Fermi potential [cf. Eq. (2.15)]. This equation shows that the plot of $(V_{th} - V_{TO})$ versus $\sqrt{2\phi_f + V_{sb}}$ will be a straight line with slope γ and intercept $\gamma\sqrt{2\phi_f}$. Thus, to calculate γ we proceed as follows:

1. Take a large geometry device and sweep the gate voltage from $V_{gs} = 0$ to V_{gs}(max), say 5 V, in steps of say 50 mV, while simultaneously monitoring the corresponding drain current I_{ds}. The drain voltage V_{ds} is fixed at 50 mV, while the back bias V_{sb} is set to zero (substrate grounded).
2. Determine V_{th} using linear extrapolation method from the point of maximum slope.
3. Repeat step 1 for several different values of V_{sb}, normally in steps of 0.1 V and determine V_{th} for the corresponding V_{sb}.
4. By initially assuming $2\phi_f = 0.6$ (a good approximation for silicon VLSI devices), plug the measured $V_{th}(V_{bs})$ data in Eq. (9.37). By using simple linear regression to $(V_{th} - V_{TO})$ versus $\sqrt{2\phi_f + V_{sb}}$ data, one can easily extract γ and ϕ_f. The new value of ϕ_f is inserted back into Eq. (9.37) in place of the initial guess value and another linear regression is performed. Regressions are iterated in this fashion until ϕ_f and γ are obtained within say 0.1%.

For a nonuniformly doped substrate γ is not constant, but becomes substrate bias dependent, as was discussed in section 5.2.1. This fact is illustrated in Figure 9.24b which is a schematic plot of V_{th} versus $\sqrt{V_{sb} + 2\phi_f}$ for implanted devices. Approximating the channel doping profile by a step profile (see Figure 9.24a), with a surface concentration N_s and substrate concentration

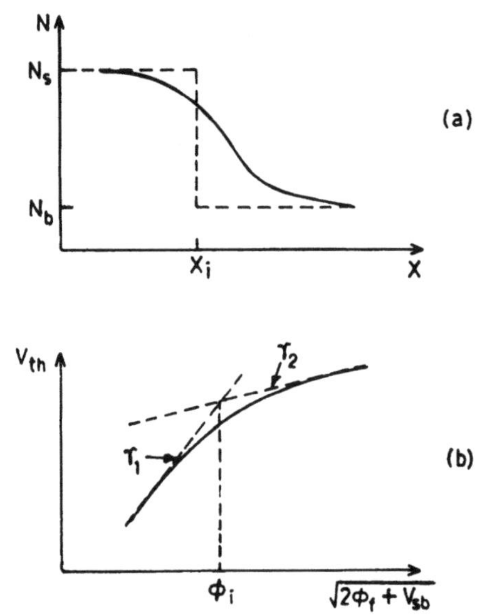

Fig. 9.24 Determination of the body factor γ for a nonuniformly doped substrate (channel implanted MOSFET). (a) Channel doping profile; dotted line shows the assumed step profile, and (b) V_{th} as a function of back bias V_{sb} showing two slopes, low V_{th} corresponds to slope γ_1, while higher V_{th} corresponds to γ_2

N_b, leads directly to the two dashed lines approximation to the curve in Figure 9.24b. The slopes γ_1 and γ_2 of the two lines represents the magnitude of the body effect corresponding to both low and high levels of V_{sb}. For γ_1 calculation, V_{th} corresponding to V_{sb} between 0–0.8 V is used, while for γ_2, V_{sb} greater than 1 V is used. In practice, often these parameters are determined by curve fitting the $V_{th} - V_{sb}$ data using nonlinear least-square optimization methods, discussed in the next chapter.

9.6 Flat Band Voltage

For a uniformly doped substrate, V_{fb} can easily be obtained from V_{th} versus V_{bs} data using the procedure discussed for determining γ (cf. section 9.5). This method yields both γ and ϕ_f. Once these two parameters are known V_{fb} can be calculated using Eq. (5.14). For non-uniformly doped substrates, V_{fb} is generally determined from $V_{th} - V_{sb}$ data at low $|V_{sb}|$ (< 1 V). Recall that V_{fb} can also be determined from an MOS capacitor (cf. section 4.7). However, it should be pointed out that V_{fb} determined from an MOS capacitor is always higher than that determined from $V_{th} - V_{sb}$ data. This is because V_{fb} determined using Eq. (5.14) is a fitting parameter rather than a true value, as obtained from an MOS capacitor (cf. section 4.7), although the two values are very close (within 10%) for uniformly doped substrates.

9.7 Drain Induced Barrier Lowering (DIBL) Parameter

Normally, the threshold voltage is measured at low drain voltages ($V_{ds} < 0.1$V). For long-channel devices, V_{th} is independent of V_{ds}. However, for short-channel devices V_{th} becomes V_{ds} dependent due to the DIBL effect (cf. section 5.3.3). The higher the V_{ds}, the lower the V_{th}, and is given by the following relation [cf. Eq. (5.96)]

$$V_{thd} \equiv V_{th}(V_{ds}) = V_{th} - \sigma V_{ds} = V_{th} - \Delta V_{th} \tag{9.38}$$

where V_{th} is the threshold voltage measured at low V_{ds}(≤ 0.1 V), called the nominal V_{th}, ΔV_{th} is the change in the threshold voltage measured at higher V_{ds} from its nominal value, and σ is the DIBL parameter. Thus, knowing ΔV_{th}, the DIBL can easily be characterized.

The measurement from which ΔV_{th} is extracted consists of the following steps:

1. The gate voltage V_{gs} is normally swept from ($V_{th} - 1$ V) to ($V_{th} + 1$ V) in steps of say 50 mV at a fixed V_{ds} and V_{sb}, while simultaneously monitoring the corresponding drain current I_{ds}.
2. Step 1 is repeated for 4 or 5 values of V_{ds} ranging from 50 mV to

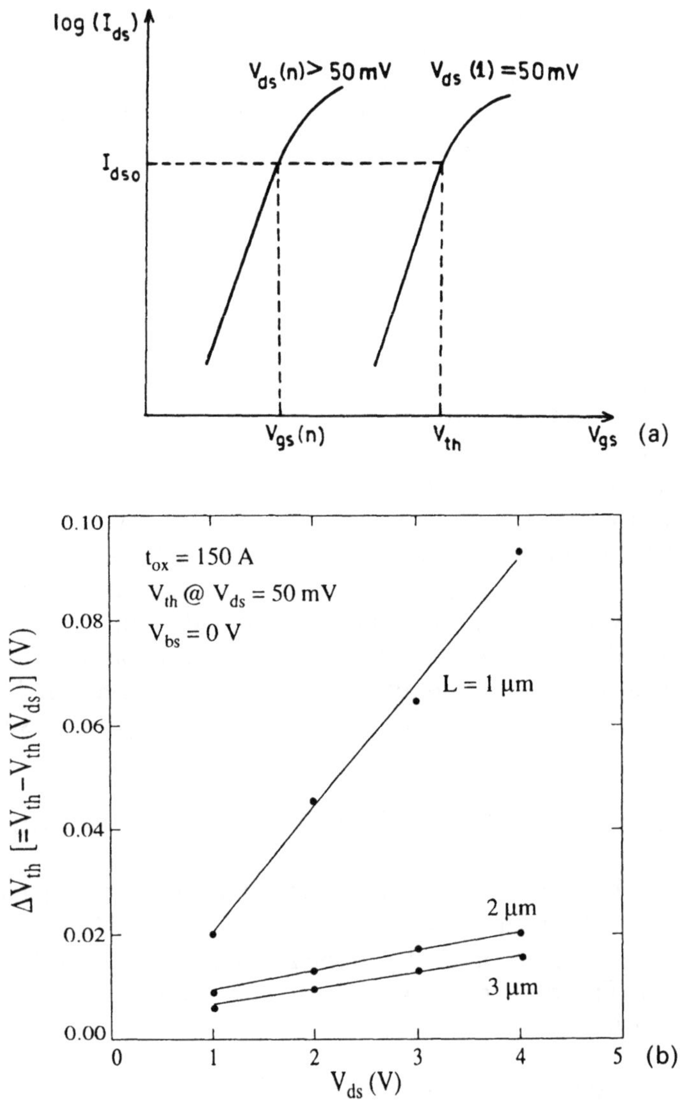

Fig. 9.25 Drain induced barrier lowering (DIBL) parameter determination. (a) Procedure for threshold voltage measurement at higher V_{ds} and (b) change ΔV_{th} in V_{th} from its nominal value ($V_{ds} < 0.1$ V) as a function of V_{ds} for 3 different channel length devices $L_m = 1$, 2 and 3 μm for nMOST with $t_{ox} = 150$ Å. The slope of these lines give value for DIBL parameter σ

V_{ds}(max) keeping V_{sb} fixed. These measurements yield I_{ds} versus V_{gs} characteristics similar to the ones shown in Figure 9.25a for a fixed value of V_{sb}.

3. Determine I_{ds} corresponding to the nominal threshold voltage V_{th}, using the linear interpolation method. This is denoted by I_{ds0} in Figure 9.25a. The gate voltage interpolated from I_{ds0} at the subsequent value of V_{ds},

denoted by $V_{gs}(n)$, is the threshold voltage $V_{th}(V_{ds})$ corresponding to that $V_{ds}(n)$.

Once $V_{th}(V_{ds})$, i.e., $V_{gs}(n)$ is determined for each V_{ds} then ΔV_{th} can be calculated for each V_{ds} as

$$\Delta V_{th}(n) = V_{th} - V_{gs}(n)$$

By repeating the steps 1–3 for different V_{sb}, one calculates ΔV_{th} as a function of V_{sb} for a given V_{ds}.
Figure 9.26b shows ΔV_{th} as a function of V_{ds} at $V_{sb} = 0$ V for different channel length devices ($L_m = 1, 2$ and $3\,\mu$m) for a typical n-channel devices fabricated using 1 μm CMOS process with $t_{ox} = 150$ Å. Note that ΔV_{th} varies linearly with V_{ds} which confirms validity of Eq. (9.38). By using linear regression to the data for different L, the DIBL parameter σ can be determined as a function of L_m. However, generally DIBL parameters are determined using the nonlinear optimization method as discussed in section 10.6.1.

9.8 Determination of Subthreshold Slope

The drain current below the threshold voltage is referred to as the subthreshold current and varies exponentially with V_{gs}. The reciprocal of the slope of the log(I_{ds}) versus V_{gs} characteristic is defined as the subthreshold slope S (see Figure 9.26). It is an important device parameter which determines how well the MOSFET functions as a switch.
To determine the subthreshold slope, the drain current I_{ds0} is first deter-

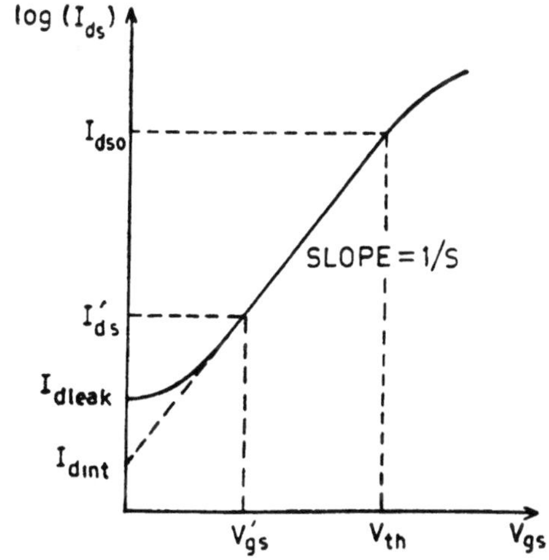

Fig. 9.26 Measurement of subthreshold slope from the device $I_{ds} - V_{gs}$ data

mined corresponding to $V_{gs} = V_{th}$, the threshold voltage at low V_{ds} (<0.1 V). Then the gate voltage V'_{gs} corresponding to the drain current I'_{ds}, which is two decades below $I_{ds0}(I'_{ds} = I_{ds0}/100)$, is obtained by interpolation. The I'_{ds} is assumed to be the boundary marking the end of subthreshold operation. Linear regression is then applied to the data between I_{ds0} and I'_{ds}. The inverse of the slope of this line yields the subthreshold slope S, given by [cf. Eq. (6.113)]

$$S = \frac{dV_{gs}}{d(\log I_{ds})} = 2.3\left[\frac{dV_{gs}}{d(\ln I_{ds})}\right] \quad \text{(V/decade)}. \tag{9.39}$$

Extrapolating the line to the I_{ds} axis gives I_{dint} which is normally compared with the measured I_{ds} at $V_{gs} = 0$ V, called the leakage current I_{dleak} (see Fig. 9.26). A large difference between I_{dleak} and I_{dint} is an indication of the bulk punchthrough or junction leakage.
If the device has different slopes in the drain current region between I_{ds0} and I'_{ds}, then S is determined from the highest value of the ratio $\Delta V_{gs}/\Delta(\log I_{ds})$ over an interval of 60 mV of V_{gs}. Recall that the minimum value of S is 60 mV. The quantity S is generally measured in both linear and saturation regions of device operation. Normally S value in the saturation region is slightly higher than the corresponding linear region value.

9.9 Carrier Inversion Layer Mobility Measurement

To measure carrier mobility in the inversion layer of MOSFETs, large devices are normally chosen in order to avoid short-channel and narrow-width effects. For a 2 μm CMOS technology with gate oxide thickness of 300 Å, a 25 × 25 μm^2 device can be considered a large device.
The quickest and traditional way to characterize the inversion layer mobility of a MOSFET is to measure drain current I_{ds} in the linear region ($V_{gs} > V_{th}$) at low V_{ds} (<0.1 V). In the linear region, the current I_{ds} at small V_{ds}(<0.1 V) can be approximated as [cf. Eq. (6.51)]

$$I_{ds} = \mu_s C_{ox} \frac{W}{L}(V_{gs} - V_{th})V_{ds} \tag{9.40}$$

and is obtained by neglecting the term $0.5\alpha V_{ds}$ in Eq. (9.35). This is a valid assumption, provided $V_{gs} - V_{th} > 0.5$ V and V_{ds} is small (<0.1 V). Differentiating Eq. (9.40) with respect to V_{gs} gives the transconductance g_m as

$$g_m = \left.\frac{\partial I_d}{\partial V_{gs}}\right|_{V_{ds}} = \mu_s C_{ox} \frac{W}{L} V_{ds}. \tag{9.41}$$

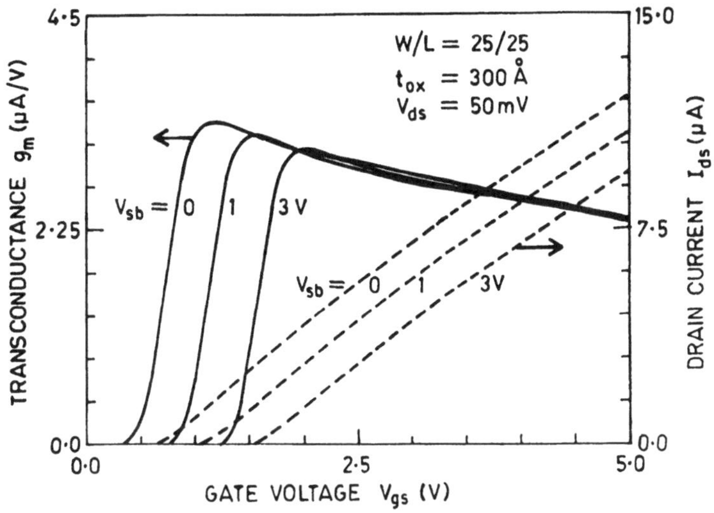

Fig. 9.27 Typical drain current $I_{ds} - V_{gs}$ (dotted lines) and transconductance $g_m - V_{gs}$ (continuous lines) characteristics of an MOSFET at three back bias $V_{sb} = 0, 1$ and 3 V. The data is from an n-channel MOSFET with W/L = 25/25 with $t_{ox} = 300$ Å

Carrier mobility is then related to the *transconductance mobility*, also called the *field-effect mobility* μ_{FE}, defined as

$$\mu_{FE} \equiv \mu_s = \frac{g_m}{C_{ox} V_{ds}} \cdot \frac{L}{W} \quad \text{(field-effect mobility).} \tag{9.42}$$

Figure 9.27 shows plots of drain current versus gate voltage at $V_{ds} = 50$ mV at three different back biases $V_{sb} = 0, 1$ and 3 V (dotted lines). The corresponding transconductance (g_m) curves are shown as continuous lines. Note that different g_m curves converge into a single curve at higher V_{gs}. Similar results are obtained for p-channel devices. Once g_m is known, μ_s can easily be calculated as a function of V_{gs} using Eq. (9.42).
Another definition of mobility, which has been used, is called the *conductivity* or *effective mobility* μ_{eff}. It is related to the DC channel conductance $g_{ds}(= \Delta I_{ds}/\Delta V_{ds})$ at constant V_{gs} as

$$\mu_{\text{eff}} \equiv \mu_s \cong \frac{g_{ds}}{C_{ox}(V_{gs} - V_{th})} \cdot \frac{L}{W} \quad \text{(conductivity mobility)} \tag{9.43}$$

and is obtained by differentiating (9.40) with respect to V_{ds}. Since V_{ds} is constant, g_{ds} is simply I_{ds}/V_{ds}. Thus, knowing g_{ds} at each V_{gs}, one can calculate μ_s. Since Eq. (9.40) is valid only for $V_{gs} > V_{th}$, the mobility measured by the g_{ds} method is good only for $V_{gs} \geq V_{th} + 0.5$; the accuracy of extracted μ_s decreases near V_{th} because the approximate relationship (9.40) becomes inaccurate. Furthermore, mobility extracted using Eq. (9.43) needs

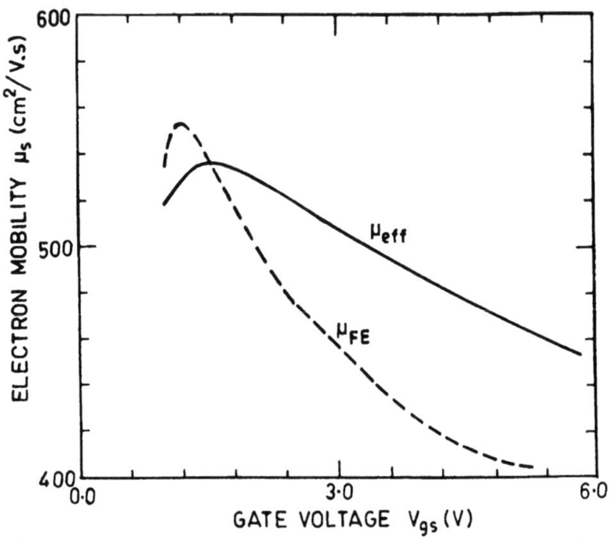

Fig. 9.28 Effective and field-effect mobility μ_{eff} (continuous line) and μ_{FE} (dotted line), respectively, as a function of gate voltage V_{gs} at $V_{sb} = 0\,V$ for a nMOST whose $I_{ds} - V_{gs}$ characteristics (dotted lines) are shown in Figure 9.27

accurate values of the threshold voltage V_{th}. Depending upon the definition of V_{th} used, μ_s could easily differ by about 10% near V_{th}.

The mobility measured using Eqs. (9.42) and (9.43) is shown in Figure 9.28, as dotted and continuous lines, respectively, for a nMOST whose $I_{ds} - V_{gs}$ characteristics are shown in Figure 9.27. Note that mobility decreases with gate voltage. This decrease has been attributed to the enhanced surface roughness scattering with increased gate voltage and to quantization effects, as was explained in section 6.5. Also note that μ_{FE} (dotted line) is lower than μ_{eff} (continuous line) over the entire gate voltage range. One may ask as to which of the two mobilities are more appropriate? To answer this question, we recalculate the drain current using Eq. (9.40) based on μ_s from Eqs. (9.42) and (9.43). Figure 9.29 shows the calculated I_{ds} with μ_s replaced by μ_{FE} (dotted line) and μ_{eff} (continuous line); circles are experimental data points. Clearly the calculated I_{ds} fits the experimental data (circles) only when μ_s is replaced by μ_{eff} [cf. Eq. (9.43)]. This shows that *μ_{eff} calculated using* Eq. (9.43) *is the appropriate method for calculating mobility* [15]. The reason that μ_{FE} is inappropriate for calculating current is that Eq. (9.42) neglects the effect of gate voltage dependence on mobility by lumping the field dependent effects, and thus is always found to underestimate the true mobility values.

Equation (9.43) gives the inversion layer mobility as a function of gate voltage V_{gs}. However, as was discussed in section 6.5, it is more appropriate to measure mobility as a function of effective (normal) electric field $\mathscr{E}_{eff}$

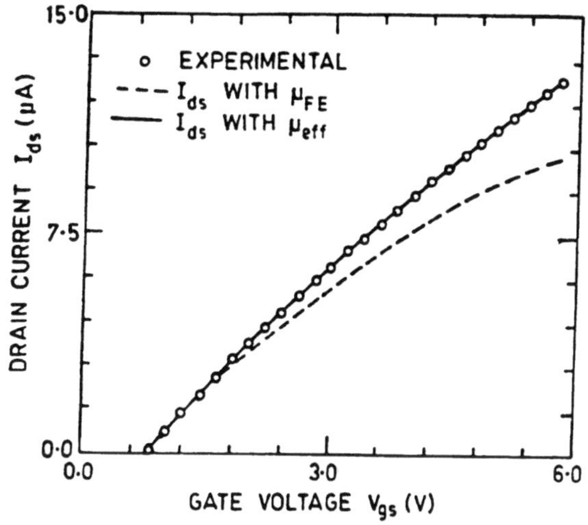

Fig. 9.29 The $I_{ds} - V_{gs}$ curve with μ_s calculated using (a) Eq. (9.42)—dotted line and (b) Eq. (9.43)—continuous line. Circles are experimental data points

defined as [cf. Eq. (6.148)]

$$\mathscr{E}_{\text{eff}} = \frac{1}{\epsilon_0 \epsilon_{si}}(Q_b + \zeta Q_i) \tag{9.44}$$

where $\zeta = 0.5$ for electrons (nMOST), 0.3 for holes (pMOST), and Q_b and Q_i are bulk and inversion charge densities, respectively. The value of ζ is such that the μ_s versus $\mathscr{E}_{\text{eff}}$ curves are independent of V_{sb} [54]. For a MOSFET with a uniformly doped substrate, Q_b can easily be calculated as [cf. Eq. (6.44)]

$$Q_b = \gamma C_{ox}\sqrt{2\phi_f + V_{sb}} \tag{9.45}$$

where γ is the body factor. However, to calculate Q_b for nonuniform doping is not a trivial task. One way to calculate Q_b is to measure the doping profile in the channel and then solve for Poisson's equation numerically. However, a somewhat simple but approximate way to calculate Q_b is to use the threshold voltage Eq. (5.7), so that

$$Q_b = C_{ox}(V_{th} - V_{fb} - 2\phi_f). \tag{9.46}$$

Assuming V_{ds} is small and $V_{gs} > V_{th}$, the inversion charge density Q_i can be calculated as [cf. Eq. (6.79)]

$$Q_i = C_{ox}(V_{gs} - V_{th}). \tag{9.47}$$

Thus, to measure mobility as a function of electric field the following steps are used:

1. The gate voltage is swept from zero to V_{gs}(max) in steps of say 0.1 V, while simultaneously monitoring the drain current I_{ds}. The drain voltage is set to 50 mV while V_{sb} is set to zero (substrate grounded).
2. Determine C_{ox} from high frequency C–V method.
3. Determine V_{th} using linear extrapolation method.
4. Calculate Q_b using Eq. (9.45) or (9.46). This requires calculation of either the body factor γ or the flat band voltage V_{fb} as discussed in sections 9.5 and 9.6, respectively.
5. Calculate Q_i as a function of V_{gs} using Eq. (9.47).
6. Use Eqs. (9.43) and (9.44) to calculate μ_s and $\mathscr{E}_{\text{eff}}$, respectively, for fixed V_{gs} and V_{sb}.
7. Repeat steps 1–6 for different V_{sb}.

Once μ_s versus $\mathscr{E}_{\text{eff}}$ data is available, the low field mobility μ_0 and mobility degradation parameter θ can easily be calculated from the following equation [cf. Eq. (6.142)]

$$\mu_s = \frac{\mu_0}{1 + \theta \mathscr{E}_{\text{eff}}} \tag{9.48}$$

using linear regression of $1/\mu_s$ versus $\mathscr{E}_{\text{eff}}$ data.
Although mobility measured using Eq. (9.43) is the appropriate mobility and is the most commonly used method, it results in an approximate value of μ_s for the following reasons:

- The inversion layer charge density $Q_i = C_{ox}(V_{gs} - V_{th})$ assumed for calculating μ_{eff} is only approximate. The deviation of Q_i from its linear approximation near threshold increases as oxide thickness decreases. This explains why mobility calculated using Eq. (9.43) becomes t_{ox} dependent, particularly near V_{th}.
- The value of the gate-to-channel capacitance, also referred to as the *inversion layer capacitance*, C_{gc} $(= dQ_i/dV_{gs})$, does not reach the value of C_{ox} until V_{gs} reaches a few tenths of a volt beyond V_{th}.
- The electric field between the drain and source does not reach the value V_{ds}/L until reasonably high gate bias.
- The diffusion current has been neglected.

To overcome the shortcomings of the above method, μ_s is more appropriately calculated by directly measuring the inversion charge density Q_i, as discussed in the next section.

9.9.1 Split-CV Method

An accurate method of carrier mobility measurement is to evaluate directly the inversion charge density Q_i, through the measurement of the gate-channel capacitance C_{gc}. This technique of measuring μ_s from Q_i is called

the *split-CV technique*. In the normal C–V measurement, the charge flowing through the gate is sensed and the measured value includes accumulation, depletion and inversion capacitances. In the split-CV method, the inversion capacitance can be measured separately from the depletion/accumulation capacitance, hence the name split-CV.
In terms of Q_i, the current I_{ds} is given by Eq. (6.33)

$$\frac{I_{ds}}{W} = Q_i \mu_s \mathscr{E} - V_t \mu_s \frac{dQ_i}{dy}. \tag{9.49}$$

If I_{ds} is measured at low $V_{ds}(< 0.1\text{ V})$, then it has been shown that [53]

$$\mathscr{E} = \frac{V_{ds}}{L} \cdot \frac{C_{gc}}{C_{ox}}$$

and

$$\frac{dQ_i}{dy} = \frac{V_{ds}}{L} C_{gc} \tag{9.50}$$

so that Eq. (9.49) becomes

$$\frac{I_{ds}}{W} = \mu_s \left(\frac{Q_i C_{gc}}{C_{ox}} - V_t C_{gc} \right) \frac{V_{ds}}{L}. \tag{9.51}$$

For $V_{gs} > V_{th}$, $C_{gc} = C_{ox}$ and $V_t \cdot C_{gc} < Q_i$, therefore Eq. (9.51) reduces to

$$\mu_s = \frac{I_{ds}}{V_{ds}} \frac{L}{W} \frac{1}{Q_i} \tag{9.52}$$

which is the same as Eq. (9.43) except that $C_{ox}(V_{gs} - V_{th})$ is replaced by Q_i. The inversion charge density Q_i can be obtained by measuring C_{gc} either using quasi-static [53]–[55] or high frequency C–V method [56]–[59]. Once C_{gc} is known, Q_i can be obtained by integrating C_{gc}, that is,

$$Q_i = \int_{-\infty}^{V_{gs}} C_{gc}(V_{gs}) dV_{gs}. \tag{9.53}$$

The lower limit of integration is normally set to V_{fb}, the flat band voltage, for convenience. This is because no inversion capacitance is measured until the gate voltage comes close to the threshold voltage. Therefore, the lower limit can be any voltage as long as it is lower than the threshold voltage. Thus, knowing C_{gc} and Q_i as a function of V_{gs}, one can calculate μ_s as a function of V_{gs} using Eq. (9.52). Remember that C_{gc} is per unit area; therefore, measured gate-to-channel capacitance C_{GC} (Farad) must be divided by WL, the effective gate area, to get C_{gc}.

The advantages of this technique are

- $Q_i(V_{gs})$ can be directly evaluated by integrating the C_{gc} versus V_{gs} curve, and the inversion charge approximation $Q_i = C_{ox}(V_{gs} - V_{th})$ can be avoided.
- The evaluation of Q_i does not require V_{th}, whose value varies depending upon what definition is used to calculate V_{th}.

The limitation of this technique is that in weak inversion it is affected by the interface states, which in turn affects the Q_i calculations. Indeed this limitation can be minimized by the use of optimum channel length devices for mobility measurements [55], [57].

In order to obtain μ_s as a function of effective field $\mathscr{E}_{eff}$, we need to calculate not only the inversion charge Q_i, but the bulk charge Q_b also, see Eq. (9.44). In the split-CV method, the bulk charge is obtained by measuring gate-to-bulk capacitance C_{gb} as a function of V_{gs} and integrating the C_{gb} curve from some reference voltage to V_{gs} as

$$Q_b = \int_{V_{fb}}^{V_{gs}} C_{gb}(V_{gs})dV_{gs}. \tag{9.54}$$

This reference voltage is normally taken as the flat band voltage V_{fb}. The Q_b calculated using C_{gb} is more accurate compared to Eq. (9.45), particularly for channel implanted devices. Once Q_i and Q_b are known, $\mathscr{E}_{eff}$ can easily be calculated from Eq. (9.44).

The results of mobility calculation using the drain conductance (conventional) method [cf. Eq. (9.43)] and the split-CV method [cf. Eq. (9.52)] for a device

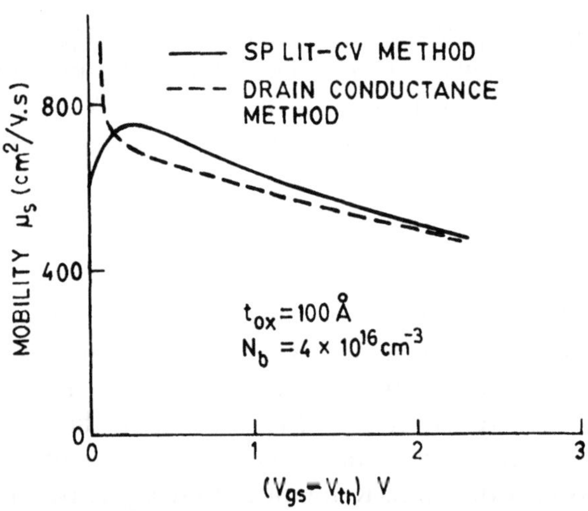

Fig. 9.30 Comparison of the mobility measured using the split-CV method (continuous line) and conventional (drain conductance) method (dotted line) (After Sodini et al. [51])

with $t_{ox} = 100$ Å is shown in Figure 9.30. The maximum error between the two curves is about 10% at $V_{gs} = V_{th} = 0.5$ V [53]. Note the trends in the two curves near the threshold voltage. The split-CV method begins to see an artificial drop in the mobility due to the surface state effects, while μ_s calculated from Eq. (9.43) diverges near V_{th}. The error in μ_s calculated using split-CV method is claimed to be negligible for $V_{gs} \gg V_{th}$ [53]. Thus, the extra complexity (as both current and capacitance measurements are required in this case) is justified by a certain increase in accuracy. However, this increase in accuracy improves only slightly over the conventional method for t_{ox} greater than 100 Å.

Although the split C–V is an accurate method for measuring mobility, it has inconsistency in the measurements of I_{ds} and Q_i. Note that, while I_{ds} is measured at a small but finite value of V_{ds} (20–50 mV), the C_{gc} required for Q_i calculation is measured at zero V_{ds}. In order to avoid this inconsistency, C_{gc} and I_{ds} should be measured at the same V_{ds}. This indeed can be done by measuring the gate-to-drain and gate-to-source capacitances, C_{gd} and C_{gs} respectively, at a desired V_{ds} and then adding the two capacitances to get C_{gc}; i.e., $C_{gc} = C_{gs} + C_{gd}$. It has been shown that this method results in a more accurate value of the inversion layer mobility [58a].

Experimental Setup for Measuring C_{gc} and C_{gb}. Figure 9.31a shows an experimental setup for measuring the gate-to-channel capacitance C_{gc} of an nMOST using an HP 4275A LCR meter. The source and drain are connected together to the capacitance meter 'Lo' terminal and the gate is connected to the 'Hi' terminal. A DC bias V_{sb} is applied to the bulk terminal. The gate voltage is stepped, from a value at which the inversion capacitance is zero, in the direction in which capacitance increases. Thus, for an n-channel MOSFET, V_{gs} is swept from $-V_{gs}$(max) to $+V_{gs}$(max) in steps of say 0.1 V. An AC signal voltage of 100 mV peak-to-peak is generally used. The frequency of the AC signal can range from 100 KHz to 1 MHz. In principle, high frequency C_{gc} measurements may be affected by the presence of interface states and channel resistance effects [57]. If present, both factors introduce a frequency dependence of the C_{gc}. However, it has been observed that C_{gc} is generally independent of frequency in the range 100 KHz to 1 MHz [58].

The value of the capacitance measured at $-V_{gs}$(max) (accumulation regime) is $C_0 = C_p + C_{ov}$, where C_p is the parasitic capacitance of the measuring system and C_{ov} is the total gate overlap capacitance. Thus, C_0 is subsequently subtracted from all previously measured values of C_{gc}. The resulting C_{gc} versus V_{gs} is shown in Figure 9.31b for an nMOST (W/L = 50/50 and $t_{ox} = 150$ Å) for three different V_{sb} (= 0,1 and 3 V). The corresponding Q_i calculated using Eq. (9.53) are shown as dotted lines. The integration for Q_i can be performed using any suitable integration routine such as the trapezoidal or Simpson method.

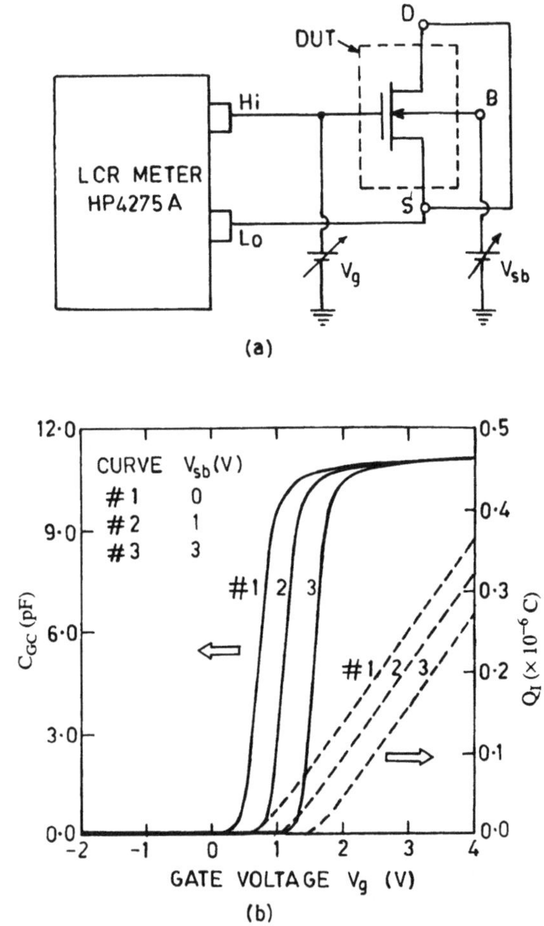

Fig. 9.31 (a) Experimental setup for measuring gate-to-channel capacitance C_{GC}. (b) C_{GC} as a function of gate voltage V_{gs} for 3 different substrate biases $V_{sb} = 0, 1$ and 3 V. The dotted lines are inversion charge density Q_I $(= Q_i/WL)$ calculated using Eq. (9.53)

To measure gate to substrate capacitance C_{gb}, the source and drain are connected to the ground, the substrate to the 'Lo' terminal of the capacitance meter and the gate to the 'Hi' terminal as shown in Figure 9.32a. Once C_{gb} as a function of V_{gs} is known, the Q_b can be obtained using Eq. (9.54) using the same procedure as for calculating Q_i. Figure 9.32b shows measured gate-to-bulk capacitance C_{gb} of an nMOST (the device whose C_{gc} is shown in Figure 9.31b) for three different V_{sb} $(= 0, 1$ and 3 V). The corresponding Q_b are shown as dotted lines.

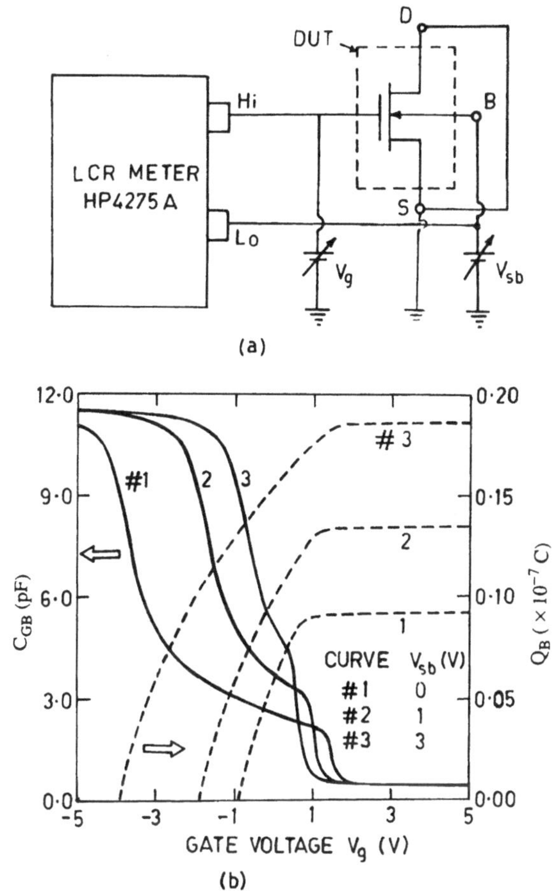

Fig. 9.32 (a) Experimental setup for measuring gate-to-bulk capacitance C_{GB}. (b) C_{GB} as a function of gate voltage V_{gs} for 3 different substrate biases $V_{sb} = 0, 1$ and 3 V. The dotted lines are bulk charge density Q_B $(= Q_b/WL)$ calculated using Eq. (9.54)

9.10 Determination of Effective Channel Length and Width

The most basic parameters of a MOSFET are those which define the effective or electrical length L and width W of the transistors. These parameters play an important role in governing the device characteristics of small geometry devices. The device L differs from the drawn channel length L_m (physical mask dimensions) by a factor ΔL such that $L = L_m - \Delta L$ [cf. Eq. (3.32)]. Similarly, the device W is generally smaller than the drawn device width W_m (physical mask dimension) by a factor ΔW such that $W = W_m - \Delta W$ [cf. Eq. (3.33)]. It is the L and W and not L_m and W_m which are used for modeling MOSFET devices (cf. section 3.7). This in turn requires ΔL and ΔW to be known. In this section we will discuss various methods of determining ΔL and ΔW of a MOSFET.

Basically, there are two methods of determining ΔL and ΔW. These are:

- *Drain current method*: [60]–[85] In this method, the drain current I_{ds} is measured as a function of gate voltage V_{gs} at a low drain voltage ($V_{ds} < 0.1$ V) and fixed back bias V_{sb}, generally zero volts. The low V_{ds} ensures device operation in the linear region.
- *Capacitance method*: [86]–[90] Here device gate-to-channel capacitance C_{GC} is measured as a function of V_{gs} at zero V_{sb} with source and drain tied together.

The drain current method is the most widely used for determining ΔL and ΔW because of its simplicity. It should be pointed out that either of the I–V or C–V methods of extracting $\Delta L(\Delta W)$ results in the effective channel $L(W)$ that is purely an electrical parameter, as discussed in section 3.7.

9.10.1 Drain Current Methods of Determining ΔL

Various drain current methods, reported in the literature, to determine ΔL are based on the following drain current equation in the linear region [cf. Eq. (6.225)]

$$I_{ds} = \frac{\beta_0 (V_{gs} - V_{th}) V_{ds}}{1 + (\theta_0 + \beta_0 R_t)(V_{gs} - V_{th})} \tag{9.55}$$

where $\beta_0 = \mu_0 C_{ox} W/L$ and we have assumed θ_b to be zero. This assumption (of $\theta_b = 0$) is made for all methods of ΔL and ΔW extraction using $I_{ds} - V_{gs}$ data in the linear region [65]. It is a good approximation provided these parameters are extracted at zero back bias, which indeed is generally the case for ΔL and ΔW extraction. However, if V_{sb} is not zero, then neglecting θ_b will cause error in the extraction. Different drain current methods use Eq. (9.55), and its variation, to determine ΔL.

Channel-Resistance Method. The one most commonly used method of ΔL extraction is the so called *channel resistance method.* In this method ΔL is extracted by measuring the response of the device channel resistance to the change in the gate voltage V_{gs} or gate drive ($V_{gs} - V_{th}$) at fixed V_{sb}. The intrinsic channel resistance R_{ch} of an MOSFET operating in the linear region is given by [cf. Eq. (6.52)]

$$R_{ch} \approx \frac{L}{\mu_s C_{ox} W (V_{gs} - V_{th})}. \tag{9.56}$$

The total resistance R_m measured between the source and drain terminals is simply

$$R_m = R_{ch} + R_t \tag{9.57}$$

where R_t is sum of the source and drain resistances. Combining Eqs. (9.56) and (9.57) we get

$$R_m = \frac{L}{\mu_s C_{ox} W(V_{gs} - V_{th})} + R_t \tag{9.58}$$

which can be written as

$$R_m = A(L_m - \Delta L) + R_t \tag{9.59}$$

where

$$A = [\mu_s C_{ox} W(V_{gs} - V_{th})]^{-1}.$$

Physically, A is the channel resistance per unit length. To unambiguously determine ΔL from Eq. (9.59) it is important that (1) R_t be independent of the external bias, and (2) A be independent of the channel length. If these two conditions are met then from Eq. (9.59) it is evident that at a given V_{gs}, the plot of the measured R_m against the drawn channel lengths L_m, for sets of adjacent transistors with the same channel widths, will be a straight line given by

$$R_m = A \cdot L_m + B \tag{9.60}$$

where the intercept B is

$$B = R_t - A \cdot \Delta L. \tag{9.61}$$

Repeating the plot for different gate voltages V_{gs} will result in a set of straight lines. In the ideal case, these different lines will all intersect at one point with the point of intersection giving R_t on the R_m axis (y-axis) and ΔL on the L_m axis(x-axis), as shown in Figure 9.33. However, often R_m versus L_m lines fail to intersect at a common point. In that case the method is carried over one step further by using second linear regression of the plot of B versus A obtained from different gate voltages V_{gs} [cf. Eq. (9.61)]. The slope and intercept of B versus A line gives ΔL and R_t, respectively. To avoid any narrow width effect wide test transistors should be used. Further, since V_{th} is channel length dependent, one should use higher gate biases (e.g. $V_{gs} > 4$–5 V) in order to minimize the effect of short channel V_{th} fall off on the parameter A. Since VLSI circuits require smaller gate biases ($V_{gs} \leq 5$ V), to minimize the effect of varying V_{th} the proper method would be to adjust V_{gs} so that the effective gate drive $V_{gt}(= V_{gs} - V_{th})$ is equal for all transistors. This way, one can use lower V_{gs} and also ensure that A is constant. The possible device to device variations of μ_s, W and C_{ox}, for the same die, are neglected in the analysis, although they can contribute to the error in the ΔL extraction.

It has been pointed out that the method of determining V_{th} also affects the extracted ΔL and R_t, particularly when using small gate drive, say $V_{gt} =$

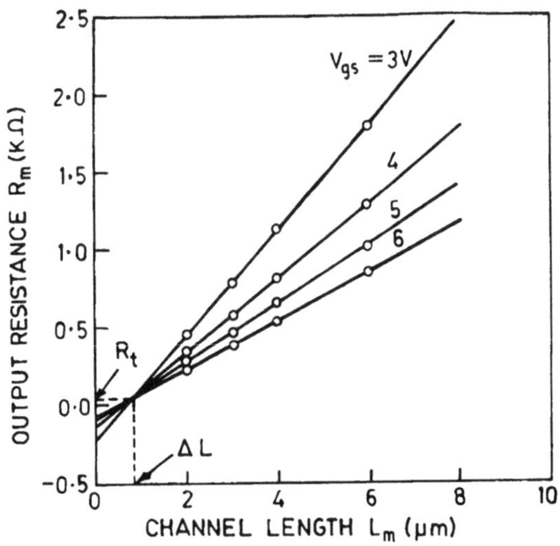

Fig. 9.33 Measured output resistance R_m versus drawn channel length L_m for nMOST with $t_{ox} = 300$ Å. Lines with different gate voltages intersect at one point from which ΔL and R_t are derived

0.5 V [67], [68]. The V_{th} determined from the constant current method, rather than the linear extrapolation method, was found to be more consistent. This is because the V_{th} measurement by linear extrapolation method is sensitive to the S/D resistance in series with the MOSFET channel resistance [67].

This method, first proposed by Terada and Muta [60], was reformulated by Chern et al. [61] and later slightly modified by many others [62]–[68]. It is the most commonly used method for determining ΔL and has become widely established as an industry standard. This is probably because of its accuracy [62] and the fact that the method also gives source/drain resistance R_t at no extra cost. The method is sensitive only to the measurement noise and does not respond to the device-to-device variation. The precision of the extracted ΔL is limited to the precision with which L_m is known [63]. Thus, in this method, ΔL and R_t are extracted using the following procedure:

1. Measure I_{ds} at low V_{ds} (typically 50 mV) and zero V_{sb}, by sweeping V_{gs} in steps of 0.1 V (or 0.05 V), for a set of transistors having the same channel width W_m but varying channel lengths L_m.
2. Determine V_{th} for each device using data from step 1.
3. For a fixed gate drive $V_{gt}(= V_{gs} - V_{th})$, determine the device output resistance $R_m(= V_{ds}/I_{ds})$ for different mask length (L_m) devices using data from steps (1) and (2). The measured R_m, at fixed V_{gt}, is plotted against

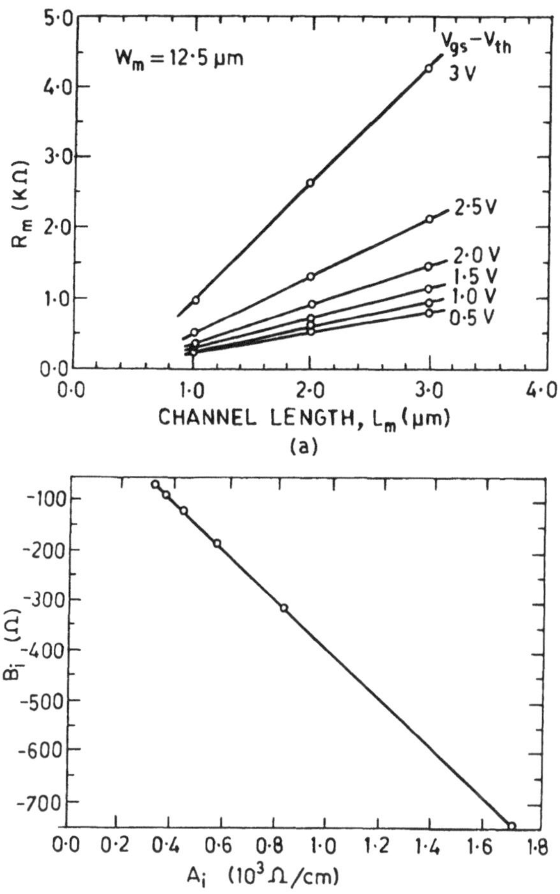

Fig. 9.34 (a) Measured output resistance R_m versus drawn channel length L_m for nMOST with $t_{ox} = 150$ Å. Lines with different gate drive voltage ($V_{gs} - V_{th}$) gives slope A_i and intercept B_i; (b) the plot of intercept B_i against slope A_i for different gate drives. The slope and intercept of this line yield ΔL and R_t

different L_m. The linear regression of this line[11] gives slope A_1 and intercept B_1.

4. Repeat step 3 for different gate drives V_{gt} in the range from 1.0 to say 5 V, in a step of, say, 0.5 V giving sets of A_i and B_i.
5. The intercepts B_i, obtained from step 4, are then plotted against the corresponding slopes A_i. The linear regression is applied again on B

[11] A linear least square regression formula based on equation $Y = AX + B$ is given in Appendix G. The regression not only gives the intercept and slope of the line, but will also yield correlation coefficient.

versus A line. The slope and intercept of this line give ΔL and R_t, respectively.

Figure 9.34a shows plots of the measured resistance R_m versus drawn channel length $L_m = 1, 1.5, 2$ and $3\,\mu$m and constant width $W_m = 12.5\,\mu$m for $V_{gt}(= V_{gs} - V_{th})$ from 0.5 V to 3 V in steps of 0.5 V. These are the n-channel conventional source/drain devices fabricated using a typical 1 μm CMOS technology, with $t_{ox} = 150\,\text{Å}$ and $V_{th} \approx 0.5$ V. The least square regression is applied to fit the straight line through the data for each specified gate drive (Figure 9.34a). A second regression is applied to B_i versus A_i data to find the slope and intercept giving ΔL and R_t, respectively (see Figure 9.34b).

Other Resistance Methods. The resistance method discussed above requires more than two devices with varying channel lengths. Various other resistance methods, based on Eq. (9.55), have been proposed which require only two devices that are identical except for the channel length. Rearranging Eq. (9.55) as [80]

$$R_m = \frac{V_{ds}}{I_{ds}} = \frac{1}{\beta_0(V_{gs} - V_{th})} + \left(\frac{\theta_0}{\beta_0} + R_t\right) \tag{9.62}$$

we plot the resistance R_m against $(V_{gs} - V_{th})^{-1}$ for each channel length as shown in Figure 9.35a. Note that R_m varies linearly with $(V_{gs} - V_{th})^{-1}$ as suggested by Eq. (9.62). The nonlinearity near $(V_{gs} - V_{th}) = 0$ results from the breakdown of the approximation used in arriving at Eq. (9.62). The V_{th} required is normally obtained by linear extrapolation of the $I_{ds} - V_{gs}$ curve. The slope of the straight line portion of the R_m versus $(V_{gs} - V_{th})^{-1}$ curve yields $1/\beta_0$ and the intercept (to the R_m axis) yields $\theta(= \theta_0/\beta_0 + R_t)$. Once θ and β_0 are obtained for devices with different L_m and fixed W_m, ΔL can be obtained using the following equation

$$\frac{1}{\beta_0} = \frac{L}{\mu_0 C_{ox} W_m} = \frac{(L_m - \Delta L)}{\mu_0 C_{ox} W_m}. \tag{9.63}$$

Figure 9.35b shows a plot of β_0^{-1} versus L_m; the ratio of the intercept to the slope of this line gives ΔL. However, if we plot intercept θ against slope $1/\beta_0$ for different L_m, then the intercept of this second regression line gives R_t.

It is important to note that in this method only two transistors are sufficient to calculate ΔL. If two transistors have the same width, the following relationship between any pair of transistors can be obtained from Eq. (9.63) that is,

$$\frac{\beta_{01}}{\beta_{02}} = \frac{L_{m2} - \Delta L}{L_{m1} - \Delta L} \tag{9.64}$$

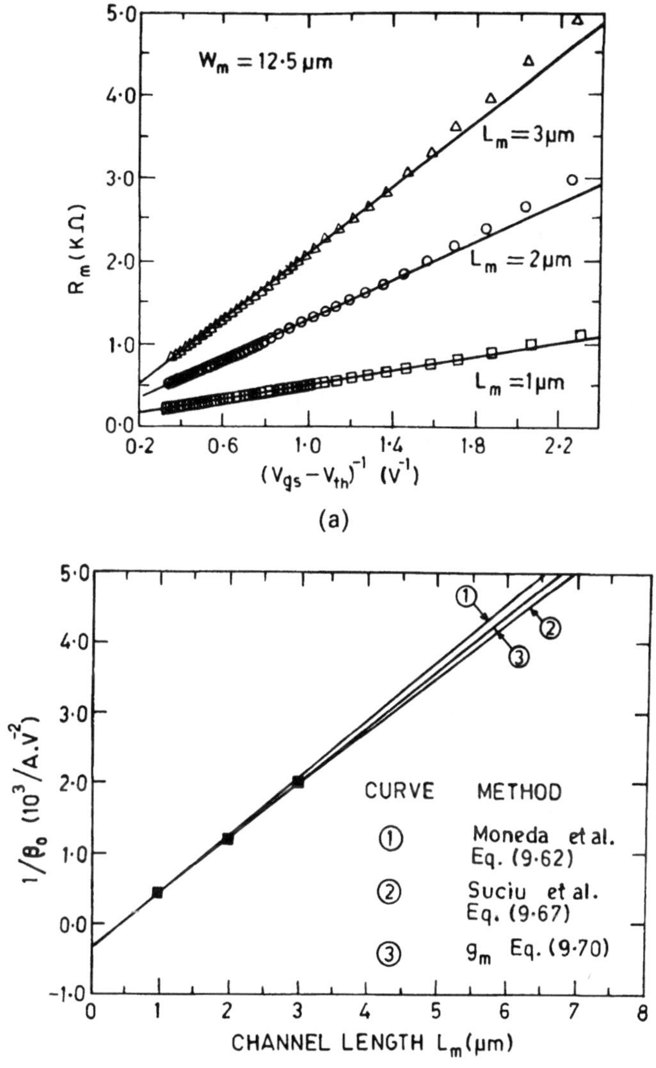

Fig. 9.35 (a) Measured output resistance R_m versus $(V_{gs} - V_{th})^{-1}$ with channel lengths L_m as a parameter for nMOST with $t_{ox} = 150$ Å ($W_m = 12.5\ \mu$m). Slope of these lines yields β_0^{-1} for each length L_m. (b) The plot of β_0^{-1} versus L_m gives ΔL

where L_{m1} and L_{m2} are drawn channel lengths for the two transistors and β_{01} and β_{02} are their corresponding β_0 values. Rearranging this equation yields

$$\Delta L = \frac{\beta_{01} L_{m1} - \beta_{02} L_{m2}}{\beta_{01} - \beta_{02}}. \tag{9.65}$$

If one of the two transistors has a considerably longer channel length, the accuracy of the ΔL extraction is substantially increased. *The advantage of this method is that it can be used to determine ΔL for a small device, provided we also have a large geometry device.*
Assuming θ_0 and R_t are independent of channel length, the series resistance R_t can be obtained from the following equation

$$R_t = \frac{\theta_1 - \theta_2}{2(\beta_{01} - \beta_{02})} \tag{9.66}$$

where θ_1 and θ_2 are the values for $(\theta_0/\beta_0 + R_t)$ for the devices with channel length L_{m1} and L_{m2}, respectively.
In a method proposed by Suciu and Johnston [79], the quantity E, obtained by rearranging Eq. (9.55), defined as

$$E \equiv \frac{(V_{gs} - V_{th})}{I_{ds}/V_{ds}} = \frac{1}{\beta_0} + \left(R_t + \frac{\theta_0}{\beta_0}\right)(V_{gs} - V_{th}) \tag{9.67}$$

is plotted against $(V_{gs} - V_{th})$ for each channel length L_m, as shown in Figure 9.36. Note that E varies linearly with $(V_{gs} - V_{th})$ as suggested by Eq. (9.67). The extrapolation of the straight line portion of the E versus $(V_{gs} - V_{th})$ curve to the E axis results in $1/\beta_0$ and the slope gives $\theta(= \theta_0/\beta_0 + R_t)$. Once β_0 for different lengths are known then ΔL can easily be calculated (see Figure 9.35b). As expected, results obtained from Eq. (9.62) and (9.67) are exactly the same as both methods are derived from the same basic equation. Note that the plot of the slope θ vs. $1/\beta_0$ for different L_m gives R_t.

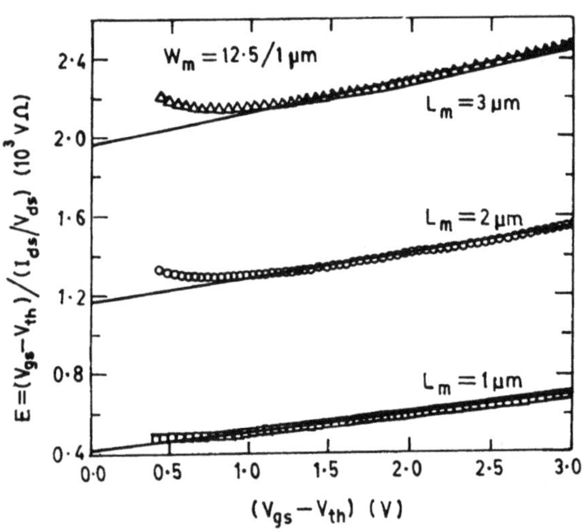

Fig. 9.36 Variation of function E versus $(V_{gs} - V_{th})$ with channel lengths L_m as a parameter for nMOST with $t_{ox} = 150$ Å ($W_m = 12.5\,\mu$m). The intercept (to E axis) yields β_0^{-1} for each length L_m. From the plot of β_0^{-1} versus L_m is derived ΔL as shown in Figure 9.35b

Transconductance Method. Another method which is often used for ΔL extraction is based on the transconductance g_m of the device. Differentiating Eq. (9.55) with respect to V_{gs} yields

$$g_m = \frac{\partial I_{ds}}{\partial V_{gs}} = \frac{\beta_0 V_{ds}}{[1 + (\theta + \beta_0 R_t)(V_{gs} - V_{th})]^2} \tag{9.68}$$

from which the maximum transconductance $g_{m,\max}$ can be seen to be

$$g_{m,\max} = \beta_0 V_{ds} = \frac{\mu_0 C_{ox} W}{L} V_{ds} \tag{9.69}$$

provided maximum g_m occurs at a point where $V_{gs} = V_{th}$. However, in practice, maximum g_m occurs at a point slightly above the threshold voltage (obtained by the linear extrapolation method). Therefore, Eq. (9.69) is accurate only when R_t can be neglected. Remembering that $L = L_m - \Delta L$, Eq. (9.69) can be rearranged as

$$g_{m,\max}^{-1} = A^{-1}(L_m - \Delta L) \tag{9.70}$$

where

$$A = \mu_0 C_{ox} W V_{ds}.$$

If we have devices with the same W_m but different drawn channel lengths L_m, then the plot of $1/g_{m,\max}$ versus L_m will be a straight-line, the extrapolation of which results in ΔL. In practice $g_{m,\max}$ is determined from the point of maximum slope of the experimental I_{ds} versus V_{gs} curve at small V_{ds}. This method is referred to as the g_m or $1/\beta$ method. Figure 9.35b shows a data plot for this method. Devices used are the same as in Figure 9.33. Note the difference between the g_m method and the one based on Eqs. (9.62) or (9.67). While the g_m method requires derivative of I_{ds} to calculate β_0, in the other methods β_0 is obtained from measured output resistance.

The main drawback of this method is that it neglects the source/drain resistance R_t. This results in data points not falling on a straight line when different device lengths are used. This nonlinearity introduces an error, which tends to underestimate ΔL. The higher the R_t, the higher the error. This method is, therefore, not suitable for LDD devices, where R_t is high. If R_t is small compared to the channel resistance, as is usually the case with standard source/drain or long-channel devices, this method yields ΔL fairly close to the resistance method [cf. Eqs. (9.62) or (9.67)]. Also note that, unlike the resistance method, this method does not yield R_t.

Transresistance Method. The g_m method requires more than two devices to determine ΔL. A method which requires only two devices and is based on both device transconductance g_m and output conductance g_d has recently been proposed [51]. Differentiating Eq. (9.55) with respect to V_{ds} gives the conductance g_d as

$$g_d \equiv \frac{dI_{ds}}{dV_{ds}} = \frac{\beta_0(V_{gs} - V_{th})}{1 + (\theta_0 + \beta_0 R_t)(V_{gs} - V_{th})}. \tag{9.71}$$

Dividing Eq. (9.71) by the square-root of Eq. (9.68) yields

$$\frac{g_d}{\sqrt{g_m}} = \sqrt{\frac{\beta_0}{V_{ds}}}(V_{gs} - V_{th}) \tag{9.72}$$

which is independent of the series resistance R_t and is a linear function of V_{gs}. The slope of $g_d/\sqrt{g_m}$ versus V_{gs} for different length devices will give β_0 from which ΔL can be determined.

Gate Bias Dependence of ΔL. All the drain current methods discussed above assumed that R_t is a constant independent of the gate voltage V_{gs}. Strictly speaking this is not true as discussed in section 3.6.1. In fact both R_t and effective channel length L depend upon V_{gs}. This is the result of the channel broadening effect where L is modulated by the gate voltage [75]. L is considered to lie between the points where the current flows from the lateral spread of the S/D diffusion layer to the inversion layer. These are the points where the conductivity of the diffusion resistance is approximately equal to the incremental inversion layer conductivity. Since the inversion layer conductivity increases with increasing gate voltage, the effective L increases. Simultaneously, the effective series resistance R_t decreases with gate voltage. The gate voltage dependence of R_t and ΔL are more pronounced for LDD devices compared to the standard S/D devices. This is due to the fact that LDD devices have an n^- regions under or near the gate which gets easily modulated by the gate voltage. This explains the failure of R_m versus L_m lines to intersect at a common point particularly for LDD devices.
Recently a method, based on Terada and Muta/Chern et al., has been proposed to calculate R_t and ΔL as a function of V_{gs} [69]. In this method, R_m is plotted against L_m for only two gate voltages that are closely spaced. The intersection of the two regression lines gives R_t and ΔL at the middle level V_{gs} as shown in Figure 9.37a. Typically the two voltages differ by less than 0.5 V so that R_t can be assumed constant over the range. With this step repeated for different V_{gs} pairs, like (1, 1.5), (1.5, 2), (2, 2.5) .... (5.5, 6), one obtains R_t and ΔL as a function of V_{gs}. For conventional S/D devices, the total variation in ΔL due to V_{gs} change is small; it is much larger for LDD devices. This is shown in Figure 9.37b, where in one process ΔL for a conventional device changes only by about 0.04 μm (from 0.53 μm at 1.25 V gate drive to 0.56 μm at 6 V gate drive), while for an LDD device the change

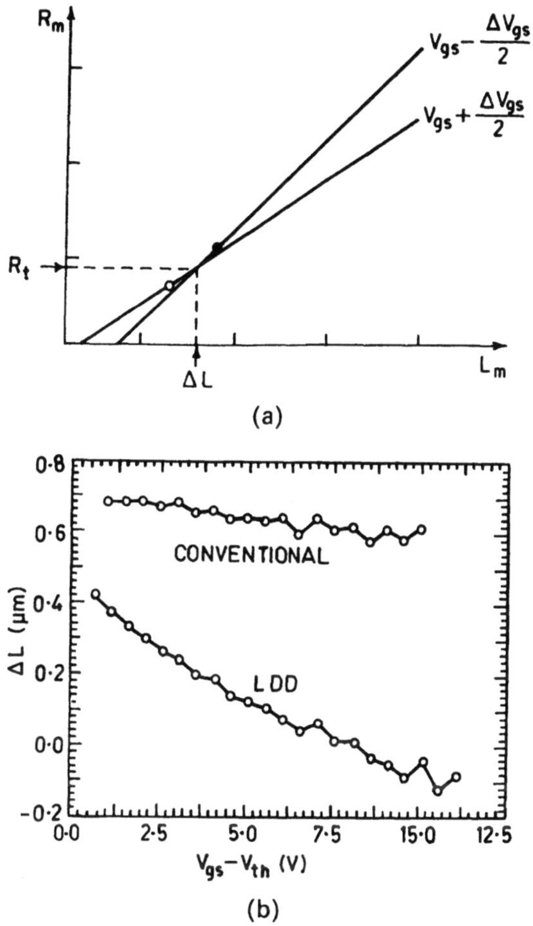

Fig. 9.37 (a) Measure output resistance R_m versus drawn channel length for two closely separated gate voltages. The intersection of the line yields ΔL. (b) Extracted ΔL for conventional and LDD n-channel MOSFET as a function of V_{gs} using procedure shown in (a). Variation of ΔL with V_{gs} is much larger for the LDD device, compared to the conventional device and depends upon doping in the n^- region. (After Hu et al. [67])

is 0.28 μm over the same gate drive voltage range. For the LDD device, this change depends upon the doping in the n^- region. The higher the doping, the smaller is the change due to V_{gs} variation.

Recently it has been reported that linear extrapolation of the effective channel length L, obtained at each pair of gate voltages, to the threshold voltage gives a value of L that corresponds to the metallurgical length for LDD devices [70].

9.10.2 *Capacitance Method of Determining* ΔL

As was pointed out earlier, the current–voltage method is the one most commonly used for determining ΔL and R_t, largely because of the measurement simplicity. Nevertheless, the capacitance method is also occasionally used because often one transistor suffices to determine ΔL [90]. Moreover, unlike in the drain current methods where R_t and ΔL cannot be separated, the capacitance method determines ΔL independent of the S/D resistance R_t.

The experimental setup for measuring ΔL using the capacitance method is shown in Figure 9.38a. The gate of the MOSFET is connected to the

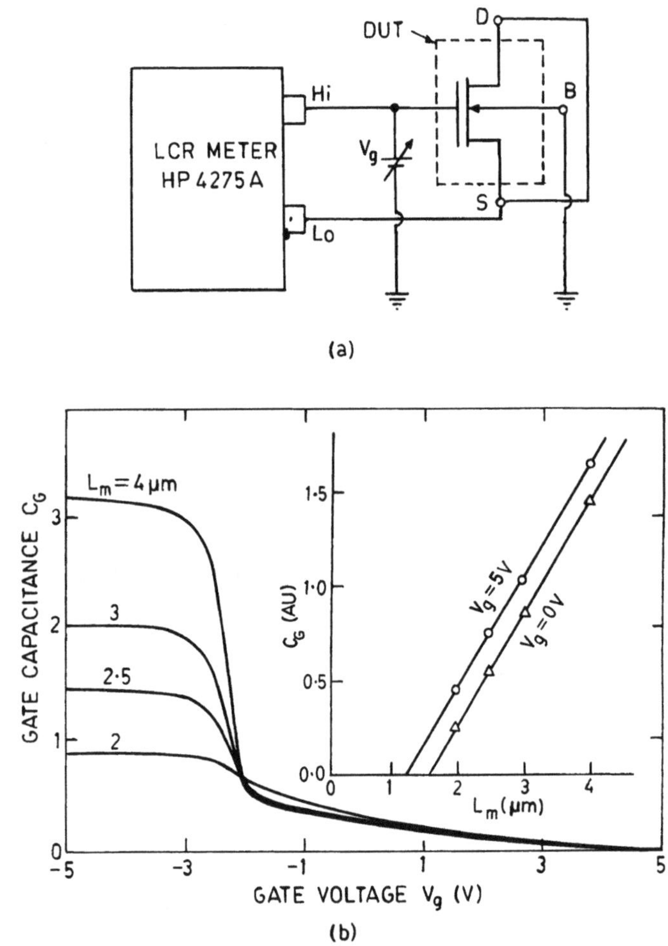

Fig. 9.38 ΔL measurement using capacitance method. (a) Experimental setup using LCR meter, (b) C–V curve for pMOST with different $L_m = 2$, 2.5, 3 and 4 μm devices. Inset shows C_G as a function of L_m for V_{gs} 5 V and $V_{gs} = 0$ V. Slope of this line yields ΔL. (After Yao [89])

'Hi' terminal of the HP4275A LCR meter, whose 'Lo' terminal is connected to the shorted source and drain. The bulk (substrate) is grounded. It should be pointed out that one can connect the 'Lo' terminal to the gate and the 'Hi' terminal to the shorted S/D. This connection has the advantage of low noise in the measurement, but has the disadvantage that one needs to track the S/D voltage by connecting a separate programable power supply to the bulk terminal [91]. A small AC voltage (20–30 mV peak-to-peak) of frequency 100 KHz is superimposed on the gate voltage, which is stepped from $V_g(\max)$ (accumulation) to $-V_g(\max)$ (inversion), and the corresponding gate capacitance C_G is measured. When the device (say p-channel) is biased in strong accumulation, a layer consisting of free electrons is formed under the gate, which is electrically disconnected from the p^+ source and drain. Therefore, the measured capacitance C_{Ga} is the sum of the overlap and parasitic capacitances C_{ovl} and C_p respectively, that is,

$$C_{Ga} = C_{ovl} W + C_p \quad \text{(F)} \quad \text{(accumulation).} \tag{9.73}$$

However, when the device is biased in strong inversion, a channel is formed under the gate, which is electrically connected to the source and drain. In this case the measured capacitance C_{Gi} is the sum of the gate-channel capacitance $C_{GC}(= C_{ox} WL)$ plus the overlap and parasitic capacitances, so that

$$C_{Gi} = C_{ox} WL + C_{ovl} W + C_p \quad \text{(F)} \quad \text{(inversion).} \tag{9.74}$$

Combining Eqs. (9.73) and (9.74) yields the gate channel capacitance C_{GC} as

$$C_{GC} = C_{Gi} - C_{Ga} = C_{ox} WL = C_{ox} W(L_m - \Delta L). \tag{9.75}$$

Thus, measuring the difference between the gate capacitance in accumulation (measured at say $V_{gs} = -5$ V) and inversion ($V_{gs} = +5$ V), we get C_{GC}. If the test transistors are wide such that $W \approx W_m$, and C_{ox} can be measured accurately, then ΔL can easily be obtained from Eq. (9.75).

The accuracy of the method depends upon the accuracy of the C_{ox} measurement. The measurement error due to C_{ox} can be eliminated by measuring C_{GC} on a series of transistors with same width W_m, but varying lengths L_m. The C_{GC} is then is plotted against L_m. The intercept to the L_m axis (x axis) gives ΔL [84]. The problem with this method is that C_G is normally very small ($\ll 1$ pF), and in practice difficult to measure accurately when the channel length becomes short.

Note that implicitly it is assumed that C_{Gi} (capacitance in inversion) and C_{Ga} (capacitance in accumulation) are constant independent of the gate bias. Indeed this is true only for long channel devices. For short channel devices (standard S/D or LDD), C_{Ga} is not constant but varies with V_{gs}. This variation is large for LDD devices compared to the standard source/drain devices. Therefore, for short channel devices C_{GC} cannot be calculated unambiguously, as it will now depend upon what gate bias in

accumulation is chosen to measure C_{Ga}. This can be seen from Figure 9.38b, where the total gate capacitance C_G is plotted as a function of the drawn channel length L_m (= 2, 2.5, 3 and 4 μm) for standard source/drain *p*-channel devices. From these measurements we get C_{GC} for each L_m; the extrapolation of C_{GC} versus L_m yields ΔL (see inset). If $V_g = V_{gs} = 0$ V is chosen for C_{GC} calculations, it yields $\Delta L = 1.54\,\mu$m, however, choosing $V_{gs} = 5$ V (accumulation) results in $\Delta L = 1.24\,\mu$m [89]. In both cases, least square fits are excellent with linear region correlation better than 0.9999. Therefore, a good least square fit alone can not be used to justify the accuracy of the measurement technique. The value of $\Delta L = 1.54\,\mu$m at $V_{gs} = 0$ V is within 4% of that determined by the resistance method. Therefore, it has been suggested that for LDD *n*-channel devices one can use $V_{gs} = 0$ rather than $V_{gs} = -5$ V (accumulation) for calculating C_{GC}. This is because in this case the n^- region is not depleted and the channel surface is not inverted [84]. However, C_{Ga} measured at $V_{gs} = 0$ contains a capacitance component due to the fringing fields between the gate and the side walls of the n^- region. This will cause C_{GC}, and hence the effective channel length L, to be smaller (ΔL to be larger) than the actual value [89].
It has been reported that there exists a fine structure (flat portion between -0.6 V to -1.1 V for *n*-channel device with W/L = 5/0.25 μm) in C–V data for short-channel devices [90]. Therefore, one can use Eq. (9.75) to determine ΔL from a single device measurement. However, others [91] have not observed such fine structures in submicron devices.

9.10.3 Methods of Determining ΔW

Most of the ΔL methods described in the previous sections can be applied to ΔW extraction. Similar to the ΔL methods, here again the drain current methods are the ones most commonly used for determining ΔW. In this case *we will need identical devices with fixed length L_m, but varying width W_m*. The length L_m should be large so that short-channel effects can be neglected.
A slight variation of the channel resistance method for ΔL extraction has been used to extract ΔW, provided R_t is redefined as source/drain series resistance per unit width [85]. Rearranging Eq. (9.58) we get

$$R_m = \frac{V_{ds}}{I_{ds}} = R_{tw}(W_m - \Delta W) + \frac{A'}{(W_m - \Delta W)(V_{gs} - V_{th})} \tag{9.76}$$

where R_{tw} is the series resistance per unit width and the constant A' is given by

$$A' = \frac{L}{\mu_s C_{ox}}.$$

For a given width device, the plot of R_m versus $(V_{gs} - V_{th})^{-1}$ should be a straight line, the slope A_w of which is given by

$$A_w(W_m - \Delta W) = A'. \tag{9.77}$$

This equation shows that a plot of $(A \cdot W_m)$ versus A_w for different width devices should be a straight line, the slope of which yields ΔW. Thus, *two linear regression steps are required to extract ΔW using this method*, namely:

1. For each device width, the device resistance R_m is plotted against gate drive $V_{gt}(= V_{gs} - V_{th})$, in the range say 1.0 V to 5.0 V. The linear regression of this data yields a slope A_w for each device width (see Figure 9.39a).
2. The slope A_w (obtained from step 1) multiplied by the drawn device

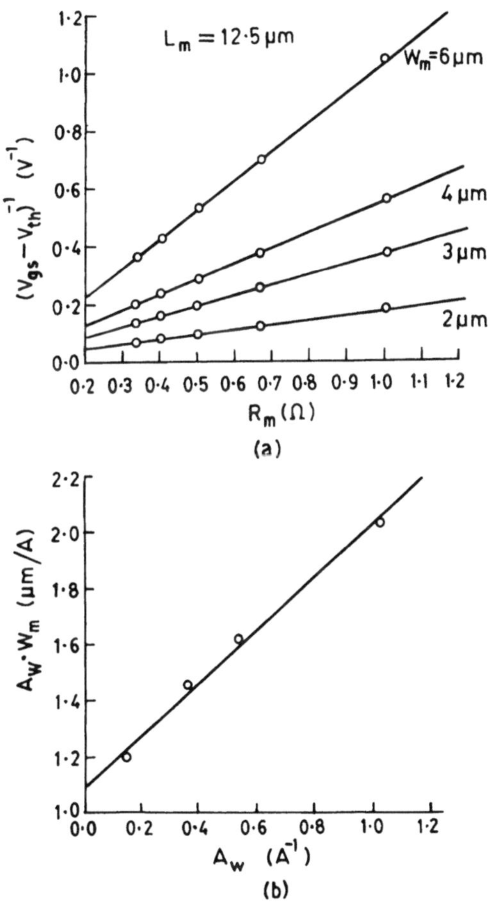

Fig. 9.39 (a) Measured MOSFET resistance R_m versus $(V_{gs} - V_{th})^{-1}$ for different devices with same channel length (12.5 µm) but different drawn widths W_m. Linear regression of these lines give slope A. (b) Plot of the slope A obtained from Figure (a) against $A \cdot W_m$. Continuous line is the least square fit to the data (circles). (After Arora et al. [85])

width is then plotted against the slope A_w. The slope of the *resulting* straight line gives ΔW (see Figure 9.39b).

Other resistance methods, discussed earlier for extracting ΔL, can also be used for ΔW extraction. The procedure is exactly the same, except that in this case we determine β_0 for devices with fixed large channel length L_m but varying widths. ΔW is then determined using linear regression of β_0 versus W_m using the following equation

$$\beta_0 = \frac{\mu_0 C_{ox}}{L}(W_m - \Delta W). \tag{9.78}$$

Alternatively, only two devices could be used as in the case of ΔL extraction. The transconductance (g_m) method for ΔL extraction, is also applicable for ΔW extraction. In this case we use the following equation

$$g_{m,\max} = \frac{\mu_s C_{ox} V_{ds}}{L}(W_m - \Delta W). \tag{9.79}$$

The extrapolation of the straight-line region of the $g_{m,\max}$ versus W_m curve yields ΔW. This method has the same drawback as ΔL extraction, that is device R_t is neglected.
The capacitance method described earlier for ΔL determination can also be used for ΔW extraction, if measurements are made on long L_m devices.

9.11 Determination of Drain Saturation Voltage

The drain saturation voltage V_{dsat} represents the boundary between the linear and saturation regions of device operation. The knowledge of the device V_{dsat} is a necessary first step for the characterization of the device in saturation. Since the transition from the linear to the saturation region is very smooth, it is difficult to determine the transition voltage V_{dsat}. Several approaches have been proposed to determine V_{dsat} [95]–[99]. Most of these approaches require iterations to a 'best fit' solution and are model dependent [95] (see Chapter 10). However, here we describe techniques which are model independent and thus require no assumptions for the parameter values [96], [99].
In one method of determining V_{dsat}, which requires device $I_{ds} - V_{ds}$ data, we define a function G such that [99]

$$\boxed{G = g_{ds}\frac{\partial}{\partial V_{ds}}\left(\frac{1}{g_{ds}}\right)} \tag{9.80}$$

where $g_{ds} = \partial I_{ds}/\partial V_{ds}$ is the output conductance of the device. The function

G is such that it increases with V_{ds} in the linear region of MOSFET operation and thus has a positive slope. In the saturation region, the function G decreases with V_{ds} and has a negative slope. Combining the linear and saturation regions of the device operation, it is realized that the function G, when plotted against V_{ds}, increases in the linear region to a peak value and then decreases in the saturation region. *The peak point, therefore, corresponds to the transition point where* $V_{ds} = V_{dsat}$. Thus, to get V_{dsat}, the following steps are used:

1. Measure I_{ds} by sweeping the drain voltage V_{ds} from zero to $V_{ds}(\max)$ at fixed V_{gs}. Repeat the process for different values of V_{gs} from say $0.2V_{gs}(\max)$ to $V_{gs}(\max)$ in equal increments. The substrate bias V_{sb} remains constant during these measurements.
2. For a given $V_{gs}(= V_{gs1})$ calculate g_{ds} from the $I_{ds} - V_{ds}$ data (see dotted lines, Figure 9.40).
3. Find the derivative of g_{ds}^{-1} with respect to V_{ds} and multiply the resulting value with the corresponding g_{ds} to obtain the function G.
4. Plot the function G against V_{ds}, the peak point gives the saturation voltage V_{dsat} at $V_{gs} = V_{gs1}$ (see continuous lines, Figure 9.40).
5. Steps 2–4 are repeated for different values of V_{gs} thus giving V_{dsat} as a function of V_{gs}.

Figure 9.40 shows a plot of the function G versus V_{ds} for various values of V_{gs}, for an n-channel MOSFET with $W/L = 12.5/1$. Also shown in this figure are values of g_{ds} versus V_{ds} for different V_{gs}. The peak points corresponding to the V_{ds} values can easily be identified as V_{dsat}. The V_{dsat} points thus calculated are shown in Figure 9.41 as circles in the measured $I_{ds} - V_{ds}$ curves, from which the function G was calculated. The triangles (Δ) show V_{dsat} calculated using Eq. (6.179).
Note that this method of calculating V_{dsat} is independent of any specific MOSFET model and requires only $I_{ds} - V_{ds}$ data at different V_{gs}. It is thus free of model parameter values. However, it involves evaluation of the derivatives, which often induce errors particularly when the value of the derivative is small. Sometimes, in the G versus V_{ds} curve, abrupt changes or multipeaks may occur (see Figure 9.40). This is because the values of g_{ds} obtained from the derivative of the measured $I_{ds} - V_{ds}$ data are not smoothly changed as V_{ds} increases, especially when V_{gs} is higher and/or V_{ds} is higher. When more than one peak point appears, the first peak with smallest V_{ds} is identified as the true peak for V_{dsat} determination. Normally a 5 point least square fit method is used to process the data and to find the first derivative [25]. Also note that the peak value of G decreases as V_{gs} increases. This can be seen as follows: In the linear region, I_{ds} is given by

$$I_{ds} = \frac{W}{L} \int_0^{V_{ds}} \mu(V) Q_i(V) dV \tag{9.81}$$

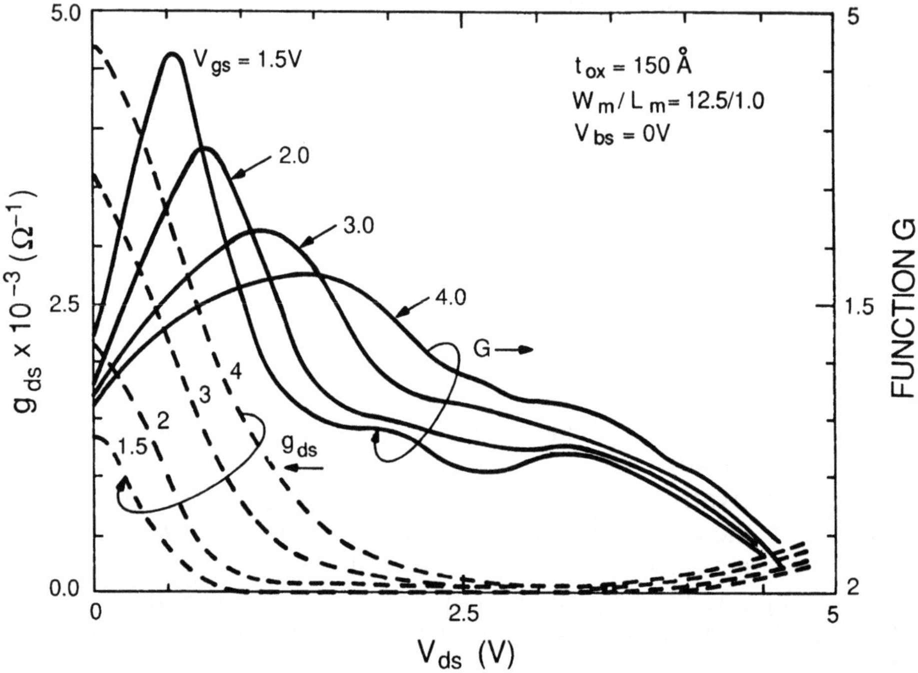

Fig. 9.40 Plots of the G function [Eq. 9.80] and output conductance g_{ds} versus V_{ds} with varying V_{gs} for a nMOST with $t_{ox} = 150\,\text{Å}$. For a given V_{gs}, the V_{ds} value at the peak corresponds to V_{dsat}

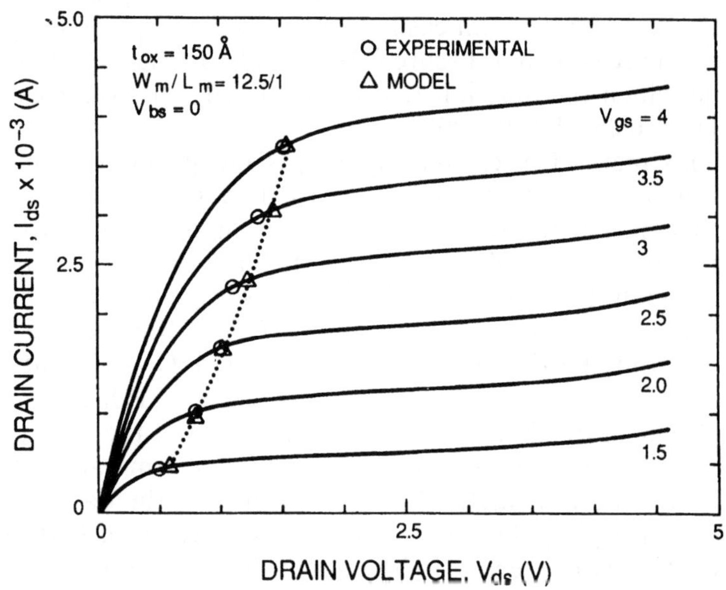

Fig. 9.41 Measured $I_{ds} - V_{ds}$ characteristics for a nMOST with $W_m/L_m = 12.5/1$ and $t_{ox} = 150\,\text{Å}$. ○ and △ represent the measured (using procedure shown in Figure 9.40) and calculated (from Eq. 6.179) V_{dsat} values, respectively

which gives

$$G = -\frac{Q_i'}{Q_i} - \frac{\mu_i'}{\mu_i} \tag{9.82}$$

where Q_i' and μ' are the first derivatives of Q_i (inversion charge) and μ (mobility) with respect to V_{ds}, respectively. As V_{gs} increases, the inversion charge density Q_i increases and therefore, the term $-Q_i'/Q_i$ of the function G (Eq. 9.82) decreases.

Another method of determining V_{dsat}, that does not require drain current derivatives, depends upon the ratio of the substrate to drain current, I_b/I_d. This method of determining V_{dsat} has been used to characterize the substrate current [96]–[97]. When the curves of constant I_b/I_d are superimposed on the plots of I_{ds} versus V_{ds}, it has been found that the loci of constant I_b/I_d are parallel to each other as shown in Figure 9.42 where loci for $I_b/I_d = 10^{-4}$, 5×10^{-4}, and 10^{-3} are shown. This is supported by the following equation for the substrate current [cf. Eq. (8.12)]

$$I_b = I_d \frac{A_i}{B_i}(V_{ds} - V_{dsat}) \exp\left[\frac{lB_i}{(V_{ds} - V_{dsat})}\right] \tag{9.83}$$

which shows that each constant I_b/I_d is determined by a constant value of $(V_{ds} - V_{dsat})$. Therefore, in the $I_{ds} - V_{ds}$ coordinate, each constant I_b/I_d locus is just a translated version of some V_{dsat} locus, where the parallel translation in the V_{ds} direction is equal to $V_{ds} - V_{dsat}$. The locus of V_{dsat} is therefore the locus of constant I_b/I_d corresponding to $V_{ds} - V_{dsat} = 0$. In other words,

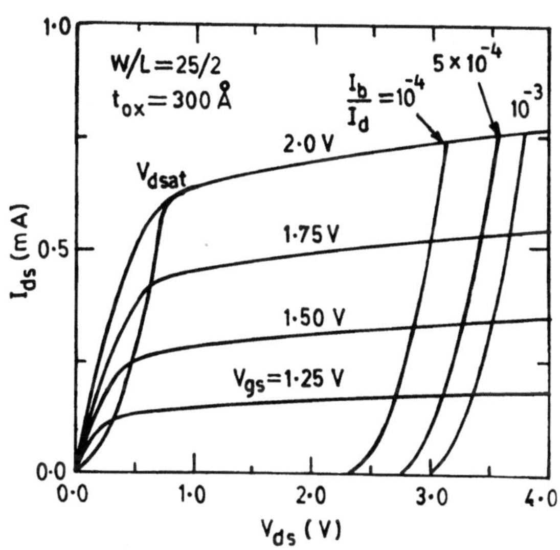

Fig. 9.42 Loci of constant I_b/I_d and V_{dsat} on the plot of $I_d - V_d$ curve illustrating V_{dsat} extraction. These curves are for nMOST with $W_m/L_m = 25/2$ and $t_{ox} = 300$ Å

the *locus of* V_{dsat} *is determined by drawing a curve through the origin which is parallel to the loci of constant* I_b/I_d.
The experimental method to determine V_{dsat} thus consist of the following steps:

1. Measure I_b and I_d by sweeping V_{ds} from zero to V_{ds}(max) at fixed V_{gs}. Repeat the process for different values of V_{gs}, from say $0.2V_{gs}$(max) to V_{gs}(max), in equal increments. The substrate bias V_{sb} remains constant during these measurements.

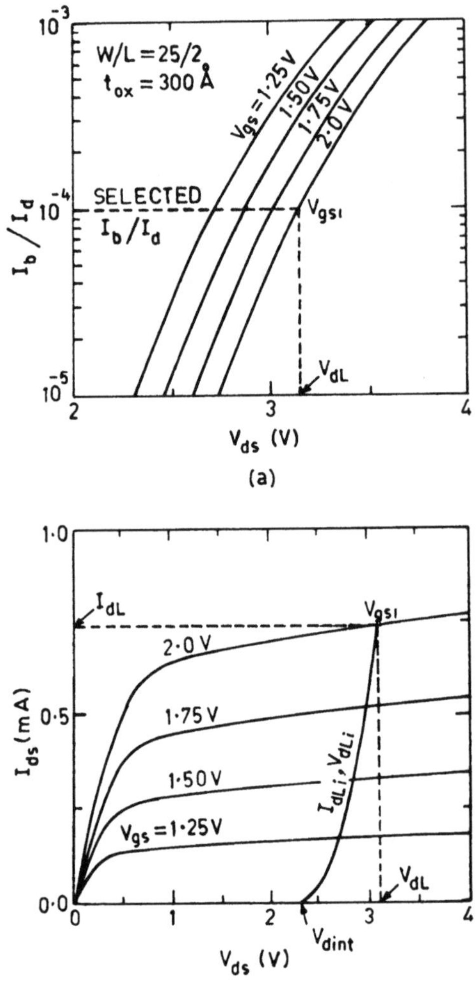

Fig. 9.43 (a) Variation of log (I_b/I_d) versus V_{ds} with V_{gs} as parameter. For selected (I_b/I_d) and $V_{gs} = V_{gs1}$, gives V_{dL} point. (b) Shifting the (I_{dLi}, V_{dLi}) curve, generated from (I_{dL}, V_{dL}) coordinate, results in locus of V_{dsat}

2. For a desired value of I_b/I_d, interpolate the $\log(I_b/I_d)$ versus V_{ds} data to find the V_{ds} value (denoted by V_{dL}) corresponding to a given V_{gs}, say V_{gs1}. Figure 9.43a shows V_{dL} for $V_{gs} = 2$ V at a specified $I_b/I_d = 10^{-4}$ in an experimental nMOST with $W/L = 25/2$ and $t_{ox} = 300$ Å.
3. Find I_{ds} corresponding to $V_{ds} = V_{dL}$ and $V_{gs} = V_{gs1}$, denoted by I_{dL}, by interpolating the $I_{ds} - V_{ds}$ data (see Figure 9.43b).
4. Repeat step 2 and 3 for different V_{gs} values. This results in a single I_b/I_d curve in $I_{ds} - V_{ds}$ coordinates shown as the (I_{dLi}, V_{dLi}) curve in Figure 9.43b.
5. Extrapolate the (I_{dLi}, V_{dLi}) curve to intersect the V_{ds} axis and calculate the V_{ds} intercept value, V_{dint}. It is the V_{dint} value by which the I_b/I_d locus curve should be shifted to obtain the locus of V_{dsat}.
6. Shift the (I_{dLi}, V_{dLi}) curve to pass through the origin by plotting the curve $(I_{dLi}, V_{dLi} - V_{dint})$, which represents the locus of V_{dsat}.
7. The intersection of this V_{dsat} locus with the $I_{ds} - V_{ds}$ curve gives V_{dsat} and I_{dsat} as a function of V_{gs}. Figure 9.42 shows the V_{dsat} locus and constant I_b/I_d loci superimposed on $I_{ds} - V_{ds}$ curve.

The I_b/I_d method is less noisy than the G function method. However, the disadvantage is that it requires substrate current data in addition to the drain current data. Moreover, it is difficult to use for *p*-channel devices because of their very low substrate current. This method is more useful when V_{dsat} is required for substrate current modeling.

9.12 Measurement of MOSFET Intrinsic Capacitances

Several methods have been developed to measure MOSFET intrinsic capacitances [100]–[103]. These methods can be broadly divided into two classes: *'on-chip' methods* [100]–[107] and *'off-chip' methods* [108]–[113]. In the so called 'on-chip' methods, a special measurement circuit is fabricated near the device whose capacitances are to be determined, in order to avoid stray capacitances and noise. The 'off-chip' method, also called the direct measurement technique, uses an external measurement system which is connected to the device under test (DUT) by probes for direct-on-wafer measurements. In what follows, we will discuss these techniques with their advantages and disadvantages. However, it should be pointed out that, for reasons discussed later, off-chip methods are the ones most commonly used for intrinsic capacitance measurements.

9.12.1 On-Chip Methods

The first on-chip technique, proposed and implemented by Iwai and Kohyama [100] and later improved by Oristian et al. [101], is based on

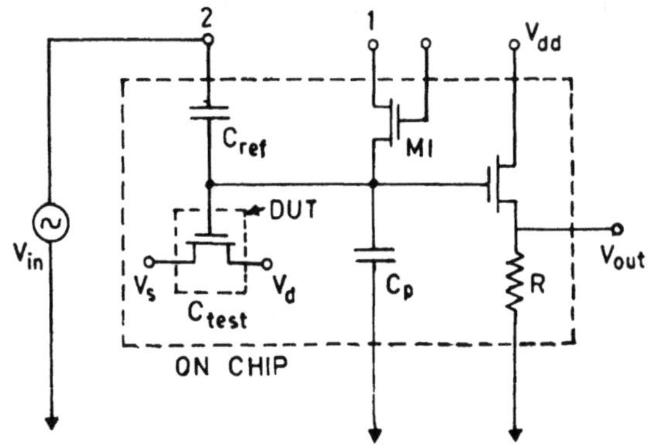

Fig. 9.44 Circuit schematic for on-chip open loop capacitance measurement technique

a capacitance voltage divider formed by series connection of a test capacitance and reference capacitance as shown in Figure 9.44. The test capacitance C_{test} is the voltage dependent gate capacitance of the DUT (both intrinsic and extrinsic capacitances). The reference capacitance is a poly-diffusion or metal–metal capacitance of known value. The output signal taken between the two capacitances is buffered by the source follower so as to isolate the DUT from measuring instruments.
A voltage V_g is applied to the gate of the DUT by pulsing M1. When M1 is turned on, the gate is charged to V_g; when M1 is turned off, the measurement is performed. A small signal is first applied to node 1 to calibrate the gain A_v of the source follower, then it is switched to node 2 to make measurements. If C_p is the stray capacitance associated with the gate node, then the output voltage V_{out} is given by

$$V_{\text{out}} = \frac{A_v C_{\text{ref}}}{C_{\text{ref}} + C_p + C_{\text{test}}} V_{in}. \tag{9.84}$$

The stray capacitance C_p includes both the interconnect capacitance of the capacitance divider output node as well as the input capacitance of the buffer amplifier. It is desirable that C_p be as small as possible, compared to the reference and test capacitances, in order to keep the gain of the divider from being reduced. By subsequently applying the AC signal to each of the source, drain and substrate terminals, three separate gate measurements are made. The gate-to source, drain and bulk capacitances C_{GS}, C_{GD} and C_{GB} respectively are then found by solving the following three voltage divider relations (one for each measurement configuration)

$$\frac{V_{o1}}{V_{in}} = \frac{C_{GS}}{C_{GS} + C_{GD} + C_{GB} + C_{\text{ref}} + C_p} \tag{9.85}$$

$$\frac{V_{o2}}{V_{in}}=\frac{C_{GD}}{C_{GS}+C_{GD}+C_{GB}+C_{\text{ref}}+C_p} \tag{9.86}$$

$$\frac{V_{o3}}{V_{in}}=\frac{C_{GB}}{C_{GS}+C_{GD}+C_{GB}+C_{\text{ref}}+C_p}. \tag{9.87}$$

Note that this method requires access to the bulk terminal of the device. This places the requirement that the DUT be in its own well. Therefore, in *n*-well CMOS processes only *n*-type devices may be used for the capacitance measurements.

This technique allows measurements of capacitance for smaller devices as process technology improves, since the reference capacitance may scale with the device capacitance being measured. However, it assumes that an accurate reference capacitor, comparable in size to the test capacitor, may be produced and that they can be accurately measured. Although this method has a simple circuit and has achieved a good resolution, it has the following disadvantages:

1. The externally applied gate voltage and the actual internal gate voltage may be slightly different because of the feedthrough due to the pulse applied to the gate of M1. The seriousness of this feedthrough effect depends upon bias voltage at the measuring node, since the charge shunted is proportional to the difference between the pulse and the bias voltage.
2. The cancellation of the parasitic input capacitance of the buffer amplifier requires the use of two test structures.

An improved on-chip capacitance measurement method was proposed by Paulous [106], which uses a switched-capacitor gain stage around the DUT, as shown in Figure 9.45a. A reference capacitance was chosen as the feedback capacitance while the DUT was used as a series capacitance. The overall gain of the circuit is given by

$$\frac{V_o}{V_{in}}=-\frac{C_{\text{test}}}{C_{\text{ref}}}. \tag{9.88}$$

This assumes that the operational amplifier (op-amp) gain and its input impedance are infinite. The frequency of operation must also be lower than the gain-bandwidth product of the op-amp, so that the gain of the circuit is frequency independent.

A small AC signal is applied to one of the terminals of the test device and a single gain measurement is made. Assuming that the reference capacitance has been accurately measured, the gain expression may be used to calculate the capacitance of the DUT. Biasing the DUT is relatively simple. Drain, source and bulk biases are directly applied to the MOSFET terminals, while the gate bias is applied through the negative terminal of the op-

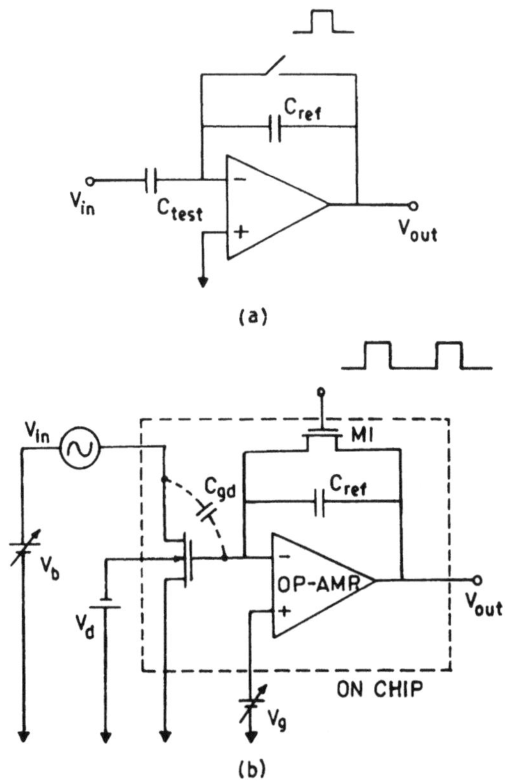

Fig. 9.45 Circuit schematic for on-chip closed loop capacitance measurement method. (a) switched capacitance AC gain stage (b) setup for measuring C_{gd}

amp. A switch (pass transistor M1) is required for DC biasing of the op-amp (see Figure 9.45b). When M1 is on, the op-amp functions as a DC voltage follower, so that the gate voltage of the DUT is at the same DC potential as the applied bias voltage. When M1 is turned off, an AC signal is applied to the DUT and the measurements are performed. Similar to the case of the voltage divider technique, the bias of the device is temporarily disturbed due to the feedback effect of transistor M1. This results in a shift in the DC output level of the op-amplifier. In theory, this would not affect the accuracy of the data, since only the AC signal is of interest.
There are several advantages of this, so called the closed-loop on-chip method, over the capacitance voltage divider (open-loop) method. Some of these are

- The calibration required to make the measurement is minimized.
- The parasitic capacitance at the gate of the DUT and at the input of the op-amp are negligible because this node is at virtual ground. Unlike

the open-loop method, this method does not require a redundant set of test structures to reduce parasitics.
- The op-amplifier is guaranteed to have higher gain than the capacitance divider structure thus resulting in better measurement resolution.

However, the main disadvantage of this method is that the test circuit becomes more complex due to fabrication of the op-amp.
Although the on-chip capacitance methods have good resolution, the requirement for the fabrication of a special on-chip circuit around the device of interest have prevented these methods from being widely adopted. The technique requires dedicated masks for the fabrication of the test structures. Moreover, a well-established fabrication process is required and the measurement circuit (op-amp) needs to be redesigned for different technologies. Another limitation of this technique is that only gate related capacitances can be measured. Measurements of the drain and source capacitances is not possible due to large DC current through the device.

9.12.2 *Off-Chip Methods*

In off-chip methods, the DUT is connected to the external measuring system without any on-chip circuit. It thus has flexibility with respect to layout of the test devices and the possibility to use commercially available measuring systems. In principle, standard test transistors can be measured with this method by careful design of the system for suppression of parasitic effects. *Since this method is more flexible and easier to realize, it is the one most commonly used for MOSFET intrinsic capacitance measurements.*
Two off-chip methods that are commonly used to measure capacitance are (1) lock-in amplifier methods [111]–[113], and (2) bridge type measurements using impedance meters such as the HP4275A [108]–[110]. Both methods measure the small signal capacitance and parallel conductance of the DUT, independent of any stray capacitance to ground. Although bridge type measurements are easier to make, they are difficult to automate. On the other hand lock-in amplifier type measurements are easier to automate but calibration is not as accurate.

Lock-In Amplifier Measuring Method. In this method shown in Figure 9.46, the switch of the on-chip closed-loop method (Figure 9.45) is replaced by a big feedback resistor ($\sim 1\,\mathrm{G\Omega}$) [113]. This is essentially a current-to-voltage (IV) converter, that converts the sensed current into a voltage, which is then detected by the lock-in amplifier such as PAR 186 or EG&G Model 5206. The op-amp in the IV converter is a commercially available high performance op-amp, such as ICH8500A or Burr Brown 3551J with very low input bias current (<0.01 pA) to reduce the noise and is thus a major

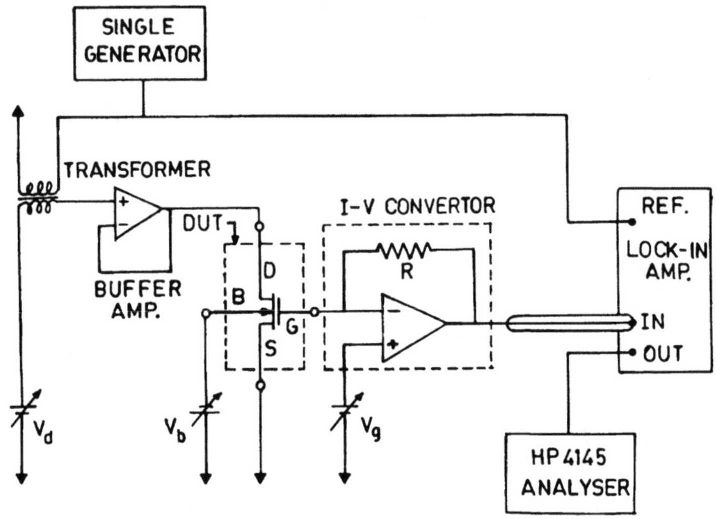

Fig. 9.46 Circuit schematic for measuring small gate capacitances using lock-in amplifier method

factor in improving the measurement resolution. The AC test signal, typically 30–50 mV (rms) is derived from the built-in oscillator of the lock-in amplifier, or can be taken from a separate signal generator, such as the HP4143. The coupling transformer and the buffer amplifier, such as the Burr Brown 3571, convert the primary test signal into a truly floating low impedance source.

Ideally the output voltage of the IV converter increases proportionally to the signal frequency f. However, f cannot be increased without limits due to the frequency limitations of the IV converter op-amplifier, buffer amplifier and stray wiring capacitance. Therefore the frequency of the test signal is generally around 1–5 KHz, depending upon the setup used. The DC bias to the DUT is supplied by an HP4145 parameter analyzer. The HP4145 also monitors the output of the lock-in amplifier, which gives capacitance information. The calibration of the measurement system is done by using a commercially available LCR meter or high precision standard capacitors. Configurations for measuring various gate capacitances can be selected with a specially designed switching matrix, such as the Kiethely model 706 scanner.

Unlike the on-chip circuitry methods, this method does not have a switch induced charge injection problem. This is because the input of the IV converter is connected to the gate terminal of the test device, which remains at virtual ground throughout the measurements. Measured results are insensitive to the parasitic capacitances on the virtual ground node as is the case for 3 or 4 terminal LCR meters.

The measurement setup shown in Figure 9.46 is to determine the gate-to-drain capacitance C_{GD}. Recall that C_{GD} is a measure of the change in the gate charge Q_G in response to the change in the gate-to-drain voltage V_{gd}. Therefore, the AC test signal is applied to the drain terminal while keeping V_{gs} and V_{gb} constant and measuring the resulting AC displacement current at the gate terminal. Similarly, for measuring C_{GS} and C_{GB}, the AC test signal is applied to the source and substrate terminals, respectively, and the AC displacement current is measured at the gate.

Bridge Type Measuring Method. MOSFET capacitances have also been successfully measured using LCR meters such as HP model 4275A which has a resolution of 0.1 fF [9]. Figure 9.47 shows the schematic diagram of the measurement system for gate-to-source capacitance C_{GS}. The gate and source are directly connected to the LCR meter; the drain and substrate are connected to the low-noise DC voltage source. While using an LCR meter, the following precautions must be taken:

- The meter has a bias supply only at the 'Hi' terminal (H_c, H_p) and the 'Lo' terminal (L_c, L_p) is at virtual ground. Therefore, the gate voltage is chosen as the reference potential, which is at virtual ground, with all other terminals referenced to the gate voltage.
- Within the LCR meter there is a resistor R_s at the 'Hi' terminal which is in series with the external bias shown in Figure 9.47. When a DC drain current flows through R_s at nonzero V_{ds}, the effective biases between the Lo and Hi terminals are reduced below the bias applied to the LCR

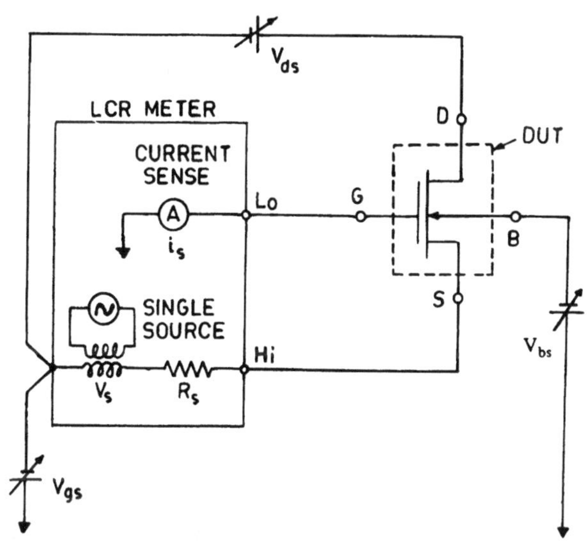

Fig. 9.47 Circuit schematic for intrinsic gate capacitance measurements

meter. Therefore, *it is necessary to correct continuously the drain(source) voltage whenever the gate bias is changed.*

For low capacitance measurement in the aF (10^{-18} F) range, it is important that a data averaging technique be used to reduce the random fluctuations of the data, thereby enhancing the resolution. Normally averaging of 20 data points (i.e., average of 20 measurements at one bias point) gives a fairly smooth curve. *Without this averaging procedure it will not be possible to achieve a resolution in the 0.1 aF range.*

The measured capacitance using the on-chip or off-chip methods consists of 3 components; the parasitic or stray capacitance, the overlap and outer fringing capacitance and the intrinsic capacitance. The parasitic and overlap capacitances have to be subtracted from raw data in order to obtain the intrinsic capacitances. As shown in Figure 9.10, the parasitic capacitance $C_p = C_{mp} + C_{ip}$. Note that in the on-chip measurements, the inter-probe capacitance C_{ip} does not exist because the small signal is buffered by the amplifier. The gate-source/gate-drain line parasitic capacitance C_{mp} is fixed and independent of bias voltage. However, test structures are used that minimize C_{mp}. It is important to use coaxial probes so that the gate electrode is shielded from the drain/source probes, thereby reducing C_{ip} to 4–5 fF. However, C_{ip} cannot be made zero as discussed in section 9.1.2.

In order to determine the overlap and fringing capacitances, the capacitance C_{GS} and C_{GD} are measured at $V_{ds} = 0$ with V_{gs} varying from strong accumulation to weak inversion. The capacitance in strong accumulation is the gate overlap capacitance plus the parasitic capacitance consisting of outer fringing capacitance and any interprobe capacitance that is present after zeroing the LCR meter. The situation becomes more complicated when overlap becomes gate bias dependent, as is discussed in the next section.

The LCR meter can be used to measure any of the 12 intrinsic capacitances for a long channel MOSFET. In general to measure capacitance C_{ij} between the terminals i and j of the MOSFET, the LCR meter is directly connected to these terminals; the remaining two terminals of the MOSFET are connected to the low-noise DC voltage source. The rest of the setup is exactly the same as described above for C_{GS} measurement.

9.13 Measurement of Gate Overlap Capacitance

The experimental setup to measure gate-to-source or gate-to-drain overlap capacitance is the same as that shown in Figure 9.38a. The gate of the MOSFET is connected to the 'Hi' terminal of the HP4275A LCR meter, whose 'Lo' terminal is connected to the shorted source and drain. The bulk (substrate) is grounded [91]. However, one can also use the lock-in-amplifier system shown in Figure 9.47. Assuming the overlap capacitance per unit width C_{gso} and C_{gdo}, at the source and drain ends, respectively, are equal, the measured capacitance in accumulation is the sum of the overlap and parasitic capacitances, that is, [cf. Eq. (9.73)]

$$C_G = 2C_{gso}W + C_p \quad \text{(F)} \quad \text{(accumulation)} \tag{9.89}$$

where the factor of 2 accounts for the source and drain side of the overlap and C_p is the stray or parasitic capacitance, i.e., $C_p = C_{mp} + C_{ip}$ (see Figure 9.10). However, in inversion the measured capacitance is the sum of the gate-channel capacitance C_{GC} plus the overlap and parasitic capacitances, so that [cf. Eq. (9.74)]

$$C_G = C_{ox}WL + 2C_{gso}W + C_p \quad \text{(F)} \quad \text{(inversion).} \tag{9.90}$$

Clearly one can calculate the overlap capacitance either from Eq. (9.89) or (9.90).

Assuming we could null out the stray capacitance C_p, then measuring the capacitance in accumulation and dividing the result by the effective width W of the device gives the total overlap capacitance per unit width $(C_{gso} + C_{gdo})$, see Eq. (9.89). A DC bias of at least $-2\,\text{V}$ (for nMOST) is applied to the gate to prevent the formation of an inversion channel and one need not have to know the full C–V curve; one data point is sufficient to calculate the overlap capacitance [114].

The overlap capacitance of a typical transistor is small, therefore, to reduce the requirement on the measuring setup, the test structure shown in Figure 9.48a can be used [86]. This structure uses multiple wide transistors connected in parallel to achieve larger, easily measurable capacitances. The metal lines connecting the source and the drain, that run parallel to the gate, should be sufficiently separated from the gate to minimize gate to metal capacitance. For such a structure, the total measured gate capacitance is

$$C_G = n(2C_{gso})W + C_p \tag{9.91}$$

where n is the number of transistors in parallel (see Figure 9.48c) and W is the width of each transistor in the test chip. For very narrow width transistors, the capacitance determined using this technique will not be accurate because of the fringing field. In order to characterize the overlap for these narrow devices, a test structure similar to that shown in Figure 9.48b can be used. Equation (9.91) may be used to calculate the overlap capacitance.

Note that measuring C_{gso} from Eq. (9.89) requires that C_p be known. However, if we have series of transistors with fixed L_m and varying W_m, then measuring the slope of the line obtained from the plot of the measured gate capacitance in accumulation as a function of device widths W_m (for fixed channel lengths) will yields the total overlap capacitance. The advantage

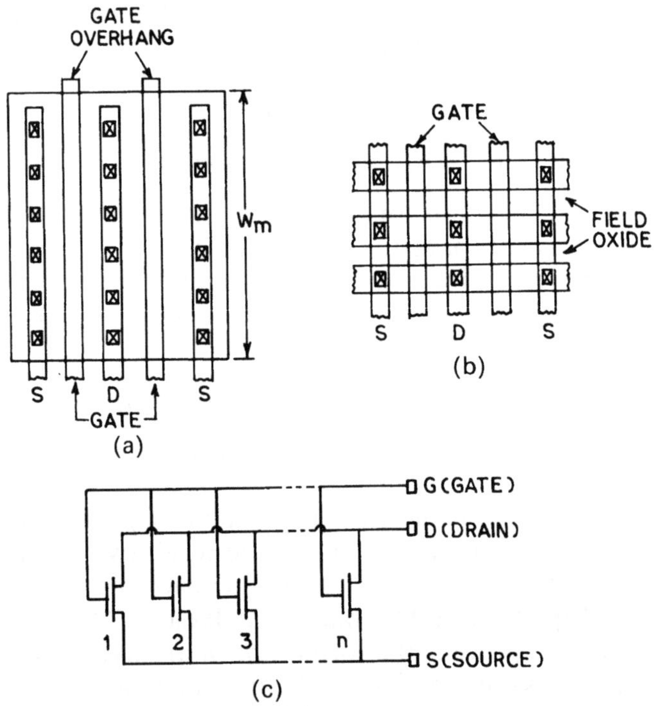

Fig. 9.48 Test structures for measuring C_{gso} and C_{gdo} including fringe effects, for transistors having small L: (a) and (b) layout, (c) schematic

of this method is that one need not know the exact value of the parasitic capacitance C_p, which is always difficult to measure. However, the disadvantage is that one now requires devices with different widths but fixed lengths.

It should be pointed out that the method of measuring overlap capacitance using Eq. (9.89) assumes that capacitance in the accumulation region is a constant, independent of bias. This indeed is true for standard source/drain junctions for large geometry devices.[12] For short channel devices or for LDD devices, the measured capacitance in accumulation becomes bias dependent, as can be seen from Figure 9.49 which is the plot of gate to S/D capacitance versus gate voltage for an n-channel ($W_m/L_m = 50/1$) LDD device (continuous line). For comparison, a device with the same W_m/L_m and C_{ox} but with standard source/drain is also shown (dashed line). Only part of the C–V curve is shown in order to highlight the change in the capacitance in accumulation and inversion. In such cases, the method

[12] Truly speaking, even for large geometry polysilicon gate devices, the gate capacitance in accumulation does change with bias, but the change is very small.

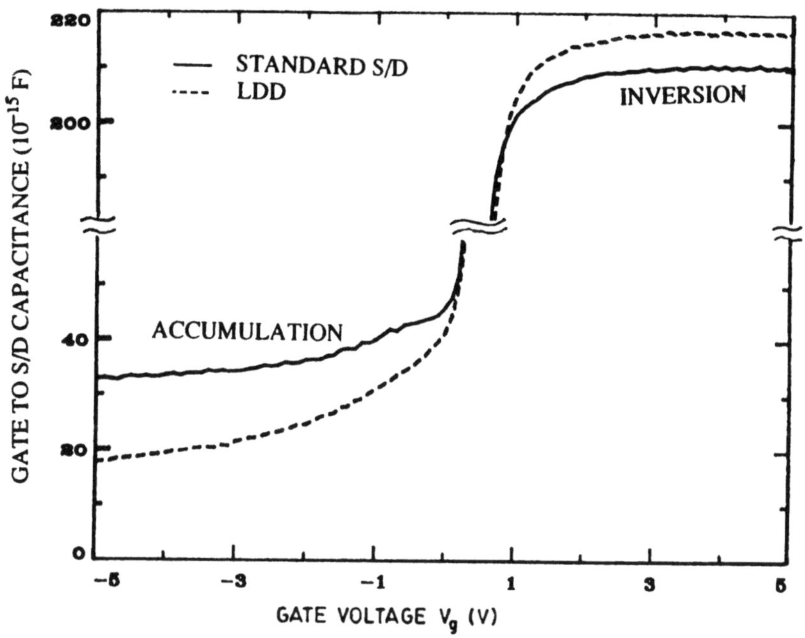

Fig. 9.49 Gate capacitance showing bias dependence of the overlap capacitance in accumulation for standard

outlined above cannot be used due to the ambiguity as to what bias point in accumulation should be chosen to calculate the gate overlap capacitance.

From Figure 9.49, note that the capacitance is constant in inversion independent of the bias, whether the device is short or is an LDD. Therefore, one can calculate C_{gso} from inversion using Eq. (9.90). However, the disadvantage of measuring C_{gso} in inversion is that we now have two extra parameters, C_{ox} and ΔL, whose values need to be known accurately. From Eq. (9.90) it is evident that the plot of the measured C_G in inversion as a function of effective channel length L will be a straight line, the intercept of which yields C_{gso} and slope yields C_{ox}, provided we could null out the stray capacitance C_p. Determining C_{gso} from such linear regression has the advantage that we could zero out the affect of C_{ox} in determining C_{gso}, although uncertainty in ΔL will still influence C_{gso}. Note that this method requires devices of the same width but different channel lengths. Further, these devices must have separate pads for all 3 device terminals (source, drain and gate), otherwise the capacitance measured may not be correct. For this reason modeling transistors in the test chips, which have their source and drain connected together, can not be used for these measurements. It should be pointed out that to determine overlap capacitance from inversion, the channel length L itself need not be known as long as a quantity proportional to L is known. A plot of C_G in inversion versus $b \cdot L$, where b is a proportionality constant, yields a line with a y-intercept of $(C_{gso} + C_{gdo})\, W$ and a slope of $c \cdot C_{ox}$. In a method proposed by Arora et al. [91], β_0 was used to extract C_{gso} from inversion. The advantage of using $\beta_0 = \mu_0 W C_{ox}/L$ instead of L is that one does not have to assume a constant ΔL for varying lengths, as is presumed by most methods of ΔL extraction (see section 9.10). Thus, to calculate overlap capacitance from inversion, the following steps are used:

1. Measure drain current I_{ds} as a function of gate voltage V_{gs} at low V_{ds} for several devices with same width and varying lengths.
2. Calculate β_0 for each device, e.g., using Eq. (9.72).
3. Measure the capacitance C_G in inversion between the gate and shorted source and drain with the substrate grounded.
4. Plot C_G in inversion versus $1/\beta_0$. One such plot for a n-channel LDD device is shown in Figure 9.50 where circles are experimental points. The devices used were $W_m/L_m = 51/0.75$, $51/1.0$ and $51/2.0$ (μm). In this plot C_G is measured at a gate voltage of 3 V (inversion). The y-intercept of this line divided by W_{eff} is $C_{gso} + C_{gdo}$.

The overlap capacitances extracted from accumulation and inversion are in general different, especially for n-channel devices [91]. However, most circuit simulators, such as SPICE, currently accept only one value for the overlap capacitance. Because the circuit delay time is very sensitive to the overlap capacitance for submicron devices, it is important that the most appropriate value of the overlap capacitance be used.

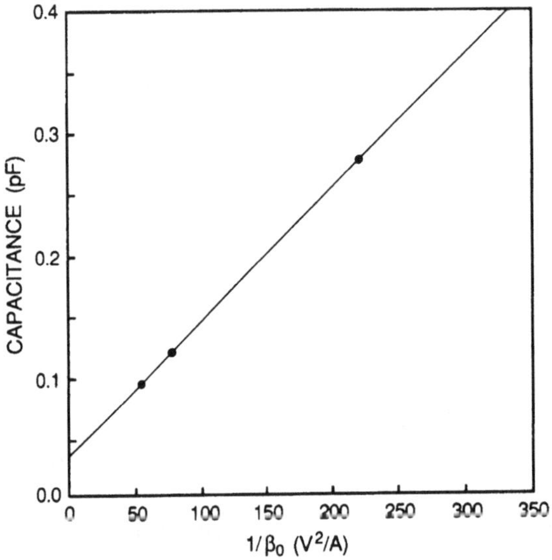

Fig. 9.50 Gate to source/drain overlap capacitance measurement using the inversion method

9.14 Measurement of MOSFET Source/Drain Diode Junction Parameters

The diode model parameters, which are important for characterizing the MOSFET source and drain *pn* junctions, are the diode reverse leakage current I_s and junction capacitance C_j. It is customary to distinguish between the area and periphery components of I_s and C_j for these diodes. As discussed in section 2.8, special test structures are fabricated to measure these components. The diode capacitance C_j is measured as a function of diode reverse voltage $V_d(=V_r)$ for both these structures. From these measurements, the area and periphery components are separated out using Eq. (2.80). Similar procedure for I_s is used by measuring I_d versus forward diode voltage $V_d(=V_f)$ and using Eq. (2.81). Once I_d versus V_d (or C_j versus V_d) data is generated for the two structures, the following procedure is then used to calculate I_s and C_j model parameters.

9.14.1 *Diode Saturation or Reverse Leakage Current I_s*

We have seen earlier that I_s is directly proportional to the active *pn* junction area [cf. Eq. (2.59)]; therefore, I_s can vary significantly from device to device. A typical value of I_s for an IC diode is $\sim 10^{-12}$ A.

Equipments required for measuring I_s are (1) a sensitive current meter capable of reading current down to 1 fA (10^{-15} A) and (2) a DC Voltage source with 10 mV resolution. A Hewlett-Packard (HP) pico Ammeter/DC voltage model 4140B or any other equivalent current meter can be used for the purpose. The advantage of the HP4140B is that it consists of both a current meter, with a resolution of 1 fA (10^{-15} A), and a DC voltage source. Due to the low level of the leakage current, the device must be well protected from noise, as discussed in section 9.1.

The leakage current I_s can be measured by sweeping the diode reverse voltage and measuring the corresponding current flowing through the diode. However, a precise method to obtain I_s is by measuring the diode current I_d as a function of diode forward voltage V_d and extrapolating the $\log(I_d)$ versus V_d curve to $V_d=0$ (see Figure 9.51). Since I_d increases exponentially with V_d it is preferable to measure the voltage V_d across the diode by forcing the current I_d through it. Recall that the diode current I_d is expressed as [cf. Eq. (2.82)]

$$I_d = I_s\left[\exp\left(\frac{V_d - I_d r_s}{\eta V_t}\right) - 1\right]. \tag{9.92}$$

From the equation above we see that if $V_d \gg I_d r_s$ and $V_d \gg V_t (\approx 0.0256\,\text{V}$ at 300 K) then

$$I_d = I_s \exp\left(\frac{V_d}{\eta V_t}\right)$$

or

$$\log(I_d) = \frac{V_d}{2.3\,\eta V_t} + \log(I_s) \tag{9.93}$$

where the factor 2.3 accounts for the conversion from "ln" (logarithm to the base e) to "log" (logarithm to the base 10). From this equation, it is evident that a plot of $\log(I_d)$ versus V_d will be a straight line whose intercept to the Y-axis gives I_s. Thus, to calculate I_s:

1. Measure I_d as a function of V_d in the forward bias region with $V_d \gg V_t$ and $V_d \gg I_d r_s$.
2. Plot $\log(I_d)$ versus V_d. One such plot for an n^+p diode with junction area $375 \times 294\,\mu\text{m}^2$ is shown in Figure 9.51 where the circles are experimental points. Note from this curve that I_d deviates from the straight line portion of the curve at high current levels due to the series resistance r_s.
3. Extrapolate the straight line portion of the $\log(I_d)$ vs. V_d curve to $V_d = 0$ resulting in I_s.

A simple least-square linear regression could be used for the purpose of extrapolating the curve to $V_d = 0$. At high V_d, the current I_d (in fact $\log(I_d)$)

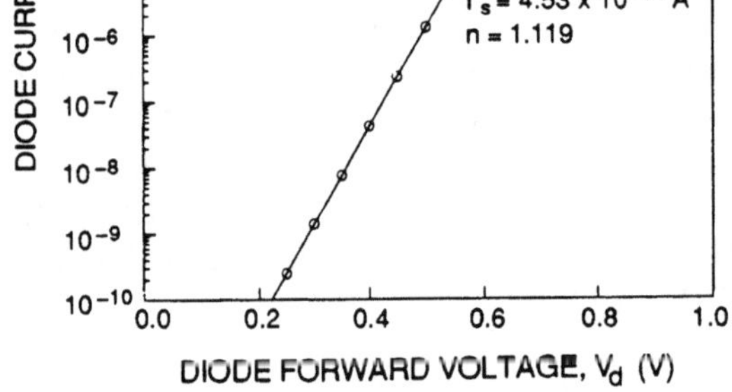

Fig. 9.51 Diode saturation current measurement procedure showing measured (circles) and calculated (continuous line) diode current

Table 9.3. *Diode parameters I_s, η and r_s*

	n^+p	p^+n
I_s	4.53×10^{-14} A	4.1×10^{-12} A
η	1.119	1.335
R_s	12.03 Ω	10.78 Ω

deviates from the straight line; therefore, while using linear regression, care must be exercised in choosing the $I_d - V_d$ data range. Generally current in the voltage range 0.2–0.5 V should be used for the purpose. Checking slopes between data points is a simple numerical method for determining linearity. The continuous line in Figure 9.51 is a linear regression of Eq. (9.93). The value of I_s thus obtained is shown in Table 9.3 for an n^+p diode (nMOST S/D junction) and p^+n diode (pMOST S/D junction) with the same junction area.

The other two parameters, the ideality factor η and the diode resistance r_s, in the diode current Eq. (9.92) are generally not needed for MOSFET S/D junction characterization. However, these parameters can be easily obtained from the same data from which I_s is determined as discussed below.

The Ideality Factor η. The ideality factor η varies in the range 1–2 depending upon the region of operation of the silicon diode or the type of the diode. However, for a silicon diode in the current range 10^{-9} to 10^{-3} A, η is very close to unity. For power diodes and SBD, η can be somewhat larger than one.

The value of η can be obtained from the slope m of the $\log(I_d)$ versus V_d curve (see Figure 9.51). According to Eq. (9.93), the slope m yields

$$\eta = \frac{1}{2.3} \cdot \frac{1}{kT} \cdot \frac{1}{(\text{slope}, m)}. \qquad (9.94)$$

Thus, knowing the temperature of measurement and slope of the $\log(I_d)$ versus V_d curve gives the ideality factor η. The value of η obtained for the n^+p diode of Figure 9.51 is shown in Table 9.3.

Diode Series Resistance r_s. The series resistance accounts for the resistance of the diode bulk region. It varies in the range 1 to 100 Ω, with a typical value of 10 Ω. Normally series resistance is significant only when the diode current is large enough (> 1 mA), so that the voltage drop across the series resistance becomes significant compared to the voltage drop across the intrinsic diode. At high current levels, as shown in Figure 9.51, the curve deviates from a straight line. This deviation is partly due to high level injection and partly due to the series resistance r_s. However, since it is difficult to distinguish between the two components, for SPICE models we generally assume that the deviation at high current levels is solely due to r_s.

One method to determine r_s is to measure the deviation of the experimental current-voltage curve from the extrapolated straight line where Eq. (9.93) is valid (see Figure 9.51), that is,

$$r_s = \frac{\Delta V_d}{I_d}. \tag{9.95}$$

Since r_s depends upon the diode current I_d, r_s is calculated at several values of I_d. The plot of I_d versus ΔV_d will be a straight line, the slope of which gives r_s.

A more accurate method of determining r_s is obtained by rearranging Eq. (9.92) in terms of V_d as

$$V_d = nV_t \ln\left(1 + \frac{I_d}{I_s}\right) + I_d r_s. \tag{9.96}$$

Assuming $I_d/I_s \gg 1$, and differentiating with respect to I_d, we get

$$I_d \frac{dV_d}{dI_d} = nV_t + I_d r_s. \tag{9.97}$$

The above equation shows that a plot of $I_d(dV_d/dI_d)$ versus I_d will be a straight line, the slope of which gives r_s. This indeed is the case, as shown in Figure 9.52, where the continuous line is the linear regression of Eq. (9.97), while circles are experimental points. Note that while plotting Eq. (9.97),

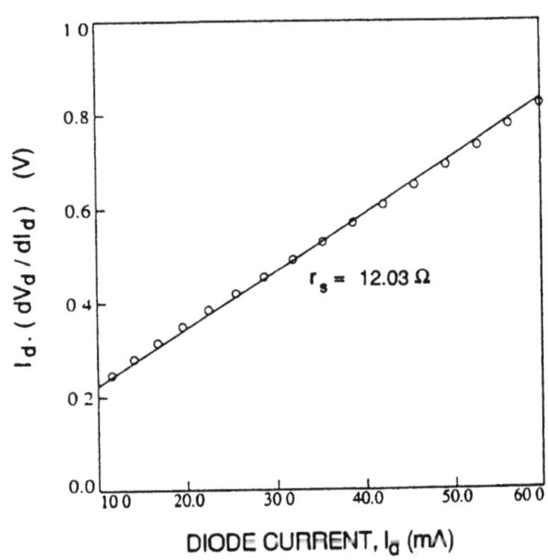

Fig. 9.52 Diode series measurement procedure. Circles are measured data, while continuous line is fit to Eq. (9.97)

we need to take I_d points outside the linear region of the $\log(I_d)$ vs. V_d plot where r_s is significant.

The methods discussed above for determining I_s, η and r_s are based on linear regression of the diode current equation. However, these parameters can also be easily determined using nonlinear least-square optimization methods, as discussed in Chapter 10.

9.14.2 Junction Capacitance

The junction capacitance, due to the space charge in the junction depletion region, as a function of diode reverse voltage V_d can easily be measured using a capacitance meter, such as HP4280A, or a general purpose LCR meter, such as the HP4275A. The signal applied to the diode has a relatively small amplitude (roughly 50 mV) in order not to affect the depletion width appreciably. However, this signal is sufficiently large to cause measurable variations of charge associated with infinitesimal variations of the depletion width. The frequency of the signal is typically 100 KHz, which is considered low enough to neglect the effects of the ohmic resistance. It is important to remember that capacitance measured by the meter will be the total junction capacitance C_J, which needs to be divided by the junction area A_d of the diode to get C_j.

The measured capacitance C_J as a function of bias is then fitted to the following equation to get the parameters C_{j0}, m and ϕ [cf. Eq. (2.74)]

$$C_J = \frac{C_{J0}}{\left(1 + \frac{V_d}{\phi}\right)^m} + C_p \tag{9.98}$$

where $C_{J0} = C_{j0} A_d$ and is the stray capacitance C_p caused by the pads, bonding wires and connecting metal lines on the chip. In practice, while it is possible to zero out the stray capacitance using an LCR meter like the HP4275, it is often not possible to eliminate the effects of stray capacitance completely. Nonetheless, it is important to make sure that C_p is negligible over the entire measurement range or is estimated accurately. Note that C_p is a redundant parameter and therefore it is recommended that C_p be obtained by other methods and not determined by curve fitting Eq. (9.98) (see section 12.1).

The parameters ϕ and m can be determined by a linear regression method using log–log data. However, the easier and commonly used method is by curve fitting Eq. (9.98) with measured data using a nonlinear least-square optimization programs (see Chapter 10). The typical values of C_{j0}, ϕ and m for area and periphery components of n^+p and p^+n diode (S/D diode of n- and p-channel MOSFET's fabricated using 1 μm CMOS technology) are shown in Table 9.4.

Table 9.4. *Junction capacitance parameters C_{j0}, ϕ and m for area and periphery components of n^+p and p^+n diodes*

	n^+p		p^+n	
	Area Component	Periphery Component	Area Component	Periphery Component
C_{j0}	1.18×10^{-8} F/cm^2	2.82×10^{-12} F/cm	4.97×10^{-8} F/cm^2	3.70×10^{-12} F/cm
ϕ	0.56 V	0.33 V	0.44 V	0.85 V
m	0.21	0.20	0.23	0.52

Note that the parameters ϕ and m are quite different in the two cases. In fact depending upon the doping profile they might take somewhat unrealistic values. Nonetheless the reverse bias capacitance data fits Eq. (9.98) fairly well (see Figure 2.19). For circuit modeling purpose, they should be taken simply as fitting parameters.

References

[1] E. H. Nicollian and J. R. Brews, *MOS (Metal Oxide Semiconductor) Physics and Technology*, John Wiley & Sons, New York, 1982.

[2] R. C. Y. Fang, R. D. Rung, and K. M. Cham, 'An improved automatic test system for VLSI parameteric testing', IEEE Trans. Instrumentation and Measurement, IM-31, pp. 198–205 (1982).

[3] B. S. Messenger, 'A fully automated MOS device characterization system for process-oriented integrated circuit design', Memorandum No. UCB/ERL M84/18, Electronic Research Laboratory, University of California, Berkeley, January 1984.

[4] O. Melstrand, E. O'Neill, G. E. Sobelman, and D. Dokos, 'A data base driven automated system for MOS device characterization, parameter optimization and modeling', IEEE Trans. Computer-Aided Design, CAD-3, pp. 47–51 (1984).

[5] E. Khalily, P. H. Decher, and D. A. Teegarden, 'TECAP2: An interactive device characterization and model development system', *Tech. Digest*, IEEE Int. Conf. on Computer-Aided Design, ICCAD-84, pp. 184–151 (1984).

[6] K. Doganis and S. Hailey, 'A unified physical device modeling environment', IEEE 1986 Custom Integrated Circuit Conference, pp. 203–207 (1986).

[7] D. Cheung, A. Clark, and R. Starr, 'The INMOS integrated parameteric test and analysis system', Proc. IEEE Int. Conf. on Microelectronic Test Structures, Vol. 2, No. 1, pp. 45–50, March 1989.

[8] *Operation and Service Manual* for Model 4145B Semiconductor Parameter Analyzer, Hewlett Packard Corporation, USA, 1986.

[9] *Operating Manual* for Model 4275A Multi-Frequency LCR Meter, Hewlett Packard Corporation, USA, 1983.

[10] *Operation and Service Manual* for Model 4140B pA Meter/DC Voltage Source, Hewlett Packard Corporation, USA, 1980.

[11] Y. Nissan-Cohen, 'A novel floating-gate method for measurement of ultra-low hole and electron gate currents in MOS transistors,' IEEE Electron Device Lett., EDL-7, pp. 561–563 (1982).

[12] N. S. Sakas, P. L. Heremans, L. Van Den Hove, H. E. Maes, R. F. De Keersmaecker, and G. J. Declerck, 'Observation of hot-hole injection in NMOS transistors using a modified floating-gate technique', IEEE Trans. Electron Devices, ED-33, pp. 1529–1533 (1986).

[13] G. B. Barbottin and A. Vapaille, Eds., *Instabilities in Silicon Devices*, Vol II (Chapter 12), North-Holland, New York, 1989.

[13a] K. R. Mistry and B. Doyle, 'AC versus DC hot-carrier degradation in n-channel MOSFETs', IEEE Trans. Electron Devices, ED-40, pp. 96–104 (1993).

[14] W. W. Lin and P. C. Chan, 'On the measurement of parasitic capacitances of device with more than two external terminals using an LCR meter,' IEEE Trans. Electron Devices, ED-38, pp. 2573–2574 (1991).

[15] D. K. Schroder, *Semiconductor Material and Device Characterization*, John Wiley & Sons Inc., New York, 1990.

[16] E. H. Nicollian and J. R. Brews, 'Instrumentation and analog implementation of Q-C method of MOS measurement', Solid-State Electron., 27, pp. 953–962 (1984). See also related papers, *ibid*, pp. 963–975 and pp. 977–988 (1984).

[17] M. Kuhn, 'A quasi-static technique for MOS C-V and surface state measurements', Solid-State Electron., 13, pp. 873–885 (1970).

[18] K. Iniewski, A. Balasinski, B. Majkusiak, R. B. Beck, and A. Jakubowski, 'Series resistance in a MOS capacitor with a thin gate oxide', Solid-State Electron., 32, pp. 137–140 (1989).

[19] K. Riedling, *Ellipsometry for Industrial Applications*, Springer-Verlag, New York, 1988.

[20] M. J. McNutt and C. T. Sah, 'Determination of the MOS Oxide capacitance', J. Appl. Phys., 46, pp. 3909–3913 (1975).

[21] D. Schmitt-Landsiedel, K. R. Hofmann, H. Oppolzer, and G. Dorda, 'Thickness determination of thin oxides in MOS structures,' in *Insulating Films on Semiconductors* (J. F. Verweij and D. R. Wolters, eds.), pp. 126, North-Holland, New York, 1983.

[22] B. Ricco, P. Olivo, T. N. Nguyen, T. S. Kuan, and G. Ferriani, 'Oxide-thickness determination in thin-insulator MOS structures', IEEE Trans. Electron Devices, ED-35, pp. 432–438 (1988).

[23] G. Sarrabayrouse, F. Campabadal, J. L. Prom, 'Oxide-thickness determination from C/V measurement in an MOS capacitor,' IEE Proceedings, 136, Pt. G, pp. 215–216 (1989).

[24] B. Majkusiak and A. Jakubowski, 'A technical formula for determining the insulator capacitance in a MOS structure', Solid-State Electron., 35, pp. 223–224 (1992).

[25] A. Savitzky and M. J. E. Golay, 'Smoothing and differentiation of data by simplified least squares procedures', Anal. Chem., 36, pp. 1627–1639 (1964).

[26] H. Maes, W. Vandervorst, and R. van Overstraeten, 'Impurity profile of implanted ions in silicon', in *Impurity Doping Processes in Silicon* (F. F. Y. Wang, Ed.), North-Holland, New York, 1981.

[27] N. D. Arora, D. J. Roulston, and S. G. Chamberlain, 'Distribution profiles of diffused layers in silicon', Solid-State Electron., 25, pp. 965–967 (1982).

[28] W. Vandervorst and T. Clarysse, 'Recent developments in the interpretation of spreading resistance profiles for VLSI technology', J. Electrochem. Soc., 137, pp. 679–683 (1990).

[29] A. Jakubowski and K. Iniewski, 'Simple formula for analysis of C–V characteristics of MIS capacitor', Solid-State Electron., 26, pp. 755–756 (1983).

[30] D. M. Brown, R. J. Conney, and P. V. Gray, 'Doping profiles by MOSFET deep depletion CV', J. Electrochem. Soc., 122, pp. 121–127 (1975).

[31] K. Ziegler, E. Klausmann, and S. Karr, 'Determination of the semiconductor doping profile right up to its surface using MIS capacitor', Solid-State Electron, 18, pp. 189–198 (1975).

[32] S. H. Lin and J. Reuter, 'The complete doping profile using MOS CV technique', Solid-State Electron., 26, pp. 343–351 (1983).
[33] G. Baccarani, H. Rudan, G. Spaini, H. Maes, W. V. Ander Vorst, and R. Van Overstraeten, 'Interpretation of C-V measurements for determining the doping profile in semiconductors', Solid-State Electron., 23, pp. 65–71 (1980).
[34] C. P. Wu, E. C. Douglas, and C. W. Mueller, 'Limitations of the C–V technique for ion-implanted profiles', IEEE Trans. Electron Devices, ED-22, pp. 319–329 (1975).
[35] B. J. Gordon, 'On-line capacitance-voltage doping profile measurement', IEEE Trans. Electron Devices, ED-27, pp. 2268–2272 (1980).
[36] K. Lehovec, 'C–V profiling of steep dopant distribution', Solid-State Electron., 27, pp. 1097–1105 (1984).
[37] I. G. McGillivray, J. M. Robertson, and A. J. Walton, 'Improved measurement of doping profile in silicon using CV techniques', IEEE Trans. Electron Devices, ED-35, pp. 174–179 (1988).
[38] K. Iniewski and C. A. T. Salama, 'A new approach to CV profiling with sub-debye-length resolution,' Solid-State Electron., 34, pp. 309–314 (1991).
[39] G. Lubberts, 'Rapid determination of semiconductor doping and flatband voltage in large MOSFETs', J. Appl. Phys., 48, pp. 5355–5356 (1977).
[40] J. A. Wikstrom and C. R. Viswanathan, 'A direct depletion capacitance measurement technique to determine the doping profile under the gate of a MOSFET', IEEE Trans. Electron Devices, ED-34, pp. 2217–2219 (1987).
[41] M. Shannon, 'DC measurement of the space charge capacitance and impurity profile beneath the gate of an MOST', Solid-State Electron., 14, pp. 1099–1106 (1971).
[42] M. G. Buchler, 'Dopant profiles determined from enhancement-mode MOSFET DC measurements', Appl. Phys. Lett., 31, pp. 848–850 (1977).
[43] M. H. Chi and C. M. Hu, 'Errors in threshold-voltage measurements of MOS transistors for dopant-profile determinations', Solid-State Electron., 24, pp. 313–316 (1981).
[44] G. P. Carver, 'Influence of short-channel effects on dopant profiles obtained from the DC MOSFET profile method', IEEE Trans. Electron Devices, ED-30, pp. 948–953 (1983).
[45] N. Kasai, N. Endo, A. Ishitani, and Y. Kurogi, 'Impurity profile measurement using $V_T - V_{SB}$ characteristics,' NEC Res. & Develop., 74, pp. 109–114 (1984).
[46] K. Iniewski and A. Jakubowski, 'A new method for the determination of channel depth and doping profile in buried-channel MOS transistors', Solid-State Electron., 31, pp. 1259–1264 (1988).
[47] D. W. Feldbaumer and D. K. Schroder, 'MOSFET doping profiling', IEEE Trans. Electron Devices, ED-18, pp. 135–139 (1991).
[48] H. G. Lee, S. Y. Oh, and G. Fuller, 'A Simple and accurate method to measure the threshold voltage of an enhancement-mode MOSFET', IEEE Trans. Electron Dev., ED-29, pp. 346–348 (1982).
[49] H. S. Wong, M. H. White, T. J. Krutsick, and R. V. Booth, 'Modeling of transconductance degradation and extraction of threshold voltage in thin oxide MOSFETs', Solid-State Electron., 30, pp. 953–968 (1987).
[50] R. V. Booth, H. S. Wong, M. H. White, and T. J. Krutsick, 'The effect of channel implants on MOS transistor characterization', IEEE Trans. Electron Devices, ED-34, pp. 2501–2508 (1987).
[51] S. Jain, 'Measurement of threshold voltage and channel length of submicron MOSFETs', Proc. IEE, Pt. I, 135, pp. 162–164 (1988).
[52] M. J. Deen and Z. X. Yan, 'A new method for measuring the threshold voltage of small-geometry MOSFETs from subthreshold conduction', Solid-State Electron., 33, pp. 503–511 (1990).

[53] C. G. Sodini, T. W. Ekstedt, and J. L. Moll, 'Charge accumulation and mobility in thin dielectric MOS transistors', Solid-State Electron., 25, pp. 833–841 (1982).

[54] N. D. Arora and G. Sh. Gildenblat, 'A semi-empirical model of the MOSFET inversion layer mobility for low-temperature operation', IEEE Trans. Electron Devices, ED-34, pp. 89–93 (1987).

[55] J. Kooman, 'Investigation of MOST channel conductance in week inversion', Solid-State Electron., 16, pp. 801–810 (1973).

[56] M. S. Liang, J. Y. Choi, P. K. Ko, and C. M. Hu, 'Inversion-layer capacitance and mobility of very thin gate-oxide MOSFETs', IEEE Trans. Electron Devices, ED-33, pp. 409–413 (1986).

[57] P.-M. D. Chow and K.-L. Wang, 'A new AC technique for accurate determination of channel charge and mobility in very thin gate MOSFETs', IEEE Trans. Electron Devices, ED-33, pp. 1299–1304 (1986).

[58] G. Sh. Gildenblat, C.-L. Huang, and N. D. Arora, 'Split C-V measurements of low temperature MOSFET inversion layer mobility,' Cryogenics, 29, pp. 1163–1166 (1989)

[58a] C. L. Huang, J. Faricelli, and N. D. Arora, 'A new technique for measuring MOSFET inversion layer mobility', IEEE Trans. Electron Devices, ED-40, pp. 1134–1139 (1993).

[59] A. Hairapetian, D. Gitlin, and C. R. Viswanathan, 'Low-temperature mobility measurements on CMOS devices', IEEE Trans. Electron Devices, ED-36, pp. 1448–1445 (1989).

[60] K. Terada and H. Muta, 'A new method to determine effective MOSFET channel length', Japanese J. Appl. Phys., 18, pp. 953–959 (1979).

[61] J. G. J. Chern, P. Chang, R. F. Motta, and N. Godinho, 'A new method to determine MOSFET channel length', IEEE Electron Device Lett., EDL-1, pp. 170–173 (1980).

[62] S. E. Laux, 'Accuracy of an effective channel length/external resistance extraction algorithm for MOSFETs', ED-31, pp. 1245–1251 (1984).

[63] J. Scarpulla and J. P. Krusius, 'Improved statistical method for extraction of MOSFET effective channel length and resistance', IEEE Trans. Electron Devices, ED-34, pp. 1354–1359 (1987).

[64] B. J. Sheu, C. Hu, P. K. Ko, and F.-C. Hsu, 'Source-and-drain series resistance of LDD MOSFETs', IEEE Electron Device Lett., EDL-5, pp. 365–367 (1984).

[65] K. K. Ng and J. R. Brews, 'Measuring the effective channel length of MOSFETs', IEEE Circuits and Devices Magazine, 6, pp. 33–38, Nov. 1990.

[66] M. R. Wordeman, J. Y.-C. Sun, and S. E. Laux, 'Geometry effects in MOSFET channel length extraction algorithms', IEEE Electron Device Lett., EDL-6, pp. 186–188 (1985).

[67] J. Y.-C. Sun, M. R. Wordeman, and S. E. Laux, 'On the accuracy of channel length characterization of LDD MOSFETs', IEEE Trans. Electron Devices, ED-33, pp. 1556–1562 (1986).

[68] D. J. Mountain, 'Application of electrical effective channel length and external resistance measurement techniques to a submicrometer CMOS process', IEEE Trans. Electron Devices, ED-36, pp. 2499–2505 (1989).

[69] G. J. Hu, C. Chang, and Y. T. Chia, 'Gate-voltage-dependent effective channel length and series resistance of LDD MOSFETs', IEEE Trans. Electron Devices, ED-34, pp. 2469–2475 (1987).

[70] J. Ida, A. Kita, and F. Ichikawa, 'Accurate characterization of gate-N^- overlapped LDD with the new Leff extraction method', IEEE IEDM, *Tech. Dig.*, pp. 219–222 (1990).

[71] K. L. Peng, and M. A. Afromowitz, 'An improved method to determine MOSFET channel length', IEEE Electron Device Lett., EDL-3, pp. 360–362 (1982).

[72] J. Whitfield, 'A modification on an improved method to determine MOSFET channel length', IEEE Electron Device Lett., EDL-6, pp. 109–110 (1985).
[73] J. H. Satter, 'Effective length and width of MOSFETs determined with three transistors', Solid-State Electron., 30, pp. 821–828 (1987).
[74] D. Takacs, W. Muller, and U. Schwabe, 'Electrical measurement of feature sizes in MOS Si-gate VLSI technology,' IEEE Trans. Electron Devices, ED-27, pp. 1368–1373 (1980).
[75] K. L. Peng, S. Y. Oh, M. A. Afromowitz, and J. L. Moll, 'Basic parameter measurement and channel broadening effect in the submicron MOSFET,' IEEE Electron Device Lett., EDL-5, pp. 473–475 (1984).
[76] C. Hao, B. Cabon-Till, S. Cristoloveanu, and G. Ghibaudo, 'Experimental determination of short-channel MOSFET parameters', Solid-State Electron., 28, pp. 1025–1030 (1985).
[77] L. Chang and J. Berg, 'A derivative method to determine a MOSFETs effective channel length and width electrically', IEEE Electron Device Lett., EDL-7, pp. 229–231 (1986).
[78] D. Takacs, W. Muller, and U. Schwabe , 'Electrical measurement of feature sizes in MOS Si-gate VLSI technology', IEEE Trans. Electron Devices, ED-27, pp. 1368–1373 (1980).
[79] P. P. Suciu and R. L. Johnston, 'Experimental derivation of the source and drain resistance of MOS transistors', IEEE Trans. Electron Devices, ED-27, pp. 1556–1162 (1980).
[80] F. H. De La Moneda, H. N. Kotecha, and M. Shatzkes, 'Measurement of MOSFET constant', IEEE Electron Device Lett., EDL-3, pp. 10–12 (1982).
[81] G. Krieger, R. Sikora, P. P. Cuevas, and M. N. Misheloff, 'Moderately doped NMOS(M-LDD)—hot electron and current drive optimization', IEEE Trans. Electron Devices, ED-38, pp. 121–127 (1991).
[82] G. Ghibaudo, 'New method for the extraction of MOSFET parameters', Electronic Letters, 24, pp. 543–545, 28th April 1988.
[83] Y. R. Ma and K. L. Wang, 'A new method to electrically determine effective MOSFET channel width', IEEE Trans. Electron Devices, ED-29, pp. 1825–1827 (1982).
[84] B. J. Sheu and P. K. Ko, 'A simple method to determine channel widths for conventional and LDD MOSFETs', IEEE Electron Device Lett., EDL-5, pp. 485–486 (1984).
[85] N. D. Arora, L. A. Bair, and L. M. Richardson, 'A new method to determine the MOSFET effective channel width', IEEE Trans. Electron Devices, ED-37, pp. 811–814 (1990).
[86] P. Vitanov, U. Schwabe, and I. Eisele, 'Electrical characterization of feature sizes and parasitic capacitances using a single structure', IEEE Trans. Electron Devices, ED-31, pp. 96–100 (1984).
[87] E. J. Korma, K. Visser, J. Snijder, and J. F. Verwey, 'Fast determination of the effective channel length and the gate oxide thickness in polycrystalline silicon MOSFETs', IEEE Electron Device Lett., EDL-5, pp. 368–370 (1984).
[88] B. J. Sheu and P. K. Ko, 'A capacitance method to determine channel lengths for conventional and LDD MOSFETs', IEEE Electron Device Lett., EDL-5, pp. 491–493 (1984).
[89] C. T. Yao, I. A. Mack, and H. C. Lin, 'Accuracy of effective channel-length extraction using the capacitance method', IEEE Electron Device Lett., EDL-7, pp. 268–270 (1986).
[90] J. Scarpulla, T. C. Mele, and J. P. Kruslus, 'Accurate criterion for MOSFET effective gate length extraction using the capacitance method', IEEE IEDM, *Tech. Dig.*, pp. 722–725 (1987).

[91] N. D. Arora, D. A. Bell, and L. A. Bair, 'An accurate method of determining MOSFET gate overlap capacitance', Solid-State Electron., 35, pp. 1817–1822 (1992).

[92] P. Antognetti, C. Lombardi, and D. Antoniadis, 'Use of process and 2-D MOS simulation in the study of doping profile influence on S/D resistance in short channel MOSFETs', IEDM, *Tech. Digest*, pp. 574–577 (1981).

[93] M. H. Seavey, 'Source and drain resistance determination for MOSFETs', IEEE Electron Device Lett., EDL-5, pp. 479–481 (1984).

[94] K. K. Ng and W. T. Lynch, 'Analysis of the gate-voltage dependent series resistance of MOSFETs', IEEE Trans. Electron Devices, ED-33, pp. 965–972 (1986).

[95] A. Vladimirescu and S. Liu, 'The simulation of MOS integrated circuits using SPICE2', Memorandum No. UCB/ERL M80/7, Electronics Research Laboratory, University of California, Berkeley, October 1980.

[96] T. Y. Chan, P. K. Ko, and C. Hu, 'A simple method to characterize substrate current in MOSFETs', IEEE Trans. Electron Device Lett., EDL-5, pp. 505–507 (1984).

[97] D. Lau, G. Gildenblat, C. G. Sodini, and D. E. Nelsen, 'Low temperature substrate current characterization of n-channel MOSFETs', IEEE—IEDM85, *Technical Digest*, pp. 565–568 (1985).

[98] R. V. H. Booth and M. H. White, 'An experimental method for determination of the saturation point of a MOSFET', IEEE Trans. Electron Devices, ED-31, pp. 247–251 (1984).

[99] W. Y. Jang, C. Y. Wu, and H. J. Wu, 'A new experimental method to determine the saturation voltage of a small-geometry MOSFET', Solid-State Electronic, 31, pp. 1421–1431 (1988).

[100] H. Iwai and S. Kohyama, 'On-chip capacitance measurement circuits in VLSI structures', IEEE Trans. Electron Devices, ED-29, pp. 1622–1626 (1982).

[101] J. Oristian, H. Iwai, J. Walker, and R. Dutton, 'Small geometry MOS transistor capacitance measurements method using simple on-chip circuit', IEEE Electron Device Lett., EDL-5, pp. 395–397 (1984).

[102] H. Iwai, J. Oristian, J. Walker, and R. Dutton, 'A scaleable technique for the measurements of intrinsic MOS capacitance with atto-Farad range', IEEE Trans. Electron Devices, ED-32, pp. 344–356 (1985).

[103] J. J. Paulous, 'Measurement of minimum-geometry MOS transistor capacitances', ED-32, pp. 357–363 (1985).

[104] C. T. Yao and H. C. Lin, 'Comments on small geometry MOS transistor capacitance measurements method using simple on-chip circuit', IEEE Electron Device Lett., EDL-6, p. 63 (1985).

[105] J. Oristian, H. Iwai, J. Walker, and R. Dutton, 'A reply to comments on "small geometry MOS transistor capacitance measurements method using simple on-chip circuit"', IEEE Electron Device Lett., EDL-6, pp. 64–67 (1985).

[106] J. J. Paulos and D. A. Antoniadis, 'Measurement of minimum geometry MOS transistor capacitances', IEEE Trans. Electron Devices, ED-32, pp. 357–363 (1985). Also see J. J. Paulos, 'Measurement and modeling of small geometry MOS transistor capacitance', *Ph.D thesis*, Massachusetts Institute of Technology, Cambridge, 1984.

[107] M. Furukawa, H. Hatano, and K. Hanihara,, 'Precision measurement technique of integrated MOS capacitor mismatching using a simple on-chip circuit', IEEE Trans. Electron Devices, ED-33, pp. 938–944 (1986).

[108] K. C. K. Weng and P. Yang, 'A direct measurement technique for small geometry MOS transistor capacitances', IEEE Electron Device Lett., EDL-6, pp. 40–42 (1985).

[109] H. Ishiuchi, Y. Matsumoto, S. Sawada, and O. Ozawa, 'Measurement of intrinsic capacitance of lightly doped drain (LDD) MOSFET's', IEEE Trans. Electron Devices, ED-32, pp. 2238–2242 (1985).

[110] Y. T. Yeow, 'Measurement and numerical modeling of short channel MOSFET gate capacitances', IEEE Trans. Electron Devices, ED-35, pp. 2510–2519 (1987).
[111] B. J. Sheu and P. K. Ko, 'Measurement and modeling of short-channel MOS transistor gate capacitances', IEEE J. Solid-State Circuits, SC-22, pp. 464–472 (1987).
[112] P. Leclaire, 'High resolution intrinsic MOS capacitance measurement system', EESDERC 1987, *Tech. Digest.*, pp. 699–702 (1987).
[113] C. T. Yao, 'Measurement and modeling of intrinsic terminal capacitances of a metal–oxide–semiconductor field effect transistor', *Ph.D. Thesis*, University of Maryland.
[114] T. Y. Chan, A. T. Wu, P. K. Ko, and C. Hu, 'A capacitance method to determine the gate-to-drain/source overlap length of MOSFET's', IEEE Electron Device Lett., EDL-8, pp. 269–271 (1987).
[115] J. Scarpulla, T. C. Mele, and J. P. Krusius, 'Accurate criterion for MOSFET effective gate length extraction using the capacitance method', IEEE IEDM, *Tech. Dig.*, pp. 722–725 (1987).
[116] C. S. Oh, W. H. Chang, B. Davari, and Y. Tur, 'Voltage dependence of the MOSFET gate-to-source/drain overlap', Solid-State Electron., 33, pp. 1650–1652 (1990).

Model Parameter Extraction Using Optimization Method 10

In the previous chapter we had discussed the experimental setup needed for acquiring the different types of data required for MOSFET model parameter measurements and/or extraction. We had also discussed linear regression methods to determine basic MOSFET parameters. In this chapter we will be concerned with the nonlinear optimization techniques for extracting the device model parameters for various DC and AC models. These techniques are general purpose model parameter extraction methods that can be used for any nonlinear physical model. There are many books devoted to the area of optimization. Our intent here is only to provide an introduction to the optimization technique as applied to the device model parameter extraction. Various optimization programs (also called optimizers), which have been reported in the literature for device model parameter extraction, differ mainly in the optimization algorithms used.
We will first discuss methods used for model parameter extraction for any MOSFET model. This will be followed by some basic definitions, which will be useful in understanding the optimization methods in general, and then discuss the optimization algorithms that are most widely used for the device model parameter extraction. The estimation of the accuracy of the extracted parameters will be discussed using confidence intervals and the confidence region approach. We will conclude this chapter with examples of extracting DC and AC model parameters.

10.1 Model Parameter Extraction

There are basically two ways to extract the model parameter values of any MOSFET model from the device I–V data or C–V data; (1) the linear regression (analytical) method, and (2) the nonlinear optimization (numerical) method.

Linear Method. In this method, the device model equations are approximated by linear functions which represents the device characteristic in a limited region of the device operation [1]–[3]. Linear regression (linear least-squares) method is then applied to those linear functions. Thus, in this method the model parameters are determined from the data local to the region of the device characteristic in which the parameter is dominant. The extracted parameter is then assumed to be known and is then used to extract further parameters. Because only few parameters are determined at one time and parameters are determined sequentially, this method is also referred to as *sequential method.* This method generally produces parameter values that have obvious physical meaning.
The linear regression methods discussed in Chapter 9 to determine parameters such as ΔL, ΔW, μ_0, θ, γ, etc., fall in this category. However, this approach is somewhat tedious and time consuming, and since each parameter value is determined by few data points, the results are not accurate over the entire data space. Also this method does not account for the interaction of the parameters among themselves and their influence in other region of operation, other than that from which it was obtained. Furthermore, as devices are scaled down it is difficult to observe linear regions of the device characteristics, and therefore special efforts are required to isolate group of parameters describing model behavior under different operating conditions.

Optimization Method. In this approach, the model parameters are extracted by curve fitting the model equations to a set of measured device data in all the regions of device operation using nonlinear least square optimization techniques [4]–[13]. Starting from the 'educated guess' values for these parameters, a complete set of optimum parameters are thus extracted using numerical methods to minimize the error between the model and the measured data. The 'educated guess' values required for the parameters are often obtained from analytical methods discussed above. The drawback of this method is that any combination of values will provide a working fit to the measured characteristics due to there being sufficient interaction between the parameters. Thus, it is not always clear as to which are the correct values. Further, parameter redundancy can lead to optimum parameter sets which are physically unrealistic. Using constraints on the parameter values and/or using sensitivity analysis on the parameters help relieve the problem [5], but does not solve it. Nonetheless, this method produces a better fit to the data over the entire data space, though at the sacrifice of some physical insight. Moreover, the whole extraction program can easily be automated so that using automatic prober units statistical distribution of the parameters can be obtained without much effort.
We have already seen that virtually all MOSFET models implemented in circuit simulators consists of different sets of equations representing different

regions of device operation. In other words, these models have separate equations for linear, saturation and subthreshold regions of the device operation with explicit formulations for threshold voltage, saturation voltage, etc. Many of the parameters are used only in a subset of these equations and therefore the approach to extract all parameters simultaneously is not a good strategy. *It turns out that it is more practical to extract the parameters by coupling the optimization technique with the approach used in the analytical method.* Thus, the parameters are extracted from one set of local data (limited part of device operating range) using optimization method in conjunction with relevant model equations. Those parameters are then frozen while determining other parameters from different local data set. Once this regional approach is completed, the data covering all regions of operation is then used to extract all the model parameters to obtain the best overall fit. This accounts for model parameter interaction as well as for the parameters which affect the device characteristics in the region of operation other than from which they were extracted earlier. Thus, in this approach, the parameters are generally split into four groups as shown in Table 10.1:

- Group I—this group of parameters are generally known from the technological process data; for example, gate oxide capacitance C_{ox}, junction depth X_j, etc. These parameters are therefore not optimized and their values are assumed known.
- Group II—the parameters determined from the I–V characteristics in the linear region of operation of the device at low V_{ds} are grouped in this category. The parameters in this group are determined from data set A (cf. section 9.1). The V_{th} model parameters that characterize the device threshold voltage fall in this group.
- Group III—the parameters in this group are mobility and electric field related model parameters and are extracted from $I_{ds} - V_{ds}$ curves with varying V_{gs} and constant V_{sb} (data set B). These characteristics are in the linear and saturation regions of device behavior.
- Group IV—the parameters determined from the I–V characteristics in the subthreshold region of device operation are grouped in this category.

Table 10.1. *Drain current model parameters grouped in four categories*

Group	Model parameters
I	C_{ox}, X_j
II	$\Delta L, \Delta W, V_{fb}, N_b, N_s, \gamma_0, \gamma_l, \gamma_w$
III	μ_0, θ, R_t, E_c
IV	m, η, σ

The procedure outlined above is one of the strategies that can be used for extracting optimum set of model parameters. However, it is possible to have any other extraction strategy coupled with the optimization technique that result in reliable parameter values. We will now discuss how an optimization method is used for parameter extraction. But before doing that, it will be instructive to discuss some basic definitions [14]–[18] which will help understand the optimization technique as used for model parameter extraction.

10.2 Basics Definitions in Optimization

Let **p** be the model parameter vector[1]

$$\mathbf{p} = \begin{bmatrix} p_1 \\ p_2 \\ \vdots \\ p_j \\ \vdots \\ p_n \end{bmatrix} \tag{10.1}$$

such that p_j is the value of the jth model parameter and n is the total number of parameters. In short, the parameter vector **p** could be written as $\mathbf{p} = [p_1, p_2, \ldots, p_n]^T$; the superscript T denotes transpose of the matrix (10.1). For example, for the SPICE Level 3 MOSFET model **p** takes the following form:[2]

$$\mathbf{p} = [V_{To}, \gamma, \mu_0 \cdots \sigma]^T.$$

This n-dimensional **p** space is usually called parameter space. Now suppose there exist a function F such that $F(\mathbf{p})$ is a measure of the modeling error incurred when the parameter **p** is used. The function $F(\mathbf{p})$ is usually called the *objective function, error criterion* or *performance measure.* Thus, an *objective function $F(\mathbf{p})$ is a measure for comparing the computed or simulated behavior (response) with that of the experimentally measured or desired behavior.* It is assumed that the function $F(\mathbf{p})$ is a real-valued function and is at least once continuously differentiable with respect to the parameter **p**.

---

[1] In this chapter we will designate vectors by a boldface lowercase letter. A matrix will be designated by boldface capital letter, while elements of the matrix (individual values in the matrix) is designated by lower case letter. In the notation for an element $[a_{ij}]$ of a matrix **A**, the first subscript refers to the row and second to the column. One may mentally visualize the subscript ij in the order $\rightarrow\downarrow$.

[2] Note that the vector **p** does not include parameters such as device channel length L and width W, and bias voltages (V_{gs}, V_{ds}, etc.) that are not varied during the optimization process.

The optimum parameter value exist at a point $\mathbf{p}^*$ when $F(\mathbf{p}^*)$ is minimum. Therefore, the problem of optimization (process of choosing the optimum set of parameters) is reduced to choosing $\mathbf{p}$ such that $F(\mathbf{p})$ is minimized. Maximization of an objective function is essentially the same problem as minimization, because maximization of $F(\mathbf{p})$ is the same as minimization of $-F(\mathbf{p})$.

A point $\mathbf{p}^*$ in the parameter space is a *global minimum* of $F(\mathbf{p})$ if $F(\mathbf{p}^*) \leq F(\mathbf{p})$ for all $\mathbf{p}$ in the region of interest. If only the strict inequality $<$ holds for $\mathbf{p}$ in the neighborhood of $\mathbf{p}^*$, we are dealing with a *local minimum* of $F(\mathbf{p})$. As an example of local and global minima, a function $F(p)$ of single parameter p given by

$$F(p) = p^4 - 11p^3 + 37p^2 - 45p + 60$$

is plotted against p (see Figure 10.1). In a given interval of p, this function has two minima (at $p = 1$ and $p = 5$) one of which is the global (at $p = 5$) minima.

Normally, we do not know the shape of the function $F(\mathbf{p})$, particularly when $\mathbf{p}$ is a function of many variables. From the minimization function we cannot conclude whether or not the minimum found is a global minimum. The possible occurrence of a local minima thus introduces an uncertainty into the solution. Since no computationally tractable algorithm is known for finding the global minima of an arbitrary function [20], *in practice minimization is carried out several times starting from different initial guess values for the parameters and observing the parameter value which gives the smallest error.*

In a device model, the objective function $F(\mathbf{p})$ is a measure of the discrepancy or error that is to be minimized between the measured response, say experimental drain current $I_{\mathbf{exp}}(i)$, and computed current (from model

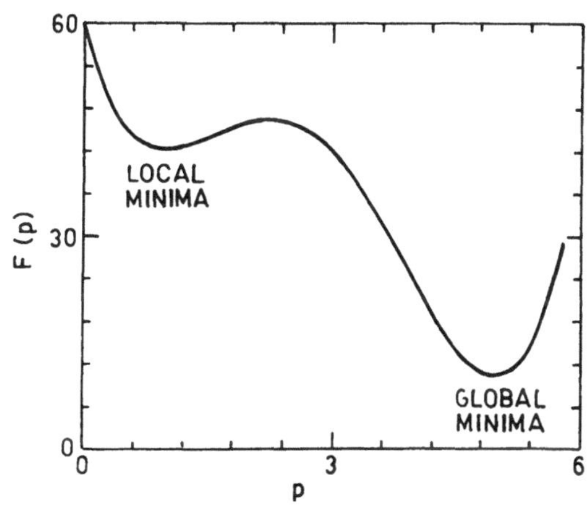

Fig. 10.1 One dimensional function $F(p)$ showing local and global minima

equations) $I_{cal}(\mathbf{p}, \mathbf{x}_i)$, where $i = 1, 2, \ldots, m$ are the data point indices and $\mathbf{x}_i$ is the set of input variables such as device L, W and bias voltages V_{ds}, V_{gs}, etc. *Selecting an objective function is the first important factor in designing a model parameter extraction program.* For many practical problems, including model parameter extraction, *a good choice of the objective function is the least-square function*, that is,

$$F(\mathbf{p}) = \sum_{i=1}^{m} w_i[r_i(\mathbf{p})]^2 \tag{10.2}$$

where r_i is the *residuals*, also called *error function*, given by

$$r_i = I_{cal}(\mathbf{p}, \mathbf{x}_i) - I_{exp}(i) \tag{10.3}$$

and w_i the *weighting function* or *weight* that assigns more weight to the specific data points in a certain region of the device characteristics than to others, so that the model is forced to fit adequately the data in those regions. In the simplest case $w_i = 1$, so that each data point is equally weighted. In general,

$$m(\text{number of data points}) > n(\text{number of model parameters}),$$

a rule of thumb is $m \geq 3n$. Sometimes the following modified form of (10.3) is used:

$$r_i = \frac{I_{cal}(\mathbf{p}, \mathbf{x}_i) - I_{exp}(i)}{\mathrm{Max}(|I_{exp}(i)|, I_{min})} \tag{10.4}$$

where I_{min} is some minimum measured value of the current, provided by the user. At current above I_{min}, the following expression for the *relative error* is used

$$r_i = \frac{I_{cal}(\mathbf{p}, \mathbf{x}_i) - I_{exp}(i)}{I_{exp}(i)} \tag{10.5}$$

otherwise the *absolute error* (scaled by I_{min})

$$r_i = \frac{I_{cal}(\mathbf{p}, \mathbf{x}_i) - I_{exp}(i)}{I_{min}} \tag{10.6}$$

is used. In general,

$$r_i(\mathbf{p}) = \frac{y(\mathbf{p}, \mathbf{x}_i) - y_m(i)}{\mathrm{Max}(|y_m(i)|, y_{min})} \tag{10.7}$$

where $y_m(i)$ is the measured response and $y(\mathbf{p}, \mathbf{x}_i)$ is the model which predicts the functional relationship between the calculated response and the input variables $\mathbf{x}_i$ and parameter vector $\mathbf{p}$. Most of the model parameter extractors [4]–[12], use the objective function given by Eq. (10.7). Once the objective function has been minimized, then the following expression is a measure of error in the model

$$\text{error} = \sqrt{\frac{F(\mathbf{p})}{m}} \tag{10.8}$$

and would be a good criterion for quantitatively evaluating agreement between the model equations and measured characteristics.
Note that in terms of error vector $\mathbf{r} = [r_1, r_2, \ldots, r_m]^T$ of size m, the objective function (10.2) can be written as

$$F(\mathbf{p}) = \mathbf{r}(\mathbf{p})^T \mathbf{W} \mathbf{r}(\mathbf{p}) \tag{10.9}$$

where $\mathbf{W}$ is a $m \times m$ *diagonal matrix*[3] whose elements w_{ii} are the weights w_i. If weights are unity, i.e., $[w_{ii}] = 1$ $(i = 1, 2, \ldots, m)$ then Eq. (10.9) becomes

$$F(\mathbf{p}) = \mathbf{r}(\mathbf{p})^T \mathbf{r}(\mathbf{p}). \tag{10.10}$$

Hessian and Jacobian. If $F(p)$ is a function of only one variable p then its Taylor series expansion is

$$F(p + \Delta p) = F(p) + \frac{dF}{dp}\Delta p + \frac{d^2F}{dp^2}\frac{(\Delta p)^2}{2} + \cdots. \tag{10.11}$$

Generalizing this equation to n dimension and retaining only the first three terms, we get the Taylor series expansion of $F(\mathbf{p})$ as

$$F(\mathbf{p} + \Delta\mathbf{p}) \approx F(\mathbf{p}) + \sum_{j=1}^{n} \frac{\partial F}{\partial p_j}\Delta p_j + \frac{1}{2}\sum_{j=1}^{n}\sum_{l=1}^{n} \frac{\partial^2 F}{\partial p_j \partial p_l}\Delta p_j \Delta p_l. \tag{10.12}$$

This equation in the vector form becomes

$$F(\mathbf{p} + \Delta\mathbf{p}) \approx F(\mathbf{p}) + [\nabla F(\mathbf{p})]^T \Delta\mathbf{p} + \tfrac{1}{2}[\Delta\mathbf{p}]^T \mathbf{H}(\mathbf{p}) \Delta\mathbf{p} \tag{10.13}$$

where $\Delta\mathbf{p}$ is a vector of the parameter increment in n dimension as

$$\Delta\mathbf{p} = [\Delta p_1, \Delta p_2, \ldots, \Delta p_n]^T, \tag{10.14}$$

and $\nabla F(\mathbf{p})$ is called the *gradient*[4] of the objective function $F(\mathbf{p})$

$$\nabla F(\mathbf{p}) = \left[\frac{\partial F}{\partial p_1}, \frac{\partial F}{\partial p_2}, \ldots, \frac{\partial F}{\partial p_n}\right]^T, \tag{10.15}$$

---

[3] A diagonal matrix is a matrix in which all the elements, except those on the principal diagonal, are zero. If the diagonal elements are unity then it is called the *unit or identity matrix*, denoted by **I**.

[4] The first derivative of a function that depends only on one parameter is called slope. At a minimum or maximum, the slope is zero. For multidimensional space, the concept of slope is generalized to define the gradient $\nabla F(\mathbf{p})$. Thus, gradient is an n-dimensional vector, the jth component of which is obtained by finding partial derivative of the function with respect to p_j.

whose jth component $\partial F/\partial p_j$ is the derivative of F with respect to p_j, and $\mathbf{H(p)}$ is a $n \times n$ symmetric matrix, called the *Hessian*, whose elements are the second derivative of $F(\mathbf{p})$ with respect to $\mathbf{p}$, defined as

$$\mathbf{H(p)} = \nabla^2 F(\mathbf{p}) = \left[\frac{\partial^2 F}{\partial p_j \partial p_l}\right]; \qquad j, l = 1, 2, \ldots, n. \tag{10.16}$$

That is, the element H_{jl} of the matrix $\mathbf{H(p)}$ in the jth row and lth column is $\partial^2 F/\partial p_j \partial p_l$.
A *necessary condition for the minimum of the objective function is that its gradient be zero*, that is

$$\nabla F(\mathbf{p}) = \frac{\partial F}{\partial \mathbf{p}} = 0. \tag{10.17}$$

Thus, finding the minimum of an objective function $F(\mathbf{p})$ is equivalent to solving n equations (10.17) in n unknown variables. An additional *sufficient condition* for a minimum of a function $F(\mathbf{p})$ is that the second derivative of $F(\mathbf{p})$, i.e., the Hessian $\mathbf{H(p)}$ be a positive definite matrix, which simply means that $\mathbf{\Delta p}^T\mathbf{H}\mathbf{\Delta p}$ must be positive for any non-zero vector $\mathbf{\Delta p}$.
We shall now calculate the gradient and Hessian of the function $F(\mathbf{p})$. We will assume that $F(\mathbf{p})$ has a quadratic form as in (10.2) as this is the most common function used for modeling work. Assuming further that $w_i = 1$, the derivative of $F(\mathbf{p})$, [cf. Eq. (10.2)], can be expressed as

$$\frac{\partial F}{\partial p_j} = 2\sum_{i=1}^{m} r_i \frac{\partial r_i}{\partial p_j}; \qquad j = 1, 2, \ldots, n \tag{10.18}$$

which in the vector form could be written as

$$\boxed{\nabla F(\mathbf{p}) = 2\mathbf{J}^T(\mathbf{p})\mathbf{r}(\mathbf{p})} \tag{10.19}$$

where $\mathbf{J(p)}$ is an $m \times n$ matrix, called a *Jacobian*, and defined as

$$\mathbf{J(p)} = \left[\frac{\partial r_i}{\partial p_j}\right]. \tag{10.20}$$

That is, the element J_{ij} of the matrix $\mathbf{J}$ in the ith row and jth column is $\partial r_i/\partial p_j$. In our example of $\mathbf{p}$ being the parameters of the drain current model, the Jacobian $\mathbf{J(p)}$ is the matrix of partial derivatives of the drain current model equation with respect to each parameter p_j; i.e., $J_{ij} = \partial I_{cal}(\mathbf{p}, \mathbf{x}_i)/\partial p_j$. Differentiating Eq. (10.18) we get the second derivative of $F(\mathbf{p})$ as

$$\frac{\partial^2 F}{\partial p_j \partial p_l} = 2\sum_{i=1}^{m}\left(\frac{\partial r_i}{\partial p_j}\frac{\partial r_i}{\partial p_l} + r_i \frac{\partial^2 r_i}{\partial p_j \partial p_l}\right) \tag{10.21}$$

which in the vector form becomes

$$\mathbf{H}(\mathbf{p}) = 2\mathbf{J}(\mathbf{p})^T\mathbf{J}(\mathbf{p}) + Q(p). \tag{10.22}$$

If the errors r_i are small then $Q(p)$ can be neglected; this is justified in most physical problems. Under this assumption, the Hessian matrix $\mathbf{H}(\mathbf{p})$ can be approximated without computing second order derivatives, that is,

$$\boxed{\mathbf{H}(\mathbf{p}) \approx 2\mathbf{J}(\mathbf{p})^T\mathbf{J}(\mathbf{p}).} \tag{10.23}$$

The error in this approximation will be small if the function $\mathbf{r}(\mathbf{p})$ is nearly linear or the function values are small.

It can easily be verfiied that the gradient [cf. Eq. (10.19)] and Hessian [cf. Eq. (10.23)] for the weighted least square objective function are given by

$$\nabla F(\mathbf{p}) = 2\mathbf{J}^T\mathbf{W}\,\mathbf{r} \tag{10.24a}$$

$$\mathbf{H}(\mathbf{p}) \approx 2\mathbf{J}^T\mathbf{W}\mathbf{J} \tag{10.24b}$$

where for the sake of brevity $\mathbf{J}(\mathbf{p})$ is simply written as $\mathbf{J}$. When $\mathbf{W} = \mathbf{I}$ (identity matrix), that is, weights are unity, Eqs. (10.24a, b) reduce to Eqs. (10.19) and (10.23), respectively.

Eigenvalues and Eigenvectors. If $\mathbf{A}$ is an $n \times n$ matrix and $\mathbf{x}$ is a nonzero n-dimensional vector such that

$$\mathbf{A}\,\mathbf{x} = \lambda\mathbf{x} \tag{10.25}$$

for some real or complex number λ, then λ is called the *eigenvalue* (or *characteristic value or latent root*) of $\mathbf{A}$ and the vector $\mathbf{x}$ that satisfies Eq. (10.25) is called the *eigenvector* of $\mathbf{A}$ associated with the eigenvalue λ. For a symmetric matrix, with which we are concerned here, all the eigenvalues are real numbers and the eigenvectors corresponding to the distinct eigenvalues are orthogonal.

The n numbers λ are eigenvalues of $n \times n$ matrix $\mathbf{A}$ if and only if the homogeneous system $(\mathbf{A} - \lambda\mathbf{I})\mathbf{x} = 0$ of n equations in n unknown has a nonzero solution $\mathbf{x}$. The eigenvalues λ are thus the roots of the characteristic equation

$$\det(\mathbf{A} - \lambda\mathbf{I}) = 0. \tag{10.26}$$

When this determinant is expanded, one obtains an algebraic equation of the nth degree whose roots λ are n eigenvalues $\lambda_1, \lambda_2, \ldots, \lambda_n$. It is common practice to normalize $\mathbf{x}$ so that it has a length of one, that is, $\mathbf{x}^T\mathbf{x} = 1$. The normalized eigenvector, generally denoted by $\mathbf{e}$, can be expressed as $\mathbf{e} = \mathbf{x}/\sqrt{\mathbf{x}^T\mathbf{x}}$ as the eigenvector corresponding to λ. The $n \times n$ matrix $\mathbf{A}$ has n pairs of eigenvalues and eigenvectors

$$\lambda_1, \mathbf{e}_1; \lambda_2, \mathbf{e}_2; \ldots; \lambda_n, \mathbf{e}_n.$$

The eigenvectors can be chosen to satisfy $\mathbf{e}_1^T\mathbf{e}_1 = \cdots \mathbf{e}_n^T\mathbf{e}_n = 1$ and be mutually perpendicular.

10.3 Optimization Methods

The problem of finding the minimum value of a function $F(\mathbf{p})$ has been extensively studied and various algorithms have been developed for this purpose. Detailed derivations of these algorithms or programming details are not given here since the emphasis is on a basic understanding of the concepts. Interested readers wishing to study these algorithms in detail are referred to the numerous books on the subject [16]–[21]. Listing of the computer programs for optimization technique, in general, can be found in various publications [21]–[25]. Software packages like SUXES [4, 5], SIMPAR [9], etc., specifically written for device model parameter extraction, are also available from universities [4], [9] and research institutions [23, 24], [27].

Most of the optimization algorithms implemented for the device model parameter extraction use *gradient methods* of optimization [4]–[12], although in some programs *direct search* optimization has also been implemented [13]. *Here we will discuss only the former method (i.e., gradient method) as it is the one most widely used for the device model parameter extraction.* It essentially consists of two steps. The first step is to select a direction of search $\mathbf{s}$ from a given point $\mathbf{p}$ (in the parameter space), while the second step is to search for the minimum of the function along the direction $\mathbf{s}$. Note that the direction $\mathbf{s}$ in n dimensional space is an n-vector $\mathbf{s} = [s_1\, s_2 \cdots s_n]^T$.

Steepest Decent Method. One of the most widely known method for minimizing a function of several variables is the method of steepest descent, often referred to as *gradient or slope-following* method. Like any other gradient method, it assumes that the objective function $F(\mathbf{p})$ is continuous and differentiable. In this method the minimum of a function is obtained by choosing the search direction $\mathbf{s}$ as the direction of the negative gradient, that is,

$$\mathbf{s} = -\nabla F(\mathbf{p}) = -\mathbf{J}^T(\mathbf{p})\mathbf{r}(\mathbf{p}) \tag{10.27}$$

while the parameter change $\Delta\mathbf{p}$ is chosen to point in the direction of the negative gradient, that is

$$\Delta\mathbf{p} = -\alpha\nabla F(\mathbf{p}) \tag{10.28}$$

where α is a positive constant. The algorithm proceeds as follows:

1. Start at some initial value of the parameter $\mathbf{p}$, which we shall designate as $\mathbf{p}^0$. This should be the best guess of the minimum being sought.

2. At the *kth iteration* ($k = 0, 1, 2, 3 \cdots$) calculate $F(\mathbf{p}^k)$ and $\nabla F(\mathbf{p}^k)$ using Eqs. (10.2) and (10.19) respectively.
3. Move in a direction $\mathbf{s}^k(= -\nabla F(\mathbf{p}^k))$. Take a step of length α along this direction such that $F(\mathbf{p}^k + \Delta\mathbf{p}^k) < F(\mathbf{p}^k)$, i.e., $F(\mathbf{p}^k + \Delta\mathbf{p}^k)$ is minimum in the direction $\mathbf{s}^k$. We can use quadratic interpolation procedure or any other method to choose the value of α^k.
4. Calculate the next step $\mathbf{p}^{k+1}$ as

$$\mathbf{p}^{k+1} = \mathbf{p}^k - \alpha\nabla F(\mathbf{p}^k). \tag{10.29}$$

5. If $|F(\mathbf{p}^k) - F(\mathbf{p}^{k+1})| > \epsilon$
   go to step 2, where ϵ is some preassigned tolerance.
6. Terminate the calculations when

$$|F(\mathbf{p}^k) - F(\mathbf{p}^{k+1})| \le \epsilon. \tag{10.30}$$

It is possible to use some other criterion to terminate the calculations in step 6, but that given by Eq. (10.30) is the one most commonly used. Various "stopping rules" have been suggested and often combination of those rules are used in practical optimization problems [5]. Some other criteria that have been proposed are

$$\text{Max}\left|\frac{\partial F(\mathbf{p})}{\partial p_j}\right| < \epsilon \tag{10.31}$$

$$\frac{|p_j^{k+1} - p_j^k|}{|p_j^k| + \delta} < \epsilon \tag{10.32}$$

where δ is set equal to some small number ($< 10^{-10}$) in the eventuality that p_j^k goes to zero. No matter what criterion is used to terminate the calculations, one needs to select the tolerance ϵ. The smaller the ϵ, the more precisely will the location of the minimum be found, though at higher computation cost as it will now require more iterations. Normally $\epsilon = 10^{-6}$ is good enough for modeling work.

This method of optimization is inherently stable and produces excellent results when $\mathbf{p}$ is away from the minimum but becomes very slow when the minimum is approached. For this reason this method is not normally used as a stand alone optimization method.

Gauss–Newton Method. In the steepest decent method, we choose the direction to move in the parameter space by considering only the first derivative term, i.e., slope. The method could be improved upon by including the second derivative term thereby taking into account both the slope and the curvature [see Eq. (10.13)]. Thus, in the new method we modify the search direction from the negative gradient to the inverse of the Hessian, that is,

$$\mathbf{s} = -\mathbf{H}^{-1}\nabla F(\mathbf{p}) \tag{10.33}$$

and the parameter change $\Delta\mathbf{p}$ is

$$\Delta\mathbf{p} = -\mathbf{H}^{-1}\nabla F(\mathbf{p}) \tag{10.34}$$

keeping the step size $\alpha = 1$ in this case. Thus, in this method the updated parameter vector $\mathbf{p}^{k+1}$ is derived from the following iterative algorithm

$$\mathbf{p}^{k+1} = \mathbf{p}^k - \mathbf{H}^{-1}\nabla F(\mathbf{p}^k) \tag{10.35}$$

so that the different steps outlined earlier still apply. This algorithm is often referred to as the Newton method for finding the minimum $F(\mathbf{p})$. The major advantage of Eq. (10.35) over Eq. (10.29) is that if the approximation is sufficiently accurate near the current parameter estimation then it gives fairly fast convergence. However, the disadvantage is that it requires prohibitively large computation effort for calculating the Hessian $\mathbf{H}$ in order to solve for $\Delta\mathbf{p}$. In general, the Hessian matrix $\mathbf{H}$ is difficult to solve with sufficient accuracy. For this reason approximations are often used for $\mathbf{H}$. The error in the approximation decreases during successive iterations as the optimization proceeds.

For the case of a quadratic $F(\mathbf{p})$ [cf. Eq. (10.2)] we have already seen that $\mathbf{H}$ could be approximated by Eq. (10.23). Substituting Eq. (10.23) for the Hessian and Eq. (10.19) for the gradient into Eq. (10.35) we get

$$\mathbf{p}^{k+1} = \mathbf{p}^k - [\mathbf{J}^{(k)T}\mathbf{J}]^{-1}[\mathbf{J}^{(k)T}\mathbf{r}^k]. \tag{10.36}$$

This algorithm is referred to as the Gauss–Newton method. Although this least square method is theoretically convergent, there are practical difficulties which hamper the convergence of the iteration process. If $\mathbf{J}^T\mathbf{J}$ is singular or nearly so, then the problem of solving $\Delta\mathbf{p}$ from Eq. (10.36) becomes ill-conditioned.

Levenberg–Marquardt Method. In order to avoid the problem of singularity of $\mathbf{J}^T\mathbf{J}$ in Eq. (10.36), Marquardt proposed an algorithm, first suggested by Levenberg, called the Levenberg–Marquardt (L–M) algorithm [26]–[28]. In this algorithm a constant diagonal matrix $\mathbf{D}$ is added to the Hessian $\mathbf{H}(\mathbf{p})$ given by Eq. (10.23). Thus, in the L–M method the updated parameter vector $\mathbf{p}^{k+1}$ is derived from the following iterative algorithm

$$\mathbf{p}^{k+1} = \mathbf{p}^k - [\mathbf{J}^{(k)T}\mathbf{J}^k + \lambda^k\mathbf{D}^k]^{-1}[\mathbf{J}^{(k)T}\mathbf{r}^k]. \tag{10.37}$$

The elements of the matrix $\mathbf{D}$ are the diagonal elements of $\mathbf{J}^T\mathbf{J}$, that is,

$$\mathbf{D}_{ii} = (\mathbf{J}^T\mathbf{J})_{ii} \tag{10.38}$$

Note that the addition of the diagonal matrix $\mathbf{D}$ ensures that the iterations matrix is nonsingular. The constant λ is called the *Marquardt parameter.*

When λ is small relative to the *norm*[5] of $\mathbf{J}^T\mathbf{J}$, the algorithm reduces to the Gauss–Newton method with its rapid convergence and when λ is large, the method becomes the steepest decent method with its inherent stability. Thus, in this method the direction $\Delta\mathbf{p}$ is intermediate between the direction of the Gauss–Newton increment ($\lambda = 0$) and direction of steepest decent ($\lambda = \infty$). Marquardt's method produces an increment $\Delta\mathbf{p}$ which is invariant under scaling transformations of the parameters. That is, if the scale for one component of the parameter vector is doubled, the increment calculated, and the corresponding component of the increment halved, the result will be the same as calculating the increment in the original scale. The algorithm proceeds as follows:

1. Start at some initial best guess value $\mathbf{p}^0$.
2. Pick a modest value of λ, say 0.01.
3. At the kth iteration ($k = 0, 1, 2, 3 \cdots$) calculate $F(\mathbf{p}^k)$.
4. Solve Eq. (10.37) for $\mathbf{p}^{k+1}$ and evaluate $F(\mathbf{p}^{k+1})$.
5. If $F(\mathbf{p}^{k+1}) \geq F(\mathbf{p}^k)$, increase λ by a factor 10 (or any other substantial factor) and go to step 4.
6. If $F(\mathbf{p}^k + \Delta\mathbf{p}^k) < F(\mathbf{p}^k)$, decrease λ by a factor 10, update the trial solution and go back to step 3.

Within the iterations λ increases until $F(\mathbf{p}^{k+1}) < F(\mathbf{p}^k)$. Between the iterations λ decreases successively so that as the minimum is reached (i.e., solution is approached) λ should tend to zero. There are other ways of incrementing λ [14–16], [32, 33] that are better than updating λ by a constant factor [12]. However, there are no rigorous approaches for choosing the best value of λ that will lead to the desired minima.

The L–M method works very well in practice and has become the standard of non-linear least square routines [22]. Various optimizers like SUXES [5], SIMPAR [9], OPTIMA [12] and most of the commercially available packages like TECAP2 [7] are based on this algorithm.

It should be pointed out that different gradient methods of optimization have been compared [17], [19], [32]. Although the L-M method is most widely used for device model parameter extraction, several modifications of the Gauss–Newton method have been found to be better than the L-M method. In fact Bard [32] appears to favor a modification of the Gauss method called interpolation–extrapolation method.

A Remark on the Calculation of Derivatives. The L–M method requires evaluation of the Jacobian $\mathbf{J}$ of the error vector $\mathbf{r}$ and solution of the n

---

[5] The norm of a vector $\mathbf{s}$ is defined as

$$\|\mathbf{s}\|^2 = \sum s_i^2.$$

normal equations at each iteration step. In our example of drain current model parameter extraction, the elements of the **J** matrix are $\partial I_{cal}(i)/\partial p_j$. Basically there are two ways to calculate these partial derivatives; (1) analytically, and (2) numerically. The analytical calculations of the partial derivatives are much more accurate and efficient when compared to the numerical methods. However, almost all optimizers use numerical methods for estimating the Jacobian. This is because the model equations are usually complex function of the model parameters, and therefore the task of deriving partial derivatives becomes tedious and cumbersome. Moreover, with numerical methods the program becomes more flexible so that any model equations could easily be implemented in the optimizer. The Jacobian is estimated numerically by using either a forward difference approximation

$$J_{ij} = \frac{\partial r_i}{\partial p_j} \approx \frac{r_i(p_1, p_2, \ldots, p_j + \delta p_j, \ldots, p_n) - r_i(\mathbf{p})}{\delta p_j} \tag{10.39}$$

or a more accurate central difference approximation

$$J_{ij} \approx \frac{r_i(p_1, p_2, \ldots, p_j + \delta p_j, \ldots, p_n) - r_i(p_1, p_2, \ldots, p_j - \delta p_j, \ldots, p_n)}{2\delta p_j} \tag{10.40}$$

where δp_j is some relatively small quantity, which could be chosen as $\delta p_j = 10^{-3} p_j$ and is frequently quite satisfactory. Bard [32] has given a brief discussion on appropriate values for δp_j other than $10^{-3} p_j$. Equation (10.40) is a more accurate estimate of the actual derivative but at the cost of the speed of evaluation of **J**. Sometimes for speed consideration, accuracy is sacrificed by using the forward difference method during the initial phase of the optimization, when the solution is still far from the optimal point, and then switching to the central difference method. When approximating **J** by the difference method, the performance normally deteriorates as the number of parameters n increases. For this reason the dynamic variable approach of approximating **J** is often used [16]–[17].

Scaling. The range of the MOSFET model parameter values are very large. For example, the substrate concentration N_b is $\sim 10^{16}\,\mathrm{cm}^{-3}$ while the difference between the drawn and effective channel length ΔL is only $\sim 10^{-5}$ cm, which results in the entries of $\mathbf{J}(\mathbf{p})$ ranging from about $\partial I_{cal}/\partial N_b = 10^{-19}$ to $\partial I_{cal}/\partial \Delta L = 10^3$. It is, therefore, very important that the entries of the Jacobian matrix should be normalized to their proper range to reduce the round-off errors. One way to achieve this normalization is to multiply each column of $\mathbf{J}(\mathbf{p})$ by a normalization factor (the current value of the corresponding variable), while each row of $\Delta \mathbf{p}^k$ is divided by the same factor so that these entries are centered at 1.

10.3.1 Constrained Optimization

During the optimization process described above, very often some physical parameter tends to take a non-physical value. To avoid this situation, generally some *constraints* are imposed on each of the parameters so that the parameters do not take unrealistic values. A common type of constraint, which is used for model parameter work, is the *box constraint* where the lower and upper bounds are given on each of the model parameter values. For example, constraint of the body factor γ might be

$$0.2 \leq \gamma \leq 3 \tag{10.41}$$

which means that the minimum value of γ can be 0.2 (lower bound) while the maximum value γ can attain is 3 (upper bound). Thus, in general the box constraint will have the following form

$$p_{j,\min} \leq p_j \leq p_{j,\max} \quad j = 1, 2, \ldots, n. \tag{10.42}$$

The box constraint given above can be expressed as a set of linear constraints for n model parameters as

$$\left[\frac{\mathbf{A}}{-\mathbf{A}}\right]\mathbf{p} \leq \mathbf{B} \tag{10.43}$$

where $\mathbf{A}$ is an $n \times n$ unit matrix and $\mathbf{B}$ is $2n \times 1$ matrix with rows consisting of upper bound ($p_{j,\max}$) and the negative value of the lower bound ($p_{j,\min}$) of the model parameter vector $\mathbf{p}$. The constraints given by (10.43), in general, could be written as

$$\mathbf{g}(\mathbf{p}) \leq 0. \tag{10.44}$$

The problem now becomes a constrained optimization problem wherein we minimize $F(\mathbf{p})$ subject to the linear constraints given by the system of equations (10.44).
The set of values of $\mathbf{p}$ satisfying the equality set of equations (10.44) forms a hypersurface, called the constraint surface, which divides the entire parameter space into two subspaces. The subspace which contains all the points that satisfy all the constraints given by Eq. (10.44) is called the *feasible region* or *region of acceptability*. By definition, no constraints are violated in the feasible region and any solution $\mathbf{p}^*$ of the constrained optimization problem must lie in the feasible region. Any point in the feasible region is called a feasible point. The constraints given by Eq. (10.44) are called *active* at the feasible point $\mathbf{p}$ if $\mathbf{g}(\mathbf{p}) = 0$ and *inactive* if $\mathbf{g}(\mathbf{p}) < 0$. The constraints at the infeasible points $\mathbf{g}(\mathbf{p}) > 0$ are also active. By convention, any equality constraint is referred as *active* and inequality constraints are active when they are violated or satisfied exactly. To illustrate this point, let us assume that the objective function $F(\mathbf{p})$ is a function of two parameters p_1 and p_2.

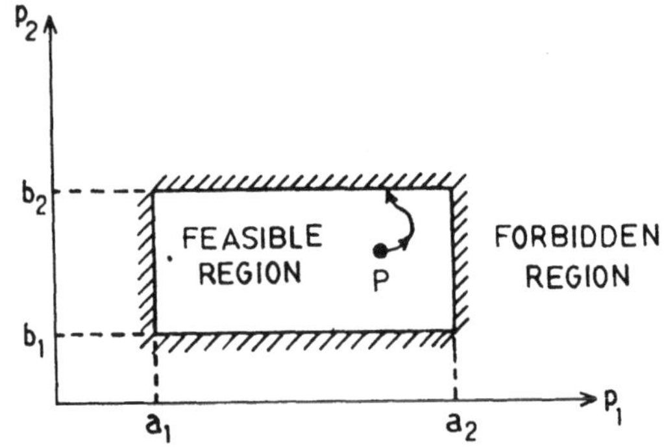

Fig. 10.2 A possible optimization path in a feasible region

Furthermore, assume that the parameters are constrained as indicated below

$$a_1 \leq p_1 \leq b_1, \quad a_2 \leq p_2 \leq b_2 \tag{10.45}$$

and shown in Figure 10.2. The region inside the shaded area is the feasible region. From the initial point $\mathbf{p}^0$ in the feasible region, the optimization procedure varied the parameters until the constraint $p_2 = b_2$ was encountered. Until that point the optimization procedure had progressed as if there were no constraint, that is, the constraints are inactive. However, when the boundary between the feasible region and the forbidden region was encountered the constraint $p_2 = b_2$ became active.

When a constraint is active, often it can be used to remove one of the parameter from the error function. One can then proceed and use an unconstrained optimization program. However, note that even though a constraint becomes active in a minimization search, it may later become inactive.

The field of optimization, or *nonlinear programming* as it is sometime called, has developed algorithms for the solution of constrained optimization problems [14]–[17]. We will not be discussing these techniques in detail because they are not widely used for the problem in hand, i.e., device model parameter extraction. However, they are common in many fields and the reader should be aware of them. Different optimization techniques use different methods to guarantee that parameters always remain in the feasible region. One way to do this is to transform the constraints using special functions so that no parameters are constrained, and thus the algorithm discussed in the previous section could be used. Once the transformed problem has been optimized, the unconstrained parameters (which are guaranteed to be in the feasible region) can be determined from the transformed parameter [29]. Still another approach is to define a new objective function $F_C(\mathbf{p})$ which is related to the original objective function $F(\mathbf{p})$ via a *penality function* [29].

$$F_C(\mathbf{p}) = F(\mathbf{p}) + \text{penality function}. \tag{10.46}$$

The introduction of the penality function makes the function $F_C(\mathbf{p})$ as a unconstrained problem. This is done by choosing the initial guess value of $\mathbf{p}$ to be in the feasible region (i.e., $\mathbf{p}$ satisfies the constraints). If the optimization procedure tries to find the minimum by going out of the feasible region, the penality function becomes larger and forces the parameter to remain in the feasible region. However, the drawback of these methods is that near the solution the problem becomes increasingly ill-conditioned. For this reason modern constrained methods are based on the *Lagrange multiplier* approach, wherein we define the Lagrange function F_L as

$$F_L(\mathbf{p}, \lambda) = F(\mathbf{p}) + \sum_{j=1}^{n} \lambda_j g_j(\mathbf{p}); \quad l = 1, 2, \ldots, n \tag{10.47}$$

where λ_j are known as Lagrange multipliers and $g_j(\mathbf{p})$ is the set of constraints given by Eq. (10.44). To solve for the optimum set of parameters $\mathbf{p}^*$, we set the gradient of (10.47) equal zero. Thus,

$$\begin{aligned} \nabla F(\mathbf{p}^*) + \sum_{j=1}^{n} \lambda_j^* \nabla g_j(\mathbf{p}^*) &= 0 \\ \lambda_j^* g_j(\mathbf{p}^*) &= 0 \\ \text{and} \quad \lambda_j \geq 0 \quad g_j(\mathbf{p}) &\leq 0. \end{aligned} \tag{10.48}$$

For more details of this method the reader is referred to reference [14]–[17], [29].

For device model parameter extraction with box constraint problems, very simple approach is often used. At every iteration, before the calculation of $F(\mathbf{p})$, the current value of $\mathbf{p}$ is subjected to Eq. (10.43) for a consistency check. If any constraint corresponding to the parameter p_j becomes active (i.e., either $p_j < p_{j,\min}$ or $p_j > p_{j,\max}$), then for that parameter p_j the component of the steepest descent direction is examined first. Thus, we determine

$$\hat{p}_j^{k+1} = p_j^k - \frac{\partial F(\mathbf{p})^k}{\partial p_j^k}. \tag{10.49}$$

If $\hat{p}_j^{k+1} < p_{j,\min}$ or $\hat{p}_j^{k+1} > p_{j,\max}$, then the constraint corresponding to this parameter remains active. In other words we insure that the parameter value is held constant during the next iteration,

$$\Delta p_j^k = p_j^{k+1} - p_j^k = 0.$$

On the other hand, if $\hat{p}_j^{k+1}$ lies within the user-specified bounds, the constraint is relaxed so that the parameter can move to the next value.

This simple approach for the box constraint can easily be implemented into the L–M algorithm for the unconstrained optimization. At each iteration, the following linear system of equations are solved [cf. Eq. (10.37)]

$$\mathbf{M}^k \Delta \mathbf{p}^k = -\mathbf{J}^{(T)k}\mathbf{r}^k$$
$$\text{where} \quad \mathbf{M}^k = \mathbf{J}^{(T)k}\mathbf{J}^k + \lambda^k \mathbf{D}^k. \tag{10.50}$$

If the jth parameter is to be constrained then we must set $\Delta p_j^k = 0$ in Eq. (10.50). This is done by zeroing the jth row and column of the matrix $\mathbf{M}^k$ and by setting the diagonal term M_{jj}^k to unity. During the next iterate p_j will be reset to the corresponding boundary values, i.e., $p_j = p_{j,\mathrm{min}}$ or $p_j = p_{j,\mathrm{max}}$. This way the algorithm discussed in the previous section 10.3 for unconstraint optimization could be used without any change [12].

10.3.2 Multiple Response Optimization

In many situations it is desirable to optimize simultaneously several objective functions; i.e., the problem is to minimize

$$\min \begin{bmatrix} F_1(\mathbf{p}) \\ F_2(\mathbf{p}) \\ \vdots \\ F_l(\mathbf{p}) \end{bmatrix} \tag{10.51}$$

subject to

$$g_j(\mathbf{p}) \leq 0 \quad j = 1, 2, \ldots, n \tag{10.52}$$

where l is the number of objective functions. For example, in some applications like analog circuit design, the small signal conductance g_{ds} is as important as the absolute value of the drain current I_{ds}. Since both of them depend on the same model parameters, it is more appropriate to extract the parameters so as to get the best fit for both the measured I_{ds} and the g_{ds} data. In other words, we now have a problem where we need simultaneous optimization of two objective functions—I_{ds} and g_{ds}.
The basic technique of finding the solution in such cases is to convert the multiple objective function into a single objective function and then solve a standard optimization problem. The key is how this conversion is actually done. The problem can be formulated in different ways [29]; however, a simple formulation assigns weights to the individual objective functions and combines these functions into a single weighted sum as the least-square function, that is,

$$F(\mathbf{p}) = \sum_{q=1}^{l} W_q F_q(\mathbf{p}) \tag{10.53}$$

where the weights W_q take into account the relative importance and the appropriate scaling associated with each objective function $F_q(\mathbf{p})$ given by

Eq. (10.2). Note that various objective functions $F_q(\mathbf{p})$ may have different units, but will depend on the same parameter vector $\mathbf{p}$. In the vector form Eq. (10.53) becomes

$$F(\mathbf{p}) = \mathbf{r}^T \mathbf{W} \mathbf{r} \tag{10.54}$$

where $\mathbf{r}(\mathbf{p})$ is given by

$$\mathbf{r}(\mathbf{p}) = [\mathbf{r}_1, \mathbf{r}_2, \ldots, \mathbf{r}_q \cdots \mathbf{r}_l]^T \quad q = 1, 2, \ldots, l \tag{10.55}$$

and

$$\mathbf{r}_q(\mathbf{p}) = [r_{q1}, r_{q2}, \ldots, r_{qi} \cdots r_{qm}]^T \quad i = 1, 2, \ldots, m \tag{10.56}$$

so that $\mathbf{r}(\mathbf{p})$ is now a vector of length $m \times l$ and $\mathbf{W}$ is the $ml \times ml$ diagonal weighting matrix. Since the multiple objective function is identical in form to that of a single objective function case (10.2), the optimization algorithms presented earlier can be used without any change. In our example of two functions I_{ds} and g_{ds}, we will have

$$F(\mathbf{p}) = W_I \sum_{i=1}^{m} w_i [r_{I_i}(\mathbf{p})]^2 + W_G \sum_{i=1}^{m} w_i [r_{G_i}(\mathbf{p})]^2 \tag{10.57}$$

where W_I and W_G are the relative weights for the current and conductance respectively, w_i is the weight for each data point (current or conductance) and $r_{I_i}(\mathbf{p})$ and $r_{G_i}(\mathbf{p})$ are the error functions for the current and conductances respectively (see Eq. 10.3). It is the algorithm given by Eq. (10.57) which is implemented in most of the device model parameter extraction programs including SUXES and OPTIMA [12].

As an example, the impact of optimizing I_{ds} and g_{ds} simultaneously is shown in Figures 10.3 and 10.4. The 'measured' g_{ds} is obtained from the $I_{ds} - V_{ds}$ data by evaluating the derivative of the $I_{ds} - V_{ds}$ curve at a given V_{ds} using the central difference method. The measured $I_{ds} - V_{ds}$ data at $V_{sb} = 0$ V for a typical 1.5 μm n-channel device (oxide thickness = 225 Å) is shown as circles in Figure 10.3. This data was fitted to the SPICE MOS Level 3 model [30] by extracting the parameters using OPTIMA. In one case, only I_{ds} was optimized (conventional approach), while in the other case, both I_{ds} and g_{ds} were optimized simultaneously. It can be seen from Figure 10.3 that while the current is modeled accurately through the conventional approach (dashed lines), the slope, especially in the saturation region, does not fit the data. This can be observed more clearly from the $g_{ds} - V_{ds}$ curves (dashed lines) in Figure 10.4. On the other hand, when both I_{ds} and g_{ds} are optimized simultaneously, the slope is modeled more accurately (solid lines in Figures 10.3 and 10.4).

Note that, in spite of optimizing the current and conductance simultaneously, the $g_{ds} - V_{ds}$ fit (Figure 10.4) does not seem to improve significantly, particularly near the saturation voltage V_{dsat}. This is because in the Level 3 model, the second derivative of the current ($\partial g_{ds}/\partial V_{ds}$) is not continuous at

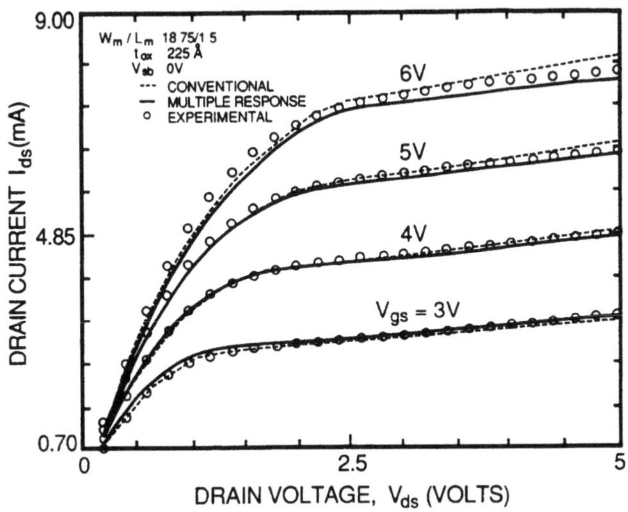

Fig. 10.3 Comparison of experimental I_{ds} vs. V_{ds} data with the MOS Level 3 model for LDD device with $W_m/L_m = 18.75/1.5$, oxide thickness = 225 Å, $V_{gs} = 3$ V, 4 V, 5 V and 6 V. Circles (○) represent experimental data, dashed lines are simulated curves from single response optimization and solid lines correspond to multiple response optimization

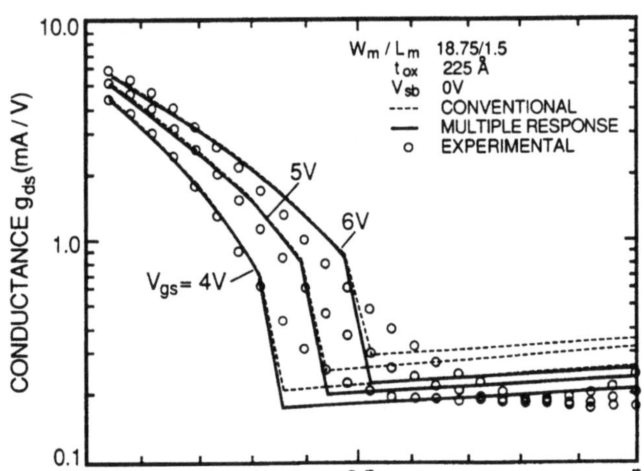

Fig. 10.4 Comparison of experimental g_{ds} vs. V_{ds} curves with the MOS Level 3 model for LDD device with $W_m/L_m = 18.75/1.5$, oxide thickness = 225 Å, $V_{gs} = 4$ V, 5 V and 6 V. Circles (○) represent experimental data, dashed lines are simulated curves from single response optimization and solid lines correspond to multiple response optimization

V_{dsat}. Indeed, the accuracy of the conductance model can be greatly improved if this condition is satisfied. In fact, the multiple response approach produces improvement in the fit only when the g_{ds} model is accurate in addition to the I_{ds} model.

10.4 Some Remarks on Parameter Extraction Using Optimization Technique

There are two potential sources of error in the optimization technique for extracting model parameters. These are:

- *Local minima* which results in non-optimum parameter set.
- *Redundancy of parameters* which produce non-unique parameter values although it does result in an optimum fit of the data.

Local Minima. The optimization algorithm always converges towards a minimum total error. The accuracy in locating the minimum is primarily determined by the convergence criteria. As was pointed out in section 10.2, the algorithms have no way to distinguish between the local and global minima. In a multidimensional error or residual plane, it is the initial guess value of the parameter which will determine to which minima the algorithm will go. More than one minima can occur in this residual plane, depending on how many parameters have to be fitted simultaneously, how complicated the model is, the type of the error function (relative or absolute error) and the data set which is used to fit the data. The optimization algorithm will then converge to that minima which is nearest to the starting guess value for different parameters. In practice, when the starting guess value is not known with any accuracy, minimization is carried out several times starting from different initial guess values for the parameters and observing the parameter value to which the algorithm converges. If the final parameter values depends on the initial starting values then one is hitting local minima. In such situations the solution which gives the smallest error should be the correct answer.

Parameter Redundancy. When a certain parameter is not sensitive to the change in the function to be fitted, that is, the function is weakly dependent on the parameter, or if the function is insensitive to the value of the parameter, then that parameter is said to be a *redundant parameter*. Thus, any value of the redundant parameter gives the same characteristics, i.e., the model parameter value is not unique. This could happen, for example, when extracting model parameters for short channel devices using data for long channel devices. Although the extracted parameter value is not unique, the fit to the data could still be optimum. Note that *non-uniqueness of the*

parameter may also be produced by convergence to a different (not desired) local minimum and therefore should not be confused with a redundant parameter. Recently it has been shown that confidence region algorithms [32, 33] when used with optimizers can be very helpful in eliminating redundant or correlated parameters, thereby increasing the predictive capability of the model [11], [34, 35]. This is discussed in the next section.
In order to obtain the model parameter values accurately and reliably, the following points must be taken into consideration:

- *The data set must be chosen carefully.* This is because if the data set does not contain data points that are sensitive to certain parameters, redundancy in the parameter set will occur. This in turn will result in a non-physical parameter set, not only for these redundant parameters but for other parameters also.
- *Good initial guess values* and boundary (minimum and maximum) values are important as discussed earlier.
- *Proper extraction methodology*: It is advisable that the parameters and data set should be divided into different regions as discussed in section 10.1. This will result in a simpler residual plane thereby reducing the possibility of multiple local minima, and thus resulting in a physical parameter set.

Clearly, though model parameter extraction programs (optimizers) are very powerful and useful tools, they must be used with care.

10.5 Confidence Limits on Estimated Model Parameter

In the previous sections we discussed how to get the optimum parameter set by fitting experimental data to the model. The data used to fit the model is not generally exact because of the inherent measurement error (sometime referred to as noise) associated with the data. This measurement error could be due to the instrument and/or statistical fluctuations [31]. Therefore, any random error in the measurement will lead to random errors in estimating the model parameter values, even when the model is exact. Clearly, knowing the model parameter value without the error estimate on the parameter, or without a statistical means of testing *goodness-of-fit* of the model, is not a very useful way of estimating a parameter.
If there were no errors associated with the measurements, then the measured data set, say $\mathbf{d}_0$, would lead to the "true" parameter set $\mathbf{p}_t$. Since measured data have random error components, one measured data set $\mathbf{d}_1$ will give parameter set $\mathbf{p}_1$. Another data set $\mathbf{d}_2$ would give a slightly different set of fitted parameters $\mathbf{p}_2$, and so on. The parameter set $\mathbf{p}_i$ therefore occurs with some probability distribution in the n-dimensional space of all possible parameter set $\mathbf{p}$. The actual extracted set $\mathbf{p}$ is one member drawn from this distribution, which in general will be different from the true parameter set

$\mathbf{p}_t$. Note that the true parameter set $\mathbf{p}_t$ is not known because we cannot get a data set without error. However, *if we know the probability distribution of the parameter set* $\mathbf{p}_i$, *we will know all about our estimate of* $\mathbf{p}$ *without the need to know the true* $\mathbf{p}_t$.

This probability distribution can be obtained from the parameter covariance matrix.[6] Assuming that the measurement errors are normally distributed and are uncorrelated, the parameter covariance matrix $\mathbf{C}$ of a parameter vector $\mathbf{p}$, for nonlinear least squares problems, is given by [32, 33]:

$$\mathbf{C} \approx \frac{F(\mathbf{p}^*)}{m-n}(\mathbf{J}^T\mathbf{W}\mathbf{J})^{-1}\mathbf{J}^T\mathbf{W}^2\mathbf{J}(\mathbf{J}^T\mathbf{W}\mathbf{J})^{-1} \tag{10.58}$$

where $\mathbf{p}^*$ is the parameter vector which minimizes the least square function $F(\mathbf{p}^*)$. Recall that during the optimization process, the Jacobian $\mathbf{J}(\mathbf{p})$ is evaluated for each nonlinear iteration. Therefore, it is readily available at the end of the optimization and can be used directly in Eq. (10.58) to evaluate $\mathbf{C}$. The covariance matrix $\mathbf{C}$ is an $n \times n$ matrix, n being the number of parameters. The diagonal terms C_{jj} of $\mathbf{C}$ represents the variance in the model parameter p_j, while the off-diagonal terms $C_{ij}(i \neq j)$ are the covariance between the parameters. For the case $\mathbf{W} = \mathbf{I}$ (unit weights), Eq. (10.58) reduces to

$$\mathbf{C} \approx \frac{F(\mathbf{p}^*)}{m-n}(\mathbf{J}^T\mathbf{J})^{-1}. \tag{10.59}$$

It should be pointed out that the above expression for the covariance matrix $\mathbf{C}$ is strictly valid only for a linear models, and hence, is only approximate for nonlinear models with the approximation being better for cases which are less nonlinear.

It is common practice to represent the probability distribution of errors in parameter estimation in the form of *confidence intervals* for individual parameters or *confidence regions* in n-dimensional parameter space. These terms are usually expressed as a percentage $100(1-\alpha)$, α being the confidence level lying between 0 and 1; e.g., 95% confidence interval for confidence level $\alpha = 0.05$. It refers to "that interval or range of values around an observed value which will in 95% of the cases include the expected value. The expected value is defined as the average of an infinite series of such determination". For example, 95% confidence interval for the normal distribution of a single parameter p is $p_t - 1.96\sigma < p < p_t + 1.96\sigma$ where σ^2 is the variance in the parameter estimate. Assuming that the measurement error is normally distributed, then each parameter p_j follows the student's t-distribution with $\nu = m - n$ degree of freedom, and has the following $100(1-\alpha)\%$ confidence

[6] These and other statistical terms are defined in Appendix H.

interval [32, 33]

$$\sqrt{C_{jj}}t_{1-\alpha/2,\nu} \leq p_j - p_{tj} \leq \sqrt{C_{jj}}t_{1-\alpha/2,\nu} \tag{10.60}$$

where p_{tj} is the 'true' value of the parameter p_j, C_{jj} (jth diagonal element of $\mathbf{C}$) is the variance of p_j and $t_{1-\alpha/2,\nu}$ is the two-sided Student's t-distribution evaluated at $1-\alpha$ probability. Thus $100(1-\alpha)\%$ confidence interval can be evaluated for each extracted parameter. Parameters with large confidence intervals are redundant. Since the device model parameters take on a wide range of values, the parameter redundancy must be characterized using an "uncertainty" value rather than the confidence interval value. The uncertainty in the parameter p_j is defined as [34]

$$\text{uncertainty } p_j = \frac{\sqrt{C_{jj}}t_{1-\alpha/2,\nu}}{p_j^*} \times 100\% \tag{10.61}$$

where p_j^* is the parameter value obtained at the minimum of $F(\mathbf{p})$. Note that the parameter uncertainties can be evaluated without any knowledge of the true parameters $\mathbf{p}_t$. If the uncertainty of a parameter is 100% or higher, the parameter is assumed to be ill-determined or redundant. In that case that parameter may be set to zero or to some fixed value with no significant loss of accuracy to the fit and optimization is then redone. Although this approach can determine redundant parameters, it provides no information on redundant parameter combinations, since the parameter covariances C_{ij} $(i \neq j)$ are not taken into account. In order to obtain redundant parameter combination, the so called *confidence region* approach, is more appropriate [35].

For zero mean (i.e., the estimated parameter values are distributed about the true value p_t) and normally distributed measurement error, the joint probability density function for the parameter vector $\mathbf{p}$ can be written as [33]

$$f(\mathbf{p}) = \frac{1}{(2\pi)^{n/2}|\mathbf{C}|^{1/2}} \exp\left[-\frac{1}{2}(\mathbf{p}-\mathbf{p}_t)^T\mathbf{C}^{-1}(\mathbf{p}-\mathbf{p}_t)\right] \tag{10.62}$$

where $|\mathbf{C}|$ is the determinant of the covariance matrix $\mathbf{C}$ of the parameter vector $\mathbf{p}$ as given by Eq. (10.58). The exponent term in Eq. (10.62) is a non-negative scalar and can be set equal to ρ^2, that is,

$$(\mathbf{p}-\mathbf{p}_t)^T\mathbf{C}^{-1}(\mathbf{p}-\mathbf{p}_t) = \rho^2 \tag{10.63}$$

which represents a hyper-ellipsoid with center at the origin and coordinates $p_1 - p_{t1}, \ldots, p_n - p_{tn}$. If a^2 represents a specific value, then $\rho^2 \leq a^2$ defines the interior of a hyper-ellipsoid in n-dimensional space and $\rho = a$ produces hypersurfaces of constant probability density. Therefore, if

$$\int_{\rho^2 \leq a_{1-\alpha}^2} \cdots \int f(\mathbf{p})dp_1 dp_2 \cdots dp_n = 1 - \alpha \quad 0 \leq \alpha \leq 1 \tag{10.64}$$

then the hyper-ellipsoidal region $\rho^2 \le a_{1-\alpha}^2$ is called the $100(1-\alpha)\%$ *confidence region.* Any parameter vector extracted from a measured data set lies within this ellipsoidal region *centered* at $\mathbf{p}_t$ with probability of $1-\alpha$.
The integral (10.64) can be evaluated more conveniently by expressing the confidence region in a coordinate system which coincides with the n axes of the hyper-ellipsoid. To obtain this transformation, we first compute the eigenvalues λ_j $(j = 1, 2, \ldots, n)$ and the corresponding normalized eigenvectors $\mathbf{e}_j$ of $\mathbf{C}^{-1}$. The matrix $\mathbf{C}^{-1}$ can be decomposed to [32]–[33]

$$\mathbf{C}^{-1} = \mathbf{Q}\mathbf{D}_\lambda\mathbf{Q}^T \tag{10.65}$$

where $\mathbf{D}_\lambda$ is the diagonal matrix with the eigenvalues $\lambda_1, \ldots, \lambda_n$ and $\mathbf{Q}$ is a matrix whose jth column is the eigenvector $\mathbf{e}_j$. By combining Eq. (10.65) with Eq. (10.63) and defining a new coordinate vector $\mathbf{h} = [h_1, h_2, \ldots, h_j, \ldots, h_n]^T$ as

$$\mathbf{h} = \mathbf{Q}^T(\mathbf{p} - \mathbf{p}_t) \tag{10.66}$$

the confidence region ellipsoid becomes

$$(\mathbf{p} - \mathbf{p}_t)^T\mathbf{C}^{-1}(\mathbf{p} - \mathbf{p}_t) = \mathbf{h}^T\mathbf{D}_\lambda\mathbf{h} = a_{1-\alpha}^2. \tag{10.67}$$

It can be easily seen that the matrix $\mathbf{D}_\lambda^{-1}$ is the covariance matrix for the transformed parameter vector $\mathbf{h}$. Since $\mathbf{D}_\lambda^{-1}$ has no off-diagonal entries, it implies that the new parameters are *uncorrelated* and correspond to each axis of the ellipsoid.
The confidence region in n-dimensional parameter space is illustrated in Figure 10.5 for the case where $n = 2$ ($\mathbf{p}$ becomes a 2-parameter vector). Then

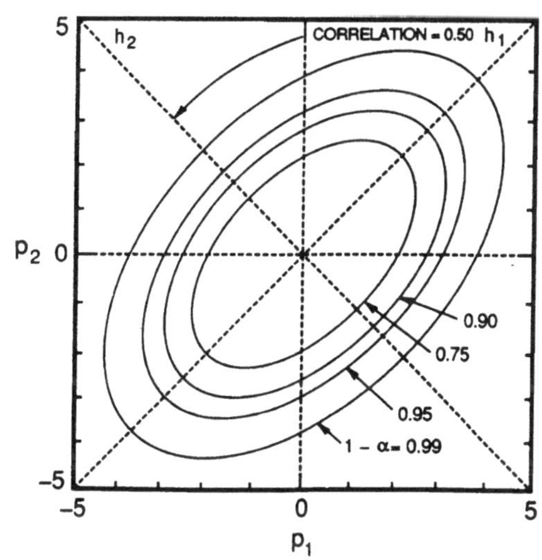

Fig. 10.5 Concentric elliptical regions of constant probability density for a normal distribution in 2D space. The rotational transformation from the original coordinates (p_1, p_2) to the new coordinates (h_1, h_2) is also indicated

the concentric ellipses correspond to different probability levels for this 2-parameter example. The integral (10.64) over each ellipse represents a fixed probability $(1-\alpha)$, so that larger ellipses correspond to larger probabilities. The rotational transformation (10.66) which relates the original parameters **p** to the new parameters **h** is also illustrated in Figure 10.5. It can be seen that h_1 and h_2 correspond to the major and minor axes, respectively.

The transformed parameters **h** are called the *principle components.* The advantage of dealing with the uncorrelated principle components **h**, rather than the correlated original parameters **p**, is that the new (transformed) parameters become statistically independent, because lack of correlation implies statistical independence. For then we can establish individual confidence interval and statistical tests for each uncorrelated parameter. By introducing another transformation

$$z_j = \sqrt{\lambda_j} h_j \tag{10.68}$$

we get a hypersphere of the following form

$$a^2 = \mathbf{z}^T \mathbf{z}. \tag{10.69}$$

In the z-coordinate system, the probability integral (10.64) reduces to

$$1 - \alpha = \int \cdots \int \frac{1}{(2\pi)^{n/2}} \exp\left[-\frac{a^2}{2} \right] dz_1 \, dz_2 \cdots dz_n. \tag{10.70}$$

The above integral can be reduced to the following standard form [33]

$$1 - \alpha = \frac{1}{\Gamma(n/2)} \int_{t=0}^{t=(1/2)a_{1-\alpha}^2} e^{-t} t^{n/2-1} dt \tag{10.71}$$

where $\Gamma(\cdot)$ represents the gamma function and $t = \frac{1}{2}a^2$. The above expression is the integral of the well known Chi-Squared probability density function. For a specified probability and number of degrees of freedom, i.e., given n and α, the integral equation (10.71) can be solved iteratively for $a_{1-\alpha}$ (see Chapter 6 of [22]). From the above value of $a_{1-\alpha}$, we can obtain the end points of each axis of the ellipsoid in terms of the new coordinates h_j:

$$\begin{array}{llll} h_1 = \pm \dfrac{a_{1-\alpha}}{\sqrt{\lambda_1}} & h_2 = 0 & \cdots \quad h_n = 0 & \text{(major axis)} \\ h_1 = 0 & h_2 = \pm \dfrac{a_{1-\alpha}}{\sqrt{\lambda_2}} & \cdots \quad h_n = 0 & \\ \vdots & & & \\ h_1 = 0 & h_2 = 0 & \cdots \quad h_n = \pm \dfrac{a_{1-\alpha}}{\sqrt{\lambda_n}}. & \end{array} \tag{10.72}$$

Consequently, individual confidence intervals can be obtained for the *uncorrelated* parameters **h**. The linear transformation Eq. (10.66) which relates h_j to the original parameters p_j is of the following form

$$h_j = e_{j1}(p_1 - p_{t1}) + e_{j2}(p_2 - p_{t2}) + \cdots + e_{jn}(p_n - p_{tn}) \tag{10.73}$$

where $e_{j1},\ldots,e_{jn}$ are the components of the jth eigenvector and $\sqrt{\sum_{q=1}^{n} e_{jq}^2} = 1$. Of particular interest is the major axis h_1, which has the largest confidence interval $\pm\frac{a_{1-\alpha}}{\sqrt{\lambda_1}}$. The dominant components of h_1 can be obtained by isolating all parameters p_j for which the magnitude of e_{1q} are high ($\geq 100\%$). These parameters (with high e_{1q}) constitute a redundant combination and must be eliminated from the extraction process.

This way we can determine the parameter combinations which are redundant. By either modifying the model equations or fixing some of these parameters to physically meaningful default values, a reduced parameter set can be re-extracted from the data. The reduced set is usually well-determined and has significantly smaller confidence regions.

The algorithm described above can easily be implemented in any nonlinear optimization program and has been used extensively for MOSFET model development and parameter extraction [35].

10.5.1 Examples of Redundant Parameters

The advantages of implementing the confidence region algorithm in the nonlinear optimization program are now discussed by taking examples from MOSFET device modeling. The first example considered is the SPICE MOS Level 2 model [30], where, surface mobility degradation is modeled as (see section 11.3.1)

$$\mu_s = \begin{cases} \mu_0(U_{crit}/\mathscr{E}_s)^{U_e} & \text{if } \mathscr{E}_s \geq U_{crit} \\ \mu_0 & \mathscr{E}_s < U_{crit} \end{cases} \tag{10.74}$$

where the transverse field $\mathscr{E}_s$ is expressed as

$$\mathscr{E}_s = \frac{C_{ox}}{\epsilon_0 \epsilon_{si}}[V_{gs} - V_{th} - U_t V_{ds}]. \tag{10.75}$$

Here μ_0, U_{crit}, U_e and U_t are the fitting parameters; the corresponding SPICE parameter names are UO, UCRIT, UTRA and UEXP, respectively. These parameters are generally extracted along with other MOS Level 2 parameters from $I_{ds} - V_{ds}$ data at small V_{ds} for several devices. The parameter values shown in Table 10.2 were obtained using $I_{ds} - V_{ds}$ data at low V_{ds} (data set A) from n-channel device ($W/L = 12.5/1$, $t_{ox} = 152$ Å and

Table 10.2. *Uncertainty estimates for the SPICE Level 2 mobility model parameters*

parameter name	initial set		reduced set	
	value	conf. interval	value	conf. interval
UO (cm^2/V-sec)	339	±3%	343	±2%
UCRIT (10^6 V/cm)	0.405	±10%	0.415	±6%
UEXP	0.476	±13%	0.524	±11%
UTRA	1.28×10^{-3}	±190%	0.0	CONSTANT
h_1	—	±320%	—	±20%

$X_j = 0.15\,\mu$m) fabricated using a typical 1 μm CMOS technology. The individual uncertainties as calculated from Eq. (10.61) are shown for these parameters, along with the uncertainty estimate for the transformed parameter h_1.

Note that the uncertainty for the parameter U_t is high as also for h_1, which has a coefficient of 0.992 along the U_t axis, indicating that it is an ill-determined parameter. Hence, we can eliminate U_t and use the following model equation for the transverse field

$$\mathscr{E}_s = \frac{C_{ox}}{\epsilon_{si}}(V_{gs} - V_{th}). \tag{10.76}$$

The parameters were re-extracted using Eqs. (10.74) and (10.76). The new parameter values (reduced set) are also shown in Table 10.2 along with the corresponding uncertainty estimates. It can be seen that the parameters in the reduced set are well determined with small uncertainties. Moreover, the average fit error of 5% did not increase when Eq. (10.76) was used in place of Eq. (10.75). Physically, this makes sense because data that was used to extract the parameter U_t is almost independent of V_{ds}.

The second example deals with the development of a model for the temperature dependence of the ionization coefficient α which for the holes and electrons can be expressed as

$$\alpha = A_i \exp\left(-\frac{B_i}{\mathscr{E}}\right) \tag{10.77}$$

where $\mathscr{E}$ is the electric field and A_i and B_i are the ionization parameters. The temperature dependence of the ionization rate has been generally modeled as [37]

$$\begin{aligned} A_i &= A_{io}[1 + \gamma_{A_i}(T - 300)] \\ B_i &= B_{io}[1 + \gamma_{B_i}(T - 300)] \end{aligned} \tag{10.78}$$

where A_{io} and B_{io} are the ionization parameter values at 300 K and γ_{A_i}

Table 10.3. *Electron ionization rate parameter uncertainties*

parameter name	initial set		reduced set	
	value	conf. interval	value	conf. interval
A_{io} (cm^{-1})	1.13×10^6	$\pm 4\%$	1.17×10^6	$\pm 3\%$
B_{io} (V/cm)	1.72×10^6	$\pm 1\%$	1.74×10^6	$\pm 1\%$
γ_{A_t} (K^{-1})	7.97×10^{-4}	$\pm 72\%$	0.0	CONSTANT
γ_{B_t} (K^{-1})	8.30×10^{-4}	$\pm 16\%$	6.69×10^{-4}	$\pm 6\%$
h_1	—	$\pm 140\%$	—	$\pm 17\%$

and γ_{B_t} are the temperature coefficients of A_i and B_i, respectively. While the confidence interval approach gives low uncertainties for all the extracted parameters ($A_{io}, B_{io}, \gamma_{A_t}$ and γ_{B_t}), we find that the major axis h_1 with dominant components along γ_{A_t} and γ_{B_t} is ill-determined. The average fit error of 3% did not increase when γ_{A_t} was set to zero and the other parameters were re-extracted. Table 10.3 gives the electron ionization parameter values and uncertainty estimates with and without γ_{A_t} as a fitting parameter. This example clearly illustrates a situation where the confidence interval approach is unable to isolate a redundant parameter combination whereas the confidence region algorithm detects this redundancy.

The value of γ_{B_t} extracted using the new approach is consistent with that determined experimentally by Grant [36]. This result implies that the ionization parameter A_i remains relatively constant, with the major variation with temperature appearing in the exponent. Similar results were obtained for the hole ionization parameters. From the confidence region results, we get the following simplified impact ionization equation:

$$\alpha = A_{io} \exp\left\{ -\frac{B_{io}[1 + \gamma_{B_t}(T - 300)]}{\mathscr{E}} \right\}. \tag{10.79}$$

The experimental and modeled electron ionization rates are compared in Figure 10.6. It can be seen that the model equation reproduces the data accurately over the entire temperature range of interest. The ionization rate Eq. (10.79) has been used to develop a temperature dependent MOSFET substrate current model in the range 273–400 K [38].

As a last example, we have considered a circuit level model for the drain-induced barrier lowering (DIBL) effect (cf. section 5.4). The most widely used circuit level model for DIBL is [cf. Eq. (5.96)]

$$V_{th}(V_{ds}) = V_{th} - \sigma V_{ds} \tag{10.80}$$

where V_{th} is the threshold voltage at low V_{ds} and σ is the DIBL parameter. Recall that this parameter depends on the effective channel length, oxide thickness, junction depth, channel doping profile and substrate bias. For a given oxide thickness, junction depth and channel doping, the following

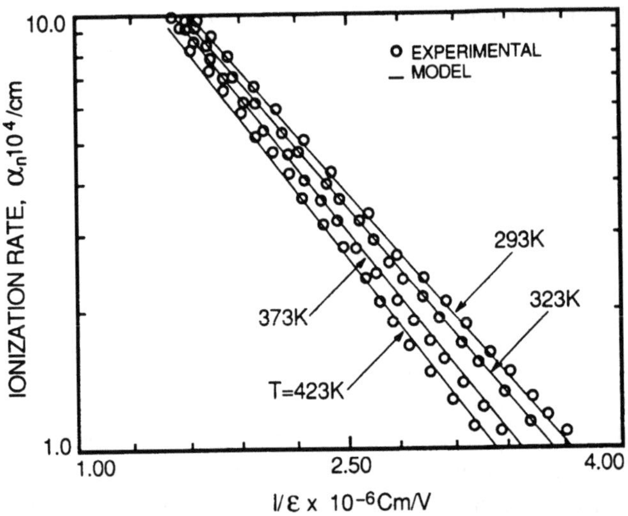

Fig. 10.6 Experimental and modeled electron ionization rate with $T = 273$ K, 323 K, 373 K and 423 K. Circles (o) represent data while curves were generated from model Eq. (10.79)

empirical expression for σ is used [cf. Eq. (5.107)]

$$\sigma = \frac{\epsilon_0 \epsilon_{si}(\sigma_0 + \sigma_1 V_{sb})}{C_{ox} L^m}. \tag{10.81}$$

We can extract model parameters σ_0, σ_1 and m from $I_{ds} - V_{ds}$ data for different channel length devices. Table 10.4 shows the confidence region results from parameter extraction done on a typical 1 μm process using n-channel enhancement type MOSFET's with $W_m/L_m = 12.5/1$, 12.5/1.5, 12.5/2 and 12.5/3. The individual uncertainties from Eq. (10.61) indicate that both σ_0 and the exponent m are ill-determined. However, this does not provide any information on the correlation between the parameters. The transformed parameter h_1 also has a large uncertainty and has components along both the σ_0 and m directions. Hence, from the confidence region result we can conclude that the (σ_0, m) pair is ill-determined. As can be seen from Table 10.4, the parameter uncertainties are greatly reduced when m

Table 10.4. *DIBL model parameter uncertainties*

parameter name	initial set		reduced set	
	value	conf. interval	value	conf. interval
σ_0	7.12×10^{-3}	±276%	0.4255	±15%
σ_1	2.39×10^{-3}	±20%	0.1541	±23%
m	1.419	±278%	1.0	CONST.
h_1	—	±544%	—	±32%

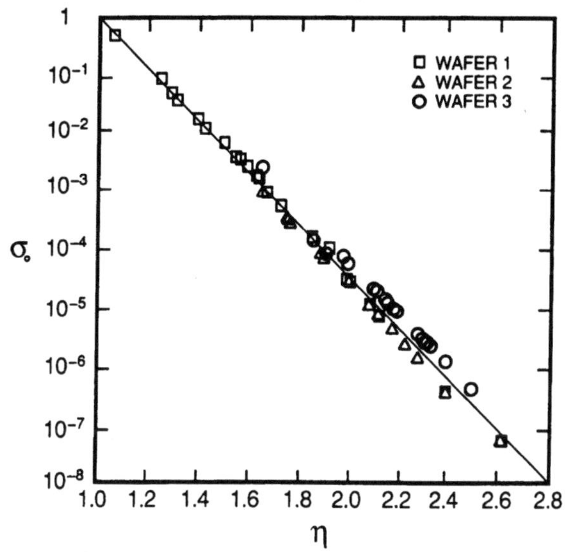

Fig. 10.7 DIBL parameter σ_0 versus m obtained from $I_{ds} - V_{ds}$ data (30 dice from 3 different wafers). The points represent extracted (σ_0, m) values corresponding to each die location. Different symbols correspond to data from different wafers

is held constant and the remaining parameters are re-extracted. The average fit error shows only a small increase from 2.2% to 2.4% when the reduced parameter set is used.

In order to confirm that indeed there is a correlation between σ_0 and m, these parameters were extracted from the $I_{ds} - V_{ds}$ data ($V_{sb} = 0$) measured at 30 die locations corresponding to 3 different wafers [12]. The $\log \sigma_0$ vs. m plot in Figure 10.7 clearly shows the strong correlation between the two parameters and confirms the confidence region results. These results imply that m must be set to an appropriate constant value for a given technology in order to get statistically meaningful parameter values.

10.6 Parameter Extraction Using Optimizer

The general purpose parameter extraction program discussed in the previous sections generally requires the following input files:

- *Estimate File*: This file gives initial values and the box constraints (minimum and maximum values) for the parameters to be extracted. Parameters not to be extracted (like C_{ox}) are also assigned values.
- *Data File*: This file contains data which is used to fit the model equations whose parameters are being extracted. For example, for DC model parameters we need I_{ds} as a function of V_{ds}, V_{gs} and V_{bs} for different

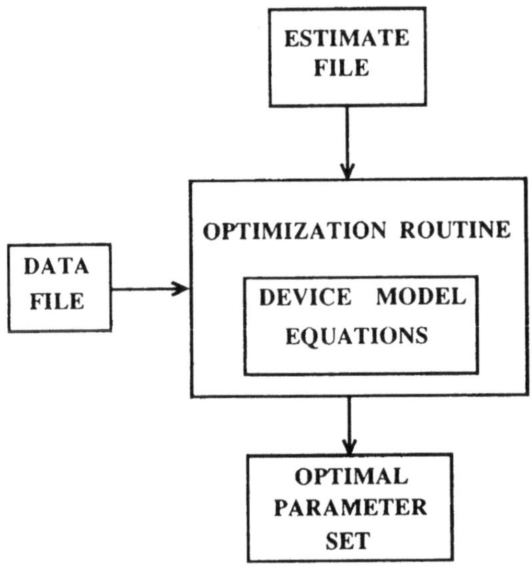

Fig. 10.8 Flow diagram for parameter extraction system

length and width devices. A simple check is normally imposed on the data which enables us to eliminate bad device measurement data. This check can easily be achieved by ensuring that I_{ds} is monotonically increasing function of bias voltages. Often this test is made outside the optimizer. The data file could consists of many segments of many different files each with different subvectors of parameters fitted.

- *Strategy File*: This file allows the user to have complete control over the extraction strategy by specifying which parameter should be optimized with respect to each data set. Normally, a flag is set to one for the parameter to be optimized and it is set to zero otherwise. In fact one need not have a separate strategy file since it could easily be combined with the estimate file as is done in SIMPAR [9] and OPTIMA [12].

The program returns the optimized parameter vector **p** and rms values of the residuals, i.e., error between the measured and calculated data points. Depending upon the optimizer, information like the percentage uncertainty in the parameter [11],[35] are also reported in the separate output file. The flow diagram for the parameter extraction is shown in Figure 10.8.

10.6.1 Drain Current Model Parameter Extraction

A large number of DC models have been described in this book. The methodology described below for extracting the model parameter values using nonlinear optimization method (optimizer) is the same irrespective of the model equations. In fact, the methodology is more important than the optimizer. Incorrect methodology can produce non-physical parameter values. Generally the model parameters, of a given model, are extracted such that a single set of parameters would give the best possible fit over a range of different device geometries. To achieve this goal, three types of $I-V$ data sets for different size transistors are required as was discussed in section 9.1. These data sets correspond to three different regions of device operation; the linear region $I_{ds} - V_{gs}$ data (data set A), the linear and saturation region $I_{ds} - V_{ds}$ data (data set B) and the subthreshold region (data set C). From these three data sets an accurate estimate of the model parameters are first determined using the linear regression method. These estimates then serve as guess values for the optimizer. A typical strategy would consist of the following steps:

1. Initially the data set A is used to determine the model parameters ΔL and ΔW (which determine the effective device dimensions) and the parasitic source/drain resistance R_t using the linear regression methods.
2. The threshold voltage related parameters, such as V_{fb}, N_b, and γ are then determined from a large reference device using V_{th} versus V_{bs} data that is obtained from data set A.
3. The initial estimate of the low field mobility μ_0 is obtained from the maximum slope of the I_{ds} versus V_{gs} curve at zero V_{bs} (data set A). The I_{ds} values at higher V_{gs} yield the mobility degradation parameter θ.
4. The drain current characteristics (data set B) is used to estimate parameters related to velocity saturation and channel length modulation. The I_{ds} at high V_{ds} is used to make sure that the device is operating in the saturation region. In practice, this step is carried out using the optimizer rather than linear regression methods.
5. After obtaining good initial estimates of the parameters in different regions of device operation, the optimizer is then used to calculate all length and width dependent V_{th} model parameters like G_l, G_w etc. from V_{th} vs V_{sb} data for different length and width devices. These parameters are then frozen, while determining all other parameters from $I_{ds} - V_{ds}$ data in the linear and saturation regions (data set B). Next we extract parameters related with the subthreshold region using data set C.

It should be emphasized that the procedure outlined above is the one commonly used, but is not necessarily the only approach.

10.6.2 MOSFET AC Model Parameter Extraction

The MOSFET AC models consists of basically the intrinsic and extrinsic capacitances. Since the intrinsic capacitances are derived from charges which are used to derive drain current, these capacitances do not contain any new parameters. However, extrinsic capacitance (source/drain junction capacitances) model parameters are almost always determined using optimization method as discussed in section 11.1.

References

[1] G. T. Wright and H. M. A. Gaffur, 'Preprocessor modeling of parameter and geometry dependence of short and narrow MOSFET for VLSI circuit simulation, optimization, and statistics with SPICE', IEEE Trans. Electron Devices, ED-32, pp. 1240–1245 (1985).

[2] M. F. Hamer, 'First-order parameter extraction on enhancement silicon MOS transistors', IEE Proc., 133, Pt. I, pp. 49–54 (1986).

[3] A. B. Bhattacharya, P. Ratnam, D. Nagchoudhuri, and S. C. Rustagi, 'On-line extraction of model parameters of a long buried-channel MOSFET', IEEE Trans. Electron Devices, ED-32, pp. 545–550 (1983).

[4] D. E. Ward and K. Doganis, 'Optimized extraction of MOS model parameters', IEEE Trans. Computer-Aided Design, CAD-1, pp. 163–168 (1982).

[5] K. Doganis and D. L. Scharfetter, 'General optimization and extraction of IC device model parameters', IEEE Trans. Electron Devices, ED-30, pp. 1219–1228 (1983).

[6] P. Yang and P. K. Chatterjee, 'An optimal parameter extraction program for MOSFET models', IEEE Trans. Electron Devices, ED-30, pp. 1214–1219 (1983).

[7] E. Khalily, P. H. Decher, and D. A. Teegarden, 'TECAP2: An interactive device characterization and model development system', *Tech. Digest*, IEEE Int. Conf. on Computer-Aided Design, ICCAD-84, pp. 184–151 (1984).

[8] S. J. Wang, J. Y. Lee, and C. Y. Chang, 'An efficient and reliable approach for semiconductor device parameter extraction', IEEE Trans. Computer-Aided Design, CAD-6, pp. 170–178 (1986).

[9] W. Maes, K. M. De Meyer, and L. H. Dupas, 'SIMPAR: A versatile technology independent parameter extraction program using new optimized fit stragegy', IEEE Trans. Computer-Aided Design, CAD-5, pp. 320–325 (1986).

[10] M. Sugimoto, 'General-purpose model parameter extraction program with initial value exploration technique', *Tech. Digest*, IEEE Custom Integrated Circuits Conference, CICC-86, pp. 624–627 (1986).

[11] B. Ankele, W. Holzl, and P. O'Levy, 'Enhanced MOS parameter extraction and SPICE modeling', Proc. IEEE Int. Conf. on Microelectronic Test Structures, Vol. 2, No. 1, pp. 73–78, March 1989.

[12] M. Sharma and N. D. Arora, 'OPTIMA: A Nonlinear Model Parameter Extraction Program with Statistical Confidence Region Algorithms', IEEE Trans. Computer-Aided Design, CAD-12, May 1993.

[13] P. Conway, C. Cahill, W. A. Lane, and S. U. Lindholm, 'Extraction of MOSFET parameters using the Simplex direct search optimization method', IEEE Trans. Computer-Aided Design, CAD-4, pp. 694–698 (1985).

[14] P. E. Gill, W. Murray, and M. H. Wright, *Practical optimization*, Academic Press, New York, 1981.

[15] J. E. Dennis Jr. and R. B. Schnabel, *Numerical Methods for Optimization and Nonlinear Equations*, Prentice-Hall, New York, 1983.

[16] D. G. Luenberger, *Linear and Nonlinear Programming*, 2nd Ed., Addison-Wesley, Reading, Mass., 1984.

[17] R. Fletcher, *Practical Methods of Optimization*, 2nd Ed., John Wiley & Sons, New York, 1987.

[18] A. L. Peressini, F. E. Sullivan, and J. J. Uhl jr., *The Mathematics of Nonlinear Programming*, Springer-Verlag, New York, 1988.
[19] F. A. Lootsma (Ed.), *Numerical Methods for Nonlinear Optimization*, Academic Press, London, 1972.
[20] L. C. W. Dixon and G. P. Szego (Eds.), *Towards Global Optimization*, North-Holland, Amsterdam, 1979.
[21] J. L. Kuester and J. H. Mize, *Optimization Techniques with Fortran*, McGraw-Hill, New York, 1973.
[22] W. H. Press, B. P. Flannery, S. A. Teukolky, and W. T. Vetterling, *Numerical Recipes: The Art of Scientific Computing*, Cambridge University Press, Cambridge, London 1986. Also by the same authors: *Numerical Recipes in C*, Cambridge University Press, 1988.
[23] Subroutines from Harwell Library, Computer Science and System Division, ARE Harwell, Oxfordshire, OX11 0RA, England.
[24] J. J. More, B. S. Garbow, and K. E. Hillstrom, '*User Guide for MINPACK-I*', Applied Mathematics Division, Argonne National Laboratory, ANL-80-74, Illinois, USA.
[25] Subroutines from International Mathematical Statistical Library, IMSL, 7500 Bellaire Blvd., Houston, Texas, USA.
[26] D. W. Marquardt, 'An algorithm for least-square estimates of non-linear parameters', SIAM Journal, 11, pp. 431–441 (1963).
[27] A. Fletcher, 'A modified Marquardt subroutine for nonlinear least squares', Harwell Report, AERE R.6799, May 1971.
[28] J. More, 'The Levenberg-Marquardt Algorithm: Implementation and Theory', *Numerical Analysis: Seventh Biannual Conference at the University of Dundee, Scotland*, Springer-Verlag, New York, 1977.
[29] R. K. Brayton and R. Spence, *Sensitivity and Optimization*, Elsevier, Amsterdam, 1980.
[30] A. Vladimirescu and S. Liu, 'The simulation of MOS integrated circuits using SPICE2', UCB/ERL Memo M80/7, Univ. of California, Berkeley, 1980.
[31] P. R. Bevington, *Data Reduction and Error Analysis for the Physical Sciences*, McGraw Hill Book Co., New York, 1967.
[32] Y. Bard, *Nonlinear Parameter Estimation*, Academic Press Inc., New York, 1974.
[33] J. V. Beck and K. J. Arnold, *Parameter Estimation in Engineering and Science*, John Wiley & Sons, New York, 1977.
[34] C. F. Machala, P. C. Pattnaik, and P. Yang, 'An efficient algorithm for the extraction of parameters with high confidence from nonlinear models', IEEE Electron Device Lett., EDL-7, pp. 214–218 (1986).
[35] M. Sharma and N. D. Arora, 'A statistical confidence region algorithm for model parameter extraction', *Proc NASECODE VII Conf.*, pp. 41–42, Front Range Press, Boulder, CO, 1991.
[36] W. N. Grant, 'Electron and hole ionization rates in epitaxial silicon at high electric fields', Solid-State Electron., Vol. 16, pp. 1189–1203 (1973).
[37] Y. Okuto and C. R. Crowell, 'Threshold energy effect on avalanche breakdown voltage in semiconductor junctions', Solid-State Electron., Vol. 18, pp. 161–168 (1975).
[38] N. D. Arora and M. Sharma, 'MOSFET substrate current model for circuit simulation', IEEE Trans. Electron Devices, Vol. ED-38, pp. 1392–1398 (1991).

11 SPICE Diode and MOSFET Models and Their Parameters

In this chapter we will discuss the *pn* junction diode and MOSFET models, as implemented in Berkeley SPICE2G and higher versions. No attempt will be made to derive the model equations, as that has already been done at appropriate places in previous chapters. Here we will only describe equations used to model different regions of device operation. Emphasis will be on model parameters required to run SPICE and how to measure them.

Berkeley SPICE has four different MOSFET models of varying complexity and accuracy [1]–[3]. These are (1) the Level 1 model—a first order model suitable only for long channel devices; (2) the Level 2 model that includes various second order effects present in small geometry devices, and is considered to be a physical model; (3) the Level 3 model—a semi-empirical model that includes most of the second order effects described in the Level 2 model; (4) the Level 4 model, called the BSIM (Berkeley Short-channel Igfet Model), that is a parameter based model. These different models can be activated by a parameter called LEVEL. We will describe all four levels of MOSFET model equations and their parameters. However, first we will describe the diode model parameters and how to determine them.

11.1 Diode Model

The SPICE diode model has been discussed in detail in section 2.9. Table 11.1 shows model parameters that determine both DC and AC characteristics of a diode.

Out of these ten parameters, the first seven (I_s, η, r_s, C_{j0}, ϕ, m and τ) are determined from diode drain current and capacitance measurements. The remaining three parameters are often not measured and default values are generally assumed for silicon *pn* junction diodes. For other type of diodes such as SBD (Schotkey Barier Diode), parameter XTI needs to be changed. In what follows we will discuss extraction for the first seven parameters.

Table 11.1. *SPICE Diode model parameters*

Parameter name in the text	SPICE parameter name	Parameter description	Default value	Units
I_s	*IS*	saturation current	$1 \cdot 10^{-14}$	A
η	*XN*	emission coefficient	1	—
r_s	*RS*	series resistance	0	Ω
C_{j0}	*CJO*	zero-bias junction capacitance	0	F
ϕ	*PB*	*pn* junction potential	1.0	V
m	*MJ*	*pn* grading coefficient	0.5	—
τ	*TT*	transit time	0	sec
V_{br}	*BV*	reverse breakdown voltage	infinite	V
E_g	*EG*	band-gap voltage	1.1	eV
p	*XTI*	IS temperature exponent	3.0	—

These parameters are entered in the MODEL statement in the SPICE input file.

Recall that SPICE calculates the diode current I_d using the following equation [cf. Eq. (2.82)]

$$I_d = I_s\left[\exp\left(\frac{V_d - I_d r_s}{\eta V_t}\right) - 1\right]$$

which after rearranging in terms of V_d (voltage across the diode) becomes

$$V_d = \eta V_t \ln\left(1 + \frac{I_d}{I_s}\right) + I_d r_s \tag{11.1}$$

where I_s, r_s and η are model parameters that can be determined either using linear regression methods, as discussed in section 9.14 or a nonlinear optimization method (cf. Chapter 10). In the latter case we fit the experimental I_d versus V_d data to model equation (11.1) such that

$$\text{error} = \sqrt{\sum_{i=1}^{l}\left[\frac{V_{\text{cal}}(i) - V_{\text{exp}}(i)}{V_{\text{exp}}(i)}\right]^2} \tag{11.2}$$

is minimum, where V_{exp} and V_{cal} are the measured and calculated V_d, respectively, and l is the number of data points. The result of this curve fitting is shown in Figure 11.1 for a typical n^+p diode fabricated using a 1 μm CMOS process. The values for the parameter I_s, η and r_s for two types of diodes (n^+p and p^+n) are shown in Table 11.2. For comparison the parameters obtained using the linear regression method (cf. section 10.14) are also shown in this table.

Note that extracted parameter values from two different methods are not exactly the same. However, for circuit simulation purposes, the parameter

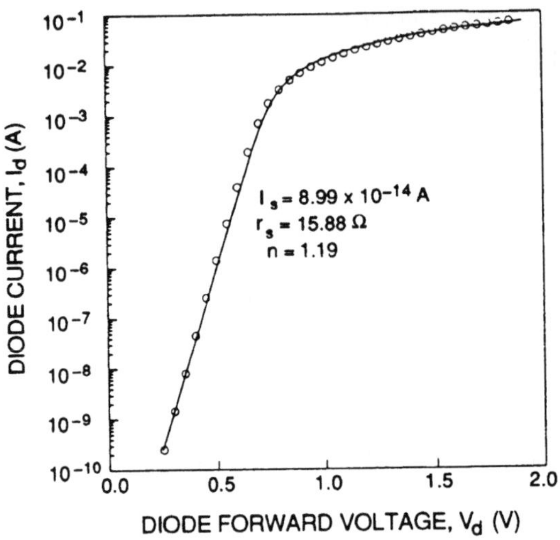

Fig. 11.1 Plot of $\log(I_d)$ versus V_d for a n^+p diode. Circles are experimental points while continuous line is nonlinear least-square fit to Eq. (11.1)

Table 11.2. *Diode parameters I_s, n and R_s*

	n^+p		p^+n	
	Linear regression	Optimization method	Linear regression	Optimization method
I_s	4.53×10^{-14} A	8.99×10^{-14} A	4.1×10^{-12} A	4.05×10^{-12} A
η	1.119	1.19	1.335	1.346
R_s	11.03 Ω	15.88 Ω	10.78 Ω	14.27 Ω

set obtained using the optimization method is more appropriate, as these values are obtained by fitting over all portion of the curve in the current range of interest. Unlike the linear regression method, the optimization method yields all three parameters simultaneously.

The parameters C_{j0}, ϕ and m describe the junction capacitance due to the space charge in the junction depletion region. When the junction reverse voltage V_d is less than $\phi/2$, the junction capacitance C_j is given by the following equation [cf. Eq. (2.74)]

$$C_j = \frac{C_{j0}}{(1 - V_d/\phi)^m} \quad (\text{F/cm}^2) \tag{11.3}$$

where C_{j0} varies from device to device, but is typically of the order of 1.0×10^{-4} pF/μm^2. The barrier potential ϕ is usually about 0.5–0.7 V and the gradient factor m is assumed to be between 0.333 (linearly graded

junction) and 0.5 (abrupt junction), although values outside this range are not uncommon.
The parameters ϕ and m are generally determined by curve fitting Eq. (11.3) with measured data using a nonlinear least square optimization program. Very often, C_{jo} is also treated as a parameter to be optimized along with ϕ and m rather than taking its value from measured data. This is because 3 parameters (C_{jo}, ϕ and m) when optimized together give better fit over the entire data range of interest (see Figure 2.19 and Table 9.4).

Transient Time τ_t. The parameter τ_t is the diode transit time and is used to calculate the diode diffusion capacitance C_{df} [cf. Eq. (2.77)] when the diode is forward biased. Typical values of τ_t range from 1 to 100 nsec. There are different electrical methods to calculate transit time τ_t, like the voltage decay method, the reverse recovery method, etc [4]. However, the simplest method of obtaining τ_t is to compute it from the reverse recovery method. In this method, we measure the diode storage time t_s by switching the diode from a forward voltage V_f to a reverse voltage V_r, and using the following equation [4]–[6]

$$\operatorname{erf}\sqrt{\frac{t_s}{\tau_t}} = \frac{1}{1 + I_r/I_f} \tag{11.4}$$

where I_f and I_r are the forward and reverse current, respectively, when the diode is switched from the forward voltage V_f to the reverse voltage V_r. Note that this equation requires evaluation of the error function, which is approximately given by [4]

$$\begin{aligned}\operatorname{erf}(x) &= \frac{2}{\sqrt{\pi}}\int_0^x \exp(-z^2)dz \\ &\approx 1 - \left[\frac{0.34802}{1 + 0.4704x} - \frac{0.095879}{(1 + 0.4704x)^2} + \frac{0.74785}{(1 + 0.4704x)^3}\right]e^{-x^2}.\end{aligned} \tag{11.5}$$

Due to the complexity of Eq. (11.4), the Newton–Raphson method is needed to compute τ_t and is thus fairly involved. However, the following simple equation is often used to calculate τ_t

$$t_s = \tau_t\left[\ln\left(1 + \frac{I_f}{I_r}\right)\right]. \tag{11.6}$$

As shown in Figure 11.2, there is a discrepancy of 30% between the τ_t calculated using Eqs. (11.4) and (11.6) even when $I_f \gg I_r$. Therefore, it is advisable to use Eq. (11.4). While using Eq. (11.6), it has been suggested that $I_f \gg I_r$ must be kept in the measurements. This way, the affect of

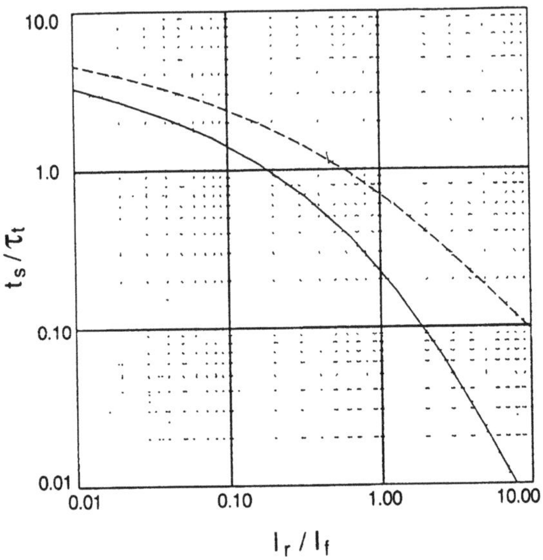

Fig. 11.2 Plot of t_s/τ_t versus I_f/I_r using Eq. (11.4) (continuous line) and (11.6) (dotted line). Continuous line predicts more exact value of τ_t

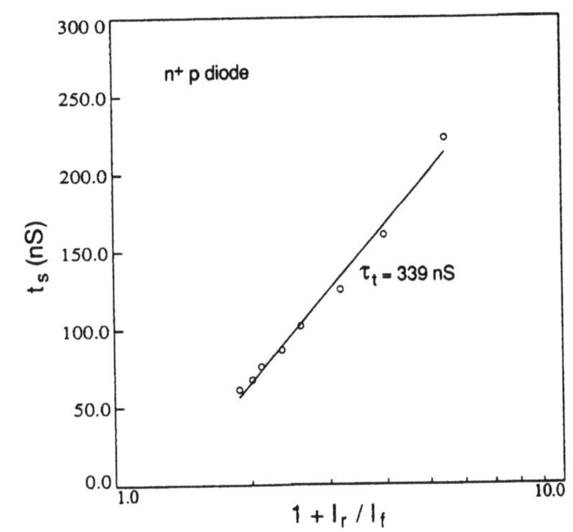

Fig. 11.3 Plot of t_s versus $(1 + I_f/I_r)$ for a n^+p diode using Eq. (11.6). Circles are experimental points while continuous line is linear regression of Eq. (11.6)

recombination in the heavily doped region is entirely eliminated [4]. Under these conditions, the plot of t_s versus $\ln(1 + I_f/I_r)$ will be a straight line (see Figure 11.3) the slope of which gives τ_t. The plot will be highly curved if the condition $I_f \gg I_r$ is not met and then a unique value of lifetime can no longer be extracted.

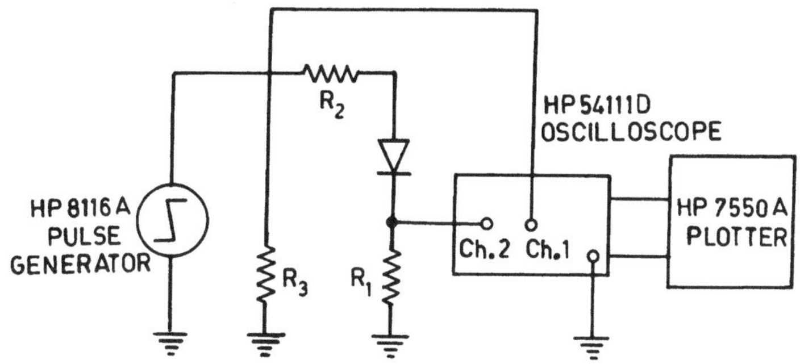

Fig. 11.4 Test setup for measuring storage time using reverse recovery method

Equipment required for measuring τ_t are (1) a fast pulse generator such as an HP8116A, (2) a fast oscilloscope, such as a Tektronix 7854 or an HP54111D with dual trace plug-in, and (3) an X-Y recorder (optional). The advantage of the HP8116A function generator is that it can supply an asymmetric pulse waveforms. However, if not available, two pulse generators are needed to adjust the voltages V_f and V_r independently. The test configuration is shown in Figure 11.4. The time delay due to connectors and series resistance in the circuit should be carefully minimized. The resistor R_2 (350 Ω) is chosen such that the DC current flowing into the diode is limited within the range of ± 15 mA for voltages between $V_f = 8$ V (forward bias) and $V_r = -3$ V (reverse bias) and the RC delay time introduced by this resistor is negligible as compared to the diode transit time.
During forward bias (at $t = 0^-$), a positive voltage ($V_f = 8$ V) at $f = 100$ Hz was applied to the circuit. The current was then calculated by dividing the

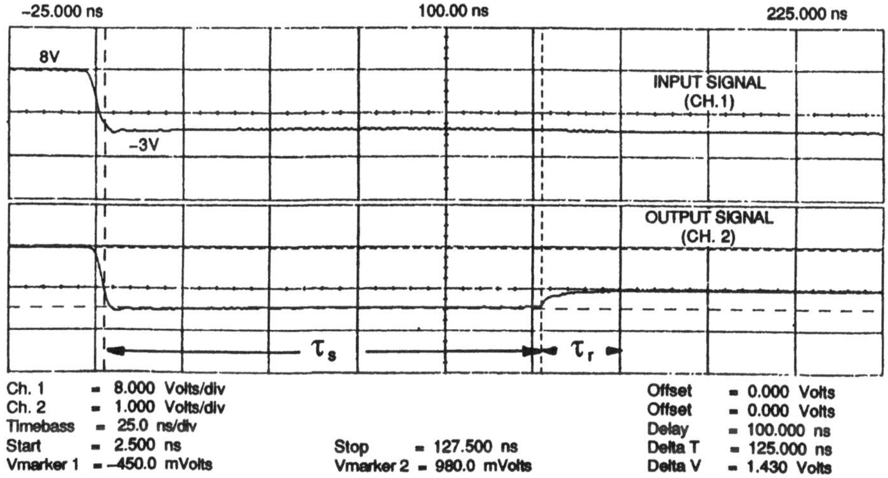

Fig. 11.5 Storage time t_s as a function of input pulse for n^+p diode

Table 11.3. *Diode transit time* τ_t

Transit time calculation using	n^+p nS	p^+n nS
Error function Eq. (11.4)	247.3	475
Log function Eq. (11.6)	339	—

voltage measured on resistor R_1 (50 Ω). At $t = 0$, a negative voltage is applied to the diode; the input pulse changes from $+8$ V to -3 V at $t = 0$. The diode storage time was measured as the time from beginning of the reverse current transition to the time when the reverse current begins to decay toward its leakage current value (see Figure 11.5).
The lifetime τ calculated using the above method for both n^+p and p^+n diodes are shown in Table 11.3. Note the difference between τ_t calculated using Eqs. (11.4) and (11.6).

11.2 MOSFET Level 1 Model

The level 1 model is often referred to as the Shichman–Hodges model. It is the simplest of the four MOSFET models in SPICE and is *accurate only for long channel devices.*

11.2.1 DC Model

The threshold voltage V_{th} for the SPICE Level 1 model is [cf. Eq. (5.16)]

$$V_{th} = V_{TO} + \gamma(\sqrt{2\phi_f + V_{sb}} - \sqrt{2\phi_f}) \tag{11.7}$$

where V_{TO} is the zero-bias ($V_{sb} = 0$ V) threshold voltage of a long channel device, γ is the body factor, and ϕ_f is the bulk Fermi potential. Note that no short channel or narrow width effects are taken into account; for details see section 5.1.
The saturation voltage V_{dsat} is calculated using the following equation [cf. Eq. (6.54)]

$$V_{dsat} = V_{gs} - V_{th}. \tag{11.8}$$

The drain current I_{ds} is calculated using the following relations [cf. Eq. (6.62)]

$$I_{ds} = \begin{cases} \beta_0[(V_{gs} - V_{th} - \frac{1}{2}V_{ds})V_{ds}](1 + \lambda V_{ds}) & \text{linear region, } V_{gs} > V_{th} \text{ and } V_{ds} \le V_{dsat} \\ 0.5\beta_0(V_{gs} - V_{th})^2(1 + \lambda V_{ds}) & \text{saturation region, } V_{ds} > V_{dsat} \\ 0 & \text{subthreshold region, } V_{gs} \le V_{th} \end{cases} \tag{11.9}$$

where

$$\beta_0 = \kappa(W/L) \quad \text{and} \quad \kappa = \mu_0 C_{ox}.$$

Note that the channel length modulation factor, λ, is included in both the linear and saturation regions, so as to make the current and its first derivative continuous, as was explained in section 6.4.1. Also note that the subthreshold current is zero.

In addition to the intrinsic MOSFET DC current equations described above, one needs to model the source/drain (S/D)-to-substrate *pn* junctions. Since in the normal operation of the device these junctions are reverse biased, the only DC parameter of the S/D junction which is of interest is the saturation (leakage) current I_s. In SPICE this is specified as J_s, the saturation current per unit area, or I_s, the total saturation current. If J_s is specified then one needs to specify the source and drain areas A_s and A_d, respectively.

11.2.2 Capacitance Model

The parameters of the dynamic model are the source/drain junction capacitances, the overlap capacitances, and the intrinsic MOSFET capacitances. The junction capacitances are the sum of both the bottom-wall (area) capacitance and side-wall (periphery) capacitance. The source diode capacitance C_{BS} is computed as follows [cf. Eq. (3.26)]

$$C_{BS} = \frac{C_{j0}A_s}{\left(1 - \dfrac{V}{\phi_{bi}}\right)^{m_j}} + \frac{C_{jsw0}P_s}{\left(1 - \dfrac{V}{\phi_{bi}}\right)^{m_{jsw}}} \tag{11.10}$$

where A_s and P_s are the area and periphery of the source-to-bulk *pn* junction, respectively, and C_{j0} and C_{jsw0} are the junction capacitance per unit area and per unit periphery, respectively, at zero back bias. A similar equation holds for the drain-to-bulk junction capacitance C_{BD}. These equations are used for all SPICE models.

The intrinsic device capacitances (also sometimes referred to as gate oxide capacitances) are based on the Meyer model (see section 7.1.1). There are only three intrinsic capacitances C_{GS}, C_{GD} and C_{GB} in the Meyer model. Their values change with bias conditions as follows:

Strong Inversion Region. In the strong inversion region when $V_{gs} > V_{th}$, the gate capacitance is calculated using the following relations:

Linear Region: In this case $V_{gs} > (V_{th} + V_{ds})$

$$C_{GS} = \frac{2}{3}C_{oxt}\left[1 - \frac{(V_{gd} - V_{th})^2}{(V_{gd} + V_{gs} - 2V_{th})^2}\right] \tag{11.11a}$$

$$C_{GD} = \frac{2}{3}C_{oxt}\left[1 - \frac{(V_{gs} - V_{th})^2}{(V_{gd} + V_{gs} - 2V_{th})^2}\right] \tag{11.11b}$$

$$C_{GB} = 0. \tag{11.11c}$$

Saturation Region: In this case $V_{th} < V_{gs} < (V_{th} + V_{ds})$

$$C_{GS} = \tfrac{2}{3} C_{oxt} \tag{11.12a}$$

$$C_{GD} = 0 \tag{11.12b}$$

$$C_{GB} = 0 \tag{11.12c}$$

where

$$C_{oxt} = WLC_{ox}. \tag{11.12d}$$

Weak Inversion Region. In SPICE this region, defined as $V_{gs} < V_{th}$, is divided into two parts. For the sake of simplicity the transition between the saturation and weak inversion regions is made linear, resulting in the following equations.
When $(V_{th} - \phi_f) < V_{gs} < V_{th}$,

$$C_{GS} = \frac{2}{3} C_{oxt} \left(\frac{V_{gs} - V_{th}}{\phi_f} + 1 \right) \tag{11.13a}$$

$$C_{GD} = 0 \tag{11.13b}$$

$$C_{GB} = C_{oxt} \left[1 + \frac{4}{\gamma^2} (V_{gb} - V_{fb}) \right]^{-1/2}. \tag{11.13c}$$

When $V_{gs} < (V_{th} - 2\phi_f)$,

$$C_{GS} = 0 \tag{11.14a}$$

$$C_{GD} = 0 \tag{11.14b}$$

$$C_{GB} = C_{oxt}. \tag{11.14c}$$

Note that these capacitances do not require any new parameters.
The overlap capacitances C_{GSO}, C_{GDO} and C_{GBO} are then added to C_{GS}, C_{GD} and C_{GB}, respectively, in different regions of device operation and are calculated from the following equations:

$$C_{GSO} = C_{gso} W \tag{11.15a}$$

$$C_{GDO} = C_{gdo} W \tag{11.15b}$$

$$C_{GBO} = C_{gbo} L. \tag{11.15c}$$

Normally $C_{gso} = C_{gdo}$, the overlap capacitance per unit width at the source and drain ends, respectively. The model parameters for the SPICE Level 1 model are shown in Table 11.4. These parameters are entered in the MODEL statement in the SPICE input file.
In addition to the model parameters shown in Table 11.4, the *device parameters* shown in Table 11.5 are also required. These device parameters

Table 11.4. *SPICE Level 1 model parameters*

Parameter name in the text	SPICE parameter name	Parameter description	Default value	Units
Level			1	
V_{To}	VTO	Zero-bias threshold voltage	0.0	V
κ	KP	Transconductance parameter	$2 \cdot 10^{-5}$	A/V^2
γ	GAMMA	Body factor	0.0	
μ_0	UO	Low field mobility	600	$cm^2/V \cdot S$
$2\phi_f$	PHI	Surface potential in strong inversion	0.1	V
λ	LAMBDA	Channel length modulation factor	0.0	V^{-1}
N_b	NSUB	Substrate Doping	0.0	m^{-3}
t_{ox}	TOX	Gate oxide thickness	10^{-7}	m
N_{ss}	NSS	Surface state density	0.0	cm^{-2}
—	TPG	Type of the gate material	1	—
I_s	IS	Bulk junction saturation current	10^{-14}	A/m^2
J_s	JS	Bulk junction saturation current per sq-meter of the junction area	10^{-14}	A
R_s	RS	Source ohmic resistance	0.0	Ω
R_d	RD	Drain ohmic resistance	0.0	Ω
ρ_s	RSH	Source, Drain diffusion sheet resistance	∞	Ω/□
—	CBS	Zero-bias B-S junction capacitance	0.0	F
—	CBD	Zero-bias B-D junction capacitance	0.0	F
C_{jo}	CJ	Zero-bias bulk junction capacitance per sq-meter of the junction area	0.0	F/m^2
m_j	MJ	Bulk junction bottom grading coefficient	0.5	—
ϕ_{bi}	PB	Bulk junction potential	0.8	V
C_{jswo}	CJSW	Zero-bias bulk junction side-wall capacitance per meter of the junction perimeter	0.0	F/m
m_{jsw}	MJSW	Bulk junction side wall grading coefficient	0.5	—
C_{gso}	CGSO	Gate-source overlap capacitance per meter channel width	0.0	F/m
C_{gdo}	CGDO	Gate-drain overlap capacitance per meter channel width	0.0	F/m
C_{gbo}	CGBO	Gate-bulk overlap capacitance per meter channel length	0.0	F/m
—	KF	Flicker noise coefficient	0.0	—
—	AF	Flicker noise exponent	1	—

are entered in the DEVICE statement in SPICE and are required for all model levels.

Note the following points, which are valid for all SPICE MOSFET models:

- Parameters *VTO* and *GAMMA* are known as the *electrical parameters* while NSUB and TOX are the *process parameters*. For all SPICE models, both kinds of parameters can be entered with the general convention that

Table 11.5. *Device parameters*

Parameter name in the text	SPICE parameter name	Parameter description	Default value	Units
L_m	L	Drawn Channel length (mask dimensions)	10^{-4}	m
W_m	W	Drawn Channel width (mask dimensions)	10^{-4}	m
A_s	AS	Source diffusion area	0.0	m^2
A_d	AD	Drain diffusion area	0.0	m^2
P_s	PS	Perimeter of the source diffusion window	0.0	m
P_d	PD	Perimeter of the drain diffusion window	0.0	m
—	NRS	Number of squares in the source diffusion	1.0	m
—	NRD	Number of squares in the drain diffusion	1.0	—

electrical parameters will always override the value computed from process parameters, if also specified. Thus, if *VTO*, *NSUB* and *TOX* are input, the threshold voltage will assume the value entered as *VTO*, while *GAMMA* will be computed from *NSUB* and *TOX*. Similarly, if *KP* is not specified but *U0* is specified, then *KP* will be computed using either the specified value of *TOX* or its default value, if not specified.

- If *VTO* is not an input parameter then one needs to specify *NSUB*, *TOX* and *TPG*, which are then used to calculate V_{T0} using Eq. (5.15). The last parameter *TPG* denotes the type of the gate and can take any of the following three values

$$TPG = \begin{cases} +1 & \text{for gate type opposite to the substrate} \\ -1 & \text{for gate type same as the substrate} \\ 0 & \text{for aluminum gate} \end{cases} \tag{11.16}$$

and is used to calculate Φ_{ms} and hence $V_{fb}(=\Phi_{ms} - qN_{ss}/C_{ox})$ [cf. Eq. (4.14)], as follows:

$$\Phi_{ms} = \begin{cases} -0.5 - 0.5E_g - 0.5\phi_f & \text{for TPG} = 0 \\ -0.5E_g - 0.5\phi_f & \text{for TPG} = 1 \\ 0.5E_g - 0.5\phi_f & \text{for TPG} = -1 \end{cases} \tag{11.17}$$

where E_g is the energy gap for silicon [cf. Eq. (2.3)].

- SPICE sets all parameters to the default values if negative values are input by the user, with the exception of *VTO*, *TPG* and *NSS*. Thus, if *GAMMA* is specified as a negative value, then SPICE assumes it to be zero, which is the default value.
- For a *p*-channel enhancement and an *n*-channel depletion device *VTO* is negative, while it is positive for *n*-channel enhancement devices. Recall that *p*-channel depletion devices are not fabricated, but if simulated, their *VTO* will be positive.

- The default value of $TOX = 10^{-7}$ m (1000 Å) is valid for the Level 2 and higher level models. If TOX is not specified for LEVEL = 1, then TOX acts as a flag and "turns off" the use of process parameters resulting in the omission of intrinsic capacitance calculations.
- The parameter LAMBDA in the Level 1 model defaults to zero if it is not specified. However, this is not the case in the Level 2 model as we will see later.
- Some parameters in the model may be specified in more than one way. For example, reverse or saturation current of the junction can be specified either as IS or JS. Whereas the first is an absolute value, the second is multiplied by AS and AD to give the saturation current of the source and drain junctions, respectively. However, the advantage of specifying JS is that the resulting value of the saturation current becomes specific to each junction of each transistor; unlike giving IS, which will result in the same value of the saturation current for all source/drain junctions. Similarly, the zero-bias depletion capacitances can be specified by CJ, which is multiplied by AS and AD, and by $CJSW$, which is multiplied by PS and PD specific to each single device. Or, they can be set by CBD and CBS, which are absolute values.
  The parasitic ohmic resistances of the source and drain junctions can be specified either by RD and RS which are the absolute values, or by RSH which is multiplied by NRS and NRD.
- If both IS and JS are specified, IS overrides JS.

Model Parameter Determination. Determination of all Level 1 parameters, except that of κ (KP) and λ (LAMBDA) have been discussed earlier. The parameter LAMBDA is a saturation region parameter and can be determined from the slope of the I_{ds} versus V_{ds} curve in the saturation region ($V_{ds} > V_{dsat}$) by dividing the slope value by the y-intercept. The slope in the saturation region is very small, and therefore care must be exercised in its determination. The parameter KP can be determined either from the slope of the linear region plot of I_{ds} versus V_{gs} at low V_{ds} or from the slope of $\sqrt{I_{ds}}$ versus V_{gs} curve with I_{ds} obtained in the saturation region. For a typical 2 μm CMOS technology, the value of KP obtained from linear region data is 27 $\mu A/V^2$, while the corresponding value obtained in saturation is 22 $\mu A/V^2$. Clearly, the value of KP obtained from the two methods is different because the mobility degradation due to the gate field is not taken into account in this model. Since SPICE allows only one value to be used for both linear and saturation regions, it is more appropriate to use an optimizer to extract KP along with other parameters.

11.3 MOSFET Level 2 Model

The Level 2 model incorporates many of the second order effects for small size devices. It can model a reasonable range of device sizes, but is computationally quite complex.

11.3.1 DC Model

The threshold voltage equation for the SPICE Level 2 model is

$$V_{th} = V_{To} - \gamma\sqrt{2\phi_f} + \gamma F_l\sqrt{2\phi_f + V_{sb}} + F_w(2\phi_f + V_{sb}) \tag{11.18}$$

where F_l is the short channel factor based on Yau's modified model as given by Eq. (5.94) and F_w is the narrow width factor based on a simplified thick field oxide model [cf. Eq. (5.91)] given by

$$F_w = \frac{\pi}{4}\frac{\epsilon_0\epsilon_{si}}{C_{ox}}\frac{G_w}{W}. \tag{11.19}$$

Linear Region Current. The drain current in the linear region is given by

$$\begin{aligned} I_{ds} = \beta_{\text{eff}}[(V_{gs} - V_{th}^* - \tfrac{1}{2}\eta V_{ds})V_{ds} - \tfrac{2}{3}\gamma F_l\{(V_{ds} + 2\phi_f + V_{sb})^{3/2} \\ - (2\phi_f + V_{sb})^{3/2}\}] \end{aligned} \tag{11.20}$$

where

$$\beta_{\text{eff}} = \kappa\frac{\mu_s}{\mu_0}\frac{L}{L_{\text{eff}}} \tag{11.21a}$$

$$V_{th}^* = V_{TO} - \gamma\sqrt{2\phi_f} + F_w(2\phi_f + V_{sb}) \tag{11.21b}$$

$$\mu_s = \mu_0\left[\frac{u_t\epsilon_0\epsilon_{si}}{C_{ox}(V_{gs} - V_{th} - u_t V_{ds})}\right]^v \tag{11.21c}$$

$$\eta = 1 + F_w \tag{11.21d}$$

$$L_{\text{eff}} = L(1 - \lambda V_{ds}) \tag{11.21e}$$

$$L = L_m - 2L_{\text{dif}} \tag{11.21f}$$

and L_m is the drawn channel length, while L_{dif} is the side diffusion [cf. Eq. (3.31)]. Note that the channel length modulation (CLM) factor λ is used for both linear and saturation regions of device operation, so as to make the current and its first derivative continuous from linear to saturation region, as was explained in Chapter 6.

Saturation Voltage. The saturation voltage $V_{d\text{sat}}$ is calculated in one of two ways. If the maximum carrier drift velocity $v_{\max}$ is assumed zero, then $V_{d\text{sat}}$

is calculated using a pinch-off model (i.e., $I_{ds}/V_{ds}=0$ at $V_{ds}=V_{dsat}$, as discussed in section 6.4.1), otherwise it is calculated using the velocity saturation model.

- V_{dsat} *using the pinch-off model*: In this case V_{dsat} is calculated from the following equation:

$$V_{dsat}=\frac{V_{gsx}-V_{th}^*}{\eta}+\frac{1}{2}\left(\frac{\gamma F_l}{\eta}\right)^2\left[1-\left[1+4\left(\frac{\eta}{\gamma F_l}\right)^2 \times\left(\frac{V_{gsx}-V_{th}^*}{\eta}+2\phi_f+V_{sb}\right)\right]^{1/2}\right] \tag{11.22}$$

  where

$$V_{gsx}=\begin{cases}V_{gs} & \text{if} \quad V_{gs}\geq V_{th}\\ V_{th} & \text{if} \quad V_{gs}<V_{th}.\end{cases}$$

- V_{dsat} *using the velocity saturation model*: In this case V_{dsat} is calculated using the Baun and Benking model from the following equation [cf. Eq. (6.173)]

$$v_{max}=\frac{I_{dsat}}{Q_{sat}W} \tag{11.23}$$

where

$$I_{dsat}=\beta_{eff}[(V_{gs}-V_{th}^*-\tfrac{1}{2}\eta V_{dsat})V_{dsat}-\tfrac{2}{3}\gamma F_l\{(V_{dsat}+2\phi_f+V_{sb})^{3/2} -(2\phi_f+V_{sb})^{3/2}\}] \tag{11.24a}$$

$$Q_{sat}=C_{ox}[V_{gs}-V_{th}^*-\eta V_{dsat}-\gamma F_l\sqrt{V_{dsat}+2\phi_f+V_{sb}}] \tag{11.24b}$$

so that

$$v_{max}=\frac{\mu_s[(V_{gs}-V_{th}^*-0.5\eta V_{dsat})V_{dsat}-\frac{2}{3}\gamma F_l\{(V_{dsat}+2\phi_f+V_{sb})^{3/2}-(2\phi_f+V_{sb})^{3/2}\}]}{L_{eff}[V_{gs}-V_{th}^*-\eta V_{dsat}-\gamma F_l\sqrt{V_{dsat}+2\phi_f+V_{sb}}]}. \tag{11.25}$$

Note that in order to solve for V_{dsat} one needs to know L_{eff}. This means that V_{dsat} calculations requires simultaneous solution of two nonlinear Eqs. (11.25) and (11.21e). However, SPICE uses the following closed form solution by making the approximation that $L_{eff}=L$ in Eq. (11.25). With this approximation one can write Eq. (11.25) in a somewhat more manageable form, if the following substitutions are made

$$V_1=\frac{V_{gs}-V_{th}^*}{\eta}+2\phi_f+V_{sb} \tag{11.26a}$$

$$V_2=2\phi_f+V_{sb} \tag{11.26b}$$

$$v = \frac{v_{max} L}{\mu_s} \tag{11.26c}$$

$$X = \sqrt{V_{dsat} + 2\phi_f + V_{sb}}. \tag{11.26d}$$

With this substitution, Eq. (11.25) becomes:

$$v = \frac{(V_1 - \frac{1}{2}V_2 - \frac{1}{2}X^2)(X^2 - V_2) - \frac{2}{3}(\gamma F_l/\eta)(X^3 - V_2^{3/2})}{V_1 - (\gamma F_l/\eta)X - X^2}. \tag{11.27}$$

It is clear that the above equation can be written as a fourth order polynomial equation in X as:

$$X^4 + AX^3 + BX^2 + CX + D = 0 \tag{11.28}$$

where the coefficients A, B, C and D are:

$$A = \frac{4}{3}\frac{\gamma F_l}{\eta}$$

$$B = -2(V_1 + v)$$

$$C = -2\frac{\gamma F_l}{\eta}v$$

$$D = 2V_1(V_2 + v) - V_2^2 - \frac{4}{3}\frac{\gamma F_l}{\eta}V_2^{3/2}.$$

Equation (11.27) is solved for X using a closed form method known as Ferrari's method. Once X is known, it is a trivial matter to obtain V_{dsat} from Eq. (11.26d). Since Eq. (11.27) is a fourth order polynomial equation, it has four possible solutions. The smallest positive solution is taken to be the valid solution. If no positive real roots are obtained, then V_{dsat} is evaluated using the pinch-off model, Eq. (11.22).

L_{eff} Calculation. The L_{eff} is calculated using Eq. (11.21e)

$$L_{eff} = L(1 - \lambda V_{ds}) \tag{11.29}$$

and depends upon whether or not the CLM term λ has a finite value. If $\lambda = 0$ is input to the model parameter file, then channel length modulation is not taken into account and $L_{eff} = L$. However, if λ is not input then it is calculated internally. Depending upon the value of v_{max}, λ is calculated from either of the following two equations:

- If $v_{max} \leq 0$, V_{dsat} is calculated using the pinch-off model, Eq. (11.22), while the effective channel length is evaluated using the following equation:

$$\lambda = \frac{X_d}{LV_{ds}}\left[\left\{\frac{V_{ds}-V_{dsat}}{4}+\sqrt{1+\left(\frac{V_{ds}-V_{dsat}}{4}\right)^2}\right\}\right]^{1/2} \text{ where } X_d = \sqrt{\frac{2\epsilon_0\epsilon_{si}}{qN_b}}. \tag{11.30}$$

- If $\upsilon_{max} > 0$, then V_{dsat} is calculated using Eq. (11.26d), and λ is given by

$$\lambda = \frac{X_d}{LV_{ds}}\left[\sqrt{\left(\frac{\upsilon_{max}X_d}{2\mu_s}\right)^2+(V_{ds}-V_{dsat})}-\frac{\upsilon_{max}X_d}{2\mu_s}\right] \text{ where } X_d = \sqrt{\frac{2\epsilon_0\epsilon_{si}}{qN_bN_{eff}}}. \tag{11.31}$$

Note that X_d used in Eqs. (11.30) and (11.31) are different. This is because Eq. (11.30) does not provide an accurate description of the output conductance in saturation and N_{eff} has to be used as an empirical factor to change the substrate doping to $N_{eff}N_b$. The larger the N_{eff}, the smaller the output conductance becomes. The range of N_{eff} is normally between 1 and 5.

Saturation Region Current. In this region, $V_{ds} > V_{dsat}$ and current is calculated using Eq. (11.20) with V_{ds} replaced by V_{dsat}.

Subthreshold Current. The current in the subthreshold region is calculated using the following equation:

$$I_{ds} = I_0 \exp\left[\frac{q}{kT}\frac{V_{gs}-V_{on}}{n}\right] \quad (\text{for } V_{gs} < V_{on}) \tag{11.32}$$

where

$$V_{on} = V_{th} + n\frac{kT}{q}$$

$$n = 1 + \frac{qN_{fs}}{C_{ox}} + \frac{C_d}{C_{ox}} + F_w$$

$$C_d = \frac{dQ_b}{dV_{bs}} = \frac{d}{dV_{sb}}[\gamma F_l\sqrt{2\phi_f + V_{sb}}]$$

and I_0 is the value of I_{ds} at $V_{gs} = V_{on}$ calculated using Eq. (11.20), N_{fs} is a curve fitting parameter and C_d is the depletion capacitance. The voltage V_{on} makes the transition from weak to strong inversion regions.

The DC parameters for Level 2 model are shown in Table 11.6. In this table only those parameters are included which are in addition to the Level 1 parameters shown in Table 11.4.

Table 11.6. *SPICE Level 2 model parameters. These are in addition to those shown in Table 11.4*

Parameter name in the text	SPICE parameter name	Parameter description	Default value	Units
Level			1	
L_{dif}	LD	Lateral diffusion	0.0	m
G_w	DELTA	Narrow width factor	0.0	—
X_j	XJ	Junction Depth	0.0	m
u_l	UCRIT	Critical field for mobility degradation	$1 \cdot 10^4$	V/cm
u_t	UTRA	Mobility transverse field coefficient	0.0	—
ν	UEXP	Exponent in mobility degradation	0.0	—
v_{max}	VMAX	Maximum carrier drift velocity	0.0	m/s
N_{eff}	NEFF	Effective substrate doping factor	1	—
N_{fs}	NFS	Fast surface state density	0.0	cm^{-2}
X_{qc}	XQC	Thin-gate oxide capacitance model flag and coefficient of channel charge share attributed to drain (0–0.5)	1.0	—

Note the following:

- The parameter *LD* accounts for the diffusion effects in the device length direction giving an effective channel length *L* as

$$L = L_m - 2L_{dif}.$$

- The UC Berkeley implementation of the Level 2 model does not have the parameter $WD(=\Delta W)$ for calculating the effective device width from drawn dimensions resulting in $W_m = W$.
- The parameter *UTRA* (cf. Eq. 11.21c) does not exist in the Berkeley version, but it is included here because it exists in most of the Level 2 models in commercially available implementations of SPICE.
- If *XJ* is not specified, the narrow channel effect is neglected.
- If *NFS* is not specified, the subthreshold current is not calculated.
- If *NFS* is not specified, $V_{on} = V_{th}$.
- If *VMAX* is not specified, the velocity saturation effect is neglected.

11.3.2 Capacitance Model

The MOSFET source and drain junction capacitance models are the same as for Level 1. However, for MOSFET intrinsic capacitances there are two models available. The first model, which is also the default model, is the Meyer model as described for Level 1; the only difference being that V_{th} is replaced by V_{on}. The second model is the charge controlled model of Ward and Dutton [9]. The parameter *XQC* is associated with partioning

Table 11.7 *Charge sharing for Level 2 capacitance model*

	Source Charge Q_s	Drain Charge Q_D
Linear Region	$Q_I/2$	$Q_I/2$
Saturation Region	$XQC \cdot Q_I$	$(1 - XQC)Q_I$

of the charge (see section 7.2). In the Level 2 model the following scheme is used to partition the channel charge Q_I into the source and drain charges, Q_S and Q_D, respectively, (see Table 11.7). The $XQC \leq 0.5$ is user input model parameter and indicates portion of the charge attributed to the drain. It also acts as a flag; $XQC = 1$ invokes the Meyer model. The partitioning scheme causes discontinuity at the boundary of the linear and saturation regions, except when $XQC = 0.5$. Note that when $XQC = 1$, then in saturation $Q_D = 0$.

Model Parameter Determination. The parameters of this model may be divided into two parts; (1) basic parameters which are basically long channel model parameters like VT0, KP, GAMMA and PHI and (2) parameters relative to second order effect not included in the basic model, and describe narrow and short channel behavior. We have already discussed parameters in the first part which are linear region parameters extracted using linear regression methods. However, the linear regression method to calculate the saturation region parameters, such as VMAX, or short-channel and narrow-width parameters, are not straight forward. It is best to determine these second order parameters using an optimizer as discussed in Chapter 10.
The presence of the parameter NFS permits calculation of the subthreshold current in the model. The parameter can be calculated using Eq. (6.102) and (6.113). For long channel device

$$N_{fs} = \frac{C_{ox}}{q}\left(\frac{S}{2.3V_t} - 1 - \frac{\gamma}{2\sqrt{2\phi_f + V_{bs}}}\right) \tag{11.33}$$

where S is the subthreshold slope. The parameters γ and ϕ_f need to be known and can be determined from V_{th} versus V_{sb} measurements. The extracted value is normally very high ($N_{fs} = 9.3 \times 10^{11}\,\mathrm{cm}^{-2}$), although fast surface states for the process are less than $10^{10}\,\mathrm{cm}^{-2}$. The NFS is treated simply as a fitting parameter. This model does not insure good correlation with measurements.
The Level 2 model, though physically based, has various drawbacks. For example, the transition from linear to saturation regions is not smooth, particularly for short-channel devices, and there is a small discontinuity in the transition from subthreshold to saturation region.

11.4 MOSFET Level 3 Model

The Level 3 is a semi-empirical model that includes second order effects due to short-channels and narrow-widths. The model is computationally efficient compared to the Level 2 model, but the empirical model parameters become geometry dependent.

11.4.1 DC Model

The threshold voltage equation for the SPICE Level 3 model is

$$V_{th} = V_{To} - \gamma\sqrt{2\phi_f} + \gamma F_l\sqrt{2\phi_f + V_{sb}} + F_w(2\phi_f + V_{sb}) - \sigma V_{ds} \tag{11.34}$$

where F_l is a short channel factor based on Dang's model, as given by Eq. (5.73), F_w is a narrow width factor as in Level 2, except that the factor of 4 is replaced by 2, and σ is the DIBL parameter given by [cf. Eq. (5.106)]

$$\sigma = \frac{8.15 \cdot 10^{-22}\eta}{C_{ox}L^3}.$$

Linear Region Current. The drain current, I_{ds}, in the linear region is given by [cf. Eq. (6.169)]

$$I_{ds} = \beta(V_{gs} - V_{th} - \tfrac{1}{2}\alpha V_{ds})V_{ds} \tag{11.35}$$

where

$$\beta = \kappa\frac{\mu_{\text{eff}}}{\mu_0}\frac{W}{L} \tag{11.36a}$$

$$\mu_{\text{eff}} = \mu_s \Bigg/ \left[1 + \left(\frac{\mu_s}{\upsilon_{\max}L}\right)V_{ds}\right] \tag{11.36b}$$

$$\mu_s = \mu_0/[1 + \theta(V_{gs} - V_{th})] \tag{11.36c}$$

$$\alpha = 1 + \frac{\gamma F_l}{4\sqrt{2\phi_f + V_{sb}}} + F_w. \tag{11.36d}$$

If the parameter $\upsilon_{\max}$ is not specified by the user, μ_{eff} is set to μ_s and the velocity saturation effect is not modeled.

Saturation Voltage. V_{dsat} is calculated from one of the following equations (see section 6.7.2)

$$V_{dsat} = \begin{cases} \dfrac{V_{gs} - V_{th}}{\alpha} + \dfrac{\upsilon_{max} L}{\mu_s} - \sqrt{\left(\dfrac{V_{gs} - V_{th}}{\alpha}\right)^2 + \left(\dfrac{\upsilon_{max} L}{\mu_s}\right)^2} & \text{(if } \upsilon_{max} \text{ specified)} \\ \dfrac{V_{gs} - V_{th}}{\alpha} & \text{(if } \upsilon_{max} \text{ not specified).} \end{cases} \tag{11.37}$$

Saturation Region Current. I_{ds} in the saturation region is calculated using the following equation

$$I_{ds} = \frac{I_{dsat}}{1 - l_d/L} \tag{11.38}$$

where

$$l_d = \sqrt{\mathscr{K} X_d^2 (V_{ds} - V_{dsat}) + \left(\frac{X_d^2 \mathscr{E}_p}{2}\right)^2} - \frac{X_d^2 \mathscr{E}_p}{2} \tag{11.39a}$$

$$X_d = \sqrt{\frac{2\epsilon_0 \epsilon_{si}}{q N_b}} \tag{11.39b}$$

$$\mathscr{E}_p = \frac{I_{dsat}}{G_{dsat} L} \tag{11.39c}$$

$$G_{dsat} = \frac{dI_{dsat}}{dV_{dsat}} \tag{11.39d}$$

and I_{dsat} is the drain current at saturation obtained by replacing V_{ds} with V_{dsat} in Eq. (11.35) and G_{dsat} is the drain conductance at saturation. The fitting parameter $\mathscr{K}$ accounts for the fact that the voltage across the depleted surface of the channel, of length l_d, is less than $V_{ds} - V_{dsat}$.

Subthreshold Region Current. It is given by the same equation as for the Level 2 model [cf. Eq. (11.32)] except that I_0 now is calculated at $V_{gs} = V_{on}$ using Eq. (11.35).

The SPICE Level 3 DC model parameters are shown in Table 11.8. These parameters are in addition to the Level 1 parameters shown in Table 11.4, except for the parameter *LAMBDA*, which is not used in Level 3.

The model parameters are generally extracted using an optimizer. Often the value of VMAX is 3–5 times higher than the physical value. To get a more realistic value, it has been suggested [13] to introduce one more

Table 11.8. *SPICE Level 3 model parameters. These are in addition to those shown in Table 11.4*

Parameter name in the text	SPICE parameter name	Parameter description	Default value	Units
Level			1	
L_{diff}	LD	Lateral diffusion		
G_w	DELTA	Narrow width factor	0.0	—
X_j	XJ	Junction Depth	0.0	m
N_{fs}	NFS	Fast surface state density	0.0	cm^{-2}
θ	THETA	Mobility degradation factor	0.0	V^{-1}
η	ETA	Static feedback factor	0.0	—
$\mathscr{K}$	KAPPA	Saturation field correlation factor	0.2	—
v_{max}	VMAX	Maximum carrier drift velocity		—

empirical parameter DEL, so that Eq. (11.36b) reads

$$\mu_{eff} = \mu_s \bigg/ \left[1 + DEL \frac{\mu_s V_{ds}}{L v_{max}} \right]. \tag{11.40}$$

Usually, the value of this parameter is less than one. Note that unlike VMAX of the Level 2 model, the VMAX parameter in level 3 is used in a very different form and is fairly easy to extract from a linear regression method.

The capacitance model (intrinsic and extrinsic) is the same as level 1 model.

11.5 MOSFET Level 4 Model

The MOSFET Level 4 model is generally known as BSIM and is in fact a modified form of CSIM (Compact Short-channel Igfet Model) [2]. This is a parameter based model whose parameters are generally extracted using automated extraction procedures using linear regression [10]. Since the model has many parameters which are bias dependent, care must be taken in extracting these parameters.

11.5.1 DC Model

In this model, threshold voltage is expressed as [cf. Eq. (5.46) and (5.96)]

$$V_{th} = V_{fb} + 2\phi_f + \gamma\sqrt{2\phi_f + V_{sb}} + K_l(2\phi_f + V_{sb}) - \sigma V_{ds}. \tag{11.41}$$

Linear and Saturation Region Current. In these regions current is given by

$$I_{ds} = \begin{cases} \beta_{\text{eff}}[(V_{gs} - V_{th} - \frac{1}{2}\alpha V_{ds})V_{ds}] & \text{linear region, } V_{gs} > V_{th} \\ \dfrac{\beta_1}{2\alpha K}(V_{gs} - V_{th})^2 & \text{saturation region, } V_{ds} > V_{\text{dsat}} \end{cases} \tag{11.42}$$

where

$$\beta_{\text{eff}} = \frac{\beta_0}{[1 + U_0(V_{gs} - V_{th})][1 + U_1 V_{ds}/L]} \tag{11.43a}$$

$$\beta_1 = \frac{\beta_0}{1 + U_0(V_{gs} - V_{th})} \tag{11.43b}$$

$$\beta_0 = \frac{\mu_0 C_{ox} W}{L} \tag{11.43c}$$

$$\alpha = 1 + \frac{\gamma}{2\sqrt{2\phi_f + V_{sb}}}\left[1 - \frac{1}{1.744 + 0.8364(2\phi_f + V_{sb})}\right]. \tag{11.43d}$$

Note that the parameter U_0 is the same as θ of Eq. (11.35) for Level 3. The saturation voltage V_{dsat} is calculated using the following equation.

$$V_{\text{dsat}} = \frac{V_{gs} - V_{th}}{\alpha\sqrt{K}} \tag{11.44}$$

where

$$K = \tfrac{1}{2}(1 + V_c + \sqrt{1 + 2V_c}) \tag{11.45a}$$

$$V_c = \frac{U_1}{L}\frac{V_{gs} - V_{th}}{\alpha}. \tag{11.45b}$$

Subthreshold Region Current. The subthreshold current is calculated using the following equation [8]

$$I'_{\text{sub}} = \frac{I_{\text{sub}} I_{dl}}{I_{\text{sub}} + I_{dl}} \tag{11.46}$$

where

$$I_{\text{sub}} = \beta_0 V_t^2 e^{1.8} \exp\left(\frac{V_{gs} - V_{th}}{nV_t}\right)\left[1 - \exp\left(-\frac{V_{ds}}{V_t}\right)\right] \tag{11.47a}$$

$$I_{dl} = \frac{\beta_0}{2}\frac{W}{L}(3V_t)^2. \tag{11.47b}$$

The factor $e^{1.8}$ is empirically chosen to achieve the best fit in the subthreshold characteristics with minimum effect on the strong inversion characteristics.
The above model has only 9 basic parameters, 5 for threshold voltage ($V_{fb}, \phi_f, \gamma, K_l$ and η) and 4 for drain current (β_0, U_0, U_1 and n). However, 5 parameters (η, β_0, U_0, U_1 and n) depend on bias voltages V_{ds} and V_{sb} as follows:

$$U_0 = U_{0z} + U_{0b}V_{bs} \tag{11.48a}$$

$$U_1 = U_{1z} + U_{1b}V_{bs} + U_{1d}(V_{ds} - V_{dd}) \tag{11.48b}$$

$$\eta_1 = \eta_{1z} + \eta_{1b}V_{bs} + \eta_{1d}(V_{ds} - V_{dd}) \tag{11.48c}$$

$$n_1 = n_0 + n_b V_{bs} + n_d V_{ds} \tag{11.48d}$$

and μ_0 (or β_0) is modeled by quadratic interpolation through 3 data points: μ_0 at $V_{ds} = 0$, μ_0 at $V_{ds} = V_{dd}$ and the slope of μ_0 with respect to V_{ds} at $V_{ds} = V_{dd}$ and can be expressed as

$$\mu_0 = \mu_0\Big|_{V_{ds}=0}\left(\frac{V_{ds}}{V_{dd}} - 1\right)^2 + \mu_0\Big|_{V_{ds}=V_{dd}}\left(2 - \frac{V_{ds}}{V_{dd}}\right)\frac{V_{ds}}{V_{dd}} + \mu_0 V_{ds}\left(\frac{V_{ds}}{V_{dd}} - 1\right) \tag{11.48e}$$

where

$$\mu_0|_{V_{ds}=0} = \mu_z + \mu_{0z}V_{bs} \tag{11.49a}$$

$$\mu_0|_{V_{ds}=V_{dd}} = \mu_s + \mu_{sb}V_{bs} \tag{11.49b}$$

$$\mu_0|_{V_{bs}=0} = \mu_s + \mu_{sd}(V_{ds} - V_{dd}) \tag{11.49c}$$

where V_{dd} is the drain voltage at which saturation region measurements are made. Thus, there are total of 20 electrical parameters including 3 subthreshold region parameters (n_0, n_b and n_d). These electrical parameters also have length and width dependence. The sensitivity of a parameter to L (effective channel length) and W (effective channel width) is denoted by adding a letter 'L' and 'W' at the start of the parameter name. For example, V_{fb} is a basic parameter with units of volts, and $LVFB$ and $WVFB$ are parameters which accounts for length and widths dependence of VFB; that is, $LVFB$ and $WVFB$ are the corresponding L and W *sensitivity factors* for VFB and have units of Volts·μm. In general a parameter P_i, which has length and width dependence, is expressed as

$$P_i = P_0 + \frac{P_l}{L} + \frac{P_w}{W}. \tag{11.50}$$

Table 11.9. *SPICE Level 4 model parameters*

Parameter name in the text	SPICE parameter name	Parameter description	Units
Level			
V_{fb}	VFB	Flat band voltage	V
$2\phi_f$	PHI	Surface potential in strong inversion	V
γ	K1	Body factor	$V^{1/2}$
K_l	K2	S/D depletion charge sharing coefficient	—
η_z	ETA	Zero-bias DIBL coefficient	—
η_b	X2E	Sens. of DIBL effect to V_{bs}	V^{-1}
η_d	X3E	Sens. of DIBL effect to V_{ds} at $V_{ds}=V_{dd}$	V^{-1}
U_{0z}	U0	Zero-bias trans. field mobility degradation	V^{-1}
U_{0b}	X2U0	Sens. of trans. field mobility degradation effect to substrate bias	V^{-1}
U_{1z}	U1	Zero-bias velocity saturation coeff.	μm/V
U_{1b}	X2U1	Sens. of velocity saturation effect to V_{bs}	$\mu m/V^2$
U_{1d}	X3U1	Sens. of velocity saturation effect to V_{ds} at $V_{ds}=V_{dd}$	$\mu m/V^2$
μ_z	MUZ	Zero-bias mobility	$cm^2/V\cdot s$
μ_{zb}	X2MZ	Sens. of mobility to V_{sb} at $V_{ds}=0$	$cm^2/V^2\cdot s$
μ_s	MUS	Mobility at $V_{bs}=0$ and at $V_{ds}=V_{dd}$	$cm^2/V^2\cdot s$
μ_{sb}	X2MS	Sens. of mobility to V_{sb} at $V_{ds}=0$	$cm^2/V^2\cdot s$
μ_{sd}	X3MS	Sens. of mobility to V_{ds} at $V_{ds}=V_{dd}$	$cm^2/V^2\cdot s$
n_0	N0	Zero-bias subthreshold slope coefficient	—
n_b	NB	Sens. of subthreshold slope to substrate bias	—
n_b	ND	Sens. of subthreshold slope to drain bias	—
ΔL	DL	channel shortening	μm
ΔW	DW	channel narrowing	μm
t_{ox}	TOX	Gate oxide thickness	μm
—	XPART	Channel charge sharing coefficient	

11.5.2 Capacitance Model

The source/drain junction capacitance model is the same as in Level 1 model but the MOSFET intrinsic capacitance model is a charge based model. The parameter *XPART* is associated with partitioning of the channel charge into drain and source components. XPART = 0 selects 60/40 partition of the channel charge to the source and drain, respectively, while XPART = 1 sets 100/0 partition in the source/drain charge in saturation. Parameters for the SPICE Level 4 model are shown in Table 11.9.

11.6 Comparison of the Four MOSFET Models

As was stated earlier, the Level 1 model is useful only for hand calculations and rough estimate of the circuit performance. The Level 2 model is more physical compared to the Level 3 model. However, Level 2 model often causes convergence problems, and also takes 25% more CPU time, compared to Level 3 model, for each model evaluation. In this respect the Level 3 model is preferable. Because of the physical nature of the Level 2 model, it is still used in spite of its drawbacks. Modifications to the Level 2 model have recently been proposed. The Level 4 model is based on the physics of the device. However, it has a large number of length and width dependent parameters, and therefore, requires large number of devices to extract the parameters.

Performance comparison of the four models have been reported recently with the aim to see how different models scale with the device length and width. For this comparison, *n*-channel MOSFETs ranging in masked channel length (L_m) and width (W_m) from 10.4 to 1.4 μm were characterized [12]. Three different size devices were used to extract the model parameters for Levels 1–3, while six W/L devices were used in order to get 34 length and width dependent parameters for Level 4. A nonlinear optimization method was used to determine the parameters. The channel length reduction parameter *LD* (due to processing effect) was about 0.35 μm, and channel width reduction parameter *WD* was 0.55 μm, resulting in an effective minimum geometry device of 0.3 by 0.7 μm. Although Levels 1–3 do not have a ΔW parameter, the parameter extraction was carried out using an effective device width obtained by subtracting the known ΔW from the drawn width.

First, the basic parameters (*MU*0, *VT*0, *GAMMA*, *NSUB*) and mobility reduction parameters (*UEXP*, *THETA*) were extracted using $I_{ds} - V_{gs}$ data. This data is measured on a large device (10.4/5.4) in the linear region of device operation. The subthreshold parameters were then extracted from the low current region of the same measurement. This is followed by the

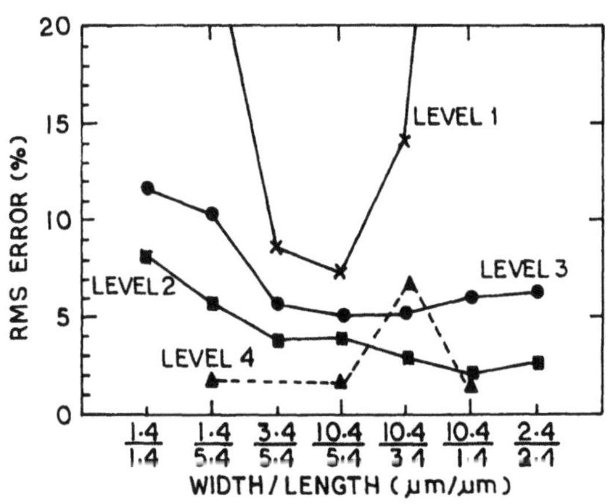

Fig. 11.6 Comparison of 4 different MOSFET models, Levels 1–4. (After Khalily et al. [12])

determination of the width and length dependent parameters using $I_{ds} - V_{gs}$ data on narrow and short devices. Finally, the velocity saturation and channel length modulation parameters were extracted using the short channel length device.
The complete set of parameters extracted from 3 different size devices are then used to simulate other geometries. Figure 11.6 shows the rms error between the simulated and measured data on different channel length and width devices. Remember that not all devices were used to extract the parameters. The Level 2 and Level 3 models show reasonable accuracy in all geometries except when the channel length is 1.4 μm (effective width of 0.3 μm). Note that the Level 1 model does not perform even for large devices. The BSIM model provides excellent accuracy near the geometries used to extract the parameter values. However, larger deviation between the measured and simulated results were encountered for different geometries. This shows that some parameters do not scale well using the $1/L$ *and* $1/W$ geometry dependence assumed in Level 4 model.

References

[1] A. Vladimirescu and S. Liu, 'The simulation of MOS integrated circuits using SPICE2', Memorandum No. UCB/ERL M80/7, Electronics Research Laboratory, University of California, Berkeley, October 1980.

[2] B. J. Sheu, D. L. Scharfetter, and H. C. Poon, 'Compact short-channel IGFET model (CSIM),' Memorandum No. UCB/ERL M84/20, Electronics Research Laboratory, University of California, Berkeley, March 1984.

[3] B. J. Sheu, D. L. Scharfetter, P. K. Ko, and M. C. Jeng, 'BSIM: Berkeley short-channel IGFET model for MOS transistors', IEEE J. Solid-State Circuits, SC-22, pp. 558–565 (1987).

[4] D. K. Schroder, *Semiconductor Material and Device Characterization*, John Wiley & Sons Inc., New York, 1990.

[5] G. W. Neudeck, *The PN Junction Diode*, Vol. II, 2nd Ed., Modular Series on Solid-State Devices, Addison-Wesley Publishing Co., Reading MA, 1987.

[6] D. J. Roulston, *Bipolar Semiconductor Devices*, McGraw-Hill Publishing Company, New York, 1990.

[7] H. J. Kuno, 'Analysis and characterization of *pn* junction diode switching', IEEE Trans. Electron Dev., ED-11, pp. 8–14 (1964).

[8] A H. C. Fung, 'A subthreshold conduction model for BSIM', Memorandum No. UCB/ERL M85/22, Electronics Research Laboratory, University of California, Berkeley, October 1985.

[9] D. Ward, 'Charge-based modeling of capacitances in MOS transistors', Stanford University Tech. Rep. G201-11, 1982.

[10] B. S. Messenger, 'A fully automated MOS device characterization system for process-oriented integrated circuit design', Memorandum No. UCB/ERL M84/18, Electronic Research Laboratory, University of California, Berkeley, January 1984.

[11] M. G. Hsu and B. J. Sheu, 'Inverse-geometry dependence of MOS transistor electrical parameters,' IEEE Trans Computer-Aided Design, CAD-6, pp. 582–585 (1987).

[12] E. Khalily, P. H. Decher, and D. A. Teegarden, 'TECAP2: An interactive device characterization and model development system', *Tech. Digest*, IEEE Int. Conf. on Computer-Aided Design, ICCAD-84, pp. 184–151 (1984).
[13] S. L. Wong and C. A. T. Salama, 'Improved simulation of p- and n-channel MOSFETs using an enhanced SPICE MOS3 model', IEEE Trans Computer-Aided Design, CAD-6, pp. 586–591 (1987).

Statistical Modeling and Worst-Case Design Parameters 12

In integrated circuit technology, the final dimensions of all structures (transistors, capacitors, interconnecting wires, etc.) on finished wafers usually differ from their drawn (intended) dimensions due to several processing effects such as lateral expansion of local oxidation, imperfect etching, mask alignment tolerances, etc. It is observed, for example, in a $2\,\mu m$ CMOS process, a polysilicon line drawn to be $2\,\mu m$ could be any where between 1.3–1.8 μm. Similarly, a $4\,\mu m$ drawn metal line would turn out to be anywhere between 3.2–4.2 μm. Further, since transistor dimensions are determined by the width of the crossing polysilicon and by the lateral diffusion of the source and drain, transistor width to length ratio (W_m/L_m) can vary appreciably from the intended value. In addition to the line-width variations, there are many other process related variations such as changes in oxide thickness, sheet resistance, threshold voltage, etc., which result in the spread in device performance. Clearly, device parameters are subject to statistical variations due to manufacturing process disturbances. These variations affect the circuit performance dramatically. For example, changes in threshold voltage due to process variations will result in changes in transistor characteristics, which in turn affect DRAM access time and refresh rate, clock speed, etc., in digital circuits and op-amp gain in analog circuits. In the worst case the circuits might cease to function.

When the chip designers design their circuits, the process variations are taken into account by simulations that use statistical MOS device models. Traditionally, the statistical models, representing process fluctuations, are simply the worst-case and best-case device model parameters which represent the worst and best case device performance [1]–[7]. *The accuracy of these sets of extreme parameters is critical for the design of the integrated circuits, as these parameters are used by the circuit designers to ensure that their chips will have acceptable parametric yield under all manufacturing process variations.* Needless to say, the creation of an accurate set of worst/best case design parameters are very important for the circuits to yield.

The statistical variations can be classified as *interdie* and *intradie* fluctuations. The interdie variability is characterized by the change in device parameters due to variations in the manufacturing process from lot to lot, wafer to wafer and chip (die) to chip (die). The intradie variability are those which produce parameter change between devices within a single chip or die. The interdie variations are much larger than the intradie variations. The intradie variations, sometimes referred to as local variations, are not considered in this book, although they might be important for some kind of circuits [8]–[10]. In this chapter we will discuss different methods of generating worst/best case design parameters that represents statistical interdie variations of device characteristics due to manufacturing process-induced fluctuations.

12.1 Methods of Generating Worst Case Parameters

The general problem of determining what combination of parameters yields worst-case conditions for an arbitrary circuit is very difficult [11]. But if the *performance function* or the circuit forms are known, then ascertaining the worst-case conditions is not that difficult. For example, in MOS digital circuit design, the transistor drive current in saturation, I_{drive} (drain current for gate and drain held at $|5.0\,\mathrm{V}|$), is a convenient transistor performance parameter. It is based on the concept that the statistical distribution of the $I-V$ characteristics represents the joint distribution of the various transitor parameters. It is this performance function that we will use to illustrate the methodology of creating worst/best case design parameters. For a given channel length and width of a device, I_{drive} will be a function of the n model parameters $p_1, p_2, \ldots, p_n$ corresponding to the adopted MOSFET model equations. The purpose is to find the set of parameters $p_i (i = 1, 2, \ldots, n)$ which will result in an optimum circuit performance (maximum yield).
It is common in industry to take account of the statistical variation in the process through worst case parameters, sometime referred as *corner design parameters* or *Worst Case Files* (WCF). There are five types of WCF, each is defined by a two letter acronym title describing the relative performance characteristics of the *p*- and *n*-channel devices, respectively, in CMOS technology, or depletion and enhancement devices, respectively, in NMOS technology. The letters describe the operation as being typical or nominal (T), fast (F), or slow (S). These are (see Figure 12.1)

- TT (Typical *p*-channel, Typical *n*-channel): This is the typical device operation case and the parameters reflect the target process.
- SS (Slow *p*-channel, Slow *n*-channel): This is the slow device operation case (worst-case) developed from parameters that produce lower I_{drive}. The parameters reflect the process variation so as to yield the slowest device operation.

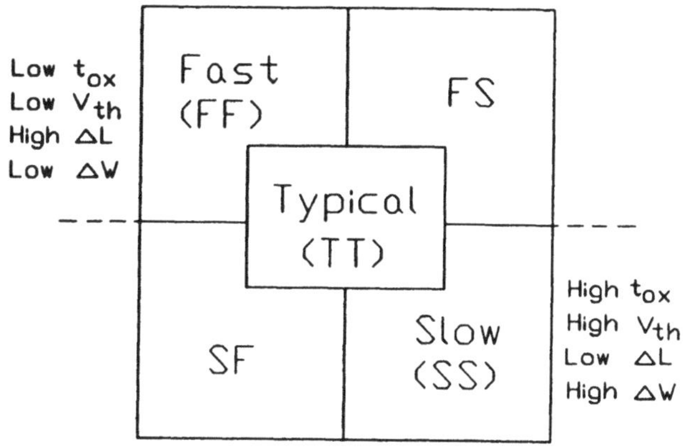

Fig. 12.1 Five different types of worst case model parameter files, commonly know as WCFs

- FF (Fast *p*-channel, Fast *n*-channel): This is the fast device operation case (best-case) developed from parameters that produce large I_{drive}. The parameters reflect the process variation shifted in such a way so as to yield the fastest device operation.
- FS (Fast *p*-channel, Slow *n*-channel): This is a mixed case in which the *p*-channel has the largest currents and *n*-channel has the lowest currents. The parameters reflect process variation with appropriate shifts to yield fast *p*-channel device operation and the *n*-channel device skewed for lower drive current.
- SF (Slow *p*-channel, Fast *n*-channel): This is a mixed case in which the *n*-channel has the largest currents and *p*-channel has the lowest currents. The parameters reflect the process variation with appropriate shifts to yield slow *p*-channel device operation and the *n*-channel device skewed for higher drive current.

Ideally these WCF should be based on parameters extracted from an extensive data base that represents the statistical variation of the process. However, with the short product time of VLSI chips, it is the common practice for chip design to be done in parallel with process development. Therefore, there is either no real data or not enough statistical meaningful data on which to base the worst/best model parameters. In such situations, one can still create WCF based on the so called *Principal Factor* model approach [2], [3] that is very conservative, or still better use statistical simulators like FABRICS II (FABrication of Integrated Circuit Simulator) [12]–[14]. FABRICS II generates samples of device parameters for a set of *process* and *layout* parameters, as well as process *disturbances* that model random fluctuations inherent in the IC fabrication process (e.g., diffusivity of impurity atoms, or linewidth variations and misalignments in the

lithography). However, incorporating suitable physically based device models is still a problem.
The creation of WCF for circuit simulation and design basically consists of two parts: (1) extracting model parameters for different length and width devices for each die and wafer, and (2) determining WCF based on the statistical variation of the extracted model parameters from part 1. In what follows we will assume that we know the model parameters which represent the device behavior in terms of device currents. Therefore, what we will discuss here is the second part of the WCF creation, namely how the interdie variation in the model parameters, due to the random variations in the manufacturing process, can be used to generate a set of parameters which represents the best and the worst case design performance. Since creating WCF requires some knowledge of statistical and probability theory, the basic theory is covered in Appendix H [15]–[17].

12.2 Model Parameter Sensitivity

The first step in generating the WCF is to obtain the *mean and standard deviation* for each of the model parameters obtained from different dice and wafers. Once the mean $\bar{p}_i$ and the standard deviation s_i for each parameter p_i is known, a test for 'outlier' data points (erroneous data points that reflect unusual or noisy data) is made. A simple method to detect outliers is to neglect points that lie outside the $\pm 3s_i$ points of the respective parameter distributions.[1] Table 12.1 shows some statistical analysis for each of the model parameters obtained from the I–V data. Note that in this table η is the subthreshold parameter, while σ_0 and m are the DIBL parameters. The results shown in the table are based on data using 55 dice from 3 different wafers (two lots) for nMOST fabricated using 1 μm CMOS process. The contribution of each parameter to the transistor drive current I_{drive} is then determined using $\pm 3s$ value. A $+3s_i$ is used if the parameters result in an increase in I_{drive}, while $-3s_i$ is used if the parameters result in a decrease in I_{drive}. In order to check whether a given parameter results in a decrease or increase in I_{drive}, one needs to calculate the sensitivity $\mathscr{S}_i$ of I_{drive} to the parameter p_i defined as

$$\mathscr{S}_i = p_i\left(\frac{dI_{\text{drive}}}{dp_i}\right) = p_i\left(\frac{\Delta I_{\text{drive}}}{\Delta p_i}\right). \tag{12.1}$$

[1] In statistical theory standard deviation is represented by σ. However, in order not to confuse this symbol with the DIBL parameter discussed earlier for the drain current modeling, here we use the symbol s for the standard deviation. There is another reason for not using the symbol σ as discussed in Appendix H. Usually there are statistical tests which could be used to test whether the data point indeed is an outlier [19].

Table 12.1. *Some statistical analysis of each of the model parameters obtained from I–V data. These results are based on data, for nMOST fabricated using 1 μm CMOS technology, using 55 sites on different wafers from two lots*

Parameter name p_i	Sensitivity of I_{drive} to p_i	Mean of p_i	σ	$p_i - 3\sigma$	$p_i + 3\sigma$	Best Case (Fast)	Worst Case (Slow)	unit
ΔL	⇑	0.26	0.03	0.14	0.37	0.33	0.18	10^{-4} cm
ΔW	⇓	0.92	0.05	0.77	1.07	0.87	.96	10^{-4} cm
C_{ox}	⇑	2.269	.0056	2.252	2.286	2.274	2.261	10^{-7} F/cm^2
V_{fb}	⇑	−0.68	0.01	−0.65	−0.72	−0.74	−0.64	V
N_b	⇓	4.11	0.09	3.83	4.39	16.618	16.615	10^{16} cm^{-3}
μ_0	⇑	522.9	10.2	485.3	569.1	536.2	510.18	cm^2/V.s
θ	⇓	0.059	0.007	0.037	0.081	0.075	0.045	cm V^{-1}
R_t	⇓	0.04	0.005	0.239	0.056	0.036	0.043	10^{-1} Ω cm
$\mathscr{E}_c$	⇑	3.24	0.182	2.69	3.78	4.55	4.46	10^4 V/cm
η	⇑	3.12	0.056	2.96	3.29	3.07	3.18	—
σ_0	⇑	0.565	0.129	0.178	0.952	0.368	0.845	cm^{-1}
m	⇑	1.021	.03	0.917	1.12	1.07	0.945	—
⋮		⋮	⋮	⋮	⋮	⋮	⋮	—

If the sensitivity $\mathcal{S}_i$ for a parameter p_i is a positive quantity then I_{drive} increases with the parameter and is shown as an arrow pointing upward in Table 12.1. Similarly, if the sensitivity is a negative quantity then I_{drive} decreases with the parameter which is shown as an arrow pointing downward. This procedure then results in individual best and worst case parameters as shown in Table 12.1, column 7 and 8, respectively.
Based on these individual best (Fast) and worst (Slow) case parameters one can calculate the spread in the transistor drive current I_{drive}. Figure 12.2 shows the spread in the drain current as a function of drain voltage V_{ds} at $V_{gs} = 3.3$ V and $V_{bs} = 0$ V for 12.5/1 nMOST. The crosses (thick line) show measured I_{drive} for 55 dice, while the continuous lines show the bounds based on Fast and Slow case results obtained from Table 12.1. It is clear from this figure that in reality the probability of occurrence of these worst case conditions is very small. In fact, such combination of device parameters may not exist. Therefore, this procedure of creating worst case design parameters normally leads to an over-estimate of the actual process spread, which when used in the design process would result in over conservative design. Thus, draw backs of this procedure for calculating WCF are:

- Although it does tell us within what limits the performance may vary, it gives no idea as to how many devices on a wafer will exhibit these performance limits.
- It assumes that the device parameters are not interdependent and they are not correlated with each other.
- It over-estimates the actual process spread.

Since we know that *all the model parameters are not independent and some of them are strongly correlated with others, statistically based methods which take into account these facts are likely to give more realistic WCF.*

12.2.1 Principal Factor Method

It has been shown [2], [3] that most of the device parameter variations can be explained by considering variations in the four parameters (1) ΔL, the difference between drawn and effective length $(L_m - L)$, (2) ΔW, the difference between drawn and effective width $(W_m - W)$, (3) C_{ox}, the gate oxide capacitance, and (4) V_{fb}, the flat band voltage. These four parameters are called the *principal factors*, probably because they are statistically independent. The other model parameters are then related to these factors by the following linear regression equation

$$p_i = a_0 + a_1 \cdot \Delta L + a_2 \cdot \Delta W + a_3 \cdot C_{ox} + a_4 V_{fb} \tag{12.2}$$

where a_0, a_1, a_2, a_3 and a_4 are called regression coefficients, to be estimated from measured data on large number of devices. Known standard methods

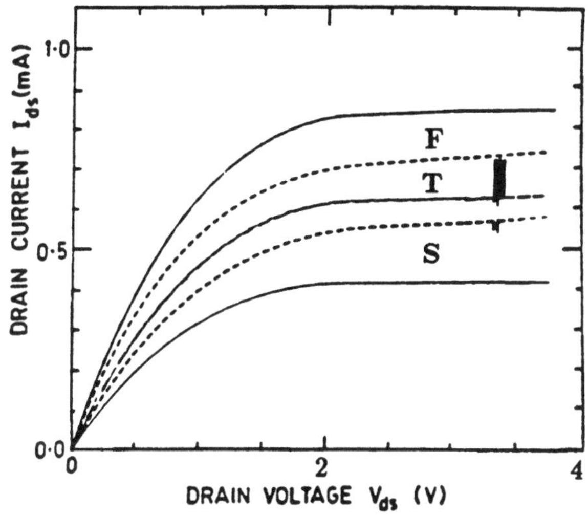

Fig. 12.2 Spread in the I–V characteristics due to random variations in the manufacturing process. Continuous lines are Fast (F) and Slow (S) bounds based on data from Table 12.1, while dashed lines are the corresponding bounds using optimization method. The line marked T corresponds to drain current for typical parameters. Thick line is experimental data points for 55 dice from different wafers

available in most of the statistical packages (see Appendix I) can be used to determine these coefficients.

The principal factor approach is more realistic compared to the one based on the sensitivity analysis. However, it is less rigorous compared to the statistical approach discussed in the next sections. Nonetheless, in the absence of large amount of statistical data, this approach of assuming ΔL, ΔW, C_{ox} and V_{fb} as the four independent parameters and setting them at their scrap limits, with all other parameters to its nominal (typical) values, will give us Slow and Fast files (parameter sets). Although this will result in a conservative design, it is the simplest approach to generate WCF. It should be pointed out that as we scale the devices towards submicron size, the ΔL variations becomes most important. In fact it has been shown that 70% of the variation in I_{drive} can be explained by simply considering variation in ΔL [21].

12.3 Statistical Analysis with Parameter Correlation

In the statistical approach, the first step is to study the distribution of each individual parameter. If there is sufficient data, often the distributions fit into either normal or log-normal type. If a distribution does not fit into either normal or log-normal type, then any convenient mathematical transformation can be used to convert a skewed distribution into a normal distribution. A useful rule of thumb is: if the deviation divided by the mean and multiplied by 100 is greater than 33%, the distribution is probably too skewed for accurate estimation of the standard deviation. Various tests exist to check for the normality of a distribution. If the number of observations is less than 50, the Shapiro-Wilks test is used, but if the number is large the Kolmogorov-Smirnov test is generally used to test for normality. A normal distribution is completely characterized by its mean and standard deviation (first two moments) as discussed in Appendix H [cf. Eq. (H.7)]. These two quantities, as well as other useful statistics of the distribution, are available in all the commercially available statistical packages.[2] The next step is to see how different model parameters are correlated with each other.

Note from Table 12.1, the parameters are measured in different units with values ranging from 10^{-7} to 10^{16}. Therefore, it is always useful to normalize the parameter, computed using the following transformation (see Appendix H)

$$z_i = \frac{p_i - \bar{p}_i}{s_i} \tag{12.3}$$

where $\bar{p}_i$ is the mean and s_i is the standard deviation of the parameter p_i. The normalized parameter z_i is unitless, and has a mean of zero and standard deviation of 1. Once all parameters are normalized, the correlation matrix for the model parameters of Table 12.1 is then generated. Test are made to see if a particular correlation coefficient r is spurious or not (see Appendix H). Using Eq. (H.15, Appendix H), it is easy to see that for m (number of data points) $= 50$, the value of r greater than 0.360 is acceptable with a 95% confidence level. Any value of r less than 0.360 has no statistical significance and its value could be set to zero. The parameters which have strong correlations (with $r > 0.7$) can easily be identified.

However, often due to the large number of parameters involved and the number of correlation coefficients which exceed the significance level (0.360 for $m = 50$), it is difficult to understand the various interrelationships between the model parameters merely by inspecting the correlation matrix. In other words, by mere inspection of the matrix, we cannot assess the *joint effects* of two parameters on another parameter, nor can we know to what extent the correlation between the two parameters is due to the third, fourth, etc., parameters. To understand the pattern of interrelationship among the model parameters and to create a WCF, two different statistical

---

[2] Various statistical packages are commercially available which calculate all the necessary statistic of our interest. Some well known are listed in Appendix 1.

techniques have been used in the literature [3,4], [22]. These are (1) *factor analysis*, and (2) techniques based on nonlinear constrained optimization theory. Both these techniques use the correlation matrix of the model parameters as their starting point, and assume that the joint effect of the different variables can be represented by a multivariate normal distribution (MVN). This necessarily implies that the parameters are individually normal. Since both these statistical techniques first need to calculate principal components of the parameter vector **p**, we will first briefly describe what it means.

12.3.1 Principal Component Analysis

The principal component method is a *technique of transforming the response (or original) variables into new, uncorrelated (independent) variables called the principal components.* Each principal component is a linear combination of the response variables. Thus if $y_1, y_2 \cdots y_n$ are the principal components for the n normalized (or standarized) response variables $z_1, z_2 \cdots z_n$, then

$$\begin{aligned} y_1 &= u_{11}z_1 + u_{12}z_2 + \cdots + u_{1n}z_n \\ y_2 &= u_{21}z_1 + u_{22}z_2 + \cdots + u_{2n}z_n \\ \vdots &\quad \vdots \\ y_n &= u_{n1}z_1 + u_{n2}z_2 + \cdots + u_{nn}z_n \end{aligned} \tag{12.2}$$

and is arranged in the order of decreasing variances. The *most informative principal component is the first, and the least informative is the last.* To know the independent variables (principal components) $y_1, y_2 \ldots$, we need to know coefficients $u_{11}, u_{12} \ldots$, etc. In terms of matrix notation, the above equations can be written as

$$\mathbf{y} = \mathbf{Uz} \tag{12.5}$$

where **y** is the independent variable vector

$$\mathbf{y} = \begin{bmatrix} y_1 \\ y_2 \\ \vdots \\ y_n \end{bmatrix} \tag{12.6}$$

U the *transformation matrix* is an $n \times n$ orthogonal matrix and **z** is the original standarized parameter vector [cf. Eq. (12.3)]. In order to determine the elements u_{ij} of the matrix **U**, the correlation matrix **R** of the normalized parameter vector **z** is diagonalized. This diagonalization is accomplished by obtaining the eigenvalues λ_i and eigenvectors $\mathbf{e}_i$ of **R** by solving the following system of equations (see section 10.1)

$$\mathbf{R}\mathbf{e}_i = \lambda_i \mathbf{e}_i$$

where $\mathbf{e}_i = (e_{i1}, e_{i2}, \ldots, e_{in})$, $i = 1, 2, \ldots, n$. By constructing the following two matrices,

$$\mathbf{E} = \begin{bmatrix} e_{11} & e_{12} \cdots & e_{1n} \\ e_{21} & e_{22} \cdots & e_{2n} \\ \vdots & \vdots & \vdots \\ e_{n1} & e_{n2} \cdots & e_{nn} \end{bmatrix} \tag{12.7}$$

$$\mathbf{D} = \begin{bmatrix} \lambda_{11} & & 0 \\ & \lambda_{22} & \\ \vdots & \vdots & \vdots \\ 0 & & \lambda_{nn} \end{bmatrix} \tag{12.8}$$

we can write $\mathbf{U} = \mathbf{D}^{-1/2}\mathbf{E}^{-1}$. Thus, substituting the matrix elements U_{lj} into (12.5), we can obtain the principal component $\mathbf{y}$ of $\mathbf{p}$. It can be shown mathematically that from the independent vector $\mathbf{y}$, the original parameter vector $\mathbf{p}$ can be obtained using

$$\mathbf{p} = \mathbf{E}\mathbf{D}^{1/2}\mathbf{y} \tag{12.9}$$

so that the rows of the first set of equation (12.5) becomes columns of the second set of equation (12.9). The principal components analysis method is available in most of the standard statistical packages. It is called PCA in BMDP, PC in SPSS-X and Prin in SAS.

12.4 Factor Analysis

Factor analysis is a technique that explains the observed relations between the numerous variables in terms of simpler relations. For a give correlation matrix of a set of variables, factor analysis enables the examination of underlying patterns so that the data can be reduced to a smaller set of *factors* that may be taken as source variables to account for the observed interrelationship in the original set of variables. *Thus, factor analysis is basically a data reduction technique which allows us to detect the most important variables out of a large set of variables.*

In factor analysis, we begin with a set of n standard variables denoted by $z_1, z_2, \ldots, z_n$. In the jargon of factor analysis the z_i's are called the original or response variables [10]. The object of the factor analysis is to represent each of these variables as a linear combination of smaller set of *common factors* $F_1, F_2, \ldots F_k$ plus a factor $g_1, g_2, \ldots, g_n$ etc., unique to each of the response variables. Thus,

$$\begin{aligned} z_1 &= l_{11}F_1 + l_{12}F_2 + \cdots + l_{1k}F_k + g_1 \\ z_2 &= l_{21}F_1 + l_{22}F_2 + \cdots + l_{2k}F_k + g_2 \\ \vdots \quad & \quad \vdots \\ z_n &= l_{n1}F_1 + l_{n2}F_2 + \cdots + l_{nk}F_k + g_n \end{aligned} \tag{12.10}$$

where k is the number of common factors which is typically much smaller than n, the coefficient l_{ij} is called the *loading* and represents the degree to which the ith variable correlates with the jth factor F_j. Note that $F_1, F_2, \ldots,$ etc., are assumed to have zero means and unit variances. Furthermore, it can be shown mathematically that

$$l_{i1}^2 + l_{i2}^2 + \cdots + l_{im}^2 = h_i^2$$

where h_i^2 is called the *communality* of variable z_i. It is the measure of the 'goodness of fit' of the variable expressed by the factors. The above equations constitute the so-called *factor model.* There are many ways available to numerically solve for these quantities and the solution process is called the initial factor extraction. A commonly used method to calculate the factors is based on Principal Component Analysis. The basic idea is to choose the first k principal components, as they explain the greatest proportion of the variance and therefore the most important, and modify them to fit the factor model described above.

To satisfy the assumption of unit variances of the factors, we divide each principal component y_i by its eigenvalue λ_i (variance of y_i). That is, if we define the jth component common factor F_j as $F_j = y_j/\lambda_j$, then we can express the ith equation for z_i from (12.5) as

$$z_1 = u_{1i}\lambda_1 F_1 + u_{2i}\lambda_2 F_2 \cdots + u_{ki}\lambda_k F_k. \tag{12.11}$$

Comparing this equation with (12.10) we can find l_{ij} and g_i. Since the response variables z_i are standardized, the factor loading l_{ij} turns out to be the correlation between p_i and F_j so that l_{ij} varies from -1 to $+1$. Notice that by factor analysis using the principal component method we extract m factors for the m response variables. If we have 10 input variables then we will extract 10 factors. The data reduction capability of the factor analysis comes from the fact that the first extracted factor accounts for the largest portion of the total variance, and each successive factor accounts for less and less. In fact the first few will typically account for more than 75% of the total variance in the data. As a rule of thumb we normally retain those factors which correspond to an eigenvalue of 1 or more. If the first four or five factors cannot explain more than 75% variance in the data, then it is fruitless to use this method as the interpretation of the components will be difficult if not impossible [25].

As a simple example we consider the diode (or source/drain) junction capacitance model parameters C_{j0}, ϕ and m [cf. Eq. (11.3)]. Factor analysis performed on these parameters measured on 55 dice show the following result:

INITIAL STATISTICS:

VARIABLE	COMMUNALITY	FACTOR	EIGENVALUE	PCT OF VAR	CUM PCT
CJO	1.000	1	2.73979	91.3	91.3
PHI	1.000	2	0.17838	5.9	97.3
M	1.000	3	0.08183	2.7	100.0

This simple example shows that 91.3% of the junction capacitance variation can be explained by taking the statistical variation of CJO. In fact, the variation of PHI and M could be ignored.

12.4.1 Factor Rotation

The factors obtained in the previous section are called initial factors. There is no simple way of deciding how many factors to be retained as the parameters tend to load heavily on more than one factor. Although the dimensionality of the given problem is reduced, the factor interpretation is not clear cut. Therefore, we find new factors whose loadings are easier to interpret. These new factors, called the *rotated factors*, are selected so that (ideally) some of the loadings are very large (near ± 1) and the remaining loadings are very small (near zero). Interpretation in terms of original variables is thus made easier. The factor rotation can be achieved in a number of different ways, but the most common technique is *varimax rotation* which is the default option in most statistical packages.

The result of this technique, when applied to the data shown in Table 12.1, indicates that most parameters are strongly related to one factor, although few parameters are strongly or moderately related to more than one factor. The parameter which has strongest dependence on each factor is then chosen as the independent (original) parameters. Remember that it is only an approximation to use the original parameters to represent factors. Therefore, the parameter with the strongest dependences on the factors should be chosen from statistical viewpoint and/or according to their availability from measurements.

12.4.2 Regression Models

Once independent parameters are chosen from factor analysis, the relationship among the other MOSFET parameters are obtained applying the regression analysis. The regression models can either be linear, like the one given by Eq. (12.2), or could be quadratic wherein nonlinear relationship between parameters can be accounted for. The quadratic model is of the form

$$y = b_0 + \sum_{i=1}^{k} b_i \cdot p_i + \sum_{i=1}^{k} \sum_{j=1}^{k} b_{ij} p_i p_j \tag{12.12}$$

where p_i's are the independent parameters, called *regressors* and y is the dependent parameter, called *regressand* and b's are the regression coefficients to be estimated. For I_{drive} as a performance parameter, a linear regression model, similar to (12.2) is good enough for statistical purposes.

12.5 Optimization Method

This technique of creating WCF affords the designer much greater flexibility since the worst/best case files are selected from the measured distribution of parameters by an optimization technique. A user defined performance function can therefore be optimized (maximized or minimized)[3], which in our case will be transistor drive current in saturation, I_{drive}. A multivariate normal distribution (MVN) is assumed whose probability density function is given by (see Appendix H)

$$f_z(\mathbf{z}) = (2\pi)^{-n/2} (\det \mathbf{R})^{-1/2} \exp[-\tfrac{1}{2}\mathbf{z}^T \mathbf{R}^{-1} \mathbf{z}] \tag{12.13}$$

where $\mathbf{z}$ is a $n \times 1$ vector representing the normalized value of the n model parameter vector $\mathbf{z} = [z_1\, z_2 \cdots z_n]^T$, and $\mathbf{R}$ is a $n \times n$ correlation matrix. To get the probability of occurrence for a given set of values of parameters, the probability density function $f_z(\mathbf{z})$ has to be integrated. Since the correlation between various parameters exists, this integration is not an easy task. It is therefore desirable to transform the parameters vector $\mathbf{z}$ to an independent parameter vector $\mathbf{y}$ as explained in the previous section. Thus, knowing $\mathbf{y}$ one can calculate the joint probability density function $f_y(\mathbf{y})$ for the independent parameter vector $\mathbf{y}$ as

$$f(\mathbf{y}) = (2\pi)^{-n/2} (\det \mathbf{D})^{-1/2} \exp[-\tfrac{1}{2}\mathbf{y}^T \mathbf{D}^{-1} \mathbf{y}] \tag{12.14}$$

where $\mathbf{D}$ is diagonal matrix whose diagonal elements are eigenvalues λ_i of the correlation matrix $\mathbf{R}$. To obtain the probability distribution function, which will yield the desired probability, the function $f_y(\mathbf{y})$ is integrated over the equidensity contours. Integration of $f(\mathbf{y})$ gives probability such that

$$\int \cdots \int f(\mathbf{y}) = \text{Probability} = 1 - \exp(-a^2/2) \tag{12.15}$$

[3] Maximization of a performance function results in best case (fast) parameters, while minimization of the function results in worst case (slow) parameters.

where a^2 is given by

$$a^2 = \frac{y_1^2}{\lambda_1} + \frac{y_2^2}{\lambda_2} + \cdots + \frac{y_n^2}{\lambda_n} \tag{12.16}$$

At this stage it is possible to optimize a user defined process function, given a probability (set by the designer). The equidensity contours (given by equation 12.13) are used as a constraint to the user defined performance function with an optimization routine which will then result in independent parameter vector **y**. By minimizing the performance function, with the equidensity contour as a constraint will yield the worst case parameter vector while maximizing the performance function will result in the best case parameter vector. In fact two optimization steps are required to get the worst/best case parameters. Once the independent parameter vector y is determined, the original parameter vector **p** can be calculated using equation

$$\mathbf{p} = \mathbf{U}^{-1}\mathbf{y} \tag{12.17}$$

The advantages with this method is that a much tighter design window will result since parameter correlations between each parameter is accounted for. The method also gives the designer the ability to customize parameter files for a given circuit by optimizing a performance function.
Thus, in this procedure the following steps are needed to obtain WCF.

1. Measure I–V and C–V data from different lots and wafers. Rule of thumb is that the number of dice should be 3 times the number of parameters.
2. Extract model parameters $p_i (i = 1, 2, \ldots, n, n$ being number of parameters) for all dice.
3. Determine statistical variation of all parameters, that is, determine mean ($\bar{p}_i$) and standard deviation (s) of each parameter.
4. Normalize each parameter so that $\bar{p}_i = 0$ and $s = 1$. The normalized parameter z_i thus corresponds to the original parameter p_i.
5. Generate correlation matrix **R** of the normalized model parameters.
6. Test if correlation coefficients are spurious. If spurious, set it to zero.
7. Calculate *eigenvalues* λ_i, and *eigenvectors* $\mathbf{e}_i$, of **R**.
8. Transform *dependent and correlated* model parameter vector **z** to *independent and uncorrelated* parameter vector **y** (principal components) using Eq. (12.5).
9. Calculate the joint probability density function $f(\mathbf{y})$ for the independent parameters vector **y** using Eq. (12.14).
10. Assuming certain value for the probability, calculate 'a' value from Eq. (12.15). For example, for 99% probability ($3s$ over n parameter space) $a = 3.035$.

11. Minimization and maximization of the performance function under the nonlinear constraint given by Eq. (12.16) gives the slow and fast **z** vectors, respectively.

The procedure outlined above was carried out using the same data as was used for generating WCF discussed in section 12.2. The results of optimization, using I_{drive} as the performance function, are shown in Table 12.2. Note that the values for only few important parameters are shown. Since I_{drive} is a function of the model parameters and device W and L, the optimization was carried out using W = 3 μm and L = 1 μm. It should be pointed out that the parameters were obtained using two steps of optimization. In the first step, parameters pertaining to the threshold voltage model were optimized. These parameters were then fixed and other parameters for I_{drive} were then optimized.
In Table 12.2, the numbers in the bracket show standard deviation times the amount of the shift in the parameter value from their respective mean (typical) values. *Note that the change in the parameter value is less than 3σ for the so called four independent parameters ΔL, ΔW, C_{ox} and V_{fb}.* This is understandable because overall change for I_{drive} is the effect of other parameters too.
The optimized DC parameter values shown in Table 12.2 are then used to calculate I_{drive} bounds. These bounds are shown as dotted line in Figure 12.2. Clearly, the optimization technique results in a more realistic WCF compared to the principal factor method.
The histrogram of I_{drive}, for $W_m/L_m = 12.5/1$ *n*-channel devices, based on data collected from 3 different lots (117 die locations from different wafers) for a typical 1 μm CMOS technology is shown in Figure 12.3. The vertical lines designated as TS and TF are the slow and fast bounds, respectively, generated by the principal factor method, while the corresponding bounds generated by the optimization method discussed above are designated as OS and OF, respectively. Note from this figure that *the bounds generated*

Table 12.2. *Mean (typical), minimum (slow) and maximum (fast) values for some important n-channel parameters obtained using the optimization method*

Parameter name	Mean (Typical)	Minimum (Slow)	Maximum (Fast)	Units
ΔL	0.36	0.315(−1.45) ⇓	0.418(1.65) ⇑	μm
ΔW	0.79	1.02(2.53) ⇑	0.570(−2.29) ⇓	μm
C_{ox}	2.25	2.23(−0.83) ⇓	2.27(0.73) ⇑	10^{-7} F/cm^2
V_{FB}	−0.65	−0.56(2.67) ⇑	−0.745(−2.79) ⇓	V
μ_0	577	600.43(1.51) ⇓	556(−1.27) ⇑	cm^2/V.s
σ_0	0.642	0.71(0.67)	0.568(−0.69)	—
⋮	⋮	⋮	⋮	—

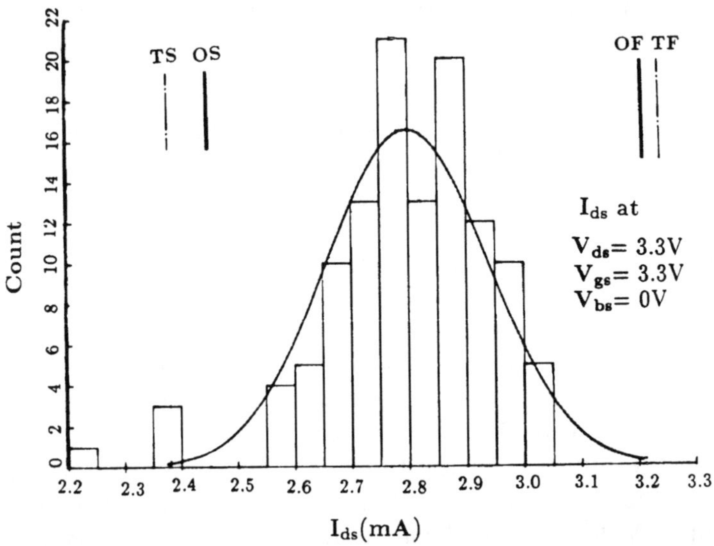

Fig. 12.3 Histogram showing variation of I_{drive} measured from different wafers and lots on *n*-channel devices. TF and TS are the Fast (best) and Slow (worst) bounds, respectively, based on principal factor method, while the corresponding bounds based on optimization method are shown as OF and OS, respectively

by the principal factor method are 8–11% higher compared to the optimization method.

Although optimization method generates realistic WCF, however, it needs large amount of statistically meaningful data that is not always available. Same is the case with the Factor rotation method. To a first approximation, WCF could be generated using principal factor method. In the latter approach, it is more appropriate to replace V_{fb} as independent parameter by N_b, the bulk concentration.

References

[1] R. Spence and R. S. Soin, *Tolerance Design of Electronic circuits*, Addison-Wesley Publishing Co., Reading, MA, 1988.

[2] P. Yang and P. Chatterjee, 'Statistical modeling of small geometry MOSFET', IEEE-IEDM82, *Tech. Digest*, pp. 286–289 (1982).

[3] P. Cox, P. Yang, S. S. Mahant-Shetti, and P. Chatterjee, 'Statistical modeling for efficient parametric yield estimation of MOS VLSI circuits', IEEE Trans. Electron Devices, ED-32, pp. 471–478 (1985).

[4] N. Herr and J. J. Barnes, 'Statistical circuit simulation modeling of CMOS VLSI', IEEE Trans. Computer Aided Design, CAD-5, pp. 15–22 (1986).

[5] J. P. Spoto, W. T. Coston, and C. P. Hernandez, 'Statistical integrated circuit design and characterization', IEEE Trans. Computer Aided Design, CAD-5, pp. 91–103 (1986).

[6] T. K. Yu, S. M. Kang, I. N. Hajj, and T. N. Trick, 'Statistical performance modeling and parametric yield estimation of MOS VLSI', IEEE Trans. Computer Aided Design, CAD-6, pp. 1013–1022 (1987).

[7] P. Tuohy, A. Gribben, A. J. Walton, and J. M. Robertson, 'Realistic worst-case parameters for circuit simulation', IEEE Proc., 134, Pt. I, pp. 137–140 (1987).

[8] S. Inohira, T. Shinmi, M. Nagata, T. Toyabe, and K. Iida, 'A statistical model including parameter matching for analog integrated circuits simulation,' IEEE Trans. Computer Aided Design, CAD-4, pp. 621–628 (1985).

[9] M. Pelgrom, A. Duinmaijer, and A. Welbers, 'Matching properties of MOS transistor', IEEE J. Solid-State Circuits, 24, pp. 1433–1439 (1989).

[10] C. Michael and M. Ismail, 'Statistical modeling of device mismatch for analog MOS integrated circuits', IEEE J. Solid-State Circuits, 27, pp. 154–166 (1992).

[11] L. A. Glasser and D. W. Doubberpuhl, *The Design and Analysis of VLSI Circuits*, Addison-Wesley Publishing Co., Reading, MA, 1985.

[12] W. Maly and A. J. Strojwas, 'Statistical simulation of the IC manufacturing process', IEEE Trans. on Computer Aided Design, CAD-1, pp. 120–131 (1982).

[13] S. R. Nassif, A. J. Strojwas, and S. W. Director, 'FABRICHSII: A statistical based IC fabrication process simulator,' IEEE Trans. on Computer Aided Design, CAD-3, pp. 40–46 (1984). Also see Report (Feb. 1990) on FABRICS II: 'A statistical simulator of the IC manufacturing process', Department of Electrical Engineering, Carnegie-Mellon University, Pittsburgh, PA, 15213.

[14] S. R. Nassif, A. J. Strojwas, and S. W. Director, 'A methodology for worst-case analysis of integrated circuits', IEEE Trans. on Computer Aided Design, CAD-5, pp. 104–113 (1986).

[15] D. G. Rees, '*Foundations of Statistics*', Chapman and Hall, New York, 1987.

[16] R. E. Walpole and R. H. Myers, *Probability and Statistics for Engineers and Scientists*, McGraw Hill, New York, 1976.

[17] C. W. Helstrom, *Probability and Stochastic Processes for Engineers*, Macmillan Publishing Company, New York, 1984.

[18] N. D. Arora and L. M. Richardson, 'MOSFET modeling for circuit simulation' in *Advanced MOS Device Physics* (N. G. Einspruch and G. Gildenblat, eds.), VLSI Electronics: Microstructure Science, Vol. 18, pp. 236–276, Academic Press Inc., New York, 1989.

[19] V. Bernett and T. Lewis, '*Outliers in Statistical Data*, John Wiley & Sons, New York, 1978.

[20] J. A. Power, A. Mathewson, and W. A. Lane, 'MOSFET statistical parameter extraction using multivariate statistics', 1991 Int. Conf. Microelectronics Test Structures, 4, pp. 209–214 (1991).

[21] M. Bolt, M. Rocchi, and J. Engel, 'Realistic statistical worst-case simulations of VLSI circuits', IEEE Trans. Semicond. Manuf., 4, pp. 193–198 (1991).

[22] D. A. Divekar, R. W. Dutton, and W. J. McCalla, 'Experimental study of Gummel-Poon Model parameter correlations for bipolar junction transistors', IEEE Journal of Solid-State Circuits, SC-12, pp. 552–559 (1977).

[23] A. A. Afifi and V. Clark, *Computer-Aided Multivariate Analysis*, Lifetime Learning Publications, Belmont, CA, 1984.

[24] J. Vlach and K. Singhal, *Computer Methods for Circuit Analysis and Design*, Van Nostrand Rienhold Company, New York, 1983.

[25] D. F. Morrison, *Multivariate Statistical Methods*, McGraw-Hill Book Company, New York, 1976.

[26] R. A. Johnson and D. W. Wichern, *Applied Multivariate Statistical Analysis*, Prentice-Hall Inc., Englewood Cliffs, New Jersey, 1982.

[27] M. E. Johnson, *Multivariate Statistical Simulation*, John Wiley & Sons, New York, 1987 (p. 52).

Appendix

Appendix A. Important Properties of Silicon, Silicon Dioxide and Silicon Nitride at 300 K

Property	Si	SiO_2	Si_3N_4	Units
Atomic number	14	—	—	
Molecular weight	28.29	60.08	140.28	g/mol
Density at 300 K	2.33	2.27	3.0	g/cm^{-3}
Relative permittivity (Dielectric constant)	11.7	3.9	7–7.5	—
Breakdown field (Dielectric strength)	3×10^5	8×10^6	1×10^7	V/cm
Refractive index	3.42	1.46	2.05	—
Thermal conductivity	1.412	0.014	—	W/cm.K
Lattice constant	5.431			Å
Energy gap E_g	1.12	8.0	5.0	eV
Intrinsic carrier concentration n_i	1.45×10^{10}	—	—	cm^{-3}
Intrinsic Debye length	2.4×10^{-5}	—	—	cm
Bulk electron mobility μ_n	1350	2–30	—	$cm^2/V.s$
Bulk hole mobility μ_p	480	—	—	$cm^2/V.s$

Appendix B. Some Important Physical Constants at 300 K

Constant	Symbol	Magnitude	Units
Electronic charge	q	1.602×10^{-19}	C
Free-electron mass	m	9.11×10^{-28}	g
Boltzmann's Constant	k	1.38×10^{-23}	J/K
		8.62×10^{-5}	eV/K
Planck's Constant	h	6.25×10^{-34}	J·s
Permittivity of free space	ϵ_0	8.854×10^{-14}	F/cm
Thermal voltage at 300 K	$V_t = kT/q$	0.02586	V
Thermal energy at 300 K	kT	0.02586	eV

Appendix C. Unit Conversion Factors

1 μm	(micrometer or micron) = 10^{-6} meters = 10^{-4} cm = 10^{4} Å
1 nm	(nanometer) = 10^{-9} meters = 10^{-7} cm = 10^{-3} μm = 10 Å
1 cm	= 10^{-8} Å
1 mil	= 25.4 μm = 10^{-3} inches
1 eV	= 1.602×10^{-19} J = 3.83×10^{-20} Cal = 1.78×10^{-36} Kg
0°C	= 273 K

Appendix D. Magnitude Prefixes

Magnitude prefix	Multiple factor	Symbol
atto	10^{-18}	a
femto	10^{-15}	f
pico	10^{-12}	p
nano	10^{-9}	n
micro	10^{-6}	μ
milli	10^{-3}	m
centi	10^{-2}	c
Kilo	10^{3}	K
Mega	10^{6}	M
Giga	10^{9}	G
Tera	10^{12}	T

Appendix E. Methods of Calculating ϕ_s from the Implicit Eq. (6.23) or (6.30)

Rearranging Eq. (6.30) for ϕ_s yields

$$\phi_s = 2\phi_f + V_{cb} + V_t \ln\left[\frac{1}{V_t}\left\{\frac{1}{\gamma^2}(V_{gb} - V_{fb} - \phi_s)^2 - \phi_s\right\}\right] \tag{E.1}$$

Assume ϕ_s^0 is an initial guess of ϕ_s then the next value of the estimate of ϕ_s is given by the Schroder series expression [1]

$$\phi_s = \phi_s^0 + K - \frac{y''}{2y'}K^2 + \frac{(3y'')^2 - y'y'''}{6(y')^2}K^3 + \frac{10y'y''y''' - (y')^2 y'''' - 15(y'')^3}{24(y')^3}K^4 \tag{E.2}$$

where only the first 5 terms in the series are shown and taken into account. The prime on y denotes the order of the derivative of the function $f(\phi_s)$ [cf. Eq. (6.31)] given by

$$y = f(\phi_s) = 0 \quad \text{and} \quad K = -\frac{y}{y'}.$$

Note that the first two terms of the series correspond to the Newton–Raphson iteration. The other 3 terms are smaller, but their contribution is significant in weak inversion.

A good initial guess for the surface potential is suggested [2]

$$\phi_s^0 = V_a + V_{cb} + V_t \ln\frac{1}{1 + \exp(V_a + V_{cb} - \phi_{ss})} \tag{E.3}$$

where

$$V_a = 2\phi_f + V_t \ln\left[\left|\frac{1}{V_t}\left\{\frac{1}{\gamma^2}(V_{gb} - V_{fb} - 2\phi_f)^2 - 2\phi_f\right\}\right| + 2\right] \tag{E.4}$$

and ϕ_{ss} is ϕ_s in weak inversion given by Eq. (6.90). That is,

$$\phi_{ss} = V_{gb} - V_{fb} + \frac{\gamma^2}{2} - \gamma\sqrt{V_{gb} - V_{fb} + \frac{\gamma^2}{4}}. \tag{E.5}$$

The semi-empirical Eq. (E.3) is such that in strong inversion $\phi_s^0 \approx V_{cb} + V_a$ and in weak inversion $\phi_s^0 \approx \phi_{ss}$; therefore, it follows the general behavior of the surface potential ϕ_s. The absolute value sign in Eq. (4) is to prevent the argument of the logarithm from becoming negative in weak inversion. With the initial guess given by Eq. (E.3), an accurate estimation of ϕ_s is obtained in all the regions of device operation using Eq. (E.2). Only one or two iterations are normally required.

Other non-iterative approaches for calculating ϕ_s, such as storing values of ϕ_s in a 2-D array [3], or approximating the potential using cubic spline functions [4], have also been proposed.

An approximate solution of Eq. (E.1) in different regimes of device operation has also been suggested. Since in strong inversion, defined as $V_{gb} > V_{gbh}$, the logarithm term varies very little, an approximate expression for ϕ_s is given by

$$\phi_s(\text{strong inversion}) \approx 2\phi_f + V_{cb} + 6V_t. \tag{E.6}$$

A better estimate (within 1% of the exact solution) for ϕ_s in strong inversion is obtained by substituting (E.3) in the right hand side of Eq. (E.1). For weak inversion region, defined as $V_{gb} < V_{gbm}$, ϕ_s is given by Eq. (E.5). A better estimate can be found by substituting ϕ_{ss} in Eq. (E.1). However, for moderate inversion, no simple relationship exist.

References

[1] A. M. Ostrowsky, *Solutions of Equations and Systems of Equations*, Academic Press, New York, 1973.

[2] C. Turchetti and G. Masetti, 'A CAD-oriented analytical MOSFET model for high-accuracy applications,' IEEE Trans. Computer-Aided Design, CAD-3, pp. 117–122 (1984).

[3] S. Yu, A. F. Franz, and T. G. Mihran, 'A physical parametric transistor model for CMOS circuit simulation', IEEE Trans. Computer-Aided Design, CAD-7, pp. 1038–1052 (1988).

[4] H. J. Park, P. K. Ko, and C. Hu, 'A charge sheet capacitance model of short channel MOSFET's for SPICE', IEEE Trans. Computer-Aided Design, CAD-10, pp. 376–389 (1991).

Appendix F. Charge Based MOSFET Intrinsic Capacitances

In this appendix, the expressions for the intrinsic capacitances for large and wide device will be presented. These capacitances are based on the charge equations given in section 7.3 and the definition of the capacitance equation (7.40). In order to write the equations in a tractable form we first define some auxiliary functions

$$
\begin{aligned}
D_{vth} &= \frac{\partial V_{th}}{\partial V_{bs}} & D_{\alpha} &= \frac{\partial \alpha}{\partial V_{bs}} \\
f_1 &= -1 + D_{vth} & f_2 &= D_{\alpha} V_{ds} \\
f_3 &= D_{\alpha} V_{ds} + \alpha & f_4 &= D_{\alpha} V_{ds}^2 \\
h_1 &= \alpha V_{ds} & h_2 &= V_{gt} - 0.5\alpha V_{ds} \\
h_3 &= \alpha(1-\alpha) V_{ds}^2 & h_4 &= 1 - \alpha \\
\mathscr{A} &= \frac{h_1}{12h_2} & \mathscr{B} &= \frac{5V_{gt} - 2h_1}{12h_2}.
\end{aligned}
\tag{F.1}
$$

Linear Region. The gate capacitances based on Eq. (7.58) can be written as

$$
\begin{aligned}
C_{GS} &= -\frac{\partial Q_G}{\partial V_s} = C_{oxt}\left[-0.5 - \frac{(f_2 V_{ds} + 2f_3)}{12h_1} - \frac{f_3 V_{ds}(f_1 + 0.5f_3)}{12h_1^2}\right] \\
C_{GB} &= -\frac{\partial Q_G}{\partial V_b} = C_{oxt}\left[\frac{f_4}{12h_2} + \frac{h_1 V_{ds}(D_{vth} + 0.5f_2)}{12h_2^2}\right] \\
C_{GD} &= -\frac{\partial Q_G}{\partial V_d} = C_{oxt}\left[-0.5 + \left(\frac{h_1}{6h_2} + \frac{h_1^2}{24h_2^2}\right)\right] \\
C_{GG} &= \frac{\partial Q_G}{\partial V_g} = C_{oxt}\left[1 - \frac{h_1 V_{ds}}{12h_2^2}\right].
\end{aligned}
\tag{F.2}
$$

Note that the sum of all the four capacitances C_{GS}, C_{GB}, C_{GD} and C_{GG} is zero. The bulk capacitances based on Eq. (7.60) can be written as

$$
\begin{aligned}
C_{BS} &= -\frac{\partial Q_B}{\partial V_s} = -C_{oxt}\left[D_{vth} + 0.5(-h_4 + f_2) + \frac{f_4(1-2\alpha) + 2h_1 h_4}{12h_2}\right. \\
&\qquad \left. + \frac{\alpha h_4(f_1 + 0.5f_3)}{12h_2^2}\right] \\
C_{BD} &= -\frac{\partial Q_B}{\partial V_d} = C_{oxt}\left[0.5\cdot(1-\alpha) - \frac{h_1(1-\alpha)}{6h_2} - \frac{h_3\alpha}{24h_2^2}\right] \\
C_{BG} &= -\frac{\partial Q_b}{\partial V_g} = C_{oxt}\left[\frac{f_2}{12h_2^2}\right] \\
C_{BB} &= \frac{\partial Q_b}{\partial V_b} = C_{oxt}\left[-D_{vth} - 0.5f_2 - \frac{f_4(1-2\alpha)}{12h_2} - \frac{h_3(D_{vth} + 0.5\cdot f_2)}{12h_2^2}\right].
\end{aligned}
\tag{F.3}
$$

The drain capacitances based on Eq. (7.55) can be written as

$$C_{DG} = -\frac{\partial Q_D}{\partial V_g} = 0.5C_{oxt}\left[1 + \frac{\mathscr{A}}{h_2}(1 - 4\mathscr{B})\right]$$

$$C_{DB} = -\frac{\partial Q_D}{\partial V_b} = C_{oxt}\left[-0.5D_{vth} - 0.5f_2 + \frac{1}{h_1^2}\right.$$

$$\left.\cdot\left\{h_1\frac{\partial h_0}{\partial V_b} + h_0 f_2 + \frac{2h_0 h_1}{h_2}(D_{vth} + 0.5f_2)\right\}\right] \tag{F.4}$$

$$C_{DS} = -\frac{\partial Q_D}{\partial V_S} = C_{oxt}\left[-0.5f_1 + 0.5f_3 + \frac{1}{h_1^2}\right.$$

$$\left.\cdot\left\{h_1\frac{\partial h_0}{\partial V_s} - h_0 f_3 - \frac{2h_0 h_1}{h_2}(f_1 + 0.5f_3)\right\}\right]$$

$$h_0 \equiv \frac{V_{gt}^2}{6} - \frac{h_1 V_{gt}}{8} + \frac{h_1^2}{40}.$$

Saturation Region. Differentiating Eq. (7.61c) with respect to V_s, V_b, V_d and V_g yields the corresponding capacitances C_{GS}, C_{GB}, C_{GD} and C_{GG}, respectively, which are shown below:

$$C_{GS} = C_{oxt}\left[-1 - \frac{f1}{3\alpha} - \frac{V_{gt}D_\alpha}{3\alpha^2}\right]$$

$$C_{GB} = C_{oxt}\left[\frac{D_{vth}}{3\alpha} + \frac{V_{gt}D_\alpha}{3\alpha^2}\right] \tag{F.5}$$

$$C_{GD} = 0$$

$$C_{GG} = C_{oxt}\left[1 - \frac{1}{3\alpha}\right].$$

Again, differentiating (7.61d) with respect to V_s, V_b, V_d and V_g we get the corresponding capacitances C_{BS}, C_{BB}, C_{BD} and C_{BD}, respectively, which are given by the following equations

$$C_{BS} = C_{oxt}\left[D_{vth} + \left(\frac{1-\alpha}{3\alpha}\right)f_1 + \frac{V_{gt}D_\alpha}{3\alpha^2}\right]$$

$$C_{BG} = C_{oxt}\left[\frac{h_4}{3\alpha}\right] \tag{F.6}$$

$$C_{BD} = 0$$

$$C_{BB} = C_{oxt}\left[-d_{vth} - \frac{f_4 D_{vth}}{3\alpha} - \frac{V_{gt}D_\alpha}{3\alpha^2}\right].$$

The transcapacitances corresponding to the drain charge will be

$$\begin{aligned} C_{DS} &= -\tfrac{4}{15}C_{oxt}f_1 \\ C_{DB} &= \tfrac{4}{15}C_{oxt}D_{vth} \\ C_{DD} &= 0 \\ C_{DG} &= -\tfrac{4}{15}C_{oxt}. \end{aligned} \tag{F.7}$$

Subthreshold Region.

$$V_1 \equiv V_{fb} + V_{bs}.$$

Differentiating Eq. (F.7) w.r.t V_s, V_b, V_d and V_g we get the corresponding capacitances C_{GS}, C_{GB}, C_{GD} and C_{GG} in the subthreshold region.

$$\begin{aligned} C_{GS} &= C_{oxt}\left[-\frac{\partial\gamma}{\partial V_{bs}}g_2\right] \\ C_{GB} &= C_{oxt}\left[\frac{\partial\gamma}{\partial V_{bs}}g_2 - \frac{\gamma}{g_1}\right] \\ C_{GD} &= 0 \\ C_{GG} &= C_{oxt}\left[\frac{\gamma}{g_1}\right] \end{aligned} \tag{F.8}$$

where

$$g_1 = \sqrt{\gamma^2 + 4(V_{gs} - V_1)}$$

and

$$g_2 = -\gamma + \frac{1}{g_1}[\gamma^2 + 2(V_{gs} - V_1)].$$

The channel charge is zero in the subthreshold region and therefore the gate and drain charges will also be zero resulting in

$$C_{DS} = 0; \quad C_{DB} = 0; \quad C_{DD} = 0; \quad C_{DG} = 0$$

and also

$$C_{SS} = 0; \quad C_{SB} = 0; \quad C_{SD} = 0; \quad C_{SG} = 0.$$

Since the channel charge is zero therefore the bulk charge Q_b is

$$Q_b = -Q_g.$$

This will result in the following capacitances

$$C_{BS} = -C_{GS}; \quad C_{BB} = -C_{GB}$$

$$C_{BD} = -C_{GD}; \quad C_{BG} = -C_{GG}$$

which were defined earlier.

Appendix G. Linear Regression

Suppose that the threshold voltage, V_{th}, of a MOSFET is measured at different temperatures T. Let us define temperature as the variable x, and V_{th} as the variable y. Clearly x (temperature) is an independent variable and y (observed V_{th}) is the dependent variable. Suppose there are m measured data points of y versus x. That is, there will be m number of data points $(y_1, x_1), (y_2, x_2), \ldots, (y_m, x_m)$, where y_i is a measured value of V_{th} at ith temperature x_i. The best fit line that relates y to x is called the *regression line*, represented by the equation

$$y_i' = a + bx_i \tag{G.1}$$

where y_i' is the predicted value of y at temperature x_i, obtained using Eq. (G.1). The constants a and b are called *regression coefficients.* Note that since Eq. (G.1) is the equation of a straight line, the parameters a and b are the intercept and slope, respectively, of the straight line.

In order to find the best fitting regresion passing through the cluster of data points, we use the so called *least squares criterion* that results in the *smallest sum of squared deviations* of the data points from the line. Stated mathematically, we determine the slope b and y-intercept a of (G.1) such that

$$\sum_{i=1}^{m} (y_i - y_i')^2 \tag{G.2}$$

is minimum, where y_i is the observed value and y_i' is the predicted value. The slope b of the best-fitting line, based on the least squares criterion, can be shown to be [1]

$$b \text{ (slope)} = \frac{m(\sum x_i y_i) - (\sum x_i)(\sum y_i)}{m(\sum x_i^2) - (\sum x_i)^2} \tag{G.3}$$

and

$$a \text{ (intercept)} = \frac{(\sum x_i y_i)}{m(\sum x_i^2) - (\sum x_i)^2} \tag{G.4}$$

where the $\sum$ (summation) is over all measurement points from $i = 1$ to m. Note that there is always an error ε_i associated with any measurement x_i. Generally this measurement error ε_i is unknown. When this is the case, the least square formulation described above is recommended. If, however, one knows that ε_i has a variance σ_i^2, some other estimate procedure might be better.

Reference

[1] N. R. Draper and H. Smith, *Applied Regression Analysis*, 2nd ed., John Wiley & Sons, New York, 1981.

Appendix H. Basic Statistical and Probability Theory

If a variable x is observed repeatedly in an experiment, the m observed values $x_1, x_2 \cdots x_m$ constitute a *sample* of size m from which the characteristics of x can be estimated.

Mean. The *mean* value of a variable x, denoted by $\bar{x}$, is defined as the sum of the observed values divided by the number of values. Thus, for the mean $\bar{x}$ of x we have

$$\bar{x} = \frac{x_1 + x_2 + \cdots + x_m}{m} = \frac{\sum_{i=1}^{m} x_i}{m}. \tag{H.1}$$

Since m represents a subset of the full set of observations, called the *population*, that might have been observed, $\bar{x}$ is called the *sample mean* in order to distinguish it from the so called *population mean*, which is generally denoted by μ in statistical theory.

Variance. A measure of spread of the value of x from its mean $\bar{x}$ is provided by the *sample variance*. It is the average of the sum of the squared deviation from the mean and is ordinarily represented by s^2. Thus,

$$s^2 = \frac{\sum_{i=1}^{m} (x_i - \bar{x})^2}{(m-1)}. \tag{H.2}$$

Note the denominator used in calculating sample variance is '(m − 1)' not 'm'. For computational purposes it is more appropriate to use the following equation

$$s^2 = \frac{\sum_{i=1}^{m} x_i^2 - (\sum_{i=1}^{m} x_i)^2/m}{(m-1)} \tag{H.3}$$

due to its better round-off error properties. Equation (H.3) can be shown to be the same as Eq. (H.2). The variance of a *population* is generally denoted by σ^2 and is expressed as

$$\sigma^2 = \frac{\sum_{i=1}^{m} (x_i - \mu)^2}{m}.$$

Standard Deviation. The square root of the sample variance is called the *sample standard deviation* denoted by s. It is a measure of the absolute variability in a data set and is a most common measure of disperson used in statistics, thus

$$s = \frac{\sqrt{\sum_{i=1}^{m} (x_i - \bar{x})^2}}{(m-1)}. \tag{H.4}$$

The *population* standard deviation is generally represented by σ and is expressed as

$$\sigma = \frac{\sqrt{\sum_{i=1}^{m} (x_i - \mu)^2}}{m}. \tag{H.5}$$

The most common measure of relative variability is the *coefficient of variation* (CV) which is simply the ratio of the standard deviation to the mean

$$\text{CV (Coefficient of Variation)} = \frac{s}{\bar{x}}. \tag{H.6}$$

Note that CV is a unitless number.

Gaussian or Normal Distribution. A quantitative measure of the frequency of occurrences of one or more specific values of a random variable[1] x is called the probability $Pr(x)$ of x. The relationship between the possible values of x and the corresponding probabilities is called the *probability distribution* of x. The probability distribution associated with a continuous random variable x is specified in terms of the *probability density function* (PDF) $f(x)$. The PDF $f(x)$ has the following properties:

- Probability must be positive, therefore, PDF must be greater than zero, that is,

$$f(x) \geq 0.$$

[1] A random variable x could be discrete or continuous. It is called continuous if it can take any numerical value in a given range. The model parameter p_i $(i = 1, 2 \cdots n)$ is an example of a continuous random variable.

- The probability that x has a value in the interval $(a \leq x \leq b)$ is given by the area under the curve $f(x)$ between a and b, that is,

$$\int_a^b f(x)dx = Pr(a \leq x \leq b).$$

- The total area under the curve $f(x)$ is unity, that is,

$$\int_{x_{min}}^{x_{max}} f(x)dx = 1.$$

Of all the probability distributions of a continuous variable x, the one most commonly used is the *normal or Gaussian distribution.* This is because (1) many practical distributions are fitted reasonably well by a normal distribution, and (2) much of probability theory takes a simple form when the distribution involved is normal. The mathematical equation representing the normal distribution of a variable x is given by

$$f(x) = \frac{1}{\sigma\sqrt{2\pi}} \exp\left[-\frac{(x-\mu)^2}{2\sigma^2}\right] \quad -\infty < x < +\infty \tag{H.7}$$

where μ is the mean value of x and is a quantitative measure of the location of the center of the curve $f(x)$ and σ is the standard deviation and is a measure of the dispersion about the mean value. Note that the symbol μ and σ used here are *population* mean and population standard deviation in order to distinguish them from the *sample* mean $\bar{x}$, and sample standard deviation s defined earlier. The constant $1/\sqrt{2\pi}$ has been chosen to ensure that the *area under this curve, obtained by integrating the density function $f(x)$, is equal to unity, or probability is* 1.0. Figure H.1 shows different curves obtained using Eq. (H.7) with mean $\mu = 3$ and standard deviation $\sigma = 0.5$ and 2. Note that the curve is symmetric about its mean μ, which locates the peak of the bell (see Figure H.1). The interval running one σ in each direction from μ has a probability of 0.683, the interval from $\mu - 2\sigma$ to $\mu + 2\sigma$ has a probability of 0.954, and the interval from $\mu - 3\sigma$ to $\mu + 3\sigma$ has a probability of 0.997. In other words,

$$Pr(\mu - \sigma \leq X \leq \mu + \sigma) = 0.683,$$

$$Pr(\mu - 2\sigma \leq X \leq \mu + 2\sigma) = 0.954,$$

$$Pr(\mu - 3\sigma \leq X \leq \mu + 3\sigma) = 0.997.$$

The curve never reaches zero for any value of x, but because the tail areas outside $(\mu - 3\sigma, \mu + 3\sigma)$ are very small, we usually terminate the curve at these points.

By convention, the probability associated with a particular variable is usually expressed as a *percentage* statement rather than as a decimal

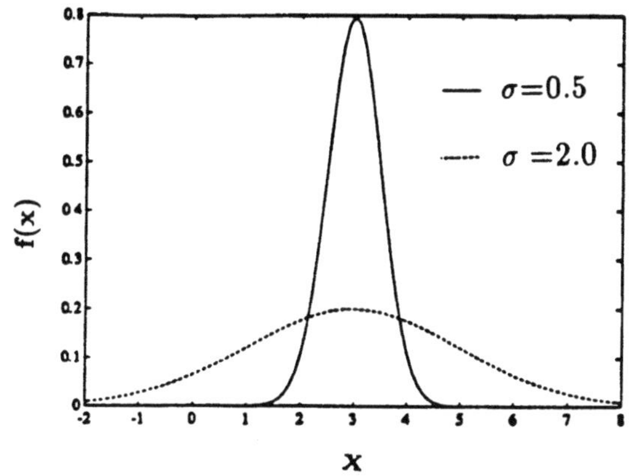

Fig. H.1 Plot of Normal or Gaussian distribution function $f(x)$ for two different values of standard deviation $\sigma = 0.5$ and 2. The mean value of x is 3

probability. Suppose the process mean of the V_{th} is 0.5 V and that the standard deviation is 0.03 V, then from the equation above the following facts would emerge.

- 68.3% of the V_{th} will lie within 0.5 ± 0.03 V ($V_{th} \pm \sigma$)
- 95.4% of the V_{th} will lie within 0.5 ± 0.06 V ($V_{th} \pm 2\sigma$)
- 99.7% of the V_{th} will lie within 0.5 ± 0.09 V ($V_{th} \pm 3\sigma$).

Integration of PDF Gives the Probability. The probability that the random variable lies in the range $(-\infty, x)$ is given by

$$D(x) = \int_{-\infty}^{x} f(x)dx. \tag{H.8}$$

The function $D(x)$ is called the *distribution function.*

Standard Normal Distribution. If we define a variable z such that

$$z = \frac{x - \mu}{\sigma} \tag{H.9}$$

then the normal distribution (H.7) becomes

$$f(z) = \frac{1}{\sqrt{2\pi}} \exp\left(-\frac{z^2}{2}\right). \tag{H.10}$$

The new transformed variable z has a mean of zero and standard deviation of 1. This particular normal distribution function $f(z)$ is called the *standard normal distribution.* The probability or the area to the left of the curve for a specified value of z has been tabulated and is usually given in most of the statistical textbooks.

Covariance. The extent of the relationship between the two variables defined in the same sample space can be determined from their scatter plot[2]. It is often difficult to judge the dependence quantitatively from such plots, except in cases when the relationship between the two parameters is very strong. *One measure of the degree of the linear association between the two parameters, which is often used, is the covariance.* Just as variance measures the spread of values of a variable around its mean, the *covariance* C_{xy} measures the joint distribution of the variables x and y around their mean. It is the sum of the deviations of the paired x_i and y_i values of the variables x and y from their respective means. Thus,

$$C_{xy} = \frac{\sum_{i=1}^{m}(x_i - \bar{x})(y_i - \bar{y})}{m-1} \tag{H.11}$$

where $\bar{x}$ and $\bar{y}$ are mean values of the variables x and y, respectively. In the special case when $x = y$, the formula for the covariance simplifies to

$$C_{xx} = \frac{\sum_{i=1}^{m}(x_i - \bar{x})^2}{m-1} = s_x^2. \tag{H.12}$$

Thus, the *covariance of a variable with itself is just the square of the standard deviation, i.e., variance.*

Correlation Coefficient. Interpretation of the covariance may not be intuitive since it is dependent on the units of measurement of the variables concerned. Therefore, to compare the interaction between different variables that have widely different units, it is necessary to standarize the variables. A variable can be converted into standarized or unitless form using the relationship given in Eq. (H.9), so that the new transformed variable z has a mean of zero and standard deviation of 1. The covariance between the standarized variables can be used to estimate the degree of interrelation between the variables, in a manner not influenced by measurement units, and is called the *correlation coefficient.*[3] When based on a sample of data (from population), the correlation coefficient is denoted by r and in turn is an estimate of the population correlation coefficient denoted by ρ. The correlation coffiicient r between the two normalized parameters z_x and z_y

---

[2] A scatter plot is simply a plot of data points between two variables x and y. A normalized scatter plot for μ_0 (low field mobility) versus θ (mobility degradation parameter) is shown in Figure H.3 (p. 597). Note that in this Figure dotted line (ellipse) is not part of the scatter plot.

[3] If the variables are reasonably commensurable, the covariance form has greater statistical appeal.

is given by

$$r = \frac{\sum_{i=1}^{m} z_x z_y}{(m-1)}. \tag{H.13}$$

The above equation when written in terms of non-normalized parameters x and y, becomes

$$r = \frac{\sum_{i=1}^{m} (x_i - \bar{x})(y_i - \bar{y})}{\sqrt{\sum_{i=1}^{m} (x_i - \bar{x})^2}\sqrt{\sum_{i=1}^{m} (x_2 - \bar{y})^2}} = \frac{C_{xy}}{s_s s_y} \tag{H.14}$$

where s_x and s_y are the standard deviation of the variables x and y, respectively, and C_{xy} are the covariance between them. Note that r is a dimensionless quantity ranging between $+1$ and -1, and gives a quantitative measure of the correlation. When the points on the scatter plot lie on a perfect straight line, the value of r is $+1$ or -1, depending on whether the line has upward or downward slope, indicating a perfect correlation. A zero value of r indicates absolutely no correlation. Intermediate values of r result when the points lie within an ellipse and in this case the interpretation of r becomes complex. The correlation between the two variables is said to be 'moderate' if the value of r is around 0.5.

It is important to note that it is possible to obtain non-zero values of the correlation coefficient even when two completely independent (uncorrelated) variables are considered. This arises because of the finite size of the sample as well as errors in the measurement. A test is normally made to check whether r is real or spurious, that is whether two variables are really related to each other in a statistical sense. In other words, it must be checked whether there is indeed a physical relationship between the pair of variables. It is to be pointed out that every statistical test is a test of the *null* hypothesis generally denoted by H_0, i.e., test of the hypothesis of zero difference or equality, in contrast to the *alternate* hypothesis, denoted by H_1 which is a statement of expected or anticipated differences. In testing a certain hypothesis we are drawing conclusions based on a few sample data, therefore, we always run the risk that we will reject, say, H_0 when H_0 is really a true hypothesis. *The risk of rejecting H_0 when H_0 is true is called the significance level of a test* and is usually represented by α. For example, $\alpha = 0.05$ (5% significance level) implies that there is a 5% chance of wrongly rejecting H_0. Frequently used standards are $\alpha = 0.05$ and $\alpha = 0.01$, but other values of α might also be chosen. If the result of the test of significance is such that the probability of the outcome is equal to or less than α, then the null hypothesis is rejected. This simply means that the probability of an error is sufficiently small that we choose to regard the null hypothesis as improbable.

Tests of significance are commonly made by transforming a statistic based on a sample into another statistic for which the probability distribution is known, given that the null hypothesis is true. When the sample size is large (as a rule of thumb, say $m \gg 30$) it is common practice to base the significance level for r and the test concerning r on the statistic

$$z_r = \frac{1}{2}\log\frac{(1+r)}{(1-r)}$$

whose distribution is approximately normal with the mean

$$z_\rho = \frac{1}{2}\log\frac{(1+\rho)}{(1-\rho)}$$

and the variance

$$\sigma_{zr}^2 = \frac{1}{n-3}.$$

Thus,

$$Z = \frac{z_r - z_\rho}{\sigma_{zr}} = \frac{\sqrt{n-3}}{2}\log\frac{(1+r)(1-\rho)}{(1-r)(1+\rho)} \tag{H.15}$$

can be looked upon as the value of a random variable which has a distribution that is approximately that of a standard normal distribution with zero mean and $\sigma = 1$ and can be evaluated in terms of the table of the standard normal distribution.

Using the above approximation we can test the null hypothesis such that $\rho = 0$ (no relationship between the variables x and y), which assumes that the sampling distribution of r values will be approximately normal in form, provided m is not too small. However, if ρ is not close to zero and if m is small, then the sampling distribution of r will not be normal in form but will be skewed instead. It has been shown that for all practical purposes the distribution of z_r is approximately normal in form, independent of the population value ρ, which means that we can test the null hypothesis for ρ other than zero. If the null hypothesis is rejected then the probability, that the sample correlation coefficient is true, is high. If not, then the correlation coefficient is spurious. For example, suppose we want to make a two-sided test of the null hypothesis that $\rho = 0$, with $\alpha = 0.01$. Also suppose we have $m = 50$ with $r = 0.5$ which gives $\sigma_{zr} = 0.1458$ and $z_r = 0.549$. From the table of the standard normal distribution, given in any statistics book, we find that the probability of obtaining $Z \geq 2.58$ (corresponding to $\alpha = 0{\cdot}01$), when the null hypothesis is true, is 0.05, and this is also the probability of obtaining $Z \leq -2.58$. Thus, with the two-sided test we will reject the null hypothesis if we obtain $-2.58 \geq Z \geq 2.58$. Substituting the value of $z_r = 0.549$ in Eq. (H.15), we have $Z = 3.7637$ and because $Z > 2.58$, the null hypothesis is rejected. We conclude that r is not spurious.

If the sample size is small then generally the t test is performed for the null hypothesis such that $\rho = 0$. All these test are available in the statistical packages (see Appendix I).
Note that the correlation coefficient r is only appropriate for measuring the degree of relationship between the two variables which are *linearly related.* A second major assumption for the use of the correlation coefficient is that the two variables have a *bivariate normal distribution* as discussed below.

Bivariate Normal Distribution. If the two variables x and y are normally distributed then their joint distribution is called a *bivariate normal distribution.* It can be imagined as a heap of sand or mount of data points, each representing a pair of values on the two variables in question. A cross-section of the mound, parallel to either the x or y axis (x and y axes are orthogonal), will result in a typical normal distribution for a single variable. This is equivalent to saying that for any given value of the variable x, the y variable is normally distributed and vice versa. The density plot for a bivariate normal distribution with standarized variables x and y for $\rho = 0.5$ and 0.9 are shown in Figure H.2. Note that the paths of x and y values yielding a constant height are ellipses. These paths are called *contours.* Thus, the contours of a bivariate normal distribution are ellipses. As the correlation coefficient between the variables increases, the mound will become narrower when viewed from above and the ellipses become more elongated (see Fig. H.2b).
The equation for the joint probability density function $f(z_1, z_2)$ of a bivariate normal distribution for two standarized variables z_1 and z_2 can be written as

$$f(z_1, z_2) = \frac{1}{2\pi\sqrt{1 - r^2}} \exp\left[-\frac{1}{2(1 - r^2)}(z_1^2 - 2rz_1z_2 + z_2^2) \right]. \tag{H.16}$$

Simplifying equation (H.16) we get

$$(z_1^2 - 2rz_1z_2 + z_2^2) = -2(1 - r^2)\ln(z_0 2\pi\sqrt{1 - r^2}) \tag{H.17}$$

which gives an equation for the contour for a given value of the function z_0 where z_0 is the value of $f(z_1, z_2)$ for a given probability. The value of z_0 will be restricted to lie in the range

$$0 < z_0 < \frac{1}{2\pi\sqrt{1 - r^2}}. \tag{H.18}$$

The probability that $a \leq z_1 \leq b$ and $b \leq z_2 \leq d$ is given by

$$\int_a^b \int_c^d f(z_1, z_2) dx_2 dz_1. \tag{H.19}$$

If the individual PDF are given for each parameter, i.e. the parameters are independent, then the joint PDF will be the product of the individual PDFs.

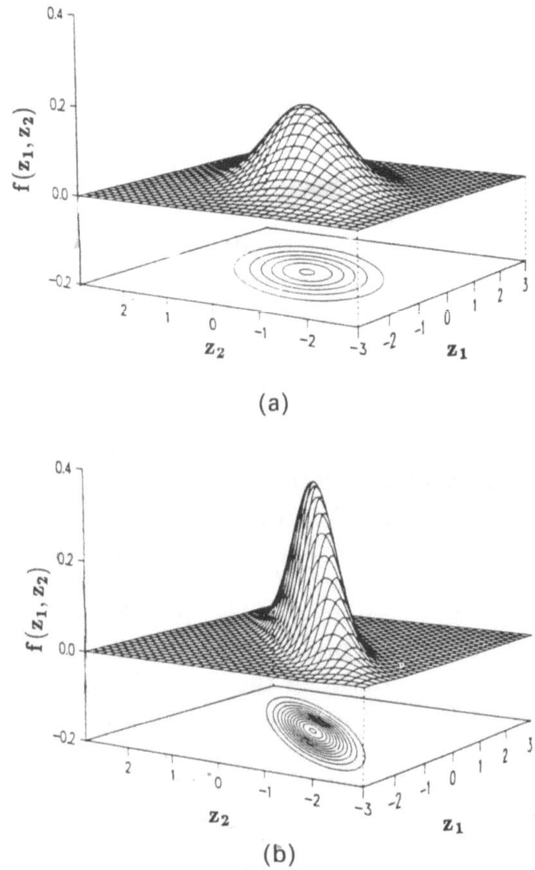

Fig. H.2 Bivariate normal density function $f(x, y)$ with correlation (a) $r = 0.5$, (b) $r = 0.9$. (After Johnson [8])

The scatter plot for the normalized model parameters μ_0 (low field mobility) and R_t is shown in Figure H.3. The measured correlation coefficient for these two parameters is 0.835 based on certain set of measured data. The equidensity contour for $z_0 = 0.006$ (which corresponds to 97% probability) is shown by dotted line in the Figure H.3. In general, for a positive value of r the ellipses are elongated with their major axes along the line $y = x$ and their minor axis along the line $y = -x$. As r approaches 1, the ellipse becomes more elongated.

Covariance Matrix. As discussed earlier, the covariance or correlation coefficient r describes the degree of relationship between two variables. In practice, usually more than two variables need to be studied. Their mutual dependence can be examined by computing the covariance of each variable with the remaining variables and expressing the results in matrix

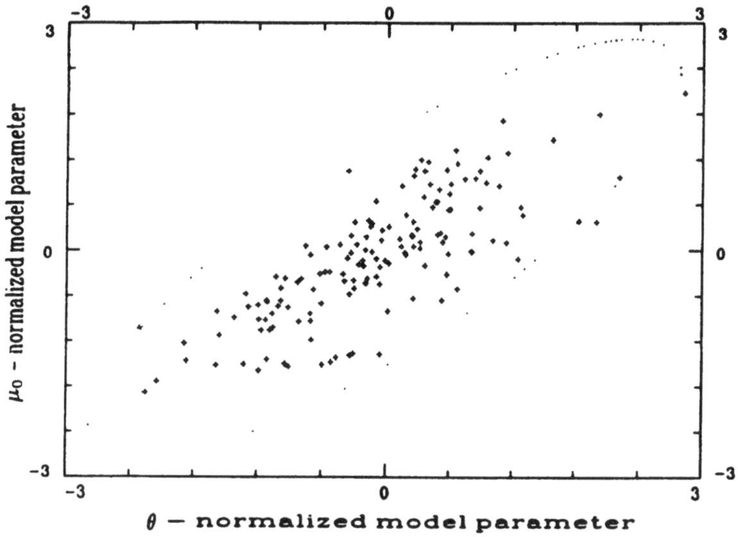

Fig. H.3 Equidensity contour for a bivariate normal distribution of two parameters μ_0 (low field mobility) and θ (mobility degradation factor). Correlation coefficient $= 0.8$, $z_0 = 0.006$

form, called the *covariance matrix*. Thus, *covariance matrix* **C** *of n variables is the matrix consisting of covariance of each variable with the remaining variables*, i.e.,

$$\mathbf{C} = \begin{bmatrix} C_{11} & C_{12} & \cdots & C_{1n} \\ C_{21} & C_{22} & \cdots & C_{2n} \\ \vdots & & \vdots & \vdots \\ C_{n1} & C_{n2} & \cdots & C_{nn} \end{bmatrix}. \tag{H.20}$$

Note the following features of the covariance matrix:

- It is a $n \times n$ square matrix; n being the number of variables. Each element of the matrix is occupied by the covariance between the variables represented by a particular row and column that the element occupies.
- The matrix is symmetrical about its diagonal, i.e., the portion above the diagonal is a mirror image of the portion below the diagonal.
- The diagonal terms of the matrix are the variance of the variables ($C_{ii} = s_i^2$).

If the variables are normalized, then the matrix is called the *correlation matrix* whose elements are the correlation coefficients between the variables represented by a row and column that the element occupies. The diagonal term equals 1 representing a perfect correlation. But this should be no surprise because they represent the correlations of each of the variables with themselves. Thus, a correlation coefficient matrix represented by **R** is

$$\mathbf{R} = \begin{bmatrix} 1 & r_{12} & \cdots & r_{1n} \\ r_{21} & 1 & \cdots & r_{2n} \\ \vdots & & \vdots & \vdots \\ r_{n1} & r_{n2} & \cdots & 1 \end{bmatrix}. \tag{H.21}$$

Visual inspection of the correlation matrix can quickly identify which two variables are highly correlated with each other or which variables correlates most highly with each of the individual variables, or even identify those which are independent. While interpreting this matrix, the physical relationship between various parameters has to be taken into account, as well as the 'significance levels' of the coefficients. Thus, each correlation coefficient is carefully studied and if spurious is set to zero in the correlation matrix.

Multivariate Normal Distribution. The normal distribution for a single variable x, given by Eq. (H.7), can be generalized to n variables and is called multivariate normal distribution (MVN) [6]–[8]. The joint probability density function for the MVN is given by,

$$f(\mathbf{x}) = (2\pi)^{-n/2}(\det \mathbf{C})^{-1/2} \exp\left[-\tfrac{1}{2}(\mathbf{x}-\mu)^T\mathbf{C}^{-1}(\mathbf{x}-\mu)\right] \tag{H.22}$$

where μ is a $n \times 1$ vector representing the mean value of the n dimensional random vector $\mathbf{x} = [x_1 x_2 \cdots x_n]^T$ and $\mathbf{C}$ is a $n \times n$ covariance matrix given by (H.20). It can be easily seen that the bivariate distribution (H.16) follows from MVN given by (H.22). For nonsingular $\mathbf{C}$ the density function f has a constant value defined by

$$(\mathbf{x}-\mu)^T\mathbf{C}^{-1}(\mathbf{x}-\mu) = a^2 \tag{H.23}$$

and thus the contours are ellipsoids defined by (H.23). These ellipsoids are centered at μ and have axes $\pm a\sqrt{\lambda_i}\mathbf{e}_i$ where

$$\mathbf{C}\mathbf{e}_i = \lambda_i \mathbf{e}_i \quad i = 1, 2, \ldots, n \tag{H.24}$$

and $(\lambda_i, \mathbf{e}_i)$ is an eigenvalue–eigenvector pair of the covariance matrix $\mathbf{C}$.

References

[1] D. G. Rees, *Foundations of Statistics*, Chapman and Hall, New York, 1987

[2] R. E. Walpole and R. H. Myers, *Probability and Statistics for Engineers and Scientists*, McGraw Hill, New York, 1976.

[3] C. W. Helstrom, *Probability and Stochastic Processes for Engineers*, Macmillan Publishing Company, New York, 1984.

[4] A. L. Edward, *An Introduction to Linear Regression and Correlation*, W. H. Freeman and Company, San Francisco, 1976.

[5] A. A. Afifi and V. Clark, *Computer-Aided Multivariate Analysis*, Lifetime Learning Publications, Belmont, CA, 1984.
[6] D. F. Morrison, *Multivariate Statistical Methods*, McGraw-Hill Book Company, New York, 1976.
[7] R. A. Johnson and D. W. Wichern, *Applied Multivariate Statistical Analysis*, Prentice-Hall, Inc., Englewood Cliffs, New Jersey, 1982.
[8] M. E. Johnson, *Multivariate Statistical Simulation*, John Wiley & Sons, New York, 1987.

Appendix I. List of Widely Used Statistical Package Programs

Name	Brief Description	Reference Manual
BMDP	General-purpose statistical package	W. J. Dixon, BMDP *Statistical Software* (1964 Westwood Blvd., Los Angles, CA 90025.)
SAS	General-purpose statistical and data analysis	SAS Institute, Inc., *SAS User's Guide*, (Cary, NC 27511.)
SPSS	General-purpose statistical package	N. H. Nie, SPSS-X (*SPSS Inc*, 444 North Michigan Ave., Chicago, IL 60611.)
RS/1	General-purpose statistical and data analysis	BBN Software Product Corp., 10 Fawcett Street, Cambridge, MA 02238

Each of these packages has special strength in one area [1]. For example, the BMDP package pays special attention to the technical aspect of the statistical procedures and emphasizes graphical output that is useful in checking assumptions. The SPSS is designed specifically to handle questionnaire and survey data arising in the social sciences. The SAS and RS/1 packages are most sophisticated in terms of data management capabilities.

Reference

[1] A. A. Afifi and V. Clark, *Computer-Aided Multivariate Analysis*, Lifetime Learning Publications, Belmont, CA, 1984.

Subject Index

Springer-Verlag Wien New York

MIX
Papier aus verantwortungsvollen Quellen
Paper from responsible sources
FSC® C105338

If you have any concerns about our products,
you can contact us on
ProductSafety@springernature.com

In case Publisher is established outside the EU,
the EU authorized representative is:
Springer Nature Customer Service Center GmbH
Europaplatz 3, 69115 Heidelberg, Germany

Printed by Libri Plureos GmbH
in Hamburg, Germany